Biology of
Microorganisms

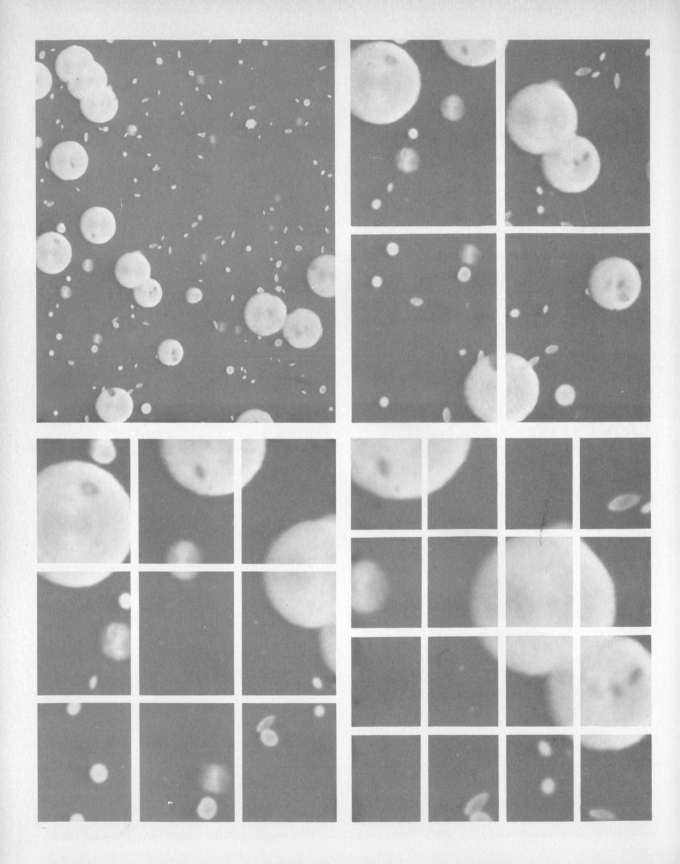

Biology of Microorganisms

Fourth Edition

Thomas D. Brock
University of Wisconsin

David W. Smith
University of Delaware

Michael T. Madigan
Southern Illinois University

Prentice-Hall, Inc., Englewood Cliffs, N.J. 07632

Library of Congress Cataloging in Publication Data

Brock, Thomas D.
 Biology of microorganisms.

 Includes bibliographies and index.
 1. Microbiology. I. Smith, David W., (date).
II. Madigan, Michael T., (date). III. Title.
QR41.2.B77 1984   576   83-24506
ISBN 0-13-078113-4

Editorial/production supervision: Karen J. Clemments
Interior design: Caliber Design Planning, Inc.
Art direction: Linda Conway
Page layout: Gail Collis
Cover design: Linda Conway and Maureen Eide
Manufacturing buyer: John Hall

Cover photograph: A fruiting body of Chondromyces crocatus, *a member
of the Myxobacterales. The size of the fruiting body is 560 μm.
Photograph by H. Reichenbach, GBF, Braunschweig, Federal Republic
of Germany; strain isolated by K. Gerth, GBF. Used by permission.*

Printed in the United States of America

10 9 8 7 6 5 4 3 2 1

ISBN 0-13-078113-4

Prentice-Hall International, Inc., *London*
Prentice-Hall of Australia Pty. Limited, *Sydney*
Editora Prentice-Hall do Brasil, Ltda., *Rio de Janeiro*
Prentice-Hall Canada Inc., *Toronto*
Prentice-Hall of India Private Limited, *New Delhi*
Prentice-Hall of Japan, Inc., *Tokyo*
Prentice-Hall of Southeast Asia Pte. Ltd., *Singapore*
Whitehall Books Limited, *Wellington, New Zealand*

Contents

6 The Autotrophic Way of Life   181

7 Growth and Its Control   213

8   The Microbe in Its Environment   239

9   Macromolecules and Molecular Genetics   271

10   Viruses   316

19 Representative Procaryotic Groups   625

Preface

We are pleased to present this fourth edition of *Biology of Microorganisms* to students and instructors of microbiology. Microbiology continues to undergo rapid changes through the impact of new developments in cell, molecular, and environmental microbiology. The revolution in genetics, as witnessed by the development of molecular cloning and genetic engineering techniques, has had profound impact on the teaching and practice of microbiology. Microbes continue to make excellent research tools for the study of many fundamental biological problems. Yet microbes are more than research tools; they are of considerable importance and interest in themselves. Basic research in such areas as ecology and evolution is advancing rapidly because of our increased understanding of fundamental microbial processes. At the same time, practical developments in industrial biotechnology, food processing, and agriculture arise from the application of microbiological principles.

The acceptance of earlier editions of this book has been gratifying. For the fourth edition, co-authors have been added for the first time. Microbiology has been advancing so rapidly that it has become virtually impossible for a single person to write a textbook that is complete, accurate, and up-to-date, and do it within a reasonable length of time. The three authors have worked closely together in the preparation of this edition, each reviewing and correcting the other's material in detail. All of us have had extensive experience teaching university-level courses in which earlier editions have been used, and were intimately familiar with the book.

As before, the book has undergone extensive changes. Also as before, the changes have been evolutionary rather than revolutionary. Naturally, the material on genetics has been extensively reworked. A whole new chapter on genetic engineering has been written, and the rest of the genetics chapters have been extensively revised. Another area that has undergone extensive revision is the chapter dealing with immunology, reflecting the current exciting research in this area. The biochemistry chapters have been modified so as to be more accessible to the beginning student. The bacterial diversity chapter has been carefully revised in line with the many new discoveries that have been made in recent years. Virtually every chapter has seen some changes, and we have taken special care to take all readers' suggestions into consideration.

As in earlier editions, particular care has been taken to keep the length of the book within bounds. We know all too well how easy it would be to simply add new material, but we believe that the ultimate user, the student, should be provided with a textbook, not a tome, and that it is the authors' responsibility to sift through the field and present those ideas and concepts that are most relevant and useful for the contemporary student.

The popular "Bit of History" topics that appeared in the third edition have been retained in the present edition, and several new entries have been added. We are especially pleased that users enjoyed these historical vignettes, which were as much fun to write as they apparently are to read.

The taxonomy of bacteria is undergoing a major revision. The new edition of *Bergey's Manual*, which will be called *Bergey's Manual of Systematic* (rather than *Determinative*) *Bacteriology,* is being published in four volumes, each of which will contain an extensive description of a group of related procaryotic genera. Because this revision is not complete, we present both the current classification of the eighth edition and the general contents of each volume of the new edition in Appendix 4.

We are most grateful to the many reviewers, mostly anonymous, who provided us with frank and constructive criticism. The following reviewers provided detailed comments: Carol Gross, John Martinko, Charles Bensen, Thomas Corner, Walter Konetzka, Jane Phillips, and Robert T. Vinopal. We trust that we have handled all criticisms and corrections properly. All errors of omission or commission are, of course, our own responsibility.

T.D.B.
D.W.S.
M.T.M.

Introduction

Microbiology is the study of microorganisms, a large and diverse group of free-living forms that exist as single cells or cell clusters. Microbial cells are thus distinct from the cells of animals and plants, which are unable to live alone in nature but can exist only as part of multicellular organisms. A single microbial cell is generally able to carry out its life processes of growth, energy generation, and reproduction independently of other cells, either of the same kind or of different kinds. Although there are exceptions, which we shall consider later, this definition provides a basis for our introduction to microorganisms.

1.1

Microorganisms as Cells

That the cell is the fundamental unit of all living matter is one of the great unifying theories of biology. A single cell is an entity, isolated from other cells by a cell wall or membrane and containing within it a variety of materials and subcellular structures. All cells contain proteins, nucleic acids, lipids, and polysaccharides. Because these chemical components are common throughout the living world, it is thought that all cells have descended from some common ancestor, a primordial cell. Through millions of years of evolution, the tremendous diversity of cell types that exist today has arisen. Cells vary enormously in size, from bacteria too small to be seen with the light microscope to the 170 by 135-mm ostrich egg, the largest single cell known. Microbial cells show a narrower but also extensive size range, some being much larger than human cells. The single-celled protozoan *Paramecium* is about 5000 times the weight of the human red blood cell.

　　　Although each kind of cell has a definite structure and size, a cell is a dynamic unit, constantly undergoing change and replacing its parts. Even when it is not growing, a cell is continually taking materials from its environment and working them into its own fabric. At the same time, it perpetually discards into its environment cellular materials and waste products. A cell is thus an open system, forever changing yet generally remaining the same.

1.2

Microbial Diversity

There are five major groups which are studied by microbiologists: fungi, protozoa, algae, bacteria, and viruses. These five may be divided further: algae, fungi, and protozoa have a form of cellular organization known as **eucaryotic,** the bacteria* are organized in a different way known as **procaryotic,** and viruses are not cells at all. The most important difference between procaryotes and eucaryotes is in the structure of the nucleus. The eucaryote has a true nucleus (*eu-* means "true"; *karyo-* is the combining form for "nucleus"), a membrane-enclosed structure within which are the chromosomes that contain hereditary material. The procaryote, on the other hand, does not have a true nucleus, and its hereditary material is contained in a single naked deoxyribonucleic acid (DNA) molecule. The many other structural differences between procaryotes and eucaryotes will be covered in some detail in Chapters 2 and 3. At present it is enough to know that these differences exist and that they are so fundamental as to make us believe they reflect an evolutionary divergence in the early history of life. Because the cells of higher animals and plants are all eucaryotic, it is likely that eucaryotic microorganisms were the forerunners of higher organisms, whereas procaryotes represent a branch that never evolved past the microbial stage.

Viruses are not cells. They lack many of the attributes of cells, of which the most important is that they are not dynamic open systems. A single virus particle is a static structure, quite stable and unable to change or replace its parts. Only when it is associated with a cell does a virus acquire attributes of a living system. Whether or not a virus is to be considered alive will depend on how life itself is defined. We shall reserve further discussion of this interesting question until Chapter 10.

Microorganisms are so diverse that it is useful to give them names, and to do this we must have ways of telling them apart. After close study of the structure, composition, and behavior of a microorganism, we can usually recognize a group of characteristics unique to a certain organism. Once an organism has been defined by such a set of characteristics, it can be given a name. Microbiologists use the binomial system of nomenclature first developed for plants and animals. The **genus** name is applied to a number of related organisms; each different type of organism within the genus has a **species** name. Genus and species names are always used together to describe a specific type of organism, whether it be a single cell or a group of such cells. Usually the names come from Latin and Greek and indicate some characteristic of the organism. For instance, *Saccharomyces*

FIGURE 1.1 Two drawings by Robert Hooke that represent one of the first microscopic descriptions of microorganisms. Top: a blue mold growing on the surface of leather; the round structures contain spores of the mold. Bottom: a mold growing on the surface of an aging and deteriorating rose leaf. (From Hooke, R. 1665. Micrographia, or some physiological descriptions of minute bodies made by magnifying glasses, with observations and inquiries thereupon. Royal Society, London.)

*This group includes the organisms which were historically called blue-green algae but are now termed **cyanobacteria,** a name which emphasizes their procaryotic form.

cerevisiae is the species of beer yeast. Yeasts convert sugar into alcohol, and *saccharo-* means "sugar." A yeast is a fungus, and the combining form *-myces* derives from the Greek word for "fungus." *Cerevisiae* is Latin for "brewer." However, many organisms are named in whole or in part after the scientists who studied them, so that it is rarely possible to break down the name of a microorganism as easily as we can this one, and even one who knows Latin and Greek would have little success in translating many species names into English. Although we shall try to keep the number of species to a minimum in this book, students should be prepared to familiarize themselves with at least the most important names (and their spellings!).

Traditionally, the living world has been divided into two kingdoms; plants and animals. Are microorganisms plants or animals? Some microorganisms, such as some of the chlorophyll-containing algae, appear to be plants, whereas many protozoa seem quite animallike. Yet we run into many difficulties. For instance, the organism *Euglena gracilis* is chlorophyll-bearing and may be considered a plant; yet after certain drug treatments it loses its chloroplasts and never regains them; thereafter its offspring live as animals. Although fungi lack chlorophyll, in many ways they are more closely related to the algae than they are to the protozoa. The procaryotic organisms, as a group, are so different from either animals or plants that it would seem foolish to try to place them in one group and some scientists classify the procaryotes in a separate kingdom called **Monera.** Some microbiologists propose the formation of a separate kingdom, called **Protista,** to include all eucaryotic microorganisms. Yet such a grouping ignores the fact that many algae clearly are more related to plants than they are to animals, and many protozoans have more in common with animals than with plants. It is likely that more than three kingdoms exist; some proposals have as many as five. As yet our knowledge of the range of microbial diversity is insufficient to develop a definitive classification.

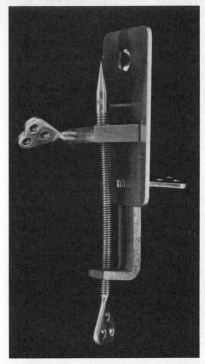

FIGURE 1.2 A replica of the kind of microscope used by Leeuwenhoek. The base was a piece of metal, with the lens inserted into the small hole at the top. The object to be viewed was placed on the small pointed tip at the end of the screw, and was moved back and forth by turning the screws.

1.3

The Discovery of Microorganisms

Although the existence of creatures too small to be seen with the eye had long been suspected, their discovery was linked to the invention of the microscope. Robert Hooke described the fruiting structures of molds in 1664 (Figure 1.1), but the first person to see microorganisms in any detail was the Dutch amateur microscope builder Anton van Leeuwenhoek, who used simple microscopes of his own construction (Figure 1.2). Leeuwenhoek's microscopes were extremely crude by today's standards, but by careful manipulation and focusing he was able to see organisms as small as bacteria. He reported his observations in a series of letters to the Royal Society of London, which published them in English translation. Drawings of some of Leeuwenhoek's "wee animalcules" are shown in Figure 1.3. His observations were confirmed by other workers, but progress in understanding the nature of these tiny organisms came slowly. Only in the nineteenth century did improved microscopes become available and widely distributed. During its history, the science of

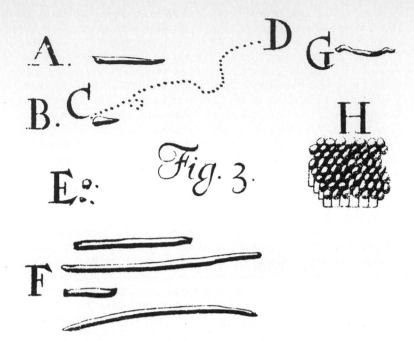

FIGURE 1.3 Leeuwenhoek's drawings of bacteria, published in 1684. Even from these crude drawings we can recognize several kinds of common bacteria. Those lettered A, C, F, and G are rod-shaped; E, spherical or coccus-shaped; H, coccus-shaped bacteria in packets. (Leeuwenhoek, A. van. 1684. Phil. Trans. Roy. Soc. London 14:568.)

microbiology has taken the greatest steps forward when better microscopes have been developed, for these enable scientists to penetrate ever deeper into the mysteries of the cell. Figure 1.4, which shows the yeast cell as seen with different microscopes, well illustrates this point.

Microbiology as a science did not develop until the latter part of the nineteenth century. This long delay occurred because, in addition to microscopy, certain basic techniques for the study of microorganisms needed to be devised. In the nineteenth century, investigation of two perplexing questions led to the development of these techniques and laid the foundation of microbiological science: (1) Does spontaneous generation occur? (2) What is the nature of contagious disease? Study of these two questions went hand in hand, and sometimes the same people worked on both. By the end of the century both questions were answered, and the science of microbiology was firmly established as a distinct and growing field.

1.4

Spontaneous Generation

The basic idea of spontaneous generation can easily be understood. If food is allowed to stand for some time, it putrefies. When the putrefied material is examined microscopically, it is found to be teeming with bacteria. Where do these bacteria come from, since they are not seen in fresh food? Some people said they developed from seeds or germs that had entered the food from the air, whereas others said that they arose spontaneously.

(a)

FIGURE 1.4 The yeast cell as it seemed to different observers. The great increase in our understanding of cell structure came with improvement in our microscopes. (a) Leeuwenhoek's drawing of yeast, dating from 1694. Note the complete absence of any cellular detail. (From Leeuwenhoek, A. van. 1694. Ondervindingen en Beschouwingen der Onsigtbar Geschapene Waarheden, 2nd ed., p. 45. Delft.) (b) Pasteur's drawings of yeast, made in 1860, showing the budding process by which yeasts grow. The contrast of the outer cell wall and inner cytoplasm is distinct. The large objects in the cytoplasm are vacuoles. (From Pasteur, L. 1860. Ann. Chim. Phys. 58:323.) (c) Drawing of the idea of a yeast cell in 1910. The greater detail inside the cell derives partly from improved microscopy and partly from the use of dyes that increase contrast and stain particular structures. However, some of the lableled structures are probably artifacts. (From Wager, H., and A. Peniston. 1910. Ann. Bot. 24:45.) (d) Photograph of a yeast cell as seen with the electron microscope. The cell is first treated with chemicals that preserve the structure and stain particular components. Magnification, 31,200×. (From Conti, S. F., and T. D. Brock, 1965. J. Bacteriol. 90:524.)

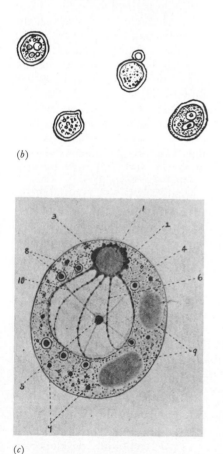

(b)

(c)

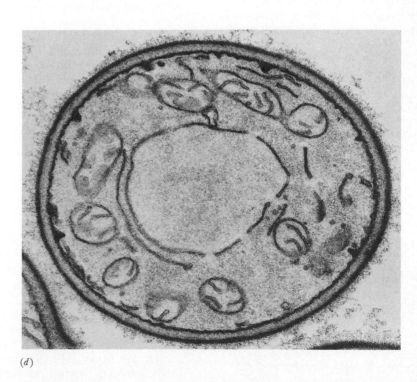

(d)

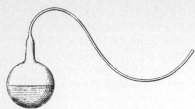

FIGURE 1.5 Pasteur's drawing of a swan-necked flask. In his own words: "In a glass flask I placed one of the following liquids which are extremely alterable through contact with ordinary air: yeast water, sugared yeast water, urine, sugar beet juice, pepper water. Then I drew out the neck of the flask under a flame, so that a number of curves were produced in it . . . . I then boiled the liquid for several minutes until steam issued freely through the extremity of the neck. This end remained open without any other precautions. The flasks were then allowed to cool. Any one who is familiar with the delicacy of experiments concerning the so-called "spontaneous" generation will be astounded to observe that the liquid treated in this casual manner remains indefinitely without alteration. The flasks can be handled in any manner, can be transported from one place to another, can be allowed to undergo all the variations in temperature of the different seasons, [yet] the liquid does not undergo the slightest alteration . . . . After one or more months in the incubator, if the neck of the flask is removed by a stroke of a file, without otherwise touching the flask, molds and infusoria begin to appear after 24, 36, or 48 hours, just as usual, or as if dust from the air had been inoculated into the flask . . . . At this moment I have in my laboratory many alterable liquids which have remained unchanged for 18 months in open vessels with curved or inclined necks." (From Pasteur, L. 1861. Ann. Sci. Nat., Ser. 4, 16:5. In Brock, T. D., ed. and trans. 1975. Milestones in microbiology. American Society for Microbiology, Washington, D.C.)

Spontaneous generation would mean that life could arise from something nonliving, and many people could not imagine something so complex as a living cell arising spontaneously from dead materials. The most powerful opponent of spontaneous generation was the French chemist Louis Pasteur, whose work on this problem was the most exacting and convincing. Pasteur first showed that structures were present in air that resembled closely the microorganisms seen in putrefying materials. He did this by passing air through guncotton filters, the fibers of which stop solid particles. After the guncotton was dissolved in a mixture of alcohol and ether, the particles that it had trapped fell to the bottom of the liquid and were examined on a microscope slide. Pasteur found that in ordinary air there exists constantly a variety of solid structures ranging in size from 0.01 mm to more than 1.0 mm. Many of these structures resemble the spores of common molds, the cysts of protozoa, and various other microbial cells. As many as 20 to 30 of them were found in 15 liters of ordinary air, and they could not be distinguished from the organisms found in much larger numbers in putrefying materials. Pasteur concluded that the organisms found in putrefying materials originated from the organized bodies present in the air. He postulated that these bodies are constantly being deposited on all objects. If this conclusion was correct, it would mean that, if food were treated to destroy all the living organisms contaminating it, then it should not putrefy.

Pasteur used heat to eliminate contaminants, since it had already been established that heat effectively kills living organisms. In fact, many workers had shown that if a nutrient solution was sealed in a glass flask and heated to boiling, it never putrefied. The proponents of spontaneous generation criticized such experiments by declaring that fresh air was necessary for spontaneous generation and that the air itself inside the sealed flask was affected in some way by heating so that it would no longer support spontaneous generation. Pasteur skirted this objection simply and brilliantly by constructing a swan-necked flask, now called the "Pasteur flask" (Figure 1.5). In such a flask putrefying materials can be heated to boiling; after the flask is cooled, air can reenter but the bends in the neck prevent particulate matter, bacteria, or other microorganisms from getting in the flask. Material sterilized in such a flask did not putrefy, and no microorganisms ever appeared as long as the neck of the flask remained intact. If the neck was broken, however, putrefaction occurred and the liquid soon teemed with living organisms. This simple experiment served to effectively settle the controversy surrounding the theory of spontaneous generation.

Killing all the bacteria or germs is a process we now call **sterilization,** and the procedures that Pasteur and others used were eventually carried over into microbiological research. Disproving the theory of spontaneous generation thus led to the development of effective sterilization procedures, without which microbiology as a science could not have developed.

It was later shown that flasks and other vessels could be protected from contamination by cotton stoppers, which still permit the exchange of air. The principles of aseptic technique, developed so effectively by Pasteur, are the first procedures learned by the novice microbiologist. Food science also owes a debt to Pasteur, as his principles are applied in the canning and preservation of many foods.

Although Pasteur was successful in sterilizing materials with simple boiling, some workers found that boiling was insufficient. We now know that the failure resulted from the presence in these materials of bacteria which formed unusually heat-resistant structures called **endospores.*** Initial work on the endospore was carried out by two men: John Tyndall in England, and Ferdinand Cohn in Germany. Both men observed that some preparations, such as the fruit juice solutions used by Pasteur, were relatively easy to sterilize, requiring only 5 minutes of boiling, whereas others were not sterilized by much longer periods of boiling, sometimes even hours. Notably difficult to sterilize were hay infusions. In addition, once hay had been brought into the laboratory, even the sugar solutions could no longer be reliably sterilized by hours of boiling. Cohn performed detailed microscopic observations and discovered an endospore forming inside each cell in old cultures of species of *Bacillus* (Figure 1.6). Cohn and another German scientist, Robert Koch, applied this observation to the study of disease, as discussed in Section 1.5. Tyndall did not observe endospores microscopically, but he did hypothesize that the hay contained bacteria which could exist in two interconvertible forms: one actively growing and heat-sensitive and the other dormant and very heat-resistant. He tested this hypothesis by boiling a hay infusion for 1 minute, presumably killing the sensitive forms. After 12 hours, he again boiled the infusion for 1 minute, under the assumption that the dormant forms (endospores) had been converted to active and were now heat sensitive. This cycle was repeated two more times with 30 seconds of boiling to ensure that all endospores had been converted to growing cells and then killed. This procedure was remarkably successful: a total of 3 minutes of boiling in this way left the infusion sterile, whereas single boilings of hours were not able to sterilize. This process of **fractional sterilization,** now referred to as **Tyndallization,** not only elegantly confirmed the existence of dormant forms, but it is also a practical procedure for the sterilization of materials which may be damaged by longer exposure to high temperatures.

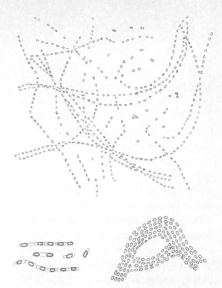

FIGURE 1.6   Drawings by Koch (top) and Cohn of bacterial endospores, published in 1876. The spores first form in long rows within the bacterial chains but are eventually liberated. The organism is *Bacillus anthracis,* the causal organism of anthrax in cattle and man. Endospores of harmless bacteria, such as *Bacillus subtilis,* are much the same. (From Koch, R. 1876. Beitr. Biol. Pflanzen 2:277. And Cohn, F. 1876. Beitr. Biol. Pflanzen 2:247.)

1.5

The Germ Theory of Disease

Proof that microorganisms could cause disease provided one of the greatest impetuses for the development of the science of microbiology. Indeed, even in the sixteenth century it was thought that something could be transmitted from a diseased person to a well person to induce in the latter the disease of the former. Many diseases seemed to spread through populations and were called **contagious;** the unknown thing that did the spreading was called the *contagion.* After the discovery of microorganisms, it was more or less widely held that these organisms might be responsible for contagious diseases, but proof was lacking.

*This structure is sometimes simply referred to as a spore, but there is danger of confusion in this terminology, since many algae and fungi form spores of a very different type, as do some complex bacteria of the *Actinomyces* group (see Section 19.24).

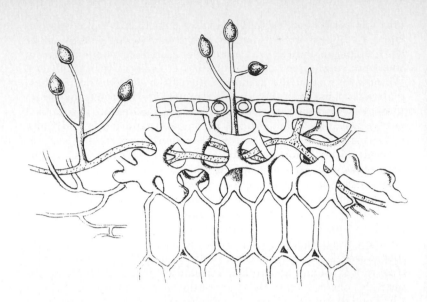

FIGURE 1.7 Berkeley's drawing of a disease-causing microorganism, done in 1846. The disease was Irish potato blight, which was responsible for the great famine in Ireland. The drawing shows the manner in which the fungus grows around and through the cells of the potato leaf. The round structures at the tips of filaments are spore bodies, which liberate the spores that transmit the fungus infection to other plants. Compare with Hooke's drawing in Figure 1.1. (From Berkeley, M. J. 1846. J. Roy. Hort. Soc. London 1:9.)

In 1845, M. J. Berkeley provided the first clear demonstration that microorganisms caused diseases by showing that a mold was responsible for Irish potato blight (Figure 1.7). Discoveries by Ignaz Semmelweis and Joseph Lister provided some evidence for the importance of microorganisms in causing human diseases, but it was not until the work of Koch, a physician, that the germ theory of disease was placed on a firm footing.

In his early work, published in 1876, Koch studied *anthrax,* a disease of cattle, which sometimes also occurs in man. Anthrax is caused by a spore-forming bacterium now called *Bacillus anthracis,* and the blood of an animal infected with anthrax teems with cells of this large bacterium. Koch established by careful microscopy that the bacteria were always present in the blood of an animal that had the disease. However, he knew that the mere association of the bacterium with the disease did not prove that it caused the disease; it might instead be a result of the disease. Therefore, Koch demonstrated that it was possible to take a small amount of blood from a diseased animal and inject it into another animal, which in turn became diseased and died. He could then take blood from this second animal, inject it into another, and again obtain the characteristic disease symptoms. By repeating this process as often as 20 times, successively transferring small amounts of blood containing bacteria from one animal to another, he proved that the bacteria did indeed cause anthrax: the twentieth animal died just as rapidly as the first; and in each case Koch could demonstrate by microscopy that the blood of the dying animal contained large numbers of the bacterium.

Koch carried this experiment further. He found that the bacteria could also be cultivated in nutrient fluids outside the animal body, and even after many transfers in culture the bacteria could still cause the disease when reinoculated into an animal. Bacteria from a diseased animal and bacteria in culture both induced the same disease symptoms upon injection. On the basis of these and other experiments Koch formulated the following criteria, now called **Koch's postulates,** for proving that a specific type of bacterium causes a specific disease:

1. The organism should always be found in animals suffering from the disease and should not be present in healthy individuals.
2. The organism must be cultivated in pure culture away from the animal body.
3. Such a culture, when inoculated into susceptible animals, should initiate the characteristic disease symptoms.
4. The organism should be reisolated from these experimental animals and cultured again in the laboratory, after which it should still be the same as the original organism.

Koch's postulates not only supplied a means of demonstrating that specific organisms cause specific diseases but also provided a tremendous impetus for the development of the science of microbiology by stressing the use of laboratory culture.

In order successfully to identify a microorganism as the cause of a disease, one must be sure that it alone is present in culture. That is, the culture must be **pure** (another word for pure is **axenic**). With material as small as microorganisms, ascertaining purity is not easy, for even a very tiny sample of blood or animal fluid may contain several kinds of organisms that may all grow together in culture. Koch realized the importance of pure cultures. He developed several ingenious methods of obtaining them, of which the most useful is that involving the isolation of single **colonies.** Koch observed that when a solid nutrient surface, such as a potato slice, was exposed to air and then incubated, bacterial colonies developed, each having a characteristic shape and color. He inferred that each colony had arisen from a single bacterial cell that fell on the surface, found suitable nutrients, and began to multiply. Because the solid surface prevented the bacteria from moving around, all of the offspring of the initial cell remained together, and when a large enough number was present, the mass of cells became visible to the naked eye. He assumed that colonies with different shapes and colors were derived from different kinds of organisms. When the cells of a single colony were spread out on a fresh surface, many colonies developed, each with the same shape and color as the original.

Koch realized that this discovery provided a simple way of obtaining pure cultures: he found that if mixed cultures were streaked on solid nutrient surfaces, the various organisms were spread so far apart that the colonies they produced did not mingle. Many organisms could not grow on potato slices; so he devised semisolid media, in which gelatin was added to a nutrient fluid such as blood serum in order to solidify it. When the gelatin-containing fluid was warmed, it liquefied and could be poured out on glass plates; upon cooling, the solidified medium could be inoculated. Later *agar* (a material derived from seaweed) was found to be a better solidifying agent than gelatin, and this substance is widely used today (Figure 1.8).

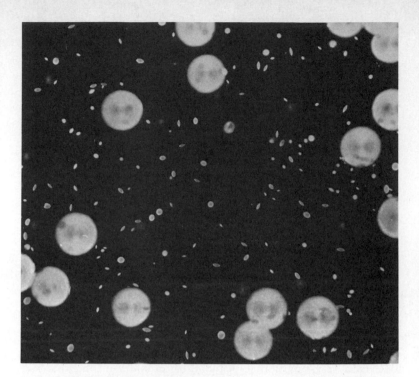

FIGURE 1.8 Bacterial colonies growing on an agar plate. The large colonies are growing on the surface, and the small ones are growing embedded in the agar gel. The large colonies are about 5 mm in diameter.

In the 20 years following the formulation of Koch's postulates, the causal agents of a wide variety of contagious diseases were isolated. These discoveries led to the development of successful treatments for the prevention and cure of contagious diseases and contributed to the development of modern medical practice. The impact of Koch's work has been felt throughout the world.

It is important to remember, however, that not all diseases are caused by microorganisms; many are inherited or are caused by deficiencies in diet or other harmful influences of the environment. Further, the microorganism is only one factor in the disease. It is a necessary cause, but it is not sufficient in itself. To produce disease, the microbe must infect a sensitive host, and not all hosts are equally susceptible. The state of health of the host, its general vigor, the presence or absence of specific immunity—all influence the outcome of the infection. Infectious disease is not a particular thing, or entity; it is a *process* in which host and microbe interact.

One might think that, because some microbes cause diseases, all microbes are harmful. This is far from the truth. Most microbes are beneficial to man or are at least harmless; it is only the rare organism that causes disease.

It is important to realize that Koch's postulates have relevance beyond identifying organisms which cause specific diseases. The essential general conclusion to be gathered is that specific organisms have specific effects. This principle that different organisms have unique activities was important in establishing microbiology as an independent biological science.

Whether life as we know it is unique to the planet earth is an unsolved question. That the earth is an excellent environment on which life has evolved and developed, however, is clear from the rich diversity of living organisms it now supports. Of all living organisms, none are more versatile than the microorganisms. In no environment where higher organisms are present are microorganisms absent, and in many environments devoid of, or hostile to, higher organisms, microorganisms exist and even flourish. Because microorganisms are usually invisible to the naked eye, their existence in an environment is often unsuspected. Yet, microorganisms carry out a number of functions vital for the life of higher organisms, so that without microbes higher organisms would quickly disappear from the earth.

The geologist divides the earth into three zones, the lithosphere, the hydrosphere, and the atmosphere; to these zones we can add the biosphere. The *lithosphere* is the solid portion of earth, composed of solid and molten rocks and soil. Microbes make up an important part of the soil. The *hydrosphere* represents the aqueous environments of the earth, such as the oceans, lakes, and rivers. Microorganisms are found throughout the oceans as well as in freshwater habitats from the tropics to the poles. The *atmosphere* is the gaseous region that surrounds the earth, relatively dense near the surface but thinning to nothing in its upper reaches. Although microbes are carried around the world on winds and other air currents, they do not actually reproduce in the atmosphere. The *biosphere* represents the mass of living organisms found in a thin belt at the earth's surface. Living organisms have had a profound influence on the earth itself, being responsible for almost all of the oxygen found in the atmosphere as well as for the enormous deposits of oil, coal, and sulfur found underground. Higher organisms and their corpses provide excellent microbial environments, and thus we find large populations of microbes associated with higher plants and animals.

Although the earth provides suitable environments for microbial growth, we do not find the same organisms everywhere. In fact, virtually every environment, no matter how slightly it differs from others, probably has its own particular complement of microorganisms that differ in major or minor ways from organisms of other environments. Because microorganisms are small, their environments are also small. Within a single handful of soil many microenvironments exist, each providing conditions suitable for the growth of a restricted range of microorganisms. When we think of microorganisms living in nature, we must learn to "think small."

New environments for microorganisms are continually being created. Some of these result from natural processes, such as the formation of a new volcanic island or the creation of a new lake after an earthquake. Most are artificial, however. Pollution of streams and lakes, clearing of forests, planting of exotic crops, introduction of new pesticides and fertilizers, and sewage treatment on a large scale all create new microbial environments. These new conditions sometimes make possible a further step in evolution and the development of a new organism.

Thus the great diversity of microbial life on planet earth should not surprise us: it only reflects the great diversity of habitats within which microbes can grow and evolve. Microorganisms are not passive inhabitants, however, and their activities affect their environments in many ways. We have already mentioned that some organisms cause diseases, effects far more harmful than we would have predicted from their small sizes. Other deleterious changes, such as food spoilage, souring of milk, deterioration of clothing and dwellings, and corrosion of metal pipes, also occur primarily through microbial action. But microorganisms play many beneficial roles in nature. They are responsible for most of the decomposition of dead animal and plant bodies, thus returning important plant nutrients to the soil. Many microorganisms that live in the intestinal tracts of animals synthesize certain vitamins, thus freeing their hosts of the need to obtain these vitamins in their diets. Without microorganisms, animals such as cows, sheep, and goats would be unable to digest the cellulose of grass and hay and hence would be unable to survive on earth. On the whole, the beneficial effects of microorganisms far outweigh their harmful ones.

1.7

The Contemporary Study of Microorganisms

Microbiology has come a long way from the days of Koch and Pasteur. Today it is one of the most sophisticated of the biological sciences, and has greatly influenced biology as a whole. Because of the special laboratory requirements for the study of microorganisms, microbiology is an independent discipline, but it is first of all a biological science.

Microorganisms have played a great role in recent years as model systems for the study of basic biological processes. Much of our understanding of molecular biology has come from studies with microbes, and thus one finds that microbiology and molecular biology are often grouped in one field. Yet there is a distinct difference between how a molecular biologist and a microbiologist studies microbes. The former is interested in microbes as models and selects for study those organisms that are simplest and easiest to examine for the purpose at hand. The microbiologist, in comparison, is interested in microbes as organisms and studies them because of the things they do, in both natural and artificial environments. Thus the microbiologist studies a wide range of microbes, both simple and complex.

As knowledge increases it becomes ever more necessary for scientists to specialize and study only a small portion of a field. No one can know everything. Specialization has led to the development of subdisciplines of microbiology, each of which has its own area of study. These various ways of dividing the field are shown in Table 1.1.

It should be pointed out that these categories are somewhat arbitrary and may overlap. The important thing to remember is that the scientists working in all of these fields are microbiologists and ultimately relate their work to the broader problems of general microbiology.

TABLE 1.1 Subdisciplines of Microbiology

Taxonomically Oriented	Habitat Oriented	Problem Oriented
Virology	Aquatic microbiology	Microbial ecology
Bacteriology	Soil microbiology	Medical microbiology
Phycology (or algology)	Marine microbiology	Agricultural microbiology
Mycology		Industrial microbiology
Protozoology		Immunology

Supplementary Readings

Abelson, J., and **E. Butz,** eds. 1980. Recombinant DNA. Science 209: 1317–1438. Articles on many new details in molecular biology.

Brock, T. D., ed. and trans. 1975. Milestones in microbiology. American Society for Microbiology, Washington, D.C. The key papers of Pasteur, Koch, and others are translated, edited, and annotated for the beginning student.

Bulloch, W. 1938. The history of bacteriology. Oxford University Press, London. This book, still in print, is the standard history of bacteriology.

Davis, B. D., R. Dulbecco, H. N. Eisen, and **H. S. Ginsberg,** 1980. Microbiology, 3rd ed. Harper & Row, Hagerstown, Md. A useful advanced reference text, which emphasizes the molecular and medical aspects of microbiology.

Dixon, B. 1976. Magnificent microbes. Atheneum, New York. Nontechnical survey of the many roles of microorganisms in daily life.

Dobell, C., ed. and trans. 1932. Anton van Leeuwenhoek and his "little animals." Constable & Co., London (1960. Dover Publications, New York.) The best introduction to the life and times of Leeuwenhoek.

Dubos, R. J. 1976. Louis Pasteur: free lance of science. Charles Scribner's Sons, New York. (Reprint of 1950 ed. Little, Brown, Boston.) A beautifully written elementary account of Pasteur's life and work.

Hooke, R. 1665. Micrographia. Royal Society, London. (1961. Dover Publications, New York.) The pioneering work in microscopy.

Miller, B. M., and **W. Litsky.** 1976. Industrial microbiology. McGraw-Hill, New York. Introductory textbook concerning areas with direct use of microbiological activities.

The Procaryotic Cell

In this chapter we present current ideas of the structure of representative procaryotic cells. We discuss the kinds of structures we can see, and their chemical nature. Our ability to see structural details of cells depends greatly upon the tools available to us. Early microscopes were crude, and knowledge of cell structure was correspondingly limited. Today, highly sophisticated microscopes have opened up new vistas to us. Although we see structural details in terms of images created by our instruments, we must remember that these structures are composed of particular chemical components. Cells, like houses, are built by connecting simple building blocks in various ways to create more complex structures. We could learn how a house is constructed by tearing it down bit by bit and examining the building blocks as they fall apart. In an analogous way, we can use this method to discover how cells are put together. Our goals are to describe the chemical building blocks of various cell structures and to discover how they are connected. Because a knowledge of structure is basic to understanding cell function, the information in this chapter lays a cornerstone on which we shall build in succeeding chapters.

2.1

Seeing the Very Small

The compound microscope The compound microscope has been of crucial importance for the development of microbiology as a science and remains a basic tool of routine microbiological research. Four types of compound microscopes are commonly used in microbiology: bright-field, phase-contrast, dark-field, and fluorescence. The **bright-field microscope** is most commonly used in elementary microbiology courses. With this microscope, objects are visualized because of the contrast differences that exist between them and the surrounding medium. Contrast differences arise because cells absorb or scatter

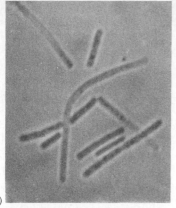
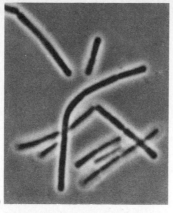
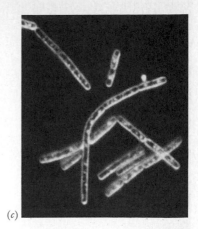

(a) (b) (c)

FIGURE 2.1 Photomicrographs of *Bacillus cereus* by (a) bright-field, (b) phase-contrast, and (c) dark-field microscopy. Magnification, 1500×.

light in varying degrees. Most bacterial cells are difficult to see well with the bright-field microscope because of their lack of contrast (Figure 2.1*a*) with the surrounding medium. Staining, which will soon be discussed, is commonly used to increase contrast. But because staining either kills or greatly modifies cells, it is desirable to avoid its use for many purposes.

The **phase-contrast microscope** was developed in order to make it possible to see small cells easily, even without staining. It is based on the principle that cells bend some of the light rays that pass through them so that they differ in refractive index from their surrounding medium. This difference can be used to create an image of much higher degree of contrast (Figure 2.1*b*) than can be obtained in the bright-field microscope. The phase-contrast microscope is now almost universally used in microbiological research laboratories. Because of its high cost and more complicated adjustments, however, it generally is not used in elementary microbiology courses. The principle of the phase-contrast microscope is shown in Appendix 5 (see Figure A5.2).

The **dark-field microscope** is an ordinary bright-field microscope in which the light system has been modified: light reaches the specimen from the sides only, and does not directly go to the lens. The only light reaching the lens is that which is scattered by the specimen; thus the specimen appears light on a dark background (Figure 2.1*c*). Dark-field microscopy makes possible the observation of living organisms or particles that are too small to be seen with bright-field or phase-contrast microscopes. However, only the outlines of objects are seen with this method, and internal structures are either invisible or may not appear accurately. Dark-field microscopy is an excellent means of observing motility of organisms and has been used widely for analyzing behavior of motile organisms (Sections 2.7 and 2.8; see also Appendix 5).

A fluorescent substance emits light of one color when light of another color shines upon it. The **fluorescence microscope** is used to visualize specimens that fluoresce, either because of the presence within them of natural fluorescent substances (for example, chlorophyll fluoresces brilliant red), or because they have been treated with fluorescent dyes. Fluorescence microscopy is widely used in microbial ecology and in immunology, as discussed in Sections 13.4 and 16.3 (see also Color Plate 7).

Staining **Positive staining** involves the use of dyes to stain cells and increase their contrast so that they can be more easily seen in the bright-field microscope. Dyes are organic compounds, and each dye used has an affinity for specific cellular materials. For example, many commonly used dyes are positively charged (cationic) and combine strongly with negatively charged cellular constituents such as nucleic acids and acidic polysaccharides. Examples of cationic dyes are methylene blue, crystal violet, and safranin. Since cell surfaces are generally negatively charged, these dyes combine with structures on the surface of cells and hence are excellent general stains. Other dyes (for example, eosin, acid fuchsin, Congo red) are negatively charged (anionic) and combine with positively charged cellular constituents, such as many proteins. Sudan black is a fat-soluble dye and combines with fatty materials in cells, thus revealing the presence and location of these substances. When a single dye is added to a specimen, the procedure is referred to as a **simple stain.**

Negative staining is the reverse of the usual staining procedure: the cells are left unstained but the background is stained, so that the cells are seen in outline. The substance used for negative staining is an opaque material that has no affinity for cellular constituents and merely surrounds the cells. Suitable negative staining materials include india ink (which is a suspension of colloidal carbon particles) and nigrosin (a black, water-insoluble dye). Negative staining is an excellent means of increasing contrast, but it is a little tricky to carry out properly, since the specimen-stain combination must be spread out in a very thin layer but must not dry out. Negative staining is most commonly used for revealing the presence of capsules around bacterial cells (see Figure 2.35a).

Differential stains are so named because they are procedures which do not stain all kinds of bacteria equally. These procedures may either be positive or negative stains. For example, there are a number of positive stain techniques which reveal endospores by making the endospore and the cell body different colors. Cells with and without endospores can therefore be differentiated. Negative stains can be differential for features on the exterior of the cell, for example, capsules, as just discussed (Figure 2.35a). One of the most important staining procedures in bacteriology is the differential **Gram stain.** On the basis of their reaction to the Gram stain, bacteria can be divided into two groups, Gram-positive and Gram-negative. This staining procedure is of considerable importance in bacterial taxonomy, and it also indicates fundamental differences in cell-wall structure of bacteria (see Section 2.5). The Gram-stain procedure is outlined in Figure 2.2. Cells are first stained with crystal violet, washed, and treated with an iodine (I_2) solution. The I_2 forms a complex with the crystal violet, which serves to fix it in the cells. Decolorization is then performed, with either alcohol or ace-

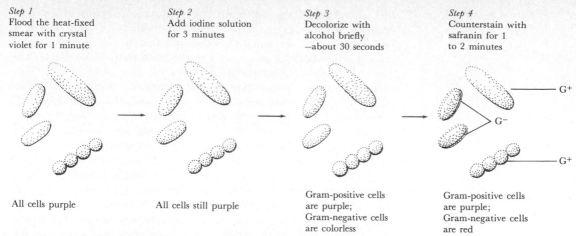

Step 1
Flood the heat-fixed smear with crystal violet for 1 minute

Step 2
Add iodine solution for 3 minutes

Step 3
Decolorize with alcohol briefly —about 30 seconds

Step 4
Counterstain with safranin for 1 to 2 minutes

All cells purple

All cells still purple

Gram-positive cells are purple; Gram-negative cells are colorless

Gram-positive cells are purple; Gram-negative cells are red

G⁺
G⁻
G⁺

FIGURE 2.2 Appearance of bacteria during steps in the Gram-stain procedure (color represents purple).

tone, substances in which the I_2-crystal violet complex is soluble. Some organisms (Gram-positive) are not decolorized, whereas others (Gram-negative) are. The essential difference between these two cell types is their comparative resistance to decolorization, the reasons for which will be discussed in Section 2.5. After decolorization, a red-colored counterstain, usually safranin, is used to render visible the now decolorized Gram-negative cells. Thus, upon microscopic examination, Gram-positive cells will appear blue, and Gram-negative cells red. It must be emphasized that the Gram stain procedure is a positive stain. The designations Gram-positive and Gram-negative refer to the ability of a specific organism to resist decolorization, not to the type of staining procedure employed.

A variety of other staining procedures have been developed over the past century of microbiology, but most are used only rarely, for special purposes, or not at all, having been rendered unnecessary due to the advent of the phase-contrast microscope.

The electron microscope To study the internal structure of pro-caryotes, an **electron microscope** is essential. With this microscope, electrons are used instead of light rays, and electromagnets function as lenses, the whole system operating in a high vacuum (see Appendix 5). The resolving power of the electron microscope is much greater than that obtained with the light microscope. Whereas with the light or phase-contrast microscope the smallest structure that can be seen is about 0.2 μm* in length, with the electron microscope

*A micrometer is the same as a *micron*, symbolized by μ. By international agreement, the word *micron* is being abandoned and replaced by *micrometer*. It should be remembered that, whenever the symbol μ precedes a unit, it means that the unit should be divided by 1 million. Thus 1 μm is one-millionth of a meter, or 10^{-6} m. or 0.001 mm. For very small dimensions, the unit *nanometer* (nm) is often used. The prefix *nano-* means that the unit following should be multiplied by 10^{-9}. The older unit *milli-micron* (mμ) is identical with the nanometer. Thus, 0.001 μm (1 mμ) equals 1 nm.

objects of 0.001 μm can readily be seen. The electron microscope enables us to see many substances of molecular size. However, electron beams do not penetrate very well and if one is interested in seeing internal structure, even a single bacterium is too thick to be viewed directly. Consequently, special techniques of thin sectioning are needed in order to prepare specimens for the electron microscope. For sectioning, cells must first be fixed, that is, treated with chemicals that prevent tissue distortion during the subsequent dehydration; which is accomplished by immersing the cells in an organic solvent. After dehydration the specimen is embedded in plastic. Thin sections are cut from this plastic with a special ultramicrotome, usually equipped with a diamond knife. A single bacterial cell, for instance, may be cut into five or more very thin slices, which are then examined individually with the electron microscope. To obtain sufficient contrast, the preparations are treated with special electron-microscope stains, such as osmic acid, permanganate, uranium, lanthanum, or lead. Because these materials are composed of atoms of high atomic weight, they scatter electrons well. Cellular structures stained with these materials give greatly increased contrast, and hence are better seen.

Fixation, embedding, and staining are potentially damaging treatments and may greatly alter cell structures. For this reason, great care must be taken in interpretation of the electron-microscope images obtained; artifacts are much more common and serious than in light microscopy. Procedures that work well for some organisms may fail with others; in general, different methods are used for procaryotic and for eucaryotic organisms.

If only the outlines of an organism need be observed, thin sections are not necessary, and whole cells can be mounted on a grid. To increase contrast of whole cells, **shadowing** is usually done. This involves coating the specimen with a thin layer of a metal such as palladium, chromium, or gold. The metal is deposited upon the specimen from one side so that a shadow is created, and the object is thus seen to have thickness and shape. Shadowing is often used to observe surface appendages on bacteria (see Figures 2.24 and 2.33).

Another way to achieve contrast with the electron microscope is by **negative staining.** The same principle applies as in negative staining with the light microscope; a substance is used that does not penetrate the structure but scatters electrons. One of the most commonly used negative stains for electron microscopy is phosphotungstic acid. Examples of negatively stained preparations are shown in several of the figures in Chapter 10.

A recently developed procedure for electron microscopy is **freeze-etching,** which was designed to prevent the formation of artifacts by eliminating chemical fixation and embedding. The specimen to be examined is frozen without chemical treatment, and the frozen block is fractured with a knife in such a way that portions of cells are exposed. Carbon is deposited on the exposed surface to make a replica, and this replica is then examined, with the net result that one sees surface or internal structures of cells (Figure 2.37). Most cellular structures seen in thin sections of chemically fixed specimens are also seen in freeze-etched material, suggesting that these structures are not artifacts.

Another instrument recently developed for examining surface structures is the **scanning electron microscope** (Appendix 5). The material to be studied is coated with a thin film of heavy metal such as gold. The electron beam is directed down on the specimen and scans back and forth across it. Electrons scattered by the heavy metal are collected and activate a viewing screen to produce an image. With the scanning electron microscope even fairly large specimens can be observed, and the depth of field is very good. A wide range of magnifications can be obtained, from as low as $15\times$ up to about $100,000\times$, but only the surface of an object can be visualized. An example of the image obtained in this manner is given in Figure 3.10.

Electron microscopy is a highly developed art, and permits us to see cellular structures that cannot be seen in any other way. The greatly increased resolution obtained fully justifies the care, time, and expense involved in the preparation of electron micrographs.

2.2

Size and Form of Procaryotes

Microorganisms are small, as the name implies. The preceding section discussed practical solutions to some of the problems raised by this small size, namely microscope techniques which allow us to see these organisms. There are other implications to the small size of microorganisms which deserve consideration as well. First, it must be realized that there are certain advantages to a cell, either procaryote or eucaryote, in being small. The accumulation of nutrients and elimination of waste products by a cell involves the cell surface, especially the cell membrane. Put in another way, the volume of the cell, where the essential metabolic activities take place, communicates with the external environment through the action of the cell membrane. We may conclude that the internal cellular reactions are controlled to a large degree by the amount of membrane available to transport materials in and out of the cell. That is, we can see a relation between cell volume and cell surface area, which is a good measure of the amount of available membrane. The relation between the volume and surface area of an object is not constant. This point may be seen most easily in the case of a sphere, in which the volume is a function of the cube of the radius ($V = \frac{4}{3}\pi r^3$) and the surface area is a function of the square of the radius ($A = 4\pi r^2$). Therefore, a smaller sphere has a higher ratio of surface area to volume than does a larger sphere, and, to return to the biological example, a smaller cell should have more efficient exchange with its surroundings than a larger cell. This pressure for a cell to be small is limited by requirements for a certain necessary minimum volume to contain all the genetic information and biochemical apparatus such as enzymes and ribosomes. Cells are not perfect spheres, of course, and many procaryotes produce more membrane without getting larger by creating infoldings or invaginations of the cell membrane. Occasionally this tendency is quite extreme, as is seen in some photosynthetic procaryotes (Figure 2.12). Eucaryotic cells, being generally

(b) *Gloeocapsa*, 5 μm

(c) *Aphanothece*, 3 μm

(d) *Thiopedia*, 2 μm

(e) *Bacillus*, 0.8 μm

(f) *Streptococcus*, 0.5 μm

FIGURE 2.3 Photomicrographs to the same magnification of procaryotic cells of various sizes. The dimensions given are the cell widths. All magnifications, 2300×. (a) Bright-field; (b) and (c) Nomarski interference contrast; (d), (e), and (f) phase contrast photomicrographs.

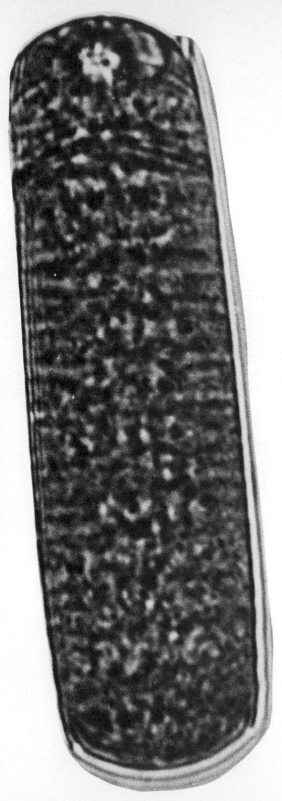

(a) *Oscillatoria*, 25 μm

larger than procaryotes, have compartmentalized cellular functions involving separate, internal membrane structures (see Chapter 3).

Although most procaryotic cells are small, there is a wide variation in size among different organisms (Figure 2.3). The procedure for measuring the size of microscopic objects is given in Appendix 5. Most bacteria have distinctive cell shapes, which remain more or less constant, although shape is influenced to some extent by the environment. Bacteria shaped like spheres are called **cocci** (singular **coccus**), whereas those shaped like cylinders are called **rods** (Figure 2.4). If a rod is many times longer than it is wide it is usually called a **filament.** Some bacteria are shaped like **spirals,** and a long spiral has the shape of a **helix** (Figure 2.4). The shape of a cell definitely affects its behavior and stability. Cocci, for instance, being round, become less distorted upon drying and thus can usually survive more severe desiccation than can rods or spirals. Rods, on the other hand, have more surface exposed per unit volume than cocci do and thus can more readily take up nutrients from dilute solutions. The spiral forms, if motile, move by a corkscrew motion, which means that they meet with less resistance from the surrounding water than do motile rods, in the same way that a screw moves into hardwood more easily than a nail. Square bacteria were recently discovered by A. Walsby. These unusual organisms are quite distinctive in their straight sides and right-angle corners (Figure 2.5). To date they have been found only in extremely salty environments, such as brines used for commercial production of salt (see Section 8.2).

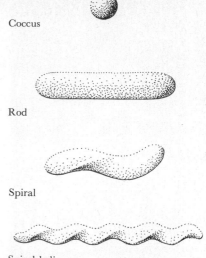

Coccus

Rod

Spiral

Spiral helix

FIGURE 2.4 Representative cell shapes in procaryotes.

FIGURE 2.5 Square bacteria. (a) Nomarski interference micrograph of cells which have formed a sheet by remaining associated through six divisions. Bar = 5.0 μm. (b) Electron micrograph of thin section of lysed cells showing that cell shape stays constant. Bar = 0.5 μm. (From Kessel, M., and Y. Cohen. 1982. J. Bacteriol. 150:851–860.)

(a)

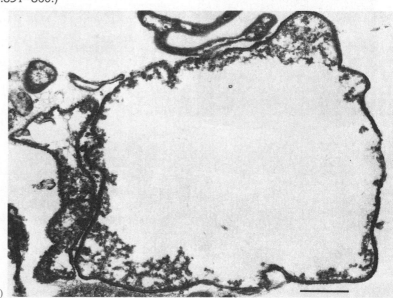

(b)

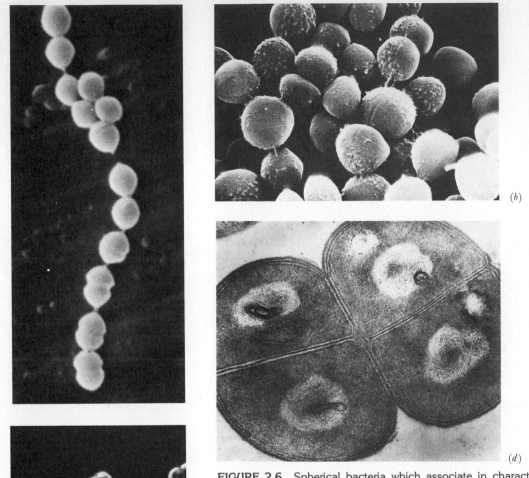

(a)

(b)

(c)

(d)

FIGURE 2.6 Spherical bacteria which associate in characteristically different ways. (a) *Streptococcus,* scanning electron micrograph. (Courtesy of Bryan Larsen, University of Iowa.) (b) *Staphylococcus,* scanning electron micrograph. Magnification, 14,800×. (From Umeda, A., et al. 1980. J. Bacteriol. 141:838–844.) (c) *Sarcina* (now *Micrococcus*) phase contrast micrograph. (From Holt, S., and E. Canale-Parola. 1967. J. Bacteriol. 93:399–410.) (d) *Sarcina* transmission electron micrograph. (From Beveridge, T. 1980. Can. J. Microbiol. 26:235–242.)

When cells divide they often remain attached to each other, and the manner of attachment is usually characteristic of both the organism and the type of division the cell has undergone. Thus, many coccus-shaped organisms form chains (*Streptoccoccus* is the best known example) by dividing always along the same axis (Figure 2.6*a*). The length of such chains may be short (2 to 4 cells) or long (over 20 cells). Some cocci divide along two axes at right angles to each other, leading to the formation of sheets of cells. If there is no pattern to the orientation of successive divisions, an irregular clump will be formed;

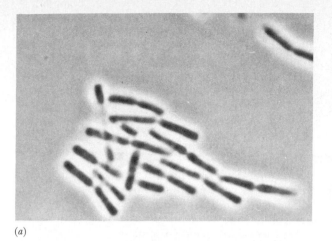

(a)

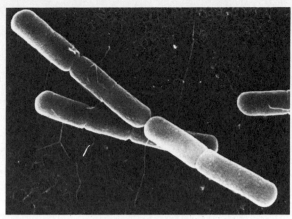

(b)

FIGURE 2.7 Chains of *Bacillus* (a) phase contrast micrograph. Magnification, 2200×. (From Brock, T. and K. Brock. 1978. Basic Microbiology with applications, 2nd ed. Prentice-Hall, Englewood Cliffs, N.J.) (b) Scanning electron micrograph. Magnification, 9600×. (From Umeda, A., and K. Amako. 1980. J. Gen. Microbiol. 118:215–221.)

such a random cluster is characteristic of *Staphyloccoccus* (Figure 2.6*b*). *Sarcina* (Figure 2.6*c*) is an example of a third group of cocci, which divide in three axes and form cube-shaped packets.

Rods always divide in only one plane and hence may form chains [such as in many *Bacillus* (Figure 2.7) species] but not more complicated arrangements. Spirally shaped organisms also divide in only one plane, but they usually separate immediately and do not form chains.

2.3

Detailed Structure of the Procaryotic Cell

The electron microscope has made possible a careful study of cell construction, and we currently have an excellent understanding of the fine structure of the cell. It has been through electron-microscopic study that we have become aware of the profound differences between procaryotic and eucaryotic cells. An electron micrograph of a thin section of a typical Gram-positive procaryotic cell is shown in Figure 2.8. Cellular components can be divided into two groups, **invariant,** found in all procaryotes, and probably essential for life, and **variant,** found in some but not all cells, and probably involved

FIGURE 2.8 Electron micrograph of a thin section of a typical bacterium, Gram-positive *Bacillus subtilis*. The cell has just divided and two membrane-containing structures are attached to the cross wall. The light region in the middle contains the DNA. Magnification, 31,200×. (Courtesy of Stanley C. Holt.)

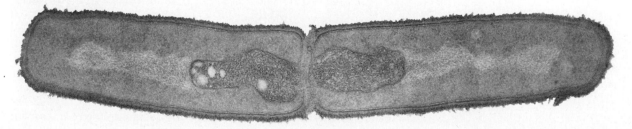

in more specialized functions. Invariant cell structures include the *cell membrane, ribosomes,* and *nuclear region.* Variant cell structures include the *cell wall* (present in most but not all organisms), *flagella, pili, capsules, slime layers, holdfasts, inclusion bodies, gas vesicles,* and *spores.* Under appropriate conditions all of these structures can be visualized in electron micrographs, and their arrangements and distribution in cells analyzed.

2.4

Cell Membranes

The plasma membrane   The **plasma membrane,** sometimes called the cell membrane, is a thin structure that completely surrounds the cell. This vital structure is the critical barrier separating the inside of the cell from its environment. If the membrane is broken, the integrity of the cell is destroyed and death usually occurs.

Thin sections of the plasma membrane can generally be visualized with the electron microscope; a representative example is seen in Figure 2.9. To prepare the plasma membrane for study with the electron microscope, the cells must first be treated with osmic acid or some other electron-dense material that combines with components of the membrane. By careful high-resolution electron microscopy, the plasma membrane appears as two thin lines separated by a lighter area (Figure 2.9). This basic and universal membrane is often called a **unit membrane** to distinguish it from membranes of more complex structure. The main components of the plasma membrane are phospholipids and proteins, the phospholipids forming the basic structure of the membrane. Phospholipid molecules (Figure 2.10) disperse themselves in water in such a way that the water-insoluble (hydrophobic) groups associate together, and the ionic (hydrophilic) groups associate together, leading almost automatically to the formation of a double-layered membrane, or bimolecular leaflet, as shown in Figure 2.11. The major proteins of the membrane are hydrophobic, and associate with and become embedded in the phos-

FIGURE 2.9   Electron micrograph of a thin section of a plasma membrane. Note the distinct double track. Magnification, 221,000×. (Courtesy of Walther Stoeckenius.)

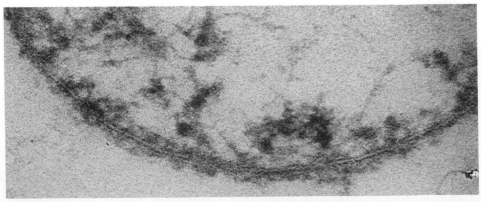

Fatty acids (e.g., palmitic acid)

$$H_2C-O-\overset{\overset{O}{\|}}{C}\begin{matrix} CH_2 \\ \\ \end{matrix}\begin{matrix} CH_2 \\ \\ \end{matrix}\begin{matrix} CH_2 \\ \\ \end{matrix}\begin{matrix} CH_2 \\ \\ \end{matrix}\begin{matrix} CH_2 \\ \\ \end{matrix}\begin{matrix} CH_2 \\ \\ \end{matrix}\begin{matrix} CH_2 \\ \\ \end{matrix}\begin{matrix} CH_3 \\ \\ \end{matrix}$$

Glycerol

$$HC-O-\overset{\overset{O}{\|}}{C}$$

$$H_2C-O-\overset{\underset{O^-}{|}}{P}-\overset{\underset{O}{\|}}{O}-CH_2-CH_2$$

Choline

$$^+N\begin{matrix}CH_3\\CH_3\\CH_3\end{matrix}$$

FIGURE 2.10 Structure of lecithin, a phospholipid. The fatty acids (in color boldface type) can be of various chain lengths and structures and are nonionic and hydrophobic (water insoluble). In phospholipids other than lecithin the choline is replaced by groups such as glycerol, ethanolamine, or serine.

pholipid matrix. Protein molecules are also bound to the ionic groups of the phospholipid, and water molecules congregate in a more or less ordered structure around the outside of the bimolecular leaflet. The structure of the plasma membrane is stabilized mainly by hydrogen and hydrophobic bonding. However, cations such as Mg^{2+} and Ca^{2+} also combine with some of the negative charges of the phospholipids and help stabilize the membrane structure.

The plasma membrane as a permeability barrier Despite its thinness, the plasma membrane functions as a tight barrier, so that the passive movement of polar solute molecules does not readily occur. Some small nonpolar and fat-soluble substances may penetrate cell membranes readily by becoming dissolved in the lipid phase of the membrane, but movement of most other molecules occurs only by means of specific transport systems, as will be described in Section 5.18. Ionized molecules, such as organic acids, amino acids, and inorganic salts do not readily bridge the membrane barrier partly because they are repelled by the electrical charge on the surface of the membrane, and partly because there are no holes in the membrane

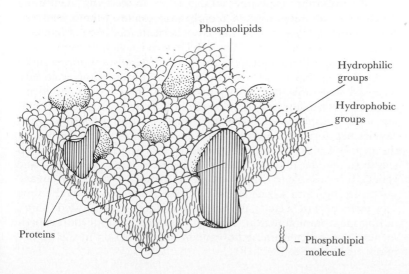

Phospholipids

Hydrophilic groups

Hydrophobic groups

Proteins

— Phospholipid molecule

FIGURE 2.11 Diagram of the construction of a cell membrane. The matrix is composed of phospholipids, with the hydrophobic groups directed inward and the hydrophilic groups toward the outside, where they associate with water. Embedded in the matrix are hydrophobic proteins. Hydrophilic proteins and other charged substances, such as metal ions, are attached to the hydrophilic surfaces. Although there are chemical differences, the overall structure shown is similar in both procaryotes and eucaryotes. (Redrawn from Singer, S. J., and G. L. Nicolson. 1972. Science 175:720–731.)

large enough for them to pass through. Even a substance as small as the hydrogen ion, H^+, does not readily breach the plasma membrane barrier passively, but must be transported (Section 4.8) because it is charged and is always hydrated, occurring as the larger hydronium ion (H_3O^+) in solution. One molecule which does freely penetrate this membrane is water itself, which is sufficiently small to pass between phospholipid molecules. Thus the plasma membrane is both a barrier to penetration of materials in general and an agent for selective transport or uptake of materials.

The interior of the cell consists of an aqueous solution of salts, sugars, amino acids, vitamins, coenzymes, and a wide variety of other soluble materials, which is called the **cell pool.** When the permeability barrier of the cell is destroyed, most of these materials are able to leak out, and only substances too large to pass through the cell-wall pores are retained. Components enter the pool either as nutrients taken up from the environment or as materials synthesized from other constituents in the cell.

Besides sustaining cell permeability, the procaryotic cell membrane plays a key role in cell respiration, as the enzymes associated with this process are part of the membrane (see Section 4.8).

Sidedness of the plasma membrane and the formation of membrane vesicles It is now well established that the two sides of the plasma membrane are different. Certain proteins embedded in the membrane project toward the inside, and others toward the outside, and of those that span the membrane, the part of the protein facing outward is different from that facing inward. This sidedness can be shown in part by careful electron microscopy, and in part by treating membranes with reagents that combine with specific chemical groups and then determining where the reagents have become attached. Thus the plasma membrane has distinctly different inner and outer surface characteristics, and this sidedness is of considerable importance in the functioning of the membrane (Sections 4.8 and 5.18).

An important experimental procedure in studying membrane function is the preparation of **membrane vesicles.** Membrane vesicles are spherical structures that have been formed from fragments of the plasma membrane. When a bacterial cell is broken under carefully controlled conditions, the plasma membrane forms small pieces that are flat at first, but immediately become converted into spherical structures, the vesicles. In essence, the edges of the membrane pieces become associated together via the hydrophobic bonding of the phospholipids. It has now been well established that when vesicles are formed under appropriate conditions the sidedness of the membrane is not altered: the surface of the membrane which was the inner face in the cell becomes the inner face of the membrane vesicle. It is possible to experimentally incorporate materials (sugars, salts, enzymes, and so on) inside the membrane vesicles at the time they form, and then study the behavior of these incorporated solutes during membrane function. This has provided an important experimental tool in determining how membranes function in the uptake of nutrients (see Section 5.18).

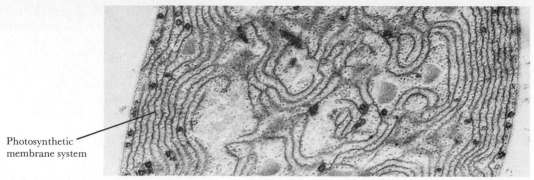

Photosynthetic
membrane system

FIGURE 2.12 Electron micrograph of a cyanobacterial cell (*Anabaena azollae*), showing the extensive array of photosynthetic membranes (thylakoids). Magnification, 14,800×. (From Lang, N. J. 1965. J. Phycol. 1:127.)

It is also possible to form structures analogous to membrane vesicles from pure phospholipids, or from phospholipid-protein mixtures. Such structures are called **liposomes,** and it is possible to synthesize functioning liposomes with a variety of proteins embedded in them. Hybrid liposomes can be made, in which proteins from several separate organisms can be incorporated into the same structure. These recent advances in the preparation and experimental manipulation of membrane structures have greatly increased our knowledge of membrane function.

Internal membranes   In addition to the membrane at the periphery of the cell, most procaryotes possess internal membranes, which can often be seen in electron micrographs. These internal membranes may be simply extensions or invaginations of the cell membrane, or they may be much more complicated. In photosynthetic procaryotes, the photosynthetic membranes often form an extensive internal membrane system (Figure 2.12; see also Figure 19.3), and the nitrifying (Figure 19.50) and the methane-oxidizing bacteria (Figure 19.42) also frequently have elaborate internal membranes (see Section 6.2 for a discussion of photosynthetic membrane systems).

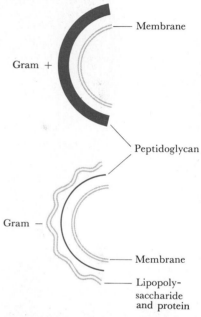

FIGURE 2.13 Comparison of the cell walls of Gram-positive and Gram-negative bacteria.

2.5

Cell Wall

One of the most important structural features of the procaryotic cell is the cell wall, which confers rigidity and shape. The procaryotic cell wall is chemically quite different from that of any eucaryotic cell, and this is one of the features distinguishing procaryotic from eucaryotic organisms. The cell wall is difficult to visualize well with the light microscope but can readily be seen in thin sections of cells with the electron microscope. Gram-positive and Gram-negative cells differ considerably in the structure of their cell walls as is shown diagrammatically in Figure 2.13 and in the electron micrographs of Figure 2.14. The Gram-negative cell wall is a multilayered structure and quite complex, while the Gram-positive cell wall consists

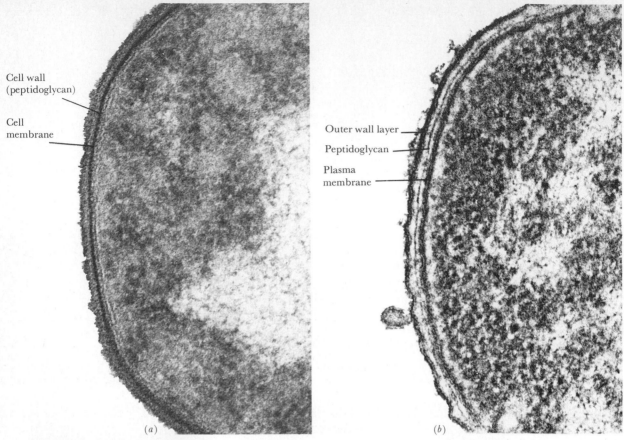

Cell wall
(peptidoglycan)

Cell
membrane

Outer wall layer

Peptidoglycan

Plasma
membrane

(a)

(b)

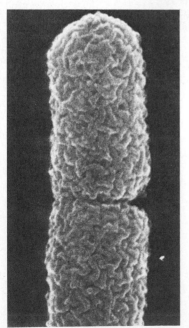

(c)

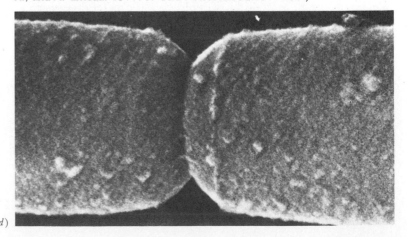

(d)

FIGURE 2.14   Cell-wall electron micrographs: (a) Gram-positive, *Arthro-bacter crystallopoietes*. Magnification, 126,000×. (Courtesy of J. L. Pate.) (b) Gram-negative, *Leucothrix mucor*. Magnification, 165,000×. (Micrograph by T. D. Brock and S. F. Conti.) (c) and (d) Scanning electron micrographs of Gram-negative (*E. coli*) and Gram-positive (*Bacillus subtilis*) bacteria. Magnification c, 56,700×; d, 78,625×. (From Amako, K., and A. Umeda. 1977. J. Gen. Microbiol. 98:297–299.)

of a single layer and is often much thicker. For most microbiologists today it is this different appearance in electron micrographs which distinguishes Gram-positive and Gram-negative cells, not the traditional staining technique described in Section 2.1. Comparison of parts *c* and *d* of Figure 2.14 shows that there is also a significant textural difference between Gram-positive and Gram-negative exteriors as revealed by negative staining procedures.

The rigid layer of both Gram-negative and Gram-positive bacteria is very similar in chemical composition. Called **peptidoglycan,*** this layer is a thin sheet composed of two sugar derivatives; *N*-acetylglucosamine and *N*-acetylmuramic acid, and a small group of amino acids, consisting of L-alanine, D-alanine, D-glutamic acid, and either lysine or diaminopimelic acid (DAP) (Figure 2.15). These constituents are connected to form a repeating structure, the *glycan tetrapeptide* (Figure 2.16). The overall structure of the peptidoglycan layer, showing the orientation of the sugars and amino acids, is given in Figure 2.17*c*.

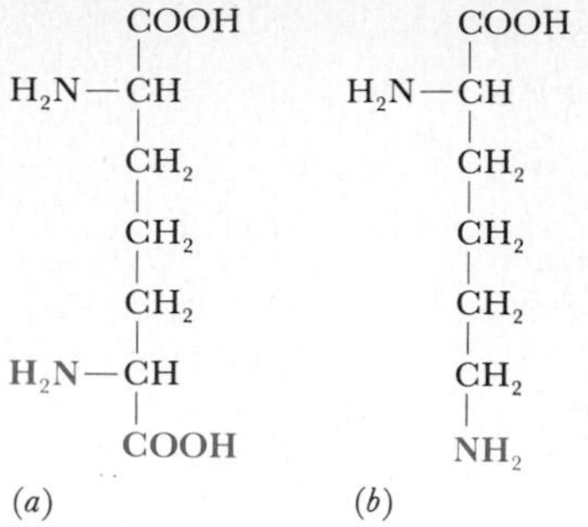

FIGURE 2.15   (a) Diaminopimelic acid. (b) Lysine.

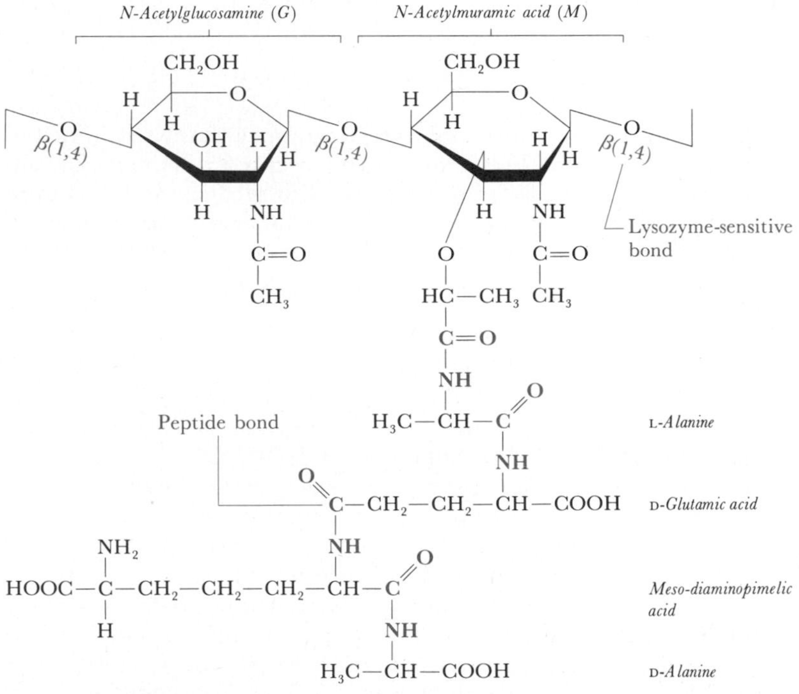

FIGURE 2.16   Structure of one of the repeating units of the peptidoglycan cell-wall structure. The structure given is that found in *Escherichia coli* and most other Gram-negative bacteria. In some bacteria, other amino acids are found.

The basic structure is in reality a thin sheet in which the *glycan* chains formed by the sugars are connected by *peptide* cross-links formed by the amino acids. The glycosidic bonds connecting the sugars in the glycan chains are very strong, but these chains alone cannot provide rigidity in all directions. The full strength of the

*The repeating structure is sometimes called *murein* or *mucopeptide* instead of peptidoglycan.

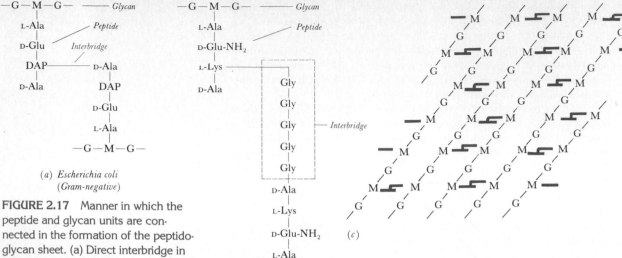

FIGURE 2.17 Manner in which the peptide and glycan units are connected in the formation of the peptidoglycan sheet. (a) Direct interbridge in Gram-negative bacteria; (b) Glycine interbridge in *Staphylococcus aureus* (Gram-positive); (c) Overall structure of the peptidoglycan. G, N-acetylglucosamine; M, N-acetylmuramic acid; heavy color lines in c, peptide crosslinks.

peptidoglycan structure is obtained when these chains are joined by peptide cross-links. This cross-linking occurs to characteristically different extents in different bacteria, with greater rigidity coming from more complete cross-linking. In Gram-negative bacteria, cross-linkage usually occurs by direct peptide linkage of the amino group of diaminopimelic acid to the carboxyl group of the terminal D-alanine (Figure 2.17a). In Gram-positive bacteria, cross-linkage is usually by a peptide interbridge, the kinds and numbers of cross-linking amino acids varying from organism to organism. In *Staphylococcus aureus,* the best studied organism, each interbridge peptide consists of five molecules of the amino acid glycine connected by peptide bonds (Figure 2.17b). Although the details are not yet known, it seems evident that the shape of a cell is determined by the lengths of the peptidoglycan chains and by the manner and extent of cross-linking of the chains.

The peptidoglycan structure is present only in procaryotes, and is found in the wall of virtually all species. The sugar *N*-acetylmuramic acid is never found in eucaryotes and the amino acid diaminopimelic acid (DAP) is also never found in eucaryotic walls. However, not all procaryotic organisms have DAP in their peptidoglycan. This amino acid is present in all Gram-negative bacteria and in some Gram-positive species, but most Gram-positive cocci have lysine instead of DAP, and a few other Gram-positive bacteria have other amino acids. Another unusual feature of the procaryotic cell wall is the presence of two amino acids that have the D configuration, D-alanine and D-glutamic acid. In proteins, amino acids are always in the L configuration. (As seen in Figure 2.16, there are two molecules of alanine in the cell-wall peptide, but only one is of the D configuration.)

The formation of the peptide cross-links involves an unusual type of peptide bond formation, called **transpeptidation,** which is also noteworthy because it is inhibited by the antibiotic *penicillin.* In *S. aureus,* the terminal amino group of the pentaglycine bridge is transferred to the carboxyl group of the subterminal D-alanine to form a peptide bond, while the terminal D-alanine is released as the free amino acid. The peptide bond between the two molecules of D-alanine serves to activate the subterminal D-alanine, thereby favoring its reaction with the glycine to form the cross-link. This reaction occurs outside the cell membrane, where ATP and the normal machinery of protein synthesis are not present, and the transpeptidation reaction replaces these requirements. It is noteworthy that the initial peptidoglycan unit has two D-alanine residues, but only one ends up in the final wall structure. Not all of the glycan chains have peptide crosslinks (see Figure 2.17). In those that do not, the terminal D-alanine is removed by a specific enzyme, D-alanine carboxypeptidase, and this reaction is also inhibited by penicillin. (However, *S. aureus* lacks a carboxypeptidase and the terminal D-alanine is retained.) Cycloserine is another antibiotic which functions by blocking transpeptidation.

Inhibition of transpeptidation by penicillin thus leads to the formation of peptidoglycan which lacks strength. The further damage to the cell, resulting in lysis and death, occurs because there are enzymes in the cell (called **autolysins**) that are involved in the opening up of the peptidoglycan structure as growth occurs. These enzymes continue to act, but because new peptidoglycan cross-links cannot occur, the cell wall becomes progressively weaker and osmotic lysis occurs. As we will see soon, lysis by penicillin can be prevented by adding an osmotic stabilizing agent such as sucrose. Under such conditions, continued growth in the presence of penicillin leads to the formation of protoplasts or spheroplasts. Pencillin-induced lysis only occurs with growing cells. In nongrowing cells, action of autolysins does not occur, so that breakdown of the cell-wall peptidoglycan is prevented.

Several generalizations regarding peptidoglycan structure can be made. The glycan portion is uniform throughout the procaryotic world, with only the sugars *N*-acetylglucosamine and *N*-acetylmuramic acid being present, and these sugars are always connected in β-1,4 linkage. The tetrapeptide of the repeating unit shows variation only in one amino acid, the lysine-diaminopimelic acid alternation. However, the D-glutamic acid at position 2 can be hydroxylated in some organisms.

The greatest variation occurs in the interbridge. Any of the amino acids present in the tetrapeptide can also occur in the interbridge, but in addition, a number of other amino acids are found in the interbridge, such as glycine, threonine, serine, and aspartic acid. However, certain amino acids are never found in the interbridge: branched-chain amino acids, aromatic amino acids, sulfur-containing amino acids, and histidine, arginine, and proline.

In Gram-positive bacteria, as much as 90 percent of the wall consists of the peptidoglycan, although another kind of constituent, teichoic acid (discussed below), is usually present in small amounts.

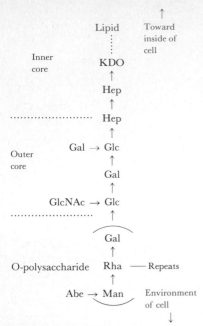

Lipid
Toward inside of cell
↑
⋮
KDO
↑
Hep
↑
Hep
↑
Gal → Glc
↑
Gal
↑
GlcNAc → Glc
↑
Gal
↑
Rha ——— Repeats
↑
Abe → Man
Environment of cell
↓

Inner core

Outer core

O-polysaccharide

FIGURE 2.18 Structure of lipopoly-saccharide of a *Salmonella.* KDO, ketodeoxyoctonate; Hep, heptose; Glc, glucose, Gal, galactose; GlcNAc, N-acetylglucosamine; Rha, rhamnose; Man, mannose; Abe, abequose.

In Gram-negative bacteria, only 5 to 20 percent of the wall is pep-tidoglycan, the rest of the wall consisting of lipid, polysaccharide, and protein, usually present in a layer outside the peptidoglycan layer. Finally, it should be noted that not all bacterial walls contain peptidoglycan. This structure has been shown to be absent from the walls of methane-producing bacteria (Section 19.20), the halobac-teria (Section 8.2), and *Sulfolobus* (Section 19.19). The cell walls of these organisms apparently have different chemical constructions.

Outer-wall (lipopolysaccharide) layer In most Gram-negative bacte-ria, the outer wall layer (illustrated in Figure 2.14*b*) exists as a true unit membrane. However, the unit membrane of the outer wall is not constructed solely of phospholipid, as is the plasma membrane, but also contains additional lipid plus polysaccharide and protein. The lipid and polysaccharide are intimately linked in the outer layer to form specific **lipopolysaccharide** (LPS) structures. Because of the presence of lipopolysaccharide, the outer layer is frequently called the lipopolysaccharide or LPS layer. Although complex, the chemical structures of some LPS layers are now understood. As seen in Figure 2.18, the polysaccharide consists of two portions, the core polysac-charide and the O-polysaccharide. In *Salmonella,* where it has been best studied, the **core polysaccharide** consists of ketodeoxyoctonate, seven-carbon sugars (heptoses), glucose, galactose, and *N*-acetylglu-cosamine. Connected to the core is the **O-polysaccharide,** which usually contains galactose, glucose, rhamnose, and mannose (all six-carbon sugars) as well as one or more unusual dideoxy sugars such as abequose, colitose, paratose, or tyvelose. These sugars are con-nected in four- or five-sugar sequences, which often are branched. When the sugar sequences are repeated, the long O-polysaccharide is formed. The structure of the **lipid** is least well known. It is not a nor-mal glycerol lipid, but instead the fatty acids are connected by ester linkage to *N*-acetylglucosamine. Fatty acids frequently found in the lipid include β-hydroxymyristic, lauric, myristic, and palmitic acids. In the outer layer, the LPS associates with phospholipids to form the outer portion of the unit membrane (Figure 2.19).

   Although we usually study the outer membrane separately, it

FIGURE 2.19 Arrangement of lipo-polysaccharide and phospholipid in the formation of the outer layer of Gram-negative bacteria. The structure formed is a true unit membrane. The phospholipid of the outer face of this membrane adheres by hydrophobic bonds to the other phospholipids, which are attached covalently to the peptidoglycan.

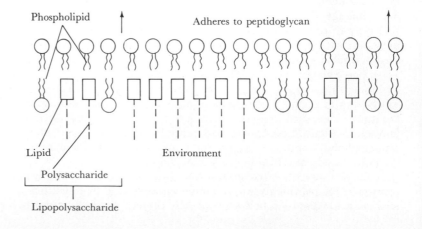

Phospholipid

Adheres to peptidoglycan

Lipid

Environment

Polysaccharide

Lipopolysaccharide

has been observed that the cell membrane and outer membrane are connected in a few places by lipoproteins, largely through hydrophobic associations. However, it is rather easy to separate the membranes experimentally, implying that this connection does not occur in many places over the surface of the cell.

In part, the functional importance of the outer layer is that it serves as an outer barrier, through which materials must penetrate if they are to reach the cell. The outer layer is permeable to small molecules, but not to enzymes or other large molecules. In fact, one of the main functions of the outer layer may be its ability to keep certain enzymes, which are present outside the peptidoglycan, from leaving the cell. These enzymes are present in an area called the **periplasmic space** (Figure 2.20); these periplasmic enzymes are probably of importance in the uptake of nutrients into the cell, as discussed in Section 5.18. Further, the outer layer in many Gram-negative bacteria possesses toxic properties and is responsible for some of the symptoms of infection (see Section 15.4). A type of toxic substance called **endotoxin** is either part of or equivalent to the LPS.

Teichoic acids Although Gram-positive bacteria do not have a lipopolysaccharide outer layer, attached to their cell wall they generally have acidic polysaccharides called **teichoic acids** (from the Greek word *teichos,* meaning "wall"). Teichoic acids contain repeating units of either glycerol or ribitol (both polyols). The *polyol units* are connected by phosphate esters and usually have other sugars and D-alanine attached. Because they are negatively charged, teichoic acids are partially responsible for the negative charge of the cell surface as a whole. Another function of certain teichoic acids is the part they play in the regulation of cell wall enlargement during growth and cell division. As mentioned above, cell-wall growth involves enzymes, called *autolysins,* which open up the existing cell wall to make room for new subunits. It is crucial for the cell that autolysin activity be regulated, because if breakdown of existing cell walls were to occur before synthesis of new cell walls, large holes would develop in the cell wall and lysis (cell rupture) would occur. Teichoic acids have been shown to regulate autolysin action, thus keeping it in balance with cell-wall synthesis. This process is discussed in more detail below.

Certain glycerol-containing acids are bound to the membrane lipid of Gram-positive bacteria; because these teichoic acids are intimately associated with lipid, they have been called *lipoteichoic acids.* Despite the fact that these teichoic acids are bound to the membrane rather than to the cell wall, they seem to extend to the outside of the cell, as shown by the fact that agents (such as antibodies) which bind to these teichoic acids can combine with them even in whole cells.

Relation of cell-wall structure to the Gram stain Are the structural differences between the cell walls of Gram-positive and Gram-negative bacteria responsible in any way for the Gram-stain reaction? Recall that in the Gram reaction an insoluble crystal violet-iodine complex is formed inside the cell and that this complex is extracted by alcohol from Gram-negative but not from Gram-positive bacteria. Gram-

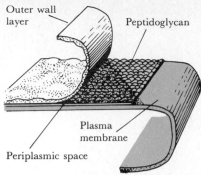

FIGURE 2.20 Diagram of the cell-wall structure of a simple Gram-negative bacterium, showing the location of the periplasmic space between the peptidoglycan layer and the outer membrane. In some Gram-negative bacteria, more layers exist. (Redrawn from Nanninga, N. 1970. J. Bacteriol. 101:297.)

positive bacteria have very thick cell walls, which become dehydrated by the alcohol. This causes the pores in the walls to close, preventing the insoluble crystal violet-iodine complex from escaping. In Gram-negative bacteria, the solvent readily dissolves in and penetrates the outer layer, and the thin peptidoglycan layer also does not prevent solvent passage. However, Gram reaction is not related directly to the bacterial cell-wall chemistry, since yeasts, which have a thick cell wall but of an entirely different chemical composition, are also Gram-positive organisms. Thus it is not the chemical constituents but the physical structure of the wall that confers Gram-positivity.

Osmosis, lysis, and protoplast formation   In addition to conferring shape, the cell wall is essential in maintaining the integrity of the cell. Most bacterial environments have solute concentrations considerably lower than the solute concentration within the cell, which is approximately 10 millimolar (mM). Water passes from regions of low solute concentration to regions of high solute concentration in a process called **osmosis.** Thus there is a constant tendency throughout the life of the cell for water to enter, and the cell would swell and burst were it not for the strength of the cell wall. This can be dramatically illustrated by treating a suspension of bacteria with the enzyme **lysozyme.** Lysozyme is found in tears, saliva, other body fluids, and egg white. It hydrolyzes the cell-wall polysaccharide (Figure 2.16), thereby weakening the wall. Water then enters, the cell swells, and bursts, a process called **lysis** (Figure 2.21).

If the proper concentration of a solute which does not penetrate the cell, such as sucrose, is added to the medium, the solute concentration outside the cell balances that inside. Under these conditions, lysozyme still digests the cell wall, but lysis does not occur, and an intact **protoplast*** is formed (Figure 2.21). If such sucrose-stabilized protoplasts are placed in water, lysis occurs immediately.

Osmotically stabilized protoplasts are always spherical in liquid medium, even if derived from rod-shaped organisms, further emphasizing that the shape of the intact cell is conferred by the cell wall. However, the wall is not a completely inelastic structure, and is able to stretch and contract to some extent. Thus it might be viewed as somewhat like the skin of a football.

If solute concentration is higher in the medium than in the cells, water flows out, the cells become dehydrated, and the protoplast collapses, a process called **plasmolysis.** This is one reason that foods can be protected from bacterial spoilage by curing them with strong salt or sugar solutions (see Section 8.2).

Although most bacteria cannot survive without their cell walls, a few organisms are able to do so; these are the mycoplasmas, a group of organisms that cause certain infectious diseases. Mycoplasmas are essentially free-living protoplasts, and are probably able to survive without cell walls either because they have unusually tough mem-

---

*Strictly defined a *protoplast* is a structure completely devoid of cell wall and consists of the cell membrane and all intracellular components. Spherical, osmotically sensitive bodies can also be formed in which the cell wall is only partially removed. Such spherical structures with some cell-wall fragments still attached are usually called *spheroplasts,* to distinguish them from true protoplasts.

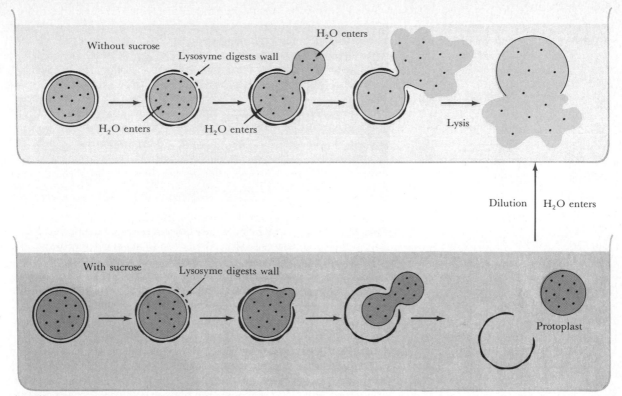

FIGURE 2.21 Lysis of protoplast in dilute solution and stabilization by sucrose.

branes or because they live in osmotically protected habitats, such as the animal body. These organisms will be discussed in Section 19.31.

Although the cell wall may resemble a smooth surface when viewed in the electron microscope, it is actually full of pores, through which water and various chemical materials pass. It is a barrier only to very large molecules such as proteins and nucleic acids, and particles such as viruses. The LPS layer acts as a molecular sieve to retain enzymes and other large molecules within the cell, and to keep large molecules outside the cell from entering. One estimate is that the LPS is a barrier to molecules larger than about 700 to 800 molecular weight. The peptidoglycan layer also acts like a molecular sieve, and may have pores similar in size to those of the LPS, or smaller. The plasma membrane, on the other hand, is passively impermeable to much smaller molecules than are either the LPS or peptidoglycan layers, as we noted in Section 2.4.

Cell-wall synthesis and cell division   When a cell enlarges during the division process, new cell-wall synthesis must take place, and this new wall material must be added in some way to the preexisting wall. This process can occur in different ways, as shown in Figure 2.22a. Small openings in the macromolecular structure of the wall are created by enzymes called autolysins, similar to lysozyme, that are produced within the cell. New wall material is then added across the openings.

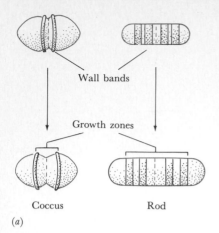

Wall bands

Growth zones

Coccus   Rod

(a)

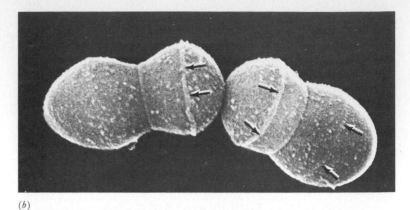

(b)

FIGURE 2.22 (a) Localization of new cell-wall synthesis during cell division. In the cocci, new cell-wall synthesis is localized at only one point, whereas in rod-shaped bacteria it occurs at several locations along the cell wall. (b) Scanning electron micrograph of *Streptococcus hemolyticus* showing division bridges (arrows). Magnification, 29,800×. (From Amako, K., and A. Umeda. 1977. J. Gen. Microbiol. 98:297–299.)

The junction between new and old peptidoglycan forms a ridge on the cell surface of Gram-positive bacteria (Figure 2.22*b*) analogous to a scar. Figure 2.22*b* also gives another example of incomplete separation of cells following division. In nongrowing cells spontaneous lysis can occur, a result of cell enzymes hydrolyzing the cell wall without concomitant new cell-wall synthesis—a process called **autolysis.** As we discussed above, certain antibiotics, such as *penicillin* and *cycloserine,* inhibit the synthesis of new cell-wall material in growing cells and induce cell lysis.

2.6

Ribosomes and Nuclear Region

Ribosomes In electron micrographs of thin sections, small dark particles can often be seen within the cytoplasm. (Note the dark granular bodies in Figure 2.14*b*, for example.) These particles, which are part of the protein-synthesizing machinery of the cell, are called **ribosomes,** because they are ribonucleic acid-containing bodies (-*some* means "body"). They are composed of about 60 percent ribonucleic acid (RNA) and 40 percent protein and are about 20 nm in diameter. Those from procaryotic microorganisms are slightly smaller and lighter in weight than those of eucaryotes. Ribosomes have characteristic sizes that are expressed by their sedimentation constants (the rate at which they fall through a liquid when subjected to high-speed centrifugation). The unit of sedimentation constant is the *Svedberg,* abbreviated "S." Procaryotic ribosomes have a sedimentation constant of 70S, while those of eucaryotes have a constant of 80S (see Chapter 3). The structure and function of ribosomes are discussed in Section 9.4.

Nuclear region: DNA Procaryotic organisms do not possess a true nucleus as do eucaryotic organisms. Instead, the DNA, which contains the genetic information for the cell, is present as a naked strand, and is not surrounded by a membrane. We discuss the structure, function, and biosynthesis of DNA in Section 9.2. Each DNA molecule consists of a single long fiber, which contains virtually all of the genetic information of the cell. The single DNA molecule of a procaryotic cell is sometimes referred to as a chromosome or genophore, but one must be careful not to confuse this structure with the more complicated eucaryotic chromosome. Eucaryotes contain their genetic information in a number of separate DNA molecules, each of which is complexed with protein to form large complex **chromosomes,** and surrounded, in the undividing cell, by the nuclear membrane.

In procaryotes the length of the DNA molecule, when stretched out, is many times the length of the cell. For instance, an *Escherichia coli* cell is about 2 μm long whereas the length of its DNA is around 1200 μm. This means that the DNA is highly folded within the cell. It is sometimes possible to visualize a portion of the procaryotic cell, called the **nuclear region,** or **nucleoid,** where the DNA is concentrated. This nuclear region is illustrated in the electron micrograph of Figure 2.8. DNA has a large number of negative charges which must be neutralized or damaging electrostatic repulsion might occur. In bacteria, this neutralization is accomplished with inorganic cations, such as Mg^{2+} or Ca^{2+}, or small organic polyamines, such as spermine or spermidine.

2.7
Flagella and Motility

Many bacteria are motile, and this ability to move independently is usually due to the presence of a special organelle of motility, the **flagellum** (plural, **flagella**). Bacterial flagella are long, thin appendages free at one end and attached to the cell at the other end. They are so thin (about 20 nm) that a single flagellum can never be seen directly with the light microscope, but only after staining with special flagella stains. One of the most common flagella stains employs the dye basic fuchsin with tannic acid as a mordant. The mordant promotes the attachment of dye molecules to the flagellum, and a crust or precipitate forms along the length of the flagellum, making it visible with the light microscope (Figure 2.23). Flagella are also readily seen with the electron microscope by shadowing (Figure 2.24).

Flagella are arranged differently on different bacteria. In **polar flagellation** the flagella are attached at one or both ends of the cell (Figure 2.23). Occasionally a tuft of flagella may arise at one end of the cell, an arrangement called "lophotrichous" (*lopho-* means "tuft"; *trichous* means "hair"). Tufts of flagella of this type can often be seen in the living state by dark-field microscopy (Figure 2.25). In **peritrichous flagellation** the flagella are not localized, but grow from many places on the cell surface (*peri-* means "around"). The type of flagellation is often used as a characteristic in the classification of bacteria (Chapters 18 and 19). Flagella are not straight but helically

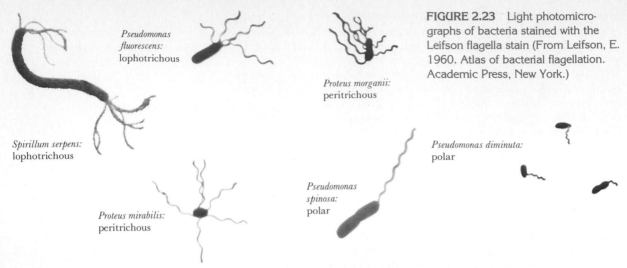

Pseudomonas fluorescens: lophotrichous

Spirillum serpens: lophotrichous

Proteus mirabilis: peritrichous

Proteus morganii: peritrichous

Pseudomonas spinosa: polar

Pseudomonas diminuta: polar

FIGURE 2.23 Light photomicrographs of bacteria stained with the Leifson flagella stain (From Leifson, E. 1960. Atlas of bacterial flagellation. Academic Press, New York.)

shaped; when flattened, they show a constant length between two adjacent curves, called the *wavelength,* and this wavelength is constant for each species.

Flagellar structure Bacterial flagella are composed of protein subunits; the protein is called **flagellin.** The amino acid composition of flagellin is somewhat atypical: there are lower amounts of sulfur-containing and aromatic amino acids than in most cellular proteins, whereas aspartic and glutamic acids occur more frequently. The shape and wavelength of the flagellum are determined by the struc-

FIGURE 2.24 Electron micrographs of metal-shadowed whole cells, showing flagella. (a) *Pseudomonas coronafaciens,* polar. Magnification, 12,800×. (b) *Erwinia carotovora,* peritrichous. Magnification, 13,000×. (Courtesy of Arthur Kelman.)

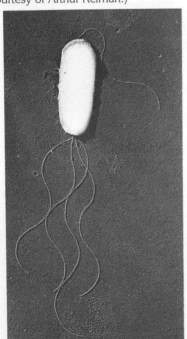

(a)

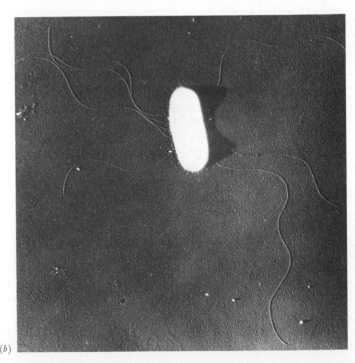

(b)

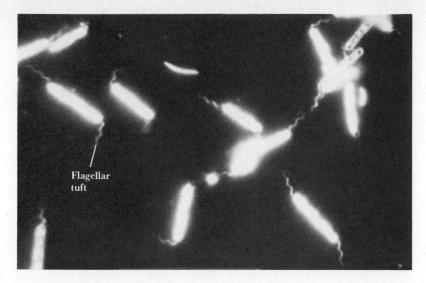

Flagellar
tuft

FIGURE 2.25 Dark-field photomicro-graph of a group of large rod-shaped bacteria with flagellar tufts at each pole. Magnification, 1900×. (From Jarosch, R. 1969. Mikroskopie 25:186–196.)

ture of the flagellin protein, and a change in the structure of the flagellin can lead to a change in the morphology of the flagellum.

The basal region of the flagellum is different in structure from the rest of the flagellum. There is a wider region at the base of the flagellum called the **hook** (Figure 2.26). Attached to the hook is the **basal body,** a complex structure involved in the connection of the flagellar apparatus to the cell envelope. The hook and basal body are composed of proteins different from those of the flagellum itself. The

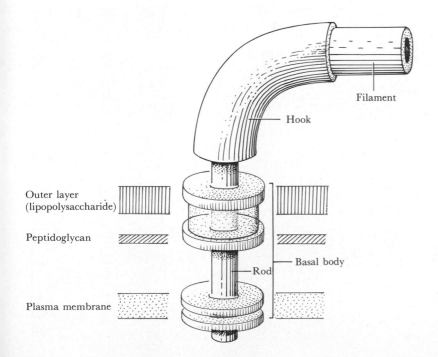

Filament

Hook

Outer layer
(lipopolysaccharide)

Peptidoglycan

Rod — Basal body

Plasma membrane

FIGURE 2.26 An interpretive drawing of the probable manner of attachment of the flagellum in a Gram-negative bacterium. (Courtesy of Julius Adler.)

basal body consists of a small central rod which passes through a system of rings. In Gram-negative bacteria, the outer pair of rings is associated with the lipopolysaccharide and peptidoglycan layers of the cell wall, and the inner pair of rings is located within or just above the plasma membrane (Figure 2.25). In Gram-positive bacteria, which lack the outer lipopolysaccharide layer, only the inner pair of rings is present.

Flagellar growth   The individual flagellum grows not from the base, as does an animal hair, but from the tip. Flagellin molecules formed in the cell apparently pass up through the hollow core of the flagellum and add on at the terminal end. The synthesis of a flagellum from its flagellin protein molecules occurs by a process called **self-assembly:** all of the information for the final structure of the flagellum resides in the protein subunits themselves. Growth of the flagellum occurs more or less continuously until a maximum length is reached; however, if a portion of the tip is broken off, it is regenerated.

When a cell divides, the two daughter cells must acquire in some way a full complement of flagella. In polarly flagellated organisms, the process of cell division probably occurs as shown in Figure 2.27*a*, the new flagellum forming at the location where cell division has just occurred. In a monopolarly flagellated cell, the two poles of

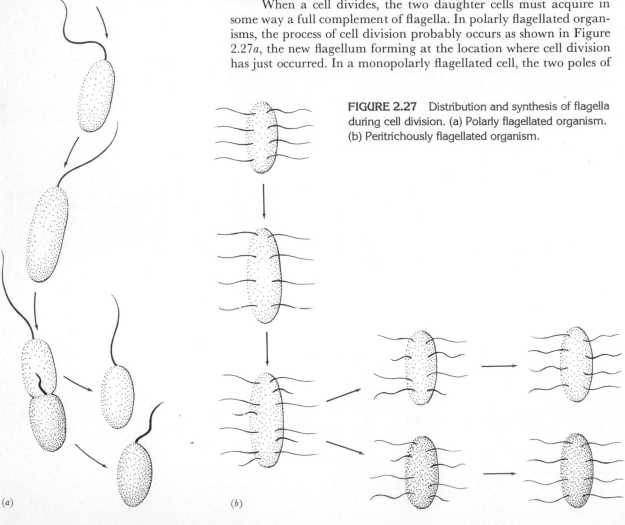

FIGURE 2.27   Distribution and synthesis of flagella during cell division. (a) Polarly flagellated organism. (b) Peritrichously flagellated organism.

(a)

(b)

the cell probably differ in some way so that the flagellum is formed at one pole and not at the other. In peritrichously flagellated organisms, the relation of cell division to flagella synthesis probably occurs as shown in Figure 2.27*b*, the preexisting flagella being distributed equally between the two daughter cells, and new flagella being synthesized and filling in the gaps.

Flagellar movement   How is motion imparted to the flagellum? For a long time it was thought that the flagellum moved in a wavelike manner from the base, in the same way that a whip is moved. However, there is now strong evidence that each individual flagellum is actually a rigid structure, which does not flex at all, but moves by rotation, in the manner of a propeller. Some evidence for this conclusion was obtained by observing the behavior of cells that were tethered by their flagella to microscope slides. It was observed that such cells rotated around the point of attachment, at rates of revolution consistent with those inferred for flagellar movement in free-swimming cells. In another experiment, mutants were used which produced flagella that were straight instead of helical; a flagellum was visualized by coating it with small latex beads. When the cells were attached to slides, the latex beads could be observed to rotate very rapidly.

It seems likely that the rotary motion of the flagellum is imparted from the basal body, which must act in some way like a motor. It is likely that the two inner rings located at the membrane rotate in relation to each other, so that (to extend the motor analogy) one of the rings could be considered to be the rotor, the other the stator.

The motions of polar and lophotrichous organisms are different from those of peritrichous organisms. Peritrichously flagellated organisms generally move and rotate in a straight line in a slow, stately fashion. Polar organisms, on the other hand, move more rapidly, spinning around and dashing from place to place. The different behavior of flagella on polar and peritrichous organisms is illustrated in Figure 2.28.

The average velocities of several bacteria are given in Table 2.1, and it can be seen that rates vary from about 20 μm to 80 μm/s. A speed of 50 μm/s is equivalent to 0.0001 mile/h, which seems slow. However, it is more reasonable to compare velocities in terms of number of cell lengths moved per second. The fastest animal, the cheetah,

FIGURE 2.28   Manner of flagellar movement in polarly and peritrichously flagellated organisms.

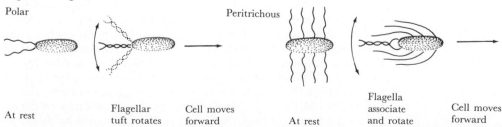

Polar

Peritrichous

At rest   Flagellar tuft rotates   Cell moves forward   At rest   Flagella associate and rotate   Cell moves forward

TABLE 2.1   Velocities of Several Common Bacteria

Organism	Flagellation	Cell length (μm)	Velocity (μm/s)	Lengths moved per second
Pseudomonas aeruginosa	Polar	1.5	55.8	37
Chromatium okenii	Lophotrichous	10	45.9	5
Thiospirillum jenense	Lophotrichous	35	86.5	2
Escherichia coli	Peritrichous	2	16.5	8
Bacillus licheniformis	Peritrichous	3	21.4	7
Sarcina ureae	Peritrichous	4	28.1	7

From Vaituzis, Z., and R. N. Doetsch. 1969. Appl. Microbiol. 17:584.

is about 4 ft long and moves at a maximum rate of 70 miles/h, or about 25 lengths per second. It can be seen from Table 2.1 that rates of bacterial movement, often around 10 or more lengths per second, are about as fast as those of higher organisms.

2.8

Chemotaxis in Bacteria

Chemotaxis is the movement of an organism toward or away from a chemical. Positive chemotaxis refers to movement toward a chemical and is usually exhibited when the chemical is of some benefit to the cell (for example, a nutrient). Negative chemotaxis is movement away from a chemical, usually one that is harmful. Chemicals that induce positive chemotaxis are called **attractants,** and chemicals that induce negative chemotaxis are called **repellents.** Although both procaryotes and eucaryotes exhibit chemotactic responses, the phenomenon has been studied most extensively in motile bacteria.

Chemotaxis is a behavioral phenomenon and suggests some kind of nervous response. There has been considerable interest in studying behavioral mechanisms in such simple organisms as bacteria, with the idea that the knowledge gained may provide insight into neural mechanisms in higher organisms. Much work has been carried out with two species, *Escherichia coli* and *Salmonella typhimurium,* and the following discussion deals only with these two organisms.

Bacterial chemotaxis can be most easily demonstrated by immersing a small glass capillary containing an attractant into a suspension of motile bacteria which does not contain the attractant. From the tip of the capillary, a chemical gradient is set up into the surrounding medium, with the concentration of chemical gradually decreasing with distance from the tip. If the capillary contains an attractant, the bacteria will move towards the capillary, forming a swarm around the open tip (Figure 2.29); and, subsequently, many of the motile bacteria will move into the capillary. Of course, some bacteria will move into the capillary even if it contains a solution of the same composition as the medium, because of random movements; but if an attractant is within, the concentration of bacteria within the capillary can be many times higher than the external concentration. On the other hand, if the capillary contains a repellent, the concentration of bacteria within the capillary will be consider-

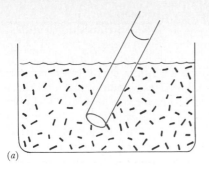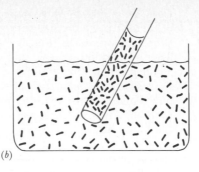

FIGURE 2.29 Capillary technique for studying chemotaxis in bacteria. (a) Insertion of capillary in a bacterial suspension. (b) Accumulation of bacteria in a capillary containing an attractant.

ably less than the concentration outside. The problem is to determine how these tiny cells are able to "sense" the chemical gradient and move towards or away from it.

To explain chemotaxis, we must first consider the behavior of a single bacterial cell when it is moving in a chemical gradient. When viewed under the microscope, the movement of single cells is extremely hard to follow with the eye, because it is so rapid. Two methods are employed. The first uses a special microscope, called a **tracking microscope,** which has been constructed to solve this problem. It automatically moves a small chamber containing the bacteria so as to keep a particular cell fixed in space, and a record is generated in three dimensions of the position of the chamber with time. An example of the kind of data obtained with this technique is shown in Figure 2.30. When movements of a large number of cells are analyzed

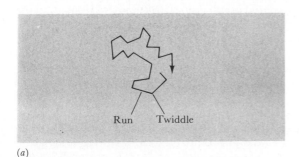

(a)

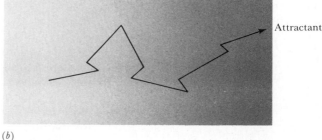

(b)

FIGURE 2.30 Diagrammatic representation of *Escherichia coli* movement, as analyzed with the tracking microscope. These drawings are two-dimensional projections of the three-dimensional movement.
(a) Random movement of a cell in a uniform chemical field. Each run is followed by a twiddle, and the twiddles occur fairly frequently. (b) Directed movement toward a chemical attractant. The runs still go off in random directions, but when the run is up the chemical gradient, the twiddles occur less frequently, and the runs are longer. When runs occur away from the chemical, twiddles occur more frequently. The net result is movement toward the chemical. (c) Directed movement away from a chemical repellent.

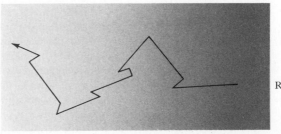

(c)

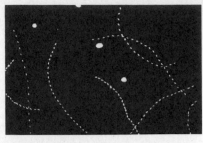

(a)

(b)

FIGURE 2.31 Chemotaxis in bacteria, as photographed with intermittent stroboscopic illumination. (a) Cells responding to a gradient with long runs and infrequent twiddles. (b) Cells in absence of gradient showing short runs and many twiddles. (From Aswad, D. and D. Koshland, Jr. 1974. J. Bacteriol. 118:640.)

with the tracking microscope, it can be concluded that bacterial behavior can be divided into two actions, called **runs** and **twiddles.** When the organism runs, it swims steadily in a gently curved path. Then it twiddles, that is, stops and jiggles in place. Then it runs off again in a new direction. A twiddle is a random event, but occurs on the average of about one per second and lasts about a tenth of a second. Following the twiddle, the direction for the next run is almost random. Thus, by means of runs and twiddles, the organism moves randomly but does not go anywhere. Now, if a chemical gradient of an attractant is present, the random movements become biased. As the organism experiences higher concentrations of the attractant, the twiddles are less frequent, and the runs are longer (Figure 2.30b). The net result of this situation is that the organism moves up the concentration gradient by increasing the length of runs that take it in a direction favoring a higher concentration of the attractant.

In the second microscopic observation method, a microscopic field is photographed continuously for a short period of time (usually less than 1 second) while illumination is provided intermittently via a rapidly flashing (stroboscopic) light. If a cell is moving during the period of photographic exposure, it will be photographed in several different places, giving the appearance of a dashed line (Figure 2.31a), a so-called *motility track*. On the other hand, if a cell is twiddling during the exposure, it will be photographed several times in basically the same place, giving the appearance of a large dot (Figure 2.31b). It is these motions which combine to cause the net motion described in Figure 2.30.

It has been well established that bacteria do not, in the ordinary sense, respond to the spatial gradient itself, but instead respond to the temporal gradient that develops as they move through the medium. In other words, if the bacterium is, by chance, moving toward the capillary, it experiences with time a progressively higher concentration of the substance, and this temporal increase in chemical concentration brings into play the mechanism to decrease twiddles, thus increasing runs.

We discussed in Section 2.7 the overall manner of movement of bacterial flagella. The specific movement of bacterial flagella in relation to the run-twiddle phenomenon also has been observed in experiments. In peritrichous organisms, such as *E. coli* and *S. typhimurium,* forward movement occurs when the flagellar bundle comes together. For forward movement, flagella within the bundle rotate counterclockwise, and the cell is propelled in a direction away from the flagella (Figure 2.28). This behavior results in a run. On the other hand, during a twiddle, the flagella rotate clockwise, and the flagellar bundle falls apart, resulting in cessation of forward motion. Bacterial flagella are rigid helices, and the direction of twist of the helix is left-handed, which results in the flagella coming together in a bundle when rotated counterclockwise, and flying apart when rotated clockwise. Thus the chemical sensing mechanism in some manner controls which direction the flagella rotate.

How do bacteria use temporal changes in chemical concentrations to control flagellar movement? Through a variety of genetic and biochemical studies, it has been shown that bacteria have specific

chemoreceptors, which sense the presence of the compound. These chemoreceptors are proteins, situated at the periphery of the cell, which bind specifically to the chemical. The binding proteins shown to be involved in chemotaxis are in at least some cases also involved in transporting those chemicals into the cell. Although the binding proteins show considerable specificity for the chemical which they sense, this specificity is not absolute. For example, the galactose chemoreceptor also recognizes glucose and fucose, and the mannose chemoreceptor recognizes glucose.

If a bacterium is presented with both an attractant and a repellent at the same time, it is forced to "choose" which compound to respond to. By carrying out experiments with pairs of such compounds at various concentrations, it has been possible to show that the nature of the response depends on the concentrations of the two agents in relation to the affinity of the chemoreceptors operating on each compound. Thus, if the concentration of *repellent* is relatively high, then the bacteria move away, whereas if the concentration of *attractant* is relatively high, they move toward the attractant even though the repellent may be harmful.

A type of chemotactic behavior occasionally observed in the laboratory is the swarming phenomenon seen on agar plates (Figure 2.32). In this figure the bacteria were inoculated as a small drop in the center of the plate. As they metabolize a nutrient in their immediate environment, they deplete the concentration of that substance and create their own chemical gradient. They thus move out from the center of the plate, following the gradient of nutrient in the form of a ring located always at the gradient. Because first one nutrient may be metabolized, then later another, more than one ring may develop, each ring "chasing" a different compound. In Figure 2.32, at least three rings can be seen. This ring formation is called a **swarm.***

FIGURE 2.32 Swarming phenomenon in a colony, due to chemotaxis. The bacteria were inoculated as a small drop in the center of the plate. As they used up a nutrient and depleted it, they moved out from the center, following the gradient of nutrient. Three rings are seen, each "chasing" a separate nutrient. (Courtesy of Julius Adler.)

Although it had been known for a long time that bacteria and blue-green algae differed from other organisms, the precise distinction between the procaryotic and eucaryotic cell did not become clear until the 1950s when the electron microscope was used for the study of cell structure. However, even with the light microscope, it was possible to make some significant observations. As early as the mid-nineteenth century, the great German microbiologist Ferdinand Cohn perceived the relationship between bacteria and blue-green algae. His observations met with considerable resistance, however, because of the obvious similarity of the photosynthetic process in blue-green algae with that in higher plants. Ernst Haeckel, an important

A Bit of History

---

*Another type of swarming phenomenon with a more complex basis is the swarming seen in highly motile strains of *Proteus vulgaris*. Depletion of a nutrient causes change in the size of the cells, from the normal, short, lightly flagellated rods, to very long, highly flagellated, and highly motile cells. The latter cells move in a wave out from the depleted region, until they arrive at an area of fresh medium, where growth resumes and the weakly motile short cells reform. After another period of depletion, swarmer cells are again formed and move. Wherever the cells stop and grow, a more dense zone is seen. The net result is a series of growth rings on the agar, which resemble superficially the chemotactic bands of *E. coli*. The rings of *Proteus* remain fixed, however, while the *E. coli* rings move.

German scholar of evolution, first pointed out the difficulty in deciding whether microorganisms were plants or animals. In 1866, he defined a separate kingdom, the Protista, which included all microorganisms. In the 1930s, H. F. Copeland separated the bacteria and blue-green algae into a separate kingdom which he called the Monera, placing the other microorganisms in a kingdom which he called the Protoctista, At about this same time, E. Chatton first coined the terms "procaryotic" and "eucaryotic," to express the organizational differences between bacteria and blue-green algae on the one hand and all other organisms on the other. The idea of a relationship between the bacteria and blue-green algae languished for many years until revived by Ernst Pringsheim, who, in 1949, pointed out not only the close relationship between these two groups, but showed that a number of the structurally more complex bacteria (for example, *Beggiatoa, Thiothrix*) could be considered to be "colorless" derivatives of blue-green algae (designated by him as apochlorotic). Electron-microscopic observations by a wide variety of investigators in the 1950s showed clearly that a major difference between procaryotic and eucaryotic cells was in nuclear organization, but that there were additional major differences, in respect to the partitioning of functions in eucaryotes into subcellular organelles, such as mitochondria and chloroplasts. At this same time, chemical studies of the cell wall of procaryotes revealed that if a wall was present, it generally contained a peculiar type of maromolecule called peptidoglycan. Another major difference between procaryotes and eucaryotes was found to be in the structures associated with cell movement, eucaryotes universally possessing flagella or cilia of complex structure with 9 + 2 fibers, whereas procaryotes possess simple flagellar structures of molecular size. As a result of these modern-day studies on cellular fine structure, R. Y. Stanier and C. B. van Niel developed a unifying concept of a bacterium in 1962: "In their totality, the bacteria cannot be clearly separated from another large microbial group, the blue-green algae. Both groups have a cellular organization, designated as procaryotic, which does not occur elsewhere in the living world. The principal distinguishing features of the procaryotic cell are: (1) absence of internal membranes which separate the resting nucleus from the cytoplasm, and isolate the enzymatic machinery of photosynthesis and of respiration in specific organelles; (2) nuclear division by fission, not by mitosis, a character possibly related to the presence of a single structure which carries all the genetic information of the cell; and (3) the presence of a cell wall which contains a specific mucopeptide [peptidoglycan] as its strengthening element."* This concept of a major chasm between procaryotic and eucaryotic cells was quickly accepted and has become a guiding light of a wide variety of studies in cell biology, molecular evolution, and biomedical research.

2.9

Other Cell and Surface Structures

Fimbriae and pili Fimbriae and pili are structures that are somewhat similar to flagella but are not involved in motility. **Fimbriae** are considerably shorter than flagella and are more numerous (Figure 2.33).† They may be chemically similar to flagella. Not all organisms

*Stanier, R. Y., and C. B. van Niel. 1962. The concept of a bacterium. Arch. Mikrobiol. 42:17–35.

†There has been considerable confusion about the terminology of fimbriae and pili. The two names have been used interchangeably by different workers, but the usage adopted here is that proposed in Ottow, J. C. G. 1975. Ecology, physiology, and genetics of fimbriae and pili. Annu. Rev. Microbiol. 29:79–108.

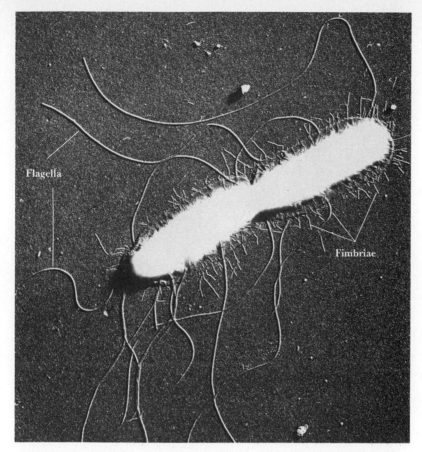

Flagella

Fimbriae

FIGURE 2.33 Electron micrograph of a metal-shadowed whole cell of *Salmonella typhosa,* showing flagella and fimbriae. Magnification, 15,100×. (From Duguid, J. P., and J. F. Wilkinson. 1961. Environmentally induced changes in bacterial morphology. In G. G. Meynell and H. Gooder, eds. Microbial reaction to environment [Eleventh symposium, Society for General Microbiology]. Cambridge University Press, London.)

have fimbriae, and the ability to produce them is an inherited trait. The functions of fimbriae are not known for certain in all cases, but there is some evidence that they enable organisms to stick to inert surfaces, or to form pellicles or scums on the surfaces of liquids.

Pili are similar structurally to fimbriae but are generally longer and only one or a few pili are present on the surface. Pili can be visualized under the electron microscope because they serve as specific receptors for certain types of virus particles, and when coated with virus can be easily seen (Figure 2.34). There is strong evidence that pili are involved in the mating process in bacteria, as will be discussed in Section 11.5. Pili are also significant in attachment to human tissues by some pathogenic bacteria.

Capsules and slime layers   Most procaryotic organisms secrete on their surfaces slimy or gummy materials, which can sometimes be seen by the use of negative stains (Figure 2.35). The terminology of this extracellular polysaccharide-containing material has undergone recent changes. The old terms **capsule** and **slime layer** are contained within the new and more general term **glycocalyx,** which is defined as that polysaccharide-containing material lying outside the cell. The

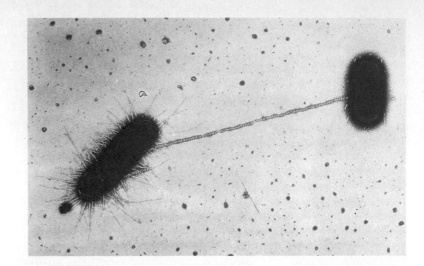

FIGURE 2.34 The presence of pili on an *E. coli* is revealed by the use of viruses which specifically adhere to the pilus. (From Brinton, C., et al. 1964. Proc. Nat. Acad. Sci. U.S. 52:776–783; courtesy of Charles Brinton.)

glycocalyx varies in different organisms, but usually contains glycoproteins and a large number of different polysaccharides, including polyalcohols and amino sugars. The glycocalyx may be thick or thin, rigid or flexible, depending on its chemical nature in a specific organism. The rigid layers are organized in a tight matrix which excludes particles, such as india ink; this form is the traditional capsule. If the glycocalyx is more easily deformed, it will not exclude particles and is more difficult to see; this arrangement is that of the traditional slime layer.

Inclusions and storage products Granules or other inclusions are often seen within the cells. Their nature differs in different organisms, but they almost always function in the storage of energy or structural building blocks. Inclusions can often be seen directly with the light microscope without special staining, but their contrast can usually be increased by using dyes. Inclusions often show up very well with the electron microscope.

FIGURE 2.35 Bacterial capsules. (a) Demonstration of the presence of a capsule by negative staining with India ink observed by phase-contrast microscopy. *Acinetobacter* sp. (Courtesy of Elliot Juni.) (b) Electron micrograph of a thin section of a *Rhizobium trifolii* cell stained with ruthenium red to reveal the capsule. Magnification, 16,500×. (Courtesy of Frank Dazzo and Richard Heinzen.)

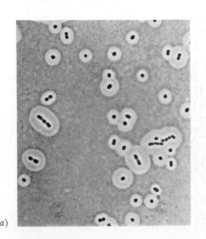

(a)

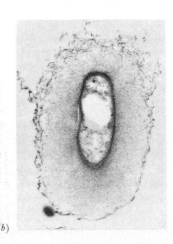

(b)

In procaryotic organisms, one of the most common inclusion bodies consists of **poly-β-hydroxybutyric acid (PHB),** a compound that is formed from β-hydroxybutyric acid units. The monomers of this acid are connected by ester linkages, forming the long PHB polymer, and these polymers aggregate into granules. With the electron microscope the positions of these granules can often be seen as light areas that do not scatter electrons, surrounded by a nonunit membrane. The granules have an affinity for fat-soluble dyes such as Sudan black and can be identified tentatively with the light microscope by staining with this compound. Poly-β-hydroxybutyric acid can be positively identified by extraction and chemical analysis. The PHB granules are a storage depot for carbon and energy.

Another storage product is **glycogen,** which is a starchlike polymer of glucose subunits. Glycogen granules are usually smaller than PHB granules, and can only be seen with the electron microscope, but the presence of glycogen in a cell can be detected in the light microscope because the cell appears a red-brown color when treated with dilute iodine, due to a glycogen-iodine reaction.

Many microorganisms accumulate large reserves of inorganic phosphate in the form of granules of **polyphosphate.** These granules are stained by many basic dyes; one of these dyes, toluidin blue, becomes reddish violet in color when combined with polyphosphate. This phenomenon is called **metachromasy** (color change), and granules that stain in this manner are often called **metachromatic granules.** *

2.10
Gas Vesicles

A number of procaryotic organisms that live a floating existence in lakes and the sea produce **gas vesicles,**† which confer buoyancy upon the cells. The most dramatic instances of flotation due to gas vesicles are seen in cyanobacteria (blue-green algae) that form massive accumulations (blooms) in lakes. Gas-vesiculate cells rise to the surface of the lake and are blown by winds into dense masses. When a sample of water containing these organisms is placed in a bottle, within minutes the organisms have floated to the surface, whereas if the gas vesicles are collapsed, the cells just as rapidly settle to the bottom (Figure 2.36). Gas vesicles are also formed by certain purple and green photosynthetic bacteria (see Section 19.1) and by some nonphotosynthetic bacteria that live in lakes and ponds.

Gas vesicles are spindle-shaped structures, hollow but rigid, that are of variable lengths but constant diameter. They are present in the cytoplasm and may number from a few to hundreds per cell. The gas vesicle membrane is an exception to the rule that membranes are composed of lipid bilayers (see Section 2.4). This membrane is composed only of protein, and consists of repeating protein subunits that are aligned to form a rigid structure. The gas vesicle membrane is impermeable to water and solutes, but permeable to gases, so that

*Another name for them is *volutin* granules.

†The former term was *gas vacuole,* but the term *gas vesicles* is now preferred. A group of gas vesicles present in a localized region of a cell is sometimes called a *gas vacuole.*

FIGURE 2.36 Flotation of cyanobacteria (blue-green algae) from a bloom in a lake, caused by the presence of gas vesicles. (a) Collapse of gas vesicles induced by hydrostatic pressure. Two identical bottles of organisms; the cork of one is struck by a hammer to increase the hydrostatic pressure. Note change in refractive index caused by collapse of the gas vesicles. (b) A few minutes later. Gas vesiculate cells (left bottle) have risen to surface, whereas cells with collapsed vesicles have sunk.

(a)

(b)

it exists as a gas-filled structure surrounded by the constituents of the cytoplasm (Figure 2.37). The rigidity of the gas vesicle membrane is essential for the structure to resist the pressures exerted on it from without; it is probably for this reason that it is composed of a protein able to form a rigid membrane rather than of lipid, which would form a fluid and a highly mobile membrane. However, even the gas vesicle membrane cannot resist high hydrostatic pressure, and can be collapsed, leading to a loss of buoyancy. The presence of gas vesicles can be determined by either bright-field or phase-contrast microscopy (Figure 2.38), but their identity is never certain unless they disappear when the cells are subjected to high hydrostatic pressure.

FIGURE 2.37 Electron micrographs of freeze-fractured cells of gas-vesiculate organisms. Magnification, 48,000×. Top, *Halobacterium* sp. Bottom left, *Amoebobacter rosea* (photosynthetic bacterium). Bottom right, *Anabaena flos-aquae*. (From Walsby, A. E. 1972. Bacteriol. Rev. 36:1–32.)

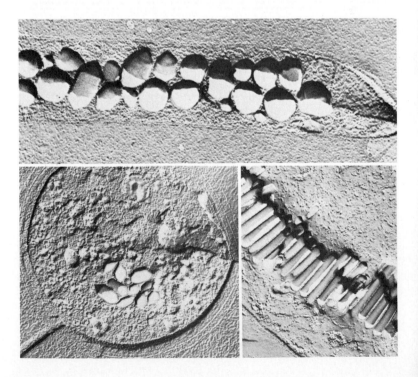

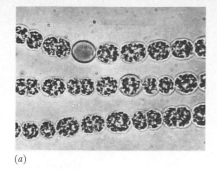
(a)

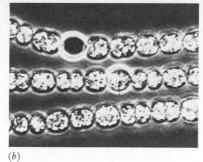
(b)

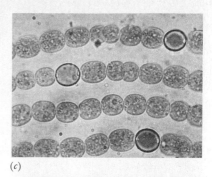
(c)

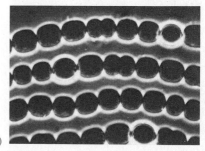
(d)

FIGURE 2.38 Gas vesicles of the cyanobacterium *Anabaena flos-aquae,* comparing gas vesiculate filaments with those whose gas vesicles have been collapsed by pressure. Vesicles intact: (a) bright-field microscopy; (b) phase-contrast microscopy. Vesicles collapsed: (c) bright-field microscopy; (d) phase-contrast microscopy. (From Walsby, A. E. 1972. Bacteriol. Rev. 36:1–32.)

2.11
Bacterial Endospores

The discovery that bacterial spores exist (noted in Chapter 1) was of immense importance to microbiology. Knowledge of such remarkably heat-resistant forms was essential for the development of adequate methods of sterilization, not only of culture media but also of foods and other perishable products. Although many organisms other than bacteria form spores, the bacterial **endospore** is unique in its degree of heat resistance. Endospores are also resistant to other harmful agents such as drying, radiation, acids, and chemical disinfectants.

Endospores (so called because the spore is formed within the cell) are readily seen under the light microscope as strongly refractile bodies (Figure 2.39). Spores are very impermeable to dyes, so that occasionally they are seen as unstained regions within cells that have been stained with basic dyes such as methylene blue. To stain spores specifically, special spore-staining procedures must be used. The structure of the spore as seen with the electron microscope is vastly different from that of the vegetative cell, as shown in Figure 2.40. The structure of the spore is much more complex than that of the vegetative cell in that it has many layers. The outermost layer is the **exosporium,** a thin, delicate covering. Within this is the **spore coat,** which is composed of a layer or layers of wall-like material. Below the spore coat is the **cortex,** and inside the cortex is the **core,** which contains the usual cell wall (core wall), cell membrane, nuclear region, and so on. Thus the spore differs structurally from the vegetative cell primarily in the kinds of structures found outside the core wall.

One chemical substance that is characteristic of spores but not of vegetative cells is **dipicolinic acid** (DPA) (Figure 2.41).* This

*One should not confuse dipicolinic acid, DPA, with diaminopimelic acid, DAP.

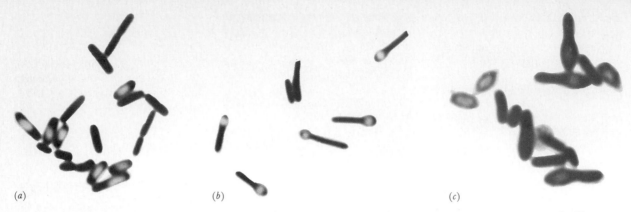

(a)　　　　　　　　　(b)　　　　　　　　　(c)

FIGURE 2.39 Light photomicrographs illustrating several types of endospore morphologies.
(a) Central spores; sporangium wall not enlarged.
(b) Terminal spores; sporangium wall enlarged.
(c) Subterminal spores; sporangium wall enlarged.
(Courtesy of The Wellcome Research Laboratories Anaerobic Bacteriology Department, Beckenham, Kent, England.)

FIGURE 2.40 Electron microscopy of the bacterial spore. (a) Formation of spores within vegetative cells of *Bacillus megaterium*. Magnification, 17,500×. (b) Mature spore of *B. megaterium*. Magnification, 35,640×. (Courtesy of H. S. Pankratz, T. C. Beaman, and Philipp Gerhardt.)

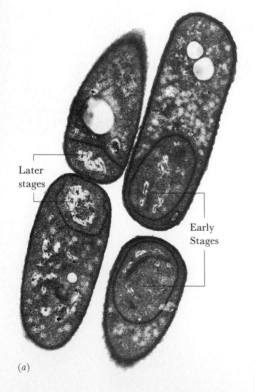

Later stages

Early Stages

(a)

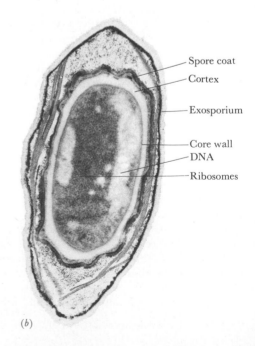

Spore coat
Cortex
Exosporium
Core wall
DNA
Ribosomes

(b)

substance has been found in all spores examined, and it is probably located primarily in the core. Spores are also high in calcium ions, most of which are also associated with the core, probably in combination with dipicolinic acid. There is good reason to believe (see Section 8.1) that the association of calcium and dipicolinic acid has some role in conferring the unusual heat resistance on bacterial spores.

The structural changes that occur during the conversion of a vegetative cell to a spore can be studied readily with the electron microscope. Under certain conditions (mainly as a result of nutrient exhaustion), instead of dividing, the cell undergoes the complex series of events leading to spore formation; these are illustrated in Figure 2.42.

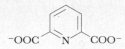

FIGURE 2.41 Dipicolinic acid (DPA). Ca²⁺ ions associate with the carboxyl groups to form a complex: Ca^{2+} DPA Ca^{2+} DPA Ca^{2+} DPA. . . .

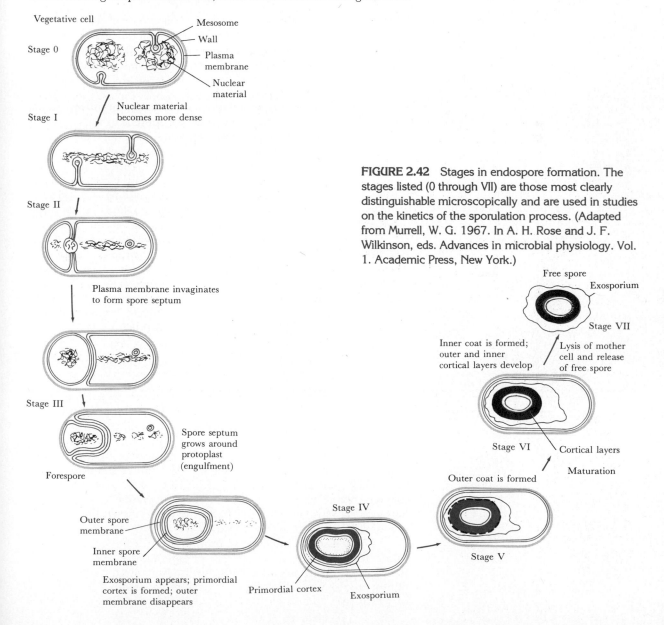

FIGURE 2.42 Stages in endospore formation. The stages listed (0 through VII) are those most clearly distinguishable microscopically and are used in studies on the kinetics of the sporulation process. (Adapted from Murrell, W. G. 1967. In A. H. Rose and J. F. Wilkinson, eds. Advances in microbial physiology. Vol. 1. Academic Press, New York.)

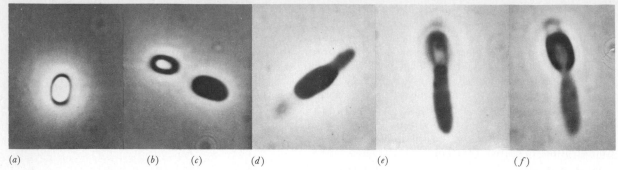

(a)   (b)   (c)   (d)   (e)   (f)

FIGURE 2.43 Photomicrographs showing the sequence of events during endospore germination in *Clostridium pectinovorum:* (a) ungerminated spore; (b, c) refractility is lost; (d–f) swelling and outgrowth. (From Hoeniger, J. F. M., and C. L. Headley. 1968. J. Bacteriol. 96:1835.)

The position of the spore within the sporangium differs among different species of bacteria. Sometimes the spore is in the center; at other times at one end **(terminal spore).** Sometimes the mature spore is larger in diameter than the sporangium, a condition called a "swollen spore." These distinctions are illustrated in Figure 2.39.

A spore is able to remain dormant for many years, but it can convert back into a vegetative cell **(spore germination)** in a matter of minutes (Figure 2.43). This process involves two steps: cessation of dormancy and outgrowth. The first is initiated by some environmental trigger such as heat. A few minutes of heat treatment at 60 to 70°C will often cause dormancy to cease. The first indications of spore germination are loss in refractility of the spore, increased stainability by dyes, and marked decrease in heat resistance. The spore visibly swells and its coat is broken. The new vegetative cell now pushes out of the spore coat, a process called **outgrowth,** and begins to divide. The spore wall and spore coat eventually disintegrate through the action of lytic enzymes.

2.12

Summary

In this chapter we have described in a general way the structure of the procaryotic cell. Procaryotic cells are usually small, and little can be learned about their structure without use of the electron microscope. Careful examination under the electron microscope by a variety of techniques reveals that procaryotic cells are rather simple in structure. They lack complex internal organelles, such as are found in eucaryotes, and their genetic material is not present in membrane-bounded nuclei. The DNA of procaryotes is present within the cell in an uncomplexed but highly folded form.

The **plasma membrane** of procaryotes is a typical unit membrane, formed by the association of phospholipid molecules in a bimolecular leaflet. Associated with this lipid membrane are a variety of proteins, some of which are embedded in the membrane, and some of which pass completely through the membrane. The **cell wall** of procaryotes is a unique feature, chemically distinct from the cell wall of eucaryotic organisms. It may consist of a single layer or it may be multilayered, but in either case it contains a rigid layer, which is responsible for the overall shape of the cell. The rigid layer consists chemically of a substance called **peptidoglycan,** which consists of a thin sheet composed of two sugar derivatives, N-acetylglucosamine and N-acetylmuramic acid, and a small group of amino acids, consisting (in most organisms) of L-alanine, D-alanine, D-glutamic acid, and either lysine or diaminopimelic acid. These constituents are connected to form a repeating structure, which is polymerized to form the peptidoglycan layer. Organisms can be divided into two groups, Gram-positive and Gram-negative, based on their reactions to the Gram-staining procedure and their appearance in electron micrographs. Both Gram-positive and Gram-negative bacteria have peptidoglycan cell walls, but Gram-positive bacteria have thicker cell walls and lack other components of Gram-negative bacteria.

Many bacteria are motile, and in most cases motility is due to the presence of **flagella.** Flagella are thin, helical structures composed of protein, which are attached in the cell membrane and extend outside the cell. Flagella appear to be rigid structures that impart motion to cells by rotating in the manner that a propeller causes a boat to move. A variety of flagellar arrangements are known, and many behavioral adaptations (chemotaxis, phototaxis) relate to flagellar function.

An important structure found in certain bacteria is the **endospore,** which is a highly heat-resistant structure. Many bacterial endospores can survive boiling temperatures for long periods of time, and thus a knowledge of spore structure and function is important in understanding how to sterilize foods and other materials that support bacterial growth. The endospore is formed internally by segmentation of a portion of the cell and contains a complex series of wall layers outside the normal cell structure. A unique chemical constituent found in spores is **dipicolinic acid,** which is involved in the unusual heat resistance of the spore. Spores are dormant and can remain stable for long periods of time. However, under proper conditions, dormancy is rapidly broken and the spores **germinate.**

Burchard, R. P. 1981. Gliding motility of prokaryotes: ultrastructure, physiology, and genetics. Annu. Rev. Microbiol. 35:497–529.

Costerton, J. W., R. T. Irvin, and K.-J. Cheng. 1981. The bacterial glycocalyx in nature and disease. Annu. Rev. Mirobiol. 35:299–324.

Doetsch, R. N., and R. D. Sjoblad. 1980. Flagellar structure and function in eubacteria. Annu. Rev. Microbiol. 34:69–108. Good summary of recent studies, pointing out a diversity of mechanisms in different organisms.

Dworkin, M. 1979. Spores, cysts, and stalks. *In* L. N. Ornston and J. R. Sokatch, eds. The bacteria, vol. 7. Academic Press, New York.

Inouye, M. ed. 1979. Bacterial outer membranes: biogenesis and functions. John Wiley, New York.

Koshland, D. E., Jr. 1979. Bacterial chemotaxis. *In* L. N. Ornston and J. R. Sokatch, eds. The bacteria, vol 7. Academic Press, New York. General review of chemotaxis and its biochemistry.

Supplementary Readings

Levinson, H., A. L. Sonenshein, and D. J. Tipper, eds. 1981. Sporulation and germination. American Society for Microbiology, Washington, D. C. Current review of specialized areas in endospore study. Useful survey of relevant literature.

Ottow, J. C. G., 1975. Ecology, physiology, and genetics of fimbriae and pili. Annu. Rev. Microbiol. 29:79–108. Useful review of fimbriae and pili in various bacteria.

Rogers, H. J. 1979. Biogenesis of the wall in bacterial morphogenesis. Adv. Microb. Physiol. 19:1–62. Bacterial cell-wall development and its relation to cell shape.

Rogers, H. J., H. R. Perkins, and J. B. Ward. 1981. Microbial cell walls and membranes. Chapman & Hall/Methuen, New York. Excellent detailed review of cell-wall and membrane structure and function.

Sutherland, I. W. 1982. Biosynthesis of microbial exopolysaccharides. Adv. Microb. Physiol. 23:79–150. Chemistry and regulation of production of a wide range of polymers.

Tipper, D. J., and A. Wright. 1979. The structure and synthesis of bacterial cell walls. *In* L. N. Ornston and J. R. Sokatch, eds. The bacteria, vol. 7. Academic Press, New York. Excellent review of different cell-wall types.

Walsby, A. E. 1977. The gas vacuoles of blue-green algae. Sci. Am. 237:90–97. A concise review of the structure and function of gas vesicles in cyanobacteria.

The Eucaryotic Cell and Eucaryotic Microorganisms

Eucaryotic cells (cells with true nuclei) are structurally much more complex than procaryotic cells. They show to varying degrees localization of cellular functions in distinct, membrane-enclosed intracellular structures called **cell organelles.** Examples include nuclei, mitochondria, and chloroplasts, which will be discussed in detail below. Cell division and sexual reproduction are also considerably more complex in eucaryotes than in procaryotes. The increased structural complexity of eucaryotes is not, however, accompanied by an increased chemical complexity.

The cells of all higher organisms, both plant and animal, are eucaryotic. Among the microorganisms, fungi, protozoa, and algae are eucaryotic. There is tremendous diversity of cell types and functions in this vast collection of eucaryotic microorganisms, but all have in common the fact that many of their cellular functions are located in or on intracellular organelles. An important point to which we shall return later in this chapter is that a cell is either procaryotic or eucaryotic—there does not seem to be any middle ground. This suggests that the evolution of the first eucaryotic cell from a procaryotic one probably involved a single large evolutionary step.

In general, the sizes of eucaryotic cells are much greater than those of procaryotes (Figure 3.1), although some eucaryotic cells are as small as the larger procaryotes. Most eucaryotic cells have fairly distinct shapes, and these are often quite complex. Cell shapes in the algae and protozoa are especially interesting (Figure 3.2), with a geometric beauty often lacking in the cells of higher plants and animals. An electron micrograph of a commonly studied eucaryotic cell, the yeast cell, is shown in Figure 3.3.

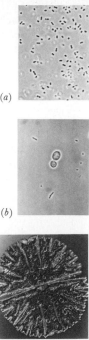

(a)

(b)

(c)

FIGURE 3.1 Photomicrographs to the same magnification of two eucaryotic cells, and a procaryotic cell for comparison. (a) Procaryote: *Acetobacter aceti.* (b) Eucaryote: yeast. (c) Eucaryote: the alga *Micrasterias.* Magnifications, 400×.

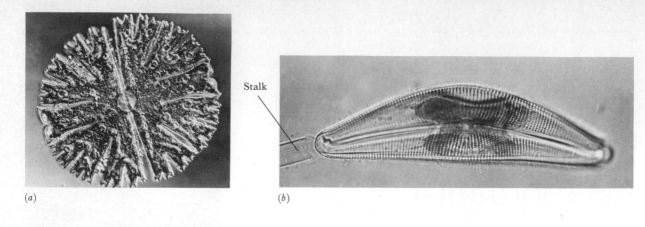

(a)

Stalk

(b)

FIGURE 3.2 Photomicrographs illustrating some of the complex cell structures of algae and protozoa. (a) The alga *Micrasterias* (a desmid). Magnification, 190×. (b) The alga *Cymbella* (a diatom). Magnification, 460×. (c) Foraminifera of several types. Magnification, 50×. (d) Radiolaria. Magnification, 150×. *c* and *d* are marine protozoa with complex shells.

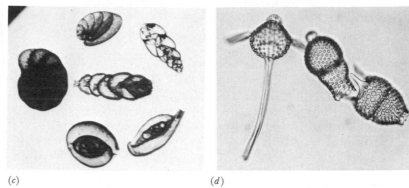

(c)

(d)

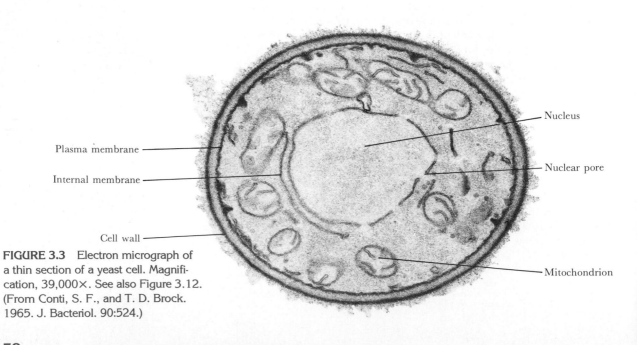

Plasma membrane

Internal membrane

Cell wall

Nucleus

Nuclear pore

Mitochondrion

FIGURE 3.3 Electron micrograph of a thin section of a yeast cell. Magnification, 39,000×. See also Figure 3.12. (From Conti, S. F., and T. D. Brock. 1965. J. Bacteriol. 90:524.)

Plasma membrane The plasma membrane of the eucaryotic cell is similar in structure to that of the procaryotic cell. The reader should refer to Section 2.4 for details of the general chemistry of the plasma membrane and its role as a permeability barrier.

From the studies at hand, it seems that the membranes of both procaryotes and eucaryotes have the same basic structure. Minor differences (which may be functionally important) arise from differences in the kinds of phospholipids and proteins of which the membrane is composed. One notable difference in chemical composition, however, is that the eucaryotes have sterols (Figure 3.4) in their membranes, whereas sterols are rare or absent in the membranes of procaryotes.*

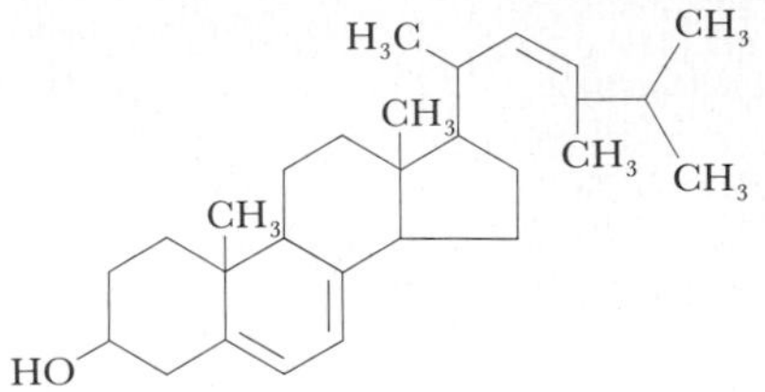

Sterols are rigid, planar molecules, whereas fatty acids are flexible. The association of sterol with the membrane works to stabilize its structure and make it less flexible. It has been shown, using artificial lipid membranes, that they are much less leaky when a sterol is part of the structure than when they are composed of pure phospholipid. Why rigidity of membrane is necessary in eucaryotes is not definitely known. One possibility, however, is that the eucaryotic cell, because it is much larger than the procaryotic cell, must endure greater physical stresses on the membrane, necessitating a more rigid membrane to keep the cell stable and functional. One group of antibiotics, the polyenes (filipin, nystatin, candicidin, for example) react with sterols and destabilize the membrane. It is of interest that these antibiotics are active against all eucaryotes, but generally do not affect procaryotes, probably because the latter lack sterols in their membranes. In addition, some bacteria of the mycoplasma group (Section 19.31), which lack a normal peptidoglycan cell wall, require sterol for growth. The sterol becomes incorporated into the plasma membrane and probably stabilizes the membrane structure. These mycoplasmas are inhibited by polyene antibiotics, in contrast to the insensitivity to these antibiotics displayed by all other procaryotes.

The variety of membranous organelles and membrane-containing structures of eucaryotes is outlined in Table 3.1; a discussion of some of the more important of these structures follows.

FIGURE 3.4 Structure of a typical sterol, yeast ergosterol.

In eucaryotic cells the processes of respiration and oxidative phosphorylation (see Chapter 4) are localized in special membrane-enclosed structures, the **mitochondria** (singular, **mitochondrion**). Mitochondria may have many shapes, but most often they are rod-shaped structures about $1 \mu m$ in diameter by 2 to 3 μm long. Figure 3.3 shows mitochondria in an electron micrograph of a thin section of a yeast cell. The mitochondrial membrane is constructed in a manner similar to other membranes: a bilayer formed of phospholipid with proteins embedded in the lipid layer. Interestingly, mito-

---

*At one time it was thought that procaryotes did not form sterols at all. However, sterols are found in small amounts in a variety of bacteria and in fairly large amounts in the methane-oxidizing bacteria.

TABLE 3.1 Membrane-Containing Structures in Eucaryotes

Structure	Characteristics	Function
Mitochondria	Bacteria-size, complex internal membrane arrays	Energy generation: respiration
Chloroplasts	Green, chlorophyll-containing, many shapes, often quite large	Photosynthesis
Endoplasmic reticulum	Not a distinct organelle, extensive array of internal membranes	Protein synthesis
Golgi bodies	Membrane aggregates of distinct structure	Secretion of enzymes and other macromolecules
Vacuoles	Round, membrane-enclosed bodies of low density	Food digestion: food vacuoles; waste product excretion: contractile vacuoles
Lysosomes	Submicroscopic membrane-enclosed particles	Contain and release digestive enzymes
Peroxisomes	Submicroscopic membrane-enclosed particles	Photorespiration in plants
Glyoxysomes	Submicroscopic membrane-enclosed particles	Enzymes of glyoxylate cycle
Nucleus	Large, generally centrally located	Contains genetic material

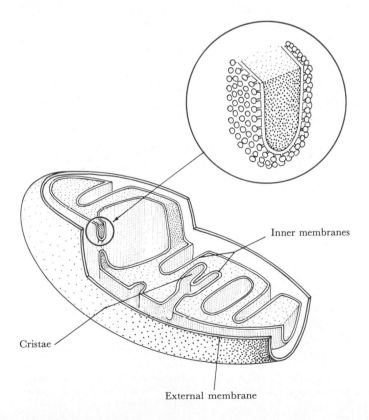

FIGURE 3.5 Structure of a mitochondrion. The structure of the inner membrane complex is shown in the insert.

Inner membranes

Cristae

External membrane

chondrial membranes seem to lack the sterols found in the plasma membrane of eucaryotes. The mitochondrial membrane is much less rigid than the plasma membrane, partly because it lacks sterols. In addition to the outer membrane, mitochondria usually possess a complex system of inner membranes; these inner membranes, called **cristae,** are unique to mitochondria and do not occur in other structures of similar size (Figure 3.5). Within the inner compartment formed by the cristae is the **matrix,** which is gellike and contains large amounts of protein. With high-resolution electron microscopy, it is possible to see that the cristae are of complex structure, consisting of small round particles attached to the basal membrane by short stalks. The various biochemical components involved in respiration, which are localized both in the membrane itself and in the small stalked particles, will be described in Chapter 4.

In recent years it has been shown that mitochondria contain DNA that probably determines the inheritance of at least some of the properties of mitochondria. Mitochondria also contain part of their own protein-synthesizing machinery in the form of ribosomes and other components; they reproduce by a division process. Hence, to a considerable extent the mitochondria retain an autonomy of function within the cell, although their growth and reproduction are precisely integrated with all other phases of cell growth and reproduction.

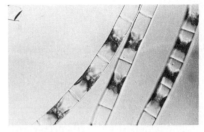

FIGURE 3.6 Photomicrograph by phase contrast of the filamentous alga *Ulothrix,* showing the centrally located chloroplast. Magnification, 230×.

3.3

Chloroplasts

Chloroplasts are green, chlorophyll-containing organelles found in all eucaryotic organisms able to carry out photosynthesis (see Chapter 6). Chloroplasts of many algae are quite large and hence are readily visible with the light microscope (Figure 3.6). The size, shape, and number of chloroplasts vary markedly throughout the different groups of algae.

Each chloroplast has an outer membrane, within which are a large number of internal membranes (called **photosynthetic lamellae,** or **thylakoids**) with which the chlorophyll is associated (Figure 3.7). In some algae the thylakoids lie separately in the chloroplast, whereas in others the thylakoids are stacked together in bands of two, three, four, or occasionally more. In the green algae (Chlorophyceae), the thylakoids are usually associated in discrete structural units called **grana,** and in this respect the green algae are similar to higher plants, where the thylakoids are also arranged in grana. Among the algae grana are found only in the Chlorophyceae, a fact which suggests that this algal group may have been the evolutionary forerunner of the higher plants.

The thylakoid membrane is constructed in a manner similar to the unit membrane described in Section 2.4, with hydrophobic protein molecules embedded in the lipid bilayer. Chlorophyll, a hydrophobic molecule, is associated with the lipid-protein membrane of the thylakoid; the exact spatial relationship between chlorophyll and the other components of the thylakoid membrane is not completely understood, although current thought is that it is combined primarily with the protein moiety.

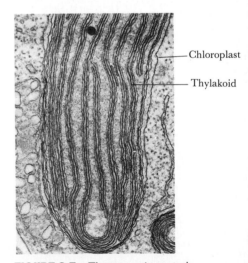

Chloroplast

Thylakoid

FIGURE 3.7 Electron micrograph showing a chloroplast of the alga *Ochromonas danica.* Note that each thylakoid consists of three parallel membranes. Magnification, 29,000×. (Courtesy of T. Slankis and S. Gibbs.)

Chloroplasts, like mitochondria, have a certain degree of autonomy in the cell. They contain ribosomes, DNA, and other components of the protein-synthesizing machinery.

On the basis of their autonomy and similarity to bacteria in many fundamental characteristics, it has been suggested that mitochondria and chloroplasts are descendants of ancient procaryotic organisms. This proposal of **endosymbiosis** (endo means "within") says that eucaryotes arose from the "invasion" of one procaryotic cell by another. The outer cell differentiated into the eucaryotic cells we see today, while the inner cell progressively lost its independence and retained selected, specialized functions, as we see in modern organelles. The endosymbiont proposal cannot be conclusively proven, but there is a large amount of evidence, biochemical, microscopic, physiological, and geological, which is consistent with it.

3.4

Movement

Almost all protozoa, most algae, and many fungi exhibit mechanical activity or movement of some kind. Two basic kinds are distinguished: **cytoplasmic streaming,** which is the movement of cytoplasm within stationary cells, and **motility,** whereby the cell moves itself through space. Many eucaryotic microorganisms exhibit motility through the activity of special organelles of motion, the flagella and cilia. The analogy between mechanical activity of microorganisms and that of muscle cells in higher animals has often been made, and it is likely that the underlying mechanisms may be similar. Yet muscle activity and mechanical activity in microorganisms differ in many details.

Flagella (singular, **flagellum**) are long filamentous structures that are attached to one end of the cell and move in a whiplike manner to impart motion to the cell. Some organisms have only a single flagellum; others have more than one. The number and arrangement of flagella are important characteristics in classifying various groups of algae, fungi, and protozoa.

It should be emphasized that flagella of eucaryotes are quite different in structure from those of procaryotes (see Section 2.7), even though the same word is used for both. Eucaryotic flagella are much more complex in structure. Without exception, flagella of eucaryotes contain two central fibers surrounded by nine peripheral fibers, each of which is doublet in nature (Figure 3.8). The whole unit is surrounded by a membrane and the fibers are embedded in an organic matrix. Each fiber is in reality a tube, called a **microtubule,** which is composed of a large number of protein molecules called **tubulin** (Figure 3.8c). Each microtubule is about the same diameter as a procaryotic flagellum, but they are composed of different proteins. Tubulin molecules are comprised of two subunits. These subunits are arranged in an alternating, helical fashion up the microtubule axis (Figure 3.8d). When observed in cross section, each tubulin microtubule appears to consist of a helix composed of exactly 13 rows of tubulin molecules (Figure 3.8c).

Attached to each of the outer doublet fibers are molecules of another protein, **dynein.** It is this protein that is involved in the con-

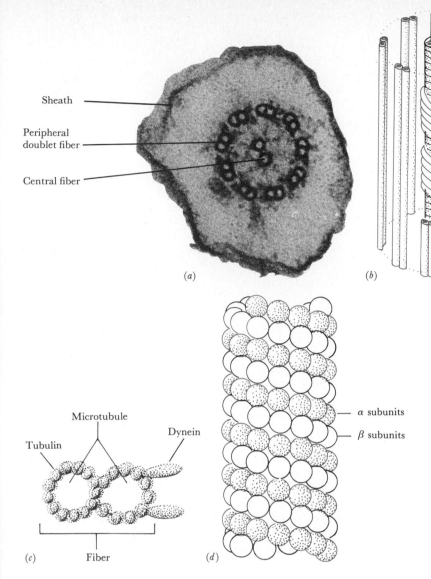

Sheath

Peripheral
doublet fiber

Central fiber

(a)

(b)

Microtubule

Tubulin

Dynein

α subunits

β subunits

(c) Fiber

(d)

FIGURE 3.8 (a) Electron micrograph of a cross section of the flagellum of the zoospore of the fungus *Blasto-cladiella emersonii* showing the outer sheath, the outer nine fibers, and the central pair of single fibers. Magnification, 132,000×. (Courtesy of Melvin S. Fuller.) (b) Interpretive diagram of the arrangement of the fibrils in a flagellum. The outer sheath is not shown. (From Paolillo, D. J., Jr. 1967. Trans. Amer. Micros. Soc. 86:428.) (c) Cross section showing construction of the outer fiber. Each microtubule is about 20 nm in diameter. (d) The tubulin proteins are actually made up of two subunits, and they are arranged in a helical fashion up the microtubule axis, as illustrated.

version of chemical energy (from adenosine triphosphate, ATP) into the mechanical energy of flagellar movement. It is thought that movement is imparted to the flagellum by coordinated sliding of the fibers toward or away from the base of the cell, in a way analogous to that in which muscle filaments slide during muscular contraction. The manner in which the flagellum moves determines the direction of cellular movement, as is illustrated in Figure 3.9. The maximum rate of movement of flagellated microorganisms ranges from 30 to 250 μm/s.

Cilia (singular, **cilium**) are similar to eucaryotic flagella in fine structure but differ in being shorter and more numerous (Figure 3.10). In microorganisms, cilia are found primarily in one group of

FIGURE 3.9 Manner in which flagella of different types impart motility to cells. (a) Wave directed away from cell pushes cell in opposite direction; type of movement found in dinoflagellates and animal spermatozoa. (b) Wave directed toward the cell pulls the cell in the direction opposite to the wave; type of movement found in trypanosomes and other flagellated protozoa. (c) Flagellum contains stiff lateral projections called mastigonemes. Wave directed away from cell pulls the cell in the same direction as the wave; type of movement found in chrysophytes. (From Jahn, T. L., and E. C. Bovee. 1967. In T. Chen, ed. Research in protozoology, vol. 1. Pergamon Press, New York.)

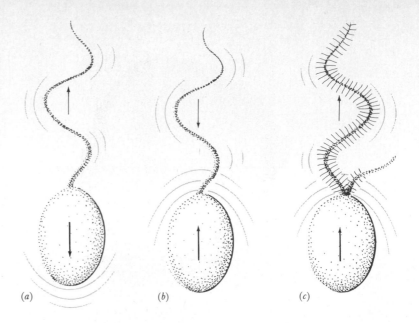

(a) (b) (c)

protozoa, appropriately called the ciliates. (Cilia are widespread among the cells of higher animals, however.) A single protozoan cell of a *Paramecium* species has between 10,000 and 14,000 cilia. These organelles are stiff and operate like oars, beating about 10 to 30 strokes per second. They do not act in unison but usually beat in a coordinated fashion, causing a wave of ciliary motion to pass over the cell, which results in cellular movement (Figure 3.11). The rate of movement of ciliates varies from 300 to 2500 μm/s for different species. Thus ciliated organisms move much more rapidly than flagellated ones.

Cytoplasmic streaming is readily detected with the light microscope in the larger eucaryotic cells by observing the behavior of various particles such as chloroplasts or mitochondria; they are seen to move en masse in a definite pattern, suggesting that they are being carried passively by the movement of the cytoplasm. Rates of cytoplasmic streaming vary greatly, depending on the organism and environmental conditions; values from 2μm/s to greater than 1000μm/s have been recorded.

In cells without walls (for example, amoebas, slime molds), cytoplasmic streaming can result in **amoeboid movement,** (Figure 3.12), so called because it is the movement characteristic of the amoebas. During movement, a temporary projection of the protoplast called a **pseudopodium** develops, into which cytoplasm streams. Cytoplasm flows forward because the tip of the pseudopodium is less contracted and hence less viscous, while the rear is contracted and viscous; thus cytoplasm takes the path of least resistance. Amoeboid motion requires a solid surface along which the protoplasm can move.

The mechanism of cytoplasmic movement is different from that of flagellar movement. Microtubules are probably not involved in cytoplasmic movement; however, solid filaments closely analogous

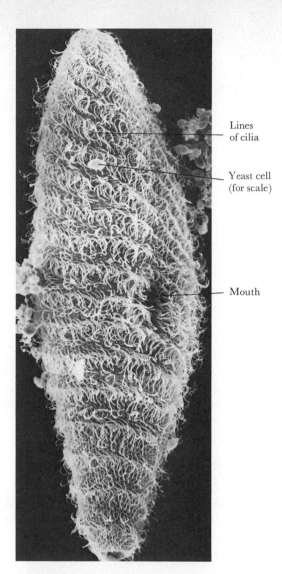

Lines
of cilia

Yeast cell
(for scale)

Mouth

FIGURE 3.10 Scanning electron micrograph of the ciliate *Paramecium*. Note the large number and orderly arrangement of the cilia. The animal is shown swimming forward (to the top), and the ciliary wave travels from posterior (bottom) to anterior. Magnification, 860×. (From Tamm, S. L. 1972. J. Cell Biol. 55:250–255.)

to the actin filaments of muscle have been shown to play a part in movement. It is possible to prepare extracts of amoebas rich in these filaments and obtain streaming-type movement when ATP is added. Microtubules definitely play a supportive role in maintaining cell shape, and are probably responsible, at least in part, for the characteristic nonspherical shapes of wall-less microbes such as the amoebas.

Motile eucaryotic cells also exhibit a variety of tactic responses such as chemotaxis, phototaxis, and geotaxis, similar to those already described for procaryotes in Section 2.8. However, the molecular mechanism of eucaryotic taxis is different and considerably more complex than that of procaryotes. The role of tactic movement in eucaryotes is analogous to that already described for procaryotes: movement towards environmental stimuli that are favorable and avoidance of environmental stimuli that are harmful.

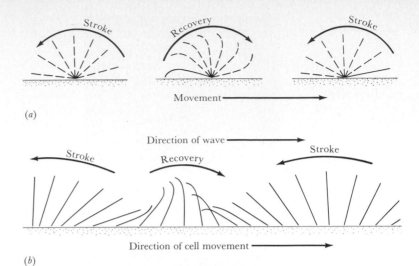

FIGURE 3.11 The beat of a single cilium (a) and the coordinated movement of a group of cilia on the surface of the cell (b).

FIGURE 3.12 Side view of a moving amoeba. *Amoeba proteus,* taken from a film, the time interval between frames being 20 seconds. The arrows point to a fixed spot on the surface. Magnification, 76×. (Courtesy of M. Haberey. The photographic technique is described in Haberey, M. 1971. Mikroskopie 27:226–234.)

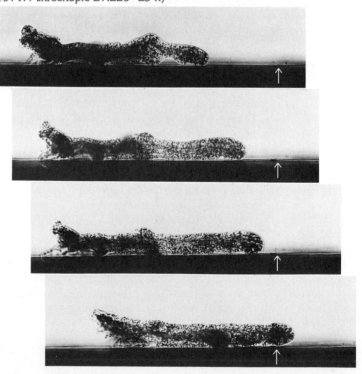

Nuclear structure One of the distinguishing characteristics of a eucaryote is that its genetic material (DNA) is organized in chromosomes located in a membrane-enclosed structure, the nucleus. In many eucaryotic cells the nucleus is a large organelle many micrometers in diameter, easily visible with the light microscope even without staining. In smaller eucaryotes, however, special staining procedures often are required to see the nucleus.

The nuclear membrane is more complex than are the membranes of other eucaryotic structures. It consists of a pair of parallel unit membranes separated by a space of variable thickness. The inner membrane is usually a simple sac, but the outer membrane is in many places continuous with the endoplasmic reticulum (mentioned in Table 3.1). The nuclear membrane possesses many round pores (Figure 3.13), which are formed from holes in both unit membranes at places where the inner and outer membranes are joined. The pores may permit the passage in and out of the nucleus of macromolecules and large particles. For instance, in eucaryotes the components of the

The Nucleus, Cell Division, and Sexual Reproduction

FIGURE 3.13 Electron micrograph of a yeast cell by the freeze-etch technique, showing a surface view of the nucleus. Note the large number of pores in the nuclear membrane. Magnification, 16,650×. (From Guth, E., T. Hashimoto, and S. F. Conti. 1972. J. Bacteriol. 109:869–880.)

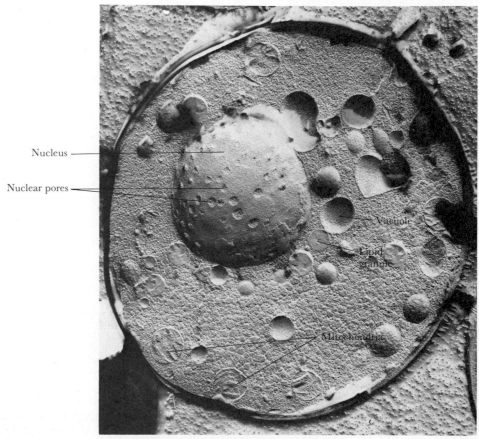

Nucleus

Nuclear pores

Vacuole

Lipid granule

Mitochondria

ribosomes are synthesized in the nucleus but function in the cytoplasm; hence pores are necessary so that these particles can leave the nucleus.

A structure often seen within the nucleus is the **nucleolus,** an area that stains differently from the rest of the nucleus because of its high content of ribonucleic acid (RNA). The nucleolus, the site of ribosomal RNA synthesis, disappears during nuclear division and re-forms when nuclear division is complete.

Chromosomes and DNA The DNA of procaryotes is contained primarily in a single molecule of free DNA (Section 2.6), whereas, in eucaryotes, the DNA is present in more complex structures, the **chromosomes.** *Chromosome* means "colored body," for chromosomes were first seen as structures colored by certain stains. Many chromosome stains involve dyes that react strongly with basic (that is, cationic) proteins called **histones,** which in eucaryotes often are attached to the DNA. Chromosomes also usually contain small amounts of RNA.

Chromosomes of cells not actively dividing are extended and quite thin and hence are difficult to see. At other times (mainly during nuclear division or mitosis, which is discussed below) they contract, making the histone density quite high and the chromosomes easy to see.

The role of the histones in chromosome contraction seems well established. Because DNA is negatively charged (due to the large number of phosphate groups present), there is a tendency for the various parts of the molecule to repel each other. Histones neutralize some of the negative charges, permitting contraction of the chromosomes. The histone content of chromosomes is the same when they are contracted as when they are highly extended; however, the histone proteins are chemically modified at the time of contraction, by phosphorylation, acetylation, and methylation. These chemical modifications change the charge on the histones, thus altering repulsive forces.

In each chromosome, the DNA is a single linear molecule, to which the histones (and other proteins) are attached. Although in higher organisms the length of the DNA molecule in a single chromosome is many times longer than the DNA of procaryotic cells, in yeast (and probably many other microorganisms), the length of the DNA in a single chromosome is actually shorter than in procaryotes. For instance, the total amount of DNA per yeast cell is only three times that of *Escherichia coli,* but yeast has 17 chromosomes, so that the average yeast DNA molecule is much shorter than the *E. coli* DNA. Seen under the electron microscope, yeast DNA molecules all appear linear, showing no evidence of the circularity seen in bacteria.

The DNA content per nucleus varies from species to species, in much the same way as does nuclear size. In addition, the chromosome number also varies greatly, from just a few to many hundreds. Another variable feature is genome size, which is the actual number of distinct genes per cell. Even with the same amount of DNA per cell, genome size can vary from organism to organism, because many genes are often present in many copies. The genome size of various eucaryotes is compared with viruses and procaryotes in Figure 3.14.

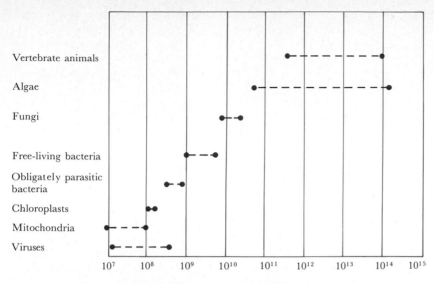

Genome size, molecular weight of DNA per cell

FIGURE 3.14 Range of genome sizes in various groups of organisms. (Data taken from Britten, R., and D. E. Kohne. 1970. Sci. American 222:24–31. Sager, R. 1972. Cytoplasmic genes and organelles. Academic Press, New York. Holliday, R., 1970. 20th Symposium, Society for General Microbiology, Cambridge University Press, New York. Kingsbury, D. T. 1969. J. Bacteriol. 98:1400–1401. Bak, A. L. et al. 1970. J. Gen Microbiol. 64:377–380.)

Cell divison   Cell division in eucaryotes is a highly ordered process in which the key events are (1) DNA synthesis, which leads to doubling of the genetic material of the cell; (2) nuclear division, by a process called **mitosis;** (3) cell division, which involves the formation of a cell membrane separating the two parts of the divided cell; and (4) cell separation, by which the two cells become detached from each other and acquire independent existences. These stages in cell division are illustrated in Figure 3.15.

Sexual reproduction   Although procaryotes show fragmentary sexual processes that are quite different from those of higher organisms (Chapters 11 and 12), eucaryotic microorganisms often exhibit sexual reproduction analogous to that of plants and animals. Sexual reproduction in eucaryotic microorganisms involves conjugation, the coming together and fusing of two cells called **gametes,** which are analogous to the sperm and egg of higher organisms. The conjugating gametes form a single cell called a **zygote,** and the nucleus of the zygote usually results from fusion of the nuclei of the two gametes. The zygote nucleus thus has twice the chromosome complement of the gametes. The chromosome number of the gametes is called the **haploid** number, and the zygote then has twice that many, the **diploid** number (Figure 3.16).

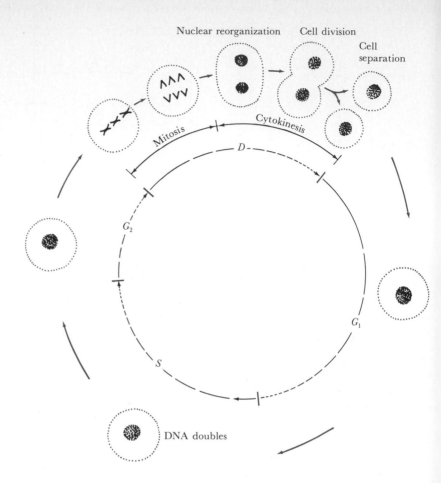

Nuclear reorganization   Cell division

Cell separation

Mitosis

Cytokinesis

D

G_2

G_1

S

DNA doubles

FIGURE 3.15 The cell-division cycle in a eucaryote. S, period during which DNA synthesis takes place; D, period of mitosis and cell division; G_1, growth period before S; G_2, growth period after S. The lengths of the various phases vary among organisms. In some organisms, G_2 is absent.

Multicellular plants and animals are usually diploid, the haploid phase being present only in the short-lived germ cells (sperm and eggs). In eucaryotic microorganisms, structures equivalent to sperm and eggs are formed, but the extent of development of the haploid and diploid phases varies. In many species the predominant growth phase is haploid, and the diploid stage is transitory; in others the diploid phase dominates; and in still others both haploid and diploid phases occur independently.

Meiosis, or "reduction division," is the process by which the change from the diploid to the haploid state is brought about. Whereas mitosis (Figure 3.15) causes duplication and doubling of the chromosome number and yields two progeny cells with the original chromosome number, the process of meiosis reduces the number of chromosomes by one-half and yields haploid cells that are precursors of germ cells (Figure 3.16). The chromosomes of diploids always occur in pairs, called **homologs,** and during meiosis one chromosome of each pair goes to each of the two gametes formed. This regular assortment occurs because the homologs pair specifically at the time of meiosis, and one of each pair is then pulled to each pole by the spindle fibers, after which division of the cell into two cells occurs.

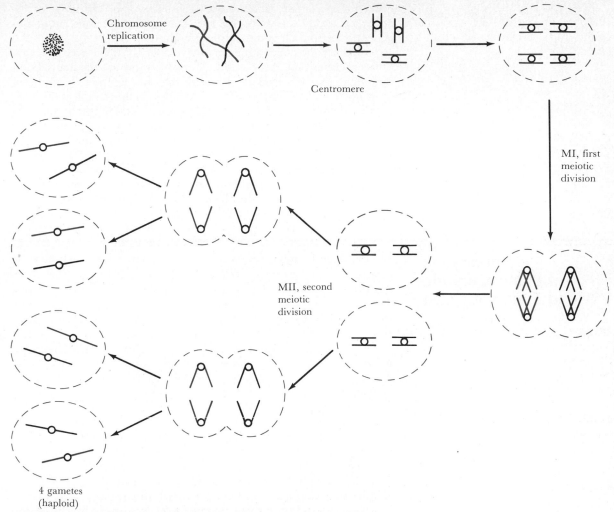

FIGURE 3.16 Diagrammatic representation of meiosis in a cell with two different chromosomes (haploid number = 2, diploid number = 4). MI and MII refer to the first and second meiotic divisions, respectively.

A second division then usually occurs resulting in the formation of four haploid gametes.

The haploid nuclei formed as a result of meiosis become incorporated into cells that are involved in the continuation of the species. These cells are called gametes if they participate directly in the fertilization process. In many microorganisms the haploid cells grow vegetatively for a time before becoming converted into gametes. In other microorganisms the initial haploid cells differentiate into **spores,** which often are dormant structures that play a role in the survival of the organism in nature. These spores germinate to become the forerunners of gametes or of haploid vegetative cell lines that later produce gametes. Great variations in these processes exist among different microorganisms; some of these are described later in this chapter. Despite variations, the result is the same—an alternation of generations from diploid to haploid to diploid.

Eucaryotic microorganisms vary widely in the degree of sexual differentiation that they exhibit. At one extreme are the many fungi and algae that show **sexual dimorphism,** that is, the production of morphologically distinct males and females. In the most general view, a *male* is defined as an organism that produces a gamete that moves in some way toward a female gamete, the latter remaining more or less passive. Sexual dimorphism is often genetically determined; a gene or group of closely linked genes controls maleness, and another set controls femaleness. Upon segregation at meiosis, the male genes are incorporated into one set of haploid nuclei and the female genes into another; each then produces the male or female plants.

3.6

Comparisons of the Procaryotic and Eucaryotic Cell

At this stage it might be useful to draw comparisons between the procaryotic and eucaryotic cell. It should be clear by now that there are profound differences in the structures of these two cell types. Table 3.2 groups these differences into several categories of which the most important are: nuclear structure and function, cytoplasmic structure and organization, and forms of motility.

Several differences listed in Table 3.2 deserve further comment. Microtubules are widespread in eucaryotes, although they seem to be absent in procaryotes. Microtubules are associated with overt motile systems such as flagella and cilia and with mechanisms of chromosome movement in the mitotic apparatus. In these cases the microtubules arise from special structures, basal bodies and centrioles, which are of similar structure. Microtubules are probably not involved in amoeboid motion or cytoplasmic streaming, but in microorganisms without walls they may be involved in maintenance of nonspherical cell shapes. Because of the universal presence of microtubules in eucaryotes, and their apparent absence in procaryotes, it may be that ability to form microtubules was an early acquisition of the evolving eucaryotic cell.

Another difference between procaryotes and eucaryotes deserving emphasis is the size of their respective ribosomes. Procaryotes have ribosomes with a sedimentation constant of 70S, which are composed of two subunits having constants of 50S and 30S. (The sum of 50 and 30 in this instance is 70 rather than 80 because the sedimentation constant is a function of the density of a particle which is mass per volume.) Cytoplasmic ribosomes of eucaryotes have a sedimentation constant of 80S, and these are composed of two subunits with sedimentation constants of 60S and 40S. The ribosomes of the mitochondria and chloroplasts of eucaryotes are 70S, however.

Although Table 3.2 emphasizes the great *structural* differences between procaryotes and eucaryotes, we should remember that these groups of organisms are chemically quite similar. Both contain proteins, nucleic acids, polysaccharides, and lipids of similar composition, and both use the same kinds of metabolic machinery. Thus it is not in their building blocks that procaryotes differ most strikingly

TABLE 3.2 Comparison of the Procaryotic and Eucaryotic Cell

	Procaryotes	Eucaryotes
Nuclear structure and function:		
Nuclear membrane	Absent	Present
Nucleolus	Absent	Present
DNA	Single molecule, not complexed with histones	Present in several or many chromosomes, usually complexed with histones
Division	No mitosis	Mitosis; mitotic apparatus with microtubular spindle
Sexual reproduction	Fragmentary process; no meiosis; usually only portions of genetic complement reassorted	Regular process; meiosis; reassortment of whole chromosome complement
Cytoplasmic structure and organization:		
Plasma membrane	Usually lacks sterols	Sterols usually present
Internal membranes	Relatively simple; mesosomes	Complex; endoplasmic reticulum; Golgi apparatus
Ribosomes	70S in size	80S, except for ribosomes of mitochondria and chloroplasts, which are 70S
Simple membranous organelles	Absent	Present in vacuoles, lysosomes, microbodies (peroxisomes)
Respiratory system	Part of plasma membrane or mesosome; mitochondria absent	In mitochondria
Photosynthetic apparatus	In organized internal membranes or vesicles; chloroplasts absent	In chloroplasts
Cell walls	Present (in most) composed of peptidoglycan	Present in plants, algae, fungi; absent in animals, protozoa; usually polysaccharide
Endospores	Present (in some), very heat resistant	Absent
Gas vesicles	Present (in some)	Absent
Forms of motility:		
Flagellar movement	Flagella of submicroscopic size; each flagellum composed of one fiber of molecular dimensions	Flagella or cilia; microscopic size; composed of microtubular elements arranged in a characteristic pattern of nine outer doublets and two central singlets
Nonflagellar movement	Gliding motility	Cytoplasmic streaming and amoeboid movement; gliding motility
Microtubules	Probably absent	Widespread; present in flagella, cilia, basal bodies, mitotic spindle apparatus, centrioles
Size	Generally small, usually <2 μm in diameter	Usually larger, 2 to >100 μm in diameter

from eucaryotes but in how these building blocks have been put together.

In the rest of this chapter we present a brief overview of the diversity of eucaryotic microorganisms.

3.7

The Algae

The term **algae** refers to a large, morphologically and physiologically diverse assemblage of organisms containing chlorophyll that carry out an oxygen-evolving type of photosynthesis. Although in the past the word *algae* was used to refer to both procaryotes and eucaryotes (the procaryotic forms were called *blue-green algae* but are now called *cyanobacteria*), currently the term is used to refer only to eucaryotic microorganisms. Although most algae are of microscopic size and hence are clearly microorganisms, a number of forms are macroscopic, some of the seaweeds growing in length to over 100 ft. It is somewhat difficult to decide how these macroscopic forms should be distinguished from such clearly nonalgal organisms as mosses and ferns. However, a characteristic of all algae is the production of spores or gametes within unicellular structures, whereas even the simplest mosses bear such reproductive cells within structures surrounded by multicellular walls.

Some algae are motile by means of flagella and seem to be related to protozoa; thus the border separating the algae from the protozoa is rather fuzzy. This need not concern us here, however, as we shall concentrate on giving some idea of the diversity of the group as a whole.

Morphology   There is considerable variability in the vegetative structures of the algae (Figure 3.17). The simplest forms are unicellular, dividing by binary fission. Among these there is a great diversity in cell shape and size. In many cases, the two daughter cells do not separate immediately upon division but remain together, forming chains or amorphous masses of cells. If the cells remain more or less attached to each other, the aggregate is called a *coenobium,* whereas if the cells have separated from each other but are held together by being embedded in mucilage, the arrangement is known as *tetrasporal.* Aggregates formed by many motile cells often show cooperative actions and can be considered more than simple colonies since they resemble in some ways multicellular organisms. Many algae are filamentous, and several types can be recognized: (1) simple, nonseptate filaments (sometimes called *siphonaceous*); (2) simple, unbranched, septate filaments; and (3) branched filamentous forms of various degrees of complexity. Some of the siphonaceous forms grow to fairly large size and can easily be seen macroscopically.

Motility   A number of algae are motile in the vegetative phase, usually because of the presence of flagella. Cilia do not occur in the algae. Simpler flagellate forms, such as *Euglena,* have a single polar flagellum, and flagellated species of the Chlorophyta have either two or four flagella of equal length, arising from one of the poles. The

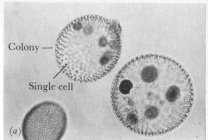

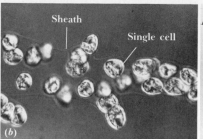

dinoflagellates have two flagella of unequal length and with different points of insertion: the transverse flagellum is attached laterally and wraps partly around the transverse groove of the cell, whereas the longitudinal flagellum originates from the lateral groove but extends lengthwise from the cell. In many cases, algae are nonmotile in the vegetative state but form motile gametes.

Although they are devoid of flagella, many diatoms are motile. The mechanism of this motility is still under study, but apparently it results from secretion of a slime through a special pore system. Some desmid algae exhibit a "creeping" kind of gliding motility, which is also thought to be due to secretion of mucilaginous slime.

Cell-wall structure Algae show considerable diversity in the structure and chemistry of their cell walls. In many cases the cell wall is composed basically of cellulose, but it is usually modified by the addition of other polysaccharides such as pectin (polygalacturonic acid), xylans, mannans, alginic acids, fucinic acid, and so on. In some algae the wall is additionally strengthened by the deposition of calcium carbonate; these are often called *calcareous* or *coralline* ("coral-like") algae. Sometimes chitin is also present in the cell wall.

In the diatoms, which are more fully discussed below, the cell wall is composed of silica, to which protein and polysaccharide are added. Even after the diatom dies and the organic materials have disappeared, the external structure remains, showing that the siliceous component is indeed responsible for the rigidity of the cell. Because of the extreme resistance of these diatom frustules to decay, they remain intact for long periods of time and constitute some of the best algal fossils.

FIGURE 3.17 Various kinds of cell shapes and arrangements in eucaryotic algae: (a) Colonial. *Volvox.* Magnification, 70×. (b) Tetrasporal. *Palmodictyon.* Magnification 250×. (c) Unbranched filament *Ulothrix.* Magnification, 250×. (d) Branched filament. *Cladophora.* Magnification, 48×.

Green algae: Chlorophyta The Chlorophyta comprise a large and diverse group of algae. Because their chlorophyll pigments are the same as those of higher plants, the Chlorophyta are often considered the likely forerunners of the higher plants. The life cycle of one representative, *Chlamydomonas,* is given in Figure 3.18. *Chlamydomonas* is the simplest form of those Chlorophyta whose vegetative cells are motile by means of flagella. In the more complex members of this group, the flagellated cells occur not singly but in colonies, and the individuality of the single cells is lost.

One of the most advanced genera of these colonial algae is *Volvox.* The number of cells in a *Volvox* colony varies from around

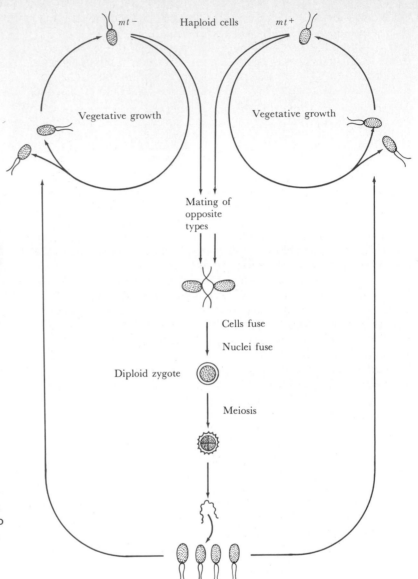

FIGURE 3.18 Life cycle of the green alga *Chlamydomonas reinhardi*. Two mating types, which look alike but are geneticaly distinct (designated mt+ and mt−), fuse to form the diploid zygote.

mt − Haploid cells *mt +*

Vegetative growth   Vegetative growth

Mating of
opposite
types

Cells fuse

Nuclei fuse

Diploid zygote

Meiosis

500 in some species to over 50,000 in others. Each cell is surrounded by a gelatinous sheath and hence is physically separated from all other cells, but in some species the cells are joined to one another by cytoplasmic strands. The arrangement of the cells in a colony is quite regular (Figure 3.17a). The cells are highly coordinated, and the *Volvox* colony moves through the water because of the integrated beating of the flagella of the individual cells.

The production of a new *Volvox* colony can occur in two ways, asexually or sexually. Most of the cells of a colony are incapable of giving rise to new colonies, but as a colony ages, a few cells in one part of the colony are converted into asexual reproductive cells called *gonidia,* each of which can be the forerunner of a new *Volvox* colony. Each gonidium is 10 or more times the diameter of a vegetative cell and lies in a gelatinous sac projecting inward from the surface of the colony. The gonidium divides a number of times in a very regular sequence, forming a group of cells called an *autocolony.* When the autocolony reaches maturity, it escapes from the parent colony and assumes an independent existence. An individual cell isolated from the colony can never initiate the formation of a new colony; thus the cells of a colony are not independent. *Volvox* cells, therefore, are more like the cells of a multicellular organism than like those of a microorganism.

Sexual reproduction in *Volvox* occurs by the formation of egg and sperm cells. In some species there is a sexual differentiation such that some colonies form male gametes and others form female gametes. Male gametes develop from enlarged cells resembling gonidia. In the female colonies, only a small number of the cells develop into eggs, each egg resembling a young gonidium. Fertilization occurs through the penetration of the egg by an individual male gamete. The zygote is thick-walled and usually contains a reddish pigment; it does not germinate immediately but remains dormant until it is liberated by death or decay of the parent colony. Under appropriate environmental conditions, dormancy is broken, meiosis occurs, and a haploid biflagellate **zoospore** is produced and becomes the forerunner of a new colony. This colony arises by a process similar to that described for asexual reproduction from gonidia.

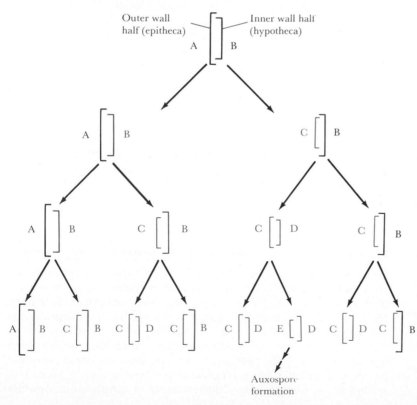

FIGURE 3.19 Cell wall formation during cell division in a diatom (color indicates new frustule formation). At division each frustule half remains intact and each half becomes the outside (epitheca) of one daughter cell. The result is two different-sized cells: one the size of the parent, the other smaller. The letters in the figure refer to relative sizes, A being larger than B which is larger than C, etc. A given frustule half is always formed as a hypotheca, but becomes an epitheca at the next and all subsequent divisions. Frustule halves remain intact indefinitely; note that after three generations the original A and B components are unchanged. The letters also demonstrate that the population will have a wide range of cell sizes after only a few generations.

FIGURE 3.20 Diatom cells (*Stauroneis anceps*) of the same culture taken from different stages of population growth, to illustrate the reduction in size and change in shape that occur during successive cell divisions. All magnifications, 3000×. (a) Initial size. (b) Intermediate. (c) Small. (From Hostetter, H. P., and R. W. Hoshaw. 1972. *J. Phycol.* 8:289–296.)

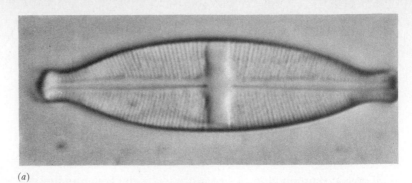

(a)

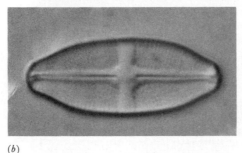

(b)

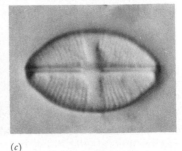

(c)

Diatoms The **diatoms** are an interesting algal group of worldwide distribution, with freshwater, marine, and soil forms. The distinctive feature of the diatom is its siliceous frustule (cell wall) composed of two overlapping halves. The frustules have distinct markings, which are characteristic of each species (see Figure 3.2*b*). Because of the presence of a rigid wall, diatoms face a special problem in cell division. The manner of cell division is illustrated in Figure 3.19, and we can see that cell size decreases with each successive division. As shown in Figure 3.20, not only does the size of the cell decrease, but its shape and proportions are also altered. Such diminution in size is eventually compensated by **auxospore** formation, in which the two frustule halves pull apart, the protoplast enlarges, and new frustule walls are formed. The old frustule is discarded, and the now enlarged diatom cell undergoes successive divisions again. In many diatoms, auxospore formation occurs as a result of a sexual process. Vegetative diatom cells, which are diploid, undergo meiosis and the haploid nuclei fuse during auxospore formation, in the process reforming the enlarged vegetative diploid cell.

3.8

The Fungi

There is no general agreement among microbiologists regarding the limits of the group of microorganisms called **fungi.** The group is often defined as comprising those eucaryotic microorganisms that have rigid cell walls and lack chlorophyll, although by this definition certain organisms that are clearly nonchlorophyllous derivatives of algae would have to be considered fungi. In addition, many fungi have affinities with the protozoa. However, the members of two great groups of fungi, the Ascomycetes and the Basidiomycetes, cannot be related in any direct way to either the algae or the protozoa. Another group of organisms, the slime molds, is sometimes included with the fungi because of formation of fruiting bodies similar to those of many fungi, but the slime molds differ in so many ways from the fungi that it is easier to consider them a separate group. Within the group classified as true fungi there are around 80,000 named species.

The habitats of the fungi are quite diverse. Some are aquatic, living primarily in fresh water, and a few marine fungi are also known; most, however, have terrestrial habitats, in soil or on dead plant matter, and these types often play crucial roles in the mineralization of organic carbon. A large number of fungi are parasites of terrestrial plants; indeed, fungi cause the majority of economically significant diseases of crop plants. A few fungi are parasitic on animals, including humans, although members of this group are much less significant as animals pathogens than are bacteria.

Vegetative structure There are two general growth forms in fungi: molds and yeasts. Generally each particular fungus fits one or the other pattern. The vegetative structure of a mold is often called a **thallus.** The most typical fungal thallus is composed of filaments, which are usually branched. Each individual filament is called a **hypha** (plural, **hyphae**), and a mass of hyphae is called a **mycelium** (Figure 3.21). Although yeasts are usually unicellular, filamentous growth is possible, depending on environmental conditions. Yeastlike growth occurs by a budding process, in which the daughter cell separates from the mother as soon as it has matured (Figure 3.22), whereas filamentous growth occurs by continuous extension of hyphal tips. Some fungi pathogenic for animals are dimorphic, growing either as mycelial or as yeastlike forms. When growing at temperatures around 37°C (such as are found in the animal body), the fungus is yeastlike, but when growing in laboratory media at lower temperatures the organism is myceliumlike.

Fungi are classified in four different groups:

Phycomycetes	Basidiomycetes
Ascomycetes	Deuteromycetes (also called *Fungi imperfecti*).

FIGURE 3.21 (a) Phase contrast micrograph of a fungal mycelium. (b) Scanning electron micrograph of a mycelium, from G. Cole. 1975. Can. J. Bot. 53:2983–3001.

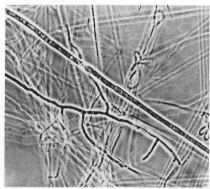
(a)

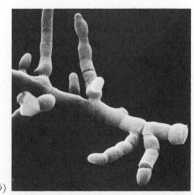
(b)

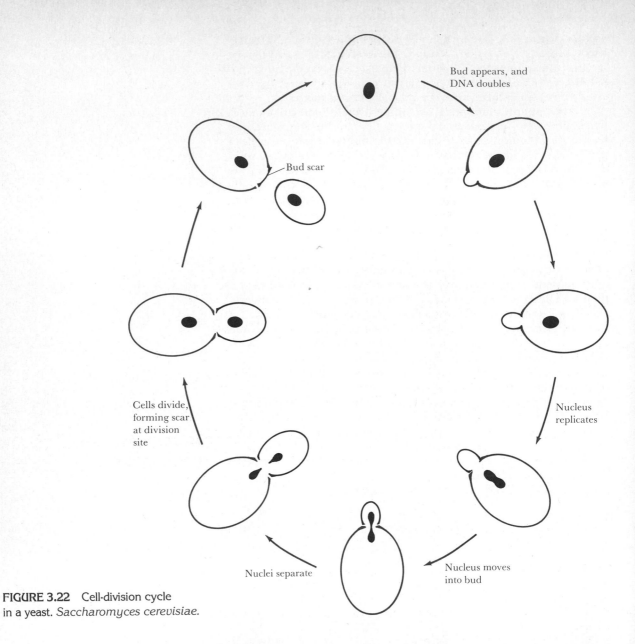

FIGURE 3.22 Cell-division cycle
in a yeast. *Saccharomyces cerevisiae.*

Classification is based on several features of which the most important is the type of reproductive spores formed, both asexual and
sexual. These spores function in dispersal of the organism to new
habitats. Asexual spores are usually resistant to drying or radiation
but are not very heat resistant and exhibit no dormancy; they are
able to germinate when moisture becomes available, often even in
the absence of nutrients. Sexual spores are usually more resistant to
heat than are asexual spores, although no fungus spore shows the
extreme heat resistance characteristic of the bacterial endospore.

Sexual spores often exhibit dormancy. Only when they have been activated in some way, such as by mild heat or by certain chemicals, do they begin to show signs of germination.

FIGURE 3.23 Phycomycetes. (a) Asexual reproduction. (b) Sexual zygospore formation.

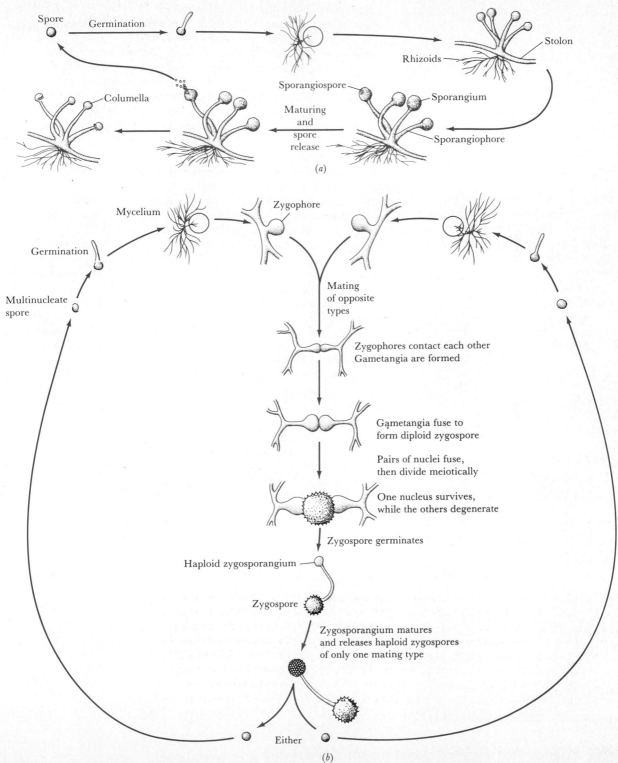

Phycomycetes are generally considered to be the most primitive. The best studied subgroup is the zygomycetes, all members of which are molds. Light-microscopic examination reveals no individual cells within the hyphal filaments. This lack of cross-walls or **septa** (singular **septum**) leads to their designation as nonseptate fungi. The hyphae contain many nuclei, however, and the organisms are **coenocytic** (meaning "without cells"). Asexual reproduction in this group occurs by the differentiation of certain parts of the hyphae, which support specialized structures called **sporangia** (singular, **sporangium**) containing **sporangiospores** (Figure 3.23a). It is characteristic of the Phycomycetes that their asexual spores are enclosed in a sac, the sporangium. When mature, the sporangia burst, releasing the spores to initiate a new cycle of growth. Sexual reproduction involves the joining of two specialized hyphae called **gametangia** with the formation of a new structure, the **zygospore,** between them (Figure 3.23b). The diploid zygospore contains a nucleus formed by the fusion of two nuclei, one coming from each of the gametangia. Often, two different mating types, designated + and −, are required and such species are said to be **heterothallic** (meaning "other body"). Some Phycomycetes are **homothallic** (meaning "same body") and have only one mating type. Although both sexual patterns are well known, a given Phycomycete species is constant, that is, either heterothallic or homothallic.

The **ascomycete** group contains both molds and yeasts. The molds appear to have septa when viewed in the light microscope (Figure 3.24a), and they are traditionally referred to as *septate* fungi. However, electron micrographs show that there are pores in these septa and the hyphae are in fact coenocytic (Figure 3.24b). Asexual reproduction consists of the formation of distinct **conidiospores** on specialized hyphal extensions called **conidiophores** (Figure 3.24c). In contrast to the Phycomycetes, the asexual spores of the Ascomycetes are always external and never enclosed. Sexual recombination is usually heterothallic and results in the formation of either four or eight spores in a sac called an **ascus,** hence the name **ascospores** (Figure 3.24d and e). Ascospore formation is observed in ascomycetous yeasts, such as *Saccharomyces cerevisiae,* as well as in molds.

Basidiomycetes are usually molds, although some basidiomycetous yeasts are known. Hyphal structure is similar to that seen

FIGURE 3.24 Ascomycetes. (a) Phase contrast micrograph of hyphae with septa. Magnification, 3500×. (b) Electron micrograph of pores in septa. Magnification, 17,000×. (a and b from Hunsley, D. and G. Gooday. 1974. Protoplasma 82:125–146). (c) Asexual reproduction.

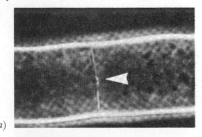

(a)

(b)

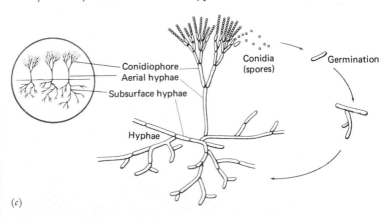

(c)

Conidiophore
Aerial hyphae
Subsurface hyphae
Conidia (spores)
Germination
Hyphae

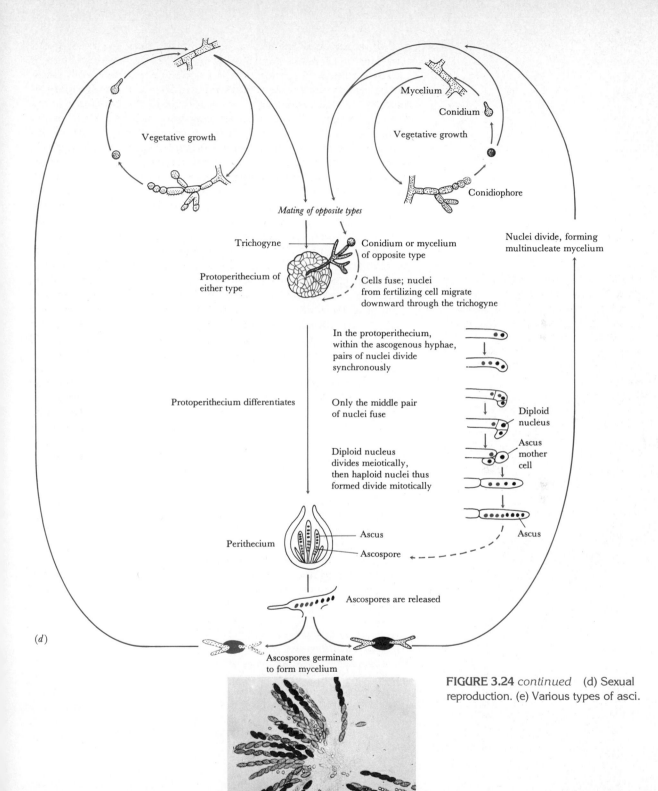

Vegetative growth

Mycelium

Conidium

Vegetative growth

Conidiophore

Mating of opposite types

Nuclei divide, forming
multinucleate mycelium

Trichogyne

Conidium or mycelium
of opposite type

Protoperithecium of
either type

Cells fuse; nuclei
from fertilizing cell migrate
downward through the trichogyne

In the protoperithecium,
within the ascogenous hyphae,
pairs of nuclei divide
synchronously

Protoperithecium differentiates

Only the middle pair
of nuclei fuse

Diploid
nucleus

Ascus
mother
cell

Diploid nucleus
divides meiotically,
then haploid nuclei thus
formed divide mitotically

Ascus

Peritherium

Ascus

Ascospore

Ascospores are released

Ascospores germinate
to form mycelium

(*d*)

(*e*)

FIGURE 3.24 *continued*   (d) Sexual
reproduction. (e) Various types of asci.

in the Ascomycetes, appearing septate in the light microscope, but revealing pores in electron micrographs. Asexual reproduction is usually confined to vegetative haploid mycelial growth, formation of asexual spores being rare. Most Basidiomycetes are heterothallic and their sexual fusion results in the formation of macroscopic structures, often impressive in size, called **fruiting bodies.** The most common examples are the mushrooms. When a group of mushrooms is observed in a given area, even of several square meters, it is found that all are connected by common diploid mycelia beneath the soil, a remarkable extent for a microorganism. The life cycle of a mushroom is shown in Figure 3.25*a*. The spores formed are called **basidiospores** because they are formed at the tip of a differentiated portion of the hyphae called a basidium (meaning "club" for its shape) (Figure 3.25*b*). Occasionally hyphae will fuse, but complete fruiting body formation does not occur immediately, development stopping with the production of small buttonlike structures, which are the mushroom primordia (the first recognizable group of cells that will form the mature mushroom). If one of these buttons is cut open the structural outlines of the mature mushroom can be seen, but these forms are usually buried beneath the litter or in the soil and are not ordinarily in view. The buttons may remain underground for considerable periods of time until favorable environmental conditions develop. After heavy rains the buttons will usually quickly enlarge into mature fruiting bodies (Figure 3.25*a*). Although hyphal growth does take place, much of the expansion at this stage is due to the uptake of water. This expansion is usually so rapid that a mature fruiting body can be formed within a few hours or a few days. In most cases a number of fruiting bodies will mature at the same time in a given area of soil, producing so-called *flushes* of fruiting bodies; one day a given locality can be apparently devoid of fruiting bodies, while the next day a large number may be present.

Deuteromycetes are septate* (at the light-microscope level) fungi for which no sexual form has been observed. Since the form of sexual spores is crucial in distinguishing Ascomycetes and Basidiomycetes, classification is impossible in the absence of sexual reproduction. The Deuteromycete group was originally set up as a provisional class for new isolates with the expectation that each member would be transferred to the Ascomycetes or Basidiomycetes once the sexual form was observed. Indeed this occurred for many isolates, but there remain many which have never reproduced sexually, even after several years in culture. It has been learned that growth medium and culture conditions are important and must be controlled carefully to encourage the sexual cycle, but many patient attempts have been unsuccessful with the Deuteromycetes.† There is another explanation for the lack of sexual activity in this group. Many of the higher fungi are known to be heterothallic. Standard microbiological practice is, of course, to maintain organisms in pure culture. The maintenance of a pure culture of a heterothallic fungus means that only one mat-

*The ability to form septa places Ascomycetes, Basidiomycetes, and Deuteromycetes together as the "higher fungi."

†In botany the "perfect" form of a plant is the sexual one, and fungi were mainly studied by botanists in the early history of microbiology, hence the term *Fungi imperfecti.*

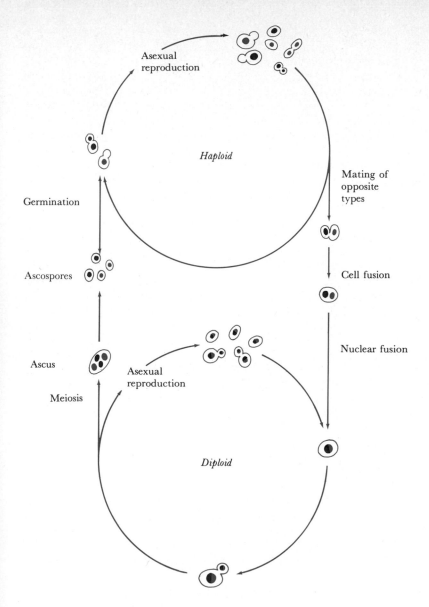

Asexual reproduction

Haploid

Germination

Mating of opposite types

Ascospores

Cell fusion

Ascus

Nuclear fusion

Meiosis

Asexual reproduction

Diploid

FIGURE 3.25 Life cycle of a typical yeast, *Saccharomyces cerevisiae.*

ing type is present. By definition this one type cannot undergo sexual reproduction by itself, and the organism therefore remains classified as a Deuteromycete. A logical conclusion is that many Deutero-mycete cultures will continue to be classified as such unless and until their appropriate mating partner is cultured with them.

Yeasts The budding cycle of a typical yeast is shown in Figure 3.22. Note that DNA synthesis and nuclear division are closely correlated with the budding process. This is in contrast to the situation in hyphal growth, where elongation of the multinucleate hypha is not synchronous with nuclear division. The sexual cycle of a typical yeast is il-

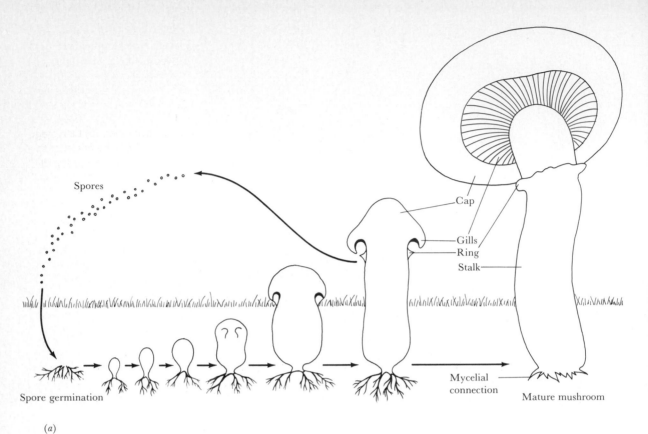

Spores

Cap

Gills
Ring
Stalk

Mycelial
connection

Mature mushroom

Spore germination

(a)

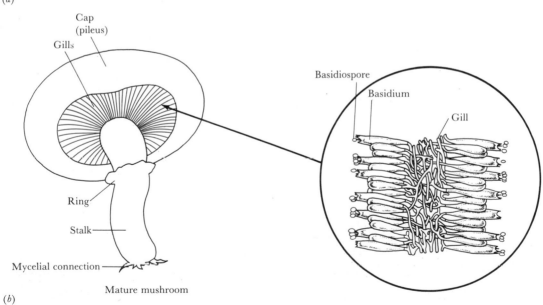

Cap
(pileus)

Gills

Basidiospore

Basidium

Gill

Ring

Stalk

Mycelial connection

Mature mushroom

(b)

FIGURE 3.26   Basidiomycetes. (a) Sexual reproduction. (b) Basidiospores in relation to mushroom structure.

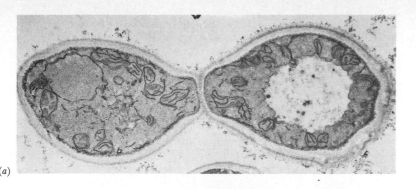

(a)

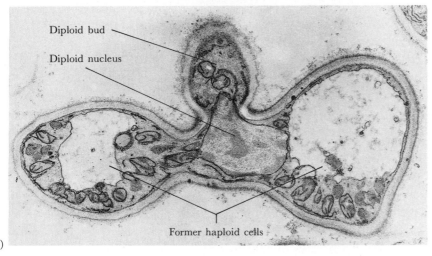

Diploid bud

Diploid nucleus

Former haploid cells

(b)

FIGURE 3.27 Electron micrographs of the conjugation process in a yeast, *Hansenula wingei*. Both magnifications, 12,000×. (a) Two conjugating cells, which have fused at the point of contact and have sent out protuberances toward each other. (b) Late stage of conjugation. The nuclei of the two cells have fused and the diploid bud has formed at a right angle to the conjugation tube. This bud will eventually separate from the conjugants and become the forerunner of a diploid cell line. (From Conti, S. F., and T. D. Brock. 1965. J. Bacteriol. 90:524.)

lustrated in Figure 3.26. Fertilization occurs through the fusion of two vegetative cells, usually of opposite mating types. The cellular events of fusion during sexual conjugation in yeast are shown in Figure 3.27. The two cells send out protuberances along their point of contact. At the same time, fusion of the cell walls takes place, so that the cells are permanently joined. After fusion is complete, the central portion of the cell wall separating the two cells dissolves, probably through the action of a lytic enzyme, and the cytoplasms of the two cells mingle. The two haploid nuclei move to the center, fuse, and form a diploid nucleus. The diploid cell forms as a bud at right angles to the fusion tube, and the diploid nucleus moves into this bud. This diploid cell now separates from the conjugation tube and begins to grow vegetatively; the parent cells usually disintegrate.

3.9

The Slime Molds

The slime molds are organisms that have at times been classified by some workers as fungi and by others as protozoa. The confusion arises because the vegetative structure of a slime mold is a protozoanlike amoeboid mass, whereas its fruiting body resembles the fruiting body of a fungus and produces spores with cell walls. Because the slime mold fruiting body was first studied by mycologists, the slime molds have traditionally been discussed in mycology rather than protozoology texts, even though the affinities of these organisms to protozoa are just as close as they are to fungi.

The slime molds can be divided into two groups, the **cellular slime molds,** whose vegetative forms are composed of single amoeba-like cells, and the **acellular slime molds,** whose vegetative forms are acellular masses of protoplasm of indefinite size and shape (called *plasmodia*). Both types live primarily on decaying plant matter, such as leaf litter, logs, and soil. Their food consists mainly of other microorganisms, especially bacteria, which they engulf and digest. One of the easiest ways of detecting and isolating slime molds is to bring into the laboratory small pieces of rotting logs and place them in moist chambers; the amoebas or plasmodia will proliferate, migrate over the surface of the wood, and eventually form fruiting bodies, which then can be observed under a dissecting microscope. The fruiting bodies of slime molds are often very ornate and colorful. Only the cellular slime molds will be discussed here.

Cellular slime molds   Few of the described species of cellular slime molds have been studied in detail. An exception is the genus *Dictyostelium,* for which the life cycle has been worked out and for which a considerable amount of information is known about the physiology and biochemistry of fruiting-body formation. The vegetative cell of *Dictyostelium discoideum* is an amoeboid cell (Figure 3.28a) that feeds on living or dead bacteria and grows and divides indefinitely. When the food supply has been exhausted, a number of amoebas aggregate to form a pseudoplasmodium, a structure in which the cells lose some of their individuality but do not fuse (Figure 3.28b and c). This aggregation is triggered by the production of *acrasin,* a substance identified as cyclic adenosine monophosphate (Section 9.5), which attracts other amoebas in a chemotactic fashion. Those cells which are the first to produce acrasin serve as centers for the attraction of other amoebas, and since each newcomer also produces acrasin, the centers increase rapidly in size. The swarm patterns around such aggregating masses are shown in Figure 3.28b and c.

Because the pseudoplasmodium formed resembles a slimy, shell-less snail in appearance and movement, it is called a "slug." Around the outside is secreted a mucoid sheath that probably protects the periphery of the slug from drying. The pseudoplasmodial slug migrates as a unit across the substratum as a result of the collective action of the amoebas (Figure 3.28d). Presumably there is some mechanism that coordinates the movements of the individual amoebas so that the slug maintains its unity, but the nature of this mechanism is unknown. The pseudoplasmodium usually migrates for a period of hours, and as it is positively phototactic, it will migrate toward the light. In nature the pseudoplasmodium is presumably formed within the dim recesses of soil or decaying bark and migrates in response to light to the surface, where fruiting-body formation takes place. Thus the formation of the pseudoplasmodium and its behavioral responses are essential features to ensure that the spores formed will be borne in the air and effectively dispersed.

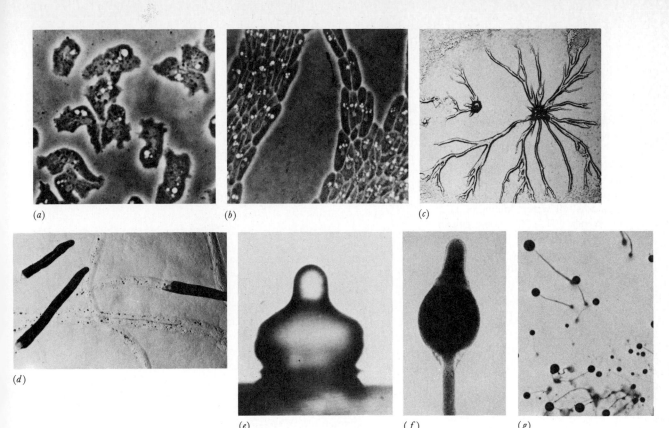

(a)   (b)   (c)

(d)   (e)   (f)   (g)

Fruiting-body formation begins when the slug ceases to migrate and becomes vertically oriented (Figures 3.28e–g and 3.29). The fruiting body is differentiated into a stalk and a head; cells in the forward end of the slug become stalk cells and those from the posterior end becomes spores. The amoebas that form the stalk cells begin to secrete cellulose, which provides the rigidity of the stalk. Other amoebas, from the rear of the slug, swarm up the stalk to the tip and form the head; most of these become differentiated into spores that are usually embedded in slime. Upon maturation of the head, the spores are released and dispersed. Each spore germinates and forms a single amoeba, which then initiates another round of vegetative growth.

The cycle of fruiting-body and spore formation described above is asexual. There is some evidence for a sexual cycle as well, but not all of the details are known and the account to be given here is therefore still somewhat preliminary. In cellular slime molds, the sexual cycle apparently involves formation of a reproductive structure, the **macrocyst.** Macrocysts are structures of multicellular origin that develop from simple aggregates of amoebas, which, when mature, are enclosed in a thick cellulose wall. Probably as a result of conjugation of two amoebas, a single large amoeba develops near the center of the cell mass and becomes actively phagocytic. This phagocytic cell continues to enlarge and engulf amoebas until eventually all surrounding cells have been engulfed. At this stage a cellulose wall de-

FIGURE 3.28 Photomicrographs of various stages in the life cycle of the cellular slime mold *Dictyostelium discoideum*. (a) Amoebas in preaggregation stage. Note irregular shape, lack of orientation. Magnification, 330×. (b) Aggregating amoebas. Notice the regular shape and orientation. The cells are moving in streams in the direction indicated by the arrow. Magnification, 330×. (c) Low-power view of aggregating amoebas. Magnification, 11×. (d) Migrating pseudoplasmodia (slugs) moving on an agar surface and leaving trails of slime in their wake. Magnification, 13×. (e) Early stage of fruiting-body formation. Magnification, 53×. (f) A late stage of a developing fruiting body. Magnification, 87×. (g) Mature fruiting bodies. Magnification, 6×. (From Raper, K. B. 1960. Proc. Am. Phil. Soc. 104:579.)

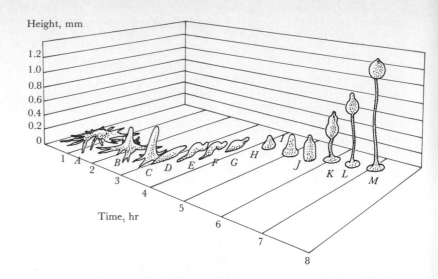

FIGURE 3.29 Stages in fruiting-body formation in the cellular slime mold *D. discoideum:* (A–C) aggregation of amoebas; (D–G) migration of the slug formed from aggregated amoeba; (H–L) culmination and formation of the fruiting body; (M) mature fruiting body. (From Bonner, J. T. 1967. The cellular slime molds, 2nd rev. ed. Princeton University Press, Princeton, N.J. Reprinted by permission of the publisher.)

velops around the limiting membrane of the now greatly enlarged amoeba, and the mature macrocyst is formed. During a period of dormancy, the putative diploid nucleus undergoes meiosis, haploid nuclei are formed, and by progressive nuclear divisions and cleavage of the cytoplasm a new generation of vegetative amoebas is formed. The macrocyst then germinates to release amoebas that reinitiate vegetative growth. There is some evidence for genetic recombination during macrocyst formation, and there is strong evidence that mating types exist, since many strains do not form macrocysts alone but do when paired with other strains. The occurrence of a sexual cycle may enable the slime mold to survive cold, drought, and other adverse conditions, to which the macrocyst is quite resistant. In comparison, the spores formed in the asexual cycle are not very resistant, and function only to disperse the organism to new locations.

3.10

The Protozoa

The simplest definition of **protozoa** states that they are unicellular animals, but as usual, such a general definition leaves many questions open, not the least of which is, What is an animal? On the one hand, some motile, chlorophyll-containing organisms (for example, *Volvox* and *Chlamydomonas*) are classified either as algae or as protozoa, depending on whether the classifier leans to botany or zoology. On the other hand, some organisms (for instance, cellular and acellular slime molds) are classified as either fungi or protozoa. Thus the lines separating the protozoa from other groups of eucaryotic microorganisms are fuzzy. In the present discussion we shall ignore these lines and concentrate instead on typical protozoa. A typical protozoan is a unicellular organism that lacks a true cell wall and obtains its food phagotrophically. Protozoa vary considerably in size and shape. Some are as small as larger bacteria, whereas others are large enough to be seen without a lens. Protozoa are found in a variety of freshwater and marine habitats; a large number are parasitic in other animals, and some are found growing in soil or in aerial habitats, such as on the surfaces of trees.

In general, the protozoa feed by ingesting particulate or macromolecular materials. The uptake of macromolecules in solution occurs by a process called **pinocytosis** (derived from a Greek word meaning "to drink"). Fluid droplets are sucked into a channel formed by the invagination of the cell membrane, and when portions of this channel are pinched off, the fluid is enclosed within a membranous vacuole. Any solutes or macromolecules dissolved in the fluid are thus taken directly into the cell. Pinocytosis differs from active transport in that it is a relatively nonspecific process. Most protozoa are also able to ingest particulate material by **phagocytosis,** a process in which bacteria-size and larger particles are ingested and digested. Phagocytosis also occurs in specialized cells of higher animals, and the process is discussed in some detail in relation to host defenses against infection in Section 15.7.

Characteristics used in separating protozoal groups As befits organisms that "catch" their own food, most protozoa are motile; indeed, their mechanisms of motility are key characteristics used to subdivide the phylum Protozoa into groups (Table 3.3). Protozoa that move by

---

TABLE 3.3 Major Taxonomic Groups of the Phylum Protozoa

Subphylum Sarcomastigophora: flagellate or amoeboid motility; some have both means of locomotion. Single type of nucleus. Typically, no spore formation:
  Superclass Mastigophora (the flagellates): flagellate motility. Cell division by longitudinal binary fission. Sexual reproduction rare. Amoeboid forms frequent in some species. Many species parasitic. See sketch (*a*).
  Superclass Sarcodina (the amoebas): amoeboid motion. Flagella, when present, restricted to developmental stages. Cell body naked or with external or internal tests or skeletons. Cell division by binary fission. Most species free-living. See sketch (*b*).
Subphylum Sporozoa: spores typically present. Single type of nucleus. Cilia and flagella absent except for flagellated gametes in some groups. All species parasitic. See sketch (*c*).
Subphylum Ciliophora (the ciliates): ciliary motility. Two types of nuclei: micronucleus and macronucleus. Cell division by transverse binary fission. Most species free-living. See sketch (*d*).

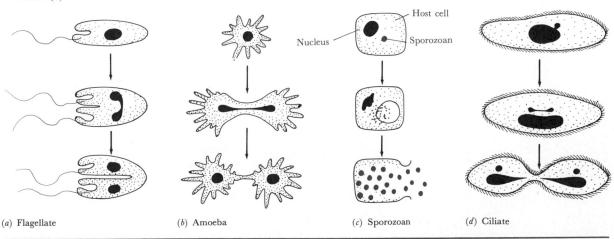

(*a*) Flagellate   (*b*) Amoeba   (*c*) Sporozoan   (*d*) Ciliate

---

(a)

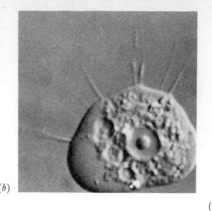

(b)

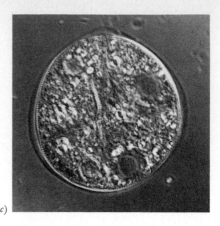

(c)

FIGURE 3.30 Photomicrographs of naked and shelled amoebas. (a) *Amoeba proteus,* a naked form. Magnification, 150×. (b) *Arcella* sp. The thin shell has openings through which pseudopodia extend. Magnification, 1500×. (c) *Actinosphaerium* sp. The rigid shell, with distinct markings, is composed of silica. Magnification, 530×.

FIGURE 3.31 Photomicrograph of the flagellate *Trypanosoma gambiense* in blood. Magnification, 1400×.

Trypanosomes

Red blood cells

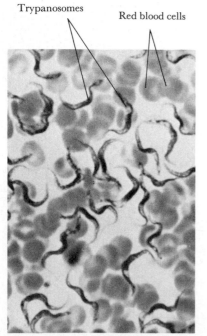

amoeboid motion are called Sarcodina; those using flagella, Flagellata or Mastigophora; and those using cilia, Ciliophora. The Sporozoa, a fourth group, are usually nonmotile and are all parasitic for higher animals. The Sporozoa often divide by multiple fission of a single cell into a number of smaller cells called "spores"—hence the name of the group. Table 3.3 describes and illustrates the representative structural characteristics of the major protozoal groups.

Although most protozoa are naked, some of the Sarcodina produce shells, or tests (Figure 3.30), which are rigid external structures either secreted by the cell or composed of foreign materials such as sand grains or diatom frustules picked up from the environment. Those sarcodines called *foraminifera* (see Figure 3.2c) produce calcium carbonate shells, whereas the *radiolaria* (Figure 3.2d) have shells of either silica or strontium sulfate. A shell is not necessarily the outermost limit of the protozoal cell; in many of the foraminifera the protoplasm can extend beyond, and may even cover, the external portion of the shell.

Some protozoa are sessile, attached by their bases with holdfasts that permit them to attach to a substratum. The structure of the holdfast is distinctive and is of use in classification. Some sessile protozoa have developed a colonial existence, in which many individuals of the same species associate and form structures of characteristic shape and size. Again, many protozoa form **cysts,** which are resting structures usually resistant to drying and occasionally resistant to other noxious agents. These enable the organism to endure a period of bad times or aid in dispersal.

Parasitic protozoa (Figure 3.31) usually show marked specificity for the animals they infect, and host range is used as one characteristic in the classification of these forms. Often these parasites show complex life cycles, alternating at different stages from one host to another (for malaria, see Section 17.12).

In contrast to those of the algae and fungi, the life cycles of most protozoa are relatively simple. Some show no sexual reproduction at all, whereas those that reproduce sexually usually exhibit a simple life cycle, without the complexities described earlier in this chapter for algae and fungi. The protozoa are a fascinating group, worthy of much more study by microbiologists.

Dustin, P. 1980. Microtubules. Sci. Am. 243:66–76. Clear description of the many functions of microtubules: good illustrations.

Fawcett, D. W. 1981. The cell. W. B. Saunders, Philadelphia. Well-illustrated survey of components of eucaryotic cells.

Gray, M. W., and **W. F. Doolittle.** 1982. Has the endosymbiont hypothesis been proven? Microbiol. Rev. 46:1–42. An up-to-date summary of endosymbiosis with molecular considerations (compare to Wallace, below).

Holwill, M. E. J. 1974. Some physical aspects of the motility of ciliated and flagellated microorganisms. Sci. Prog. 61:63–80. An advanced but understandable review of how flagella and cilia move.

Lehninger, A. L. 1975. Biochemistry, 2nd ed. Worth, New York. Chapter 19 has a good discussion of mitochondrial structure, Chapter 22 has a brief discussion of chloroplast structure.

Margulis, L. 1981. Symbiosis in cell evolution. W. H. Freeman, San Francisco. A current summary in favor of the endosymbiosis hypothesis.

Morré, D. J. 1975. Membrane biogenesis. Annu. Rev. Plant Physiol. 26:441–481. Relates membrane biochemistry to membrane structure in eucaryotes. Good discussion of the interrelationships between different membranous structures in eucaryotes.

Richardson, M. 1974. Microbodies (glyoxysomes and peroxisomes) in plants. Sci. Prog. 61:41–61. An excellent brief review of the structure, function, and methods of study of microbodies.

Stanier, R. Y., and **C. B. van Niel.** 1962. The concept of a bacterium. Arch. Mikrobiol. 42:17–35. Now classic paper pointing out the profound differences between procaryotes and eucaryotes.

Wallace, D. C., 1982. Structure and evolution of organelle genomes. Microbiol. Rev. 46:208–240. Molecular biology of chloroplasts and mitochondria with discussion of endosymbiosis (compare to Gray and Doolittle, above).

Bold, H. C., and **M. J. Wynne.** 1978. Introduction to the algae: structure and reproduction. Prentice-Hall, Englewood Cliffs, N.J. An up-to-date, well-illustrated textbook on the algae.

Fritsch, F. E. 1935, 1945. The structure and reproduction of the algae, vols. I and II. Cambridge University Press, New York. (Reprinted, 1965.) Although quite dated and with very unsatisfactory illustrations, these two volumes still provide the only detailed overview of all algal groups, and they are an indispensable guide to the older literature.

Pringsheim, E. G. 1949. Pure cultures of algae; their preparation and maintenance. Cambridge University Press, New York. (1967. Hafner, New York.) A good introduction for those beginning the laboratory study of algae.

Rosowski, J. R., and **B. C. Parker.** 1982. Selected papers in phycology II. Phycological Society of America, Lawrence, Kans. Updated version with key papers from virtually all areas of algal physiology and taxonomy.

Smith, G. M. 1950. The fresh-water algae of the United States, 2nd ed. McGraw-Hill, New York. Out of date but still the best treatment of the freshwater algae.

Stein, J. R., ed. 1973. Handbook of phycological methods: culture methods and growth measurements. Cambridge University Press, New York. Thirty specialists write individual chapters on the practical aspects of growth and culture of marine and freshwater algae. Recommended.

Stewart, W. D. P., ed. 1975. Algal physiology and biochemistry. Blackwell Scientific, Oxford. A series of articles providing an up-to-date picture of algal metabolism.

Supplementary Readings
Eucaryotic Cell Structure

The Algae

The Fungi

Alexopoulos, C. J., and **C. W. Mims.** 1979. Introductory mycology, 3rd ed. John Wiley, New York. Revised edition of a widely used book on all aspects of fungi.

Cole, G. T., and **R. A. Samson.** 1979. Pattern of development in conidial fungi. Pitman, London. Excellent illustrations of a specialized aspect of mycology.

Dahlberg, K. R., and **J. L. Van Etten.** 1982. Physiology and biochemistry of fungal sporulation. Annu. Rev. Phytopathol. 20:281–301.

Smith, J. E., and **D. R. Berry,** eds. 1978. The filamentous fungi, vol. 3. Developmental mycology. John Wiley, New York. A book with 22 well-referenced chapters written by experts on separate features of mold growth.

The Slime Molds

Erdos, G. W., A. W. Nickerson, and **K. B. Raper.** 1973. The fine structure of macrocyst germination in *Dictyostelium mucoroides.* Dev. Biol. 32:321–330. Research paper describing aspects of the sexual process.

Gray, W. D., and **C. J. Alexopoulos.** 1968. Biology of the myxomycetes. Ronald Press, New York. An extensive, advanced text on the acellular slime molds.

Loomis, W. F., ed. 1982. The development of *Dictyostelium discoideum.* Academic Press, New York. Eleven chapters by experts on various aspects of cellular slime mold physiology.

Raper, K. B. 1960. Levels of cellular interaction in amoeboid populations. Proc. Am. Philos. Soc. 104:579–604. The classic paper on the cellular slime molds.

Godfrey, S. and **M. Sussman.** 1982. The Genetics of Development in *Dictyostelium discoideum.* Annu. Rev. Genet. 16:385–404.

The Protozoa

Levandowsky, M., and **S. H. Hutner,** eds. 1980. Biochemistry and physiology of protoza, Academic Press, New York. An update of a definitive work on a variety of protozoa.

Lynn, D. H. 1981. The organization and evolution of microtubular organelles in ciliated protozoa. Biol. Rev. 56:243–292. Emphasis on evolution and structure of cilia, with consideration of function.

Patterson, D. J. 1980. Contractile vacuoles and associated structures: their organization and function. Biol. Rev. 55:1–46. Summary of the many roles of these important eucaryotic structures.

Energetics

4

Research on the chemical nature of life in the last 100 years has shown that the details of biological processes are completely understandable in terms of fundamental chemical principles. There is no special kind of chemistry unique to life. The key feature of living systems is their ability to organize molecules and reactions into carefully controlled, systematic sequences. The ultimate expression of this organization is the ability of a living organism to replicate itself.

The discipline of **biochemistry** deals with the study of the chemical processes in living cells. An important term used to describe collectively all the chemical processes occurring in living organisms is **metabolism.** When we speak of metabolic reactions, we mean chemical reactions occurring in living organisms.

Chemical systems, if left to themselves, go downhill with time and ultimately become completely random. Living organisms, on the other hand, are highly ordered and nonrandom. Since a living organism is a chemical system, how is it possible to maintain a nonrandom condition? It takes energy, and the ability of organisms to use and transform energy is a central facet of the life process. Organisms must continuously consume energy in order to stay alive. It is the purpose of this chapter to present the general principles of energy transformation in microorganisms.

We can distinguish two general kinds of metabolic activities: the consumption of energy from the environment, and the utilization of energy in creating the ordered structure of the organism. These two activities are generally referred to as *catabolism* and *anabolism,* respectively. Chapter 4 presents the major processes of catabolism used by organisms to obtain energy from the oxidation of organic compounds. Chapter 5 analyzes important processes of anabolism in which energy is used to synthesize essential cellular components. Chapter 6 considers generation of energy from sources other than organic compounds, namely inorganic compounds and light.

Energy

In order to examine metabolic processes in detail we must understand that energy exists in several different forms, some of which are much more relevant biologically than others. **Mechanical** energy is developed, for example, during cellular movement, beating of flagella and cilia, cytoplasmic streaming, movement and reorganization of intracellular structures such as mitochondria and chloroplasts, and alteration of cell shape. **Electrical** energy is produced when electrons move from one place to another; it is usually expressed as a flow of current between two points due to a difference in voltage. All living organisms produce electrical energy as a part of their cellular activity. **Electromagnetic** energy occurs in the form of **radiations,** and in biology the most significant is that from visible or near-visible light, such as radiations from the sun. Light is the primary energy source for photosynthetic organisms, and can also be used for certain functions by many nonphotosynthetic organisms. Some organisms, called **bioluminescent,** produce light energy. **Chemical** energy is the energy that can be released from organic or inorganic compounds by chemical reactions, and is the primary source of energy for nonphotosynthetic organisms. Cells also store chemical energy in the form of certain storage products; this energy can later be released. **Thermal** energy or **heat** is energy caused by molecular agitation: the faster the molecules in a system are moving, the hotter the system is. All organisms produce heat as a part of their normal energy-transformation processes. Since heat cannot be used as a primary energy source by living organisms, that energy released by an organism as heat represents wasted energy. **Atomic** energy is contained within the structure of atoms themselves and is released in the form of atomic radiations. It is not utilizable by living organisms, and on the contrary it may cause damage to them.

Three forms of energy are useful biologically: chemical, electrical, and light energy. In keeping with the idea that life is basically a chemical process, we find that the large majority of energy reactions within a cell involve chemical energy. Energy can be obtained from outside the cell in the form of chemical energy or light energy. When light is used as an energy source, it is converted to chemical energy before it is actually of value to the cell. Electrical energy is not directly utilizable as an energy source, but it is very important as an intermediate form in energy conversions and in the transport of some materials across the cell membrane. Energy can be converted from one form to another, but can neither be created nor destroyed. Because its many forms are interconvertible, energy is most conveniently expressed by a single unit. Although a variety of units exist, in biology the most commonly used energy unit is the **kilocalorie** (kcal), which is defined as the quantity of heat energy necessary to raise the temperature of 1 kilogram of water $1°C$.*

Chemical reactions are accompanied by changes in energy. The amount of energy involved in a chemical reaction is expressed in terms of the gain or loss of energy during the reaction. There are two

*An international commission has recommended that energy be expressed in *joules* or *kilojoules* (kj). One kilocalorie is equivalent to 4.184 kJ.

types of expression of the amount of energy released during a chemical reaction, abbreviated H and G. H, called **enthalpy,** expresses the total amount of energy released during a chemical reaction. However, some of the energy released is not available to do useful work. G, called **free energy,** is used to express the energy released that is available to do useful work. The change in free energy during a reaction is expressed as $\Delta G^{0\prime}$, where the symbol "Δ" should be read to mean "change in" (see Appendix 1 for definitions). Reactant and product concentrations and pH (when H^+ is a reactant or product) will affect the observed free energy changes. The superscripts "0" and "$\prime$" mean that a given free-energy value was obtained under "standard" conditions: pH 7 and all reactants and products initially at 1 molar (M). If $\Delta G^{0\prime}$ is negative, then free energy is released, and the reaction will occur spontaneously; such reactions are called **exergonic.** If $\Delta G^{0\prime}$ is positive, the reaction will not occur spontaneously, but the reverse reaction can occur spontaneously; such reactions are called **endergonic.**

In an **exergonic** reaction, the reaction proceeds until the concentration of products builds up, and then the reverse reaction, the conversion of products back to reactants, increases. An equilibrium is eventually reached, in which the forward and reverse reactions are exactly balanced. This balanced condition does not mean that reactant and product occur in equal concentrations. The concentration of products and reactants at equilibrium is related to the free energy of the reaction. If the reaction proceeds with a large negative $\Delta G^{0\prime}$, then the equilibrium is far toward the products and very little of the reactants will remain. In comparison, if the reaction proceeds with a small negative $\Delta G^{0\prime}$, then at equilibrium there are more nearly equal amounts of products and reactants. By determining the concentrations of products and reactants at equilibrium, it is possible to calculate the free-energy yield of the reaction (see Appendix 1).

This analysis can be extended to reveal the relative likelihood of occurrence of different reactions. Reactions which are more exergonic have a lower ratio of substrate molecules to product molecules at equilibrium, have more negative $\Delta G^{0\prime}$ values, and are more likely to occur.

In addition to speaking of free energy yield of reactions, it is also possible to talk about the free energy of individual substances. This is the free energy of formation, the energy yield or energy requirement necessary for the formation of a given molecule from the elements. By convention, the free energy of formation of the elements is zero.

If the formation of a compound from the elements proceeds exergonically, then the free energy of formation of the compound is negative, whereas if the reaction is endergonic, then the free energy of formation of the compound is positive. A few examples of free energies of formation: H_2O (water), -57 kcal/mole; CO_2 (carbon dioxide), -94.5 kcal/mole; H_2 (hydrogen), 0 kcal/mole (by definition); NH_4^+ (aqueous ammonia), -19 kcal/mole; N_2O (nitrous oxide), $+24.7$ kcal/mole; glucose -219 kcal/mole; CH_4 (methane), -12 kcal/mole. For most compounds the free energy of formation is negative, reflecting the fact that compounds form spontaneously from the elements. Again, the relative probabilities of different reactions (formations in this case) can be derived from comparison of the respective energies of formation. Thus we see that glucose (energy of formation of -219 kcal/mole) is more likely to form from carbon, hydrogen, and oxygen (that is, have a high ratio of product to reactants) than is methane (energy of formation of -12 kcal/mole) to form from carbon and hydrogen. This greater tendency to formation may also be viewed as greater stability. The positive energy of formation for nitrous oxide ($+24.7$ kcal/mole) tells us that this molecule is unstable and will spontaneously decompose to nitrogen and oxygen. The free energies of formation of a number of compounds of microbiological interest are given in Appendix 1, Table A1.1.

4.2

Activation Energy, Catalysis, and Enzymes

Free-energy calculations tell us only what conditions will prevail when the reaction or system is at equilibrium; they do not tell us how long it will take for equilibrium to be reached. The formation of water from gaseous oxygen and hydrogen is a good example. We have already seen that this reaction is quite favorable (energy of formation of -57 kcal/mole). However, if we were to mix O_2 and H_2 together, no reaction would occur within our lifetime. The explanation is that the rearrangement of oxygen and hydrogen atoms to form water requires that the molecular bonds of the reactants be broken first. The breaking of bonds requires energy, just as the formation of bonds releases energy. The required energy is referred to as **activation energy.** With O_2 and H_2, an electrical spark can provide activation energy, initiating the reaction, and once the reaction has begun it continues spontaneously since heat (energy) is released and serves to activate more molecules, leading to an explosion.

This example leads us to consider the phenomenon of **catalysis.** A **catalyst** is a substance which serves to lower the activation energy of a reaction, thereby reducing or even eliminating the need for an external supply of energy. A catalyst increases the rate of a reaction even though it is itself not changed. It is important to note that catalysts do not affect the free energy change of a reaction. An energy-requiring reaction cannot be made to occur simply by the addition of a catalyst. Free energy (an equilibrium concept) tells us what reactions are possible; catalysts affect the speed at which reactions proceed.

Most reactions in living organisms will not occur at appreciable rates without catalysis. In fact, the requirement of catalysis is important for the development of the living state, because if reactions occurred without catalysis, all reactants in cells would quickly proceed to the lowest energy levels of the products. It is only because of the presence of temporarily energy-rich compounds in cells that cell function is possible.

The catalysts of biological reactions are proteins called **enzymes.** Enzymes are highly specific in the reactions which they catalyze. That is, each enzyme catalyzes only a single reaction or a group of closely related reactions. This specificity and the action in lowering activation energy are related to the precise three-dimensional structure of the enzyme molecule. In an enzyme-catalyzed reaction, the enzyme temporarily combines with one or more reactant molecules, each of which is then termed a **substrate** of the enzyme. The enzyme is generally much larger than the substrate and the combination of enzyme and substrate depends on so-called *weak bonds* such as hydrogen bonds and van der Waals' forces to join the enzyme to the substrate. These weak bonds are effective only over very short distances and therefore the enzyme and its substrate must fit closely together in a **complementary** fashion. The small area on the enzyme to which the substrate binds is referred to as the **active site** of the enzyme. A useful analogy is the complementarity between a lock and a key. The requirements for fairly precise complementarity explain why each enzyme is so restricted in its range of action. There are several possible mechanisms for catalytic activity of an enzyme, all of which are related to the precise matching of shapes and charges. The enzyme may concentrate a substrate from the dilute surroundings, hold it in a particular position for reaction with other parts of the enzyme, or hold two substrates in proper orientation for reaction with each other.

Enzymes are polymers of subunits called amino acids. Each enzyme has its specific three-dimensional shape as a consequence of the linear sequence of amino acids in the polymer. Figure 4.1 is a diagrammatic example showing the manner in which amino acid sequence determines folding of the linear polymer into final protein

FIGURE 4.1 Protein structure. (a) A short segment of the polypeptide chain. The amino acids are connected by peptide bonds. A single polypeptide may have several hundred amino acid units of up to 20 different kinds. (b) The peptide chain twists into a specific shape. Here the chain is shown in a helix, although other configurations are also possible. In the helix, the $C{=}O$ from one turn forms a hydrogen bond with the N—H of the turn below. (c) The helix folds into a globular configuration, the various parts held together by ionic, hydrophobic, and hydrogen bonds and by disulfide bridges. (d) Several polypeptides may combine to form the final protein. These subunits may be the same as each other or may be different. In many multimeric proteins there are only two to four polypeptide monomers, while a few proteins are composed of as many as 16 separate protein subunits. If the multimeric protein were an enzyme, the active site often involves portions of more than one of the subunits acting in a cooperative fashion.

glycine—alanine—histidine—valine—cysteine

phenylalanine—glycine—lysine—glutamic acid

arginine—

(*a*) *Primary structure*

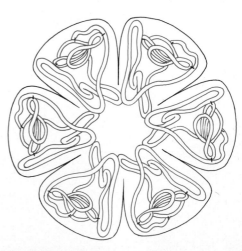

Disulfide bridge

(*b*) *Secondary structure*   (*c*) *Tertiary structure*   (*d*) *Quarternary structure*

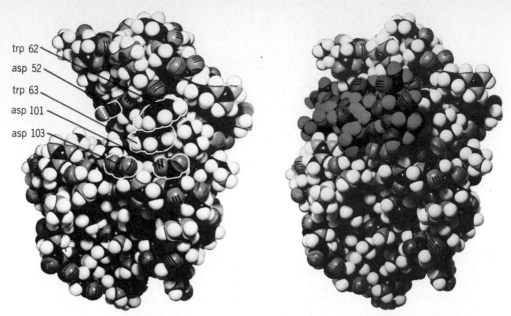

FIGURE 4.2 Space-filling model of the enzyme lysozyme. Left: enzyme without substrate, showing active site crevice. Right: enzyme–substrate complex, with substrate in color. The numbers refer to the position of the relevant amino acids in the polypeptide chain. Abbreviations are as follows: trp = tryptophan; asp = aspartate. (Reprinted by permission from The Structure and Action of Proteins by R. E. Dickerson and I. Geis. Benjamin/Cummings, Publishers, Menlo Park, California, 1969.)

shape. The precise enzyme-substrate complementarity may be seen more easily in a space-filling model of an enzyme with and without its substrate (Figure 4.2). In this example, the key roles of several amino acids (numbered on photograph) are clear in binding the enzyme to its substrate.

Many enzymes contain small nonprotein molecules which assist the catalytic function. These small molecules may be generally divided into two categories on the basis of the nature of the association with the enzyme, although the classifications may overlap somewhat. **Prosthetic groups** are bound very tightly to the enzyme, usually permanently. If an enzyme contains a prosthetic group, the protein part alone is called an **apoenzyme** and the complete enzyme, formed when the prosthetic group is attached, is called a **holoenzyme.** The heme group of many cytochromes is an example of a prosthetic group; cytochromes will be described in detail later in this chapter. **Coenzymes** are bound rather loosely to enzymes and each coenzyme may be associated with a number of different enzymes at different times during growth. Coenzymes serve as intermediate carriers of small molecules from one enzyme to another. Most coenzymes are synthesized from vitamins, the small organic molecules required in trace amounts for the growth of many organisms.

Enzymes are generally named for their substrates or for the reaction which they catalyze, by means of the combining form *-ase.*

Thus cellul*ase* is an enzyme that attacks cellulose, glucose oxid*ase* is an enzyme that catalyzes the oxidation of glucose, and ribonucle*ase* is an enzyme that decomposes ribonucleic acid.

Oxidation-Reduction

The utilization of chemical energy in living organisms generally involves what are called oxidation-reduction reactions. Although some oxidation-reduction reactions involve oxygen, most do not; instead of oxygen transfer, the real basis of an oxidation-reduction reaction is electron transfer. For example, when *ferrous* iron loses an electron, $Fe^{2+} \rightarrow Fe^{3+} + e^-$, it is converted into *ferric* iron. The release of an electron is called **oxidation**; we may say that ferrous iron has been **oxidized** to ferric iron. In real solutions, electrons cannot exist alone; they are always associated with some atom or molecule. The equation shown above for the oxidation of ferrous iron is referred to as a **formal** reaction. The equation gives us chemical information but does not represent a real reaction by itself. This formal equation is also known as a **half-reaction,** a term which implies the need for some other reactant. The balanced chemical process is completed when we identify a half-reaction which consumes electrons. For example, oxygen is an excellent electron acceptor: $O_2 + e^- \rightarrow O_2^-$. This gain of an electron is called a **reduction** and oxygen has been **reduced** in the given half-reaction. The coupling of the half-reactions will result in an overall reaction of **electron transfer** in which one molecule (Fe^{2+} in our example) is an **electron donor** and another molecule (O_2) is an **electron acceptor** (Figure 4.3). The key to understanding biological electrochemistry is the proper association of half-reactions. It must be remembered that neither oxidations nor reductions ever occur alone in real reactions; they must always be coupled, since electrons do not exist alone in solution.

Reduction potentials   Substances vary in their tendencies to give up electrons and become oxidized. This tendency is expressed as the **reduction potential.** The word "potential" is used here in two senses. First, it refers to the possibility of a reaction; and second, since the transfer of electrons is the definition of electric current, there is an electrical potential created. This potential is measured electrically in reference to a standard substance, hydrogen (H_2). By relating all reduction potentials to a standard, it is possible to express potentials for various half-reactions on a single scale, making possible ready comparison between various reactions. By convention, reduction potentials of half-reactions are written with the oxidant on the left; that is, as reductions, so we speak of **reduction** potential, thus: oxidant $+ e^- \rightarrow$ reductant. If hydrogen ions are involved in an oxidation-reduction reaction, as is often the case, then the reduction potential will be influenced by hydrogen ion concentration (usually expressed as pH). By convention in biology, reduction potentials are given for neutrality (pH 7). Using these conventions, the reduction potential of the half-reaction $Fe^{3+} + e^- \rightarrow Fe^{2+}$ is $+0.77$ volt (V), that of $\frac{1}{2} O_2 + 2H^+ + 2e^- \rightarrow H_2O$ is $+0.816$ V, and that of $2H^+ + 2e^- \rightarrow H_2$ is -0.421 V. Note that this potential for forming H_2 is not 0 volts, be-

Electron donating half-reaction

$$Fe^{2+} \longrightarrow Fe^{3+} + e^-$$

Electron accepting half-reaction

$$O + e^- \longrightarrow O^-$$

Complete (coupled) reaction may be written two ways

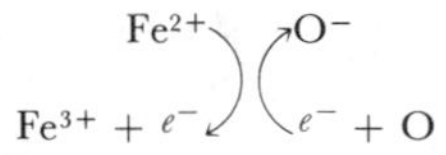

$$Fe^{2+} \quad O^-$$
$$Fe^{3+} + e^- \quad e^- + O$$

or

$$Fe^{2+} \longrightarrow Fe^{3+} + e^-$$
$$O + e^- \longrightarrow O^-$$

Net reaction $Fe^{2+} + O \longrightarrow Fe^{3+} + O^-$

Fe^{2+} is the reductant or electron donor; it becomes oxidized.

O is the oxidant or electron acceptor; it becomes reduced.

FIGURE 4.3   Example of a coupled oxidation–reduction reaction, the oxidation of ferrous iron (Fe^{2+}) by oxygen.

cause we make biological calculations at pH 7. The standard H_2 value of 0 is obtained at pH 0 (1 N H^+).

Most molecules can be either electron donors (reductants) or electron acceptors (oxidants) at different times, depending upon what other materials they react with. The term *oxidation-reduction* is often shortened to **redox** for use in such phrases as *redox potential* or *redox reaction*. A list of reduction potentials for a number of substances is given in Appendix 1, Table A1.2.

Redox pairs and coupled reactions The two substances on each side of the arrow in the half reaction, oxidant and reductant, can be thought of as a "redox pair," such as $2H^+/H_2$, Fe^{3+}/Fe^{2+}, $O_2/2O^-$. When writing a redox pair, the oxidized form will always be placed first.

In constructing oxidation-reduction couples from half-reactions, it is simplest to remember that the *reduced* substance of a redox pair whose potential is more *negative* donates electrons to the *oxidized* substance of a redox pair with the more *positive* potential. Thus, in the redox pair $2H^+/H_2$ with a potential of -0.42 V, H_2 has a great tendency to donate electrons. On the other hand, in the redox pair $\frac{1}{2}O_2/H_2O$, with a potential of $+0.82$ V, H_2O has a very slight tendency to donate electrons, but O_2 has a great tendency to accept electrons. It follows then that in the coupled reaction of H_2 and O_2, hydrogen will serve as the electron donor, and become oxidized, and oxygen will serve as the electron acceptor and become reduced:

$$H_2 \longrightarrow 2H^+ + 2e^-$$
$$\tfrac{1}{2}O_2 + 2e^- \longrightarrow 2O^-$$
$$2H^+ + 2O^- \longrightarrow H_2O$$

Even though both half-reactions are written formally as reductions by convention, in a real redox reaction one of the two half-reactions must be an oxidation and therefore proceeds in the reverse direction from the formal presentation. In the example above, the oxidation of H_2 to $2H^+ + 2e^-$ is reversed from the formal half-reaction, which is a reduction.

The electron tower A convenient way of viewing electron transfer in oxidation-reduction reactions is to imagine a vertical tower (Figure 4.4). The tower represents the range of reduction potentials for redox pairs, from the most negative at the top to the most positive at the bottom. If we make an analogy to potential energy, the reduced substance in the pair at the top of the tower has the greatest amount of potential energy (roughly, the energy that it took to lift the substance to the top), and the reduced substance at the bottom of the tower has the least amount of potential energy. On the other hand, the oxidized substance in the pair at the top of the tower has the least tendency to *accept* electrons, whereas the oxidized substance in the pair at the bottom of the tower has the *greatest* tendency to accept electrons.

As the electrons from the electron donor at the top of the tower fall, they can be caught by acceptors at various levels. The farther the electrons drop before they are caught, the greater amount of energy released. O_2, at the bottom of the tower, is the final acceptor (or put in other terms, is the most powerful oxidizing agent). In the middle

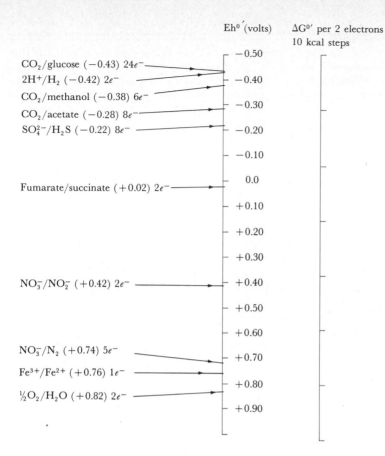

$Eh^{0'}$ (volts)

$\Delta G^{0'}$ per 2 electrons
10 kcal steps

CO$_2$/glucose (-0.43) $24e^-$

2H$^+$/H$_2$ (-0.42) $2e^-$

CO$_2$/methanol (-0.38) $6e^-$

CO$_2$/acetate (-0.28) $8e^-$

SO$_4^{2-}$/H$_2$S (-0.22) $8e^-$

Fumarate/succinate ($+0.02$) $2e^-$

NO$_3^-$/NO$_2^-$ ($+0.42$) $2e^-$

NO$_3^-$/N$_2$ ($+0.74$) $5e^-$

Fe^{3+}/Fe^{2+} ($+0.76$) $1e^-$

½O$_2$/H$_2$O ($+0.82$) $2e^-$

-0.50
-0.40
-0.30
-0.20
-0.10
0.0
$+0.10$
$+0.20$
$+0.30$
$+0.40$
$+0.50$
$+0.60$
$+0.70$
$+0.80$
$+0.90$

FIGURE 4.4 The electron tower. Redox pairs are arranged from the strongest reductants (negative reduction potentials) at the top to the strongest oxidants (positive reduction potentials) at the bottom. As electrons are removed from the top of the tower, they can be caught by acceptors at various levels. The farther the electrons fall before they are caught, the greater the difference in reduction potentials between electron donor and electron acceptor, the more energy is released. On the right, the energy released is given in 10 kcal/mol steps, assuming 2 electron transfers in each case. Some of the half-reactions indicated involve the transfer of several electrons, for example the CO$_2$/glucose couple. This reaction is included as a biologically important example of an overall process. The actual reduction of CO$_2$ to glucose involves several smaller redox reactions.

of the tower, the redox pairs can generally act as either electron acceptors or electron donors. For instance, in Figure 4.4, the SO$_4^{2-}$/H$_2$S couple has an intermediate reduction potential of -0.22 V. In the electron-donating mode, H$_2$S can donate electrons to acceptors more positive, such as NO$_3^-$ or O$_2$. When this happens, H$_2$S becomes oxidized to SO$_4^{2-}$. On the other hand, in the electron-accepting mode, SO$_4^{2-}$ can accept electrons from H$_2$ and become reduced to H$_2$S. Indeed, both types of reactions are known in microbiology, the H$_2$S oxidation being carried out by *Thiobacillus* and the SO$_4^{2-}$ reduction by *Desulfovibrio*.

In biological catabolism the electron donor is often referred to as an **energy source.** We must be careful, however, not to interpret this term to mean that all the energy is contained in the electron donor and is released upon oxidation. It is necessary to remember that it is the coupled redox process which releases energy. As discussed in the context of the electron tower, the amount of energy released in a redox reaction depends on the identities of both the electron donor and the electron acceptor: the greater the difference between reduction potentials of the two half-reactions, the more energy there will be released upon their coupling.

Electron Carriers

In most biochemical redox reactions the transfer of electrons from donor to acceptor is not direct, but rather involves one or more intermediates, referred to as **carriers.** When such carriers are used, we refer to the starting compound as the **primary** or **initial** electron donor and to the final acceptor as the **terminal** or **ultimate** acceptor. The net energy change of the complete reaction sequence is determined by the difference in reduction potentials between the initial donor and the ultimate acceptor. The transfer of electrons through the intermediates is a series of redox reactions, but the energy change from these individual steps must add up to the value obtained by considering only the starting and ending compounds.

The intermediate electron carriers may be divided into two general classes: some may move about the cell, associating with different enzymes; others are attached to the cell membrane (usually permanently). The fixed carriers function in respiration and are discussed in Section 4.7. The coenzymes NAD (nicotinamide-adenine dinucleotide) and NADP (NAD phosphate) (structures in Figure 4.5) are the most common of the diffusible carriers. In order to be chemically stable, NAD and NADP must transport hydrogen ions at the same time they transport electrons. The ratio is two electrons to one hydrogen ion for each coenzyme in the reduction reaction: $NAD^+ + 2H^+ + 2e^- \rightarrow NADH + H^+$. Thus dual function of NAD in carrying both electrons and protons is important in many catabolic oxidation

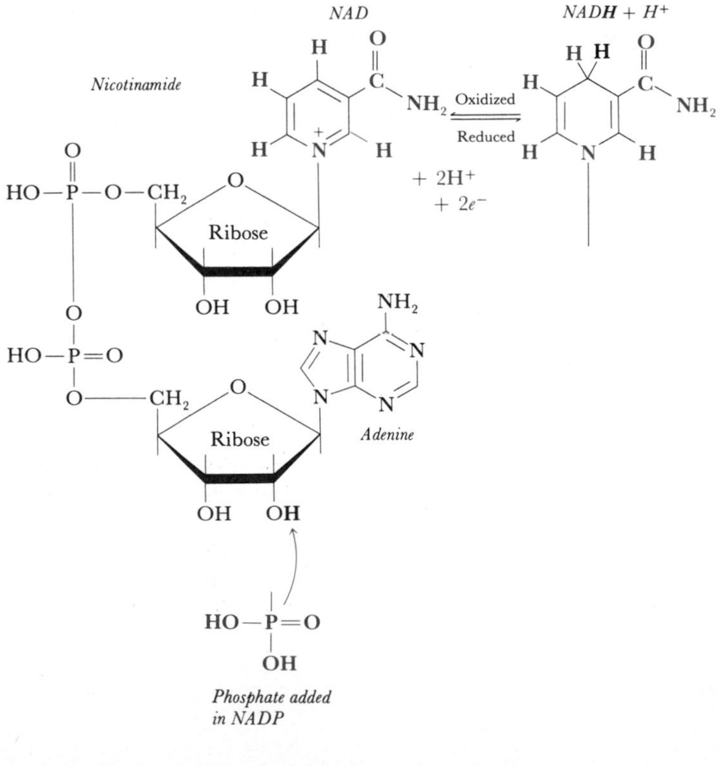

FIGURE 4.5 Structure of the oxidation–reduction coenzyme nicotinamide adenine dinucleotide (NAD). In NADP, a phosphate group is present, as indicated. Both NAD and NADP undergo oxidation–reduction as shown.

reactions in which protons are removed with electrons. This process is referred to as **dehydrogenation**; and one example is

$$\underset{\textit{Ethanol}}{\text{H}_3\text{C}-\overset{\displaystyle\overset{\text{H}_2}{|}}{\text{C}}-\text{OH}} \longrightarrow \underset{\textit{Acetaldehyde}}{\text{H}_3\text{C}-\overset{\displaystyle\overset{\text{H}}{|}}{\text{C}}=\text{O}} + 2\text{H}^+ + 2e^-$$

Both the electrons and protons are accepted by NAD.

The reduction potential of the NAD/NADH pair is -0.320 V, and that of the NADP/NADPH pair is -0.324 V, which places these redox pairs fairly high on the electron tower. Although both NAD and NADP redox pairs have similar potentials, they serve different purposes in the cell. NAD is directly involved in energy-generating reactions, as we will see later in this chapter, whereas NADP is involved primarily in biosynthetic reactions, as discussed in Chapter 5.

These coenzymes increase the diversity of possible redox reactions by making it possible for chemically dissimilar molecules to be coupled as initial electron donor and ultimate acceptor. As we have discussed, most biological reactions are catalyzed by specific enzymes which can only react with a limited range of substrates. Redox reactions may be considered to proceed in two separate stages: removal of electrons from the donor and addition of electrons to the acceptor. These two stages are usually catalyzed by different enzymes, each of which binds to its substrate and the coenzyme. Figure 4.6 is a generalized description of this functioning of a coenzyme. Specific biological sequences will be developed in Sections 4.6 and 4.7.

High-Energy Phosphate Compounds and Adenosine Triphosphate (ATP)

It is of great importance in energy generation that the energy released as a result of oxidation-reduction reactions be conserved so that it can be used for cell functions. If chemical energy is converted into heat during oxidation-reduction, it is no longer available and has been wasted. In living organisms, chemical energy released in redox reactions is most commonly transferred to a variety of phosphate compounds in the form of high-energy phosphate bonds, which serve as intermediaries in the conversion of energy into useful work.*

In most phosphate compounds, phosphate groups are attached via oxygen atoms in **ester linkage,** as illustrated in Figure 4.7. Not all phosphate ester linkages are high-energy bonds. As a means of expressing the energy of phosphate bonds, the free energy released when water is added and the bond is hydrolyzed can be given. As seen in Figure 4.7, the $\Delta G^{0\prime}$ of hydrolysis of the phosphate bond in glucose-6-phosphate is only -3.3 kcal/mole, whereas the $\Delta G^{0\prime}$ of hydrolysis of the phosphate bond in phosphoenolpyruvate is -14.8 kcal/mole, four times that of glucose-6-phosphate. (Free energies of hydrolysis for a number of other high-energy phosphate compounds are given in Appendix 3, Table A3.1.)

*It is not exactly accurate, chemically, to speak of "bond energy" and "high-energy phosphate bonds," but this simplifying usage is close enough to reality for present purposes.

Part I	Part II	Overall
$A_{red} \longrightarrow A_{ox} + 2e^-$	$C_{red} \longrightarrow C_{ox} + 2e^-$	$A_{red} + C_{ox} \longrightarrow A_{ox} + C_{red}$ (I)
$C_{ox} + 2e^- \longrightarrow C_{red}$	$B_{ox} + 2e^- \longrightarrow B_{red}$	$C_{red} + B_{ox} \longrightarrow C_{ox} + B_{red}$ (II)
Net: $A_{red} + C_{ox} \longrightarrow A_{ox} + C_{red}$	Net: $C_{red} + B_{ox} \longrightarrow C_{ox} + B_{red}$	Net: $A_{red} + B_{ox} \longrightarrow A_{ox} + B_{red}$

(a)

Part I: Enzyme I reacts with electron donor
   substrate and oxidized coenzyme

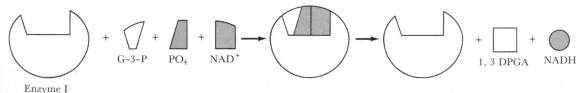

Part II: Enzyme II reacts with electron acceptor
   substrate and reduced coenzyme

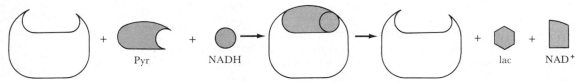

(b)

FIGURE 4.6 (a) Generalized example of a redox process involving a coenzyme. Compound A is the initial electron donor, compound B is the ultimate electron acceptor and compound C is the intermediate electron carrier. The subscripts red and ox refer to the reduced and oxidized forms, respectively, of the indicated molecules. The process is presented in two portions to highlight the function of compound C as an intermediate. Note that in the overall reaction, compound C is unchanged. (b) Diagrammatic interpretation of reactions shown in part a using an actual catabolic pair of reactions. Abbreviations: G-3-P, glyceraldehyde-3-phosphate; PO_4, inorganic phosphate; NAD^+, oxidized coenzyme NAD; 1,3 DPGA, 1,3 diphosphoglyceric acid; NADH, reduced coenzyme NAD; pyr, pyruvate; lac, lactic acid. Enzyme I is glyceraldehyde-3-phosphate dehydrogenase and enzyme II is lactic acid dehydrogenase. Geometrical shapes of substrates and enzyme active sites do not reflect actual geometry of these compounds; they are presented schematically for convenience and clarity. The meanings of Part I and Part II are the same as for part a of this figure.

Adenosine triphosphate (ATP) The most important high-energy phosphate compound in living organisms is adenosine triphosphate, ATP. ATP serves as the prime energy carrier in living organisms, being generated during certain oxidation-reduction reactions, and being used during biosynthetic reactions. The structure of ATP is shown in Figure 4.7, where it is seen that two of the phosphate bonds of ATP are high-energy bonds.

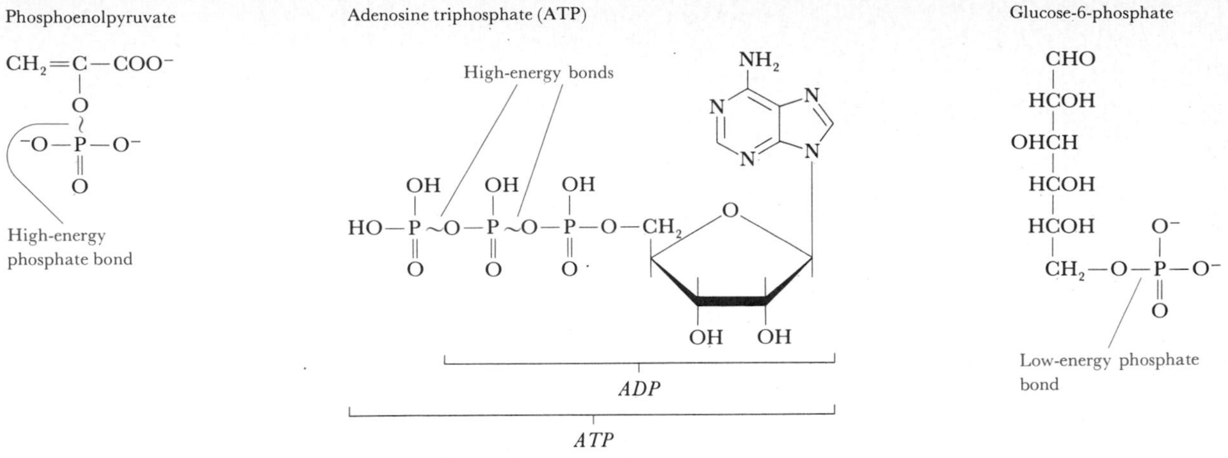

Phosphoenolpyruvate

Adenosine triphosphate (ATP)

Glucose-6-phosphate

$$CH_2=C-COO^-$$
$$|$$
$$O$$
$$|$$
$$-O-P-O^-$$
$$||$$
$$O$$

High-energy phosphate bond

High-energy bonds

$$NH_2$$

$$OH \quad OH \quad OH$$
$$HO-P\sim O-P\sim O-P-O-CH_2$$
$$|| \qquad || \qquad ||$$
$$O \qquad O \qquad O$$

OH   OH

ADP

ATP

CHO
|
HCOH
|
OHCH
|
HCOH
|
HCOH   O⁻
| |
CH₂—O—P—O⁻
           ||
           O

Low-energy phosphate bond

FIGURE 4.7 High-energy phosphate bonds. The list below the figure shows the free energy of hydrolysis of some of the key phosphate esters, indicating that some of the phosphate ester bonds are of higher energy than others. Structures of three of the compounds are given to indicate the position of low-energy and high-energy bonds.

Compound	$G^{0'}$ kcal/mole
Phosphoenolpyruvate	−14.8
1,3-diphosphoglycerate	−11.8
Acetyl phosphate	−10.1
ATP	− 7.3
ADP	− 7.3
Glucose-1-phosphate	− 5.0
Fructose-6-phosphate	− 3.8
AMP	− 3.4
Glucose-6-phosphate	− 3.3

It should be emphasized that although we express the energy of the high-energy phosphate bonds in terms of the free energy of hydrolysis, in actuality it is undesirable for these bonds to hydrolyze in cells if no other reaction takes place as well. If simple hydrolysis takes place, heat is generated and the energy of the high-energy bond is lost. The free energy of the high-energy phosphate bonds[†] is generally used to drive biosynthetic reactions and other aspects of cell function through carefully regulated processes in which the energy released from ATP hydrolysis is coupled to energy-requiring reactions.

4.6

Energy Release in Biological Systems

Organisms use a wide variety of energy sources and mechanisms to synthesize high-energy phosphate bonds in ATP. Both chemical energy and light energy are used to synthesize ATP. Among chemical substances, both organic and inorganic compounds may serve as electron donors. Similarly, a number of different electron acceptors can be used in coupled redox reactions, including both organic and inorganic compounds.

In the rest of this chapter we will be concerned with the mechanisms by which ATP is synthesized as a result of oxidation-reduction

[†]The free energy of hydrolysis is influenced by pH, the ionic environment, and the concentrations of reactants and products. The values given in Figure 4.7 are for standard conditions, for purposes of comparison. It has been calculated that within cells, the free energy of hydrolysis of the terminal phosphate on ATP may be considerably higher than − 7 kcal. Values have been calculated of − 10.0 kcal/mole at an intracellular pH of 6, − 10.5 kcal/mole at pH 7, and − 12.0 kcal/mole at pH 8.5. In general, the intracellular pH of cells is neutral or slightly acidic.

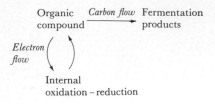

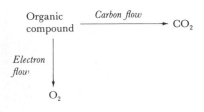

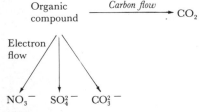

FIGURE 4.8 Contrasts in electron and carbon flow in energy-yielding oxidation of an organic compound.

reactions involving organic compounds. Metabolism of organic compounds containing reduced forms of carbon is the source of energy for all animals and for the vast majority of microorganisms, including all fungi, protozoa, and most bacteria.

The series of reactions involving the oxidation of a single compound is called a *biochemical pathway.* The pathways for oxidation of organic compounds and conservation of energy in ATP can be divided into three major groups (Figure 4.8): (1) **fermentation,** in which the redox process occurs in the absence of any added electron acceptor; (2) **respiration,** in which molecular oxygen serves as the electron acceptor; and (3) **anaerobic respiration,** in which an electron acceptor other than O_2 is involved, such as NO_3^-, SO_4^{2-}, or CO_3^{2-}.

Substrate-level and electron-transport phosphorylation The processes by which ATP is synthesized during oxidation-reduction reactions can be grouped into two major categories, called substrate-level and electron-transport phosphorylation. In **substrate-level phosphorylation (SLP),** ATP is synthesized directly during a specific enzymatic step in the oxidation pathway of the electron donor. The specific phosphorylating reaction involves the transfer of a phosphate group from an intermediate organic-phosphate compound to ADP, forming ATP. **Electron-transport phosphorylation (ETP),** also sometimes called oxidative phosphorylation, occurs during the transfer of electrons through an electron-transport system in a membrane. A single electron-transport system can be involved in the final oxidation-reduction reactions of a wide variety of compounds, so that ATP synthesis via electron-transport phosphorylation is a much more general process than substrate-level phosphorylation. However, electron-transport phosphorylation requires the presence of an external electron acceptor, so that it only occurs during respiration or anaerobic respiration. Substrate-level phosphorylation, on the other hand, may occur during oxidation-reduction reactions in all three types of processes: fermentation, respiration, and anaerobic respiration. Although only a very few enzyme reactions exist in which substrate-level phosphorylation can occur, organisms are able to convert usable chemical compounds into one or the other of these few substrates, thus enabling substrate-level phosphorylation to occur under a wider range of circumstances. In the following section we discuss one of the major series of reactions by which ATP is synthesized via substrate-level phosphorylation.

4.7

Fermentation

In the absence of an added electron acceptor, many organisms perform balanced redox reactions of some organic compounds with the release of energy, a process called **fermentation.** Under these conditions only partial oxidation of the carbon atoms of the organic compound occurs and therefore only a small amount of the energy is released. Some atoms of the compound are oxidized and others are reduced. As an example, the catabolism of glucose by yeast in the absence of oxygen:

$$C_6H_{12}O_6 \longrightarrow 2CH_3CH_2OH + 2CO_2 \qquad \Delta G^{0\prime} - 57 \text{ kcal/mole}$$

Glucose
(intermediate
oxidation
level)

Ethanol
(reduced
product)

Carbon
dioxide
(oxidized
product)

Energy

Note that some of the carbon atoms end up in CO_2, a more oxidized form than the carbon atoms in the starting glucose, while other carbon atoms end up in ethanol, which is more reduced (that is, it has more hydrogens and electrons per carbon atom) than glucose.* The energy generated in this fermentation (57 kcal/mole) is not all released as heat; some is conserved in the form of high-energy phosphate bonds in ATP, with a net production of two such bonds. We discuss now the biochemical steps involved in the fermentation of glucose to ethanol and CO_2, and the manner in which some of the energy released is conserved in high-energy phosphate bonds.

Glucose fermentation   The biochemical pathway for the breakdown of glucose can be divided into three major parts (Figure 4.9). The first part is a series of preparatory rearrangement reactions that do not involve oxidation-reduction, and lead to the production of the key intermediate, *glyceraldehyde-3-phosphate.* In the second part, an oxidation-reduction reaction occurs, high-energy phosphate bond energy is produced in the form of ATP, and pyruvate is formed. In the third part, a second oxidation-reduction reaction occurs and the fermentation products ethanol and CO_2 are released. The biochemical pathway to pyruvate is called **glycolysis,** and it is also sometimes called the Embden–Meyerhof pathway, after two of its discoverers. For details of the biochemistry involved, see Appendix 3, Figure A3.1.

    Initially, the glucose is *phosphorylated* by ATP, yielding glucose-6-phosphate. Phosphorylation reactions of this sort often occur preliminary to oxidation. When ATP is converted to adenosine diphosphate (ADP), energy is utilized because the organic phosphate bond in glucose-6-phosphate is at a lower energy level than was the phosphate bond of ATP. (The energy lost at this step will be regained later in the reaction sequence.) The initial phosphorylation of glucose *activates* the molecule for the subsequent reactions. Glucose-6-phosphate is converted into its isomer fructose-6-phosphate and another phosphorylation leads to the production of *fructose-1,6-diphosphate,* which is a key intermediate product in the breakdown process. The enzyme *aldolase* now catalyzes the splitting of fructose-1,6-diphosphate into two three-carbon molecules, glyceraldehyde-3-phosphate and dihydroxyacetone phosphate.† Note that as yet there has been no redox change

*The degree of oxidation or reduction of a carbon compound is expressed by its oxidation state, as discussed in Appendix 1, Part B. In all oxidation-reduction reactions there must be a balance between oxidized and reduced products. In the present reaction, the oxidation state of glucose is 0, that of ethanol is -4 and that of CO_2 is $+4$. Because equal amounts of ethanol and CO_2 are formed, an oxidation-reduction balance results. Calculation of an oxidation-reduction balance provides a check to ascertain that the reaction written is correct.

†There is an enzyme that catalyzes the interconversion of dihydroxyacetone phosphate and glyceraldehyde phosphate. For simplicity, we write only glyceraldehyde-3-phosphate in Figure 4.9 since that is the compound which undergoes subsequent reaction.

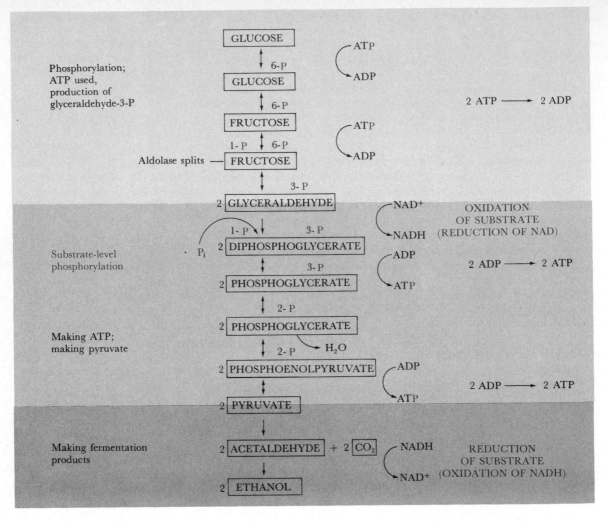

Phosphorylation;
ATP used,
production of
glyceraldehyde-3-P

Aldolase splits ──

Substrate-level
phosphorylation

Making ATP;
making pyruvate

Making fermentation
products

OVERALL REACTION: GLUCOSE ⟶ 2 ETHANOL + 2CO₂

Note that there is no
change in the proportions
of NAD⁺ and NADH.

Free energy yield: 57 kcal/mole
glucose fermented. Assuming an
energy value for the high-energy
phosphate bond in ATP of −7.4
kcal/mole, −14.8 kcal are
conserved in ATP, for an efficiency
of 26%.

FIGURE 4.9 Ethanol fermentation, the sequence of
enzymatic reactions in the conversion of glucose to
alcohol and CO₂ by yeast under anaerobic conditions.
The steps from glucose to pyruvate are sometimes
collectively called glycolysis, and are also called the
Embden-Meyerhof pathway. For a more detailed version
of this pathway, see Appendix 3, Figure A3.1.

since all of the reactions proceeded without any electron transfer, although two high-energy phosphate bonds from ATP have been used.

The first oxidation reaction occurs in the conversion of glyceraldehyde-3-phosphate to 1,3-diphosphoglyceric acid. In this reaction, the coenzyme NAD accepts two electrons and is converted into NADH, while inorganic phosphate is converted into an organic form. This energetically favorable reaction in which an inorganic phosphate molecule is converted into an organic form (chemically we say that the phosphate has been esterified) sets the stage for the nxt process, the substrate-level phosphorylation step in which ATP is formed. The SLP reaction is possible because one of the phosphates represents a newly synthesized high-energy phosphate bond, unlike the sugar-phosphate bonds. The synthesis of ATP at this stage represents the successful conservation of some of the energy released during the oxidation of glyceraldehyde-3-phosphate. This reaction is probably the most important substrate-level phosphorylation in living organisms. Further rearrangements shown in Figure 4.9 convert the remaining phosphate bond to a high-energy bond, leading ultimately to the synthesis of *pyruvate* and to transfer of the energy of the high-energy phosphate bonds to ADP, forming ATP.

In the complete fermentation, two ATP molecules were used initially to phosphorylate the sugar, and four ATP molecules were synthesized (two from each three-carbon fragment), so that the net gain is two ATP molecules per molecule of glucose fermented. Since the energy value of a high-energy bond of ATP is about 7.4 kcal/mole, and 57 kcal of energy is released per mole of glucose during the alcoholic fermentation, about 26 percent of the energy released during the oxidation of glucose is retained in the high-energy bonds of ATP (14.8 kcal in the 2 moles of ATP), and the rest is lost as heat.

In the oxidation step of the glycolytic pathway, NAD was reduced to NADH. The cell has only a limited supply of NAD, and if all of it were converted to NADH, the oxidation of glucose would stop, since the oxidation of each molecule of glyceraldehyde-3-phosphate can proceed only if there is a molecule of NAD to accept the released electrons. This "roadblock" is overcome in the complete fermentation by the oxidation of NADH back to NAD through reactions involving the conversion of pyruvate to ethanol and CO_2. The first step is the conversion (decarboxylation) of pyruvate to acetaldehyde and CO_2; electrons are then transferred to acetaldehyde (the ultimate electron acceptor) from NADH, leading to the formation of ethanol and NAD. The NADH that had been produced earlier is thus oxidized back to NAD.

In any energy-yielding process, oxidation must balance reduction, and there must be an electron acceptor for each electron removed. In this case, the reduction of NAD at one enzymatic step is coupled with its oxidation at another. The final products, CO_2 and ethanol, also are in oxidation-reduction balance (see footnote, page 109). Compare the function of the coenzyme NAD to the generalized use of intermediate electron carriers described in Figure 4.6.

The ultimate result of this series of reactions is the net synthesis of two high-energy phosphate bonds, two molecules of ethanol, and two molecules of CO_2. For the yeast cell the crucial product is ATP, which is used in a wide variety of energy-requiring reactions, and ethanol and CO_2 are merely waste products. The latter substances would hardly be considered waste products by the distiller or brewer, however, for the anaerobic fermentation of glucose by yeast is a means of producing ethanol, the crucial product in alcoholic beverages. For the baker the desired product is CO_2, which is essential in the rising of bread dough.

The reactions proceeding from glucose to pyruvate, described above, occur in a wide variety of microorganisms, but the resulting pyruvate may be processed further in a number of different ways. Many bacteria, as well as higher animals, carry out the reaction *pyruvate + NADH → lactic acid + NAD,* with pyruvate serving as the ultimate electron acceptor to form the end product lactic acid instead of alcohol and CO_2. Other bacteria form acetic, succinic, or other organic acids; alcohols such as butanol; and ketones such as acetone (see Chapter 19).

Other fermentations Many compounds other than glucose can be fermented. These include most sugars, many amino acids, certain organic acids, purines, pyrimidines, and a variety of miscellaneous products (Figure A3.8). The biochemical reactions involved in the fermentation of some of these compounds are discussed in Chapters 5 and 19. For a compound to be fermentable it must be neither too oxidized nor too reduced. If it is too oxidized, it will be a poor electron donor and therefore be unable to complete very favorable redox reactions. Similarly, if it is too reduced, it will be a poor electron acceptor and only slightly favorable redox couples will be possible.

Another requirement is that the compound be convertible to an intermediate that can partake in substrate-level phosphorylation. Nonfermentable compounds include hydrocarbons, fatty acids, and other highly reduced compounds. In some cases, an organic compound cannot be fermented unless another organic compound is present as the electron acceptor. Mixed fermentations of this type are common in the genus *Clostridium,* a large group of spore-forming, anaerobic, rod-shaped bacteria, and are discussed in Section 19.22 (see especially Figure 19.59).

Phosphoroclastic reaction and H_2 production The generation of energy from the anaerobic breakdown of pyruvate by clostridia can be taken as an example of another process resulting in substrate-level phosphorylation. Pyruvate is formed by clostridia from a variety of fermentable substrates, including amino acids, organic acids, and purines and pyrimidines. Pyruvate is then further degraded in a series of steps called the **phosphoroclastic reaction,** in which inorganic phosphate is used:

$$\text{Pyruvate} + \text{coenzyme A} \longrightarrow \text{acetylcoenzyme A} + CO_2 + H_2$$
$$\text{Acetylcoenzyme A} + PO_4^{3-} \longrightarrow \text{acetylphosphate} + \text{coenzyme A}$$

In this reaction, the phosphate linkage in acetylphosphate is a high-energy one (see Table A3.1). (Coenzyme A is an important coenzyme in acetyl transfer reactions and will be discussed in more detail below.) The high-energy phosphate bond of acetylphosphate can now be used to synthesize ATP in a substrate-level phosphorylation:

$$\text{Acetylphosphate} + \text{ADP} \longrightarrow \text{acetate} + \text{ATP}$$

In the phosphoroclastic reaction, the oxidation-reduction balance is maintained in an interesting way, through the production of molecular hydrogen (H_2). Molecular hydrogen is produced by a variety of anaerobic bacteria, and serves to dispose of excess electrons. In the clostridia, production of molecular hydrogen is closely associated with the presence of **ferredoxin,** an electron carrier of low redox potential (-0.430 V). Ferredoxins are small proteins containing iron (see the next section) and are involved not only in hydrogen production but also in photosynthesis and other biochemical reactions described in Chapters 5 and 6. Electrons from pyruvate are transferred first to ferredoxin, and from ferredoxin to H^+ in a reaction catalyzed by the enzyme hydrogenase (Figure 4.10). The ability to produce H_2 is a characteristic of most clostridia, and the two gasses H_2 and CO_2 produced by clostridia during anaerobic fermentation of canned goods are usually responsible for the bursting of cans in some types of canned food spoilage.

$$
\begin{array}{c}
H_2 \\
\uparrow \\
2H^+ \\
\uparrow \; \textit{Hydrogenase} \\
\text{Ferredoxin} \\
\uparrow \\
2e^- \\
\uparrow \\
\text{Pyruvate}
\end{array}
$$

FIGURE 4.10 Production of molecular hydrogen from pyruvate.

4.8
Respiration

We have discussed above the metabolism of glucose molecules by a fermentative pathway that functions in the absence of an external electron acceptor. A relatively small amount of energy is released in this process (and few ATP molecules synthesized) since the carbon atoms in these fermentation products have the same net oxidation state as those in the starting glucose, as explained in Section 4.7. This small energy release may be understood in terms of the electron tower and the formal principles of redox reactions. Fermentation processes yield little energy for two reasons: (1) the carbon atoms in the starting glucose are only partially oxidized, and (2) the difference in reduction potentials between the initial electron donor and ultimate electron acceptor is small. However, if O_2 is present as an electron acceptor, respiration can occur, all the substrate molecules can be oxidized completely to CO_2, and a yield of *thirty-eight* ATP molecules per glucose unit is theoretically possible. The greater energy release during respiration occurs because respiring cells surmount the two limitations just listed for fermentation: (1) the carbon atoms in the starting glucose are completely oxidized to CO_2; and (2) the ultimate electron acceptor (oxygen) has a very positive reduction potential, leading to a large net difference in potentials between the initial donor and ultimate acceptor and therefore the synthesis of much ATP. Our discussion of respiration will deal with the mechanisms of these two processes: (1) The biochemical pathway used to remove all the available electrons from the carbon in glucose, and (2) the way these electrons are transferred to oxygen while ATP is synthesized.

Tricarboxylic Acid cycle The pathway by which the organic product of glycolysis (pyruvate) is completely oxidized to CO_2 is called the

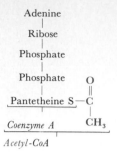

Adenine
|
Ribose
|
Phosphate
|
Phosphate   O
| ‖
Pantetheine S — C
| |
Coenzyme A   CH₃
_____
Acetyl-CoA

FIGURE 4.11 Structure of CoA and acetyl-CoA.

tricarboxylic acid cycle (TCA cycle*). Pyruvate, a key metabolic intermediate, is the starting compound in the TCA cycle. Note that the TCA cycle does not begin with the fermentation products, such as ethanol and lactic acid, which are formed by specific enzymatic reactions of pyruvate.

Pyruvate is first decarboxylated, leading to the production of one molecule of NADH and an acetyl radical coupled to coenzyme A. Acetylcoenzyme A (abbreviated acetyl-CoA) (Figure 4.11) is an activated form of acetate, the acetyl-CoA bond being a high-energy bond. In addition to being a key intermediate in the tricarboxylic acid cycle, acetyl-CoA also plays many important biosynthetic roles (see Chapter 5). The acetyl group of acetyl-CoA combines with the four-carbon compound oxalacetate, leading to the formation of *citric acid,* a six-carbon organic acid, the energy of the high-energy acetyl-CoA bond being used to drive this synthesis (Figure 4.12). Dehydration, decarboxylation, and oxidation reactions follow, and two CO_2 molecules are released. Ultimately oxalacetate is regenerated, and can serve again as an acetyl acceptor, thus completing the cycle.

For each pyruvate molecule entering the cycle, three CO_2 molecules are released, one during the formation of acetyl-CoA, one by the decarboxylation of isocitrate, and one by the decarboxylation of α-ketoglutarate. As in fermentation, the electrons released during the TCA cycle oxidation of the substrate are usually transferred initially to the coenzyme NAD. Respiration differs from fermentation specifically in the manner in which the reduced NADH is oxidized. The electrons from NADH, instead of being transferred to an intermediate such as pyruvate, are transferred to oxygen through the mediation of an electron-transport system, forming oxidized NAD and H_2O.

Electron-transport systems   Electron transport systems are composed of the membrane-associated electron carriers mentioned before. These systems have two basic functions: (1) to accept electrons from the electron donor and transfer them to the electron acceptor (in this case, O_2), and (2) to conserve some of the energy that is released during electron transfer by the synthesis of ATP.

Several types of oxidation-reduction enzymes and electron-transport proteins are involved in electron transport: (1) *NAD and NADP dehydrogenases,* which transfer electrons from NADH or NADPH; (2) riboflavin-containing electron carriers, generally called *flavoproteins;* (3) *iron-sulfur proteins,* similar to the ferredoxins but of higher reduction potential; and (4) *cytochromes,* which are proteins containing an iron-porphyrin ring called *heme.* In addition, one class of nonprotein electron carriers is known, the lipid-soluble *quinones,* sometimes called *coenzymes Q.* These electron-transport components are embedded in the membrane in an ordered arrangement which permits energy conservation as illustrated in Figures 4.11 and 4.12.

Flavoproteins are proteins containing a derivative of riboflavin (Figure 4.13); the flavin portion, which is bound to a protein, is the

---

*It is sometimes also called the *citric acid cycle* or the *Krebs cycle,* after one of its discoverers, Sir Hans Krebs.

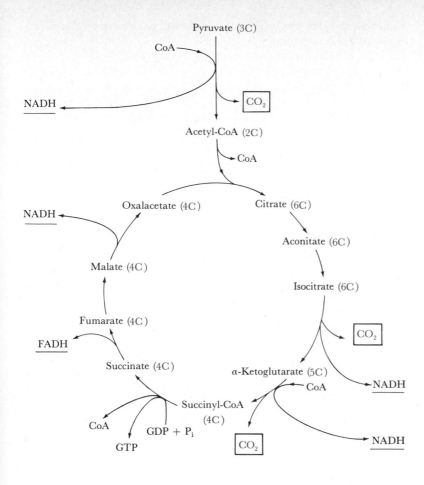

Pyruvate (3C)

CoA

NADH

CO_2

Acetyl-CoA (2C)

CoA

Oxalacetate (4C)

Citrate (6C)

NADH

Aconitate (6C)

Malate (4C)

Isocitrate (6C)

Fumarate (4C)

CO_2

FADH

Succinate (4C)

α-Ketoglutarate (5C)

NADH

CoA

Succinyl-CoA (4C)

CoA

GDP + P_i

CO_2

NADH

GTP

Overall reaction: Pyruvate + 4NAD + FAD $\longrightarrow$ 3CO_2 + 4NADH + FADH
GDP + P_i $\longrightarrow$ GTP
GTP + ADP $\longrightarrow$ GDP + ATP

15 ATP

Electron-transport   4 NADH $\equiv$ 12 ATP
phosphorylation   FADH $\equiv$ 2 ATP

FIGURE 4.12 The tricarboxylic acid cycle. The three-carbon compound, pyruvate, is oxidized to CO_2, with the electrons being transferred to NADH and FADH. Electron-transport phosphorylation with these electron carriers leads to the synthesis of ATP. The cycle actually begins when the 2-carbon compound acetyl-CoA condenses with the 4-carbon compound oxalacetate, to form the 6-carbon compound citrate. Through a series of oxidations and transformations, this 6-carbon compound is ultimately converted back to the 4-carbon compound oxalacetate, which then passes through another cycle with the next molecule of acetyl-CoA. The overall balance sheet is shown at the bottom. The reducing equivalents formed (NADH and FADH) are shown on the left and right sides of the cycle. For chemical and enzymatic details of the tricarboxylic acid cycle, see Appendix 3, Figure A3.2.

prosthetic group, which is alternately reduced as it accepts electrons and oxidized when the electrons are passed on. Two flavins are known, flavin mononucleotide (FMN) (Figure 4.13) and flavin-adenine dinucleotide (FAD), in which FMN is linked to ribose and adenine through a second phosphate. The riboflavin portion of these molecules consists of an isoalloxazine ring connected to ribitol, an alcohol derivative of ribose. Riboflavin, also called *vitamin B_2*, is a required growth factor for some organisms.

The **cytochromes** are proteins with iron-containing porphyrin rings attached to them (Figure 4.14). They undergo oxidation and

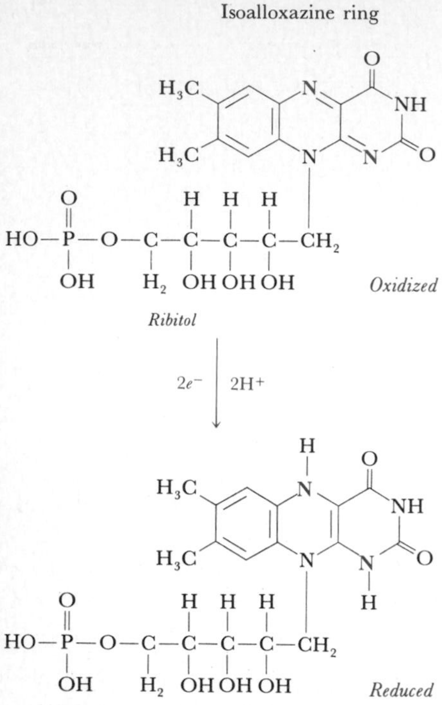

Isoalloxazine ring

FIGURE 4.13 Flavin mononucleotide (FMN) (riboflavin phosphate). Note that $2H^+$ are taken up when the flavin becomes reduced, and $2H^+$ are given off when the flavin becomes oxidized.

reduction through loss or gain of an electron by the iron atom at the center of the cytochrome:

$$\text{Cytochrome-Fe}^{2+} \rightleftharpoons \text{Cytochrome-Fe}^{3+} + e^-$$

$$\textit{(Reduced)} \qquad \textit{(Oxidized)}$$

Several cytochromes are known, differing in their reduction potentials. One cytochrome can transfer electrons to another that has a higher reduction potential and can accept electrons from a more reduced cytochrome. The different cytochromes are designated by letters, such as cytochrome *a*, cytochrome *b*, cytochrome *c*. The cytochromes of one organism may differ slightly from those of another, so that there are designations such as cytochrome a_1 and cytochrome a_2.

Iron-sulfur proteins are associated with the electron-transport chain in several places. Because the iron in these proteins is not in a cytochrome, they are sometimes called *nonheme iron proteins*. The standard potentials of these iron-sulfur proteins vary over a wide range, so that they can serve at several steps in the electron-transport process. One nonheme iron protein already mentioned is ferredoxin, an iron-sulfur protein of very low reduction potential that is present in anaerobic bacteria.

The **quinones** (Figure 4.15) are lipid-soluble substances involved in the electron-transport chain. One of these, called *coenzyme Q*, is a benzoquinone derivative with a long hydrophobic side chain. Others called menaquinones and naphthoquinones are found in bacteria and are related to vitamin K, a growth factor for higher animals.

FIGURE 4.14 Structure of the porphyrin (heme) prosthetic group of cytochrome c, indicating the manner in which the porphyrin is connected to the protein.

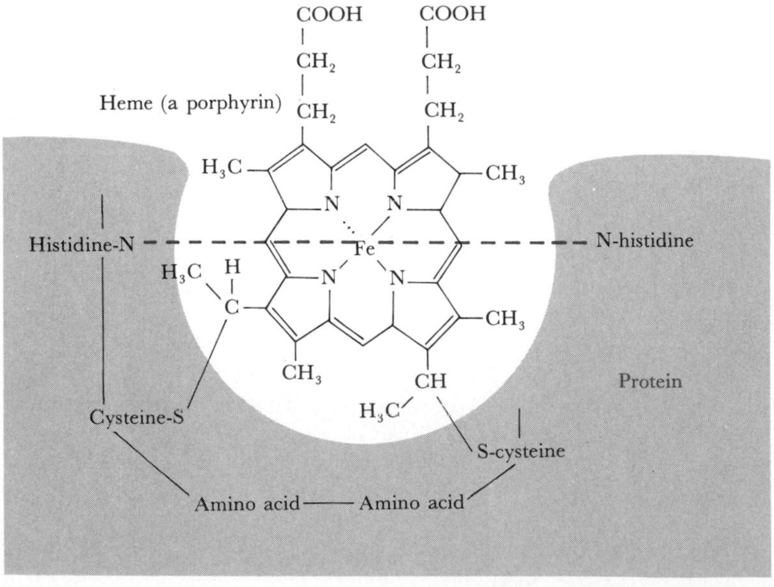

Energy conservation in the electron-transport system The overall process of electron transport in the electron-transport chain is shown in Figure 4.16. From the reduction potentials of the various electron carriers, it is possible to calculate the free energy release at each step.

Electron-Transport (Oxidative) Phosphorylation

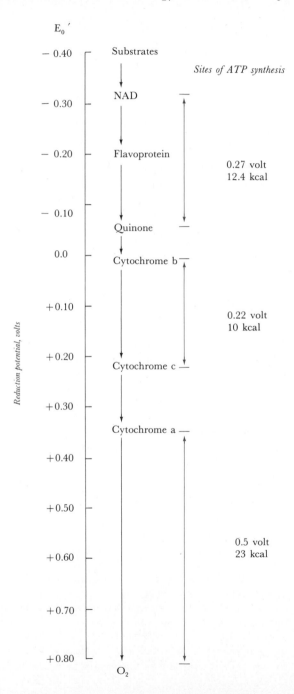

$E_0{}'$

Reduction potential, volts

− 0.40	Substrates
	Sites of ATP synthesis
− 0.30	NAD
− 0.20	Flavoprotein
	0.27 volt 12.4 kcal
− 0.10	Quinone
0.0	Cytochrome b
+0.10	
	0.22 volt 10 kcal
+0.20	Cytochrome c
+0.30	Cytochrome a
+0.40	
+0.50	
	0.5 volt 23 kcal
+0.60	
+0.70	
+0.80	O_2

FIGURE 4.15 Structure of oxidized and reduced forms of coenzyme Q, a quinone. The five-carbon unit in the side chain (an isoprenoid) occurs in a number of multiples. In bacteria, the most common number is n = 6; in higher organisms, n = 10. Note that $2H^+$ are taken up when the quinone becomes reduced, and $2H^+$ are given off when the quinone becomes oxidized.

FIGURE 4.16 One complete electron-transport system, leading to the transfer of electrons from substrate to O_2. By breaking up the complete oxidation into a series of discrete steps, energy conservation is possible, and ATP synthesis can occur. Experimental studies and calculations of differences in reduction potentials show that there are three places in the chain where sufficient drop in potential occurs to permit ATP synthesis. These are shown in color on the right.

As seen, there are three places in the chain where large amounts of free energy are released: between NAD and flavin, between cytochrome b and c, and between cytochrome a and O_2. At each of these locations, sufficient energy is released to synthesize a high-energy phosphate bond in ATP although, as we shall see, the actual synthesis of ATP is physically and chemically separated from these very favorable reactions in the membrane. Note that electrons are not transferred from NADH to O_2 directly, but rather go through a number of intermediate redox reactions. This multiple step process is essential in the mechanism by which ATP is synthesized.

Of considerable interest is the manner in which ATP is synthesized during electron transport, a process called *electron-transport phosphorylation* or oxidative phosphorylation. The former term is now preferred because there are many processes, especially in bacteria, in which electron transport leading to ATP synthesis can occur in the absence of O_2.

Generation of a proton-motive force   To understand the process of electron-transport phosphorylation, we must first discuss the manner in which the electron-transport system is oriented in the cell membrane. The overall structure of the membrane was outlined in Section 2.4 (see Figure 2.11). It was shown that proteins or protein clusters are embedded in the lipid bilayer, but that the orientation of components in the membrane is asymmetric, some components being accessible from the outside and others accessible from the inside. The electron-transport carriers are oriented in the membrane in a series of loops, in such a way that there is a separation of the movement of hydrogen atoms and electrons. The hydrogen atoms, removed from hydrogen carriers (that is, NADH) or substrates on the inside of the membrane, are carried to the outside of the membrane; the electrons removed from these hydrogen atoms return to the cytoplasmic side of the membrane through other electron carriers of the redox loop (Figure 4.17). At the termination of the electron-transport system, the electrons are passed to the final electron acceptor (in the case of aerobic respiration, this is O_2) and reduce it. When O_2 is reduced it combines with H^+ from the cytoplasm, causing a net formation of OH^- on the inside. The membrane is not freely permeable to either H^+ or OH^-, so that equilibrium is not spontaneously restored. While electron transport to O_2 does produce water, it also forms the components of water, H^+ and OH^-, on opposite sides of the membrane. The net result is the generation of a pH gradient and an electrical potential across the membrane, with the inside of the cytoplasm electrically negative and alkaline, and the outside of the membrane electrically positive and acidic. This pH gradient, ΔpH, and electrical potential, $\Delta\psi$, represent an energized state of the membrane, and can be used by the cell to do useful work. In the same way that the energized state of a battery is expressed as its electromotive force, the energized state of a membrane can be expressed as a **proton-motive force** (H^+ are protons). The energized state of the membrane induced as a result of electron-transport processes can be used directly to do useful work in such processes as active transport (see Section 5.18) or flagellar rotation, or it can be used indirectly to do useful work through the

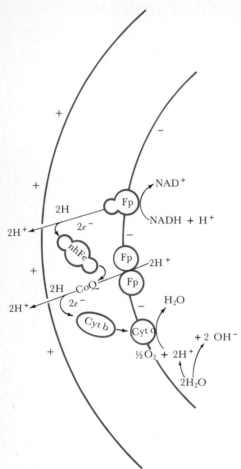

FIGURE 4.17   Orientation of the electron-transport system in the bacterial membrane, showing the manner in which electron transport through the respiratory loops can lead to charge separation and the extrusion of H^+. Note that the consumption of H^+ in the final transfer of electrons to O_2 causes the net production of OH^- ions. A tentative scheme for *Escherichia coli*. Fp, flavoprotein; nhFe, non-heme iron; CoQ, coenzyme Q; cyt b, cytochrome b; cyt O, cytochrome o. Cytochrome o here has a similar function to cytochrome a in Figure 4.16.

formation of high-energy phosphate bonds in ATP, as will be described below.*

The key substances involved in this transfer of H+ across the membrane are the coenzymes NAD, flavin and quinone. The structures of these substances were given in Figures 4.5, 4.13, and 4.15. As seen, each of these substances takes up 2H+ when becoming reduced, and discharges 2H+ when becoming oxidized again. Apparently, when protein-bound in the membrane, these substances are oriented in the membrane in such a way that they take up H+ from the inside, and discharge H+ to the outside, as they alternately become reduced and oxidized. In this way, a gradient of pH is constructed, in which the inside is alkaline with respect to outside.

This series of redox reactions may be analyzed by examining each pair of carriers sequentially, as we did in a generalized way in Figure 4.6. The transfer of two electrons through this system causes a net imbalance of six proton equivalents. The two protons associated with the starting NADH are extruded from the cell together with two more protons transported via the association of the pair of flavoproteins and CoQ. Two hydroxyl ions are formed on the inside as $\frac{1}{2}O_2$ is reduced. The net charge movement is therefore 6: four protons extruded and two hydroxyl ions formed internally. Many different electron transport sequences are possible and are indeed observed in different organisms. These sequences differ in both the number and the identity of the membrane-bound carriers.

ATP synthesis How is the proton-motive force used to synthesize ATP? An important component of this process is a membrane-bound enzyme, ATPase, which contains two parts, a headpiece present on the inside of the membrane, and a proton-conducting tailpiece that spans the membrane (Figure 4.18). This enzyme catalyzes a reversible reaction between ATP and ADP + P_i (inorganic phosphate) as shown in Figure 4.18. Operating in one direction (Figure 4.18a), this enzyme catalyzes the formation of ATP by allowing the controlled reentry of protons across the energized membrane. Just as the formation of the proton gradient required energy, the carefully controlled dissipation of the proton motive force releases energy, some of which is used to synthesize ATP. The proton motive force is more than just a supply of energy. As Figure 4.18a shows, protons are also reactants in this synthesis of ATP. So we can see that the mechanism by which the energy contained in the proton motive force is converted into ATP depends on basic principles of chemical mass action.

*The strength of the proton-motive force (that is, the amount of work done by a single proton going once around the circuit) is Δp (in units of millivolts, mV) and can be calculated from the following formula:

$$\Delta p = \Delta \psi - \frac{2.3RT}{F} \Delta pH$$

where $\Delta \psi$ is the membrane potential, ΔpH the pH gradient, both in mV, and R, T, and F are the gas constant, the absolute temperature, and the faraday, respectively. In *Escherichia coli*, ΔpH has been measured in one set of experiments as -100 mV, $\Delta \psi$ as -137 mV, and Δp as -237 mV. Similar values for Δp have been found in other organisms. This is just about enough energy to synthesize a high-energy phosphate bond.

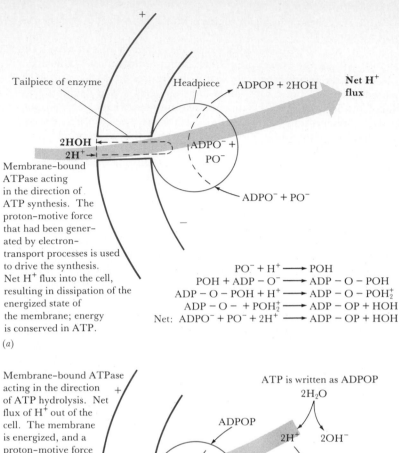

Tailpiece of enzyme

Headpiece

ADPOP + 2HOH

Net H⁺ flux

2HOH
2H⁺

ADPO⁻ +
PO⁻

+

ADPO⁻ + PO⁻

−

Membrane-bound ATPase acting in the direction of ATP synthesis. The proton-motive force that had been generated by electron-transport processes is used to drive the synthesis. Net H⁺ flux into the cell, resulting in dissipation of the energized state of the membrane; energy is conserved in ATP.

$$PO^- + H^+ \longrightarrow POH$$
$$POH + ADP - O^- \longrightarrow ADP - O - POH$$
$$ADP - O - POH + H^+ \longrightarrow ADP - O - POH_2^+$$
$$ADP - O - + POH_2^+ \longrightarrow ADP - OP + HOH$$
$$Net: ADPO^- + PO^- + 2H^+ \longrightarrow ADP - OP + HOH$$

(a)

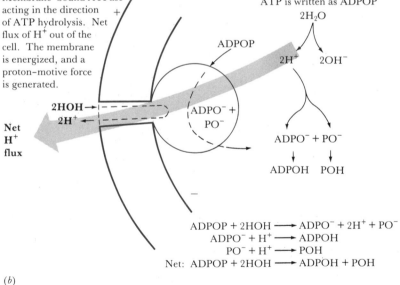

Membrane-bound ATPase acting in the direction of ATP hydrolysis. Net flux of H⁺ out of the cell. The membrane is energized, and a proton-motive force is generated.

+

ATP is written as ADPOP

$2H_2O$

ADPOP

2HOH
2H⁺

ADPO⁻ +
PO⁻

2H⁺   2OH⁻

Net H⁺ flux

ADPO⁻ + PO⁻

−

ADPOH   POH

$$ADPOP + 2HOH \longrightarrow ADPO^- + 2H^+ + PO^-$$
$$ADPO^- + H^+ \longrightarrow ADPOH$$
$$PO^- + H^+ \longrightarrow POH$$
$$Net: ADPOP + 2HOH \longrightarrow ADPOH + POH$$

(b)

FIGURE 4.18 Membrane-bound ATPase acts as a proton channel between cytoplasm and cell exterior. When the net H⁺ flux is to the interior the energized state of the membrane is dissipated and ATP is synthesized. When the net H⁺ flux is to the outside ATP is consumed and the membrane is energized. To follow the precise position of hydrogen and hydroxyl ions on ATP, this diagram should be related to that showing the complete chemical structure of ATP, Figure 4.7.

ATPase may also catalyze the reverse reaction, that is, the hydrolysis of ATP and the translocation of 2H⁺ to the outer portion of the membrane (Figure 4.18b). This results in the conservation of energy of the high-energy phosphate bond through the formation of a proton-motive force.* Operating in the other direction, the proton-motive force generated by electron transport brings about the *synthesis* of ATP, transferring energy from the energized state of the membrane to high-energy phosphate bond energy. Thus, even though the synthesis of a high-energy phosphate bond in ATP is energetically unfavorable, the reaction can be driven through the operation of the proton-motive force.

Uncouplers and inhibitors of electron-transport phosphorylation   A variety of chemical agents, called *uncouplers,* inhibit the synthesis of ATP during electron transport without inhibiting the electron-transport process itself. Examples of such uncoupling agents are dinitrophenol, dicumarol, carbonylcyanide-*m*-chlorophenylhydrazone, and salicylanilide. All of these agents are lipid-soluble substances that are acidic and can pass through the lipid matrix of the membrane when combined with H atoms. They thus promote the passage of H^+ ions across the membrane, causing dissipation of the proton-motive force. Characteristically, uncouplers actually stimulate respiration, while completely inhibiting ATP synthesis, thus resulting in wasted energy.

Various chemicals inhibit electron transport by interfering with the action of electron carriers. Carbon monoxide combines directly with the terminal cytochrome, cytochrome oxidase, and prevents the attachment of oxygen. Cyanide (CN^-) and azide (N_3^-) bind tightly to the iron of the porphyrin ring of the cytochromes and prevent its oxidation and reduction. The antibiotic antimycin A inhibits electron transport between cytochrome *b* and *c*. All of these inhibitors and uncouplers are powerful poisons for cells, inhibiting growth and other functions.

4.10

The Balance Sheet of Aerobic Oxidation

The net result of the tricarboxylic acid cycle is the complete oxidation of pyruvic acid to CO_2 with the production of four molecules of NADH. Each of the NADH molecules can be oxidized back to NAD through the electron-transport system, producing three ATP molecules per molecule of NADH oxidized. In addition, the oxidation of α-ketoglutarate to succinate involves substrate-level phosphorylation, producing guanosine-5′-triphosphate (GTP; see Figure A3.16), which is later converted to ATP, and this oxidation also involves donation of electrons to the flavin of an electron-transport particle without the mediation of NAD, producing two more molecules of ATP, for a total in the two reactions of three ATP. Thus a total of 15 ATP molecules are synthesized for each turn of the cycle (Figure 4.12) and since in the oxidation of glucose, two molecules of pyruvic acid are formed from each glucose molecule, 30 molecules of ATP can be synthesized in the citric acid cycle. Also, the two NADH mol-

*We will discuss in Section 6.3 how this sort of process can be used to form reducing power for biosynthetic reactions in purple and green bacteria.

TABLE 4.1   ATP Yields from the Fermentation and Respiration of Glucose

	Glycolysis	Citric Acid Cycle	Total
Fermenting organisms	2 ATP	Not operative	2 ATP
Respiring organisms	2 ATP (substrate-level phosphorylation)	2 ATP (substrate-level phosphorylation: succinyl CoA)	
	6 ATP* (electron-transport phosphorylation)	28 ATP (electron-transport phosphorylation)	38 ATP

*In eucaryotes, only 4 ATP may be produced by electron-transport phosphorylation of the NADH produced during glycolysis due to the fact that NADH generated by glycolysis in the cytoplasm is not directly available to the mitochondrion, where electron-transport phosphorylation occurs, due to mitochondrial impermeability to NADH.

ecules produced during glycolysis can be reoxidized by the electron-transport system, yielding six more molecules of ATP. Finally, two molecules of ATP are produced by substrate-level phosphorylation during the conversion of glucose to pyruvic acid, so that aerobes can form 38 ATP molecules from glucose breakdown, in contrast to the two molecules of ATP produced anaerobically (Table 4.1). If we again assume that the high-energy phosphate bond of ATP is about 7 kcal/mole, this means that 266 kcal of energy could be converted to high-energy phosphate bonds in ATP by the complete oxidation of glucose to CO_2 and H_2O. Since the total amount of energy available from the complete oxidation of glucose by oxygen is about 688 kcal/mole, respiration is about 39 percent efficient, the rest of the energy being lost as heat.

In addition to its function as an energy-yielding mechanism, the tricarboxylic acid cycle provides key intermediates for biosynthesis. Oxalacetate and α-ketoglutarate lead to several amino acids (see Section 5.8), succinyl-CoA is the starting point for porphyrin synthesis and acetyl-CoA provides the material for fatty acid synthesis.

The amount of ATP that an organism can produce from the oxidation or fermentation of a compound determines the amount of growth that the organism can achieve. Thus, yeast should be able to form 19 times more cell mass from a given amount of glucose when growing aerobically than when growing anaerobically. This is of considerable practical importance when yeast is being produced commercially for baking purposes. We shall discuss growth yields in relation to ATP yields further in Section 7.6.

Turnover of ATP and the role of energy-storage compounds   It can be calculated (see Section 7.6) that for the synthesis of 1 gram (g) of cell material (wet weight), about 20 millimoles (mmoles) of ATP would be consumed. Since the intracellular concentration of ATP is only 2 mM, ATP obviously has only a catalytic role during growth. It has been calculated that during a doubling of the cell mass, ATP must turn over about 10,000 times.

Related to this short life of ATP is the fact that if ATP is not immediately used for growth and biosynthesis, it is hydrolyzed by reactions not yielding energy. For long-term storage of energy, most organisms produce organic polymers that can later be oxidized for the production of ATP. The glucose polymers starch and glycogen are produced by many microorganisms, both procaryotic and eucaryotic, and poly-β-hydroxybutyrate (PHB) is produced by many procaryotes. These polymers often are deposited within the cells in large granules that can be seen with the light or electron microscope (see Section 2.9). In the absence of an external energy source, the cell may then oxidize its energy-storage material and thus be able to maintain itself even under starvation conditions.

Polymer formation has a twofold advantage to the cell. Not only is energy stored in a stable form, but also polymers have little effect on the internal osmotic pressure of cells, whereas the same number of units present as monomers in the cell would markedly increase the cellular osmotic pressure, resulting in inflow of water and possible swelling and lysis. A certain amount of energy is lost when a polymer is formed from monomers, but this disadvantage is more than offset by the benefits to the cell.

4.11

Anaerobic Respiration

Although O_2 is the most common electron acceptor in the oxidation of NADH in electron-transport systems, some organisms can use other electron acceptors (Table 4.2). Oxidation with these alternative electron acceptors is called **anaerobic respiration.** One of the most common of these alternative acceptors is nitrate, NO_3^-, which is converted into more reduced forms of nitrogen, N_2O and N_2, a process also called **denitrification.** Nitrate is first reduced to nitrite by nitrate reductase; this enzyme is then reduced by cytochrome b. Thus the electron-transport chain is shortened from that involving O_2 and only two rather than three ATP molecules are generated by electron-transport phosphorylation. This is consistent with the fact that the reduction potential for the reduction of nitrate to nitrite is $+0.4$ V, as opposed to the reduction potential of $+0.8$ V for the reduction of O_2. Thus growth on nitrate is less efficient than growth on O_2 and in most organisms nitrate reduction is strongly inhibited by O_2.

All bacteria that reduce nitrate are called facultative anaerobes since they will transfer electrons to O_2 if it is present and to NO_3^- only when O_2 is absent. The nitrite formed is further reduced to the gases N_2O and N_2, but the biochemistry of these reactions is not well understood. The inorganic anion chlorate (ClO_3^-) is a specific inhibitor of nitrate reduction: it will inhibit growth of bacteria using nitrate as electron acceptor but not of the same bacteria using O_2. The gaseous products of denitrification, if produced in soil, escape to the atmosphere. Hence denitrification is considered a detrimental process agriculturally, as it results in loss of nitrogen from the soil (see Section 13.8).

Another electron acceptor used by some bacteria is sulfate (SO_4^{2-}), which is reduced to H_2S. Sulfate-reducing bacteria are usually strict anaerobes, unable to grow on or use O_2, and often they are even killed by O_2. The reduction potential for the reduction of sulfate is even lower than that for nitrate reduction, and efficiency of growth is also lower. The biochemistry of sulfate reduction and the ecological importance of sulfate-reducing bacteria are discussed in Sections 5.14, 13.10, and 19.16.

The organic compound fumarate is used by a wide variety of bacteria as an electron acceptor; fumarate is reduced to succinate. This is the reverse reaction of that found in the tricarboxylic acid cycle (Figure 4.12), in which succinate is oxidized to fumarate. The reduction potential of the fumarate/succinate pair is relatively high ($+0.030$ V), which allows coupling of fumarate with NADH and a variety of other electron donors. Fumarate is thus readily available as an electron acceptor. The energy yield is sufficient for the synthesis of 1 ATP. Bacteria able to use fumarate as an electron acceptor include *Vibrio succinogenes* (which can grow on H_2 as sole energy source and use fumarate as electron acceptor), *Desulfovibrio gigas* (a sulfate-reducing bacterium), some clostridia, *Escherichia coli,* and *Proteus rettgeri.* Another bacterium, *Streptococcus faecalis,* can use fumarate as an electron acceptor but does not couple this with electron-transport phosphorylation. In the case of the latter organism, use of fumarate merely serves to reoxidize NADH formed during glycolysis.

A number of other electron acceptors are known to be reduced by bacteria, but it has not been definitely shown that they are coupled to electron-transport phosphorylation. These are listed in Table 4.2. The reduction of CO_2 to methane (CH_4) is an extremely important process in the anaerobic carbon cycle, as will be outlined in Sections 13.7 and 19.20. The organisms that carry out this process, called the methanogenic bacteria, are a diverse group of extremely oxygen-sensitive anaerobes. They are widespread in anaerobic muds, the intestinal tract, the rumen of cows and other ruminants, and in anaerobic sewage treatment installations. Most of the methanogenic bacteria can grow on H_2 as sole electron donor and CO_2 as sole electron acceptor. Because there is no known mechanism by which the utilization of H_2 can be coupled to substrate-level phosphorylation, it is hypothesized that the methane bacteria carry out electron-transport phosphorylation. Neither quinones nor cytochromes are present in the methane bacteria, but they contain a new type of electron carrier called F_{420} (see Section 19.20), which may be involved in electron-transport phosphorylation.

The ferric ion (Fe^{3+}) can be used by a variety of bacteria as an electron acceptor, being reduced to the ferrous ion (Fe^{2+}). This process is found in many of the organisms that reduce nitrate, and there is some suggestion that perhaps the same enzyme system is involved in both reductions. Ferric ion is present in soils and rocks, often as the insoluble ferric hydroxide $Fe(OH)_3$, and when conditions become anaerobic, reduction to the ferrous state can occur. The ferrous ion is much more soluble than the ferric ion, and reduction by microorganisms thus leads to the solubilization of iron, an important first step in the formation of the type of ore deposit called *bog iron* (see Color Plate 3c and Section 13.11). However, although the reduction of ferric iron occurs in many nitrate-reducing bacteria, it has not yet been shown that this is coupled to electron-transport phosphorylation.

The other electron acceptors listed in Table 4.2 are definitely reduced by a variety of bacteria, but further work is necessary to show that their reductions are coupled with electron-transport phosphorylation.

TABLE 4.2 Electron Acceptors Used or Possibly Used in Electron-Transport Phosphorylation

Acceptor	Reduced Products
Acceptors definitely linked to electron-transport phosphorylation:	
O_2	H_2O
Nitrate (NO_3^-)	Nitrite (NO_2^-), nitrous oxide (N_2O), nitrogen (N_2)
Nitrite (NO_2^-)	Nitrous oxide, nitrogen
Sulfate (SO_4^{2-})	Sulfide (H_2S)
Fumarate ($^-OOC-CH=CH-COO^-$)	Succinate ($^-OOC-CH_2-CH_2-COO^-$)
Dimethyl sulfoxide (CH_3-S-CH_3) $\overset{\|}{\underset{}{O}}$	Dimethylsulfide (CH_3-S-CH_3)
Elemental sulfur (S^0)	Sulfide
Acceptors possibly linked to electron-transport phosphorylation:	
Carbon dioxide (CO_2)	Methane (CH_4)
Ferric (Fe^{3+})	Ferrous (Fe^{2+})
Tetrathionate ($S_4O_6^{2-}$)	Thiosulfate ($S_2O_3^{2-}$)
Glycine (H_2C-COO^-) $\underset{NH_2}{\|}$	Acetate (CH_2COO^-) + NH_4^+
Manganic oxide (MnO_4)	Manganous oxide (MnO_2)

4.12

Summary

We have learned in this chapter that the utilization of chemical energy in living organisms involves oxidation-reduction reactions. Although some oxidation-reduction reactions in the cell involve oxygen, most do not. Electron transfer is the actual basis of oxidation-reduction. A primary electron donor is required. This donor loses electrons, becoming oxidized. An electron acceptor is needed for every oxidation-reduction reaction; it accepts the electrons and becomes reduced. The tendency of a substance to give up electrons is expressed as its reduction potential, which is measured in reference to the reduction potential of hydrogen.

Electrons never exist in the free state; they must always be carried by specific substances called *electron carriers*. In living organisms, a variety of electron carriers exist, some freely diffusing, others bound to electron-transport particles. Electron-transport particles are generally incorporated into membranes, either the plasma membrane, in procaryotes, or mitochondrial membranes, in eucaryotes. The most important diffusible electron carriers are nicotinamide-adenine dinucleotide (NAD) and nicotinamide-adenine dinucleotide phosphate (NADP). The most important bound-electron carriers are the flavoproteins, cytochromes, and quinones. Another group of electron carriers comprises the iron-sulfur proteins, of which ferredoxin is the best known.

Coupled with electron transfer reactions is the synthesis of high-energy phosphate bonds in adenosine triphosphate (ATP). ATP serves as the prime energy carrier in living organisms, being generated during or following oxidation-reduction reactions and being used during biosynthesis reactions. One of the interesting things about ATP is that the high-energy phosphate bond in this molecule can be synthesized from many types of oxidation-reduction reactions, irrespective of the reduction potentials of the systems concerned. It is because the synthesis of, and the release of energy from, ATP is independent of the redox scale that ATP can be used as an energy carrier in a wide variety of biochemical reactions. This fact was surely of major importance in the evolution of ATP as the prime energy carrier of living organisms. However, ATP is only a short-lived energy reserve. For long-term energy storage, organic polymeric compounds such as starch, glycogen, or PHB are formed.

To be utilized as an energy source, the organic compound must be able to give up electrons and become oxidized, and since each oxidation reaction must be accompanied by a reduction, there must be an electron acceptor to take up the electrons from the energy source. The most widely occurring terminal electron acceptor is oxygen, which is activated by means of a cytochrome system present in an electron-transport particle. When O_2 accepts electrons it becomes reduced to H_2O. The utilization of O_2 as an electron acceptor is called **respiration.** Other electron acceptors that can replace O_2 are the inorganic compounds nitrate, ferric iron, sulfate, and CO_2. The utilization of these electron acceptors in place of O_2 is called **anaerobic respiration.** Organic compounds can also be utilized as energy sources in the absence of an added electron acceptor by a process called **fermentation.** In this process, organic compounds serve as both electron donors and electron acceptors. In the fermentation of glucose by yeast, for instance, some carbon atoms of the glucose are oxidized to CO_2, whereas other carbons are reduced to alcohol. Although NAD is involved in fermentation, a cytochrome system is not. Whereas in respiration all of the potential chemical energy of an organic compound can be released, in fermentation it cannot. Thus energy and ATP yields in fermentation are much lower than in respiration.

The alternative flow patterns of electrons in the various oxidation-reduction reactions in living organisms are diagrammed in Figure 4.19. There are two distinct means by which ATP is synthesized during oxidation-reduction reactions, termed **substrate-level and electron-transport phosphorylation.** In substrate-level phosphorylation, ATP is synthesized directly during one of the enzymatic reactions involved in oxidation of the energy source. Substrate-level phosphorylations involve the incorporation of inorganic phosphate into an activated organic compound, thus generating a high-energy phosphate bond. In a subsequent reaction, this high-energy phosphate bond is transferred to ATP. There is only a restricted number of energy-rich compounds known in biochemistry to be coupled to ATP synthesis, so that if substrate-level phosphorylation is to occur, it must be through one of these compounds. Substrate-level phosphorylation is the only way in which ATP is synthesized in fermentation. Thus, if an organic compound is to serve as an energy source via fermentation, it must be capable of being converted into one of these few energy-rich compounds.

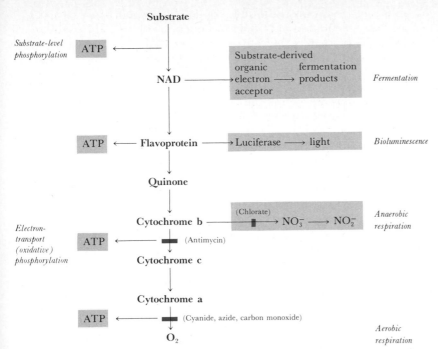

FIGURE 4.19 Electron flow in fermentation and aerobic and anaerobic respiration. Electron flow in aerobic respiration proceeds vertically down the page from substrate to O_2. In the absence of O_2 some bacteria use NO_3^- (anaerobic respiration), whereas in the absence of any external electron acceptor a substrate-derived electron acceptor may be used (fermentation). Sites of ATP formation and sites of inhibitor action are shown. Bioluminescence is an alternate pathway and leads to production of light rather than ATP (see Section 19.14). All alternatives shown do not occur in any single organism.

Electron-transport phosphorylation, sometimes called oxidative phosphorylation, occurs during the transfer of electrons through a membrane-bound electron-transport particle. This can occur only if there is an external electron acceptor, such as O_2, NO_3^- or SO_4^{2-}. Key components of electron-transport particles are the electron carriers NAD, flavoprotein, cytochromes, and quinones. These carriers have different reduction potentials and are positioned in the electron-transport particle in order of their potentials. Further, the electron carriers in the particle are oriented in the membrane in a series of loops, in such a way that there is a separation of the movement of hydrogen atoms and electrons. A proton-motive force is set up across the membrane, and this energized state of the membrane either can be used directly to do useful work (as in active transport and flagellar rotation) or indirectly to do useful work through the formation of ATP. A complex, membrane-bound enzyme, ATPase, is involved in this latter process, and not only serves as a means for ATP synthesis, but also as a channel for conducting protons across the membrane. There are three sites in the electron-transport chain where there is sufficient energy difference for ATP synthesis to occur; thus during the transfer of two electrons from NAD to O_2, three molecules of ATP can be synthesized.

When a sugar such as glucose is used as an energy source via fermentation, it undergoes a series of biochemical steps called **glycolysis**, which lead to the synthesis of pyruvate as a key substance in the eventual synthesis of ATP. During the steps leading from glucose to pyruvate, substrate-level phosphorylation occurs and a net of two

ATP molecules are synthesized. At the same time, two molecules of NAD are reduced to NADH. To maintain the electron balance, the two NADH must be oxidized back to NAD. This is done by reactions in which pyruvate is converted into a more reduced product, such as alcohol or lactate.

When a sugar such as glucose is oxidized completely in respiration, 38 molecules of ATP can be synthesized. The early steps in glucose oxidation may follow glycolysis, with the synthesis of two ATP; however, the pyruvate formed is not converted into fermentation products, but is completely oxidized to CO_2, through a series of biochemical reactions called the **tricarboxylic acid cycle.** During the tricarboxylic acid cycle, a total of 30 molecules of ATP can be synthesized. Because the NADH produced during initial conversion of glucose to pyruvate can also be reoxidized in the electron-transport system, six more molecules of ATP are made, for a total of 38 molecules. Thus aerobic utilization of glucose via respiratory processes is much more favorable energetically than anaerobic utilization via fermentation.

Supplementary Readings

Chou, P. Y., and **C. D. Fasman.** 1978. Empirical prediction of protein conformation. Annu. Rev. Biochem. 47:251–276. Examination of the relation between amino acid sequence and protein shape.

Crow, D. R. 1979. Principles and applications of electrochemistry, 2nd ed. Chapman & Hall, London. Excellent summary of basic equations and variations at other than standard conditions.

Dickerson, R. E., R. Timkovich, and **R. J. Almassy.** 1976. The cytochrome fold and the evolution of bacterial energy metabolism. J. Mol. Biol. 100:473–491. Excellent review of general role of cytochromes, with good illustrations and intriguing speculation about relationships between procaryotes and eucaryotes.

Fillingame, R. H. 1980. The proton-translocating pumps of oxidative phosphorylation. Annu. Rev. Biochem. 49:1079–1113. Comparison of chemiosmosis in bacterial and mitochondrial systems.

Haddock, B. A., and **W. A. Hamilton,** eds. 1977. Microbial energetics. Symposium 27, Society for General Microbiology. Cambridge University Press, New York. Excellent series of articles on microbial energetics.

Harold, F. M. 1977. Ion currents and physiological functions in microorganisms. Annu. Rev. Microbiol. 30: 181–203. A brief review of the generation and utilization of the proton-motive force, with emphasis on eucaryotic microorganisms.

Harold, F. M. 1978. Vectorial metabolism. *In* L. N. Ornston and J. R. Sokatch, eds. The bacteria, vol. 6. Academic Press, New York. Role of chemiosmosis in active transport by bacteria.

Khan, S., and **R. M. Macnab.** 1980. Protein chemical potential, proton electrical potential and bacterial motility. J. Mol. Biol. 138:599–614. The energy for bacterial motility is derived from chemiosmotic processes.

Knowles, R. 1982. Denitrification. Microbiol. Rev. 46:43–70. Thorough coverage of biochemistry and ecology of denitrification.

Mitchell, P. 1979. Keilin's respiratory chain concept and its chemiosmotic consequences. Science 206: 1148–1159. The development of the chemiosmotic hypothesis in the Nobel lecture by its originator.

Nichols, D. G. 1982. Bioenergetics: an introduction to the chemiosmotic theory. Academic Press, New York. Excellent up-to-date summary of chemiosmosis.

Postgate, J. R. 1979. The sulphate-reducing bacteria. Cambridge University Press, New York. A thorough survey of sulfate reduction by a leader in the field.

Quayle, J. R., ed. 1979. Microbial biochemistry (Vol. 21 of International review of biochemistry). University Park Press, Baltimore. Chapter 7 contains a good summary of bacterial methane production and consumption.

Stryer, L. 1981. Biochemistry, 2nd ed. W. H. Freeman, San Francisco. An excellent coverage of basic energetics and biochemistry. Beautifully illustrated and easy to read. See especially Chapters 11 through 14.

Thauer, R. K., K. Jungermann, and K. Decker. 1977. Energy conservation in chemotrophic anaerobic bacteria. Bacteriol. Rev. 41:100–180. A thorough discussion of energy-yielding reactions in bacteria, well grounded in fundamental principles of free energy. Excellent discussion of electron-transport phosphorylation during anaerobic respiration. Useful tables of reduction potentials and free-energy yields. Also provides a useful review of biochemical pathways in fermentative bacteria.

Walsh, C. 1979. Enzymatic reaction mechanisms. W. H. Freeman, San Francisco. Description of basic chemical principles in enzyme function.

5

Biosynthesis and Nutrition

Anabolism and Catabolism

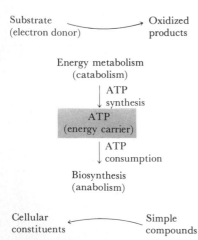

FIGURE 5.1 Schematic representation of anabolism and catabolism showing key role of ATP in relating the processes.

In the preceding chapter we have seen that many microbes generate energy during the metabolism of organic compounds. These reactions involve oxidation-reduction processes accompanied by the release of energy, some of which is conserved in high-energy phosphate bonds in adenosine triphosphate (ATP). The compound may be oxidized either completely to CO_2 or only partially, but in either case the compound is degraded into smaller and simpler products. The processes involved in this breakdown of compounds in cells are collectively called **catabolism,** and the enzymatic reactions involved in breakdown are called **catabolic reactions.**

In this chapter we are concerned with the reactions by which microorganisms build up the vast array of chemical substances of which they are composed. These substances, often quite complex, are synthesized from simpler molecules by processes collectively called **anabolism.** The enzymatic reactions involved in anabolism are often called **biosynthetic reactions.** Biosynthetic reactions are often energy-requiring, and ATP formed in catabolic (energy-generating) reactions is used up in anabolic reactions. Collectively, catabolic and anabolic reactions are called **metabolic reactions; metabolism** refers to the whole array of degradative and biosynthetic reactions taking place within cells. Although catabolism and anabolism are in some senses opposites, it must be understood that the detailed reactions of each are not merely the reverse of each other. For example, the enzymes used in degrading glucose are quite different from those which function in glucose synthesis. The relationship between catabolism and anabolism, illustrating the central role of ATP as an energy carrier, is shown in Figure 5.1.

Biosynthesis Although energy is required in certain key biosynthetic reactions, the focus of biosynthesis is not on energy but on carbon and on the intermediates occurring in the buildup of cell constituents from simple starting materials. In Chapter 4 we described the intermediates in the oxidation of glucose in both the glycolytic and the tricarboxylic acid cycles, but our main concern there was not with the ultimate fate of these intermediates but with how their transformations led to the formation of ATP. In the present chapter, we shall describe biochemical pathways of the key biosynthetic reactions for the synthesis of sugars, amino acids, fatty acids, purines, pyrimidines, and other important cell constituents. In many of these pathways, anabolism and catabolism share common intermediates: compounds such as pyruvate, acetyl-CoA, oxalacetate, glucose-6-phosphate, and so on. These intermediates are formed as a result of catabolism, but also serve as starting materials in biosynthetic reactions. Some of the starting compounds for biosynthesis are formed by simple depolymerizations of large molecules. For example, proteins may be hydrolyzed to amino acids, which can be used directly in new protein synthesis. Similarly, nucleic acid monomers (bases and nucleosides) may be formed by nuclease action and used for synthesis (Figure 5.2).

The relationships between catabolism and anabolism occur not only at the common intermediates shown in Figure 5.2. Certain biochemical pathways also play dual roles, functioning in both anabolism and catabolism. For instance, the tricarboxylic acid cycle is involved not only in the oxidation of pyruvate and acetyl-CoA, but also in the generation of succinyl-CoA, oxalacetate, and α-ketoglutarate, which serve as intermediates in the normal pathways for the synthesis of amino acids, porphyrins, and other compounds. A pathway that serves the dual function of catabolism and anabolism is called an **amphibolic pathway.** Many pathways are amphibolic. For example, glycolysis forms dihydroxyacetone phosphate as an intermediate, which may either be further metabolized to pyruvate during energy generation or converted to glycerol for initiation of phospholipid synthesis. The TCA cycle is also clearly amphibolic, as discussed in Section 4.9. For example α-ketoglutarate may be oxidized to succinate, leading to energy generation, or it may be converted to the amino acid glutamate for use in protein synthesis. In these examples there are obviously choices to be made. A given molecule at a given moment may go to catabolic or to anabolic sequences; it cannot do both. However, it should be clear from a study of the tricarboxylic acid pathway in Figure 4.12 that if any compound is removed from the cycle by being used up in biosynthesis, the cycle cannot continue to function, as links in the cycle would be deleted. This deficiency is overcome by replacement of oxalacetate by ancillary synthetic reactions that are out of the main cycle. Reactions of this type have been grouped under the term **anaplerotic** (meaning "filling up" or "replenishing"). Several anaplerotic reactions will be described later in this chapter.

One key point regarding the contrasts between anabolic and catabolic reactions, as mentioned above, is that even where the same

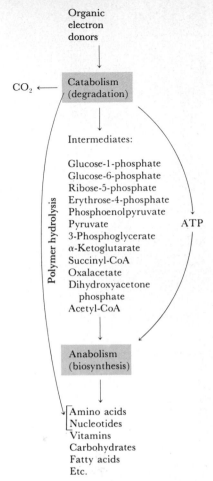

FIGURE 5.2 Interrelation of catabolism and anabolism via a key group of common intermediates.

reactants are involved, the enzymes acting in anabolic reactions are often different from those involved in catabolic reactions. For instance, a sugar or amino acid may be broken down to pyruvate in a catabolic reaction, and pyruvate may be the precursor of the same sugar or amino acid in an anabolic reaction, but the anabolic sequence is not necessarily a mere reversal of the catabolic one. Different steps are usually involved, with quite different intermediates. Even if exactly the same intermediates are involved, the catabolic and anabolic enzymes may be different. The reasons for this are two-fold: (1) catabolic reactions are energy-yielding degradations, whereas anabolic reactions often require the input of energy and by using different enzymes, the proper energy relations can be maintained; and (2) if different reaction sequences and enzymes are involved, effective control of the relative rates of catabolism and anabolism can be maintained, through control of the reaction rates of the two enzyme systems.

In eucaryotes, another difference between anabolism and catabolism is in the cellular localization of the processes. Many catabolic reactions are localized in the mitochondria and microbodies, whereas anabolic reactions are primarily cytoplasmic. At least one advantage of having separate sites for catabolic and anabolic reactions is that both can occur at the same time without confusion arising. In procaryotes, where compartmentation is much less structured, control of reactions occurs mainly at the level of single enzymes, as will be described later in this chapter.

5.2

Nutrition and Biosynthesis

Chemical composition of a cell During our discussion of cellular structure in Chapters 2 and 3, some aspects of the chemical makeup of various cell constituents were presented. It may be useful at this point to summarize what is known about the overall chemical composition of a cell (Table 5.1). Cells are made up of large numbers of small molecules, such as water, inorganic ions, carbohydrates, and amino acids, but contain a much smaller number of large molecules, the polymers of the cell, the most important of which are the proteins and nucleic acids. The cell may obtain most of the small molecules from the environment in preformed condition, whereas the large molecules are synthesized in the cell. There are complex interrelationships between the compounds taken into the cell and the compounds synthesized by the cell, and these relationships will be the main topic of discussion in this chapter.

Nutrition Substances in the environment used by organisms for catabolism and anabolism are called **nutrients.** Nutrients can be divided into two classes: (1) necessary nutrients, without which a cell cannot grow; and (2) useful but dispensable nutrients, which are used if present but are not essential. Some nutrients are the building blocks from which the cell makes macromolecules and other structures, whereas other nutrients serve only for energy generation without being incorporated directly into the cellular material; sometimes a nutrient can play both roles. The required substances can be divided

TABLE 5.1   Overall Macromolecular Composition of an Average *E. coli* B/r Cell*

Macromolecule	Percentage of Total Dry Weight	Weight per Cell ($10^{15} \times$ Weight, Grams)		Molecular Weight	Number of Molecules per Cell	Different Kinds of Molecules
Protein	55.0	155.0		4.0×10^4	2,360,000	1050
RNA	20.5	59.0				
23 S rRNA			31.0	1.0×10^6	18,700	1
16 S rRNA			16.0	5.0×10^5	18,700	1
5 S rRNA			1.0	3.9×10^4	18,700	1
transfer			8.6	2.5×10^4	205,000	60
messenger			2.4	1.0×10^6	1,380	400
DNA	3.1	9.0		2.5×10^9	2.13	1
Lipid	9.1	26.0		705	22,000,000	4†
Lipopolysaccharide	3.4	10.0		4346	1,200,000	1
Peptidoglycan	2.5	7.0		$(904)_n$	1	1
Glycogen	2.5	7.0		1.0×10^6	4,360	1
Total macromolecules	96.1	273.0				
Soluble pool	2.9	8.0				
building blocks			7.0			
metabolites, vitamins			1.0			
Inorganic ions	1.0	3.0				
Total dry weight	100.0	284.0				
Total dry weight/cell		2.8×10^{-13}g				
Water (at 70% of cell)		6.7×10^{-13}g				
Total weight of one cell		9.5×10^{-13}g				

*In balanced growth at 37°C in glucose minimal medium, mass doubling time, g, of 40 minutes.

†There are four classes of phospholipids, each of which exists in many varieties as a result of variable fatty acyl residues.

From Ingraham, J. L., O. Maaløe, and F. C. Neidhardt. 1983. Growth of the Bacterial Cell. (Sunderland, Mass.: Sinauer Associates, Inc.), p. 3. Used by permission. The data have been compiled from a number of sources.

into two groups, macronutrients and micronutrients, depending on whether they are required in large or small amounts. It is easy to detect when a macronutrient is required, merely because so much of it is needed. Often micronutrients are required in such small amounts that it is impossible to measure exactly how much is required; indeed, one may not even suspect that the particular micronutrient is present in the medium in which an organism is growing, because medium components may contain these micronutrients as trace contaminants.

5.3

Sugar Metabolism

Hexose utilization   Sugars with six carbon atoms, called **hexoses,** are the most important electron donors for many microorganisms, and are also important structural components of microbial cell walls, capsules, slimes, and storage products. The most common hexose sources in nature are listed in Table 5.2, from which it can be seen that most are polysaccharides, although a few are disaccharides.

TABLE 5.2 Sources of Hexose Sugars in Nature*

Substance	Composition	Sources	Enzymes Breaking Down
Cellulose	Glucose polymer (β-1,4-)	Plants (leaves, stems)	Cellulases (β,1-4 glucanases)
Starch	Glucose polymer (α-1,4-)	Plants (leaves, seeds)	Amylase
Glycogen	Glucose polymer (α-1,4- and α-1,6)	Animals (muscle)	Amylase, phosphorylase
Laminarin	Glucose polymer (β-1,3-)	Marine algae (Phaeophyta)	β-1,3-Glucanase (laminarinase)
Paramylon	Glucose polymer (β-1,3-)	Algae (Euglenophyta and Xanthophyta)	β-1,3-Glucanase
Agar	Galactose and galacturonic acid polymer	Marine algae (Rhodophyta)	Agarase
Pectin	Galacturonic acid polymer (from galactose)	Plants (leaves, seeds)	Pectinase (polygalacturonase)
Sucrose	Glucose-fructose disaccharide	Plants (fruits, vegetables)	Invertase
Lactose	Glucose-galactose disaccharide	Milk	β-Galactosidase

*Each of these is subject to degradation by microorganisms.

Utilization involves enzymatic breakdown to monosaccharides, conversion to glucose-6-phosphate, and catabolism via the glycolytic pathway or other hexose oxidation pathways (see top part of Figure 5.3). One of these pathways, the pentose-phosphate (or hexose-monophosphate) pathway is especially important in biosynthesis as well. As shown in Appendix 3, Figure A3.6, this pathway involves the oxidative decarboxylation of glucose-6-phosphate, yielding CO_2 and ribulose-5-phosphate while reducing two molecules of coenzyme, usually NADP. Ribulose-5-phosphate molecules formed in this way are converted to two other pentose phosphates, xylulose-5-phosphate and ribose-5-phosphate. These two pentoses then undergo a series of rearrangements brought about by the enzymes transketolase and transaldolase (Figure 5.4). As summarized in Figure A3.6, the net products of these transfer reactions, fructose-6-phosphate and glyceraldehyde-3-phosphate, may be further metabolized in a variety of energy-yielding processes.

As mentioned above, the pentose-phosphate pathway has important biosynthetic activities. The pentoses are more than just intermediates; they are essential precursors for nucleic acid synthesis, not only the ribose and deoxyribose moieties, but also in purine synthesis (see Section 5.9). In addition, ribose is present in many other compounds, including ATP, NAD, and other coenzymes. The erythrose-4-phosphate formed by transaldolase is a precursor in the synthesis of the aromatic amino acids (see Figure A3.14). Therefore, the actual functioning of the pentose-phosphate pathway cannot be drawn in simple schematic form. Key compounds are constantly being re-

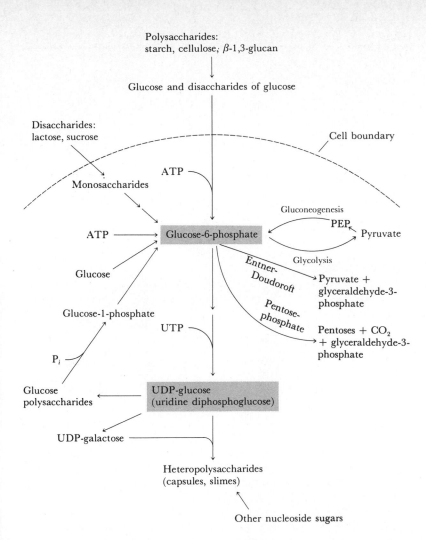

Polysaccharides:
starch, cellulose, β-1,3-glucan

Glucose and disaccharides of glucose

Disaccharides:
lactose, sucrose

Cell boundary

ATP

Monosaccharides

Gluconeogenesis

PEP

Pyruvate

ATP ———→ Glucose-6-phosphate

Glucose

Glycolysis

Entner-Doudoroft

Pyruvate +
glyceraldehyde-3-
phosphate

Glucose-1-phosphate

Pentose-phosphate

UTP

Pentoses + CO_2
+ glyceraldehyde-3-
phosphate

P_i

Glucose
polysaccharides

UDP-glucose
(uridine diphosphoglucose)

UDP-galactose

Heteropolysaccharides
(capsules, slimes)

Other nucleoside sugars

FIGURE 5.3 Hexose metabolism:
summary of main routes.

moved, requiring introduction of more hexose molecules at the beginning to keep the pathway going.

A final important feature of the pentose-phosphate reactions is that the hexose oxidations are usually linked to the reduction of NADP. The generation of reduced coenzyme (NADPH) in this fashion is extremely important, since it is a general observation that biosynthetic reactions which require reducing power proceed with NADPH as the immediate electron donor. This separate use of NADPH in biosynthesis (see also Section 6.7) and NAD in energy generation (Sections 4.6 and 4.7) allows finer control of the balance between anabolism and catabolism. The pentose-phosphate reactions are a major source of generating NADPH. It is proper to conclude that this pathway is clearly amphibolic, but the biosynthetic function, both direct and indirect, seems to predominate over the rather small ATP yield in most organisms.

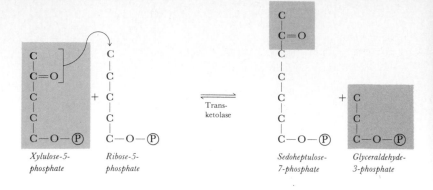

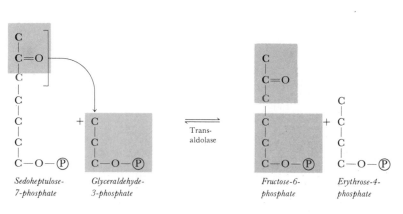

FIGURE 5.4 Reactions of transketolase and transaldolase. Transketolase transfers two-carbon fragments from keto sugars (xylulose, sedoheptulose) to aldose acceptors (ribose, erythrose). Transaldolase transfers three-carbon fragments from keto sugars (sedoheptulose, xylulose) to aldose acceptors. Functioning of these enzymes in pentose-phosphate pathway is shown in Figure A3.6.

Pentose utilization   In addition to their central biosynthetic roles pentoses may serve as electron donors in fermentation and respiration. Xylose is a common pentose in wood and straw, present in the form of a polymer called *xylan.* Arabinose is the sugar in many plant gums. Hydrolytic enzymes degrade these polymers to single pentose units that then pass into the cell. Before catabolism, pentoses are phosphorylated and converted into *xylulose-5-phosphate,* which is a central intermediate of pentose metabolism (Figure 5.5). In the fermentation of pentose, two pathways are possible, depending on the enzymes present in the organism. The direct breakdown of xylulose-5-phosphate involves a substrate-level phosphorylation with the enzyme *phosphoketolase,* leading to the formation of glyceraldehyde-3-phosphate and acetylphosphate. Glyceraldehyde phosphate is converted to lactate via the normal glycolytic pathway (see Figure 4.9) with the synthesis of two ATP molecules, and another molecule of ATP is generated from acetylphosphate. Since one ATP molecule was used in the initial phosphorylation of pentose, the net ATP yield from the fermentation of pentose is two ATP molecules, the same resulting from the fermentation of hexose (see Section 4.7). The end products are lactate and acetate. This pentose fermentation is carried out by various lactobacilli and enteric bacteria. At least two bacteria, *Thiobacillus novellus* and *Thiobacillus A2,* couple the phosphoketolase cleavage to respiration by entering the TCA cycle with pyruvate and acetyl-CoA, as shown in color in the bottom part of Figure 5.5.

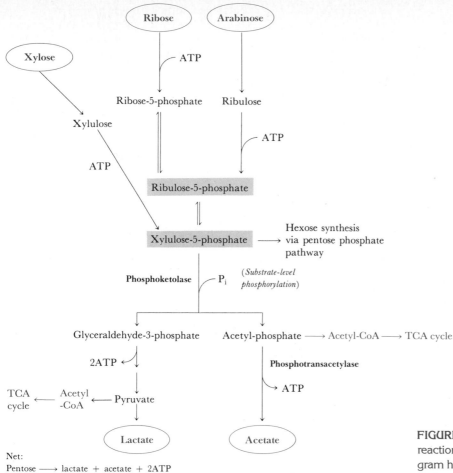

FIGURE 5.5 Pentose catabolism. The reactions in the lower part of the diagram have been studied primarily in the lactic acid bacteria and *Thiobacillus*.

If the enzyme phosphoketolase is absent, xylulose-5-phosphate cannot be broken down directly but is converted first to hexose phosphate via reactions of the pentose phosphate pathway (described above), and the hexose phosphate may then be metabolized via glycolysis, leading to fermentation or respiration.

Polysaccharide utilization *Polysaccharides,* usually insoluble and too large to pass through cell membranes, are hydrolyzed by enzymes that are excreted outside the cell. Two general kinds of enzymes are involved, those hydrolyzing hexose units from the ends of polysaccharide chains, called *exohydrolases,* and those attacking internal units, called *endohydrolases.* Although both kinds of enzymes act independently, the endohydrolases, by creating more chain ends, greatly increase the rate at which the exohydrolases can act.

Although both starch and cellulose are composed of glucose units, they are connected differently (Table 5.2), and this profoundly affects their properties. Cellulose is much more insoluble than starch and is usually less rapidly digested. Cellulose forms long fibrils, and

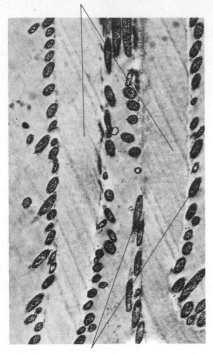

Cellulose fibers

FIGURE 5.6 Attachment of cellulose-digesting bacteria, *Sporocytophaga myxococcoides,* to cellulose fibers. Magnification, 7280×. (From Berg, B., B. V. Hofsten, and G. Pettersson. 1972. J. Appl. Bacteriol. 35:215–219.)

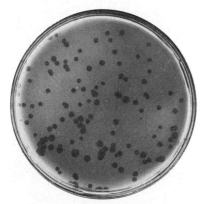

FIGURE 5.7 *Cytophaga hutchinsoni* colonies on cellulose-agar plate. Clear areas are where cellulose has been digested. (Courtesy Katherine M. Brock.)

organisms that digest cellulose are often found closely associated with them (Figure 5.6). Many fungi are able to digest cellulose and these are mainly responsible for decomposition of plant materials on the forest floor. Among bacteria, however, cellulose digestion is restricted to only a few groups, of which the gliding bacteria (Figure 5.7), clostridia, and actinomycetes are the most common. Anaerobic digestion of cellulose is carried out by a few *Clostridium* species, which are common in lake sediments, animal intestinal tracts, and systems for anaerobic sewage digestion. Starch is digestible by many fungi and bacteria; this is illustrated for a laboratory culture in Figure 5.8. Starch-digesting enzymes, called amylases, are of considerable practical utility in many industrial situations where starch must be digested, such as the textile, laundry, paper, and food industries, and fungi and bacteria are the commercial sources of these enzymes.

Agar is another polysaccharide that is of considerable interest because of its frequent use in solid media for culturing microorganisms. It occurs naturally in many seaweeds but is not found in nonmarine plants. This polysaccharide, composed of galactose and galacturonic acid (a galactose derivative) units, can be digested by some marine microorganisms. When a colony of agar-digesting organisms grows on an agar plate, a small depression is produced around the colony, which then slowly sinks into the agar (see Figure 5.9 for an analogous phenomenon with pectin).

All of the polysaccharides occurring extracellularly and utilized as substrates are broken down to monomer sugar units by hydrolysis. In contrast, the polysaccharides formed within cells as storage products are broken down not by hydrolysis but by **phosphorolysis.** This process, involving the addition of inorganic phosphate, results in the formation of hexose phosphate rather than the free hexose and may be summarized as follows for the degradation of starch, an α-1,4 polymer of glucose:

$$(C_6H_{12}O_6)_n + P_i \longrightarrow (C_6H_{12}O_6)_{n-1} + \text{glucose-1-phosphate}$$

There are two advantages to this phosphorylation. First, the cell membrane is rather impermeable to phosphorylated sugars and they are therefore much less likely to be lost from the cell by diffusion. Second, there is a net energy savings from this phosphorolysis. As we saw in Section 4.7, an early step in glycolysis is the phosphorylation of glucose at the expense of a molecule of ATP. When glucose-1-phosphate is produced directly by phosphorolytic cleavage, the expenditure of one ATP is saved. This phosphorolytic cleavage represents the trapping of energy released from a very favorable reaction (starch phosphorolysis) in the form of hexose phosphate. Glucose-1-phosphate formed in this way may also be converted to glucose-6-phosphate by the enzyme phosphoglucomutase and then metabolized via the pentose-phosphate pathway (Figure A3.6).

Many microorganisms can use *disaccharides* for growth (Table 5.2). *Lactose* utilization by microorganisms is of considerable economic importance because milk-souring organisms produce lactic acid from lactose. The utilization of lactose commonly involves the enzyme **β-galactosidase,** an enzyme of considerable interest in molecular biology because its genetics and mechanism of synthesis have been widely studied in *Escherichia coli* (see Chapters 9 and 12). Many bacteria living in the mammalian intestine form β-galactosidase, which enables them to metabolize some of the lactose that reaches the intestinal tract. *Sucrose,* the common disaccharide of higher plants, is usually first hydrolyzed to its component monosaccharides (glucose and fructose) by the enzyme *invertase,* and the monomers are then metabolized by normal pathways. Another way of hydrolyzing sucrose, which actually serves a biosynthetic function, is the formation of dextran, a glucose polymer. Dextran formation will be described later in this section.

Pectin   Another important polysaccharide, pectin, is a polymer of the sugar acid *galacturonic acid* (this is derived from galactose by oxidation of the 6-carbon sugar to a carboxylic acid). Pectin occurs in the cell walls and intercellular layers of higher plants and plays a major role in the formation of plant tissues. Pectin-digesting bacteria are very common in soil, and many plant pathogenic bacteria also possess the ability to digest this polymer. The bacteria that cause *soft rot* of carrots, potatoes, and other root vegetables are highly pectinolytic, and it is the bacterial digestion of pectin between the cell walls that leads to the maceration of the tissue. An interesting practical process involving pectin-digesting bacteria is *retting,* a process used in the manufacture of linen from flax. The flax stems are tied in bundles and laid in moist fields or submerged at the bottom of a slow-flowing river or a pond. The retting bacteria colonize the flax, hydrolyze the pectin, and cause a loosening of the fibers, so that they can be peeled away and processed into linen. Demonstration of pectinolytic activity in cultures can be done by solidifying a culture medium with pectin instead of agar; the colonies producing the enzyme form depressions in the gel surface which gradually deepen (Figure 5.9).

Hexose and polysaccharide biosynthesis   We have already seen in Chapter 4 that catabolism of glucose is a widespread and important activity of many organisms. It should also be understood that hexoses in general have other important cellular functions as well. These sugars are key precursors for the synthesis of polysaccharides, including storage compounds (glycogen, starch) and cell walls (peptidoglycan). Many compounds other than glucose can be utilized as electron donors and every organism must thus have the ability to synthesize hexoses. The synthesis of glucose is called *gluconeogenesis,* which means the creation of new glucose from noncarbohydrate precursors. The key compound in this synthesis is phosphoenolpyruvate (PEP). Gluconeogenesis begins with PEP and uses reactions of the Embden–Meyerhof pathway (Figure 5.10). Note that we describe gluconeogenesis as starting with PEP and not pyruvate. Pyruvate is a major intermediate formed by many different reactions and it is appropriate to consider it here. However, it is not possible to convert pyruvate directly to PEP. In the Embden–Meyerhof pathway, the conversion of PEP to pyruvate by pyruvate kinase is essentially irreversible. On those occasions when pyruvate is to be used in gluconeogenesis, it must first be converted to PEP through two separate reactions:

FIGURE 5.8   Demonstration of hydrolysis of starch by colonies of *Bacillus subtilis.* After incubation, the plate was flooded with Lugol's iodine solution. Where starch hydrolysis occurred, the characteristic purple color of the starch-iodine complex is absent. Hydrolysis of starch occurs at some distance from the bacterial colonies because of the production of extracellular amylase, which diffuses into the surrounding medium.

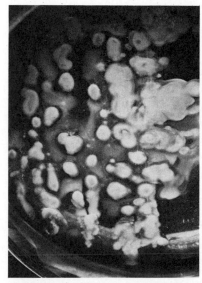

FIGURE 5.9   Colonies of *Erwinia carotovora,* a pectin-digesting bacterium, growing on a medium solidified with pectin. Note the depression which the colonies have formed as they digest the pectin.

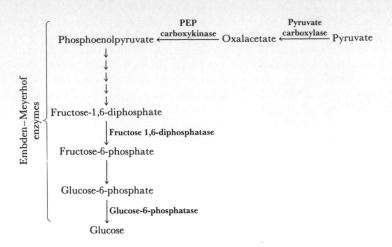

FIGURE 5.10 Gluconeogenesis starting with phosphoenolpyruvate (PEP). The two enzymes in this synthesis which are not found in the Embden–Meyerhof pathway are indicated. Pyruvate may be converted to PEP to begin gluconeogenesis by the indicated enzymes.

$$\text{Pyruvate} + CO_2 + ATP + H_2O \longrightarrow$$
$$\text{Oxalacetate} + ADP + P_i + 2H^+$$

and $\text{Oxalacetate} + ATP \longrightarrow PEP + ADP + CO_2$

Net: $\text{Pyruvate} + 2\,ATP + H_2O \longrightarrow PEP + 2ADP + 2H^+$

Pyruvate carboxylase catalyzes oxalacetate formation and phosphoenolpyruvate carboxykinase the formation of PEP. This reaction sequence might seem inefficient, but it must be remembered that a living cell is a dynamic system and it is necessary to keep anabolism and catabolism in careful balance with each other. These reactions are essential in regulating the balance.

The Embden–Meyerhof reactions between fructose-1,6-diphosphate and PEP are freely reversible so that under the right conditions hexose diphosphate synthesis will proceed from PEP. Fructose-1,6-diphosphate is then converted to fructose-6-phosphate in an irreversible reaction catalyzed by an enzyme (fructose-1,6-diphosphatase) which is not part of the catabolic Embden–Meyerhof pathway. Gluconeogenesis is completed by conversion of fructose-6-phosphate to glucose-6-phosphate followed by formation of glucose by the action of glucose-6-phosphatase, another enzyme not in the Embden–Meyerhof pathway. Although many of the enzymes in gluconeogenesis are the same ones which function in catabolism via the Embden–Meyerhof pathway, the overall process is not the simple reverse of glycolysis since two key enzymes (the phosphatases) are unique to gluconeogenesis and serve to pull the overall reaction sequence toward hexose synthesis when this synthesis is required by the cell.

As shown in the lower part of Figure 5.3, a key compound in the biosynthesis of other hexoses from glucose is uridine diphosphoglucose (UDPG), which has the structure uracil-ribose-phosphate-phosphate-glucose. This is called a *nucleoside diphosphate sugar* (see Section 5.9 for the definition and structure of nucleosides), since it contains the nucleoside uridine (uracil-ribose). UDPG is an activated form of glucose and serves both as a starting material for the synthesis of many other nucleoside diphosphate sugars and as the glucose precursor of cellular polysaccharides. Thus, while glucose-6-

phosphate is the central intermediate in glucose catabolism, UDPG is the central intermediate in glucose anabolism. Other nucleoside diphosphate sugars derived from UTP include UDP-galactose, UDP-*N*-acetylglucosamine, and UDP-*N*-acetylmuramic acid. The latter two are intermediates in the biosynthesis of the cell wall in procaryotes, as will be described in Section 5.11. Other pyrimidines or purines occur in the activated forms of other sugars, such as thymidine diphosphoriboribose, guanosine diphosphomannose, and cytidine diphosphoribitol. At least 20 nucleoside diphosphate sugars are known in microorganisms alone.

The synthesis of UDPG and its conversion to UDP-galactose can be used as a representative of the kinds of reactions involved in nucleoside diphosphate sugar transformations:

Glucose + ATP $\longrightarrow$ glucose-6-phosphate $\longrightarrow$
glucose-1-phosphate + UTP (uridine triphosphate) $\longrightarrow$
$$\text{UDP-glucose} + PP_i$$

The synthesis of galactose involves the conversion of UDP-glucose to UDP-galactose by an *epimerase:*

$$\text{UDP-glucose} \xrightleftharpoons{\text{NAD}} \text{UDP-galactose}$$

UDP-galactose may then be converted to galactose-1-phosphate:

UDP-galactose $\rightleftharpoons$ galactose-1-phosphate
+ UMP (uridine monophosphate)

If galactose is used as a carbon source and an electron donor, it is first converted to galactose-1-phosphate (using ATP), and this reacts with UTP to form UDP-galactose (plus PP_i). The UDP-galactose is converted to UDP-glucose by the epimerase reaction, and the glucose subsequently ends up as glucose-6-phosphate, which is metabolized by the regular glycolytic route.

The *synthesis* of polysaccharides is not merely a reversal of the reactions involved in breakdown. A nucleoside sugar (for example, UDP-glucose) is the starting material, and sugar units are added stepwise to the end of a polysaccharide chain:

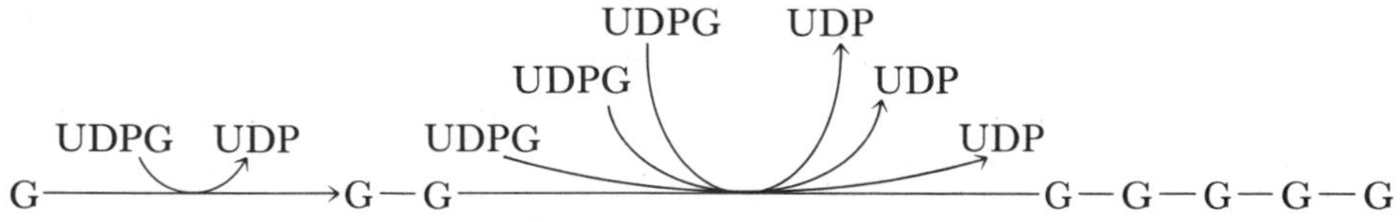

Energy to drive the synthesis is obtained by hydrolysis of the high-energy sugar-phosphate bond in the nucleoside sugar.

The microbial polysaccharide *dextran* is synthesized in a different manner, the enzyme *dextransucrase* being involved. The starting material is the disaccharide sucrose; an α-1,6-glucose polymer is formed from the glucose moiety, and fructose is liberated in the free state:

$$n\ \text{Sucrose} \longrightarrow (\text{glucose})_n + n\ \text{fructose}$$

(Glucose-   *Dextran*
α-1,4-fructose)

Hydrolysis of the glycosidic bond connecting the two sugars of sucrose is energetically very favorable and the energy released is used to

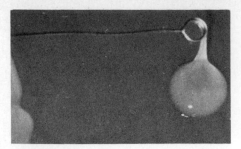

FIGURE 5.11 Slimy colony formed by dextran-producing bacterium, *Leuconostoc mesenteroides,* growing on a sucrose-containing medium. When the same organism is grown on glucose, the colonies are small and not slimy.

drive dextran synthesis. In this way the requirement for a nucleoside sugar with its high-energy phosphate bond is eliminated. Dextran is formed in this way by the bacterium *Leuconostoc mesenteroides* and a few others, and the polymer formed accumulates around the cells as a massive slime or capsule (Figure 5.11). Since sucrose is required for dextran formation, no dextran is formed when the bacterium is cultured on a medium with glucose or fructose. The dextran formed by *L. mesenteroides* has been used medically as a blood plasma substitute, and can be produced commercially by allowing the purified enzyme to react with sucrose, the length of the polymer chains being determined by how long the reaction is allowed to proceed. Dextran is also one substance that enables bacteria causing tooth decay to adhere to the teeth and form what is called dental plaque (see Section 14.4). Since dextran is formed only when sucrose is present, bacterial adherence should occur only when sucrose is present in the diet, and at least one way of controlling tooth decay is to reduce the sucrose intake by eliminating candies and other sweets.

5.4

Organic Acid Metabolism

A variety of organic acids can be utilized by microorganisms as carbon sources and electron donors. The acids of the tricarboxylic acid cycle such as citrate, malate, and succinate are common natural products formed by plants and are also fermentation products of microorganisms. The tricarboxylic acid cycle is common in microbes; thus it is not surprising that many microorganisms are able to utilize these acids as electron donors and carbon sources. Aerobic utilization of four-, five-, and six-carbon acids can be accomplished by means of enzymes of the tricarboxylic acid cycle, with ATP formation by electron-transport phosphorylation. However, some microorganisms that have a normal tricarboxylic acid cycle are impermeable to organic acids and cannot take these up from the environment. Such microorganisms, even though they contain the proper enzymes for the metabolism of organic acids, are unable to use them as electron donors. Permeability barriers to the utilization of nutrients are discussed in detail in Section 5.18.

Anaerobic utilization of organic acids usually involves conversion to pyruvate and ATP formation via the phosphoroclastic reaction (see Section 4.6).

The glyoxylate cycle Utilization of two- or three-carbon acids as carbon sources cannot occur by means of the tricarboxylic acid cycle alone. This cycle can continue to operate only if the four-carbon acid oxalacetate is regenerated at each turn of the cycle (see Figure 4.12), and any removal of carbon compounds for biosynthetic reactions would prevent completion of the cycle. When acetate is utilized, the oxalacetate needed to continue is produced through the **glyoxylate cycle** (Figure 5.12), so called because glyoxylate is a key intermediate. This cycle is composed of most of the TCA cycle reactions plus two additional enzymes: **isocitrate lyase,** which splits isocitrate to succi-

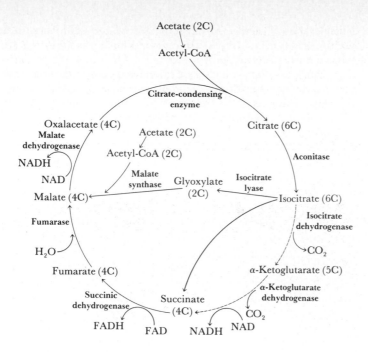

FIGURE 5.12 The glyoxylate cycle, leading to the synthesis of oxalacetate from acetate. Two unique reactions, isocitrate lyase and malate synthase (shown in color), operate with a majority of the TCA cycle reactions. Two TCA reactions not involved in the glyoxylate cycle are shown with dashed lines.

nate and glyoxylate, and **malate synthase,** which converts glyoxylate and acetyl-CoA to malate.

The net biosynthetic action of this cycle occurs as follows. One molecule of acetate is used in the formation of succinate from isocitrate via isocitrate lyase and a second molecule of acetate combines with glyoxylate (formed as the other half of the isocitrate lyase reaction) to form malate. These molecules of succinate and malate are converted to two molecules of oxalacetate via normal TCA cycle reactions. One of these oxalacetate molecules represents the return of starting material to the cycle and the other represents a net gain of a key four-carbon compound for biosynthesis. The extra oxalacetate formed this way may be used in amino acid synthesis (see Section 5.8), or it may be converted to PEP by the enzyme PEP carboxykinase as described above. Phosphoenolpyruvate may then be used for gluconeogenesis as described in Section 5.3.

Pyruvate utilization   Three-carbon compounds such as pyruvate or compounds converted to pyruvate (for example, lactate or carbohydrates) also cannot be utilized as carbon sources through the tricarboxylic acid cycle alone. The oxalacetate needed to keep the cycle going is synthesized from pyruvate by the addition of a carbon atom from CO_2. In some organisms this step is catalyzed by the enzyme pyruvate carboxylase:

$$\text{Pyruvate} + \text{ATP} + CO_2 \longrightarrow \text{oxalacetate} + \text{ADP} + P_i$$

whereas in others it is catalyzed by phosphoenolpyruvate carboxylase:

$$\text{Phosphoenolpyruvate} + CO_2 \longrightarrow \text{oxalacetate} + P_i$$

These reactions replace oxalacetate that is lost when compounds of the tricarboxylic acid cycle are removed for use in biosynthesis, and the cycle can continue to function.

Such reaction sequences as the glyoxylate cycle and those involving the carboxylating enzymes, which serve to supplement the catabolic reactions of the tricarboxylic acid cycle, are good examples of anaplerotic reactions (see Section 5.1).

5.5

Molecular Oxygen (O_2) as a Reactant in Biochemical Processes

We have discussed in some detail in Chapter 4 the role of O_2 as an electron acceptor in energy-generating reactions. Although this is by far the most important role of O_2 in cellular metabolism, O_2 also plays an interesting and important role in certain types of biosynthetic reactions. Even though O_2 has a very positive reduction potential, it is not a very reactive molecule because of its peculiar chemical properties. In particular, O_2 exists in a triplet ground state with two unpaired electron spins (see Section 8.5). Carbon, on the other hand, exists in organic compounds in the single state, so that concerted reactions between singlet carbon and triplet oxygen are not possible. In order for oxygen to react with carbon, it must first be activated and converted into the singlet state. In biological systems this is generally done through interaction of O_2 with an enzyme containing a transition element such as iron, Fe, or copper, Cu. Such transition elements have one or more unpaired electrons and thus direct reaction between these metals and O_2 is possible. Oxygen and some of its derivatives are also potentially dangerous to organisms. The mechanisms of the danger and protection against it are described in detail in Section 8.5.

Oxygenases are enzymes that catalyze the incorporation of O_2 into organic compounds. There are two kinds of oxygenases: *dioxygenases* catalyze the incorporation of *both* atoms of O_2 into the molecule; and *monooxygenases* catalyze the transfer of only one of the two O_2 atoms to an organic compound as a hydroxyl (OH) group, with the second atom of O_2 ending up as water, H_2O. Because monooxygenases catalyze the formation of hydroxyl groups (OH) in organic compounds, they are sometimes called *hydroxylases*. Because monooxygenases require a second electron donor to reduce the second oxygen atom to water, they are also sometimes called *mixed-function oxygenases*. In most monooxygenases, the second electron donor is NADH or NADPH, although the direct coupling to O_2 is through a flavin which becomes reduced by the NADH or NADPH donor.

As we will see, there are several types of reactions in living organisms that require O_2 as a reactant. One of the best examples is the involvement of O_2 in steroid biosynthesis. Such a reaction can obviously not take place under anaerobic conditions, so that organisms which grow anaerobically must either dispense with this reaction or obtain the required substance (steroid) preformed from their environment. The requirement of O_2 as a reactant in biosynthesis is of considerable evolutionary significance, as molecular O_2 was originally absent from the atmosphere of the earth when life evolved and only became available after the evolution of cyanobacteria, the first photosynthetic organisms to produce O_2 (see Chapter 6).

Fat and phospholipid hydrolysis   Fats are esters of glycerol and fatty acids. Microbes utilize fats only after hydrolysis of the ester bond, and extracellular enzymes called **lipases** are responsible for the reaction. The end result is formation of glycerol and free fatty acids (Figure 5.13*a*). Lipases are not highly specific, and attack fats containing fatty acids of various chain lengths. However, a different enzyme is usually involved in hydrolyzing the fatty acid at the internal position of the glycerol molecule than is involved at the two outside fatty acids. Microbial attack on food lipids is of considerable practical significance, since many of the fatty acids liberated by lipase activity produce undesirable odors. Milk fat is high in fats containing esters of butyric acid, and liberation of this odoriferous substance by microbial lipases may be responsible for the development of rancid odors in spoiled milk and dairy products. Microbial lipase production can be detected using an agar medium to which glycerol tributyrate is added: when this insoluble fat is hydrolyzed the products formed are soluble, and the opalescence of the medium is cleared in the area surrounding the microbial growth. Lipase-producing organisms are thus indicated by clear zones around colonies.

Phospholipids are mixed glycerol esters of fatty acids and a phosphate ester of a non-fatty acid unit, sometimes called the X-group. Their structure and role in the formation of membranes were discussed in Section 2.4. Substances forming the X-group include ethanolamine, choline, serine, inositol, and glycerol. These X-groups vary in charge, and this variation affects the physical properties of the phospholipid molecule. Although phospholipids are mainly found in cell membranes, they are also found in large amounts in egg yolk and in blood serum. Phospholipids are hydrolyzed by specific enzymes called **phospholipases,** given different designations depending on which ester bond they cleave (Figure 5.13*b*). Phospholipases A and B cleave fatty acid esters and thus resemble the lipases described above, but phospholipases C and D cleave phosphate ester linkages and hence are quite different types of enzymes. Attack by the phospholipase C of *Clostridium perfringens* on the phospholipid of the red blood cell membrane results in lysis

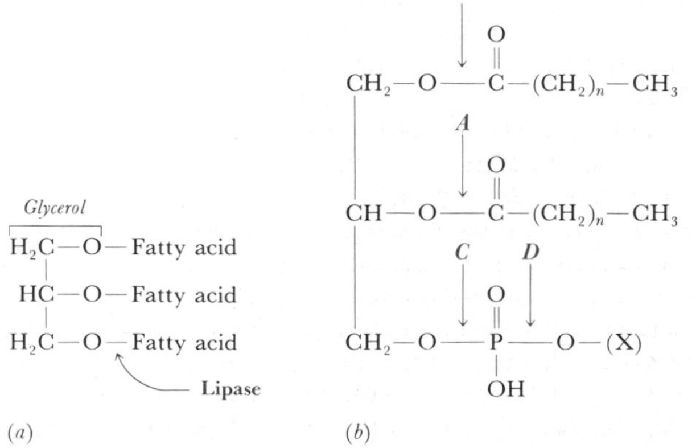

FIGURE 5.13   (a) Action of lipase on fat; (b) phospholipase action on phospholipid. The sites of action of the four distinct phospholipases *A, B, C,* and *D* are shown. X refers to a number of small organic molecules which may be at this position in different phospholipids. Compare this diagram to the more complete figure of a phospholipid in Figure 2.10.

Clostridium perfringens

Phospholipase action: Egg–yolk precipitation

Antibody against enzyme added: phospholipase action inhibited: no precipitation of egg yolk

FIGURE 5.14 Action of phospholipase around streak of *Clostridium perfringens,* growing on an agar medium containing egg yolk. On half of the plate, antibody was added, preventing action of the enzyme. (Courtesy G. Hobbs, Torry Research Station, Aberdeen, Scotland.)

(**hemolysis**), and this enzyme is partly responsible for the symptoms of gas gangrene induced by this bacterium (see Section 15.3). A nutrient agar medium to which about 4 percent of sterile egg yolk has been added is used for general detection of microbial phospholipase production. The resulting medium is clear, since the polar groups of the phospholipid make it water soluble. If the X-group is hydrolyzed by the enzyme, the resulting fat no longer contains polar groups and is water insoluble. Enzyme action is thus recognized by the zone of opalescence that develops in the medium surrounding the colony of a phospholipase-producing organism (Figure 5.14).

Fatty acid oxidation The fatty acids released by action of lipases and phospholipases are oxidized by a process called **beta oxidation,** in which two carbons of the fatty acid are split off at a time (Figure 5.15). In eucaryotes the enzymes are in the mitochondria, whereas in procaryotes, they have not been localized. The fatty acid is first activated with coenzyme A; oxidation results in the release of acetyl-CoA and the formation of a fatty acid shorter by two carbons. The process of beta oxidation is then repeated and another acetyl-CoA molecule is released. Two separate dehydrogenation reactions occur. In the first, electrons are transferred to flavin-adenine dinucleotide (FAD), whereas in the second they are transferred to NAD. Both of these coenzymes are reoxidized through electron transport phosphorylation with the synthesis of ATP. Most fatty acids have an even number of carbon atoms, and complete oxidation yields only acetyl-CoA. The acetyl-CoA formed is then oxidized by way of the TCA cycle or is converted into hexose and other cell constituents via the glyoxylate cycle. Fatty acids are good electron donors in respiration. For example, the complete oxidation of the 16-carbon fatty acid palmitic acid results in the net synthesis of 129 ATP molecules from electron-transport phosphorylation of reducing equivalents generated during the formation of acetyl-CoA and from the oxidation of acetyl-CoA through the TCA cycle.

Fatty acid synthesis Synthesis of these essential polymers is a good example of the earlier observation that catabolism and anabolism of large molecules are not simply the reverse of each other. Fatty acid synthesis consists of the stepwise buildup of long chains from two-carbon fragments of acetyl-CoA, but involves an entirely different set of enzymes from those in fatty acid breakdown. The reaction sequence of fatty acid synthesis is fundamentally the same in all organisms (Figure 5.16). In the bacterium *Escherichia coli,* the acetyl group of acetyl-CoA is transferred to a small protein which serves as a carrier, called *acyl carrier protein* (ACP). The acetyl-ACP is attached to the enzyme *fatty acid synthetase* throughout the series of reactions in which successive two-carbon fragments are added and reduced. Interestingly, although the fatty acid chain is increased two carbons at a time, the immediate precursor is a three-carbon compound, *malonyl-ACP*. Malonyl-ACP itself is synthesized from malonyl-CoA; the latter is derived from acetyl-CoA and carbon dioxide in an enzymatic reaction dependent on the vitamin *biotin* (Figure 5.17). Malonyl-ACP can be viewed as an activated form of acetyl-ACP

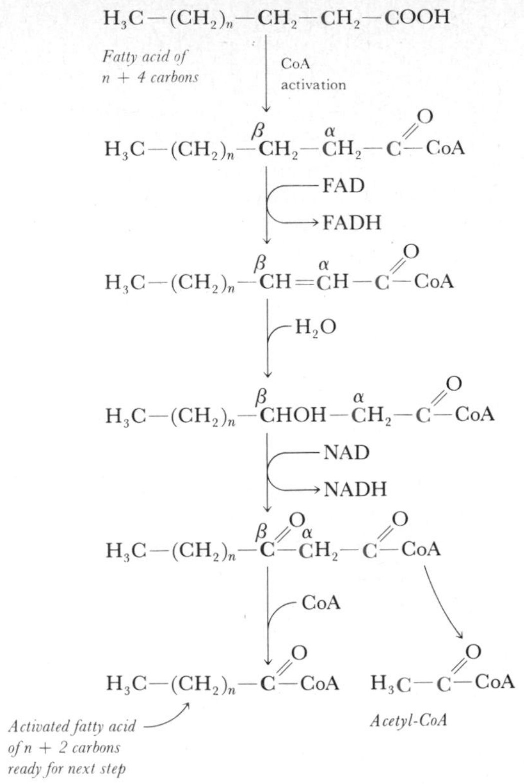

$$H_3C-(CH_2)_n-CH_2-CH_2-COOH$$

Fatty acid of n + 4 carbons

CoA activation

$$H_3C-(CH_2)_n-\overset{\beta}{C}H_2-\overset{\alpha}{C}H_2-\overset{O}{\overset{\|}{C}}-CoA$$

FAD
FADH

$$H_3C-(CH_2)_n-\overset{\beta}{C}H=\overset{\alpha}{C}H-\overset{O}{\overset{\|}{C}}-CoA$$

H_2O

$$H_3C-(CH_2)_n-\overset{\beta}{C}HOH-\overset{\alpha}{C}H_2-\overset{O}{\overset{\|}{C}}-CoA$$

NAD
NADH

$$H_3C-(CH_2)_n-\overset{\beta}{\overset{O}{\overset{\|}{C}}}-\overset{\alpha}{C}H_2-\overset{O}{\overset{\|}{C}}-CoA$$

CoA

$$H_3C-(CH_2)_n-\overset{O}{\overset{\|}{C}}-CoA \qquad H_3C-\overset{O}{\overset{\|}{C}}-CoA$$

Activated fatty acid of n + 2 carbons ready for next step

Acetyl-CoA

FIGURE 5.15 Mechanism of beta oxidation of a fatty acid, which leads to successive formation of two-carbon fragments of acetyl-CoA.

since the presence of another carboxyl group makes the methyl group of acetyl-ACP more reactive. The buildup of a fatty acid proceeds by stepwise addition of two of the carbons of malonyl-ACP, with the third carbon being released as CO_2 (Figure 5.16). Each acetyl fragment added to the end of the chain is then reduced in a series of reactions involving NADPH. In the first turn of the cycle, butyryl-ACP is formed from two starting acetyl-CoA molecules. Butyryl-CoA may then accept additional acetyl units from malonyl-ACP. The fatty acid chain is thus lengthened two carbons at a time. It may be for this reason that the most common fatty acids found in organisms are those with even numbers of carbon atoms, such as palmitic acid ($C_{16}H_{32}O_2$), stearic acid ($C_{18}H_{36}O_2$), and oleic acid ($C_{18}H_{34}O_2$). There are three general biosynthetic conclusions to be made from this pathway of fatty acid synthesis. First, there is often a need to activate compounds before reaction is possible; in this case the acyl carrier protein and acetyl-CoA are both involved. Second, this process consumes ATP; directly in the synthesis of malonyl-CoA and indirectly in the synthesis of the activated CoA and ACP molecules. Third, there is a substantial requirement for reducing power in the form of NADPH, which is formed during energetically favorable redox reactions, such as those of the pentose-phosphate pathway (see Section 5.4).

It is of interest that CO_2, through its involvement in malonyl-CoA synthesis, is an essential component in fatty acid biosynthesis

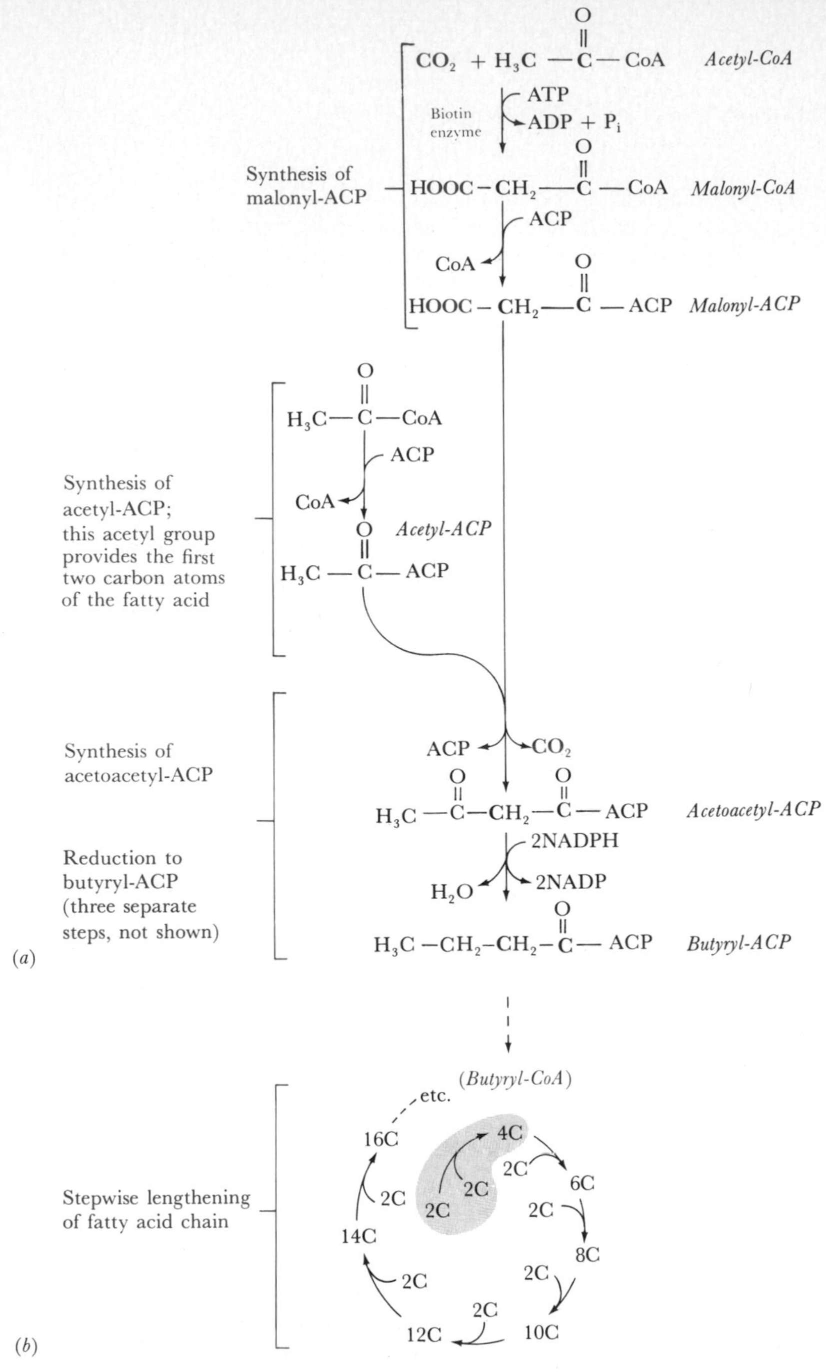

FIGURE 5.16 Synthesis of fatty acids by stepwise addition of two-carbon units from malonyl-CoA. ACP stands for acyl carrier protein. The two carbon atoms of malonyl-CoA that end up in the fatty acid are shown in boldface; the carboxyl derived from CO_2 is shown in color. (a) Synthesis of butyryl-ACP, the key acceptor, from two molecules of CO_2. (b) Schematic representation of complete synthesis of a fatty acid from acetyl-CoA precursors. The compounds in color represent the reactions of part a; the four-carbon compound is butyryl-CoA.

even though carbon from CO_2 does not appear in the final product. This explains a peculiar observation that was made years ago in studying the nutrition of microorganisms: CO_2 was found to be required for growth of certain nonphotosynthetic organisms, but this requirement could be replaced by a fatty acid, such as oleic acid.

Also, biotin, which is a required growth factor for some microorganisms, and which is involved in the CO_2 reaction, could also be replaced by a fatty acid. Since there is a small amount of CO_2 in normal air (about 0.05 percent), it is usually unnecessary to provide CO_2 for laboratory cultures. However, some pathogenic bacteria, such as *Neisseria meningitidis,* require higher concentrations of CO_2 than are present in normal air, and one must take this into account when attempting to isolate such organisms.

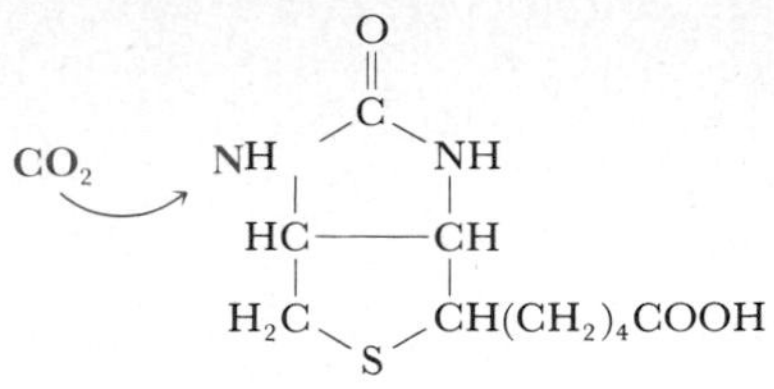

FIGURE 5.17 Biotin, the vitamin involved in activation of CO_2 in fatty acid synthesis.

Unsaturated fatty acids contain one or more double bonds. In most aerobic organisms the formation of the double bond involves a reaction requiring molecular oxygen; the requirement for O_2 explains why, when microorganisms such as yeast are cultured under anaerobic conditions, an unsaturated fatty acid, such as oleic acid, must be added to the medium as a growth factor. In anaerobic and some aerobic bacteria unsaturated fatty acids are synthesized through dehydration of a hydroxy acid, a process not involving O_2. This difference in the manner in which unsaturated fatty acids are synthesized is thought to be of considerable evolutionary significance.

Some microorganisms produce fatty acids with branched chains. These are made by using as the initiating molecule a branched-chain fatty acid (isobutyryl-CoA or isovaleryl-CoA) instead of acetyl-CoA. All subsequent two-carbon fragments are added to make a straight chain, so that the branch is only at one end.

Both branched and unsaturated fatty acids play roles in maintaining the fluidity of the cell membrane. Fluidity is important in cell membrane function (Section 2.4) and is controlled by the degree to which the side chains of the fatty acids are stacked together. The unsaturated and branched side chains do not stack as tightly as those of saturated fatty acids, so that by controlling the proportion of these fatty acids, the precise degree of membrane fluidity can be established. In most organisms, fluidity is regulated by the proportion of unsaturated fatty acids, but a number of bacteria use branched chain fatty acids for this purpose.

Fats are formed by the successive esterification of glycerol phosphate by CoA or ACP derivatives of two fatty acids to form a diacyl glycerol phosphate (also called phosphatidic acid). The phosphate group is then split off and the third fatty acid added. In the synthesis of *phospholipids,* after the initial esterifications, the phosphatidic acid in converted into a nucleotide derivative, cytidine diphosphate diglyceride, and the X-group is added in place of cytidine monophosphate. Some of the X-groups of phospholipids are not added directly but are formed by conversion of other X-groups.

Isoprenoids There is a class of lipids which are polymers of the five-carbon branched hydrocarbon **isoprene** subunit:

$$CH_2{=}CH-\overset{\displaystyle \overset{CH_3}{|}}{C}{=}CH_2$$

Isoprene

These compounds serve a variety of important biochemical roles. Called isoprenoids, they include carotenoids, coenzyme Q, and the bactoprenols, which function as carriers in bacterial cell-wall synthesis. The sterols, ring compounds present in eucaryotes (see Section 3.1), are derivatives of isoprene. The side chain of chlorophyll is phytol, an isoprenoid. Other isoprenoids not found in microorganisms but common in higher plants are rubber and terpenes.

5.7

Hydrocarbons

Hydrocarbons are organic carbon compounds containing only carbon and hydrogen and are highly insoluble in water. Low-molecular-weight hydrocarbons are gases, whereas those of higher molecular weight are liquids or solids at room temperature. Some hydrocarbons are aliphatic compounds, a class of carbon compounds in which the carbon atoms are joined in open chains. There is a tremendous variation among aliphatic hydrocarbons in chain length, degree of branching, and number of double bonds. Another important group of hydrocarbons contains the aromatic ring and can be viewed as derivatives of benzene.

Aliphatic hydrocarbons   Only relatively few kinds of microorganisms (for example, *Nocardia, Pseudomonas, Mycobacterium,* and certain yeasts and molds) can utilize hydrocarbons for growth. Utilization of aliphatic hydrocarbons is strictly an aerobic process: in the absence of O_2, hydrocarbons are completely unaffected by microbes. This accounts for the fact that hydrocarbons in petroleum deposits have remained unchanged for millions of years. As soon as petroleum resources are brought to the surface and exposed to air, oxidation of the hydrocarbons begins and the materials are eventually decomposed (see Section 13.13). The steps in hydrocarbon utilization are not completely known, but the ultimate products probably are acetate units, which are then oxidized through the tricarboxylic acid cycle.

The initial oxidation step of aliphatic hydrocarbons involves molecular oxygen (O_2) as a reactant, and one of the atoms of the oxygen molecule is incorporated into the oxidized hydrocarbon. In the oxidation of aliphatic hydrocarbons, monooxygenases are most generally involved, and a typical reaction sequence is that shown in Figure 5.18. The end product of the reaction sequence is acetyl-CoA. However, the initial oxidation is not at the terminal carbon in all cases. Oxidation may sometimes occur at the second carbon and then quite different subsequent reactions occur.

Aromatic hydrocarbons   There is a very large variety of aromatic hydrocarbons which can be used as electron donors by microorganisms, of which bacteria of the genus *Pseudomonas* are the best studied. It has been demonstrated that the metabolism of these compounds, some of which are quite complex, has as its initial stage the formation of either of two molecules, protocatechuate and catechol:

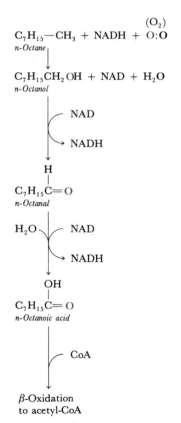

FIGURE 5.18   Steps in oxidation of an aliphatic hydrocarbon, catalyzed by a monooxygenase.

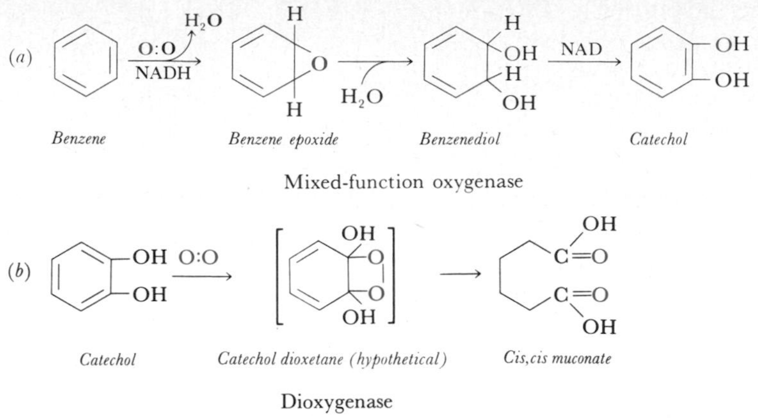

(a) Benzene → Benzene epoxide → Benzenediol → Catechol

Mixed-function oxygenase

(b) Catechol → Catechol dioxetane (hypothetical) → Cis,cis muconate

Dioxygenase

FIGURE 5.19 Roles of oxygenases in catabolism of aromatic compounds. (a) Hydroxylation of benzene to catechol by a mixed-function oxygenase in which NADH is the second electron donor. (b) Cleavage of catechol to *cis,cis*-muconate by a dioxygenase. Reactant oxygen atoms are shown in color in both reactions to demonstrate the different mechanisms.

These single ring compounds are referred to as **starting substrates,** since oxidative catabolism proceeds only after the complex aromatic molecules have been converted to these more simple forms. Protocatechuate and catechol may then be further degraded to compounds which can enter the TCA cycle: succinate, acetyl-CoA, pyruvate. Several steps in the catabolism of aromatic compounds require oxygenases. Figure 5.19 shows two different oxygenase-catalyzed reactions, one using a mixed-function oxygenase and the other a dioxygenase (see Section 5.5 for terminology).

5.8

Amino Acid Metabolism

Amino acid synthesis   There are 20 amino acids common to proteins, and those organisms that cannot obtain some or all amino acids preformed from the environment must synthesize them from other sources. The amino acids in proteins are α-amino acids in L configuration. An amino acid can be represented schematically as

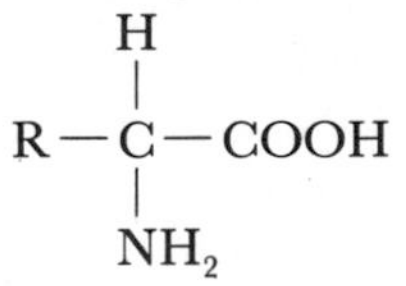

$$R-\overset{\displaystyle H}{\underset{\displaystyle NH_2}{C}}-COOH$$

with the R group shown varying from one amino acid to another. See Appendix 3, Figure A3.9, for the structures of the amino acid R groups. Two problems are involved in the synthesis of the amino acids: the synthesis of the "carbon skeleton," and the synthesis and attachment of the amino group to the skeleton. The synthesis of the carbon skeleton is usually straightforward, as there are only a few organic compounds which are important precursors. The various amino acids can be grouped into families based on the identity of these precursors (Figure 5.20). Note that the precursor organic molecules are ones which we have encountered before in glycolysis and the TCA cycle.

The attachment of the amino group is a crucial step since nitrogen is often a limiting factor for the growth of many organisms. Two general enzymatic methods of incorporating the amino group may

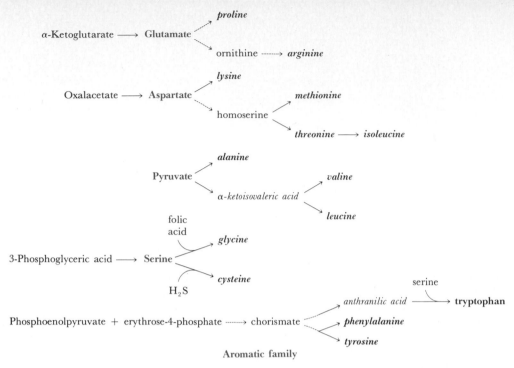

FIGURE 5.20 Amino acid families. The amino acids on the right are derived from the starting materials on the left.

be distinguished: *transaminases,* which exchange an amino group on one organic compound for the keto group of another; and *amino-acid dehydrogenases,* which catalyze the reductive assimilation of ammonia from the environment. Transaminase reactions involve the coenzyme pyridoxal phosphate, which activates the amino group so that it is readily transferred. Two examples of transamination are shown in Figure 5.21. The incorporation of ammonia from the environment by glutamic dehydrogenase is presented in Figure 5.22. The equilibrium of the reaction as shown favors the formation of glutamic acid ($\Delta G^{0\prime}$ of -7.5 kcal/mole). However, this synthesis requires reducing power in the form of NADH, which was generated in a separate redox reaction. The complete biosynthetic process of the amino acid may therefore be seen as the coupling of two redox reactions, with the energy released during the first being used to drive the second. The reduced coenzyme NADH functions as an intermediary in this transfer.

The detailed pathways for amino acid biosynthesis are given in Appendix 3, Figures A3.10 through A3.15. One of these, the synthesis of histidine, is probably the most complicated, partly because of the nature of the intermediates, and partly because it overlaps part of the pathway for the synthesis of the purine ring (Section 5.9). As shown in Figure 5.23 (see also Figure A3.15), five carbon atoms of histidine are derived from ribose, a nitrogen from glutamine, and a carbon and a nitrogen from ATP. The latter is accomplished by

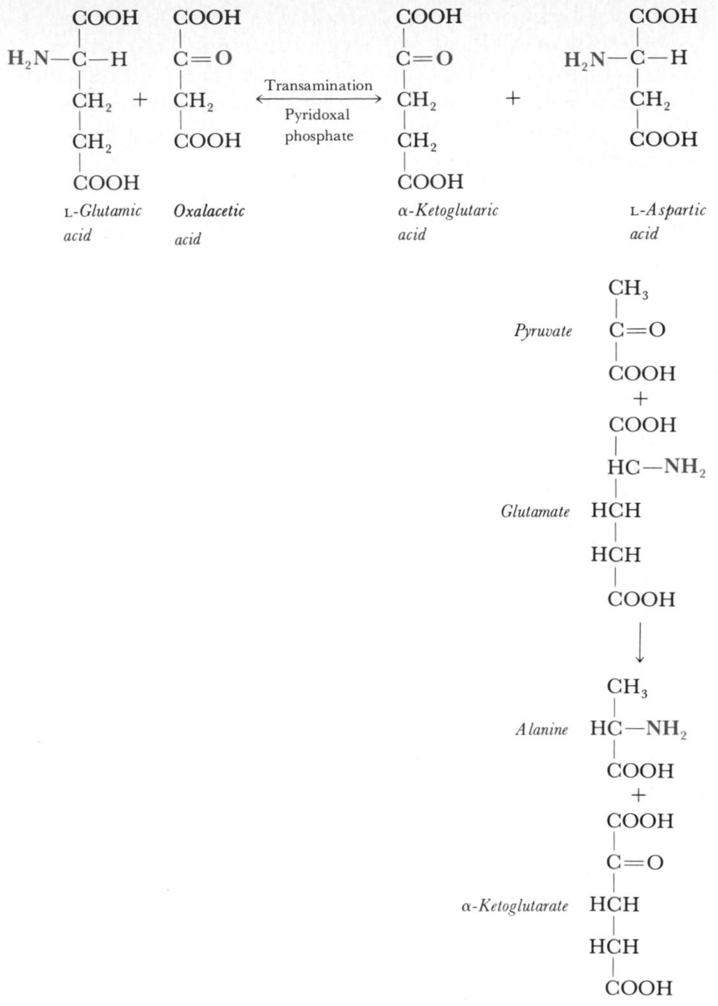

COOH   COOH   COOH   COOH

H_2N-C-H $C=O$ Transamination $C=O$ H_2N-C-H

CH₂ + CH₂ ⟷ Pyridoxal   CH₂ + CH₂

CH₂   COOH   phosphate   CH₂   COOH

COOH   COOH

L-*Glutamic*   *Oxalacetic*   *α-Ketoglutaric*   L-*Aspartic*
acid   *acid*   *acid*   *acid*

Pyruvate

CH₃

$C=O$

COOH

+

COOH

$HC-NH_2$

Glutamate   HCH

HCH

COOH

↓

CH₃

Alanine $HC-NH_2$

COOH

+

COOH

$C=O$

α-Ketoglutarate   HCH

HCH

COOH

FIGURE 5.21   Synthesis of amino acids by transamination.

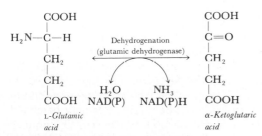

COOH   COOH

H_2N-C-H Dehydrogenation $C=O$

CH₂   (glutamic dehydrogenase)   CH₂

CH₂   CH₂

COOH H_2O NH_3 COOH

NAD(P)   NAD(P)H

L-*Glutamic*   *α-Ketoglutaric*
acid   *acid*

FIGURE 5.22   Synthesis of an amino acid, glutamate, by an amino acid dehydrogenase.

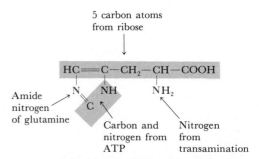

FIGURE 5.23   Origin of the carbon and nitrogen atoms of the amino acid histidine.

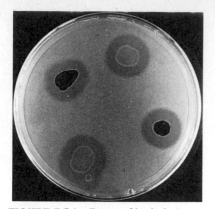

FIGURE 5.24 Zones of hydrolysis around colonies of protease-producing bacteria. The protein casein was incorporated into the agar. As it is digested by the protease, clearing occurs because the insoluble casein disappears.

actually splitting the adenine ring of ATP. The remainder of the adenine ring of ATP can be recycled through the pathway involved in the synthesis of purine (Section 5.9), and the ATP reformed.

This section has dealt with synthesis of amino acids from simpler compounds. The use of these amino acids in synthesis of proteins will be discussed in Chapter 9.

Amino acid utilization When amino acids are used as carbon sources, usually the first transformation to take place is the removal of the amino group, converting the amino acid into an organic acid. This can involve either a transaminase or an amino-acid dehydrogenase operating in the reverse direction of its synthesis of the amino acid (see Figure 5.22). In the latter case energy is generated in the formation of NADH. Many microorganisms have **amino-acid decarboxylases,** which convert the carboxyl group of the amino acid to CO_2 and the amino acid to an amine, a derivative of ammonia. By any of the reactions, an organic acid is ultimately formed, which is then utilized via the TCA cycle, as described earlier in this chapter.

Protein utilization Proteins can be metabolized only after they are broken down to small peptides or individual amino acids through the activity of enzymes called **proteases.** Some proteases specifically hydrolyze the peptide linkage between particular amino acids, whereas other proteases are nonspecific and hydrolyze any peptide bond. The amino acids released by protease action are utilized as carbon sources by one of the mechanisms described above. Many proteins are insoluble, and the production of extracellular proteases by cultures can be easily detected by incorporating an insoluble protein into an agar medium and looking for zones of clearing around the colonies after growth (Figure 5.24).

Not all microorganisms can utilize proteins for growth. Among those that can are many pathogenic microorganisms such as streptococci and staphylococci, soil microorganisms such as bacilli, and food-spoilage organisms such as certain species of *Pseudomonas.* Some fungi can digest and grow on keratin, a specialized, highly insoluble protein that is the main component of animal hair; this property is often found among fungi that infect the skin.

5.9

Purine and Pyrimidine Metabolism

Purines and pyrimidines are components of nucleic acids, as well as of many vitamins and coenzymes (for example, ATP, NAD, thiamine). The terminology of purine and pyrimidine derivatives is cumbersome, but important to understand. Figure A3.16 gives a diagrammatic presentation of the following summary. The individual purine and pyrimidine rings are called the *free bases.* The base with sugar attached is called a *nucleoside;* if the sugar is ribose the compound may be called a *riboside,* and if the sugar is deoxyribose it may be called a *deoxyriboside.* The nucleoside with a phosphate attached to the sugar is called a *nucleotide,* or more specifically a *ribotide* or *deoxyribotide.* If the nucleotide has two phosphates, it is called a *nucleoside diphosphate;* if it has three phosphates, the term *nucleoside triphosphate* is used. Terminology also exists for the various nucleosides and nucleotides. Thus, if uracil is the free base, its nucleoside is called *uridine* and its nucleoside monophosphate is called *uridylic acid.* The other common bases and their derivatives are adenine, adenosine, adenylic acid; guanine, guanosine, guanylic acid; cytosine, cytidine, cytidylic acid; thymine, thymidine, thymidylic acid.

Purine synthesis   The purine ring is built up almost atom by atom, using carbons and nitrogens derived from amino acids, CO_2, and formyl groups (themselves derived from serine). These formyl groups are added stepwise to a ribose-phosphate starting material, forming inosinic acid after 11 reactions (Figure 5.25; for details, see Figure A3.17).

The two carbon atoms in the purine ring derived from formyl groups are added in reactions involving a coenzyme derivative of the vitamin *folic acid.* This coenzyme is involved in a number of reactions involving addition of one-carbon units, and is of additional interest because a number of microorganisms cannot synthesize their own folic acid but must have it provided from the environment (see Section 5.17). Additionally, certain antimicrobial drugs called *sulfonamides* are specific inhibitors of folic acid synthesis. Sulfonamide inhibition of growth can be mostly overcome by providing the end products of pathways in which folic acid serves as a coenzyme, such as the purine synthesis pathway.

As shown in Figure 5.25, the first purine ring formed is that of inosinic acid, which serves as an intermediate for the formation of the two key purine derivatives, adenylic acid and guanylic acid.

Pyrimidine synthesis   In contrast to purine ring synthesis, a complete pyrimidine ring (orotic acid) is built up before the sugar is added. After the addition of ribose and phosphate to orotic acid, the other pyrimidines are successively synthesized (Figure 5.26). Complete structures and details of pyrimidine biosynthesis are given in Figure A3.20.

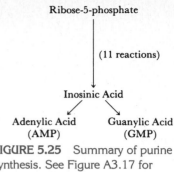

FIGURE 5.25   Summary of purine synthesis. See Figure A3.17 for chemical structures.

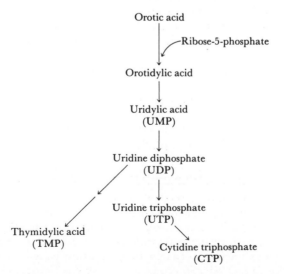

FIGURE 5.26   Summary of pyrimidine synthesis. See Figure A3.20 for chemical structures.

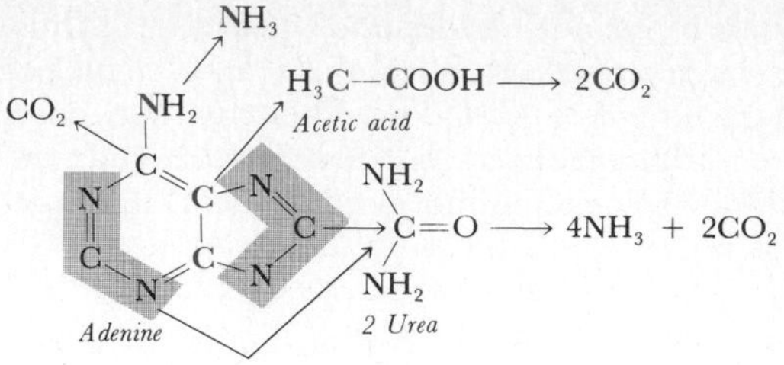

FIGURE 5.27 Degradation of a purine ring.

Summation:
Adenine $\longrightarrow$ $5NH_3$ + $5CO_2$

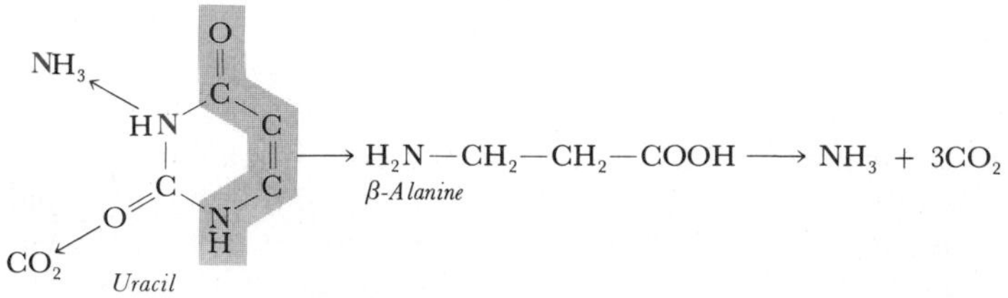

FIGURE 5.28 Degradation of a pyrimidine ring.

Summation:
Uracil $\longrightarrow$ $2NH_3$ + $4CO_2$

The pyrimidine thymine is present in DNA but not in RNA, and the sugar unit of its nucleotides is deoxyribose rather than ribose. Conversion of ribose to deoxyribose occurs by reduction of the ribose of a pyrimidine nucleotide, and the addition of the methyl group of thymidine involves participation of folic acid. The pyrimidine cytosine and the purines adenine and guanine are present in both RNA and DNA. Their deoxyribose derivatives are formed by reduction of the nucleoside diphosphates, followed by addition of the final phosphate from ATP to form the deoxyribonucleotide triphosphates that are the precursors of DNA (see Figure A3.21 for details).

Utilization There are a number of microorganisms that can utilize nucleic acids as sources of carbon and nitrogen and as electron donors. Initially, the nucleic acids are hydrolyzed to nucleotides by nucleases. The purine and pyrimidine bases liberated by hydrolysis of nucleic acids can be used directly by certain bacteria. The degradation of the bases occurs not by reversal of the biosynthetic pathways but by entirely different routes. The end products of purine degradation are NH_3 and CO_2; urea and acetic acid are intermediates (Figure 5.27). The end products of pyrimidine degradation are also NH_3 and CO_2, but the intermediate is β-alanine (Figure 5.28), an amino acid not found in proteins.

5.10

Porphyrin Ring

The porphyrin ring is found in pigments involved in energy-generating systems: the cytochromes (see Figure 4.14) and the chlorophylls (see Figure 6.2). Porphyrin is built up of four pyrrole units connected in a ring (the *tetrapyrrole ring*). Different porphyrins are formed by modifications of the side chains of the ring.

The biosynthesis of pyrrole and the tetrapyrrole ring is shown in Figure 5.29. The starting materials, succinyl-CoA and glycine, condense to form δ-aminolevulinic acid, two molecules of which condense, making porphobilinogen, a pyrrole. Four molecules of porphobilinogen condense into the tetrapyrrole ring system, and subsequent steps add side chains. The addition of a metal ion to the center of the ring (Mg^{2+} for chlorophyll and Fe^{2+} for heme) occurs near the end of the sequence. As far as is known, the early steps in porphyrin biosynthesis are the same in all organisms. The synthesis of the porphyrin ring was a significant event in early evolution and, by making photosynthesis and electron-transport phosphorylation possible, has had profound effects on subsequent evolutionary events.

A few microorganisms are unable to synthesize the porphyrin ring and require it preformed, usually as heme, the iron-containing porphyrin moiety of hemoglobin. One of the richest sources of heme is blood, and these heme-requiring microorganisms were first detected by observing a requirement for a growth factor from blood. Bacteria of the genus *Haemophilus* (meaning literally, "blood-loving") and a number of protozoan blood parasites require heme or a heme derivative for growth.

5.11

Bacterial Cell-Wall Synthesis

Although complex, the biosynthetic pathway for synthesis of the peptidoglycan of the bacterial cell wall is of great interest and importance. As seen in Figures 2.16 and 2.17, the peptidoglycan is a mixed polymer of amino sugars and amino acids. The amino sugars, *N*-acetylmuramic acid and *N*-acetylglucosamine, are connected in glycosidic linkage, and form the backbone of the polymer. The amino acids, connected in peptide linkage, form the cross-links. Most information is known about the synthesis of the cell wall of *Staphylococcus aureus,* and the discussion here will be restricted to that organism. In *S. aureus* the peptide chains are connected by a bridge composed of five glycine units.

Two carrier molecules participate in bacterial cell-wall synthysis, uridine diphosphate and a lipid carrier. As we have seen earlier in this chapter, uridine diphosphate is a sugar carrier for the formation of various glycan polymers. The lipid carrier, called *bactoprenol,* is a C_{55} isoprenoid alcohol which is connected via two phosphate esters to *N*-acetylmuramic acid:

$$
\begin{array}{ccc}
CH_3 & CH_3 & CH_3 \\
| & | & | \\
\end{array}
$$

CH$_3$C=CHCH$_2$(CH$_2$C=CHCH$_2$)$_9$CH$_2$C=CHCH$_2$
|
O
|
O=P—O$^-$
|
O
|
O=P—O$^-$
|
O—[*N*-acetyl-
muramic acid]

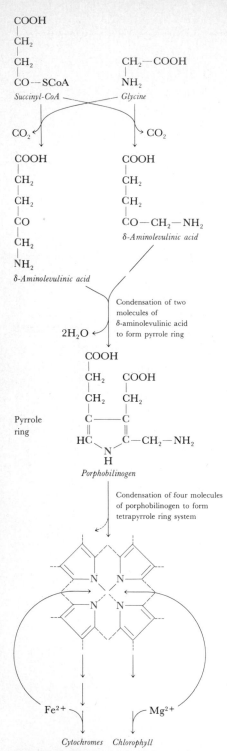

COOH
|
CH_2
|
CH_2
|
CO — SCoA
Succinyl-CoA

CH_2 — COOH
|
NH_2
Glycine

CO_2

CO_2

COOH
|
CH_2
|
CH_2
|
CO
|
CH_2
|
NH_2
δ-Aminolevulinic acid

COOH
|
CH_2
|
CH_2
|
CO — CH_2 — NH_2
δ-Aminolevulinic acid

Condensation of two
molecules of
δ-aminolevulinic acid
to form pyrrole ring

$2H_2O$

COOH
|
CH_2   COOH
| |
CH_2   CH_2
| |
C — — C
‖   ‖
HC   C — CH_2 — NH_2
   \ N /
     |
     H
Porphobilinogen

Pyrrole
ring

Condensation of four molecules
of porphobilinogen to form
tetrapyrrole ring system

Fe^{2+} Mg^{2+}

Cytochromes   Chlorophyll

FIGURE 5.29 Steps in the synthesis of
the pyrrole ring and the porphyrin ring.

The assembly of polymers outside the cell membrane presents special problems of transport and control. Bactoprenol is involved in the transport of the peptidoglycan building block across the cell membrane, where the peptidoglycan is then inserted into a growing point of the cell wall. As in the uridine diphosphate derivatives, the sugars it carries are usually activated by pyrophosphate linkage. The same or a similar lipid carrier is also involved in transport of the intermediates for synthesis of the lipopolysaccharide components of the outer wall layer of Gram-negative bacteria, and also for teichoic acid and extracellular mannan synthesis in certain Gram-positive bacteria. The function of the bactoprenol lipid carrier is apparently to render sugar intermediates sufficiently hydrophobic so that they will pass through the hydrophobic cell membrane.

Synthesis of the peptidoglycan can be divided into five stages (Figure 5.30):

1. Synthesis of UDP-muramic acid pentapeptide in the cytoplasm
2. Transfer of the muramic acid pentapeptide to the lipid carrier at the bacterial membrane
3. Addition of the second amino sugar, *N*-acetylglucosamine, from a nucleotide derivative, addition of the pentaglycine bridge, and addition of a residue of ammonia, events that involve cytoplasmic components and presumably take place on the inner side of the cytoplasmic membrane
4. Movement of the completed peptidoglycan unit through the membrane
5. Linkage of the peptidoglycan unit at a growing point of the cell wall and peptide cross-link formation

We discuss here only the final stages of the process, involving the formation of the glycan chain and the cross-linking peptide. As seen in Figure 5.30, the lipid carrier inserts the disaccharide into the glycan backbone and then moves back inside the cell to pick up another peptidoglycan unit. The final step in cell-wall synthesis is the formation of the peptide cross-links between adjacent glycan chains through the process of transpeptidation as discussed in Section 2.5. The energy to drive this peptide bond formation is released during the hydrolysis of the terminal D-alanine. An energy conservation scheme of this sort is necessary to support a biosynthetic process outside the cell membrane, where ATP is not supplied.

The cross-linking reaction inhibited by penicillin involves peptide formation with one of several different amino acids, depending on the bacterium involved. The one illustrated here is that found in *S. aureus,* which involves the formation of a pentaglycine bridge. In Gram-negative bacteria, such as *Escherichia coli,* there is no bridge, but a direct link between diaminopimelic acid on one peptide and D-alanine on an adjacent peptide (Figure 2.17a). This linkage also occurs by transpeptidation and is inhibited by penicillin. It is fascinating to consider that one of the key developments in human medcine, the discovery of penicillin, is linked to a specific biochemical reaction involved in cell-wall synthesis, transpeptidation.

Control of Biosynthetic versus Degradative Processes

In the preceding pages we have discussed both biosynthesis and utilization of carbohydrates, organic acids, fatty acids, amino acids, purines, and pyrimidines. It has been emphasized that in most cases the pathway for the degradation of each substance was different from a mere reversal of the pathway for biosynthesis. Different intermediates and different enzymes were involved. This is perhaps one of the most surprising and interesting discoveries to arise from studies on microbial metabolism. By utilizing different pathways for synthesis and for breakdown, an organism is able to keep its signals straight. A further means of control is through the use of two different coenzymes, NAD and NADP, the former being involved in degradative reactions and the latter in biosynthetic reactions.

Another problem that an organism has is controlling the rate of synthesis of the various amino acids, purines, pyrimidines, and other constituents that serve as the building blocks of cellular structures. Some of these building blocks are required in larger amounts than others, and the rate of synthesis balances the rate of utilization, so that excess amounts of these materials are not built up. One mechanism of control is **feedback inhibition:** the amino acid or other end product of a biosynthetic pathway inhibits the activity of the first enzyme in this pathway. Thus, as the end product builds up in the cell, its further synthesis is inhibited. If the end product is used up, however, synthesis can resume (Figure 5.31).

How is it possible for the end product to inhibit the activity of an enzyme that acts on a compound quite unrelated to it? This occurs because of a property of the inhibited enzyme known as **allostery.** An allosteric enzyme has two important sites, the active site, where the substrate binds, and the allosteric site, where the inhibitor (sometimes called an "effector") binds reversibly. When an inhibitor binds at the allosteric site, the structure or conformation of the enzyme molecule changes so that the substrate no longer binds efficiently at the active site (Figure 5.32). When the concentration of the inhibitor falls, equilibrium favors the dissociation of the inhibitor from the allosteric site, returning the active site to its catalytic shape. Allosteric enzymes are very common in both biosynthetic and degradative pathways, and are especially important in branched pathways. For example, the amino acids proline and arginine are both synthesized from glutamic acid. Figure 5.33 shows that these two amino acids can control the first enzyme unique to their own synthesis without affecting the other, so that a surplus of proline, for example, will not cause the organism to be starved for arginine.

In addition, some biosynthetic pathways are regulated by the use of **isozymes** (short for isofunctional enzymes: *iso* means same or constant). These enzymes catalyze the same reaction, but are subject to different regulatory control. An example is the synthesis of the aromatic amino acids as shown in Figure 5.34. The three different isozymes which catalyze the first condensation reaction are regulated independently by the three different end product amino acids. Unlike the earlier examples of feedback inhibition where inhibitors completely stopped an enzyme activity, in this case the total amount of

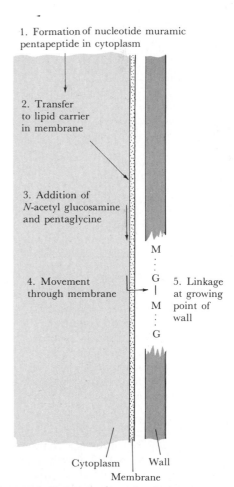

1. Formation of nucleotide muramic pentapeptide in cytoplasm

2. Transfer to lipid carrier in membrane

3. Addition of *N*-acetyl glucosamine and pentaglycine

4. Movement through membrane

5. Linkage at growing point of wall

Cytoplasm　Wall

Membrane

FIGURE 5.30 Five stages in biosynthesis of the bacterial cell-wall peptidoglycan.

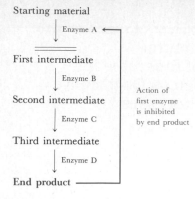

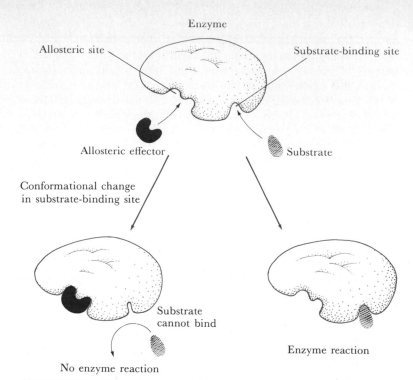

FIGURE 5.31 Feedback inhibition. The activity of the first enzyme of the pathway is inhibited by the end product, thus controlling production of end product.

FIGURE 5.32 Mechanism of enzyme inhibition by allosteric effector. When the effector combines with the allosteric site, the conformation of the enzyme is altered so that the substrate can no longer bind.

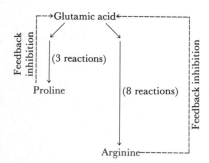

FIGURE 5.33 Feedback inhibition (dashed lines) in a branched biosynthetic pathway. Intermediates are labeled arbitrarily for convenient reference. The end products, proline and arginine, are shown in color to emphasize their role as regulatory molecules. Reaction details may be found in Figure A3.10.

FIGURE 5.34 Regulation of a complex biosynthetic pathway involving feedback inhibition (dashed lines) at branch points and the use of isozymes. Roman numerals refer to alternate forms of enzyme distinguished on basis of regulating compound. See Figure A3.14 for details.

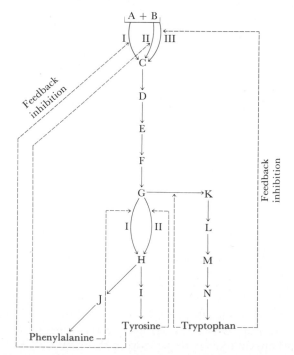

the initial enzyme activity is diminished in a stepwise fashion and falls to zero only when all three products are present in excess.

These examples are representative of the major regulatory patterns observed; there are other, rather elegant examples of regulation in Appendix 3. It is assumed that such enzymes have developed because efficient control of the rate of enzyme activity enables an organism to adapt to changing environments. Examples of how the rates of synthesis of various end products are regulated can be found in some of the detailed pathways given in Appendix 3.

5.13

Inorganic Nitrogen Metabolism

Nitrogen, which is needed for amino acid, purine, and pyrimidine biosynthesis, can be obtained by microorganisms from either inorganic or organic forms. The most common inorganic nitrogen sources are nitrate and ammonia, but other inorganic sources used by certain microbes include cyanide (CN^-), cyanate (OCN^-), thiocyanate (SCN^-), cyanamide (NCN^{2-}), nitrite (NO_2^-), and hydroxylamine (NH_2OH). Free nitrogen gas (N_2) is also used by a variety of bacteria. Ecological aspects of N transformations are discussed in Section 13.8.

Ammonia Although ammonia has the same oxidation state as an amino group, its assimilation into amino acids still requires expenditure of energy. Many bacteria have two pathways for ammonia assimilation, and which one functions depends on the external ammonia concentration. When ammonia concentrations are high, the **glutamate dehydrogenase** pathway operates, leading to the synthesis of glutamic acid from ammonia and α-ketoglutarate, as illustrated in Figure 5.22. As seen, when operating in the direction of glutamate synthesis, this reaction requires reducing power in the form of NADPH (or occasionally NADH).

When ammonia concentrations are low, a more complex reaction sequence comes into play which involves *glutamine* (the amide of glutamic acid), which is synthesized from glutamate and ammonia in a reaction requiring ATP and catalyzed by the enzyme **glutamine synthetase**:

$$HO-\overset{\overset{O}{\|}}{C}-(CH_2)_2-\underset{\underset{NH_2}{|}}{CH}-COO^- + ATP + NH_3 \longrightarrow H_2N-\overset{\overset{O}{\|}}{C}-(CH_2)_2-\underset{\underset{NH_2}{|}}{CH}-COO^- + ADP + P_i$$

Glutamate   *Glutamine*

In a second reaction, the enzyme **glutamate synthase** catalyzes the transfer of the amide group of glutamine to α-ketoglutarate, forming two molecules of glutamate:

Glutamine + α-ketoglutarate $\rightleftharpoons$ 2 glutamate

When these two reactions are summed, the net reaction is

α-Ketoglutarate + NH_3 + ATP $\longrightarrow$ glutamate + ADP + P_i

Once the N atom has been incorporated into glutamate, it can be transferred to other carbon skeletons by means of transaminases, as described in Section 5.8. Glutamine itself also functions as an amino donor for a variety of reactions, most importantly in the synthesis of the purine ring (Figures 5.28 and A3.17).

A rather interesting control is exerted on the activity of the enzyme glutamine synthetase. Glutamine synthetase exists in two forms, the active form and an inactive form, which results when adenyl residues from ATP are attached to the molecule, a process called adenylylation. Adenylylation of glutamine synthetase is promoted by high concentrations of ammonia, leading to inactivation of the enzyme. However, since glutamate dehydrogenase functions when ammonia concentrations are high, ammonia assimilation is shifted from the glutamine pathway to the glutamate pathway. When ammonia concentrations fall, glutamine synthetase becomes deadenylylated and becomes active again. Control via adenylylation is different from allosteric control described in Section 5.12, as adenylylation requires a covalent chemical modification of the enzyme rather than merely a change in its conformation.*

Nitrate   When nitrate is utilized as a nitrogen source, it is reduced to ammonia. The first step in this process is the reduction of nitrate to nitrite by the enzyme *nitrate reductase,* a molybdenum-containing flavoprotein. A second enzyme, *nitrite reductase,* reduces nitrite further (possibly via hydroxylamine, NH_2OH) but the subsequent steps to ammonia are not completely understood. The ammonia formed is converted into organic form via one of the reactions just described above. The presence of the metal molybdenum in nitrate reductase explains an observation made many years ago when studying the nutrition of certain fungi: it was found that molybdenum was required for growth when using nitrate as a nitrogen source but not when using ammonia. As we will see, the enzyme involved in nitrogen fixation, nitrogenase, also has a molybdenum-containing component.

The reduction of nitrate to ammonia discussed here is called **assimilatory** nitrate reduction, in contrast to **dissimilatory** nitrate reduction (Section 4.10 and Figure 13.13), the process occurring in denitrifying bacteria, in which the function of nitrate is that of an alternate electron acceptor to O_2 in electron transport phosphorylation. In assimilatory nitrate reduction, ATP synthesis does not occur. Different enzymes are involved in assimilatory and dissimilatory reductions.

Nitrogen fixation   The utilization of nitrogen gas (N_2) as a source of nitrogen is called **nitrogen fixation** and is a property of only certain bacteria and cyanobacteria. An abbreviated list of nitrogen-fixing organisms is given in Table 5.3, from which it can be seen that a variety of bacteria, both anaerobic and aerobic, fix nitrogen. In addition, there are some bacteria, called *symbiotic,* that fix nitrogen only when present in nodules or on roots of specific host plants. As far as is currently known, no eucaryotic organisms fix nitrogen. Symbiotic nitrogen fixation will be discussed in Section 14.10.

*Regulation of this system has several more aspects not covered in this simplified discussion. For a brief review of the complete system, see the article by B. Magasanik cited in the Supplementary Readings for this chapter.

TABLE 5.3 Some Nitrogen-Fixing Organisms

Free-living				Symbiotic	
Aerobes		**Anaerobes**		**Leguminous Plants**	**Nonleguminous Plants**
Heterotrophs	Photosynthetic	Heterotrophs	Photosynthetic		
Bacteria: 　*Azotobacter* spp. 　*Klebsiella* 　*Beijerinckia* 　*Bacillus polymyxa* 　*Mycobacterium flavum* 　*Spirillum lipoferum*	Cyanobacteria （various, but not all）	Bacteria: 　*Clostridium* spp. 　*Desulfovibrio*	Bacteria: 　*Chromatium* 　*Chlorobium* 　*Rhodospirillum* 　*Rhodopseudomonas* 　*Rhodomicrobium*	Soybeans, peas, clover, locust, etc., in association with a bacterium of the genus *Rhizobium*	*Alnus, Myrica, Ceanothus, Comptonia, Casuarina;* in association with actinomyces of the genus *Frankia* Grasses: *Azospirillum lipoferum* (true symbiont?)

Although biological nitrogen fixation has been known for a long time, it has only been in the past two decades that any type of real understanding of the process has developed. The first nitrogen-fixing bacterium was isolated by Sergei Winogradsky in the 1890s; this was the anaerobe *Clostridium pasteurianum.* In 1901, Martinus Beijerinck isolated *Azotobacter chroococcum,* an aerobic nitrogen fixer. Perhaps because aerobes are easier to cultivate than anaerobes, most of the biochemical work was done on *Azotobacter* for many years. The most critical problem in any attempt to study the biochemistry of a process is to obtain an active enzyme system free of whole cells. All attempts to obtain such a cell-free system that would fix nitrogen failed, and long after the main pathways of carbon metabolism were well worked out, virtually nothing was known of the pathway by which N_2 was converted into amino acids. The inability to prepare active cell-free extracts was almost certainly due to the extreme O_2 sensitivity of nitrogenase, even from aerobes such as *Azotobacter,* so that the enzyme was inactivated before it could be assayed. The breakthrough came in 1960, when a group at the E. I. DuPont Company (J. E. Carnahan, L. E. Mortenson, H. F. Mower, and J. E. Castle) developed a cell-free system with *Clostridium pasteurianum.* The use of this anaerobe was probably the critical factor, for it was natural, when working with an anaerobe, to avoid the presence of O_2. Another key to success was that a large amount of pyruvate was added as a source of energy, and *C. pasteurianum* has an active phosphoroclastic pathway for generating reducing power and ATP from pyruvate. Thus the three items necessary for obtaining active nitrogen fixation were present: reducing power, ATP, and absence of O_2. In 1962, the DuPont group (Mortenson, Carnahan, and R. C. Valentine) discovered ferredoxin and showed that it was involved in electron transfer to N_2. It was later found that ferredoxin is involved in many other electron transfer reactions in bacteria.

A Bit of History

In the fixation process, N_2 is reduced to ammonium and the ammonium converted into organic form. The reduction process is catalyzed by the enzyme *nitrogenase,* which consists of two separate protein components, called components I and II. Both components contain iron, and component I contains molybdenum as well. The

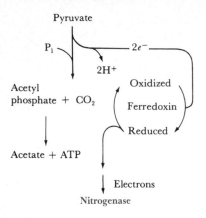

I. Generation of reducing power and ATP (*Clostridium;* other bacteria may use different pathways)

Pyruvate

P_i

$2e^-$

$2H^+$

Acetyl phosphate $+ CO_2$

Oxidized

Ferredoxin

Reduced

Acetate $+$ ATP

Electrons

Nitrogenase

II. Reduction of N_2

Reduced ferredoxin

ATP —— Component II

$P_i + $ ADP ←— Component I

N_2

III. Progressive six-electron reduction of N_2 (none of the intermediates accumulate)

$N \equiv N$
$\downarrow 2H$
$HN = NH$
$\downarrow 2H$
$H_2N - NH_2$
$\downarrow 2H$
$H_3N \quad NH_3$

IV. Overall reaction

$$6H^+ + 6e^- + N_2 \longrightarrow 2NH_3$$
$$18\text{–}24\ \text{ATP} \longrightarrow 18\text{–}24\ \text{ADP} + 18\text{–}24\ P_i$$

FIGURE 5.35 Steps in nitrogen fixation: reduction of N_2 to $2NH_3$.

nitrogenase components are inactivated by oxygen, even if isolated from an aerobe. Owing to the stability of the $N \equiv N$ triple bond, N_2 is extremely inert and its activation is a very energy-demanding process. Six electrons must be transferred to reduce N_2 to $2NH_3$, and several intermediate steps might be visualized; but since no intermediates have ever been isolated, it is now assumed that the three successive reduction steps occur with the intermediates firmly bound to the enzyme. Nitrogen fixation is highly reductive in nature and the process is inhibited by oxygen since component I is rapidly and irreversibly inactivated by O_2. In aerobic bacteria, N_2 fixation occurs in the presence of O_2 in whole cells, but not in purified enzyme preparations, and it is thought that the nitrogenase within the cell is in an O_2-protected microenvironment. Some bacteria and cyanobacteria able to grow both aerobically and anaerobically fix N_2 only under anaerobic conditions.

The electrons for nitrogen reduction are transferred to the nitrogenase from ferredoxin, the low-redox-potential electron carrier already described in Section 4.6. (In fact, ferredoxin was first discovered in nitrogen-fixing bacteria through studies on the nature of the electron carrier in nitrogen fixation, and was then found to be present in non-nitrogen-fixing organisms as well.) In *Clostridium pasteurianum*, ferredoxin is reduced by phosphoroclastic splitting of pyruvate (Figure 5.35; see also Section 4.6), and ATP is synthesized at the same time. In addition to reduced ferredoxin, ATP is required for N_2 fixation in all organisms studied. The ATP requirement for nitrogen fixation is very high, about 4 to 5 ATP being split to ADP + P_i for each $2\ e^-$ transferred. ATP is apparently required in order to lower the reduction potential of the system sufficiently so that N_2 may be reduced. The reduction potential of component II is -0.294 V and this is lowered to -0.402 V when the enzyme combines with ATP. Electrons are first transferred from ferredoxin to component II (Figure 5.35), after which ATP binds and lowers the reduction potential. This complex then combines with component I, and component I becomes reduced. Reduced component I can now convert N_2 to NH_3, but the details of the six-electron reduction are not known.

Components I and II have been purified from a large number of nitrogen-fixing organisms. It is of considerable evolutionary interest that component I from one organism will function with component II from another organism. Even when the components are taken from organisms thought to be distantly related (for example, an aerobe and an anaerobe; a photosynthetic and a nonphotosynthetic organism) the hybrid complex retains some activity. This can be interpreted to mean that the structures of the components have not changed markedly during evolution, suggesting that the molecular requirements for the nitrogenase proteins to be active in N_2 reduction are fairly specific.

Nitrogenase is not specific for N_2 but will also reduce cyanide (CN^-), acetylene ($HC \equiv CH$), and several other compounds. The reduction of acetylene is only a two-electron process and ethylene ($H_2C = CH_2$) is produced. The reduction of acetylene probably serves no useful purpose to the cell but provides the experimenter with a simple way of measuring the activity of nitrogen-fixing systems, since it is fairly easy to measure the reduction of acetylene to ethylene ($H_2C = CH_2$). This technique is now used to detect nitrogen fixation in unknown systems. Previously, it was not easy to prove that an organism fixed N_2; indeed, many claims for nitrogen fixation in microorganisms have been erroneous. The growth of an organism in a medium to which no nitrogen compounds have been added does not mean that the organism is fixing nitrogen from the air since traces of nitrogen compounds often occur as contaminants in the ingredients of culture media or drift into the media in gaseous form or as dust particles. Even distilled water may be contaminated with ammonia. One method of proving nitrogen fixation is to show a net increase in the total nitrogenous content of the medium plus its organisms after incubation; the assumption would be that the increased nitrogen could come only from N_2 from the air. A more sensitive procedure is to use an isotope of nitrogen, ^{15}N, as a tracer.* The gas phase of a culture is enriched with ^{15}N, and after incubation the cells and medium are digested, the ammonia produced being distilled off and assayed for its ^{15}N content. If there has been a significant production of ^{15}N-labeled NH_3, it is proof of nitrogen fixation. However, the acetylene-reduction method is an even more sensitive way of measuring nitrogen fixation and is rapidly replacing the more difficult ^{15}N method. The sample, which may be soil, water, a culture, or a cell extract is incubated with acetylene, and the reaction mixture is later analyzed by gas chromatography for production of the gaseous substance ethylene. This method is far simpler and faster than other methods.

5.14

Sulfur Metabolism

Although the two sulfur-containing amino acids cysteine and methionine are utilized as sulfur sources by many microorganisms, most microorganisms are also able to use inorganic sulfate (SO_4^{2-}) as the sole source of sulfur to synthesize not only these amino acids but also the sulfur-containing vitamins (thiamin, biotin, and lipoic acid). The assimilation of sulfate first involves its activation by reaction with ATP in two steps to form phosphoadenosine phosphosulfate (PAPS) (Figure 5.36). Subsequently, the sulfate radical attached to PAPS is reduced to sulfite (SO_3^{2-}) by an enzyme that uses NADPH as the electron donor. Sulfite is then reduced to hydrogen sulfide (H_2S) by another enzyme using NADPH. The conversion of H_2S to organic sulfur occurs by reaction with the amino acid serine (Figure 5.36). Other organic sulfur compounds can later be synthesized from the reduced sulfur of cysteine.

Sulfate-reducing bacteria (for example, *Desulfovibrio*), which use sulfate as a terminal electron acceptor (Section 4.10), carry out dissimilatory sulfate reduction, producing H_2S as the end product. In these bacteria the reduction of sulfate occurs with adenosine phosphosulfate (APS) rather than PAPS. This is another example illus-

---

*^{15}N is not a radioactive isotope but a stable isotope, and its presence must be detected with the mass spectrometer, an expensive and rather cumbersome instrument.

1 Synthesis of active sulfate:

$$ATP + SO_4^{2-} \rightleftharpoons \text{adenosine phosphosulfate} + PP_i$$
$$ATP + \text{adenosine phosphosulfate} \longrightarrow \text{phosphoadenosine phosphosulfate} + ADP$$

2 Reduction of active sulfate to sulfite:

$$\text{Phosphoadenosine phosphosulfate} \xrightarrow{\quad NADPH \quad NADP \quad} \text{phosphoadenosine phosphate} + SO_3^-$$

3 Reduction of sulfite to hydrogen sulfide:

$$SO_3^- \xrightarrow{\quad NADPH \quad NADP \quad} H_2S + H_2O$$

4 Formation of amino acid:

$$H_2S + H_2C \overset{OH}{\underset{}{-}} \overset{NH_2}{\underset{H}{C}} -COOH \rightleftharpoons H_2C \overset{SH}{\underset{}{-}} \overset{NH_2}{\underset{H}{C}} -COOH + H_2O$$

FIGURE 5.36 Steps in activation and reduction of sulfate to H_2S and formation of a sulfur-containing amino acid.

trating that synthetic and degradative processes in microorganisms proceed by different pathways.

5.15

Mineral Nutrition

A variety of minerals are required by organisms for growth. These can be separated into two groups, *macronutrient minerals* and *micronutrient minerals* or *trace elements*.

Macronutrients *Phosphorus* occurs in nature in the form of organic and inorganic phosphates and is utilized by microorganisms primarily to synthesize phospholipids and nucleic acids. Probably most or all microorganisms utilize inorganic phosphate for growth. One of the most important reactions in the conversion of inorganic to organic phosphate is the glyceraldehyde phosphate dehydrogenase reaction of the Embden–Meyerhof pathway, which was discussed in Section 4.7 (see also Figure A3.1).

Organic phosphate compounds occur very often in nature, and they are utilized as phosphate sources through the action of *phosphatases,* which hydrolyze the organic phosphate ester. Two types of enzymes are known, acid and alkaline phosphatases, distinguished by the pH optima. Phosphatases are present in nearly all organisms. In Gram-negative organisms they are often localized between the cell wall and membrane in the periplasmic space (see Section 2.5). Thus situated, they are in an excellent position to act on external phosphates.

Potassium is universally required. A variety of enzymes, including some of those involved in protein synthesis, are specifically activated by potassium. In laboratory cultures, potassium often can be replaced by rubidium, its heavier relative in the periodic table, but not by sodium, its lighter relative.

Magnesium functions to stabilize ribosomes, cell membranes, and nucleic acids, and this element is also required for the activity of many enzymes, especially those involving phosphate transfer. Thus relatively large amounts of magnesium are required for growth. Interestingly, Gram-positive bacteria require about 10 times more magnesium than do Gram-negative bacteria. When cells are cultured under magnesium-deficient conditions, growth may continue for a while as stored magnesium is diluted out, but it eventually ceases. Those structures with the greatest magnesium requirement, ribosomes and cell membranes, are most strongly affected by magnesium deficiency and break down first under conditions of magnesium deficiency.

Calcium ions play a key role in the heat stability of bacterial spores and may also be involved in the stability of the cell wall. Even though calcium and magnesium are closely related in the periodic table, calcium cannot replace magnesium in many of its roles in the cell.

Sodium is required by some but not all organisms, and its need may reflect the environment; for example, seawater has a high sodium content, and marine microorganisms generally require sodium for growth, whereas closely related freshwater forms may be able to grow in the complete absence of sodium. For some reason, all cyanobacteria, even those found in freshwater, require Na^+.

The element *silicon* is required by those organisms that have silicon-containing cell walls, the diatoms (algae; see Section 3.7) and radiolaria (protozoa, Section 3.10). Although silicon is one of the most plentiful elements on earth, it is often in short supply in the environment because it forms highly insoluble structures. In laboratory culture, silicon is frequently present in small amounts as a contaminant, since the vessels used for culture are generally composed of glass made from silica sand, but optimal growth of silicon-requiring organisms usually requires addition of small amounts of a soluble form of silicon, such as sodium silicate, to the culture medium. Germanium, which is related to silicon in the periodic table, is a specific inhibitor of silicon metabolism and hence inhibits growth of silicon-requiring organisms. Germanium dioxide (GeO_2) is sometimes added to culture media in order to isolate algae other than diatoms from nature by inhibiting growth of the rather weedlike diatoms.

Trace elements   The trace-element, or micronutrient, requirements of microorganisms are difficult to determine experimentally. Even though the concentration of an element in the culture medium may have been reduced to such an extent that it can no longer be detected by chemical assay, it may still be present in sufficient quantity to meet a requirement for growth. Contaminating amounts of trace elements may come from the glassware, distilled water, culture ingredients, and even cotton plugs. Often, only after the glassware has been scrupulously cleaned and the culture ingredients highly purified is it possible to demonstrate a trace-element requirement. The trace elements commonly required by most microorganisms are zinc, copper, cobalt, manganese, and molybdenum. These metals function in enzymes or coenzymes. The trace element *cobalt* is needed only for the formation of vitamin B_{12}, and if this vitamin is added to the medium, cobalt may no longer be needed. *Zinc* plays a structural role in certain enzymes in that it helps hold together protein subunits in the proper configuration for enzyme activity.

Molybdenum is present in certain flavins called molybdoflavoproteins, involved in assimilatory nitrate reduction. It is also present in nitrogenase, the enzyme involved in N_2 reduction (N_2 fixation).

Copper plays a role in certain oxidation-reduction enzymes. *Manganese* is an activator of many enzymes that act on phosphate-containing compounds, substituting in these enzymes for magnesium. In *Bacillus* spp., it is required in considerably larger amounts for sporulation than for vegetative growth.

5.16

Iron Nutrition and Biochemistry

Iron is a rather special case, as it is required in fairly large amounts by most organisms (although not at the level of a macronutrient), yet is normally present in the environment in very insoluble form, so that organisms must have special mechanisms for obtaining iron from their habitats. Iron has two oxidation states, ferrous, Fe^{2+}, and ferric, Fe^{3+}. Ferrous iron is generally more soluble than ferric iron, so that in the anaerobic environments in which it is present, iron availability may not be a problem. In fact, iron can to some extent be solubilized by reducing it from the ferric to the ferrous state.

Iron forms compounds with a wide variety of organic compounds. Compounds that complex specifically with metals such as iron are called **chelators,** and chelators play a special role in transport of iron from place to place. The atoms in organic compounds with which metals combine are called **ligands,** and the main iron ligands are of two types, oxygen and nitrogen. In general, Fe^{3+} forms strongest chelates via O ligands, whereas Fe^{2+} forms chelates via N ligands. Organisms which live in low-iron environments, or where iron is tied up in insoluble precipitates, usually produce iron chelators that enable them to obtain iron. The oceans and hard-water lakes are generally low in available iron. Interestingly, despite the fact that animals have large amounts of iron present in their tissues and fluids, the animal body is low in available iron, because iron is tied up in the body in substances such as hemoglobin, and in iron-carrying proteins called **transferrins.** Thus, blood and milk are quite low in available iron and at least in part because of this, they are not good microbial habitats.

Iron enzymes and proteins We have discussed earlier a number of iron-containing enzymes or proteins. A brief summary of some of the key iron proteins is given in Table 5.4, which shows the critical importance of iron in cellular function. Almost all microorganisms require iron. Aerobes need iron for the synthesis of cytochromes and other iron-containing components of the electron-transport chain, whereas anaerobes need iron for the synthesis of ferredoxin and other nonheme iron proteins of low reduction potential. There are a few bacteria which may have no iron requirement at all. These are members of the lactic acid bacteria. This group of bacteria (see Section 19.23), which are important in milk fermentations and as pathogens, do not contain any cytochromes or ferredoxins, and certain species

TABLE 5.4 Iron-Containing Proteins and Enzymes

Protein	Function
Heme-containing proteins:	
Cytochromes	Electron-transport systems
Hemoglobins (leghemoglobin in root nodules)	Oxygen-binding
Catalase	$H_2O_2 \longrightarrow H_2O + O_2$
Peroxidase	$NADH + H_2O_2 \longrightarrow NAD + 2H_2O$
Oxygenases (some)	$R + O_2 \longrightarrow RO_2$
Iron-sulfur (nonheme) proteins:	
Rubredoxins	High-reduction-potential electron carriers
Ferredoxins	Low-reduction-potential electron carriers
Nitrogenase (components I and II)	$N_2 \longrightarrow 2NH_3$
Hydrogenase	$H_2 \rightleftharpoons 2H^+ + 2e^-$

seem to lack other iron-containing proteins as well. They obtain their energy exclusively via glycolysis, forming primarily lactic acid as an end product. Interestingly, these bacteria live in milk and body fluids where all of the iron present is tied up and hence unavailable, and it may be because of this that these bacteria evolved a metabolism which does not involve iron proteins.

Iron transport into the cell   Many organisms produce specific iron-binding organic compounds, called **ironophores,** which solubilize ferric iron and transport it into the cell. One major group of irono-phores are derivatives of **hydroxamic acid,** which bind iron very strongly, as shown below.

$$
\begin{array}{c}
R-N\!\!-\!\!-\!\!-\!\!C-R + Fe^{3+} \longrightarrow R-N\!\!-\!\!-\!\!-\!\!C-R \\
\;\;\;|\quad\quad\; \| \quad\quad\quad\quad\quad\quad\quad\quad |\quad\quad\; | \\
\;\;\underline{OH\quad O}\quad\quad\quad\quad\quad\quad\quad\quad O\quad\quad O \\
\quad\textit{Hydroxamate}\quad\quad\quad\quad\quad\quad\quad\quad\;\searrow\;\swarrow \\
\quad\quad\textit{group}\quad\quad\quad\quad\quad\quad\quad\quad\quad\quad Fe^{3+}
\end{array}
$$

Some microbially produced hydroxamic acids have been called *siderochromes* or *ferrichromes,* and act as growth factors when organisms are cultured in media with very low iron concentrations. It is thought that once the iron-hydroxamate complex has passed into the cell, the iron is reduced to the Fe^{2+} state, and since the complex with Fe^{2+} is less strong, the iron is released and can be used in synthesis of porphyrins and other iron-containing compounds (Figure 5.37). Some microorganisms cannot synthesize hydroxamates and must have them preformed in their culture media, the hydroxamates thus acting as vitamins (see Section 5.17 and Table 5.5). Names for some of the hydroxamates that function as vitamins are *terregens factor, coprogen,* and *mycobactin.* A number of microbially produced hydroxamates are antibiotics rather than growth factors, apparently because they tie up iron irreversibly and create an iron deficiency. Names of some of these antibiotics are albomycin, aspergillic acid, grisein, and fer-rimycin. In at least some cases, the antibiotic action can be nullified by simultaneously adding one of the vitaminlike hydroxamates.

In some bacteria, the iron-binding compounds are not hydrox-

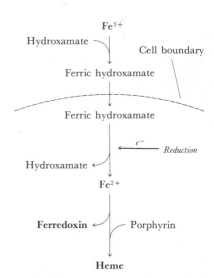

FIGURE 5.37 Postulated function of iron-binding hydroxamate in the iron nutrition of an organism living in an environment low in iron.

TABLE 5.5 Commonly Required Vitamins

Vitamin	Function
p-Aminobenzoic acid	Precursor of folic acid
Folic acid	One-carbon metabolism; methyl group transfer
Biotin	Fatty acid biosynthesis; β-decarboxylations; CO_2 fixation
Cobalamin (B_{12})	Reduction of and transfer of single carbon fragments; synthesis of deoxyribose
Lipoic acid	Transfer of acyl groups in decarboxylation of pyruvate and α-ketoglutarate
Nicotinic acid (niacin)	Precursor of NAD; electron transfer in oxidation-reduction reactions
Pantothenic acid	Precursor of coenzyme A; activation of acetyl and other acyl derivatives
Riboflavin	Precursor of FMN, FAD in flavoproteins involved in electron transport
Thiamin (B_1)	α-Decarboxylations; transketolase
Vitamin B_6 (pyridoxal-pyridoxamine group)	Amino acid and keto acid transformations
Vitamin K group; quinones	Electron transport
Hydroxamates (terregens factor, coprogen, mycobactin, etc.)	Iron-binding compounds; solubilization of iron and transport into cell

amates but phenolic acids. Enteric bacteria such as *Escherichia coli* and *Salmonella* spp. produce complex phenolic acid derivatives called **enterobactins** or **enterochelins**.

Heme itself will serve as a source of cellular iron for some bacteria and either is incorporated directly into porphyrin-containing proteins or releases iron within the cell for subsequent use by iron-requiring systems.

Consequences of iron deficiency Iron deficiencies can result in dramatic alterations in cell metabolism. The most noteworthy example is the effect of iron deficiency on the synthesis of diphtheria toxin by *Corynebacterium diphtheriae*. In media with sufficient iron little or no toxin is formed, whereas in iron-deficient media toxin production is high. The iron concentration of the tissues in which *C. diphtheriae* lives thus controls toxin production and the resultant disease symptoms.

Because of the general unavailability of iron in the animal body, the iron nutrition of pathogenic organisms plays a significant role in host-parasite relationships. Most pathogens produce iron chelators that enable them to obtain iron in the body. These chelators bind iron more strongly than it is bound in hemoglobin or transferrins, and therefore are able to displace iron from these proteins and make it available to the pathogen. In certain types of infections, a considerable increase in severity may result if soluble iron is injected into or fed to the host.

The synthesis by microorganisms of ironophores is controlled by the iron concentration of their environments. When available iron concentration is high, the ironophores are not synthesized, but when the available iron drops to a low level, large amounts of the iron-chelating compounds are synthesized and excreted into the surroundings.

In culture media, iron is rendered available by providing it in chelated form with a synthetic chelating agent, such as ethylene-diamine-tetraacetic acid (EDTA) or nitrilotriacetic acid (NTA). A concentration of iron of 10 micrograms per milliliter (μg/ml) would be an excess for virtually any microbial culture. Providing iron in chelated form is especially important when organisms are being grown in media that lack natural iron chelators such as amino acids. In such media, iron deficiencies are dramatically evident when small amounts of inoculants are used, for even if an organism can produce an iron chelator, the amount produced by a single cell will be vanishingly small, and if the chelator diffuses away from the organism, it will be lost and growth cannot commence. Several of the known ironophores were first discovered because they were essential for the growth of an organism when a small inoculum was used but were not essential for growth from a large inoculum. Soil extracts often contain natural iron chelators and commonly have been used to obtain growth of algae on mineral media. In current practice, however, soil extracts, which are ill-defined components, are no longer used because synthetic iron chelators are available.

5.17

Organic Growth Factors

Growth factors are specific organic compounds that are required in very small amounts and cannot be synthesized by the cell. Substances frequently serving as growth factors are vitamins, amino acids, purines, and pyrimidines. Some organisms are able to synthesize all of these compounds, whereas others require the addition of one or more to the culture medium.

Vitamins   The first growth factors to be discovered and studied in any detail were those that we now call **vitamins.** Table 5.5 lists commonly needed vitamins and their functions. In living organisms most vitamins function in forming coenzymes; for instance, the vitamin nicotinic acid is a part of the coenzyme NAD. Some microorganisms have very complex vitamin requirements, whereas others require only a few vitamins. The lactic acid bacteria, which include the genera *Streptococcus* and *Lactobacillus,* are renowned for their complex vitamin requirements, which are even greater than those of humans. Cobalamin (vitamin B_{12}) is often required by aquatic organisms, bacteria as well as algae. Over half of the algae tested, both freshwater and marine, require some form of vitamin B_{12}, and often thiamin and biotin as well. Fungi, on the other hand, never require vitamin B_{12}.

Amino acids   Many microorganisms require specific amino acids. Inability to synthesize an amino acid is related to lack of the enzymes needed for its synthesis. The required amino acids can usually be supplied either as the free amino acids or in small peptides. When the peptide enters the cell, it is hydrolyzed by a peptidase to the component amino acids. In some cases, the peptide can be used, whereas the free amino acid cannot. This is usually due to impermeability of the cell to the free amino acid. The small peptide enters more readily and is then hydrolyzed, liberating the needed amino acid. In some cases, determination of amino acid requirements is complicated by the fact that the presence of one amino acid in higher than appropriate amounts leads to inhibition of uptake of another required amino acid. This is called **amino acid imbalance,** and arises because several amino acids are taken into the cell by the same transport system. Amino acid imbalance in culture media can be overcome by taking care to keep the concentration of all amino acids at equivalent and low levels, or in many cases by providing the amino acids in small peptides. Amino acid imbalance can also affect growth because of complexities of the feedback inhibition process, as outlined in Figures A3.11 and 12.

Purines and pyrimidines   Purines and pyrimidines, which are the building blocks of nucleic acids and coenzymes, are growth factors for a number of microorganisms. If they are provided as the free bases (simple purines or pyrimidines), they must be converted inside the cell into nucleosides and nucleotides before they can serve their proper functions. Two pathways exist for formation of nucleotides. The first, the direct route, employs PRPP transferase:

Guanine + phosphoribosyl pyrophosphate (PRPP) $\longrightarrow$

guanine monophosphate (GMP) + pyrophosphate

The second is an indirect route using nucleoside kinase:

(a) Guanine + ribose-1-phosphate $\longrightarrow$

guanine-ribose + phosphate

(Guanosine)

(b) Guanosine + ATP $\longrightarrow$ GMP + ADP

If the second pathway is present, the organism can use both nucleosides and free bases as the added growth factor, whereas if just the first occurs, only the free bases can be used. The nucleotides usually cannot be used as growth factors because the presence of the phosphate group makes them ionized and hence less able to penetrate the cell.

Other growth factors   Some organisms require the porphyrin ring or one of its derivatives. Recall that chlorophyll, the cytochromes, and the hemoglobin of animals all contain porphyrin rings. A porphyrin requirement explains why certain pathogenic organisms (for example, *Haemophilus influenzae*) are cultured on media containing red blood cells, since one of the major constituents of red cells is hemoglobin, which contains the porphyrin heme. Heme-requiring bacteria lack the ability to synthesize porphyrins and hence are restricted to environments where blood is present. Since a porphyrin is needed for synthesis of the cytochrome system, the respiratory apparatus of these bacteria becomes deranged when they are cultured in the absence of heme.

Another interesting example of a growth-factor requirement concerns the bacteria that live in the rumen of cows, sheep, and other ruminants (see Section 14.8). Many of these organisms were difficult to cultivate in the laboratory until it was found that they required specific growth factors present in rumen fluid. These growth factors were subsequently identified as four-carbon to six-carbon branched- and straight-chain fatty acids. As far as is known, these growth factors are required only by rumen organisms.

Transport and Nutrition

As was discussed in Chapters 2 and 3, the cell membrane is a barrier through which solutes pass in either direction. Since nutrients must enter the cell through the membrane, the permeability properties of the membrane are important in cell nutrition. A compound can penetrate the membrane in either of two ways, by **passive diffusion** or by **carrier-mediated transport** (Figure 5.38). Passive diffusion is quite nonspecific and generally slow, and the compound usually distributes itself in such a way that its concentration inside the cell is the same as outside. Carrier-mediated transport is highly specific, very rapid, and in many cases allows the cells to accumulate nutrients from the medium even against large concentration gradients. The latter process is generally called **active transport.**

The necessity for active-transport mechanisms in microorganisms can readily be seen. Most microbes live in environments whose concentrations of salts and other nutrients are many times lower than concentrations within the cells. If passive uptake were the only type to occur, these cells would not be able to acquire the proper concentrations of solutes. Active-transport mechanisms overcome this problem by enabling the cell to accumulate solutes against a concentration gradient. With low external concentrations, a solute accumulation against a concentration gradient can occur only through the expenditure of energy derived from the cell's metabolic machinery. In diffusion-limited uptake both the rate of uptake and the intracellular level are proportional to the external concentration. On the other hand, as shown in Figure 5.38, carrier-mediated transport shows a saturation effect: if the concentration of substrate in the medium is high enough to saturate the carrier, which is likely to be the case even at quite low substrate concentrations, the rate of uptake (and often the internal level as well) becomes maximal and can be increased no further.

Active-transport mechanisms show marked specificity. In a group of very closely related substances some may be transported rapidly, some slowly, and others not at all. Further, there may be competition for uptake; for instance, L-alanine, L-serine, and glycine, all closely related amino acids, compete with each other for uptake, as do L-valine, L-leucine, and L-isoleucine. This is one reason that amino acid imbalance in a culture medium leads to poor growth of organisms requiring them—an excess of one amino acid prevents the uptake of the other. A similar kind of competition for uptake is also shown among related cations, such as Na^+, K^+, and Rb^+.

Cells have evolved several distinct mechanisms to extract nu-

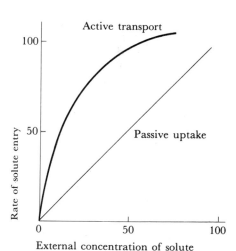

FIGURE 5.38 Relationship between uptake rate and external concentration in passive uptake and active transport. Note that in active uptake the uptake rate shows saturation at high external concentrations.

trients from the growth medium, which fall into two fundamental classes. (1) The substance is converted into a derivative as part of the process of translocation across the membrane. It is thus a chemical group derived from a molecule, rather than the whole molecule, which passes across the membrane, and the process is called **group translocation.** (2) The substance combines with a membrane carrier, which probably undergoes a conformational change, leading to release of the substance inside the cell without chemical alteration. When the substance is brought inside to the same concentration as outside, we speak of **facilitated diffusion.** Very often, however, the substance accumulates in the cytoplasm, chemically unaltered but to a much higher concentration than that outside. This process of **active transport** requires the performance of work, and thus involves coupling of the membrane carriers to the metabolic machinery of the cell. In many cases, energy coupling is apparently linked to the generation of ion gradients across the membrane.

Group translocation These are processes in which the substance is chemically altered in the course of passage across the membrane. Since the product that appears inside the cell is chemically different from the external substrate, no concentration gradient is produced across the membrane. The best studied cases of group translocation involve transport of the sugars glucose, mannose, fructose, N-acetylglucosamine, and β-glucosides, which are phosphorylated during transport by the phosphotransferase system. The phosphotransferase system is composed of at least two distinct enzymes, one of which is specific for each sugar and the other of which is nonspecific. A third protein component of the system is a small heat-stable protein designated HPr, which acts as a carrier of high-energy phosphate:

$$\text{Phosphoenolpyruvate} + \text{HPr} \longrightarrow \text{HPr} - \textcircled{P} + \text{pyruvate}$$
$$\text{HPr} - \textcircled{P} + \text{sugar} \longrightarrow \text{sugar} - \textcircled{P} + \text{HPr}$$

The phosphorylation of HPr by phophoenolpyruvate is carried out by an enzyme present in the cytoplasm, and HPr-phosphate then moves to the membrane. The sugar being transported is then phosphorylated by HPr-phosphate, the reaction catalyzed by an enzyme within the membrane specific for that sugar. Finally, the phosphorylated sugar moves into the cell, where it can be metabolized by normal catabolic pathways.

One of the best pieces of evidence for the involvement of this system in transport comes from studies on mutants unable to utilize sugars. Two kinds of mutants can be isolated, those unable to utilize specific sugars, due to a block in synthesis of a specific membrane-bound enzyme, and those unable to utilize all of the sugars transported by the phosphotransferase system, due to a block in synthesis of either HPr or the enzyme involved in its phosphorylation, both of which are nonspecific. Another ingenious experiment showed that phosphorylation occurred only during transport and not after the sugar had entered the cytoplasm. This involved preparation of **membrane vesicles,** small closed spheres derived from the cell membrane by fragmentation followed by closure. These vesicles will carry out active transport and have the virtue that substances can be placed inside the vesicles at the time of closure. A sugar placed inside the vesicles along with the enzymes of the phosphotransferase system was not phosphorylated, whereas a sugar applied externally appeared within the vesicles in phosphorylated form, thus showing that translocation occurred by unidirectional phosphorylation.

Other substances thought to be transported by group translocation include purines, pyrimidines, and fatty acids. However, many substances are not taken up by group translocation, including several sugars, which seem to be accumulated instead by active transport.

Facilitated diffusion and active transport   These are processes in which the substance being transported combines with a membrane-bound carrier protein which then releases the substances unchanged on the inside. If coupling to energy-generating systems occurs, the substance may accumulate in the cell in concentrations much higher than the external concentrations, whereas in the absence of energy coupling, transport may still occur but accumulation against a concentration gradient may not. Since the substance is not altered inside the cell, if metabolism does not remove it, its concentration may reach many times the external concentration; but eventually a limit is reached. At this point, further transport may cease, or else exit may occur at a rate sufficient to balance entry. The exit process also involves the carrier. The exact manner in which energy coupling occurs is not known, but many investigators believe that the carrier is altered in such a way that its affinity for the substrate on the inside is reduced, whereas its affinity for external substrate remains high. Thus as long as energy coupling exists, transport will occur more readily inward than outward, but if energy coupling is blocked (or energy generation stops), entry and exit occur at the same rate and concentrative uptake no longer occurs. This model is diagrammed in Figure 5.39.

Substances probably transported by facilitated diffusion and active transport include β-galactosides (such as lactose), galactose, arabinose, sulfate, phosphate, amino acids, and organic acids (such as citrate). However, a substance may be transported by different systems in different organisms. For example, lactose is transported by facilitated diffusion in *Escherichia coli,* but by a phosphotransferase system (group translocation) in *Staphylococcus aureus.* An experimental distinction between group translocation and other carrier-mediated transport processes rests on determining whether or not the substance is altered chemically in the course of transport. A distinction between facilitated diffusion and simple (passive) diffusion rests on showing the involvement of a carrier in the former. Carrier involvement can be shown by three types of experiments: (1) transport via a carrier will show a saturation effect whereas simple diffusion will not (Figure 5.38); (2) related substances will compete with each other for uptake if a carrier is involved (for example, glycine uptake is often inhibited by alanine and vice versa); and (3) mutants may be isolated that lack the carrier protein and hence no longer carry out facilitated diffusion, or active transport, but passive diffusion continues.

Role of ion gradients in active transport   Active transport designates processes in which nutrients are "pumped" across the membrane, so

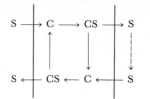

(a) *Facilitated diffusion*

(b) *Facilitated diffusion with energy coupling*

C = carrier
C* = energized carrier
S = substance transported

FIGURE 5.39   Model for functions of carrier in facilitated diffusion. (a) At equilibrium, exit balances entrance. Transport occurs but accumulation against a concentration gradient does not. (b) Energized carrier has greatly reduced affinity for substance on the cytoplasmic side of the membrane. Exit is much reduced over entrance. Accumulation against a concentration gradient occurs. In neither case does the carrier move laterally in the membrane, but remains in one location and undergoes changes in conformation.

as to accumulate in the cytoplasm chemically unchanged but at a higher concentration. As in any other pump, this requires the performance of work, apparently at the level of the membrane carriers. Evidence now indicates that the proton-motive force generated as a result of cellular metabolism (Section 4.8), is used to transport ions and other molecules across the plasma membrane. Cellular metabolism primarily serves to use the energy released from catabolic redox reactions to establish gradients of ion concentration across the membrane, accompanied by an electrical potential. That is, energy is transformed from its original chemical form to the more physical form of the gradient. It is the electrochemical potential gradient that does the work of transport, as follows. Each membrane carrier has specific sites for both its substrate (for example, galactose or sulfate) and the coupling ion. Thus the force exerted on the ion, primarily the difference in electrical potential, is used by the cells to drive the movement of a substrate which may be uncharged. In principle, various ions such as Na^+, K^+, and H^+ could serve as coupling ions, but it appears that bacteria employ predominantly H^+.

There are at least four independent mechanisms by which bacteria generate gradients of H^+ activity. One is the hydrolysis of ATP to ADP + P_i by a membrane-bound ATPase, which brings about the extrusion of protons from the cell (Figure 4.18); this route occurs in anaerobic bacteria and in the plasma membrane of eucaryotes. Aerobic organisms can extrude H^+ more directly, as a result of the operation of the respiratory chain. As discussed in Section 4.8, the respiratory chain appears to be spatially oriented within the membrane in such a way as to separate protons (H^+ ions) from electrons. The protons are ejected on the outer surface of the membrane; the electrons are passed to oxygen and, together with H^+ from the cytoplasm, reduce it to H_2O, causing a net formation of OH^- in the cytoplasm. The membrane is impermeable to both H^+ and OH^-, so that equilibrium is not spontaneously restored. In consequence, respiration sets up a gradient of H^+ activity across the membrane with the cytoplasm alkaline and electrically negative compared to the other surface (Figure 5.40). Two other mechanisms depend on light; they are the photosynthetic apparatus (Figure 6.8) and bacteriorhodopsin, a pigment unique to halophilic bacteria (page 256).

The gradient of pH and of electrical potential, whether established by the ATPase or by respiration, is apparently used by the cell to transport both ions and uncharged molecules. Let us recall that the transport process involves membrane carriers, specific for each substrate; the H^+ gradient serves as a link between these membrane carriers and the metabolic machinery, making it possible for the carriers to "pump" nutrients inward. Cations, such as K^+, may accumulate in the cytoplasm in response to the electrical gradient, since the interior of the cell is negative. Uptake of anions probably occurs together with protons so that it is effectively the undissociated acid that enters the cell. Even transport of uncharged molecules such as sugars or amino acids may be linked to the electrical gradient: the carrier transports both the substrate and one or more protons, and the association of the carrier with H^+ modifies its affinity for the substrate (Figure 5.40). There is now substantial evidence both for

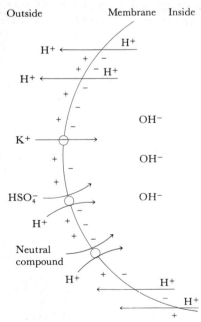

FIGURE 5.40 Explanation of how an ion gradient (OH^- inside, H^+ outside) could lead to transport of either cations or anions (see text for details).

the generation of ion gradients and electrical potentials and for the role of such gradients in energy coupling of active transport.

Periplasmic binding proteins   In Gram-negative bacteria the cell envelope is complex, with an outer lipopolysaccharide layer enclosing the peptidoglycan and plasma membrane. The outer lipopolysaccharide functions as an external membrane and between this layer and the plasma membrane is a definite space called the **periplasmic space.** In addition to enzymes such as the phosphatases involved in phosphate nutrition (Section 5.15), the periplasmic space holds a number of specific proteins involved in transport. These proteins are called **binding proteins** because, although not enzymes, they bind specifically the substances that they transport. These binding proteins are readily released from cells by treatments that disrupt the outer lipopolysaccharide layer, and because they leave the cell so readily they are not thought to be an integral part of the plasma membrane, as are the carriers described above. Their function may be in the initial stages of transport, binding the substance and bringing it to the membrane-bound carrier; these transport systems are probably not linked to the H^+ gradient but may use ATP as an energy source. Binding proteins seem to be absent in Gram-positive bacteria, which also lack the outer wall layer and a defined periplasmic space.

5.19
Summary

This chapter is summarized in Figure 5.41. We begin with the nutrients available to the organism from its environment. Small molecules can be taken up directly but polymeric macromolecules are first hydrolyzed by extracellular enzymes. Catabolic processes convert this vast array of materials into a small group of intermediates that serve as the starting materials for biosynthetic reactions. Catabolism also leads to the formation of ATP, the carrier of energy for biosynthetic reactions, and NADPH, the carrier of reducing power for biosynthesis. Inorganic substances are either assimilated directly and used in biosynthetic reactions, as are the metals shown in the diagram, or, as shown for sulfur and nitrogen, they are converted to simple reduced compounds which are used in biosynthesis. Phosphate is interesting in that it is really assimilated by catabolic rather than anabolic reactions, being converted into organic form either by substrate-level or electron-transport phosphorylation. Biosynthesis leads to the production of the cell constituents involved in growth and macromolecular synthesis.

Space on Figure 5.41 does not permit listing the key intermediates situated at the intersection between catabolism and anabolism, but these are listed in Table 5.6. Interestingly, they are only 12 in number, and most of these have been discussed in this chapter in relation to biosynthetic pathways. Study of Table 5.6, in relation to the material presented in this chapter, will provide the reader with a review of biosynthetic pathways, and may help to clarify the intricacies of cellular metabolism.

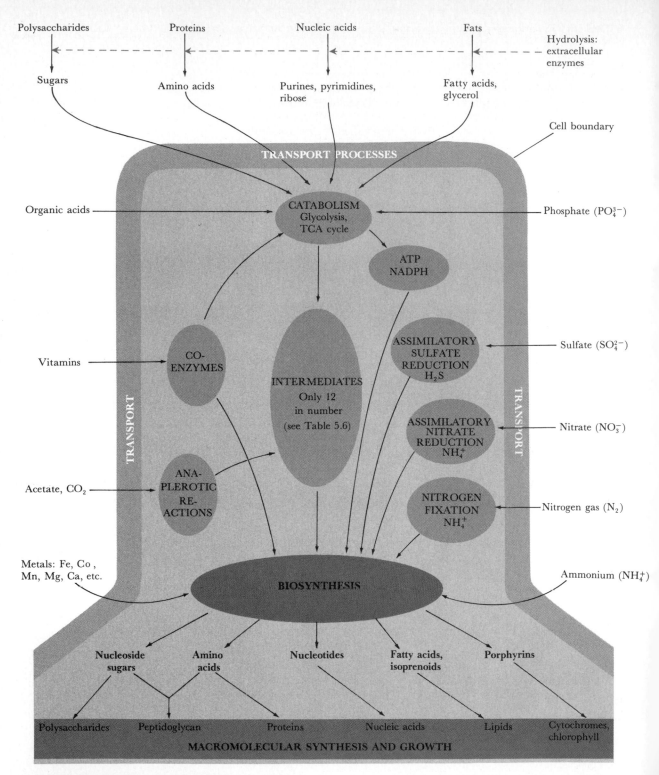

FIGURE 5.41 Nutrition and biosynthesis: a summary.

TABLE 5.6   Intermediates at the Intersection between Catabolism and Biosynthesis

Intermediate	Catabolic Origin	Role in Biosynthesis
Glucose-1-phosphate	Galactose, polysaccharides	Nucleoside sugars
Glucose-6-phosphate	Glycolysis	Pentose, storage polysaccharides
Ribose-5-phosphate	Pentose-phosphate pathway	Nucleotides, deoxyribonucleotides
Erythrose-4-phosphate	Pentose-phosphate pathway	Aromatic amino acids
Phosphoenolpyruvate	Glycolysis	Phosphotransferase system (sugar transport) aromatic amino acids, anaplerotic reactions (CO_2 fixation), muramic acid synthesis
Pyruvate	Glycolysis, phosphoketolase (pentose oxidation)	Alanine, valine, leucine, anaplerotic reactions (CO_2 fixation)
3-Phosphoglyceric acid	Glycolysis	Serine, glycine, cysteine
α-Ketoglutarate	Tricarboxylic acid cycle	Glutamate, proline, arginine, lysine
Succinyl-CoA	Tricarboxylic acid cycle	Methionine, porphyrins
Oxalacetate	Tricarboxylic acid cycle, anaplerotic reactions	Aspartic acid, lysine, methionine, threonine, isoleucine
Dihydroxyacetone phosphate	Glycolysis	Glycerol (fats, phospholipids)
Acetyl-CoA	Pyruvate decarboxylation, fatty acid oxidation, pyrimidine breakdown	Fatty acids, isoprenoids, sterols, lysine (two carbons), leucine (two carbons)

Supplementary Readings

Dagley, S., and **D. E. Nicholson.** 1970. An introduction to metabolic pathways. John Wiley, New York. An excellent reference book for further study of the pathways given in this chapter. Dozens of pathways are given in outline form, and structures, enzymes, and mechanisms of action are shown. Especially good for unique pathways found in bacteria.

Doelle, H. W. 1975. Bacterial metabolism, 2nd ed. Academic Press, New York. Although dated, this textbook is still useful as a centralized source of information regarding the biochemical pathways of bacteria.

Eddy, A. A. 1982. Mechanisms of solute transport in selected eucaryotic microorganisms. Adv. Microb. Physiol. 23:1–78. Detailed account of transport processes in yeasts, with comparison to bacterial processes.

Gottschalk, G. 1979. Bacterial metabolism. Springer-Verlag, New York. A brief but clearly written summary of major metabolic reactions of bacteria.

Harold, F. M. 1978. Vectorial metabolism. *In* L. N. Ornston and J. R. Sokatch, eds. The bacteria, vol. 6. Academic Press, New York. Role of chemiosmosis in active transport by bacteria.

Hobbs, A. S., and **R. W. Albers.** 1980. The structure of proteins involved in active membrane transport. Annu. Rev. Biophys. Bioeng. 9:259–291. Thorough review of transport mechanisms in bacteria and eucaryotes.

Ingraham, J. L., O. Maaløe and **F. C. Neidhardt.** 1983. Growth of the bacterial cell. Sinauer Associates, Inc., Sunderland, Mass. Good presentation of molecular aspects of cell growth.

Kaback, H. R. 1974. Transport studies in bacterial membrane vesicles. Science 186:882–892. Bacterial membrane vesicles have provided some of the best experimental material for studies on transport. This article by the discoverer of membrane vesicles reviews the preparation, structure, and function of these interesting objects.

Magasanik, B. 1982. Control of nitrogen assimilation in bacteria. Annu. Rev. Genet. 16:135–168.

Roberts, G. P., and **W. J. Brill.** 1981. Genetics and regulation of nitrogen fixation. Annu. Rev. Microbiol. 35:207–235. Current details in a rapidly changing field.

Rosen, B. P. 1978. Bacterial transport. Marcel Dekker, New York. Series of chapters by experts covering structure and function of membranes as well as energetics of transport.

Stryer, L. 1981. Biochemistry, 2nd ed. W. H. Freeman, San Francisco. The best general biochemistry textbook, with excellent illustrations.

Thauer, R. K., K. Jungermann, and **K. Decker.** 1977. Energy conservation in chemotrophic anaerobic bacteria. Bacteriol. Rev. 41:100–180. Deals primarily with energy generation but relates these processes to biosynthesis.

Tipper, D. J., and **A. Wright.** 1979. The structure and synthesis of bacterial cell walls. *In* L. N. Ornston and J. R. Sokatch, eds. The bacteria, vol. 7. Academic Press, New York.

Umbarger, H. E. 1978. Amino acid biosynthesis and its regulation. Annu. Rev. Biochem. 47:533–606. Detailed review of amino acid synthesis in bacteria.

The Autotrophic Way of Life

Our discussion of biosynthesis so far has dealt with metabolism and rearrangement of organic compounds. It has been assumed that the organisms in question obtain organic compounds for growth from their surroundings. Organisms which require a constant supply of organic compounds for the majority of their biosynthetic reactions are called **heterotrophs.** This group includes all animals, fungi, protozoa, and most bacteria. **Autotrophs,** on the other hand, are organisms that are able to obtain the carbon they need for cellular biosynthesis from carbon dioxide (CO_2), and can live and grow in the complete absence of organic matter.* Autotrophs thus are capable of synthesizing organic material from CO_2, a process sometimes called *CO_2 fixation* (in analogy to N_2 fixation, Section 5.13). Autotrophs, however, do perform many of the same reactions as heterotrophs. Their use of carbon dioxide should be viewed as an alternative way to obtain biosynthetic precursors. Autotrophy does not replace standard biosynthetic reactions; rather, it augments them by allowing biosynthesis to begin with a much simpler compound, CO_2. In this chapter, for purposes of simplification, we shall first discuss the energy metabolism of autotrophs and then go on to study the manner in which they are able to use CO_2 as a major carbon source. It will be seen that there are several different processes used for energy generation by autotrophs, but that virtually all use the same pathway for converting CO_2 to organic compounds.

One large group of autotrophs obtains energy for cellular metabolism from light; these organisms are called **phototrophic** or

*Some autotrophs have vitamin requirements and hence do not grow in the complete absence of organic matter, but such requirements constitute a trivial level of organic carbon, as far as biosynthetic processes are concerned. Similarly, all heterotrophs also have an absolute requirement for some CO_2 in the anaplerotic reactions discussed in Chapter 5. It is clear, therefore, that the terms *autotroph* and *heterotroph* refer to an organism's major mode of obtaining carbon for growth; neither designation is absolute.

photosynthetic. Other autotrophs, called **lithotrophic,** get their energy from the oxidation of inorganic compounds, such as elemental sulfur (S^0), ammonium ion (NH_4^+), ferrous iron (Fe^{2+}), or hydrogen (H_2). Both groups are able to grow on completely inorganic media, and hence their designation as autotrophic, meaning literally "self-nourishing." Phototrophs are a diverse group, including both eucaryotic organisms (algae and higher plants) and procaryotic forms. Lithotrophic autotrophs, on the other hand, are a much more restricted group, including relatively few kinds of bacteria and no eucaryotes. The ability to transform inorganic compounds makes lithotrophs ecologically important in nutrient cycles (see Sections 13.8 and 13.10) and economically significant in certain problems of metal corrosion.

6.1

Photosynthesis

One of the most important biological processes on earth is **photosynthesis,** which is the conversion of light energy into chemical energy that can then be used for the energy-requiring reactions of biosynthesis, including the reduction of CO_2 to organic compounds. The ability to photosynthesize is dependent on the presence of special light-gathering pigments, the **chlorophylls,** which are found in plants, algae, and some bacteria. Photosynthesis has been studied for many years in higher plants, but in recent years our knowledge of this process has been greatly advanced through biochemical studies on algae and photosynthetic bacteria. It has been found that the growth of a phototrophic autotroph can be analyzed as two more or less distinct sets of reactions: the **light reactions,** in which light energy is converted into chemical energy, and the **dark reactions,** in which this chemical energy is used to reduce CO_2 to organic compounds. The conversion of CO_2 to organic compounds is an energy-requiring reduction. Energy is needed in the form of ATP; the immediate source of electrons for the reduction is NADPH.

The function of the light reactions is to convert light energy into chemical energy in the form of ATP and NADPH. The purple and green bacteria use light primarily to form ATP; they obtain NADPH (also referred to as reducing power) in one of two ways: either from constituents of their environment, such as H_2S and organic compounds; or by using the energy in ATP to reduce NADP in a process called *reversed electron transport*. In the anaerobic environments in which these bacteria live, reduced compounds are usually available to them. Green plants, algae, and cyanobacteria, however, do not generally use H_2S and other compounds to acquire reducing power. Instead, they obtain electrons from the oxidation of H_2O molecules and then use these electrons ultimately to reduce NADP to NADPH while at the same time producing O_2 (Figure 6.1). As we shall see, this important difference between the purple and green bacteria on the one hand, and the cyanobacteria, algae, and plants on the other, has its basis in significant differences in the functions of the photosynthetic apparatus.

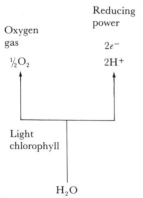

Oxygen gas

$\frac{1}{2}O_2$

Reducing power

$2e^-$
$2H^+$

Light chlorophyll

H_2O

FIGURE 6.1 Production of O_2 and reducing power from H_2O.

The Role of Chlorophyll in Photosynthesis

As we have noted, photosynthesis occurs only in organisms that possess some type of chlorophyll, and it seems reasonable therefore to conclude that chlorophyll is related to the assimilation of light energy. Chlorophyll is a porphyrin, as is the heme group of the cytochromes, but chlorophyll contains a *magnesium* atom instead of an iron atom at the center of the porphyrin ring. Chlorophyll also contains a long hydrophobic side chain, and because of this side chain, chlorophyll associates with lipid and hydrophobic proteins of photosynthetic membranes.

The structure of chlorophyll *a*, the principal chlorophyll of higher plants, of most algae and of the cyanobacteria, is shown in Figure 6.2. Chlorophyll *a* is green in color because it absorbs red and blue light preferentially and transmits green light. The spectral properties of any pigment can be best expressed by its **absorption spectrum** which indicates the degree to which the pigment absorbs light of different wavelengths. The absorption spectrum of an ether extract of chlorophyll *a* shows strong absorption of red light (maximum absorption at a wavelength of 665 nm) and blue light (maximum at 430 nm).

There are a number of chemically different chlorophylls that are distinguished by their absorption spectra. Chlorophyll *b*, for instance, absorbs maximally in the red region at 645 nm rather than at 665 nm. Many plants have more than one chlorophyll, but the most common are chlorophylls *a* and *b*. Among the procaryotes, the cyanobacteria have chlorophyll *a*, but the purple and green bacteria have chlorophylls of different structure, which are called *bacteriochlorophylls*. Bacteriochlorophyll *a* (Figure 6.2) from purple photosynthetic bacteria absorbs maximally at 770 nm; other bacteriochlorophylls absorb maximally at 660 nm and 870 nm (see also Figure 19.1).

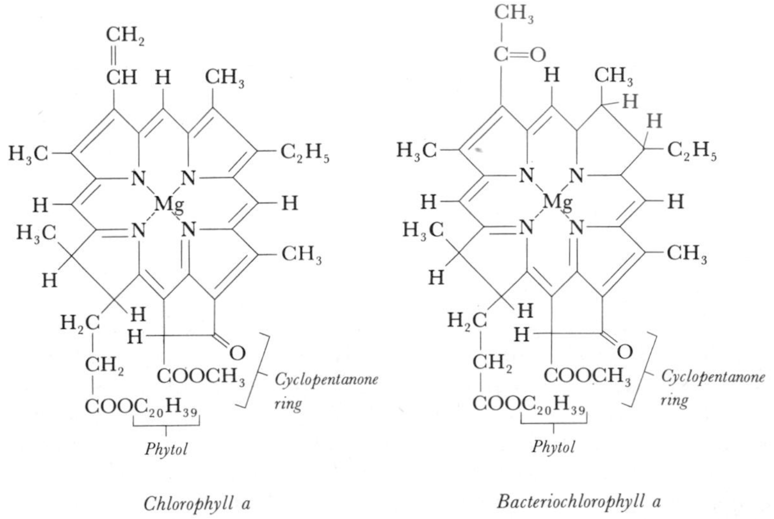

FIGURE 6.2 Structures of chlorophyll *a* and bacteriochlorophyll *a*. The two molecules are almost identical except for the atoms shown in color on the bacteriochlorophyll *a* structure.

Chlorophyll a

Bacteriochlorophyll a

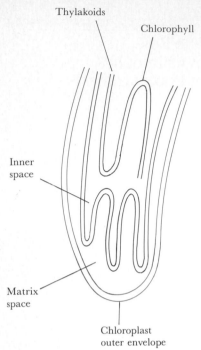

Thylakoids

Chlorophyll

Inner space

Matrix space

Chloroplast outer envelope

FIGURE 6.3 Details of chloroplast structure, showing how the convolutions of the thylakoid membranes define an inner space and a matrix space within the overall chloroplast envelope.

The absorption spectra mentioned above for the chlorophylls were measured on chlorophyll dissolved in ether. It is also possible to measure the absorption spectrum of chlorophyll in its natural state inside cells. The results indicate that the absorption maximum of chlorophyll within cells is at a higher wavelength than it is in an extract. A physical chemist would tell us that this observation means that the chlorophyll molecules within the cell are not in simple solution but are associated in complex aggregates or arrays.

Just where is the chlorophyll located inside of the cell? We noted in Section 3.3 that eucaryotic algae have special chlorophyll-containing intracellular organelles, the chloroplasts. Even here, the chlorophyll is not found uniformly distributed throughout the chloroplast but appears only in association with the sheetlike (lamellar) membrane structures of the chloroplast (Figure 6.3). These photosynthetic membrane systems are sometimes called **thylakoids.** The thylakoids are so arranged that the chloroplast is divided into two regions, the **matrix space,** which surrounds the thylakoids, and the **inner space** within the thylakoid array. This arrangement makes possible the development of a pH gradient that can be used to synthesize ATP, as described below. Within the thylakoid membrane, the chlorophyll molecules are associated in groups consisting of about 200 to 300 molecules. It is now apparent that only some of the chlorophyll molecules participate directly in the conversion of light energy to ATP. This chlorophyll is referred to as **reaction center** chlorophyll and receives light energy by transfer from the more numerous **light-gathering** (or *antenna*) chlorophyll molecules (Figure 6.4). The chlorophyll molecules are closely associated with proteins which precisely control their orientation in the membrane so that energy absorbed by one chlorophyll (initially as light) can be efficiently transferred to another. In photosynthetic procaryotes, chloroplasts are absent, and instead chlorophyll is found in extensive internal membrane systems (see Figures 2.12, 19.3, and 19.11). In the purple and green bacteria, a single reaction center probably comprises about 40 chlorophyll molecules.

6.3

Light Reactions in the Purple and Green Bacteria

The photosynthetic bacteria differ from the cyanobacteria, algae, and plants in several important respects. They live primarily under anaerobic conditions and do not produce O_2. Their photosynthesis is therefore said to be **anoxygenic.** Instead of water, they use sulfur or organic compounds from their environment as electron donors. A brief overview of the characteristics of the purple and green bacteria is given in Table 6.1. Many of the purple and green bacteria are strict anaerobes, but one subgroup of the purple bacteria, the Rhodospirillaccac, and the genus *Chloroflexus* of the green bacteria, are able to grow aerobically in the dark as heterotrophs. Photosynthetic growth in the purple and green bacteria is always strictly anaerobic, however, even in those organisms capable of aerobic heterotrophic growth. Photosynthetic growth by these anaerobes also depends on the availability in the environment of a reduced compound to serve as electron donor, such as a reduced sulfur compound (for example, H_2S, ele-

TABLE 6.1 Characteristics of the Purple and Green Bacteria

Purple Bacteria	Green Bacteria
Bacteriochlorophyll *a* or *b*	Bacteriochlorophyll *c*, *d*, or *e*
Purple- or red-colored carotenoids	Yellow- or brown-colored carotenoids
Photosynthetic membrane system in a series of lamellae or tubes, continuous with the plasma membrane	Photosynthetic membrane system in a series of vesicles, located just beneath the plasma membrane

mental sulfur, thiosulfate) or an organic compound (for example, succinate, malate, butyrate). During phototrophic growth, the bacteria oxidize the external electron donor in order to reduce NADP to NADPH. For instance, their growth on hydrogen sulfide often leads to production of elemental sulfur which is deposited either outside the cells, as in the green bacteria (Figure 6.5*a*), or inside the cells, as in some of the purple bacteria (Figure 6.5*b*).

The conversion of light energy to ATP involves a series of redox reactions with membrane-bound carriers. This process will be examined first in the photosynthetic bacteria, which have a relatively simple reaction series.

An overall scheme of electron flow in photosynthesis in the purple and green bacteria is shown in Figure 6.6. Reaction-center chlorophyll receives the energy from photons with wavelengths in the infrared region of the light spectrum and is designated P870. The P870 molecule is unstable due to the extra energy it has absorbed and returns to a stable configuration by ejecting an electron, thereby becoming positively charged. This loss of an electron is an oxidation, and the chlorophyll is said to have been **photooxidized.** All oxidations must be accompanied by reductions; in this case the chlorophyll passes its electron to an unidentified electron acceptor called X. As shown in Figure 6.6, the reduction potential of X is much more negative than that of the starting P870 chlorophyll molecule. Remember from the concept of the electron tower in Figure 4.4 that electrons spontaneously proceed from compounds with negative reduction potentials to those with positive reduction potentials, a transfer accompanied by the release of energy. The transfer of an electron from P870 to acceptor X is essentially a transfer from the bottom of the electron tower to the top and is made possible by the external supply of energy to P870 in the form of the photon it absorbed. From acceptor X the electron proceeds through an electron-transport system involving ubiquinone, an iron-sulfur protein, and cytochromes. Synthesis of ATP occurs probably as a result of the formation of a proton gradient generated through the function of ubiquinone in a fashion analogous to that seen in electron-transport phosphorylation in Section 4.8. (See Figures 4.15 and 4.17 for a description of the way quinones act as carriers of protons.) The redox series is completed when oxidized P870 receives the electron and is thereby re-reduced to its original form. It is then capable of absorbing the energy from another photon and reinitiating the process. This method of making ATP is called **cyclic photophosphorylation** since the electron is repeatedly moved around a closed circle. There is no external electron donor or acceptor.

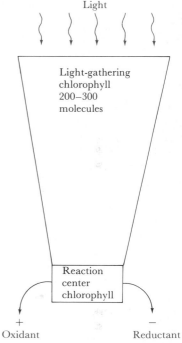

FIGURE 6.4 The photosynthetic unit and its associated reaction center. A packet of energy, absorbed by chlorophyll in the light-harvesting system, migrates to the reaction center, where it promotes the separation of oxidant and reductant.

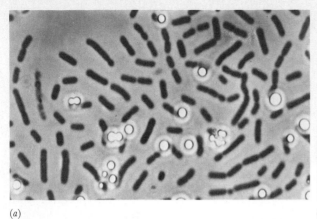

(a)

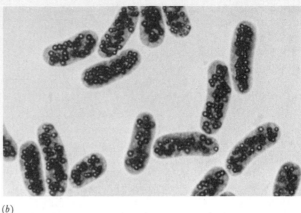

(b)

FIGURE 6.5 Photomicrographs of photosynthetic bacteria. (a) Green bacterium: *Chlorobium limicola*. The refractile bodies are sulfur granules deposited outside the cell. Magnification, 2000×. (b) Purple bacterium: *Chromatium okenii*. Notice the sulfur granules deposited inside the cell. Magnification, 1200×. (Photographs courtesy of Norbert Pfennig.)

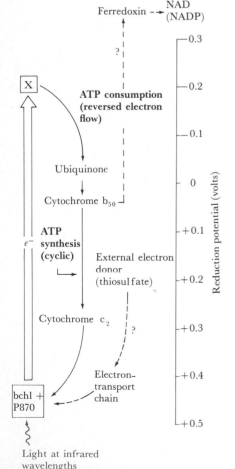

FIGURE 6.6 Electron flow in bacterial photosynthesis. Only a single light reaction, equivalent to photosystem I, is involved. Note that the potential of X is less negative than that in green plants. The dashed lines indicate pathways for which there is still some doubt.

Considerable uncertainty still exists about the manner in which photosynthetic bacteria create reducing power (NADPH). It is clear that they use reduced compounds from the environment, but there are two ways which may be distinguished for transfer of electrons from the external donor to NADP. The first involves direct transfer from a highly reduced compound to NADP. This direct reaction is possible in the presence of the appropriate enzymes, if the donor has a reduction potential more negative than that of NADP. Hydrogen gas (H_2) is an example of such a compound: its reduction potential is -0.42 V, whereas that of NADP is about -0.32 V. Thus, with H_2, light need only be used to generate ATP by cyclic phosphorylation, all reducing power coming from the environment. Further, one group, the so-called nonsulfur purple bacteria (Rhodospirillaceae), grow primarily photoheterotrophically. This means they use organic compounds as carbon sources for growth, rather than CO_2, and use light energy only to generate ATP. Provided the organic compounds used are sufficiently reduced, they can be assimilated without the need for a photosynthetically generated reducing power.

However, many purple and green bacteria will grow anaerobically using inorganic substances such as H_2S, thiosulfate ($S_2O_3^{2-}$), or elemental sulfur (S^0), which have higher reduction potentials than NADP. During growth on these substances, the bacteria oxidize them anaerobically in light-mediated reactions. The probable mechanism involves some components of the photosynthetic electron transport chain in a noncyclic manner, as shown by the dashed lines in Figure

6.6. The situation is not straightforward, however, as the reduction potential of the primary electron acceptor X is not sufficiently negative to be able to reduce ferredoxin directly. One postulated mechanism by which reducing power can be generated under such conditions is a process called **reversed electron transport,** which functions through the proton-conducting property of the membrane-bound ATPase described in Figure 4.18*b*. As shown on that figure, when ATP is broken down to ADP + P$_i$, H$^+$ ions are extruded, setting up an energized state of the membrane. This energized state makes possible a reversed electron transport, with electrons flowing toward NADP (the reverse of the electron-transport system outlined in Figure 4.16) from the photosynthetic electron donor (that is, thiosulfate, elemental sulfur). Reversed electron transport is more firmly established in lithotrophic bacteria (see Section 6.10), but there is some evidence that it occurs in purple and green bacteria.

6.4

Electron Flow in Green-Plant Photosynthesis

Green plants, algae, and cyanobacteria also use light to generate the compounds ATP and NADPH. However both compounds are made as direct consequences of light reactions with the electrons for the reduction of NADP to NADPH coming from the splitting of water molecules. The overall scheme of electron flow in green-plant photosynthesis is shown in Figure 6.7. This is sometimes called the *Z scheme* of photosynthesis, because of the Z-shaped pathway of the electrons' movement. There are two systems of light reactions, called *photosystem I* and *photosystem II,* associated with the flow of electrons. Each photosystem has its own type of chlorophyll; each chlorophyll is distinguished by the fact that it absorbs light best at a certain wavelength. Photosystem I chlorophyll, called P700, absorbs light best at long wavelengths (far-red light), whereas photosystem II chlorophyll, called P680, absorbs light best at shorter wavelengths (blue and near-red light). (Both P700 and P680 are chlorophyll *a*, with the absorption spectrum modified as a result of association of the chlorophyll *a* molecules with the particular part of the reaction center.) In effect, these two photosystems operate cooperatively to create a sufficient charge separation so that reducing power can be obtained by the splitting of water. As discussed in Section 4.3, the potential of the O$_2$/H$_2$O redox pair is +0.81 V; through the action of photosystem II, a positively charged chlorophyll is formed that has a reduction potential sufficiently positive so that it is more oxidizing than the O$_2$/H$_2$O pair.

Although some of the electron carriers in the photosynthetic scheme are familiar (cytochromes, ferredoxin), others are unique to photosynthesis and have not been previously described. The pathway of electron transport is ultimately from H$_2$O to NADP. The first step involves the metal *manganese,* which is directly involved in the initial splitting of water and the liberation of molecular oxygen (O$_2$). The electron then passes to chlorophyll P680, which has been previously photooxidized. The primary electron acceptor in photosystem II is designated Q, which stands for *quencher*. This is a component that has not been isolated biochemically but has been demonstrated spectro-

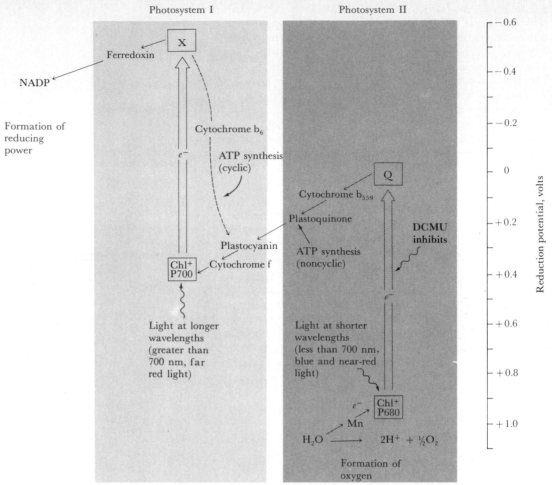

FIGURE 6.7 Electron flow in green-plant photosynthesis, the "Z" scheme. Two photosystems are involved, photosystems I and II. Water is split to form electrons and O_2 in photosystem II. Q and X are primary electron acceptors of photosystems I and II, respectively, identified by spectroscopic means only. ATP synthesis occurs at the locations shown.

scopically. From the quencher Q the electron passes through an electron-transport chain, which includes two cytochromes (cytochrome b_{559} and cytochrome f), a quinone called *plastoquinone* (so-called because it is found in the chloro*plast*), and a copper-containing protein called *plastocyanin*. The electron is now accepted by the reaction-center chlorophyll of photosystem I, chlorophyll P700, which has been previously photooxidized. When P700 receives its photon the electron is donated to an acceptor with a very negative reduction potential. The primary acceptor in photosystem I is designated X. The acceptor X also has not been isolated biochemically but has been identified spectroscopically. When reduced, X is at a reduction potential sufficiently negative so that it can reduce ferredoxin. The electron is then passed from ferredoxin via an NADP reductase to NADP, which is reduced to NADPH. The joint action of photosystems I and II is termed **noncyclic photophosphorylation** since the electrons are moved in a linear sequence and do not end up in the starting chlorophyll. At the conclusion of this series of reactions both chlorophylls, P700 and P680, have been re-reduced and are ready to absorb more photons. The net reaction is the transfer of two electrons from H_2O to NADP, accompanied by the synthesis of ATP.

As noted, in addition to producing reducing power, ATP is synthesized during electron flow. In the Z scheme, the site of ATP synthesis has been identified at the step between plastoquinone and plastocyanin. The synthesis of ATP proceeds by a proton gradient mechanism similar to that already described for electron-transport phosphorylation in respiration (Section 4.8) and in the cyclic photophosphorylation process of the purple and green bacteria. In chloroplasts the pH gradient is created across the thylakoid membrane. The components of the photosynthetic electron-transport chain are oriented in the thylakoid membrane in such a way that plastoquinone can act to shuttle H^+ from the chloroplast matrix to the chloroplast inner space (Figure 6.8).

A second process of ATP synthesis occurs in photosystem I under conditions when reducing power is not being synthesized (dashed line

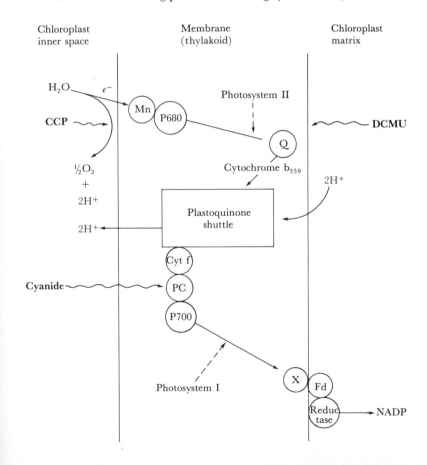

FIGURE 6.8 Mechanism for generation of a pH gradient across the thylakoid membrane. As a result of this gradient, ATP can be synthesized in the manner illustrated in Figure 4.18. See Figure 6.7 for the identity of the various electron carriers. Abbreviations: CCP, carbonylcyanide-phenylhydrazone; DCMU, 3-(3,4-dichlorophenyl)-1,1-dimethylurea.

in Figure 6.7). This is a type of **cyclic phosphorylation,** because the electron goes around a cycle from P700 to acceptor X and back to P700 again, without actually leaving the system. The normal process is noncyclic phosphorylation; cyclic phosphorylation occurs primarily when photosystem II is blocked in some way, or when light of only long wavelengths is available. Cyclic phosphorylation may also occur when algae are growing under conditions of very low intensity, where there is insufficient energy to initiate the function of both photosystems. Cyclic photophosphorylation is discussed in more detail in Section 6.5. Note that this form of cyclic photophosphorylation is entirely separate from the cyclic photophosphorylation performed by the photosynthetic bacteria, especially in terms of the chlorophylls involved.

That two light reactions occur in green-plant photosynthesis was one of the more surprising results of work on photosynthesis in the past generation. The discovery arose from studies on the effect of monochromatic (single-wavelength) light on the rate of photosynthesis. It was found that far-red light (greater than 680 nm) was quite inefficient in bringing about oxygen evolution in algae, but if a second beam of shorter wavelength light were superimposed on the first, the efficiency of the far-red light was enhanced. It seemed as if light of shorter wavelength was actually making it possible for the algae to use the longer-wavelength light more efficiently. As might be expected, this enhancement phenomenon has not been found in the purple and green bacteria, which possess only one light system.

The differences between photosynthesis in the purple and green bacteria and other photosynthetic organisms are summarized in Table 6.2. In Chapter 2 we emphasized the structural similarities between cyanobacteria and other procaryotes, but now we see that, as far as the photosynthetic process is concerned, cyanobacteria resemble the eucaryotes more than they do the other photosynthetic bacteria.

TABLE 6.2   Differences Between Plant-Type and Bacterial-Type Photosynthesis

	Plant-Type Photosynthesis	Bacterial-Type Photosynthesis
Organisms	Cyanobacteria, eucaryotic algae, higher plants	Purple bacteria, green bacteria
Chlorophyll type	Chlorophyll a (absorbs in red); some have chlorophyll b, c, d, or e	Bacteriochlorophylls (some absorb in far red)
Cyclic photophosphorylation	Present (photosystem I)	Present
Photosystem II (noncyclic photophosphorylation)	Present	Absent
Produce O_2	Usually (when performing noncyclic photophosphorylation)	No
Source of reducing power	Usually H_2O (when performing noncyclic photophosphorylation)	H_2, H_2S, other sulfur compounds, organic compounds

Most of the early work on photosynthesis was carried out with higher plants, and the microbe had little to contribute to the research. By the mid-1930s, the overall process of CO_2 fixation and O_2 production was generally understood, but the source of O_2 was not known. One hypothesis said that oxygen came from CO_2 and the other said that it came from H_2O. A strong case for the latter developed from studies on purple and green bacteria by Cornelius B. Van Niel, a Dutch microbiologist working at Stanford University. It is interesting to note that the first clear idea of the source of O_2 in green-plant photosynthesis came from studies on photosynthetic organisms which do not themselves produce O_2.

The study of the purple and green bacteria, especially the sulfur forms, has a long history. These organisms were first studied by Sergei Winogradsky, who in the 1880s had also studied the lithotrophic sulfur bacteria. Winogradsky showed that the purple and green sulfur bacteria oxidized H_2S and produced sulfur granules, and that the sulfur granules were gradually converted to sulfate. He thought that these organisms were just another group of autotrophic sulfur bacteria. About the same time, T. W. Engelmann, a German microbiologist, proposed the idea that these bacteria were photosynthetic. This was based on his experiments showing that they were phototactic and tended to accumulate at that part of a light spectrum at which their pigments absorbed (see Figure 6.16 for the modern equivalent of Engelmann's experiment). Engelmann attempted to show that these bacteria produced O_2 but failed to obtain a clear-cut answer. Later H. Buder, another German microbiologist, showed that O_2 did not form; he believed that it was produced during the photosynthetic process but immediately reacted in dark reactions with H_2S to form sulfur or sulfate. This idea seemed reasonable since at that time it was thought that O_2 formation was one of the hallmarks of the photosynthetic process.

The proof that photosynthetic sulfur bacteria do not produce O_2 but obtain reducing power for CO_2 fixation directly from H_2S was provided by the brilliant experiments of Van Niel published in 1931. Van Niel isolated pure cultures of these bacteria and grew them on completely mineral medium, so that he could control substrates and analyze products. He did careful quantitative analyses of the amount of H_2S oxidized and the amount of CO_2 fixed. In the case of the green sulfur bacteria, which showed the simplest situation, the following stoichiometry was found (where CH_2O represents carbon at the oxidation level of carbohydrate):

$$CO_2 + 2H_2S \longrightarrow (CH_2O) + H_2O + 2S$$

That is, a ratio of $1:2:2$ was found between CO_2, H_2S, and S. This simple stoichiometry was at variance with the stoichiometry found for the lithotrophic sulfur bacteria; with the latter, as many as 40 molecules of H_2S were required to fix one molecule of CO_2. Thus it seemed unlikely that the photosynthetic sulfur bacteria were first liberating O_2 by light reactions and then using it to oxidize H_2S by dark reactions. Now, green plant photosynthesis can be represented as follows:

$$CO_2 + H_2O \longrightarrow (CH_2O) + O_2$$

If we add H_2O to both sides of this equation, we obtain

$$CO_2 + 2H_2O \longrightarrow (CH_2O) + H_2O + O_2$$

This equation is strikingly similar to that for the green sulfur bacteria given above, with only O replacing S. If we assume that H_2S in the green sulfur bacteria and H_2O in green plants have the same function, as sources of reducing power, we can express this in a generalized way using H_2A, where A is the compound from which the reducing power is derived:

A Bit of History

$$CO_2 + 2H_2A \longrightarrow (CH_2O) + H_2O + 2A$$

From these considerations, Van Niel hypothesized that the source of O_2 in green plants is H_2O and not CO_2. This was later clearly shown to be the case by use of isotopically labeled water, $H_2^{18}O$. Further, it can be considered that photosynthetic bacteria that use reducing substances other than H_2S use them in a similar way, as a source of electrons for reduction of CO_2. This could be true even of the organic compounds used by many purple and green bacteria, although when organic compounds are used the situation is complicated by the fact that organic materials can be assimilated directly as carbon sources by many purple and green bacteria.

The overall scheme of photosynthesis advanced by Van Niel—$CO_2 + 2H_2A \rightarrow (CH_2O) + H_2O + 2A$—has provided a useful framework for all of the research that has subsequently taken place on the process in green plants. The proof that green plants use water as a source of electrons for the formation of reducing power came from a large number of studies on chloroplasts isolated from green plants, and microbes have not played a significant role in this research. However, it was the original studies on purple and green bacteria that placed the whole problem in a clear light and provided the impetus for later work. This brief history illustrates an important point about science: it is sometimes better not to study directly a process of interest, but to study an analogous process more favorable for research, with the hope that the definitive studies on the simpler system will shed light on the more complicated process.

6.5

Cyclic Photophosphorylation in Photosystem I

As we have seen, photosystems I and II normally function together in a noncyclic, oxygenic process. However, under certain conditions many algae and some cyanobacteria are able to carry out cyclic photophosphorylation using only photosystem I. This alteration requires the presence of anaerobic conditions as well as a reducing substance such as H_2 or H_2S. Under these conditions, the electrons for CO_2 reduction come not from water but from the reducing substance. In the eucaryotic algae, H_2 is generally the reducing substance. About 50 percent of eucaryotic algae have the enzyme hydrogenase, which catalyzes H_2 oxidation. When the alga is adapted for a short time to anaerobic conditions, the hydrogenase comes into function, and if light is present an overall reaction of the following kind is carried out: $2H_2 + CO_2 \rightarrow (CH_2O) + H_2O$. This process occurs under natural conditions only when light intensities are low, apparently because under such circumstances there is insufficient energy available for photosystem II activity. At higher light intensities, even under anaerobic conditions, photosystem II acts, and water is split, leading to O_2 production, and the system gradually becomes aerobic as O_2 builds up. In the laboratory, anoxygenic photosynthesis can be made to occur even at high light intensities if an inhibitor of photosystem II, such as the plant herbicide 3-(3,4-dichlorophenyl)-1,1-dimethylurea (DCMU) is present (see Figure 6.7).

A number of cyanobacteria can use H_2S as a reducing agent in anoxygenic photosynthesis. When H_2S is used, it is oxidized to elemental sulfur (S^0), and sulfur granules are deposited outside the cells. This process is thus quite similar to the usual photosynthesis carried out by the purple and green sulfur bacteria. Again, it occurs primarily at low light intensities, or when only far-red or infrared radiation is available so that photosystem II cannot function. It has also been found that high concentrations of H_2S inhibit photosystem II; thus even at higher light intensities cyclic photophosphorylation may occur if the H_2S concentration is proper. In one cyanobacterium, *Oscillatoria limnetica,* anoxygenic photosynthesis has been shown to occur under natural conditions using H_2S.

From an evolutionary point of view, the existence of cyclic photophosphorylation indicates a close relationship between green-plant and bacterial photosynthesis. Although organisms carrying out oxygenic photosynthesis have acquired photosystem II, and hence the ability to split H_2O, they still retain the ability under certain conditions to use photosystem I alone.

6.6

Additional Aspects of Photosynthetic Light Reactions

Accessory pigments   Although chlorophyll is obligatory for photosynthesis, most photosynthetic organisms have other pigments that are involved, at least indirectly, in the capture of light energy. The most widespread accessory pigments are the **carotenoids,** which are almost always found in photosynthetic organisms. Carotenoids are insoluble in water but soluble in organic solvents; the structure of a typical carotenoid is shown in Figure 6.9. They have long hydrocarbon chains with alternating $C-C$ and $C=C$ bonds, an arrangement called a conjugated double-bond system. As a rule, carotenoids are yellow in color and absorb light in the blue region of the spectrum. The carotenoids are usually closely associated with chlorophyll in the photosynthetic apparatus, and there are approximately the same number of carotenoid as there are chlorophyll molecules. Carotenoids do not act directly in photosynthetic reactions, but they may transfer the light energy they capture to chlorophyll, and this energy may thus be used in photophosphorylation in the same way as is the light energy captured directly by chlorophyll.

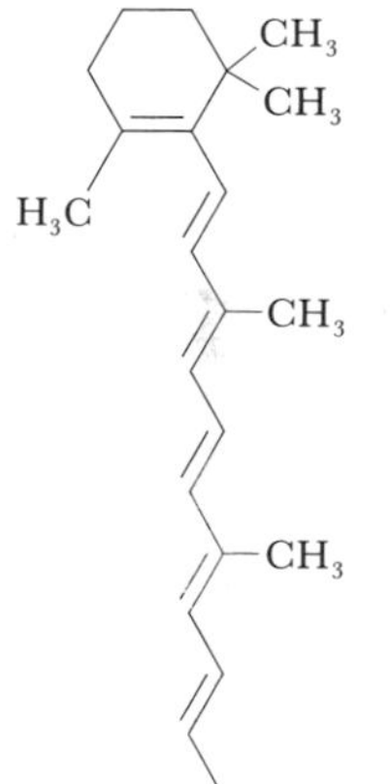

Cyanobacteria, red algae, and a few others contain **biliproteins,** which are accessory pigments that are red or blue in color. The red pigment, called *phycoerythrin,* absorbs light most strongly in the wavelengths around 550 nm, whereas the blue pigment, *phycocyanin,* absorbs most strongly at about 620 to 640 nm. These biliproteins contain open-chain tetrapyrrole rings called *phycobilins* (Figure 6.10), which are coupled to proteins. The biliproteins occur as high-molecular-weight aggregates called *phycobilisomes,* attached to the photosynthetic membranes. They are closely linked to the chlorophyll-containing system, which makes for very efficient energy transfer, approaching 100 percent, from biliprotein to chlorophyll.

The light-gathering function of the accessory pigments seems to be of obvious advantage to the organism. Light from the sun is distributed over the whole visible range; yet chlorophylls absorb well in only a part of this spectrum. By having accessory pigments, the organism is able to capture more of the available light. Another function of accessory pigments, especially of the carotenoids, is as photoprotective agents. Bright light can often be harmful to cells, in that

FIGURE 6.9 A typical carotenoid, β-carotene.

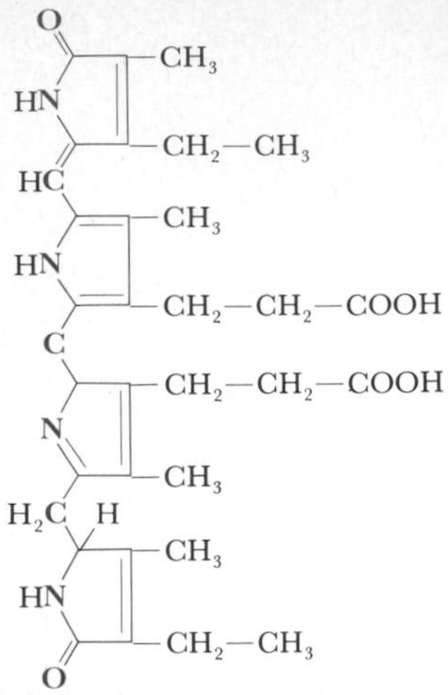

FIGURE 6.10 A typical phycobilin. This compound is an open-chain tetrapyrrole, derived biosynthetically from a closed porphyrin ring by loss of one carbon atom as carbon monoxide. The structure shown is the prosthetic group of phycocyanin, a proteinaceous pigment found in cyanobacteria and red algae.

it causes various photooxidation reactions that can actually lead to destruction of chlorophyll and of the photosynthetic apparatus itself. The accessory pigments absorb much of this harmful light and thus provide a shield for the light-sensitive chlorophyll. Since photosynthetic organisms must by their nature live in the light, the photoprotective role of the accessory pigments is an obvious advantage.

Experimental dissection of photosynthesis with inhibitors and mutants
Among the many techniques used to understand the overall scheme of light reactions in photosynthesis, an important approach has been to use procedures that block certain steps of the process without affecting other steps. One means by which this has been done has been to isolate mutants of algae that are unable to carry out photosynthesis and then to determine which steps of the process do not occur. Mutants have been isolated that are blocked in photosystem II, and some of these mutants have been shown to be deficient in cytochrome b_{559}, which is involved in electron transport in this pathway (Figure 6.8). Mutants have also been isolated that are blocked in photosystem I. One such mutant is deficient in plastocyanin, another in cytochrome b, and another seems to be unable to carry out cyclic phosphorylation because of a block in the membrane ATPase involved in ATP synthesis.

Several inhibitors have been shown to specifically affect steps in photosynthetic light reactions. The plant herbicide 3-(3,4-dichlorophenyl)-1,1-dimethylurea (DCMU) is a specific inhibitor of photosystem II. In the presence of this agent, O_2 evolution and noncyclic photophosphorylation do not occur and NADPH is not formed. Cyclic photophosphorylation occurs normally, and some organisms are able to grow photoheterotrophically in the presence of DCMU, using organic compounds as carbon sources and light as an energy source in cyclic phosphorylation. In the presence of suitable organic compounds, CO_2 fixation is not necessary, thus reducing power is not required and growth can occur in the absence of photosystem II activity. As noted above, purple and green bacteria are only capable of cyclic photophosphorylation with bacteriochlorophylls. These organisms are not inhibited by DCMU, whether growing on inorganic or organic substrates. The position at which DCMU affects photosystem II is on the matrix side of the thylakoid membrane, as shown in Figure 6.8. Another inhibitor, carbonylcyanide-phenylhydrazone (CCP) inhibits the formation of O_2 from water, and hence also specifically inhibits photosystem II activity. Another way to prevent photosystem II activity is to grow organisms under a manganese deficiency, since manganese is specifically required for the O_2-producing step (Figure 6.7).

In mutants, artificial electron carriers, such as certain dyes, can be added to bypass the block. In this way, details of the electron flow can be analyzed. Another important approach involves the use of spectroscopic techniques, which permit observation of changes in light absorption by chlorophylls, cytochromes, or other light-absorbing components of the system. There is usually a spectral shift (change in peak wavelength absorbed) when a component is oxidized or reduced, and light-induced spectral shifts indicate the flow of electrons through specific carriers.

Energetics of photosynthesis and the quantum yield   Energy from light does not come in continuous fluxes but in small discrete packets, each of which is called a **quantum.** A fundamental law of physics relates the energy in a quantum to the wavelength of light (see Appendix 1). If we take light of 680 nm, which is the wavelength absorbed by reaction-center chlorophyll in photosystem II, it can be calculated from the equation relating energy to wavelength (Appendix 1) that a mole of quanta has 41 kcal (a mole of quanta is Avogadro's number of quanta).

From free energy data (Section 4.1 and Appendix 1, Table A1.1), the **quantum yield** (number of quanta needed to fix one molecule of CO_2) can be calculated. The synthesis of a mole of glucose from CO_2 and H_2O has a $\Delta G^{0\prime}$ of $+686$ kcal. Taking one-sixth of this, we can calculate that the fixation of 1 mole of CO_2 requires $+114$ kcal, and since a mole of quanta of red light has 41 kcal, it would take at least 3 quanta (114/41) to fix one molecule of CO_2. This value of 3 quanta per one molecule of CO_2 fixed is the maximum possible quantum yield of photosynthesis. Experimental studies using uniformly illuminated suspensions of algae and careful measurement of light absorbed and CO_2 fixed indicate that in practice it takes about 8 quanta to fix one molecule of CO_2, which is about 38 percent of the theoretical value of 3 quanta.

Carrying this analysis further, we know from the balance sheet for photosynthetic CO_2 fixation (Section 6.7) that for each molecule of CO_2 fixed, two NADPH and three ATP are required. In noncyclic photophosphorylation (Figure 6.7), 2 quanta are required to eject each electron (one for each photosystem), but two electrons are required to reduce one NADP. Thus, to synthesize the two NADPH necessary for CO_2 fixation, four electrons must be ejected, which is equivalent to 8 quanta. It also requires two electrons to synthesize one ATP, so that when 8 quanta are absorbed, two molecules of ATP can be synthesized. Thus 8 quanta must be absorbed to synthesize two NADPH and two ATP, and a quantum yield of 8 has been measured experimentally. This does not quite add up, since as we have just noted, three ATP are required to reduce one molecule of CO_2, and we have only made two ATP. Since the quantum yield of 8 has been measured experimentally, it seems reasonable to conclude that extra ATP is made somewhere during the process; perhaps in some manner by cyclic photophosphorylation.

If we consider the energetics of anoxygenic photosynthesis carried out by the photosynthetic bacteria, we find that much less energy is required to fix CO_2 into organic compounds than is required by oxygenic photosynthesis. This is because reducing power does not come from the splitting of water but from stronger reductants such as H_2 or H_2S. If we write the simplest equation for CO_2 fixation by photosynthetic bacteria using cyclic photophosphorylation:

$$6CO_2 + 12H_2S \longrightarrow C_6H_{12}O_6 + 6H_2O + 12S$$

and calculate the free energy from the values given in Table A1.1, we find that the $\Delta G^{0\prime}$ is $+97$ kcal/mole, or $+16$ kcal per mole of CO_2 fixed. This is considerably less than the $+114$ kcal per mole of CO_2 required in green-plant photosynthesis. It is perhaps because of the small energy demand of anoxygenic photosynthesis that purple and green bacteria can exist in environments where light intensities are very low. It may also explain how these bacteria are able to function using only cyclic photophosphorylation to generate ATP.

6.7

Autotrophic CO$_2$ Fixation

In the first half of this chapter, we have discussed the energy metabolism and light reactions of phototrophic autotrophs. We now discuss the biochemistry of the process by which autotrophs, both phototrophic and lithotrophic, convert CO$_2$ into organic matter.

Although all organisms require some of their carbon for cellular biosynthesis to come from CO$_2$, autotrophs can obtain *all* of their carbon from CO$_2$. The overall process is called CO$_2$ fixation, and all of the reactions of CO$_2$ fixation will occur in complete darkness, using ATP and reducing power (NADPH) generated either during the light reactions of photosynthesis or during oxidation of inorganic compounds. Most autotrophs so far examined have a special pathway for CO$_2$ reduction, the **Calvin cycle,** also referred to as the C$_3$ or photosynthetic carbon cycle. This cycle is summarized later, but first let us consider some of the key individual reactions.

Enzyme reactions of the Calvin cycle   The first step in CO$_2$ reduction is the reaction catalyzed by the enzyme *ribulose diphosphate carboxylase,* which involves a reaction between CO$_2$ and ribulose diphosphate (Figure 6.11) leading to formation of two molecules of 3-phosphoglyceric acid (PGA), one of which contains the carbon atom from CO$_2$. PGA constitutes the first identifiable intermediate in the CO$_2$ reductive process, and it is because of the formation of this three-carbon compound that the photosynthetic carbon cycle is sometimes called the C$_3$ cycle. The carbon atom in PGA is still at the same oxidation level as it was in CO$_2$; the next two steps involve reduction of PGA to the oxidation level of carbohydrate (Figure 6.12). In these steps, *both* ATP and NADPH are required: the former is involved in the phosphorylation reaction that activates the carboxyl group, the latter in the reduction itself: the carbon atom from CO$_2$ is now at the reduction level of carbohydrate (CH$_2$O), but only one of the carbon atoms of glyceraldehyde phosphate has been derived from CO$_2$, the other two having arisen from the ribulose diphosphate. Since autotrophs can grow on CO$_2$ as a sole carbon source, there must be a way by which carbon from CO$_2$ can become incorporated into the other positions in glyceraldehyde phosphate. Further, the ribulose diphosphate that was used up in the ribulose diphosphate carboxylase step must be regenerated. The remainder of the reactions of the C$_3$ cycle are concerned with these matters.

The Calvin cycle   The series of enzyme reactions leading to synthesis of ribulose diphosphate is shown in summary form in Figure 6.13. In this series of reactions five molecules of glyceraldehyde phosphate (or its equivalent, dihydroxyacetone phosphate) are used up, and three molecules of ribulose diphosphate are formed; three molecules of ATP are also required, but no NADPH, as there are no reductive steps. Notice that the reactions in the middle part of Figure 6.13 are

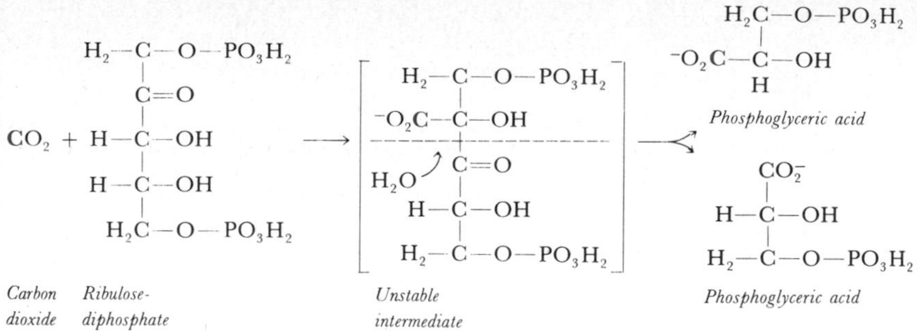

$$CO_2 + \begin{array}{c} H_2-C-O-PO_3H_2 \\ | \\ C=O \\ | \\ H-C-OH \\ | \\ H-C-OH \\ | \\ H_2C-O-PO_3H_2 \end{array} \longrightarrow \left[\begin{array}{c} H_2-C-O-PO_3H_2 \\ | \\ {}^-O_2C-C-OH \\ \text{------}|\text{------} \\ C=O \\ H_2O \nearrow | \\ H-C-OH \\ | \\ H_2-C-O-PO_3H_2 \end{array} \right] \rightarrow \begin{array}{c} H_2C-O-PO_3H_2 \\ | \\ {}^-O_2C-C-OH \\ H \\ \textit{Phosphoglyceric acid} \\[2mm] CO_2^- \\ | \\ H-C-OH \\ | \\ H_2-C-O-PO_3H_2 \\ \textit{Phosphoglyceric acid} \end{array}$$

Carbon Ribulose-   Unstable
dioxide diphosphate   intermediate

FIGURE 6.11 Reaction of the enzyme ribulose diphosphate carboxylase.

$$\begin{array}{c} H_2-C-O-PO_3H_2 \\ | \\ HO-C-H \\ | \\ CO_2^- \\ \textit{Phosphoglyceric acid} \end{array} + ATP \longrightarrow \begin{array}{c} H_2-C-O-PO_3H_2 \\ | \\ HO-C-H \\ | \\ H_2O_3P-O-C=O \\ \textit{1,3-Diphosphoglyceric acid} \end{array} + ADP$$

$$\begin{array}{c} H_2-C-O-PO_3H_2 \\ | \\ HO-C-H \\ | \\ H_2O_3P-O-C=O \\ \textit{1,3-Diphosphoglyceric acid} \end{array} + NADPH \longrightarrow \begin{array}{c} H_2-C-O-PO_3H_2 \\ | \\ HO-C-H \\ | \\ H-C=O \\ \textit{Glyceraldehyde-3-phosphate} \end{array} + P_i + NADP$$

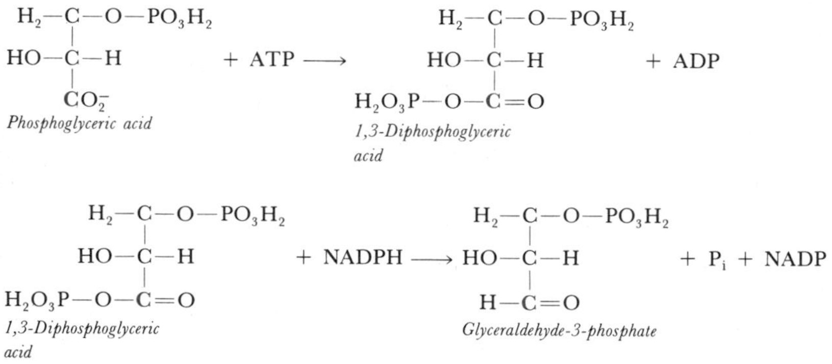

FIGURE 6.12 Steps in the conversion of 3-phosphoglyceric acid (PGA) to glyceraldehyde-3-phosphate. Note that both ATP and NADPH are required.

the same as those of the pentose phosphate cycle described in Figure 5.4, and involve the same enzymes, transaldolase and transketolase. The Calvin cycle is sometimes called the **reductive pentose cycle,** to distinguish it from the pentose cycle described in Section 5.2.

The final step in the regeneration of ribulose diphosphate is the phosphorylation of ribulose-5-phosphate with ATP by the enzyme *phosphoribulokinase.* This enzyme is not involved in the normal pentose phosphate cycle but is unique to the photosynthetic carbon cycle.

In a single carboxylation reaction, one CO_2 (1C) and one ribulose diphosphate (5C) yield two molecules of PGA ($2 \times 3C$), which are converted to two molecules of glyceraldehyde phosphate. Since we need five molecules of glyceraldehyde phosphate (or their equivalent, five molecules of dihydroxyacetone phosphate) to complete one turn of the cycle, this means that three carboxylation reactions must take place for each turn of the cycle. Since each carboxylation yields two glyceraldehyde phosphates, three carboxylations would yield six, but we need only five for regeneration of ribulose diphosphate. The sixth glyceraldehyde phosphate thus represents a net gain: three CO_2 molecules have been in effect converted into one glyceraldehyde phosphate.

Some of the enzymes of the Calvin cycle were already known from studies on pentose metabolism (Chapter 5), but the key enzyme, ribulose diphosphate carboxylase, which is involved in the initial CO_2 fixation reaction, does not function in normal pentose metabolism. This enzyme has been found in virtually all photosynthetic organisms examined—plants, algae, and bacteria. It is also found

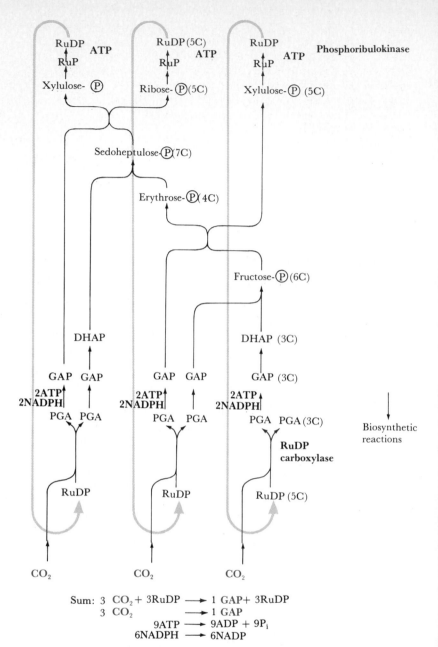

FIGURE 6.13 The Calvin cycle. Three turns of the cycle are necessary to synthesize one glyceraldehyde-3-phosphate from three CO_2 molecules. RuDP stands for ribulose-1,5-diphosphate; DHAP, for dihydroxyacetone phosphate; GAP, for glyceraldehyde-3-phosphate; PGA, for 3-phosphoglyceric acid; RuP, for ribulose-5-phosphate. The number of carbon atoms in each compound is shown in parentheses.

$$\text{Sum:}\ 3\ CO_2 + 3RuDP \longrightarrow 1\ GAP + 3RuDP$$
$$3\ CO_2 \longrightarrow 1\ GAP$$
$$9ATP \longrightarrow 9ADP + 9P_i$$
$$6NADPH \longrightarrow 6NADP$$

in lithotrophic autotrophic bacteria, such as the sulfur, iron, and nitrifying bacteria.* The second key enzyme is glyceraldehyde phosphate dehydrogenase, which catalyzes the reductive step of the Calvin

*Two groups of autotrophs, the green bacteria and the methanogenic bacteria (Section 19.20), may use a different pathway for CO_2 fixation. In both of these groups, it has not been possible to unequivocally demonstrate the presence of the enzyme ribulose diphosphate carboxylase. The precise mechanism by which these bacteria fix CO_2 is still under study.

cycle (Figure 6.12). It should be noted that a glyceraldehyde phosphate dehydrogenase is crucial to the fermentation of glucose in the glycolytic pathway (Figure 4.9). In glycolysis the reaction proceeds from glyceraldehyde-3-P to 1,3-diphosphoglyceric acid, involving a substrate-level phosphorylation and a reduction of NAD to NADH. In the Calvin cycle, the reaction proceeds in the opposite direction from 1,3-diphosphoglyceric acid to glyceraldehyde-3-P, a high-energy phosphate bond (that in 1,3-diphosphoglyceric acid) is *lost*, and NADPH is oxidized. One way in which the two reactions are separately controlled is through the use of NAD in the energy-generating reaction and NADP in the energy-requiring reaction.

The enzyme phosphoribulokinase, which converts ribulose-5-phosphate to ribulose-1,5-diphosphate, is also unique to the Calvin cycle. Two additional enzymes that may be specific for the Calvin cycle are the phosphatases that convert the diphosphates of sedoheptulose and fructose to the monophosphates (not shown in the abbreviated scheme of Figure 6.13). All of the other enzymes of the Calvin cycle are the same as those for reactions of the normal pentose phosphate cycle, which operates in nonautotrophic organisms.

Biosynthetic reactions   Most photosynthetic organisms live in nature in alternating light and dark regimes, but the primary products of the light reactions, ATP and NADPH, are short-lived and must be quickly used up since they cannot be stored. Photosynthetic organisms circumvent this difficulty by converting CO_2 into energy-rich storage products during the light cycle, then using these during the dark cycle. In algae and cyanobacteria, storage products are usually carbohydrates such as sucrose or starch, whereas in purple and green bacteria the main storage product is poly-β-hydroxybutyric acid.

The balance sheet of the Calvin cycle   Let us consider now the complete balance sheet for conversion of three molecules of CO_2 into one molecule of glyceraldehyde phosphate. Six ATP molecules and six NADPH molecules are required for the reduction of six molecules of PGA to glyceraldehyde phosphate, and three ATP molecules are required for conversion of ribulose phosphate to ribulose diphosphate. Thus six NADPH and nine ATP are required to manufacture one glyceraldehyde phosphate from CO_2. Since two glyceraldehyde phosphates can give rise to one hexose sugar molecule, 12 NADPH and 18 ATP are required to make one hexose molecule from $6CO_2$.

6.8

Variants of the Photosynthetic Carbon Cycle

The C_4 pathway   Although the C_3 cycle is well established as the primary pathway for net CO_2 reduction in autotrophic organisms, other pathways of CO_2 fixation do exist. In Section 5.4 we discussed the fixation of CO_2 by the enzymes pyruvate carboxylase and phosphoenolpyruvate carboxylase, which in both cases leads to synthesis of the four-carbon compound oxalacetate as the first product of CO_2 fixation. Oxalacetate then enters the tricarboxylic acid cycle, or is converted to sugars via 3-phosphoglyceric acid. This pathway of CO_2 fixation is called the **C_4-dicarboxylic acid pathway,** to contrast it

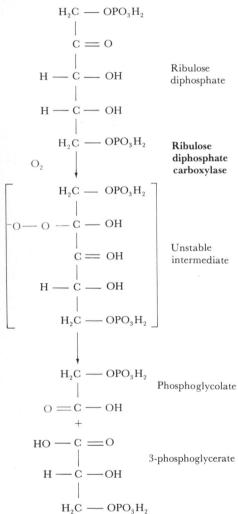

H_2C — OPO_3H_2

|

$C = O$

|

H — C — OH Ribulose diphosphate

|

H — C — OH

|

H_2C — OPO_3H_2 **Ribulose diphosphate carboxylase**

O_2 ↓

H_2C — OPO_3H_2

|

O — O — C — OH

|

$C = OH$ Unstable intermediate

|

H — C — OH

|

H_2C — OPO_3H_2

↓

H_2C — OPO_3H_2

| Phosphoglycolate

$O = C$ — OH

+

HO — $C = O$

| 3-phosphoglycerate

H — C — OH

|

H_2C — OPO_3H_2

FIGURE 6.14 Action of ribulose diphosphate carboxylase as an oxygenase under high O_2 and low CO_2 conditions. Compare with Figure 6.11, which shows the normal reaction of this enzyme. When acting as an oxygenase, RuDP carboxylase catalyzes the synthesis of phosphoglycolate and only one molecule of phosphoglycerate.

with the C_3 cycle, where the first stable product of CO_2 fixation is the C_3 compound 3-phosphoglyceric acid. A number of higher plants (especially grasses) fix CO_2 initially by this pathway, and the alga *Chlorella* also fixes CO_2 primarily by this method when grown under an atmosphere rich in CO_2. The C_4 pathway functions to trap and store CO_2 but does not result in net CO_2 fixation or sugar synthesis by itself. Rather it operates by efficiently supplying CO_2 to the C_3 cycle for net photosynthesis, since all organisms using the C_4 pathway also have the C_3 cycle.

Photorespiration   Many higher plants and eucaryotic algae have been shown to carry out a light-stimulated production of CO_2 from organic matter. This process, called *photorespiration,* can occur simultaneously with photosynthesis, so that some of the CO_2 converted to organic matter is released again. Photorespiration should be distinguished from normal dark respiratory processes, which also occur in these organisms. In the light, dark respiration is restricted and photorespiration is probably the predominant process of CO_2 release. The two respiratory processes occur in different parts of the cell and involve different biochemical pathways. Dark respiration occurs predominantly in the mitochondria, whereas photorespiration occurs predominantly in microbodies called **peroxisomes** (see Table 3.1) as well as in chloroplasts and mitochondria. A key intermediate in photorespiration is glycolate, which is formed in the chloroplast as a result of photosynthetic CO_2 reduction and is transferred to the peroxisome where oxidation occurs.

Although all the details are not known, there is strong evidence that photorespiration occurs because of an additional activity of the enzyme ribulose diphosphate carboxylase (RuDP carboxylase) under conditions of high O_2 and low CO_2 concentration. As described in Figure 6.11, RuDP carboxylase catalyzes the addition of CO_2 to RuDP and the immediate splitting of the unstable intermediate formed into two molecules of phosphoglyceric acid. RuDP carboxylase will also act as an oxygenase and catalyze the addition of molecular oxygen (O_2) onto RuDP, and the unstable intermediate formed splits into only one molecule of phosphoglycerate and one molecule of phosphoglycolate (Figure 6.14). The phosphoglycolate formed is converted into glycolate (CH_2OH—$COOH$) by a phosphatase. Glycolate is then either excreted or is oxidized further. Because RuDP is split in the oxygenase reaction and thus taken out of the photosynthetic carbon cycle, CO_2 fixation is diminished. Good evidence of the involvement of the oxygenase reaction in photorespiration is shown by the fact that glycolate is also formed in purple bacteria in the presence of O_2. It is also formed in lithotrophic sulfur bacteria, and hydrogen-oxidizing bacteria, which also use the RuDP carboxylase reaction to fix CO_2 (see Section 6.10).

Although glycolate can be oxidized to glyoxylate and then used in the biosynthesis of serine and glycine, much of the glycolate in aquatic microorganisms is probably excreted into the environment, and thus represents a wastage of energy. It thus appears that photorespiration, an apparently inevitable consequence of the oxygenase activity of RuDP carboxylase, is primarily a detrimental process to the autotrophic organism.

Photosynthetic organisms are found in nature almost exclusively in areas where light is available. Thus caves, deep ocean waters, turbid rivers, the interiors of animals, and other permanently dark habitats are usually devoid of photosynthetic organisms, although these places are usually well colonized by nonphotosynthesizers. Photosynthetic microorganisms are most abundant in aquatic environments where light penetration is good, but they are also found on the surfaces of soils and rocks. In a lake or ocean the depth to which light will penetrate in sufficient amounts to allow for the growth of photosynthetic organisms will vary with the turbidity of the water; in very clear waters light penetration sufficient for photosynthesis to exceed respiration may be 100 m. The light intensity at which photosynthesis occurs at a rate just sufficient to balance respiration (both light and dark) is called the **compensation point.**

Light quality   The spectral quality of light changes with depth, water absorbing red light more effectively than blue light. This selective absorption of red light puts a special limitation on photosynthetic bacteria, whose chlorophyll absorption maxima are in the far-red region. In fact, at wavelengths beyond 900 nm, the absorption of light by water is so strong that none of it ever reaches the organisms.

All species of photosynthetic organisms absorb light selectively at certain wavelengths, and the light absorbed by one organism is of course not available to another. Figure 6.15 shows the absorption spectra of several different microorganisms measured directly in vivo, without extraction of the pigments, and we can see that some organisms absorb light well in regions where other organisms absorb light poorly. These differences are of ecological significance. For instance, since algae are aerobes, they grow in surface waters and absorb much of the light in the blue and red regions. Purple and green bacteria are anaerobes, and these organisms must grow in deep-lying waters and on the surface of muds, where anaerobic conditions prevail. If they are to survive, they must use the light the algae allow to pass—light mainly in the far-red and infrared regions of the spectrum. Thus the different absorption spectra of various groups of photosynthetic microorganisms have an ecological basis.

Global significance   The global significance of autotrophs is emphasized by the fact that they are the organisms which convert CO_2 into organic compounds that then become available as food for heterotrophic organisms, such as animals, fungi, and most bacteria (see Chapter 13). Of equally great importance is the fact that O_2-producing photosynthetic organisms are responsible for virtually all of the O_2 present in the atmosphere. On the primitive earth the atmosphere was virtually devoid of O_2. Purple and green bacteria probably existed, but since they perform only cyclic photophosphorylation,

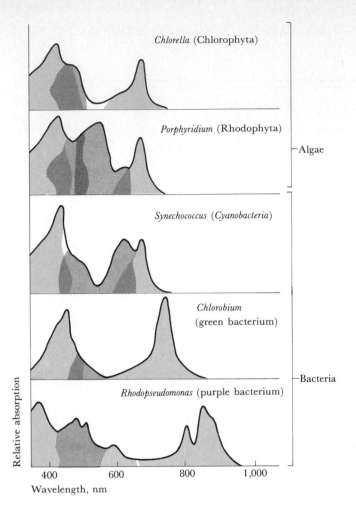

Chlorella (Chlorophyta)

Porphyridium (Rhodophyta)

}Algae

Synechococcus (Cyanobacteria)

Chlorobium (green bacterium)

}Bacteria

Rhodopseudomonas (purple bacterium)

Relative absorption

400   600   800   1,000

Wavelength, nm

FIGURE 6.15 Absorption spectra of whole cells, resulting from the presence of chlorophylls and accessory pigments. (The chlorophylls are in color, accessory pigments, light and dark gray.) Note that the photosynthetic purple and green bacteria absorb at longer wavelengths than the algae and cyanobacteria and are thus able to utilize light that the other organisms have not absorbed. (From Stanier, R. Y., and G. Cohen-Bazire. 1957. In Microbial ecology [Seventh Symposium, Society for General Microbiology]. Cambridge University Press, Cambridge, England.)

they do not produce O_2. Upon the evolution of noncyclic photophosphorylation, probably in the cyanobacteria, O_2 production by photosynthesis could occur. Fossil evidence suggests that cyanobacteria arose between 1 and 2 billion years ago (see Section 19.33), and geological evidence suggests that the atmosphere became oxidizing around 1 billion years ago. Since all higher organisms require O_2, the evolutionary significance of photosynthetic O_2 production is clear.

Phototaxis Many photosynthetic microorganisms move toward light, a process called **phototaxis** (Section 2.8). The advantage of phototaxis to a photosynthetic organism is that it allows the organism to orient itself for most efficient photosynthesis; indeed, if a light spectrum is spread across a microscope slide on which there are motile photosynthetic bacteria, the bacteria accumulate at those wavelengths at which their pigments absorb (Figure 6.16).

We discussed in Section 2.8 the details of chemotaxis, the movement of motile bacteria toward and away from chemicals. The mech-

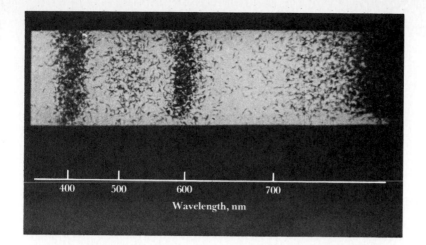

FIGURE 6.16 Phototactic accumulation of the photosynthetic bacterium *Thiospirillum jenense* at light wavelengths at which its pigments absorb. A light spectrum was displayed on a microscope slide containing a dense suspension of the bacteria; after a period of time, the bacteria had accumulated selectively and the photomicrograph was taken. The wavelengths at which accumulations occur are those at which bacteriochlorophyll a absorbs. (Courtesy of Norbert Pfennig.)

anism of phototaxis is somewhat different. The bacteria exhibit a kind of response called a **shock movement** (from the German word "Schreckbewegung"). When a bacterial cell moves out of the light into the dark it stops suddenly (as if shocked) and reverses direction. This shock movement is the only response the bacterial cell makes to changing light conditions, but it is sufficient to keep the bacterial cell within light. Imagine a photosynthetic bacterium moving in a spot of light. As it courses around through the liquid, it may by accident wander out of the light field. Immediately, the cell stops, reverses its direction, and thus reenters the light. Suppose now that we examine a cell which is completely removed from the light spot. It moves back and forth through the dark liquid, showing no special attraction to the light. (After all, it has no way of knowing the light spot is even there.) But if, in the course of its movement, it should wander into the light spot, it would remain in the light, since whenever it leaves the light spot, it reverses direction. Thus, even though bacteria do not swim *toward* the light, most of the cells end up in the light spot, and only a few remain in the dark. This accumulation of photosynthetic cells in the light is therefore a biased random walk, as is the chemotaxis discussed in Section 2.8. In addition to being able to distinguish light from dark, these bacteria can distinguish bright light from dimmer light. It is not the absolute amount of light to which they respond, but to differences in light intensity; phototactic bacteria can distinguish between two light sources that differ in intensity by only 5 percent. The accumulation of bacteria at different wavelengths of light, as illustrated in Figure 6.16, results because the pigments of the bacteria absorb light at certain wavelengths better than at others, so that the bacteria "perceive" these wavelengths as brighter. It is of historical interest that T. W. Engelmann, the first person to suggest that certain bacteria were photosynthetic, was led to this idea by observing that the purple bacteria he was studying accumulated preferentially at specific regions of the light spectrum (see "A Bit of History," page 191).

6.10

Lithotrophy: Energy from the Oxidation of Inorganic Electron Donors

Bacteria that can obtain their energy from the oxidation of inorganic compounds are called **lithotrophs.*** Most of them can also obtain all of their carbon from CO_2 and they are therefore also autotrophs. Most of these organisms reduce CO_2 to organic carbon by processes very similar to those used by photosynthetic organisms (see Section 6.7), but the mechanism by which they produce ATP is quite different from that of the photosynthesizers. In lithotrophs ATP is generated by electron-transport phosphorylation during oxidation of the inorganic electron donor.

A Bit of History

The discovery of autotrophy in lithotrophic bacteria was of major significance in the advance of our understanding of cell physiology, since it showed that CO_2 could be converted into organic carbon without the intervention of chlorophyll. Previously, it had been thought that only green plants converted CO_2 into organic form. The idea of lithotrophic autotrophy was first developed by the great Russian microbiologist Sergei Winogradsky, who did most of this work while he was still rather a junior scientist. Winogradsky began by studying iron bacteria, but soon turned to studies on sulfur bacteria because certain of the colorless sulfur bacteria (*Beggiatoa*, *Thiothrix*) are very large and hence are easy to study even in the absence of pure cultures. Springs with waters rich in H_2S are fairly common around the world, and Winogradsky studied several such springs in the Bernese Oberland district of Switzerland. In the outflow channels of sulfur springs, vast populations of *Beggiatoa* and *Thiothrix* develop, and suitable material for microscopic and physiological studies can be obtained by merely lifting up the white filamentous masses. Pure cultures are not needed for many studies. As Winogradsky noted, "This state of demipurity would be poor in an ordinary culture but is sufficient for a culture under the microscope, since it is possible to observe the development from day to day, almost from hour to hour, and see easily the presence of contaminants." Winogradsky first showed that the colorless sulfur bacteria were present only in water containing H_2S. As the water flowed away from the source, the H_2S gradually dissipated. Finally the water contained no H_2S, and sulfur bacteria were no longer present. This suggested that their development was dependent on the presence of H_2S. Winogradsky then showed that by starving *Beggiatoa* filaments for a while, they lost their sulfur granules; he found, however, that the granules were rapidly restored if a small amount of H_2S was added. He thus concluded that H_2S was being oxidized to elemental sulfur. But what happened to the sulfur granules when the filaments were starved of H_2S? Winogradsky showed by some clever microchemical tests that when the sulfur granules disappeared, sulfate appeared in the medium. Thus he formulated the idea that *Beggiatoa* (and by inference other colorless sulfur bacteria) oxidize H_2S to elemental sulfur and subsequently to sulfate. Because they seemed to require H_2S for development in the springs, he postulated that this oxidation was the principal source of energy for these organisms.

Studies on *Beggiatoa* provided the first evidence that an organism could oxidize an inorganic substance as a possible energy source, and this was the origin of the concept of lithotrophy. However, the sulfur bacteria proved difficult to work with, primarily because there are a number of spontaneous

*Lithotroph means "rock-eating."

chemical changes in sulfur compounds, which can also occur and confuse the study. Winogradsky thus turned to a study of the nitrifying bacteria, and it was with this group that he clearly was able to show that autotrophic fixation of CO_2 was coupled to the oxidation of an inorganic compound in the complete absence of light and chlorophyll. The process of nitrification had been known before Winogradsky's work from studies on the fate of sewage when added to soil. Two Frenchmen, T. Schloesing and A. Muntz, had shown that the process was due to living organisms. Winogradsky proceeded to isolate some of the bacteria, using completely mineral media in which CO_2 was the sole carbon source and ammonia the sole electron donor. Because ammonia is chemically stable, it was easy to show that the oxidation of ammonia to nitrite, and subsequently to nitrate, is a strictly bacterial process. As no organic materials were present in the medium, it was also possible to show that organic matter (the bacterial cell material) was formed only from CO_2. If the ammonia was left out of the medium, no growth occurred. Winogradsky concluded, "This [process] is contradictory to that fundamental doctrine of physiology which states that a complete synthesis of organic matter cannot take place in nature except through chlorophyll-containing plants by the action of light." His basic conclusion has been confirmed by a large number of subsequent studies. At least in one way, however, autotrophy in lithotrophs and phototrophs is similar, in that in both processes, the pathway of CO_2 fixation follows the same biochemical steps (the Calvin cycle) involving the enzyme ribulose diphosphate carboxylase.

Hydrogen   Some bacteria can use hydrogen gas an an electron donor in respiration: $2H_2 + O_2 \rightarrow 2H_2O$. This reaction is catalyzed by the enzyme *hydrogenase,* and the hydrogens are transferred to NAD. The NADH thus formed donates electrons to an electron-transport particle, and ATP is synthesized by electron-transport phosphorylation. All such hydrogen bacteria are also able to use a certain number of organic compounds as electron donors in respiration.

Sulfur   Many reduced sulfur compounds can be used as electron donors by a variety of *colorless sulfur bacteria* (called "colorless" to distinguish them from the chlorophyll-containing sulfur bacteria). The most common sulfur compounds used as electron donors are H_2S, elemental sulfur (S^0), and thiosulfate ($S_2O_3^{2-}$), which are electron donors for electron-transport phosphorylation. The first oxidation product of H_2S is sulfur, a highly insoluble material. Some bacteria deposit sulfur inside the cell (Figure 6.17a), whereas others deposit it extracellularly. The sulfur deposited as a result of the initial oxidation is an energy reserve, and when the supply of H_2S has been depleted, additional energy can be obtained from oxidation of sulfur to SO_4^{2-}. Production of SO_4^{2-} can lead to highly acidic conditions, with pH values sometimes below 2.0.

One species of sulfur-oxidizing bacteria, *Thiobacillus thiooxidans,* is unusually resistant to these acidic conditions and is found in nature in highly acidic environments, the acidity of which results from the metabolism of these bacteria. The ability of sulfur-oxidizing bacteria to produce sulfuric acid is sometimes employed in agricultural practice in alkaline soils; powdered sulfur is plowed into the soil, and sulfur bacteria naturally present in the soil oxidize it and reduce the soil pH to values more suitable for agricultural crops. The biochemistry

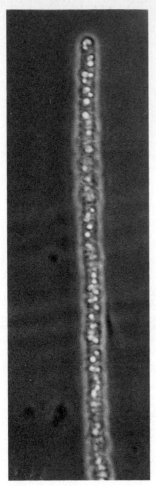

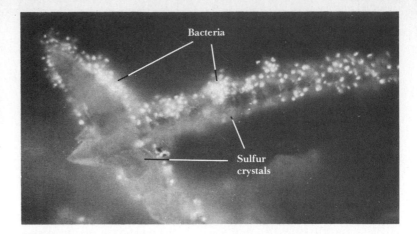

(b)

FIGURE 6.17 Sulfur bacteria. (a) Deposition of internal sulfur granules by *Beggiatoa*. Magnification, 2300×. (b) Attachment of the sulfur-oxidizing bacterium *Sulfolobus acidocaldarius* to a crystal of elemental sulfur. Visualized by fluorescence microscopy after staining the cells with the dye acridine orange. The sulfur crystal does not fluoresce. Magnification, 630×.

(a)

of sulfur oxidation is outlined in Figure 19.52. Although most thiobacilli are aerobes, at least two types can use nitrate as their terminal electron acceptor and are thus denitrifiers (see Section 4.11 on anaerobic respiration).

When elemental sulfur is used as an electron donor by an organism, the organism must grow attached to the sulfur particle because of the extreme insolubility of the elemental sulfur (Figure 6.17b). By adhering to the particle, the organism can efficiently obtain the few atoms of sulfur which go spontaneously into solution, and as these are oxidized, more dissolve; gradually the particle is consumed.

Iron The aerobic oxidation of iron from the ferrous to the ferric state is an energy-yielding reaction for a few bacteria. Only a small amount of energy is available from this oxidation, and for this reason the bacteria must oxidize large amounts of iron in order to grow. Ferric iron forms a very insoluble hydroxide [$Fe(OH)_3$] in water, and this insoluble iron precipitates as oxidation proceeds (see Color Plates 3b and c).

Ferrous iron also oxidizes spontaneously in air at neutral pH values. Since the spontaneous reaction is rapid, ferrous iron does not accumulate in amounts sufficient for the growth of iron-oxidizing bacteria, and autotrophic iron bacteria living at neutral pH have not been demonstrated. Under acidic conditions, ferrous iron is not oxidized spontaneously, and one species of acid-tolerant iron bacteria (*Thiobacillus ferrooxidans*) can carry out iron oxidation and grow. The water that drains coal-mining dumps is often acidic due to bacterial sulfuric acid production and contains ferrous iron, and it is in these acid mine waters that *T. ferrooxidans* proliferates. This species is also able to oxidize elemental sulfur and is closely related to *T. thiooxidans*. Another iron-oxidizing bacterium is *Sulfolobus acidocaldarius,* which lives in hot, acid springs. This organism is extremely versatile, as it not only oxidizes ferrous iron, but H_2S, elemental sulfur, and a variety of organic compounds. It can grow heterotrophically by using organic compounds as electron donors.

Nitrogen   The most common inorganic nitrogen compounds used as electron donors are *ammonia* (NH_3) and *nitrite* (NO_2^-), which are aerobically oxidized by certain bacteria called the **nitrifying bacteria.** One group of organisms (*Nitrosomonas* is one genus) oxidizes ammonia to nitrite, and another group (*Nitrobacter*) oxidizes nitrite to nitrate; the complete oxidation of ammonia to nitrate is carried out by members of these two groups of organisms acting in sequence. It is puzzling that the complete oxidation of ammonia to nitrate requires cooperation of two separate organisms, whereas the oxidation of H_2S to sulfate can be done by a single organism. No explanation for this difference exists. An additional point is that nitrite is itself rather toxic, and under mildly acidic conditions it is a powerful mutagenic agent (Table 11.3). Organisms producing or oxidizing nitrite must deal in some way with this toxicity.

   Nitrifying bacteria are widespread in the soil, and their significance in soil fertility and in the nitrogen cycle is discussed in Section 13.8; their taxonomy, physiology, and structure are discussed in Section 19.18.

Energy yields in lithotrophs   Table 6.3 summarizes the energy yields from aerobic oxidations of various inorganic electron donors. The amount of ATP formed is directly proportional to the amount of energy released in a given oxidation. Since reactions that provide little energy yield little ATP, the organisms using these reactions are in turn able to synthesize only small amounts of cell substance per mole of substrate oxidized. However, since these organisms utilize electron donors that are not available to other organisms, they are able to survive in nature.

   When growing autotrophically, lithotrophs must also produce reducing power in addition to ATP if they are to grow with CO_2 as a sole carbon source. Although NAD can be reduced by H_2, all of the

TABLE 6.3   **Energy Yields from the Oxidation of Various Inorganic Electron Donors**

Reaction	$\Delta G^{0'}$ (kcal/mole)
$HS^- + H^+ + \frac{1}{2}O_2 \longrightarrow S^0 + H_2O$	-48.5
$S^0 + 1\frac{1}{2}O_2 + H_2O \longrightarrow SO_4^{2-} + 2H^+$	-140.6
$NH_4^+ + 1\frac{1}{2}O_2 \longrightarrow NO_2^- + 2H^+ + H_2O$	-62.1
$NO_2^- + \frac{1}{2}O_2 \longrightarrow NO_3^-$	-18.1
$H_2 + \frac{1}{2}O_2 \longrightarrow H_2O$	-54.6
$Fe^{2+} + H^+ + \frac{1}{4}O_2 \longrightarrow Fe^{3+} + \frac{1}{2}H_2O$	-17.0

Data for Fe^{2+} taken from Lees, H., S. C. Kwok, and I. Suzuki. 1969. Can. J. Microbiol. 15:43–46. Other data calculated from Appendix 1, Table A1.1. Values for Fe^{2+} are for pH 2–3, others, for pH 7. The energy yield depends on both pH and concentrations of reactants, and the values given should be viewed only as average values, suitable primarily for comparative purposes.

TABLE 6.4 Reduction Potentials of Several Redox Pairs Involved in Lithotrophic Growth

Redox Pair	Reduction Potential (V)
$2H^+/H_2$	-0.41
NAD/NADH	-0.32
S^0/HS^-	-0.27
SO_4^{2-}/HS^-	-0.22
NO_3^-/NO_2^-	$+0.43$
Fe^{3+}/Fe^{2+}	$+0.77$

other inorganic electron donors have reduction potentials higher than that of NAD (Table 6.4). If the reduction potential of an electron donor is higher than that of NADH, there is no way in which its oxidation can be *directly* coupled to the reduction of NAD to NADH. There is good evidence that the NAD is reduced through the process of reversed electron transport, already discussed for the purple and green bacteria in Section 6.4. In the lithotrophic bacteria, reversed electron transport would lead to the reduction of NAD (or NADP) using energy from ATP that had been generated during the oxidation of the inorganic electron donor.

Note that the transformations of inorganic elements discussed here are essentially the reverse of those discussed for anaerobic respiration in Section 4.11. In anaerobic respiration oxidized forms of the elements are reduced using energy usually derived from organic compounds, whereas in lithotrophic oxidation the reduced forms of the elements are oxidized, usually using O_2 as an electron acceptor. By the combined action of lithotrophs and organisms carrying out anaerobic respiration, *cycles* of transformation can occur. Thus

$$H_2S \longrightarrow S^0 \longrightarrow SO_4^{2-} \quad \text{(lithotrophic)}$$
$$SO_4^{2-} \longrightarrow H_2S \quad \text{(anaerobic respiration)}$$

These cycles of key inorganic elements will be discussed in detail in Chapter 13.

With most lithotrophic autotrophs it has been possible to demonstrate the presence of ribulose diphosphate carboxylase, as well as the other enzymes of the "photosynthetic" carbon cycle, that is, the Calvin cycle (Section 6.7).

6.11

Comparison of Autotrophs and Heterotrophs

The energy metabolism and the terminology used in describing the energy relationships of autotrophs and heterotrophs are indicated in Table 6.5, which emphasizes the similarities and differences among various groups. The two factors to be considered in comparing these organisms are the nature of their energy metabolism and the nature of their carbon source.

Autotrophs obtain all of their carbon from CO_2, but their energy comes either from light or respiration of inorganic chemicals. Organisms that use inorganic chemicals are called **lithotrophs,** those that photosynthesize are called **phototrophs.** A number of phototrophs can use organic compounds as carbon sources and obtain some or all of their energy from light; such organisms are called **photoheterotrophs. Conventional heterotrophs** belong to the large group of organisms that use organic compounds for carbon as well as electron donors.

Early investigators concluded that some autotrophic bacteria were not able to grow on organic media; these bacteria were termed *obligate autotrophs.* It is now clear that this terminology is inaccurate since all of these supposedly deficient organisms have been shown to assimilate organic matter from their surroundings. These studies are made difficult by the very specific nutritional requirements of the bacteria and their relatively slow growth rates, with doubling times often being on the order of days or weeks. Although all of the bacteria

TABLE 6.5 Comparison and Classification of Autotrophs and Heterotrophs Based on Sources of Energy and Carbon

	Terminology
Method of energy generation:	
1. Photosynthesis	Phototroph
2. Oxidation of inorganic compounds	Lithotroph
Carbon source:	
1. Carbon dioxide	Autotroph
2. Organic compounds	Heterotroph

Another class of organisms, conventional heterotrophs, use organic compounds as both carbon source and electron donor.

Examples of organisms using different modes of energy generation and sources of carbon:

Photoautotrophs:
Green plants
Most algae
Cyanobacteria
Some purple and green bacteria

Photoheterotrophs:
Few algae
Most purple and green bacteria
Some cyanobacteria

Lithotrophic autotrophs:
Hydrogen bacteria
Colorless sulfur bacteria
Nitrifying bacteria
Iron bacteria

Lithotrophic heterotrophs:
Beggiatoa; few other colorless sulfur bacteria

Conventional heterotrophs:
Animals
Most bacteria
Fungi
Protozoa

in question can use organic compounds for carbon sources, several of them are restricted to using inorganic compounds as electron donors in their energy metabolism. These bacteria may correctly be referred to as obligate lithotrophs. Examples include some (but not all) members of the genus *Thiobacillus* and the nitrifying bacteria.

The ability of lithotrophs to assimilate organic compounds as carbon sources while using inorganic oxidations for energy makes possible the growth of these organisms as lithotrophic heterotrophs. This type of growth has also been referred to as **mixotrophic.** Some lithotrophs grow best under mixotrophic conditions, because the electron donor need be used only for generation of ATP, reducing power either not being needed or coming from the organic compound. At least one sulfur bacterium, *Beggiatoa,* seems to require mixotrophic conditions to use reduced sulfur compounds as electron donors. *Beggiatoa* is unable to grow on a completely inorganic medium with reduced sulfur compounds and CO_2, but will use reduced sulfur compounds as electron donors when acetate is present. It will also grow on acetate as sole source of carbon and energy in the absence of reduced sulfur compounds. A number of flagellated algae (eucaryotes) also require organic compounds such as acetate for carbon, but will use light as an energy source. One bacterium, *Thiobacillus perometabolis,* is apparently deficient in CO_2 fixation ability while still being capable of lithotrophic oxidation of reduced sulfur compounds. This organism is not able to make ribulose diphosphate carboxylase and may be described as an obligate mixotroph.

The ability of an alga to use organic compounds as carbon sources can be easily tested experimentally by culturing it in the presence of DCMU [3-(3,4-dichlorophenyl)-1,1-dimethylurea]. As we have noted, this inhibitor blocks photosystem II but not photosystem I, so that cyclic photophosphorylation and ATP generation can occur in its presence, but neither noncyclic photophosphorylation nor generation of reducing power can occur. In the presence of DCMU, photoautotrophic growth will not occur, but photoheterotrophic growth may take place, if an assimilable carbon source is present.

One group of photosynthetic bacteria, the Rhodospirillaceae (nonsulfur purple bacteria), shows either photoheterotrophic or conventional heterotrophic metabolism, depending on environmental conditions. They are photosynthetic under anaerobic conditions and heterotrophic under aerobic conditions. In these organisms, O_2 inhibits formation of the photosynthetic pigments, so that under aerobic conditions the organisms become colorless and carry on a normal heterotrophic respiratory metabolism. Under anaerobic conditions, O_2-mediated respiration cannot occur, of course, and chlorophyll is made and cyclic photophosphorylation can take place, although organic compounds are still used as sources of carbon. Under anaerobic conditions, some of these organisms can also grow slowly in the dark and then energy generation is by fermentation. The nonsulfur purple bacteria are thus one of the most versatile groups of organisms from the viewpoint of energy-generating mechanisms.

6.12

Summary

We have seen in this chapter that autotrophs are organisms that can obtain all of their carbon for growth from inorganic sources, carrying out a process called CO_2 fixation. Two broad groups of autotrophs are recognized, depending on their sources of energy: phototrophic autotrophs, which use light energy; and lithotrophic autotrophs, which oxidize certain inorganic chemicals for energy. Both groups of autotrophs use the same biochemical pathway for the fixation of CO_2 into carbon, a process called the photosynthetic carbon cycle or **Calvin cycle.** A key enzyme of the Calvin cycle is **ribulose diphosphate carboxylase,** which catalyzes the initial step of CO_2 fixation. The utilization of light energy to form organic matter from CO_2 is called **photosynthesis.**

The reduction of CO_2 to the oxidation level of organic matter requires energy in the form of ATP and reducing power in the form of NADPH. In phototrophs, the utilization of light energy to form ATP and NADPH requires the participation of a key photosynthetic pigment, chlorophyll. Chlorophyll is present in the cell as part of a photosynthetic membrane system, where it is positioned so that it can absorb light and cause the flow of electrons and the formation of a proton-motive force. Green plants, algae, and cyanobacteria carry out **oxygenic photosynthesis,** in which water molecules are split and molecular oxygen (O_2) is formed. The electrons given up when the water molecules are split are transferred through the photosynthetic electron-transport system and are used to reduce NADP to NADPH. The proton-motive force, which develops during photosynthetic electron transport, is used to synthesize ATP by a process similar to that already described (Section 4.9) for the synthesis of ATP during respiratory electron transport.

The purple and green bacteria (and some algae and cyanobacteria under certain special conditions) carry out a type of photosynthesis called **anoxygenic,** because water is not split and molecular oxygen is not formed. In the purple and green bacteria, the electrons for the generation of reducing power are not obtained from water. Reducing power comes instead either from a constituent of the environment (for example, H_2) or by a process called **reversed electron transport,** in which the membrane-bound ATPase is used to create a proton-motive force through the consumption of energy from ATP.

The lithotrophs obtain their energy from the oxidation of inorganic compounds. Some inorganic compounds known to serve as electron donors for lithotrophs include H_2, H_2S, NH_4^+, nitrite (NO_2^-), and ferrous iron (Fe^{2+}). Oxidation of these compounds occurs through electron-transport systems, with the generation of a proton-motive force which can be used to synthesize ATP. Reducing power for CO_2 fixation can either come from the compound itself (in the case of H_2), or can be formed from ATP by reversed electron transport.

The autotrophs as a whole are of great importance for the economy of nature, because they synthesize organic matter from CO_2, thus providing the food base for all heterotrophs. Additionally, oxygenic photoautotrophs play another major role in nature, because they are responsible for the formation of the O_2 of the atmosphere, thus creating the aerobic environment needed by animals. Without the activity of the oxygenic photoautotrophs, the earth would soon become anaerobic and unfavorable for the existence of higher forms of life.

Supplementary Readings

Bogorad, L. 1975. Phycobiliproteins and complementary chromatic adaptation. Annu. Rev. Plant Physiol. 26:369–401. Reviews function and biosynthesis of phycobilins.

Clayton, R. K., and **W. R. Sistrom,** eds. 1978. The photosynthetic bacteria. Plenum Press, New York. Comprehensive survey of this group of bacteria.

Drews, G., and **J. Oelze.** 1981. Organization and differentiation of membranes of phototrophic bacteria. Adv. Microb. Physiol. 22:1–92. Photosynthetic electron transport chains are compartmentalized differently.

Lehninger, A. L. 1975. Biochemistry, 2nd ed. Worth, New York. Chapters 22 and 23 provide an excellent treatment of photosynthesis and photosynthetic CO_2 fixation.

Miller, K. R. 1979. The photosynthetic membrane. Sci. Am. 241:102–113. Excellent photographs and drawings showing the asymmetry of chloroplast membranes and the relation to proton gradients.

Padan, E. 1979. Facultative photosynthesis in cyanobacteria. Annu. Rev. Plant Physiol. 30:27–40. A summary of alternative electron donors in the cyanobacteria.

Parson, W. W. 1982. Photosynthetic bacterial reaction centers: interactions among the bacteriochlorophylls and bacteriopheophytins. Annu. Rev. Biophys. Bioeng. 11:57–80. Physical chemistry of energy transfer, emphasizing importance of spatial relations of chlorophyll molecules.

Stewart, W. D. P. 1980. Some aspects of structure and function in N_2-fixing cyanobacteria. Annu. Rev. Microbiol. 34:497–536.

Stoeckenius, W., and **R. A. Bogomolni.** 1982. Bacteriorhodopsin and related pigments of halobacteria. Annu. Rev. Biochem. 51:587–616. Function of bacteriorhodopsin in chemiosmotic terms.

Whittenbury, R., and **D. P. Kelly.** 1977. Autotrophy: a conceptual phoenix. Pages 121–150 *in* B. A. Haddock and W. A. Hamilton, eds. Microbial energetics. 27th Symposium, Society for General Microbiology. Cambridge University Press, New York. One definition of an autotroph and a brief overview of autotrophs as a group.

Zelitch, I. 1975. Pathways of carbon fixation in green plants. Annu. Rev. Biochem. 44:123–145. A good review of photorespiration.

Growth and Its Control

7

Growth is defined as an orderly increase in all the cellular constituents and structures of an organism. In most microorganisms this increase continues until the cell divides into two new cells, a process termed **binary fission** (binary refers to the fact of *two* new cells, and fission means separation). Microbial growth therefore usually results in an increase in cell number, the new cells formed eventually attaining the same size as the original cell.

It is important to distinguish between the growth of individual cells and the growth of populations of cells. The growth of a cell is an increase in its size and weight and is usually a prelude to cell division. Population growth, on the other hand, is an increase in the *number* of cells as a consequence of cell growth and division. It is extremely difficult to study the growth of individual cells quantitatively, especially in microorganisms, because analytical techniques are not sensitive enough for use with these small structures. Therefore, almost all growth studies of microorganisms are population studies involving very large numbers of organisms. Although such studies have proven very valuable, it must always be remembered that the answers obtained reflect the average condition of all cells in the population; individual variations between cells cannot be detected in this way.

7.1

Population Growth

Under ideal conditions unicellular organisms are growing and dividing continuously. After each cell division, the two new daughter cells immediately begin a new cycle of growth leading to division. In our idealized example all cells are growing at the same rate, probably their fastest possible, or *maximum*, rate. Therefore, there is a constant time interval required for one new cell to complete a growth and division cycle, giving rise to two daughter cells. The amount of time

Elapsed Time (hours)	Cell Number	Log$_2$ (cell no.)	Log$_{10}$ (cell no.)
0	1	0	0
0.5	2	1	0.301
1	4	2	0.602
1.5	8	3	0.903
2	16	4	1.204
2.5	32	5	1.505
3	64	6	1.806
3.5	128	7	2.107
4	256	8	2.408
4.5	512	9	2.709
5	1,024	10	3.010
⋮	⋮		⋮
10	1,048,576	20	6.021

required for the complete division cycle is called the **doubling time** or the **generation time,** since, for unicellular organisms, each doubling represents a new generation. A hypothetical growth experiment beginning with a single cell having a doubling time of 30 minutes is presented in Table 7.1. This pattern of population increase, where the number of cells increases by a constant factor during each unit time period, is referred to as **exponential growth** (see Equation 1 in Appendix 2). When the cell number is graphed on arithmetic coordinates as a function of elapsed time, one obtains a curve with a constantly increasing slope (dashed line, Figure 7.1). Examination of curved lines is not convenient and population results are usually *transformed* by taking the logarithm of each data point. The last two columns in Table 7.1 are the logarithm base 2 (log$_2$) and logarithm base 10 (log$_{10}$), respectively, of the cell number values at each time point. The logarithm values show a simple linear increase; the difference between the two bases is a constant factor. The log$_{10}$ values are presented in Figure 7.1 (solid line, right-hand axis) in a semilogarithmic graph which results in a straight line. This semilogarithmic graph is convenient and simple to use for calculating generation time from a set of results. The doubling time may be read directly from the graph (Figure 7.2).* For some purposes it is useful to express growth mathematically. This is discussed in Appendix 2.

*Populations of microorganisms are expressed in terms of cell concentration, that is, number of cells per milliliter of suspension. These cell concentrations (or cell densities) are often of such magnitude that it is difficult to express them easily. To simplify the handling of these large numbers, the microbiologist uses exponents of 10. Thus 10^6 is equivalent to 1,000,000, 10^8 to 100,000,000, and so on. Cell numbers can then be expressed as 2×10^8/ml (200,000,000/ml), 5×10^7/ml (50,000,000/ml), 6.5×10^3/ml (6500/ml), and so on, where the exponent of 10 represents the number of places to the right of the decimal point. When adding two figures expressed in exponents of 10, we must first write out the numbers or reduce them to the same power of 10:

$$2.5 \times 10^7/ml = 25,000,000/ml = 25 \times 10^6/ml$$
$$3.0 \times 10^6/ml = 3,000,000/ml = 3.0 \times 10^6/ml$$
$$2.8 \times 10^7/ml = 28,000,000/ml = 28 \times 10^6/ml$$

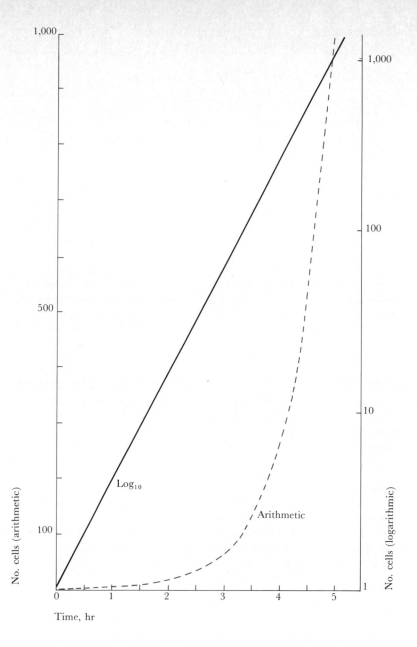

FIGURE 7.1 Growth rates as plotted on arithmetic and logarithmic scales.

One of the characteristics of exponential growth is that the rate of increase in cell number is slow initially but increases at an ever faster rate. This results, in the later stages, in an explosive increase in cell numbers. A practical implication is that during the early stages of exponential growth, a nonsterile product such as milk may be

When making graphs of growth curves, it is convenient to use five-cycle semilogarithmic paper, since this makes it possible to represent on one sheet of paper an increase in cell density of 10^5 times, a 100,000-fold increase.

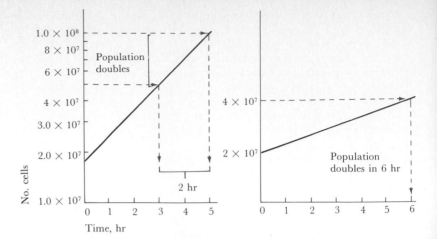

FIGURE 7.2 Method of estimating the generation times of exponentially growing populations.

allowed to stand under conditions conducive to microbial growth for several hours without detriment. However, letting it stand for the same length of time during the later stages of exponential growth would be disastrous.

7.2

Measuring Microbial Number and Weight

Let us consider the procedures used to obtain the data in Table 7.1. That is, how does the experimenter actually determine what the population size is at any particular time? There are a number of different methods which are used, each with its own advantages and disadvantages.

Direct count   The size of a microbial population may be measured by counting the number of individual cells with a microscope, a procedure called the **direct microscopic count.** The small size of most microbes and their large population densities make it necessary to use special chambers such as the Petroff–Hausser or hemocytometer chambers to count the number of cells in a sample (Figure 7.3). There are three aspects of direct microscopic counting which are limitations in its use. First, it is tedious and therefore impractical for large numbers of samples. Second, it is not very sensitive, because at least 10^6 bacterial cells per milliliter must be present before a single cell will be seen in the microscope field. Third, living cells cannot be distinguished from dead cells.

Viable count   In many cases we are interested in enumerating only living cells. A living cell is defined as one that is able to divide, and viable cell counting is usually done by determining the number of cells in a population capable of dividing and forming colonies. The **plate count** method is most widely used, and two versions of this technique are the spread plate and the pour plate. In the **spread plate,** a small sample (usually 0.1 ml) is aseptically spread over the surface of an agar plate containing an appropriate medium, and the plate incubated until the colonies are visible to the eye (Figure 7.4a). It is inferred that each colony arose from a single cell, and by counting the number of colonies one can calculate the number of viable cells in the sample. With the **pour plate** procedure, the sample is mixed with melted agar and the mixture poured into a sterile plate. The organisms are thus fixed within the agar gel, and form colonies (Figure 7.4b). With pour plates, larger sample volumes can be used, 1.0 ml or more, and heavy slurries or suspensions, such as from homogenized food or soil can be counted. A major disadvantage of the pour plate is that organisms affected by the brief heating in the melted agar cannot be counted.

The plate count is very sensitive because in principle any viable cell, when placed on an appropriate medium, will give rise to a colony. In addition to sensitivity, plate counting also allows for the positive identification of the organism counted, as the colony that is formed can serve as the inoculum for a pure culture, which can be identified taxonomically. Furthermore, since different organisms often produce colonies of different shape, size, texture, and color, several kinds of organisms can be counted in a mixture. A major assumption in plate count procedures is that each colony has arisen from a single cell. If, by chance, two cells are placed very close to each other during inoculation, they may give rise to a single fused colony which cannot be distinguished from a colony which came from one cell. This systematic underestimation of population size will of course be more serious with larger population sizes. Practical experience has determined that there should be fewer than 300 colonies per plate to minimize this problem. Thus, for dense suspensions, viable counting requires considerable dilution of the sample before plating, a condition that increases the possibility of errors due to extra handling.

An additional practical limitation is that there is no single agar medium which will allow the growth of all bacteria since different bacteria have different nutritional requirements. The most extreme example of this restriction occurs with bacteria which will not grow on any agar medium, requiring liquid culture for growth. It is possible to perform a viable count determination using only liquid cultures, with the most-probable-number (MPN) technique. In this technique, the sample to be counted is diluted until the cell density is less than one per milliliter. At these low levels, it is no longer appropriate to speak of cell concentrations, but rather of the probability of finding a cell. For example, if there are five cells in 10 ml of dilution buffer, the concentration could be expressed as 5×10^{-1} cells/ml. However, viable cells are of course individuals. Therefore, we say that in this example there is a probability of 0.5 for each 1-ml portion in the tube that it will contain a viable cell. If the tube were emptied by the withdrawal of 10 samples of 1 ml each, five of them would be expected to contain one cell each, and five of them would be expected to contain no cells. This concept is the basis of the MPN procedure. A number of replicate small portions (usually 1 ml) are taken from a diluted sample and each portion is inoculated into a separate tube of fresh growth medium, which is then incubated to allow growth. Those tubes which received a cell will show growth and those which did not receive a cell will not show growth. The proportion of inoculated tubes showing growth is a measure of the probability just dis-

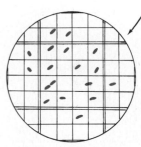

Ridges which support cover slip

Sample added here; care must be taken not to allow overflow; space between cover slip and slide is 0.02 mm ($\frac{1}{50}$ mm.). Whole grid has 25 large squares, a total area of 1 sq mm and a total volume of 0.02 mm³.

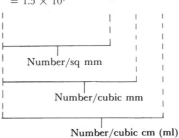

Microscopic observation; all cells are counted in large square: 12 cells (in practice, several squares are counted and the numbers averaged)

To calculate number per milliliter of sample:
12 cells $\times$ 25 squares $\times$ 50 $\times$ 10³
= 1.5 $\times$ 10⁷

Number/sq mm

Number/cubic mm

Number/cubic cm (ml)

FIGURE 7.3 Direct microscopic counting procedure using the Petroff–Hausser counting chamber.

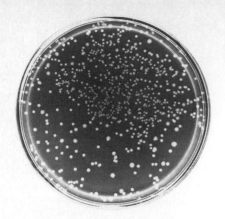

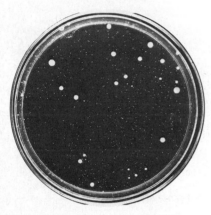

FIGURE 7.4 (a) Colony size and distribution on a spread plate used for viable counting. (b) Colony size and distribution on a pour plate. Both plates *Escherichia coli.*

cussed. This probability can be converted back to cell concentration in the undiluted sample by consideration of the number of dilutions and the use of the MPN table (Appendix 2, Table A2.3). This table is a statistical derivation of the most expected or *most probable* number of cells which would result in the observed growth pattern. In order for the MPN procedure to be useful, the sample must be diluted to the correct extent. If the sample has not been diluted sufficiently, all aliquots will contain viable cells and all inoculated tubes will show growth upon incubation. If the sample has been diluted too far, none of the inoculated tubes will show growth. Since it is not possible in advance to know exactly the proper dilution, it is usual practice to inoculate growth tubes with samples from each of several different dilutions. Statistical tables have been constructed for use with inoculations of three, five, or ten replicate growth tubes. Greater accuracy is obtained by using larger numbers of tubes, but at the expense of more time and materials in conducting the measurement.

Measurement of microbial mass For many studies, especially those dealing with the biochemistry of growth processes, we are interested in determining the mass of the population rather than the number of cells present. Mass can be measured directly by determining either the dry or the wet weight of a sample taken from the population. Dry weight is usually about 20 to 25 percent of the wet weight.

The average weight of a single cell can be calculated by measuring the weight of a large number of cells and dividing by the total number of cells present, as measured by direct microscopic count. Procaryotic cells range in dry weight from less than 10^{-15} g to greater than 10^{-11} g, and the cells of eucaryotic microorganisms range from about 10^{-11} g to about 10^{-7} g.

In dense suspensions, turbidity measurements can be used to quantify growth, and a variety of commercial instruments are available for this purpose. Turbidity is more closely related to cell mass than cell number. Although turbidity measurements can be made quickly and accurately without disturbing a culture, they are not very sensitive, for a suspension of bacteria does not show visible turbidity until the density is about 10^7 bacteria per milliliter.

No methods of mass determination can distinguish between the contribution of live and dead cells to the total mass.

7.3

The Growth Cycle of Populations

The hypothetical culture examined in Table 7.1 and Figure 7.1 reflects only part of the growth cycle of a microbial population. A typical growth curve for a population of cells is illustrated in Figure 7.5. This growth curve can be divided into several distinct phases, called the **lag phase, exponential phase, stationary phase,** and **death phase.**

Lag phase When a microbial population is inoculated into a fresh medium, growth usually does not begin immediately, but only after a period of time called the *lag phase,* which may be brief or extended,

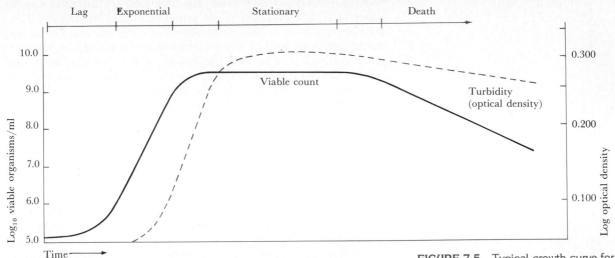

Growth phases

FIGURE 7.5 Typical growth curve for a bacterial population.

depending on conditions. If an exponentially growing culture is inoculated into the same medium under the same conditions of growth, a lag is not seen, and exponential growth continues at the same rate. However, if the inoculum is taken from an old (stationary phase) culture and inoculated into the same medium, a lag usually occurs even if all of the cells in the inoculum are alive. This is because the cells are usually depleted of various essential coenzymes or other cell constituents, and time is required for resynthesis. A lag also ensues when the inoculum consists of cells that have been damaged (but not killed) by treatment with heat, radiation, or toxic chemicals, due to the time required for the cells to repair the damage.

A lag is also observed when a population is transferred from a rich culture medium to a poorer one. This happens because for growth to occur in a particular culture medium the cells must have a complete complement of enzymes for the synthesis of the essential metabolites not present in that medium. On transfer to a new medium, time is required for synthesis of the new enzymes.

Exponential phase   The *exponential phase* has already been discussed and was shown to be a consequence of the fact that each cell in a population divides into two cells. Under optimal conditions, growth rates of different organisms vary widely, apparently owing to inherent genetic limitations. In general, procaryotes grow faster than eucaryotes, although there are some notably slow-growing procaryotes. For a given organism, growth rate varies with environmental conditions and with the composition of the culture medium. In a rich medium with many preformed cell constituents, growth rate is usually faster than in a poor medium, in which the cell must synthesize these things for itself.

The relationship between growth rate and macromolecular synthesis is discussed in Chapter 9.

Stationary phase In a closed culture vessel, a population cannot grow indefinitely at an exponential rate. Growth limitation occurs either because the supply of an essential nutrient is exhausted or because some toxic metabolic product has accumulated. The period during which growth of a population ceases is called the *stationary phase* (Figure 7.5). Thus it is the activity of the organism itself that induces the changes in the medium leading to the stationary phase. A mathematical expression of the growth-rate equation, which includes the self-crowding effects that lead to the stationary phase, is given in Appendix 2.

Death phase If incubation continues after a population reaches the stationary phase, the cells may remain alive and continue to metabolize, but often they die. If the latter occurs, the population is said to be in the *death phase.* During this death phase the direct microscopic count may remain constant but the viable count slowly decreases. In some cases death is accompanied by cell lysis, leading to a decrease in the direct microscopic count concurrent with the drop in viable count. A discussion of the effects of germicidal chemicals on viability will be found later in this chapter.

It must be emphasized that these phases are reflections of the events in a population, not of individual cells. The terms lag phase, exponential phase, stationary phase, or death phase do not apply to individual cells, but only to populations of cells.

7.4

Effect of Nutrient Concentration on Growth

Nutrient concentration can affect either growth rate or total growth, as shown in Figure 7.6. At very low concentrations of the nutrient, the rate of growth is reduced, whereas at moderate and higher levels of nutrient, growth rates are identical but the total growth (sometimes called the **total** or **maximum crop**) is limited. As the nutrient concentration is increased still further, a concentration will be reached that is no longer limiting, and further increases then no longer lead to increases in total crop. At this point, some nutrient or environmental factor other than the one being varied is limiting to further increases in growth.

The effect of nutrient concentration on *total growth* is easy to understand, since much of the nutrient is converted into cell material, and if the amount of nutrient is limited, the amount of cell material will also be limited. The reason for the effect of very low nutrient concentrations on *growth rate* is less certain. One idea is that at these low nutrient concentrations the nutrient cannot be transported into the cell at sufficiently rapid rates to satisfy all of the metabolic demands for the nutrient. Conceivably not all carrier sites are fully occupied (see Section 5.18 for a discussion of the role of carriers in transport). The shape of the curve relating growth rate to nutrient concentration (Figure 7.6*b*) resembles a saturation process of the kind also seen for active transport (Figure 5.38). This ability to alter the growth rate by nutrient concentration is used in the operation of the type of continuous-culture apparatus called a **chemostat,** described in Section 7.5.

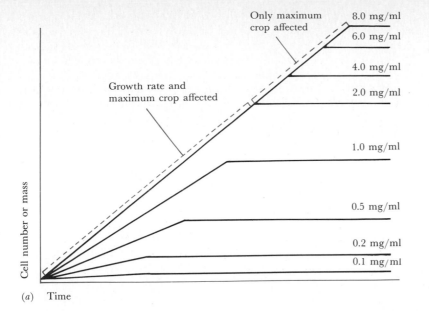

Only maximum
crop affected

Growth rate and
maximum crop affected

8.0 mg/ml
6.0 mg/ml
4.0 mg/ml
2.0 mg/ml
1.0 mg/ml
0.5 mg/ml
0.2 mg/ml
0.1 mg/ml

Cell number or mass

(a)   Time

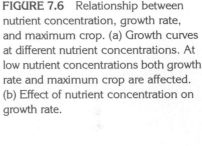

(b)   Nutrient concentration, mg/ml

FIGURE 7.6   Relationship between nutrient concentration, growth rate, and maximum crop. (a) Growth curves at different nutrient concentrations. At low nutrient concentrations both growth rate and maximum crop are affected. (b) Effect of nutrient concentration on growth rate.

Microbiological assay of growth factors   Microorganisms are very useful tools for the measurement, or assay, of small amounts of vitamins, amino acids, and other growth factors and hence have been widely employed for examination of foods, pharmaceuticals, and other preparations. A microbiological assay has the virtues of specificity, sensitivity, and simplicity. It also makes possible the assay of compounds whose exact chemical structure is unknown. To perform the assay it is necessary to have a culture of a bacterium which has a requirement for the substance to be assayed. Several species of *Lactobacillus* (see Section 19.23) have multiple requirements and are good candidates for this assay procedure. A series of culture media is prepared in which all materials needed by the microorganism are supplied, except for the substance to be assayed. This substance is then added at different concentrations to each preparation of growth medium. If the concentration range is low enough, there will be a linear relation between the maximum crop of cells and added substance. This relation is presented in Figure 7.7, which is derived from the data of Figure 7.6a. After a relation such as this has been determined, it is possible to measure the concentration of the substance in a sample by adding a small amount of the sample to growth medium, inoculating it with the same strain of test bacteria, and measuring the amount of bacterial growth which results. Reference to a graph such as that in Figure 7.7 will give a direct determination of the concentration of the substance in the sample.

   Microbiological assays were formerly used for amino acids, but modern chemical methods are as sensitive and are more rapid. Unlike microbiological assays, however, chemical assays do not distinguish between the D and L optical isomers of an amino acid. Some vitamins (for example, thiamin, riboflavin) can also be assayed more conve-

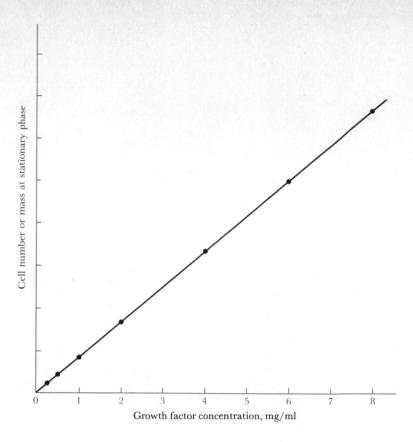

Cell number or mass at stationary phase

Growth factor concentration, mg/ml

FIGURE 7.7 Microbiological assay of growth factors. Maximum crop of bacterial culture is directly proportional to growth factor concentration over the range tested.

niently by chemical means, but vitamins such as B_{12} and biotin, which are active at extremely low concentrations, must still be assayed microbiologically.

7.5

Continuous Culture

Our discussion of population growth has been confined to closed or *batch cultures,* growth occurring in a fixed volume of a culture medium that is eventually altered by the actions of the growing organisms so that it is no longer suitable for growth. In the early stages of exponential growth in batch cultures, conditions may remain relatively constant but in later stages drastic changes usually occur. For many studies, it is desirable to keep cultures in constant environments for long periods, and this is done by employing *continuous cultures.* A continuous culture is essentially a flow system of constant volume to which medium is added continuously and from which a device allows continuous removal of any overflow. Once such a system is in equilibrium, cell number and nutrient status remain constant, and the system is said to be in **steady state.**

The most common type of continuous-culture device used is called a **chemostat** (Figure 7.8), which permits control of both the population density and the growth rate of the culture. Two elements are used in the control of a chemostat—the flow rate and the concentration of a limiting nutrient, such as a carbon, energy or nitrogen source, or growth factor. As described in Figure 7.6b, at low concentrations of an essential nutrient, growth rate is proportional to the nutrient concentration. However, at these low nutrient concentrations, the nutrient is quickly used up by growth. In batch cultures growth ceases at the time the nutrient is used up, but in the chemostat continuous addition of fresh medium containing the limiting nutrient permits continued growth. Since at the low nutrient levels used, the limiting nutrient is quickly assimilated, its concentration in the chemostat vessel itself is always virtually zero.

   Effects of varying dilution rate and concentration of the inflowing growth-limiting substrate are given in Figure 7.9. As seen, there are rather wide limits over which dilution rate will control growth rate, although at both very low and very high dilution rates the steady state breaks down. At high dilution rates, the organism cannot grow fast enough to keep up with its dilution, and the culture is

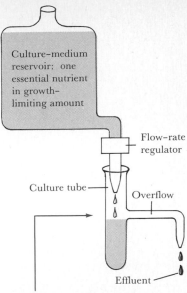

Chemostat. In such a device the population density is controlled by the concentration of the limiting nutrient in the reservoir and the growth rate is controlled by the flow rate, which can be set by the experimenter.

FIGURE 7.8 Schematic for a continuous-culture device (chemostat).

FIGURE 7.9 Steady-state relationships in the chemostat. The dilution rate is determined from the flow rate and the volume of the culture vessel. Thus, with a vessel of 1000 ml and a flow rate through the vessel of 500 ml/hr, the dilution rate would be 0.5 hr⁻¹. Note that at high dilution rates, growth cannot balance dilution, the population washes out, and the substrate concentration rises to a maximum (since there are no bacteria to use the inflowing substrate). However, throughout most of the range of dilution rates shown, the population density remains constant and the substrate concentration remains at a very low value. Note that although the population density remains constant, the growth rate (doubling time) varies over a wide range. Thus, the experimenter can obtain populations with widely varying growth rates, without affecting population density. See Appendix 2 for a mathematical treatment. (Based on Herbert, D., R. Elsworth, and R. C. Telling. 1956. The continuous culture of bacteria: a theoretical and experimental study. J. Gen. Microbiol. 14:601. By permission of Cambridge University Press.)

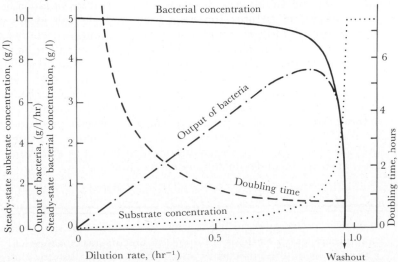

washed out of the chemostat. At the other extreme, at very low dilution rates a large fraction of the cells may die from starvation, since the limiting nutrient is not being added fast enough to permit maintenance of cell metabolism. There is probably a minimum amount of energy necessary to maintain cell structure and integrity, called **maintenance energy,** and this nutrient used for maintenance energy is not available for biosynthesis and cell growth. Thus at very low dilution rates steady-state conditions will probably not be maintained, and the population will slowly wash out (not shown in Figure 7.9).

The cell density in the chemostat is controlled by the level of the limiting nutrient. If the concentration of this nutrient in the incoming medium is raised, with dilution rate remaining constant, cell density will increase although growth rate will remain the same and the steady-state concentration of the nutrient in the culture vessel will still be virtually zero. Thus, by adjusting dilution rate and nutrient level, the experimenter can obtain at will a variety of population densities at a variety of growth rates. The actual shape of the curve for bacterial concentration given in Figure 7.9 will depend on the organism, the environmental conditions, and the limiting nutrient used. Mathematical relationships for chemostats, expressing the parameters of Figure 7.9, are given in Appendix 2.

Chemostats have been quite useful in genetic, physiological, and ecological studies with microorganisms. For genetic studies, chemostats have been especially useful in studying the mutation process (see Chapter 9), since they permit maintenance of populations of constant size growing at constant rates for long periods of time. In physiological studies, chemostats have been useful in studying the relationships between growth and protein and nucleic acid synthesis, and the control of enzyme synthesis. Another physiological use is that an organism can be cultured under a variety of different environmental conditions but at the same growth rate in each, so that the effect of the environment on the properties of the organism can be studied without concern that the environment may merely be affecting growth rate. In this way, the effects of temperature variation on the properties of an organism have been studied, with identical growth rates at all temperatures. In ecological studies, chemostats have served as useful models for microbial growth in nature under conditions of low nutrient concentration, and have found especial use in studying bacterial growth in lakes, rivers, and oceans, where concentrations of organic substances are often low. Finally, chemostats are useful in the selection, from either natural populations or laboratory cultures, of strains able to carry out specific processes of interest. The selective properties of a chemostat should be clear, since the strain best adapted to the conditions used should be able to outgrow other strains and soon dominate the culture.

7.6

Growth Yields

Growth is an energy-requiring process. In Chapter 5 we saw several examples in which synthesis of important precursors consumes ATP. The assembly of these precursors into macromolecules also consumes

ATP (which will be seen in Chapter 9), as do active transport, motility, and other cellular activities. Growth may be examined in more detail by measurement of cell yield (that is, biomass produced) in comparison to amount of ATP generated during catabolism. The parameter calculated is called Y_{ATP}, for cell yield per ATP generated. Experimental determination of these values is done most easily with fermentative organisms in which the amount of ATP produced per molecule of substrate metabolized can be deduced from knowledge of the catabolic pathway used by the organism under study. For example, the fermentation of glucose to lactic acid via the glycolytic pathway leads to the formation of two ATP from each glucose (Section 4.7). The actual experiment consists of culturing the organism anaerobically in a rich medium which contains all necessary monomers, and with a single fermentable electron donor such as glucose. At the end of the experiment, the amount of glucose consumed is determined and the dry weight of new cell material is measured. It should be noted that this procedure mainly measures the energetic cost of assembling macromolecules, which is only part of the overall growth process. Table 7.2 presents yield data from some fermenting organisms.

Note that the Y_{ATP} values in Table 7.2 are quite similar, between 9 and 10, even though the ATP yield per mole of substrate varies from 1 to 3. It can be calculated (Table A1.3) that 1 mole of ATP has the potential for forming 32 g of polymers in a typical bacterial cell. Therefore, fermenting bacterial cells appear to synthesize only one-third of their potential. This difference probably reflects (1) our lack of knowledge of some energy-requiring steps; *and* (2) wastage of energy by the cell. For example, active transport and motility are energy-requiring activities for which precise energy costs cannot be accurately estimated. One must also consider the concept of **maintenance energy,** an expenditure necessary to repair damage to vital structures such as the cell membrane and cell wall.

Growth yield studies are another area in which the chemostat is very useful. As noted in Section 7.5 and Appendix 2, when the

TABLE 7.2   Molar Growth Yields for Anaerobic Growth of Fermentative Organisms Using Glucose as Electron Donor

Organism	$Y_{substrate}$ (grams dry weight per mole substrate used)	ATP yield (moles ATP per mole of substrate)*	Y_{ATP} (grams dry weight per mole ATP)
Streptococcus faecalis	20	2	10
Streptococcus lactis	19.5	2	9.8
Lactobacillus plantarum	18.8	2	9.4
Saccharomyces cerevisiae	18.8	2	9.4
Zymomonas mobilis	9	1	9
Aerobacter aerogenes	29	3	9.6
Escherichia coli	26	3	8.6

*ATP yield is based on a knowledge of the pathway by which glucose is fermented in the different organisms.

Data from Forrest, W. W., and D. J. Walker. 1971. Adv. Microb. Physiol. 5:250–251.

dilution rate and substrate level in a chemostat are properly controlled, then all the substrate will be consumed and growth yield will be directly proportional to the concentration of substrate in the incoming medium. Calculation of $Y_{substrate}$ is then straightforward.

7.7

Antimicrobial Agents

An antimicrobial agent is a chemical that kills or inhibits the growth of microorganisms. Such a substance may be either a synthetic chemical or a natural product. Natural products that are antimicrobial agents are called **antibiotics**; most antibiotics are produced by microorganisms.

Agents that kill organisms are often called "cidal" agents, with a prefix indicating the kind of organism killed. Thus we have **bactericidal, fungicidal,** and **algicidal** agents. A bactericidal agent, or bactericide, kills bacteria. It may or may not kill other kinds of microorganisms. Cidal agents with a broad spectrum of target organisms are usually called **germicides.** Germicides are sometimes conveniently divided into two groups, **antiseptics** and **disinfectants.** An antiseptic is a germicide that is sufficiently harmless so that it can be applied to the skin or mucous membranes, although it is not necessarily safe enough to be taken internally. A disinfectant is an agent that kills microorganisms (but not necessarily their spores) and is distinguishable from an antiseptic by the fact that it is not safe for application to living tissue, and its use is restricted to inanimate objects such as tables, floors, or dishes.

Agents that do not kill but only inhibit growth are called "static" agents, and we can speak of **bacteriostatic, fungistatic,** and **algistatic agents.** The distinction between a static and a cidal agent is often arbitrary, since an agent that is cidal at high concentrations may only be static at lower concentrations. To be effective, a static agent must be continuously present with the product, for if it is removed or its activity neutralized the organisms present in the product could initiate growth. Static agents are often used as food preservatives, and since they must be present continuously to be effective they remain in the food when eaten and hence must be nontoxic. Many drugs used in treatment of microbial infections are static agents and must be kept present for a period of time long enough for body defenses to destroy the infecting organism. Such drugs obviously must be nontoxic to the body. Although there is a wide variety of chemicals that are static agents, most of them are too toxic for use in foods or as drugs.

Many antimicrobial agents are effective in low concentrations, 1 to 10 parts per million (1 to 10 μg/ml). This is of practical significance since it means that the agent will be active even after it is highly diluted, making it possible to apply the agent in effective concentration to animals by injection or to plants by spraying. Some agents not active at low concentrations can still have practical use, however, especially when applied to inanimate objects where high concentrations can be more effectively employed.

An important feature of many antimicrobial agents is **selective** toxicity, which means that the agent is more active against the microbe than against the animal or plant host. Agents that act selectively on disease-causing organisms without affecting human tissue are of medical value. Some antimicrobial agents act selectively against procaryotic organisms and are relatively harmless to eucaryotes. Since humans are eucaryotes and many disease-causing microorganisms are procaryotes, agents selective against procaryotes potentially have wide medical uses. A wide variety of such selective agents is known. There are also a few agents that act selectively on eucaryotic microorganisms (fungi, protozoa, algae) without affecting the eucaryotic cells of higher animals, including humans, and such agents have important medical applications. A substance that selectively attacks microorganisms without harming human cells is called a **chemotherapeutic agent.** The use of chemotherapeutic agents to control infectious disease is discussed in Chapter 15.

7.8

Effect of Antimicrobial Agents on Growth

Antimicrobial agents affect growth in a variety of ways, and a study of the action of these agents in relation to the growth curve is of considerable aid in understanding their modes of action. The most accurate way of observing the effect of a chemical is to add it at an inhibitory concentration to an exponentially growing culture and continue to sample the culture for growth and viability. For measurement of growth, turbidity is most convenient, although direct microscopic counting may also be useful in some cases. For measurement of viability, some procedure is necessary by which the inhibitory activity of the chemical is nullified, since if growth-inhibitory concentrations are carried over into the medium used for viable counting, even viable cells will not be able to grow and form colonies. In many cases, dilution for viable counting is sufficient to eliminate growth-inhibitory properties of the agent, but in some cases the organisms may have to be removed from the chemical by filtration or centrifugation.

Three distinct kinds of effects are observed when an antibiotic is added to an exponentially growing bacterial culture, bacteriostatic, bactericidal, and bacteriolytic. As described in the preceding section, a **bacteriostatic** effect is observed when growth is inhibited, but no killing occurs (Figure 7.10a). Bacteriostatic agents are frequently inhibitors of protein synthesis and act by binding to ribosomes. The binding, however, is not tight, and when the drug concentration is lowered the agent becomes free from the ribosome and growth is resumed. The mode of action of protein synthesis inhibitors is discussed in Section 9.5. **Bactericidal** agents prevent growth and induce killing, but lysis or cell rupture does not occur (Figure 7.10b). Bactericidal agents generally bind tightly to their cellular targets and are not removed by dilution. **Bacteriolytic** agents induce killing by cell lysis, which is observed as a decrease in cell numbers or in turbidity after the agent is added (Figure 7.10c). Bacteriolytic agents include antibiotics that inhibit cell-wall synthesis, such as penicillin (see Section 2.5), as well as agents that act on the cell membrane.

Quantification of Antimicrobial Action

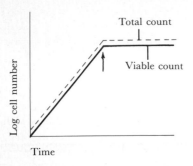

(a) Bacteriostatic

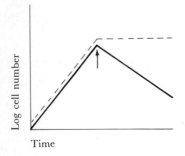

(b) Bactericidal

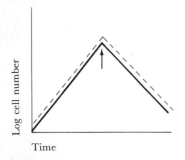

(c) Bacteriolytic

FIGURE 7.10 Three types of action of antimicrobial agents. At the time indicated by the arrow, a growth-inhibitory concentration was added to the exponentially growing culture. Note the different behaviors of the viable and the total counts.

To compare the relative activities of a group of antimicrobial agents, the minimum concentration of each agent that completely inhibits the growth of a test organism in culture tubes under defined growth conditions is determined. This value, the **minimum inhibitory concentration** (MIC), will be lowest for the most potent agent. Another commonly used method of studying the quantitative aspects of antimicrobial action is the **agar diffusion method.** A molten agar culture medium is evenly inoculated with a suspension of the test organism and poured into petri plates. Known amounts of the antimicrobial agent are placed on filter-paper disks on the surface of the plates or in metal wells embedded in the agar (Figure 7.11). During incubation, the agent diffuses into the surrounding zone, a concentration gradient is set up, and at some distance the MIC is reached. As the organism grows it forms a turbid layer, except in the region where the concentration of agent is above the MIC; here a zone of inhibition is seen. The size of the zone of inhibition is determined by the sensitivity of the organism, the nature of the culture medium, incubation conditions, the rate of diffusion of the agent, and the concentration of the agent. Some antibiotics diffuse much more readily in agar than do others, and hence it is not always possible to conclude that the antibiotic that gives the widest zone is the most active one against the test organism. For instance, substances of larger molecular weight diffuse less rapidly than do those of lower molecular weight, and some substances may diffuse poorly because they become bound to the agar; in both cases smaller zones of inhibition are produced.

Quantification of killing To measure the lethal effect of an antimicrobial agent, viable counts must be made. From the data obtained, time-survivor curves can be plotted, which will show the effect of the agent upon the population over a period of time. Two basic types of time-survivor curves are obtained, the sigmoid and the exponential (Figure 7.12). The *sigmoid curve* results if the population contains cells with various degrees of sensitivity to the agent, a small proportion of the cells being unusually susceptible, most of the cells having average sensitivity, and another small proportion of cells being unusually resistant. This variation in sensitivity of cells is called a *normal distribution* by the statistician and is the type of distribution to be expected for the usual population of microorganisms. Interestingly, although commonly seen with animals and plants, sigmoid time-survivor curves are rare for microbial disinfection. In fact, the *exponential curve* is much more common. The explanation of the exponential curve is not straightforward. One attempt to explain it draws an analogy between microbial killing and the way in which chemical reactions obey the Mass Law. In chemical reactions, the reaction velocity (degree of change occurring in unit time) is proportional to the concentration of reacting substances. In the case of disinfection, the reagents are cells and antimicrobial molecules, but the antimicrobial chemical is usually present in so great an excess compared to the weight of the bacteria acted upon that its concentration remains constant throughout the reaction. The rate of killing will therefore depend only on the concentration of surviving bacteria. The type of reaction which depends directly on the concentration of a single reactant is called a *first-order relationship* and is quite common in chemical kinetics. To explain chemical reactions, the chemist hypothesizes that an activated state is necessary if the chemical is to undergo reaction, with molecules constantly fluctuating between activated and

FIGURE 7.11 Antibiotic inhibition zones on an agar plate inoculated with a culture of *Staphylococcus aureus*. Metal cups placed in the agar were filled with solutions of the antibiotic. The zone size is proportional to the concentration of antibiotic. (Courtesy of Merck Sharp and Dohme Research Laboratories.)

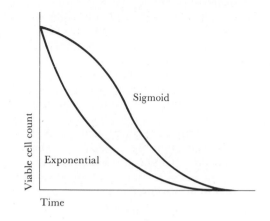

(a)

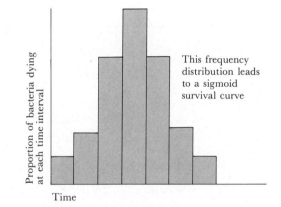

This frequency distribution leads to a sigmoid survival curve

(b)

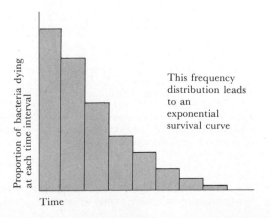

This frequency distribution leads to an exponential survival curve

(c)

FIGURE 7.12 Time-survivor curves for disinfection and their interpretation. (a) Time-survivor curves plotted on arithmetic graph paper, contrasting the sigmoid and the exponential curves. (b) The frequency distribution which leads to a sigmoid curve. This is the "normal" distribution of the statistician. (c) The distribution leading to an exponential survival curve.

normal states. When there is a large number of molecules, the number in the activated state will be large, so that many molecules will undergo reaction, whereas as the number of molecules is reduced, the number of activated molecules will be smaller. In such a situation, the rate of reaction depends on the total number of molecules present, and an exponential reaction rate is obtained. Returning to cell killing, an exponential rate of disinfection might then also be explained by the assumption that cells undergo fluctuations between a sensitive and a resistant state, and that these fluctuations are of a random nature. The sensitive state might be one in which either the cell absorbs the chemical agent more effectively or in which the target becomes accessible to attack.

The correct interpretation of time-survivor curves has some theoretical importance in an analysis of how chemical agents kill cells; from a practical point of view, however, the important observation is that the disinfection process is a gradual one, and that in any situation there will be a minority of cells that will survive very much longer than the majority. To secure complete control of the population, attention must be focused on these unusually resistant cells, and sufficient time must be allowed to permit the disinfection process to go to completion.* A mathematical description of death rates is given in Appendix 2.

7.10

Growth-Factor Analogs

In Section 5.17 we discussed growth factors and defined them as specific chemical substances required in the medium because the organism cannot synthesize them. Substances exist that are related to growth factors and that act to block the utilization of the growth factor. These "growth-factor analogs" usually are structurally similar to the growth factors in question but are sufficiently different so that they cannot duplicate the work of the growth factor in the cell. The first of these to be discovered were the *sulfa drugs,* the first modern chemotherapeutic agents to specifically inhibit the growth of bacteria; they have since proved highly successful in the treatment of certain diseases. The simplest sulfa drug is sulfanilamide (Figure 7.13a). Sulfanilamide acts as an analog of *p*-aminobenzoic acid (Figure 7.13b), which is itself a part of the vitamin folic acid (Figure 7.13c). In organisms that synthesize their own folic acid, sulfanilamide acts by blocking the synthesis of folic acid. Sulfanilamide is active against bacteria but not against higher animals because bacteria synthesize

*This is extremely important in the preparation of virus vaccines, used in the immunization process (Section 16.8). Because the viruses used are potentially harmful to people, it is necessary that all active virus particles be killed in the preparation of the vaccine. To this end, chemical agents are usually used, such as formaldehyde, and it is essential that the inactivation process be allowed to proceed long enough so that all particles are destroyed. Although the exponential killing curve in theory never reaches zero, in practice it should be possible to sterilize any virus preparation. An understanding of the quantitative aspects of chemical disinfection is of great value to workers in the virus vaccine field. (Some early batches of the Salk vaccine, used to immunize for polio, contained live virus particles and caused disease in recipients. The live virus remained probably because insufficient formaldehyde was used to cause complete inactivation during the time period used.)

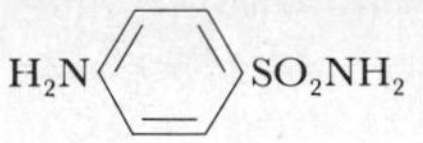

H₂N⟨ ⟩SO₂NH₂   H₂N⟨ ⟩COOH

(a) Sulfanilamide 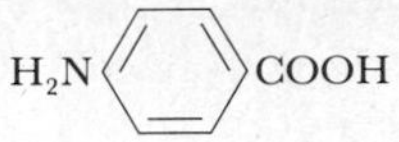   (b) p-Aminobenzoic acid

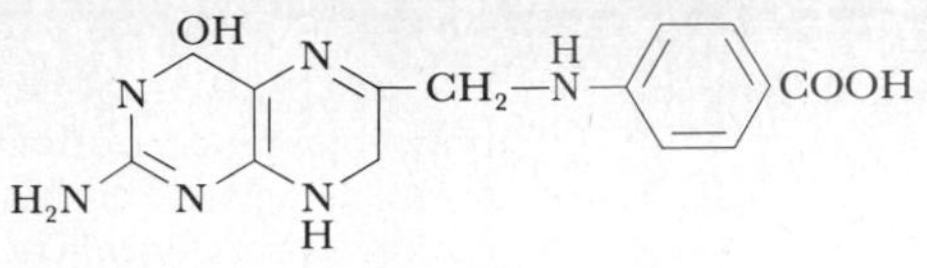

(c) Folic acid

FIGURE 7.13 (a) The simplest sulfa drug, sulfanilamide. It is an analog of *p*-aminobenzoic acid (b), which itself is part of the growth factor folic acid (c).

their own folic acid, whereas higher animals obtain folic acid preformed in their diet.

The concept that a chemical substance can act as a competitive inhibitor of an essential growth factor has had far-reaching effects on chemotherapeutic research, and today analogs are known for various vitamins, amino acids, purines, pyrimidines, and other compounds. A few examples are given in Figure 7.14. In these examples, the analog has been formed by addition of fluorine or bromine. Fluorine is a relatively small atom and does not alter the overall shape of the molecule, but it changes the chemical properties sufficiently so that the compound does not act normally in cell metabolism. Fluorouracil resembles the nucleic-acid base uracil; bromouracil resembles another base, thymine. This is because the bromine atom is considerably larger than the fluorine atom and thus resembles the methyl group of thymine.

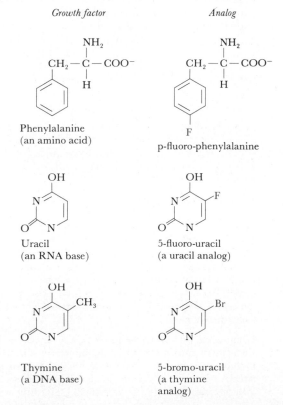

Growth factor   *Analog*

Phenylalanine
(an amino acid)

p-fluoro-phenylalanine

Uracil
(an RNA base)

5-fluoro-uracil
(a uracil analog)

Thymine
(a DNA base)

5-bromo-uracil
(a thymine
analog)

FIGURE 7.14 Examples of growth-factor analogs.

A Bit of History

The development of chemotherapeutic agents probably has had a greater impact on medicine than any other discovery. Although a variety of natural chemical agents had been used earlier, the real advances in work with chemotherapuetic agents began with the German scientist Paul Ehrlich. In the early 1900s, Ehrlich developed the concept of selective toxicity. He began his work by studying the staining of microorganisms by dyes and observed that some dyes stain microorganisms but not animal tissue. He assumed that if a dye did not stain a tissue, the dye molecules were unable to combine with any of the cell constituents. He then reasoned that if such a dye had toxic properties it should not affect the animal cells because it could not combine with them, but it should attack the microbial cells. In an infected animal, chemicals of this sort should behave like "magic bullets," striking the pathogen but missing the host. It was of course not necessary that the chemical be a dye; it only need be selective in its binding properties. Ehrlich proceeded to test large numbers of chemicals for selectivity and discovered the first chemotherapeutic agents, of which Salvarsan, a drug for the cure of syphilis, was the most famous. However, no chemical agents were discovered that were able to affect the vast majority of infectious agents until the 1930s, when Gerhard Domagk discovered the sulfa drugs. The discovery of the sulfas came about as a direct offshoot of the approach used by Ehrlich: the large-scale screening of chemicals for activity in infectious diseases in experimental animals. Domagk, at the Bayer Chemical Company in Germany, tested a large variety of synthetic organic chemicals, mainly dyes, for their ability to cure streptococcal infections in mice. The first active compound was Prontosil, which had the intriguing property that it was active in the mouse but had no activity against streptococci in the test tube. It was then discovered that in the animal body, Prontosil broke down to sulfanilamide, which was the actual active agent. With this discovery, it was possible to embark on a program of synthesis based on the sulfanilamide structure, which yielded a large number of active drugs with various medical uses. D. D. Woods in England then showed that p-aminobenzoic acid specifically counteracted the inhibitory action of sulfanilamide, and he also showed that the streptococci required p-aminobenzoic acid for growth. This led to the concept of the growth-factor analog, which enabled chemists to pursue the synthesis of a wide variety of chemotherapeutic agents.

However, despite the successes of the sulfa drugs, most infectious diseases still were not under chemical control. It took the discovery of penicillin by Alexander Fleming, a Scottish physician engaged in research at St. Mary's Hospital in London, to point the direction. Fleming's first paper on penicillin, published in 1929, begins as follows:

While working with staphylococcus variants a number of culture plates were set aside on the laboratory bench and examined from time to time. In the examinations these plates were necessarily exposed to the air and they became contaminated with various micro-organisms. It was noticed that around a large colony of contaminating mould the staphylococcus colonies became transparent and were obviously undergoing lysis. Subcultures of this mould were made and experiments conducted with a view of ascertaining something of the properties of the bacteriolytic substance which had evidently been formed in the mould culture and which had diffused into the surrounding medium.

Fleming characterized the product, and since it was produced by a fungus of the genus *Penicillium,* gave it the name *penicillin.* His work, however, did not include a process for large-scale production, or show that penicillin was effective in the treatment of infectious disease. This was later done by a group of British scientists at Oxford University, headed by Howard Florey, who began their work in 1939, motivated in part by the impending World War II and the knowledge that infectious disease was at that time the leading cause of death among soldiers on the battlefield. Florey and his colleagues developed methods for the analysis and testing of penicillin and for its production in large quantities. They then proceeded to test penicillin against bacterial infections in humans. Penicillin was dramatically effective in controlling staphylococcal and pneumococcal infections and was also more effective for streptococcal infections than the sulfa drugs. With the effectiveness of penicillin demonstrated and the war in Europe becoming more intense, Florey brought cultures of the penicillin-producing fungus to the United States in 1941. He persuaded the U.S. government to create a large-scale research program and a joint effort of the pharmaceutical industry, the U.S. Department of Agriculture at its laboratory in Peoria, Illinois, and several universities was mounted. By the end of World War II, penicillin was available in large amounts, for civilian as well as military use. As soon as the war was over, pharmaceutical companies entered into commercial production of penicillin on a competitive basis and began to look for other antibiotics. Success was quick and dramatic, and the impact on medicine has been close to phenomenal. Infant and child mortality have been greatly reduced, and many diseases that at one time had high fatality rates are now no more than medical curiosities.

7.11

Antibiotics

An antibiotic is a chemical substance produced by one microorganism that is able to kill or inhibit the growth of other microorganisms. Thousands of antibiotics have been discovered, but only a relative few have turned out to be of great practical value in medicine. However, the most important of these antibiotics have found widespread and revolutionary use in the treatment of many infectious diseases.

The production of antibiotics by microorganisms is a common occurrence, and a variety of microorganisms produce antibiotics acting against other microorganisms. Antibiotic-producing microorganisms are especially common in the soil. Three groups of microorganism are responsible for the production of most of the antibiotics used in medicine: (1) fungi, especially those of the genus *Penicillium,* which produce antibiotics such as penicillin and griseofulvin; (2) bacteria of the genus *Bacillus,* which produce antibiotics such as bacitracin and polymyxin; and (3) actinomycetes of the genus *Streptomyces* (Section 19.28), which produce antibiotics such as streptomycin, chloramphenicol, tetracycline, and erythromycin. It is among members of the genus *Streptomyces,* a genus of organisms widespread in soil, that most of the antibiotics have been discovered.

Antibiotics are a diverse group of compounds, but they can be grouped in families with similar chemical structures. The antibiotics of one group usually have similar types of activity and find similar uses in practice. For instance, there is a large group of penicillins, most of which have similar activity, although certain penicillins are more active against Gram-negative bacteria and others are more active against Gram-positive bacteria.

The sensitivity of microorganisms to antibiotics varies. Gram-positive bacteria are usually more sensitive to antibiotics than are Gram-negative bacteria, although conversely some antibiotics act

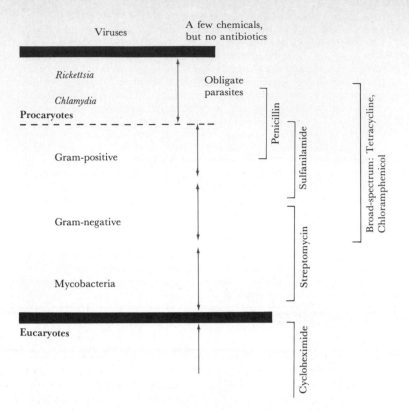

Viruses · A few chemicals, but no antibiotics

Rickettsia

Chlamydia

Procaryotes

Obligate parasites

Gram-positive

Gram-negative

Mycobacteria

Eucaryotes

Penicillin

Sulfanilamide

Streptomycin

Broad-spectrum: Tetracycline, Chloramphenicol

Cycloheximide

FIGURE 7.15 Range of activity of several antibiotics against various microbial groups, illustrating the idea of broad-spectrum antibiotics.

only on Gram-negative bacteria. An antibiotic that acts on both Gram-positive and Gram-negative bacteria is called a **broad-spectrum antibiotic.** In general, a broad-spectrum antibiotic will find wider medical usage than a narrow-spectrum one, although the latter may be quite valuable for the control of certain kinds of diseases (Figure 7.15). We discuss the molecular mechanisms of antibiotic action in Chapter 9 and the development of antibiotic resistance in Chapters 9, 11, and 12.

The search for new antibiotics Seeking new antibiotics has occupied the time of a large number of microbiologists in pharmaceutical companies for the past generation. The most widely used procedure is what is called the *screening approach.* A large number of possible antibiotic-producing organisms are isolated from natural environments, and each is tested against a variety of organisms to see if antagonisms exist. The battery of test organisms usually is large and includes Gram-positive, Gram-negative, and acid-fast bacteria, yeasts and other fungi, and perhaps protozoa and algae. Antitumor agents are sought using tumor tissue cultures as test organisms, and virus-infected animal-cell cultures are used to screen for antiviral agents.

A convenient procedure in examining for antibacterial agents is to streak the possible antibiotic producer along one chord of an agar plate and incubate the plate for a few days to allow the organism to grow and produce antibiotic. Then a series of test bacteria are streaked at right angles to the chord, and the plate is reincubated and examined for zones of inhibition (Figure 7.16). Since different antibiotics attack different groups of organisms, it is desirable that any antibiotic screening program use a wide range of test organisms to make certain that few antibiotic-producing organisms are missed. It should be obvious that appropriate medium and culture conditions permitting the production of the antibiotic are necessary since environment can greatly affect antibiotic production.

Once antibiotic activity is detected in a new isolate, the microbiologist will determine if the agent is new or old. With so many antibiotics already known, the chances are good that the antibiotic is already known. Often simple chemical methods will permit the characterization and identification of the antibiotic. If the agent appears new, larger amounts are produced and the antibiotic purified and sometimes crystallized. Finally, the antibiotic would be tested for therapeutic activity, first in infected animals, later perhaps in humans (see Section 15.12).

7.12

Germicides, Disinfectants, and Antiseptics

Earlier, we defined germicides as chemical agents that kill microorganisms (Section 7.7). We have differentiated antiseptics, which can be used on the skin, from disinfectants, which are used only on inanimate objects. The quantitative aspects of the killing of microorganisms with chemical agents have been discussed in Section 7.8. Germicides have wide use in situations where it is impractical to use heat for sterilization. Hospitals find it frequently necessary to sterilize heat-sensitive materials, such as surgical instruments, thermometers, lensed instruments, polyethylene tubing and catheters, and inhalation and anaesthesia equipment. In the food industry, floors, walls, and surfaces of equipment must often by treated with germicides to reduce the load of microorganisms. Drinking water is commonly disinfected to reduce or eliminate any potentially harmful organisms, and treated wastewater is generally disinfected before it is discharged into the environment.

Although the testing of germicides in laboratory situations is relatively straightforward, in practical cases, the determination of efficacy is often very difficult. This is because many germicides are neutralized by organic materials, so that germicidal concentrations are not maintained for sufficient time. Further, bacteria and other microorganisms are often encased in particles, and the penetration of a chemical agent to the viable cells may be slow or absent. Also, bacterial spores are much more resistant to many germicides than are vegetative cells. Thus germicide effectiveness must ultimately be determined under the intended conditions of use. It should be emphasized that germicidal treatments do not necessarily sterilize. **Sterility** is defined as the complete absence of living organisms, and sterilization with chemicals often requires long contact periods under special conditions. In most cases, the use of germicides ensures only that the microbial load is reduced significantly, although perhaps with the hope that pathogenic organisms are completely eliminated. However, bacterial endospores as well as vegetative cells such as those

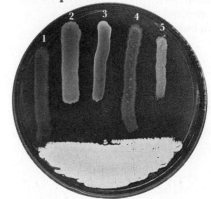

FIGURE 7.16 Method of testing a microbial isolate for antibiotic production. The streptomycete was streaked across one chord of the plate. After several days, to permit growth and antibiotic production, the test bacteria were streaked at right angles. Several of the bacteria are sensitive to the antibiotic produced by this streptomycete. Test organisms: (1) *Escherichia coli;* (2) *Bacillus subtilis;* (3) *Staphylococcus epidermidis;* (4) *Klebsiella pneumoniae;* and (5) *Mycobacterium smegmatis.*

TABLE 7.3 Classes of Antiseptics and Disinfectants and Their Uses

Heavy metals:	
Mercury: $HgCl_2$	Combines with SH groups in proteins; very toxic to humans; activity neutralized by SH compounds and organic matter; used as disinfectant
Silver: $AgNO_3$	Protein precipitant; activity neutralized by organic matter; used in eyes of newborns to prevent gonorrhea
Copper: $CuSO_4$	Algicide in swimming pools and water supplies; fungicide for plant diseases
Halogens:	
Iodine: Tincture of iodine (0.2% I_2 in 70% ethanol), iodophors (I_2 complexes with detergents)	Iodinates proteins containing tyrosine residues; antiseptic on skin; disinfectant of medical instruments; small-scale water purification; relatively nontoxic on skin but toxic internally
Chlorine: Cl_2 gas, $Ca(OCl)_2$, $NaOCl$; active ingredient is HOCl, formed at neutral or acidic conditions	Oxidizing agent; activity neutralized by organic matter, NH_3; used in water purification, general disinfectant in food and dairy industries
Alcohols:	
Ethanol, isopropanol; most active with some water present (50–70% alcohol in water)	Lipid solvents and protein denaturants; antiseptic on skin; disinfectant for hospital items
Aldehydes:	
Formaldehyde, HCHO; available as 37% solution (formalin); used as 2% aqueous solution	Alkylating agent; combines with NH_2, COOH, and SH groups in nucleic acids and proteins; neutralized by organic matter; used for embalming of corpses; small amounts from wood smoke participate in meat-smoking process
Glutaraldehyde, $OHCCH_2CH_2CH_2CHO$; used as 2% aqueous solution	Less toxic than formaldehyde; cold sterilization of hospital goods; neutralized by organic matter
Ethylene oxide: Used as a gas	Alkylating agent; very toxic, used in special gas sterilization units for heat-sensitive goods
Phenols:	
Phenol, carbolic acid, C_6H_5OH; used as aqueous solution (5%); Many phenol derivatives used: with halogen, alkyl, OH groups (cresols, thymol, etc.)	Protein denaturant; disrupts cell membrane at low concentrations; very toxic, activity increased by soaps, not affected by organic matter; disinfectant for large, dirty surfaces (floors, walls, etc.)
Cationic detergents: where one of the R groups is a long-chain alkyl and the other three are methyl, benzyl, etc.	Affect cell membrane through charge interactions with phospholipids; neutralized by phospholipids, metal ions, low pH, soaps, organic materials; skin antiseptics; disinfectants for medical instruments, food and dairy equipment
Ozone, O_3: gas; must be generated at the site of use (high-voltage electric discharge)	Oxidizing agent; very toxic to humans; water purification; action neutralized by organic matter

of *Mycobacterium tuberculosis,* the causal agent of tuberculosis, are very resistant to the action of many germicides, so that even the complete elimination of pathogens by germicidal treatment may not occur.

A summary of the most widely used antiseptics and disinfectants, and their modes of action, is given in Table 7.3.

7.13
Summary

In this chapter we have discussed some of the processes involved in microbial growth and its control. Populations of unicellular microorganisms increase exponentially, and because of this, very large populations of cells can develop quickly. However, growth does not continue indefinitely. Eventually a stationary phase is reached, owing to either exhaustion from the medium of an essential nutrient or accumulation of a toxic waste product. The stationary phase is often followed by a death phase.

A number of processes or agents can be used to control microbial growth. When all organisms are killed, this is generally called **sterilization.** A wide variety of chemicals, called **antimicrobial agents,** can be used to control microbial growth. Cidal agents kill microorganisms, resulting in sterilization, whereas static agents inhibit growth without causing death. An important feature of many antimicrobial agents is their **selective toxicity,** the ability to act more effectively against some organisms than others. Agents that are selective for microorganisms without affecting humans can be used in control of infectious diseases.

In the last part of this chapter we have discussed a variety of chemicals that are used as antiseptics and disinfectants. Most of these agents are much less specific in their action than the antibiotics and other chemotherapeutic agents, but they find wide use in many practical situations, in hospitals, food service, and the water supply industry. In the next chapter we shall discuss in some detail those methods of destroying microorganisms that employ physical methods, such as heat and radiation.

Supplementary Readings

Albert, A. 1965. Selective toxicity, 3rd ed. John Wiley, New York. Although somewhat dated, this book is still a useful discussion of general principles of selective toxicity of chemotherapeutic agents.

American Public Health Association. 1980. Standard methods for the examination of water and wastewater, 15th ed. American Public Health Association, Washington, D.C. This is the standard reference on the determination of numbers of indicator bacteria in water; provides excellent detailed methods for most-probable-number and membrane-filtration methods of viable counting.

Block, S. S., ed. 1977. Disinfection, sterilization, and preservation, 2nd ed. Lea & Febiger, Philadelphia. The standard reference on chemical disinfection, with separate chapters on each of the different groups of agents.

Davis, B. D., R. Dulbecco, H. N. Eisen, and **H. S. Ginsberg.** 1980. Microbiology, 3rd ed. Harper & Row, Hagerstown, Md. Chapter 5 has a good summary of the features of the bacterial growth curve. Chapter 64 has an excellent brief treatment of sterilization and disinfection.

Gerhardt, P. ed. 1981. Manual of methods for general bacteriology. Amer. Soc. Microbiol., Washington. Detailed, up-to-date presentation of bacteriological procedures with extensive references.

Lancini, G., and **F. Parenti.** 1982. Antibiotics, an integrated view. Springer-Verlag, New York. Summary of all aspects of antibiotics, from discovery to production and mechanism of action.

Mendelson, N. H. 1982. Bacterial growth and division: genes, structures, forces, and clocks. Microbiol. Rev. 46:341–375. Analysis of cell growth activities in evolutionary and theoretical terms.

Monod, J. 1949. The growth of bacterial cultures. Annu. Rev. Microbiol. 3:371–394. The classic review of the bacterial growth curve, still current.

Norris, J. R., and **D. W. Ribbons,** eds. 1969. Methods in microbiology, vol. 1. Academic Press, New York. This volume has a number of excellent articles on methods for studying microbial growth. Especially valuable are the following: **C. E. Helmstetter** on methods for studying the microbial division cycle; **M. F. Mallette** on evaluation of growth by physical and chemical means; **J. R. Postgate** on viable counts and viability.

Sargent, M. G. 1978. Surface extension and the cell cycle in prokaryotes. Adv. Microb. Physiol. 18:105–176. Integration of overall physiological activities with details of cell expansion in bacteria.

Tempest, D. W., and **O. M. Neijssel.** 1978. Eco-physiological aspects of microbial growth in aerobic nutrient-limited environments. *In* M. Alexander, ed. Advances in microbial ecology, vol. 2. Plenum Press, New York.

The Microbe in Its Environment

8

The preceding chapter described growth of microorganisms under hypothetical, essentially ideal, conditions. That analysis provides a useful introduction, but experimental work with actual organisms requires consideration of many factors. It is to be expected that the activities of microorganisms are greatly affected by the chemical and physical conditions of their environments. Understanding environmental influences helps us to explain the distribution of microorganisms in nature and makes it possible for us to devise methods for controlling microbial activities and destroying undesirable organisms. Not all organisms respond equally to a given environmental factor. In fact, an environmental condition may be harmful to one organism, and actually beneficial to another. However, organisms can tolerate some adverse conditions under which they cannot grow, and hence we must distinguish between the effects of environmental conditions on the viability of an organism and effects on growth, differentiation, and reproduction.

8.1

Temperature

Temperature is one of the most important environmental factors influencing the growth and survival of organisms. It can affect living organisms in either of two opposing ways. As temperature rises, chemical and enzymatic reactions in the cell proceed at more rapid rates and growth becomes faster. However, above a certain temperature, proteins, nucleic acids, and other cellular components are sensitive to high temperatures and may be irreversibly inactivated. Usually, therefore, as the temperature is increased within a given range, growth and metabolic function increase up to a point where inactivation reactions set in. Above this point cell functions fall sharply to zero. Thus we find that for every organism there is a **minimum temperature** below which growth no longer occurs, an **optimum temperature** at

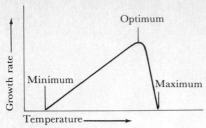

FIGURE 8.1 Effect of temperature on growth rate.

which growth is most rapid, and a **maximum temperature** above which growth is not possible (Figure 8.1). The optimum temperature is always nearer the maximum than the minimum. These three temperatures, often called the **cardinal temperatures,** are generally characteristic for each type of organism, but are not completely fixed, as they can be modified by other factors of the environment. The maximum temperature most likely reflects the inactivation discussed above. However, the factors controlling an organism's minimum temperature are not as clear. As mentioned earlier (Sections 2.5 and 5.6), the cell membrane must be in a fluid state for proper functioning. Perhaps the minimum temperature of an organism results from "freezing" of the cell membrane so that it no longer functions properly in nutrient transport or proton gradient formation. This explanation is supported by experiments in which the minimum temperature for an organism can be altered to some extent by causing it to incorporate different fatty acids into its phospholipids, thereby changing the fluidity. It is also observed that the cardinal temperatures of different microorganisms differ widely; some microbes have temperature optima as low as 5 to 10°C and some as high as 90 to 100°C. The temperature range throughout which growth occurs is even wider than this, from below freezing (-12°C) up to boiling (100°C). No single organism will grow over this whole temperature range, however; the usual range for a given organism is about 30 to 40 degrees, although some have a much broader temperature range than others. Those organisms with narrow temperature ranges are called **stenothermal,** and are generally found in habitats of relatively constant temperature. **Eurythermal** organisms have a wider temperature range and are usually found in environments where the temperature varies considerably. Both stenothermal and eurythermal organisms have defined minimum, optimum, and maximum temperatures.

Although there is a continuum of organisms, from those with very low temperature optima to those with high temperature optima, it is possible to broadly distinguish three groups of organisms: **psychrophiles,** with low-temperature optima, **mesophiles,** with mid-range-temperature optima, and **thermophiles,** with high-temperature optima (Figure 8.2). These temperature distinctions are made for convenience and the precise numbers should not be taken as absolutes. Mesophiles are found in warm-blooded animals and in terrestrial and aquatic environments in temperate and tropical latitudes. Psychrophiles and thermophiles are found in unusually cold or unusually hot environments, respectively. The latter two groups are discussed in some detail below.

Cold environments   Much of the world has fairly low temperatures. The oceans, which make up over half of the earth's surface, have an average temperature of 5°C, and the depths of the open oceans have temperatures of around 1 to 2°C. Vast land areas of the Arctic and Antarctic are permanently frozen, or are unfrozen only for a few weeks in summer. These cold environments are rarely sterile, and some microorganisms can be found alive and growing at any low temperature at which liquid water still exists. Even in many frozen materials there are usually microscopic pockets of liquid water pres-

Plate 1

(a) Snowbank in the Beartooth Mountains, Montana, with red coloration caused by the presence of snow algae. Pink snow such as this is common on summer snowbanks at high altitudes throughout the world.

(a)

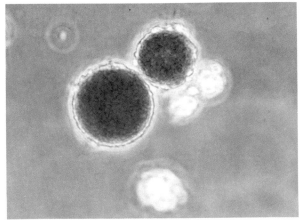

(b) Photomicrograph of red-pigmented cells of the snow alga Chlamydomonas nivalis. (Photomicrograph by J. L. Mosser and T. D. Brock.)

(c) Extensive development of pink halophilic bacteria and red Dunaliella salina, a eucaryotic green alga, in a saltern where solar salt is being prepared by the evaporation of sea water. Sodium chloride is precipitating out. The red coloration generally develops only when the brine reaches saturation for sodium chloride. The phenomenon is found in salterns in all parts of the world.

(c)

Plate 2

(a)

(a) A very large hot spring, Grand Prismatic Spring, in Yellowstone National Park, showing the development of extensive mats of photosynthetic bacteria (primarily cyanobacteria and green bacteria). Where the water overflows the edge of the pool, it cools sufficiently so that the highly pigmented bacteria develop. The orange color predominates because of the rich carotenoid pigments in these organisms.

(b) Close-up of a small bacterial mat, in Octopus Spring, Yellowstone National Park. The temperature where the mat shows the most extensive development is about 55°C.

(b)

Plate 3

(a)

(a) A large leaching dump where low-grade copper ore is extracted using the bacterium Thiobacillus ferrooxidans. Acid water is pumped in at the top of the dump, and the copper-rich water from the bottom is collected so that the copper can be recovered. Bingham Canyon, Utah.

(b)

(b) Acid mine drainage, showing the confluence of a normal river and a creek draining a coal-mining area. The acidic creek is very high in ferrous iron. At low pH values, ferrous iron does not oxidize spontaneously in air, but Thiobacillus ferrooxidans carries out the oxidation. Insoluble ferric hydroxide and complex ferric salts precipitate, forming the precipitate called "yellow boy" by coal miners. (Courtesy of Bill Strode and the Courier-Journal and Louisville Times.)

(c)

(c) Extensive development of insoluble ferric hydroxide in a small pool draining a bog in Iceland. Iron deposits such as this are widespread in cooler parts of the world and are modern counterparts of the extensive bog-iron deposits of earlier geological eras. These ancient deposits are now the sources of much commercially mined iron ore. In the water-saturated bog soil, facultatively anaerobic bacteria reduce ferric iron to the more soluble ferrous state. The ferrous iron leaches into the drainage area surrounding the bog, where oxidation occurs, either spontaneously or through the agency of iron-oxidizing bacteria (Gallionella and Leptothrix), and the insoluble ferric hydroxide deposit is formed.

Plate 4

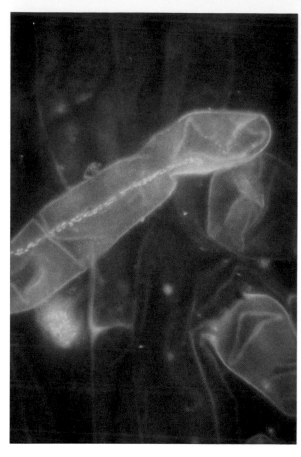

(a)

(a) The infection thread formed in a root hair of white clover by the bacterium Rhizobium trifolii. The preparation was stained with the fluorochrome dye acridine orange and observed with the epifluorescence microscope. Magnification, 2300×. (Courtesy of Ben B. Bohlool, University of Hawaii.)

(b)

(b) The sea anemone (a coelenterate) is green because of the presence of intracellular symbiotic algal cells. The fish is a clownfish. (Photograph by Paul A. Zahl, taken on a coral reef of the Great Barrier Reef of Australia. © 1957 by National Geographic Society.)

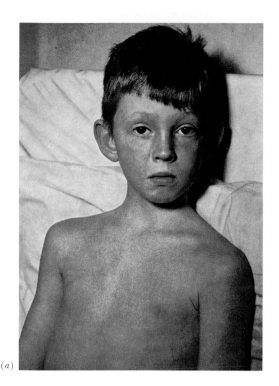

(a)

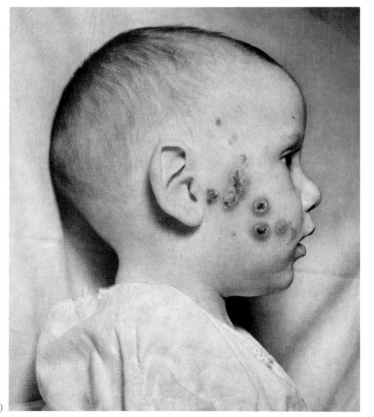

(b)

(a) The typical rash of scarlet fever, due to the action of erythrogenic toxin. Causal organism, Streptococcus pyogenes.

Plate 5

(b) Typical lesions of impetigo, commonly caused by Staphylococcus aureus. (Both plates courtesy of Parke Davis Co. From Franklin H. Top, Communicable and infectious diseases, 6th ed. C. V. Mosby Co., St. Louis, 1968.)

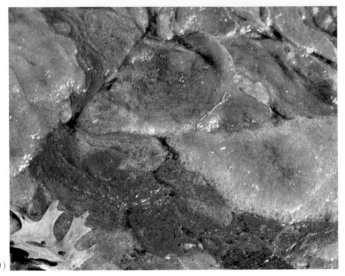

(a)

(a) Massive accumulation of purple sulfur bacteria, Thiopedia sp., in a spring in Madison, Wisconsin. The bacteria grow on the bottom of the spring pool and float to the top when disturbed. The green alga is Spirogyra.

(b)

(b) Water sample from a depth of 7 meters in a stratified Norwegian lake. The color is caused by the presence of large numbers of Pelodictyon luteolum, a green sulfur bacterium. (Courtesy of Norbert Pfennig.)

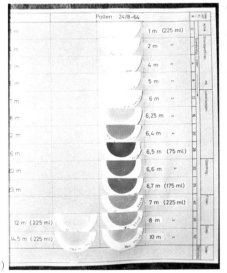

(c)

(c) Series of membrane filters through which were passed water samples taken from varying depths in a stratified Norwegian lake. At 6.4 to 6.6 meters, a heavy cyanobacterial bloom is present, and at 6.7 to 7 meters the green sulfur bacterium Pelodictyon luteolum reaches its highest population density. (Courtesy of John Ormerod and the Norwegian Institute for Water Research.)

Plate 6

(d) Enrichment cultures of purple and green photosynthetic bacteria. Water from the anaerobic zone of a stratified lake was enriched with hydrogen sulfide and incubated at low light intensities. Left, green sulfur bacterium. Center, purple sulfur bacterium. Right, a green sulfur bacterium, which forms extensive amounts of brown carotenoid pigments.

(d)

Plate 7

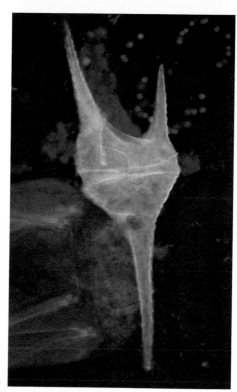

(a)

(a) Many microorganisms contain fluorescent pigments and can be visualized with a fluorescence microscope. To make this photomicrograph, a water sample from a lake was filtered through a membrane filter, and the filter was observed with an incident-light fluorescence microscope. The large organism is Ceratium, a eucaryotic alga. This organism has an outer shell (theca), which fluoresces green, and chloroplasts within the organism fluoresce red due to the presence of chlorophyll. The smaller organisms are cyanobacteria, which fluoresce red due to the presence of chlorophyll.

(b) In this photomicrograph, the organisms were rendered fluorescent by reaction with fluorescent antibody. Cells of Clostridium septicum were treated with antibody conjugated with fluorescein isothiocyanate, which fluoresces green. Cells of Clostridium chauvei were stained with antibody conjugated with rhodamine B, which fluoresces red. (Courtesy of the Anaerobic Department, Wellcome Research Laboratories, Beckenham, Kent, England.)

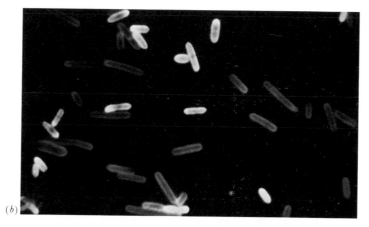

(b)

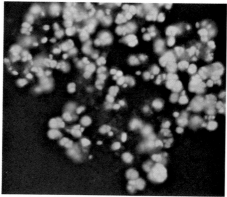

(c)

(c) Autofluorescence of methane bacterium, due to the presence of the unique electron carrier, F_{420}. The organisms were visualized with blue light in a fluorescence microscope. Methanosarcina barkeri.

(d) F_{420} fluorescence in the methane bacterium Methanobacterium formicicum.

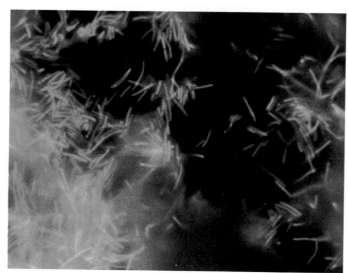

(d)

Plate 8

Series of photographs of myxobacterial fruiting bodies, presented to give an idea of the complexity of structures formed. (a) *Myxococcus fulvus*. The fruiting bodies are resting on filter paper. 75 to 150 μm in diameter. (b) *Stigmatella aurantiaca*. 150 μm high. (c) *Podangium (Melittangium) erectum*. 50 to 60 μm high. (d) *Myxococcus stipitatus*, resting on a soil crumb. 170 μm high. (Photographs courtesy of Hans Reichenbach and Martin Dworkin.)

(a)

(b)

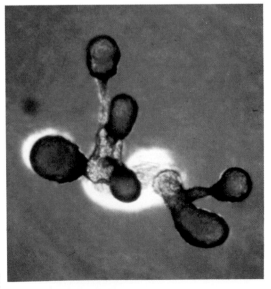

(c)

(d)

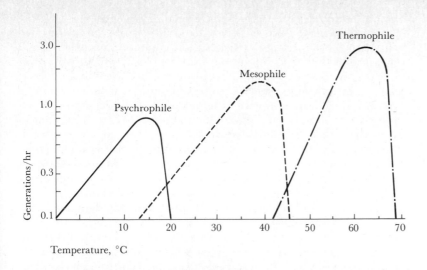

FIGURE 8.2 Relation of temperature to growth rates of a psychrophile, a mesophile, and a thermophile.

ent where microbes can grow. It is important to distinguish between environments that are cold throughout the year and those that are cold only in winter. The latter, characteristic of continental temperate climates, may have summer temperatures as high as 30°C, and winter temperatures far below 0°C. Such highly variable environments are much less favorable for cold-adapted organisms than are the constantly cold environments found in polar regions, at high altitudes, and in the depths of the oceans.

As noted earlier, organisms that are able to grow at low temperatures are called **psychrophiles** or **cryophiles**. A number of definitions of psychrophile exist, the most common being that a psychrophile is any organism able to grow at 0°C. This is a rather imprecise definition because many organisms with temperature optima in the 20s or 30s grow, albeit slowly, at 0°C. A current definition is that a true or strict psychrophile is an organism with an optimal temperature for growth of 15°C or lower, and a minimal temperature for growth at 0°C or lower. Organisms which grow at 0°C but have optima of 25 to 30°C have often been called facultative psychrophiles, but a more compact term is *psychrotroph*. True psychrophiles (also called *obligate* psychrophiles) are generally stenothermal, whereas psychrotrophs are commonly eurythermal.

Strict psychrophiles are found in environments that are constantly cold, and they are usually rapidly killed even by brief warming to room temperature. For this reason, their laboratory study is unusually difficult, as great care must be taken to ensure that they never warm up during sampling, transport to the laboratory, plating, pipetting, or other manipulations. All media and equipment must be precooled before use, and the work must be carried out in a cold room or in a refrigerator chest. Because of these technical problems, strict psychrophiles have not been well studied. However, enough work has been done for us to know that strict psychrophiles exist and that they are an interesting and diverse group, including bacteria, fungi, and algae from a variety of genera and species. The best studied psychrophiles have been the eucaryotic algae that grow in dense masses within and under the ice in polar regions. Polar explorers first observed these algae from ships breaking through ice floes. Upturned chunks of ice showed a brown coloration, commonly visible in bands, due primarily to algae of various diatom genera which had grown under and become frozen in the ice. Because polar seas in the area of sea ice are generally just at the freezing point, the diatoms grow in a constantly cold environment. Although light intensities are low under the ice, there is sufficient light for net photosynthesis to occur. One of these diatoms, *Fragillaria sublinearis,* has been shown in laboratory culture to have a temperature optimum for growth of 7°C, and a maximum around 12°C. A bacterium isolated from Antarctic seawater was also found to have an extremely low temperature optimum, 4°C, and was unable to grow at all at 10°C. This organism also grew readily at −2.5°C, a temperature at which seawater does not freeze.

Psychrophilic algae are often seen on the surfaces of snowfields and glaciers (see Color Plate 1) in such large numbers that they impart a distinctive red or green coloration to the surface. The most common snow alga is the eucaryote *Chlamydomonas nivalis;* its brilliant red spores (see Color Plate 1) are responsible for the red color. The alga probably grows within the snow as a green-pigmented vegetative cell, and then sporulates; as the snow dissipates by melting, erosion, and vaporization, the spores become concentrated on the surface. Snow algae are most commonly seen on permanent snowfields in mid- to late summer, and are especially common in sunny dry areas, probably because in more rainy areas they are washed away from the snowfields.

Another interesting organism with a low temperature optimum is the snow mold, a fungus of the genus *Typhula.* This fungus is pathogenic on grasses and other plants (for example, winter wheat), which overwinter (pass the winter) in the vegetative state. The fungus cannot grow at temperatures over 15°C and remains dormant through the summer (*estivate* would be the proper term for "oversummering") in a resting form called a *sclerotium.* When the plants become covered with snow the sclerotia germinate, and the fungus grows. Snow mold is much more common in areas where fairly deep snows occur, because the thick layer of snow effectively insulates the plants from the winter cold, and the temperature at the snow-plant interface remains around 0°C all winter. Thus a constantly cold habitat is created, which provides the proper conditions for growth of the fungus.

Psychrotrophs (facultative psychrophiles) are much more widely distributed than strict psychrophiles and can be isolated from soils and water in temperate climates. As we noted, they grow best at a temperature between 25 and 30°C, with a maximum of about 35°. Since temperate environments do warm in summer it is understandable that they cannot support the very sensitive obligate psychrophiles, the warming essentially providing a selective force favoring the facultative psychrophiles and excluding the obligate forms. It should be emphasized that although psychrotrophs do grow at 0°C, they do not grow very well, and one must often wait several weeks before visible growth is seen in culture media. Various genera of bacteria, fungi, algae, and protozoa have members that are psychrotrophic.

Meat, milk and other dairy products, cider, vegetables, and

fruits, when stored in refrigerated areas, provide excellent habitats for the growth of psychrotrophic organisms. Growth of bacteria and fungi in foods at low temperatures can lead to changes in the quality of the food and eventually to spoilage. The lower the temperature, the less rapidly does spoilage occur, but only when food is solidly frozen is microbial growth impossible. The frozen food industry owes its great development in recent years to the much greater keeping qualities of frozen foods over merely refrigerated foods. Fortunately, the organisms that colonize and grow in cold foods are psychrotrophs rather than psychrophiles, since these are the kinds of organisms present in the environments where human civilization is developed. Strict psychrophiles would probably grow much faster in foods at low temperature, but would be unlikely to survive dispersal from polar or alpine regions.

It is not completely understood how some organisms can grow well at low temperatures that prevent the growth of other organisms. At least one factor is that psychrophiles, as opposed to other organisms, have enzymes that are able to catalyze reactions more efficiently at low temperatures. Perhaps as a consequence of this, the enzymes of psychrophiles are very sensitive to higher temperatures, being rapidly inactivated at temperatures of 30 to 40°C. Another feature of psychrophiles is that their active-transport processes function well at low temperatures, thus making it possible for the organisms to effectively concentrate essential nutrients. In the same sense as we discussed minimum growth temperatures earlier in this section, it is thought that this efficient low-temperature transport mechanism is due to peculiarities of the cell membrane. Psychrophiles have a higher content of unsaturated fatty acids in the cell membrane than do other organisms, and a consequence of this is that the membrane remains semifluid at low temperatures. Membranes with predominantly saturated fatty acids would be expected to be solid and nonfunctional at low temperatures and would become semifluid and functional only at higher temperatures.

Freezing Despite the ability of some organisms to grow at low temperatures, there is a lower limit below which reproduction is impossible. Pure water freezes at 0°C and seawater at −2.5°C, but freezing is not continuous and microscopic pockets of water continue to exist at much lower temperatures. Although cellular growth ceases at about −30°C, enzymatic reactions can occur, albeit slowly, at even lower temperatures, and the lower limit for biochemical reactions in aqueous systems is probably the glass transition point of water at −140°C.* (Activity at even lower temperatures might occur if the freezing point were lowered by use of solvents such as alcohol or glycerol.) Although freezing prevents microbial growth it does not always cause microbial death. In fact, freezing is one of the best ways of keeping many microbial cultures for later study and is used quite extensively in research laboratories.

The molecular events during freezing are now reasonably well

---

*The glass transition point is the temperature at which a solid changes from a somewhat viscous form to a hard and relatively brittle one.

understood. When cells are subjected to freezing temperatures, their cytoplasm does not freeze as fast as the surrounding medium and the cell contents become supercooled. Eventually the cells do develop equilibrium with their frozen surroundings, and how this occurs depends partly on whether the cells are cooled slowly or quickly, and partly on how permeable they are to water. Since the water-vapor pressure in a supercooled cell is higher than in ice, if they are cooled slowly or are highly permeable to water the cells begin to equilibrate by losing water, and become dehydrated. If cooling is rapid or if the permeability to water is low, equilibration occurs by intracellular freezing, and ice crystals form in the cytoplasm. The transformation of cellular water into ice crystals results in an increase in the concentration of solutes in the water left within the cell, effectively causing dehydration. Many cells are quite sensitive to dehydration and do not survive freezing for this reason. However, the main damage from intracellular freezing probably results from the physical effect of the ice crystals on cellular structures, especially on the plasma membrane. If the crystal size is kept small by very rapid freezing, damage may be minimized, although crystal growth can also occur during warming, so that this process should also be carried out rapidly. At sufficiently low temperatures, the whole cytoplasm freezes.

The medium in which the cells are suspended considerably affects sensitivity to freezing. Water-miscible liquids such as glycerol and dimethylsulfoxide, when added at about 0.5 molar (M) to the suspending medium, penetrate the cells and protect by reducing the severity of dehydration effects. Macromolecules such as serum albumin, dextran, and polyvinylpyrrolidone, added at 10^{-5} to 10^{-3} M, do not penetrate the cells and probably protect by combining with the cell surface and preventing freezing damage to the cell membrane. The practical significance of this is that sensitive cells can often be preserved in the frozen state by selecting the proper suspending medium.

High-temperature environments As noted earlier, organisms that grow at temperatures above 45 to 50°C are called **thermophiles.** Temperatures as high as these are found in nature only in certain restricted areas. For example, soils subject to full sunlight are often heated to temperatures above 50°C at midday, and darker soils may become warmed even to 70°C, although a few inches under the surface the temperature is much lower. Fermenting materials such as compost piles and silage usually reach temperatures of 60 to 65°C. However, the most extensive and extreme high-temperature environments are found in nature in association with volcanic phenomena.

Many hot springs have temperatures around boiling, and steam vents (fumaroles) may reach 150 to 500°C. Molten lava has temperatures over 1000°C. Hot springs occur throughout the world but are especially concentrated in western United States, New Zealand, Iceland, Japan, the Mediterranean region, Indonesia, Central America, and central Africa. The area with the largest single concentration of hot springs in the world is Yellowstone National Park, Wyoming. Although some springs vary in temperature, others are very constant, not varying more than 1 to 2°C over many years.

Many of these springs are at the boiling point for the altitude (92 to 93°C at Yellowstone, 99 to 100°C at locations where the springs are close to sea level). As the water overflows the edges of the spring and flows away from the source, it gradually cools, setting up a thermal gradient (see Color Plate 2). Along this gradient, microorganisms develop (Figure 8.3), with different species growing in the different temperature ranges. By studying the species distribution along such thermal gradients and by examining hot springs and other thermal habitats at different temperatures around the world it is possible to determine the upper temperature limits for each kind of microorganism (Table 8.1). From this information we conclude that (1) procaryotic organisms in general are able to grow at temperatures higher than those at which eucaryotes can grow, (2) nonphotosynthetic organisms are able to grow at higher temperatures than can photosynthetic forms, and (3) structurally less complex organisms can grow at higher temperatures than can more complex organisms. However, it should be emphasized that not all organisms from a group are able to grow near the upper limits for that group. Usually only a relatively few species or genera are able to function successfully near the upper temperature limit.

As mentioned above, the precise temperatures defining each type of microorganism are not absolute. For example, the lower temperature limit for thermophily is quite arbitrary and depends somewhat on the kind of organism. Green algae that grow at 35°C are often called thermophiles, whereas thermophilic protozoa grow above 40°C (range, 40 to 60°C) and thermophilic fungi above 55°C (range, 55 to 60°C). Only cyanobacteria that grow above 50°C would probably be called thermophiles, whereas for other bacteria, 55°C would be near the lower limit. The important point is that the different groups can be generally distinguished on a consistent basis.

Bacteria can grow over the complete range of temperatures in which life is possible, but no one organism can grow over this whole range. As mentioned above, each organism is limited to a restricted range of perhaps 30°C, and it can grow well only within a still narrower range. Even within the thermophilic group of bacteria there are differences. Some thermophilic bacteria have temperature optima for growth at 55°C, others at 70°C, and still others at 100°C. In most boiling springs (Figure 8.4) it is found that a variety of bacteria grow, often surprisingly rapidly. The growth of such bacteria can be easily studied by immersing microscope slides into the spring and retrieving

FIGURE 8.3 Algal growth in a hot spring. Yellowstone National Park. As the water flows away from the source, it cools, and the temperature becomes favorable for algal growth. The upper temperature limit is about 70° to 73°C. In the background, Great Fountain Geyser is erupting.

TABLE 8.1 Approximate Upper Temperature Limits for Different Kinds of Microorganisms

Organism	Temperature (°C)
Protozoa	45–50
Eucaryotic algae	56
Fungi	60
Photosynthetic bacteria (including cyanobacteria)	70–73
Bacteria	>99

Data from Brock, T. D. 1967. Science 158:1012; and from Tansey, M. R., and T. D. Brock. 1972. Proc. Natl. Acad. Sci. USA 69:2426.

FIGURE 8.4 A typical small boiling spring in Yellowstone National Park. This spring is superheated, having a temperature 1 to 2 degrees above the boiling point. The mineral deposits around the spring consist mainly of silica. The log pole holds a rope containing a microscope slide rack used to incubate slides upon which bacteria attach and grow in the spring.

them after a few days. Microscopic examination of the slides generally reveals small or large colonies of bacteria (Figure 8.5) that have developed from single bacterial cells which attached to and grew on the glass surface.

Ecological studies of organisms living in boiling springs have shown that growth rates are surprisingly rapid, doubling times of 2 to 7 hours having been found. By use of radioactively labeled substrates it has been possible to show that bacteria living at these high temperatures are functional and have optimum temperatures near those of their environments. Both aerobic and anaerobic bacteria have been found living at high temperatures, and many morphological types exist. These organisms have been studied with the electron microscope and show unusual fine structure (Figure 8.6). The classification of many of these high-temperature bacteria is uncertain. Bacteria living in the temperature range of 55 to 70°C are members of such diverse genera as *Bacillus, Clostridium, Thermoactinomyces,* and *Methanobacterium;* and a wide variety of others probably exist. At temperatures above 75°C, only a few bacteria have been cultured, among which are many new genera. One anaerobe, isolated from an undersea thermal vent, has an optimum of 105°C.

Thermophilic bacteria related to those living in hot springs have also been found in artificial thermal environments. The hot-water heater, domestic or industrial, usually has a temperature of 55 to 80°C and is a favorable habitat for the growth of thermophilic bacteria. Organisms resembling *Thermus aquaticus,* a common hot spring organism, have been isolated from the hot water of many installations. Electric power plants, hot industrial process water, and other artificial thermal sources probably also provide sites where thermophiles can grow. Presumably, hot springs are the natural reservoirs of these organisms, and inoculation of the hot water occurs via air or water dispersal.

How can organisms survive and grow at these high temperatures? First, their enzymes and other proteins are much more resistant to heat than are those of mesophiles, just as the proteins of mesophiles are more stable to heat than are those of psychrophiles. Furthermore, the protein-synthesizing machinery (that is, ribosomes and other constituents) of the thermophiles, as well as such structures as the cell membrane, are likewise more resistant. We mentioned earlier that psychrophiles have membrane lipids rich in unsaturated fatty acids which make them functional at low temperatures but unstable at moderate temperatures; conversely, thermophiles have membrane lipids rich in saturated fatty acids, which enable the membranes to remain stable and functional at high temperatures. Saturated fatty acids form much stronger hydrophobic bonds than do unsaturated fatty acids, which accounts for the membrane stability.

Why is it that some groups of microorganisms have representatives that will grow at high temperatures, whereas others do not? The absence of eucaryotes at temperatures above 60°C may be due to the inherent thermolability of the organellar membranes of these organisms, especially those present in the nucleus, mitochondria, and chloroplast. The mitochondrial membrane seems especially sensitive to heat, and mitochondria disappear when cells are heated to a few degrees above their maximum temperature. The organellar membranes of eucaryotes probably must be fairly fluid to permit the passage of high-molecular-weight components. Messenger RNA and ribosomes, for instance, are made in the nucleus and must move through the nuclear membrane to the cytoplasm. It is conceivable that eucaryotes cannot synthesize a functional nuclear membrane that is still stable at high temperatures.

The absence of photosynthetic procaryotes (mainly cyanobacteria) at temperatures above 70 to 73°C may be related to the thermolability of the photosynthetic membrane system, a structure absent from heterotrophs and lithotrophs.

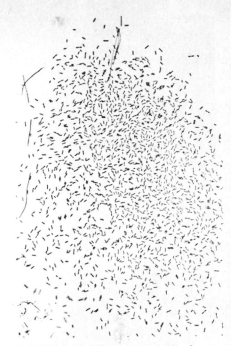

FIGURE 8.5 Photomicrograph of a bacterial microcolony that developed on a microscope slide immersed in a boiling spring such as that shown in Figure 8.4. Magnification, 380×.

Destruction of microorganisms by heat   Let us now consider the effects of temperature on viability. As temperature rises past the maximum temperature for growth, lethal effects become apparent. As shown in Figure 8.7, death from heating is an exponential or first-order function and occurs more rapidly as temperature is raised. These facts have important practical consequences. If we wish to kill every cell, that is to *sterilize* a population, it will take longer at lower temperatures than at higher temperatures. It is thus necessary to select the time and temperature that will sterilize under stated conditions.

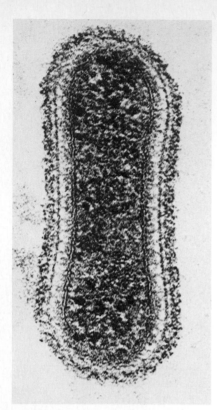

FIGURE 8.6 Electron micrograph of a bacterium living in a boiling hot spring. Magnification, 160,000×. (Electron micrograph by T. D. Brock and Mercedes Edwards.)

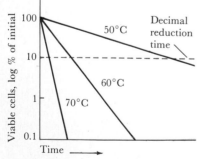

FIGURE 8.7 Effect of temperature on viability of a mesophilic bacterium.

The first-order relationship shown in Figure 8.7 means that the rate of death is proportional at any instant only to the concentration of organisms at that instant and that the time taken for a definite fraction (for example, 90 percent) of the cells to be killed is independent of the initial concentration. This means that heat treatment has no cumulative effect on a cell, and that the length of time a cell has been exposed to heat prior to death has no perceivable effect on it. First-order death of this type is found among unicellular organisms, but not among multicellular organisms, where heat treatment seems to have a cumulative effect. (For a further discussion of the meaning of first-order kinetics of cell killing, see Section 7.8.)

The rate of inactivation may deviate from a simple first-order reaction for a number of reasons (Figure 8.8). If the cell suspension is clumped, every cell in the clump must be inactivated before the colony-forming ability of the clump will be inactivated, so that the rate of death during the early stages of heating will be slower than later (Figure 8.8a). Second, if the suspension is composed of two populations with different heat resistances, the more heat-sensitive cells will be inactivated first, and then the inactivation rate will slow to a new rate during the time when the more resistant organism is being inactivated (Figure 8.8b). If inactivation of spores is being studied, germination of the spore is usually activated by heat, which sometimes results in an initial rise in the viable count before a subsequent first-order inactivation (Figure 8.8c).

Not all organisms are equally susceptible to heat. As might be expected, psychrophiles are the most heat sensitive, followed by mesophiles; thermophiles are the most heat resistant. Some thermophiles actually grow best under conditions that rapidly kill mesophiles and psychrophiles. Bacterial endospores are much more heat resistant than are vegetative cells of the same species. Since endospores are nearly always present in foods and other substances to be sterilized, heating times must be long enough to kill all spores present. Spores of thermophiles are more heat resistant than are spores of mesophiles.

The most useful way of characterizing heat inactivation is to determine the time at which a defined fraction of the population is killed. For a variety of reasons the time required for a tenfold reduction in the population density at a given temperature, called the **decimal reduction time** or D, is the most useful parameter. Over the range of temperatures usually used in food sterilization, the relationship between D and temperature is essentially exponential, so that when the logarithm of D is plotted against temperature a straight line is obtained (Figure 8.9). The slope of the line provides a quantitative measure of the sensitivity of the organism to heat under the conditions used, and the graph can be used in calculating process times for sterilization, such as in canning operations. Organisms and spores vary considerably in heat resistance. For instance, in the autoclave (a device used for sterilizing by means of superheated steam under pressure), a temperature of $121°C$ may be reached. Under these conditions, spores from a very heat-resistant organism may require 4 to 5 minutes for a decimal reduction, whereas spores of other types show a decimal reduction in 0.1 to 0.2 minute at this temperature. Vegetative cells may require only 0.1 to 0.5 minute at 65°C for a decimal reduction in numbers.

Although accurate, determination of decimal reduction times is a fairly lengthy procedure, since it requires making a number of viable count measurements. A less satisfactory but easier way of characterizing the heat sensitivity of an organism is to determine the **thermal death time,** the time at which all cells in a suspension are killed at a given temperature. This is done simply by heating samples of this suspension for different times, mixing the heated suspensions with culture medium, and incubating. When all cells are killed, no growth will be evident in the incubated samples. Of course, the thermal death time determined in this way will depend on the size of the population tested, since a longer time will be required to kill all cells in a large population than in a small one. If the number of cells is standardized, then it is possible to compare the heat sensitivities of different organisms by comparing their thermal death times. When the logarithm of the thermal death time is graphed versus temperature, a straight line similar to that shown in Figure 8.9 is obtained.

The nature of the medium in which heating takes place influences the killing of vegetative cells or spores. Microbial death is more rapid at acidic pH values, and for this reason acid foods such as tomatoes, fruits, and pickles are much easier to sterilize than more neutral foods such as corn and beans. High concentrations of sugars, proteins, and fats usually increase the resistance of organisms to heat, while high salt concentrations may either increase or decrease heat resistance, depending on the organism. Dry cells (and spores) are more heat resistant than moist ones; for this reason, heat sterilization of dry objects always requires much higher temperatures and longer times than does sterilization of moist objects. (Even some multicellular desert animals can survive brief heating to 100°C if they are dry.)

Bacterial endospores (Section 2.11) are the most heat-resistant structures known: they are able to survive conditions of heating that would rapidly kill vegetative cells of the same species. Although it is not completely certain, it seems likely that a major factor in heat resistance is the amount and state of water within the endospore, since dry cells are much more resistant to heat than wet cells. During endospore formation, the protoplasm is reduced to a minimum volume as a result of the accumulation of metal ions, such as Ca^{2+} (high in spores), and dipicolinic acid (present only in spores), which lead to formation of a gel-like structure. At this stage, the thick cortex forms around the protoplast (see Figure 2.40b), and contraction of the cortex results in a shrunken protoplast low in water. The water content of the protoplast determines the degree of heat resistance of the spore. If the spores of a bacterial species have high water content, they have low heat resistance, and the heat resistance of spores can be varied by altering the water content of the spores. Water moves freely in and out of spores, so that it is not the impermeability of the spore wall that affects water content, but the physical state of the spore protoplast, that is, the degree of gel-like structure. The tightly fitting cortex prevents the protoplast from imbibing water and swelling. The importance of an intact spore cortex in heat resistance is shown by the fact

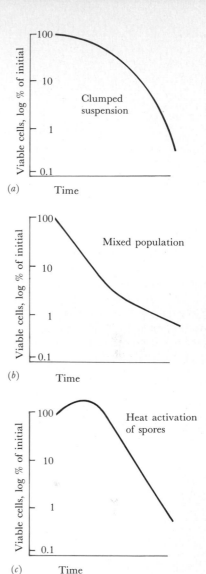

FIGURE 8.8 Deviations from first-order heat inactivation. Curves shown represent responses at single temperatures. Compare with Figure 8.7.

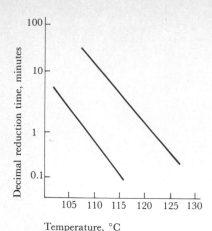

FIGURE 8.9 Relationship between temperature and rate of killing as indicated by the decimal reduction time. Data for such a graph are obtained from curves such as those given in Figure 8.7. The upper line in the figure represents data for a very heat-resistant organism.

that, when germination occurs, loss of refractility (which is due to the dissolution of the cortex) and decrease in heat resistance parallel each other.

Pasteurization and sterilization Pasteurization is a process using mild heat to reduce the microbial populations in milk and other foods that are exceptionally heat sensitive. It is named for Louis Pasteur, who first used heat for controlling the spoilage of wine. It is not synonymous with sterilization since not all organisms are killed. Originally, pasteurization of milk was used to kill pathogenic bacteria, especially the organisms causing tuberculosis, brucellosis, and typhoid, but it was discovered that the keeping qualities of the milk were also improved. Today, since milk rarely comes from cows infected with the pathogens mentioned above, pasteurization is used primarily because it improves the keeping qualities.

Pasteurization of milk is usually achieved by passing the milk continuously through a heat exchanger where its temperature is raised quickly to 71°C and held there for 15 seconds, and then is quickly cooled, a process called **flash pasteurization.** Occasionally pasteurization is done by heating milk in bulk at 63 to 66°C for 30 minutes and then quickly cooling. Flash pasteurization is more satisfactory in that it alters the flavor less and can be carried out on a continuous-flow basis, thus making it adaptable to large dairy operations.

Sterilization by heat involves treatment that results in the complete destruction of all organisms, and since bacterial endospores are ubiquitous, sterilization procedures are designed to eliminate them. This requires heating at temperatures above boiling and the use of steam under pressure. The usual procedure is to heat at 1.1 kg/cm² (15 lb/in²) steam pressure, which yields a temperature of 121°C. Heating is usually done in an autoclave, although if only small batches of material need be sterilized, a home pressure cooker is quite satisfactory and much less expensive. At 121°C, the time of autoclaving to achieve sterilization is generally considered to be 10 to 15 minutes. If bulky objects are being sterilized, heat transfer to the interior will be slow, and the heating time must be sufficiently long so that the total material is at 121°C for 10 to 15 minutes.

8.2

Water and Water Activity

All organisms require water for life. The amount of water in different environments varies. However, water availability does not depend only on water content: it is a complex function of adsorptive and solution factors. Water adsorbed to surfaces may or may not be available, depending on how tightly it is adsorbed and how effective the organism is in removing it. Also, when solutes are dissolved in water, they become more or less hydrated, and the degree to which a solute becomes hydrated affects the availability of water to organisms. There are a number of ways in which water availability as influenced by adsorption and solution factors can be expressed, but the simplest to understand is **water activity.**

Water activity   Water activity, abbreviated a_w, is related to the vapor pressure of water in the air over a solution or substance and is estimated by measuring the relative humidity of the vapor phase. Relative humidity, a familiar meteorological concept, expresses the ratio of the amount of water present in air at a given temperature to that which the air could hold if it were saturated with water. In meteorological usage, relative humidity is usually expressed as a percentage, but water activity is given as the fraction. Thus a_w 0.75 equals 75 percent relative humidity. To measure the water activity of a substance or solution, it is placed in an enclosed space and allowed to come to equilibrium with the air. Measurement of the relative humidity in the air (usually measured by sensors called *psychrometers*) then gives the water activity of the material (Figure 8.10). Because temperature greatly affects the amount of water that air can hold, temperature must be specified (and held constant during equilibration). Table 8.2 gives the water activities for several solutes as a function of solute concentration. As can be seen, water activity of freshwater and marine habitats is relatively high, and it is only in solutions with high concentrations of solute that water activity is markedly lowered. Different solutes affect water activity to differing degrees, depending on how they dissociate and hydrate when dissolved in water.

In soil microbiology, the concept of **water potential** is often used instead of water activity. Water potential also expresses the availability of water, but is an energy term, defined as the difference in free energy between the system under study (for example, soil, food) and a pool of pure water at the same temperature. Water potential can be expressed in a number of different units, but the most widely used unit is the *bar,* which is equivalent to 10^6 dynes/cm^2 or 0.986 atmospheres (atm). The values are always negative (similar to suction pressures), as shown in Table 8.2. Water activity and water potential are not exactly proportional, because water activity does not take into account the temperature of the system, but as a rough rule, a decrease of 0.01 a_w in the range 0.75 to 1.0 a_w is equivalent to a decrease of potential of about 15 bar.

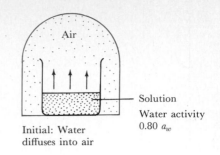

Initial: Water diffuses into air

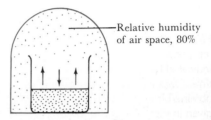

At equilibrium: Measurement of relative humidity of air space gives water activity of solution

FIGURE 8.10   Measurement of water activity of a solution by measuring relative humidity (RH) of the vapor in equilibrium with it: $a_w \times 100 = $ RH.

TABLE 8.2   Water Activity for Several Solutes at 25°C

Water Activity a_w	Approximate Water potential (bar)	NaCl (g/100 ml water)	Sucrose (g/100 ml water)	Glycerol (g/100 ml water)
0.995	−7	0.87	9.2	2.6
0.980	−28	3.5 (seawater)	34.2	10.1
0.960	−56	7	64.9	20.2
0.900	−145	16.2	140 (maple syrup)	51.5
0.850	−210	23 (Great Salt Lake)	205 (saturated)	78.2
0.800	−307	30 (saturated)	—	105.8
0.700	−491	—	—	168.4
0.650	−593	—	—	202.4

Water activity and microbial growth When an organism grows in a medium with low water activity, due to addition of a solute, it must perform work to extract water from the solution. This usually results in lessened growth yield or a lower growth rate. In many cases, the same result will be seen with any solute that lowers the water activity. However, there is always the possibility that the solute added may have a specific toxic effect on the cells, so that two media at the same water activity may affect growth to different degrees. In Section 2.5 we learned that the cytoplasm usually has a higher solute concentration than the medium, so that water tends to flow into the cell. If the external solute concentration is raised to a value higher than internal, water will flow out, and an organism can only obtain water and grow under these conditions by increasing its internal solute concentration. This increase in internal solute concentration is an energy-requiring process and the organisms perform work to bring about the concentration increase. Some organisms are better at this effort than others and are thus able to grow in media at lower water activities. The minimum a_w values for growth of different kinds of microorganisms are given in Table 8.3, and it can be seen that certain fungi and yeasts are best able to grow at low water activities. Although the effect of solute concentration on growth is often considered to be due to its effect on osmotic pressure, it is really due to the effect on water activity. Organisms such as the yeasts and fungi listed in Table 8.3, which can grow in media with high concentrations of solute, are often called **osmotolerant,** indicating their ability to withstand high osmotic pressures.

TABLE 8.3 Approximate Limiting Water Activities for Microbial Growth

Water Activity, a_w	Aquatic Environments	Foods	Bacteria	Yeasts	Fungi	Eucaryotic Algae
1.00	Blood, seawater	Vegetables, meat, fruits	*Caulobacter* *Spirillum* spp.			
0.95		Bread	Most Gram-positive rods	Basidiomycetous yeasts	Basidiomycetes	Most algal groups
0.90		Ham	Most cocci *Lactobacillus* *Bacillus*	Ascomycetous yeasts	*Fusarium* *Mucor*	
0.85		Salami	*Staphylococcus*	*Saccharomyces rouxii* (in salt) *Debaryomyces* (in salt)		
0.80		Fruit cake preserves		*S. bailii* (in sugars)	*Penicillium*	
0.75	Salt lake	Salt fish	*Halobacterium* *Halococcus*		*Wallemia* *Aspergillus* *Chrysosporum*	*Dunaliella*
0.70		Cereals, confectionery, dried fruit				
0.65					*Eurotium*	
0.60				*S. rouxii* (in sugars)	*Xeromyces bisporus*	

Based on Brown, A. D. 1976. Microbial water stress. Bacteriol. Rev. 40:803–846.

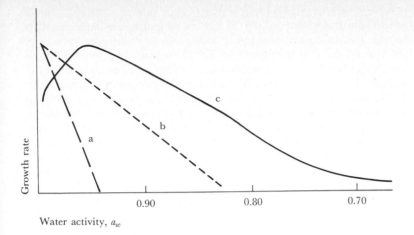

FIGURE 8.11 Effect of variation in water activity on the growth of micro-organisms. (a) Normal organism, growing best at high water activities and being markedly inhibited by reduced activities. (b) Osmotolerant organism, growing best at high water activities but growing reasonably well at somewhat reduced activities. (c) Osmophilic organism, showing optimal growth at reduced water activity.

A few fungi, such as the *Xeromyces* listed in Table 8.3 are actually **osmophilic**, requiring media with increased solute concentration (the optimum a_w for *Xeromyces* is 0.90), but most of these organisms grow best at a_w values of 0.95 to 0.99. A comparison of the effect of water activity on the growth of normal, osmotolerant, and osmophilic microorganisms is given in Figure 8.11.

The *Halobacterium* listed in Table 8.3 is a clear example that the nature of the solute influences the results. It actually *requires* sodium ions for growth and grows optimally in media to which sodium chloride has been added to obtain a_w values less than 0.80, that is, approximately saturated NaCl (see Table 8.2). It is thus proper to call this organism **halophilic** (salt-requiring).

Matric and osmotic water activity   As we have noted, water activity can be affected either by adsorption or by solute interaction. Effect of adsorption on water activity is often called a *matric effect*, the matrix of substances or materials adsorbing the water really being responsible for reducing water availability (Figure 8.12). Effect of solute interactions on water activity is then called an *osmotic effect*. Solid materials such as soil, food, wood, metal, glass, and so on adsorb water, and the avidity of adsorption is determined by the chemical and physical properties of the material. However, if such a material is allowed to equilibrate with an atmosphere of known relative humidity (water activity), at equilibrium the activity of adsorbed water will have the same a_w as did the atmosphere. If an organism now grows on the material, it will be subjected to exactly the same water activity as it would be subjected to if it were growing in a solution whose water activity had been controlled by a solute. In both cases the organism must extract water from its environment, in one case water bound to a surface and in the other case water associated with solute molecules. How does a microorganism obtain water from an environment of low a_w? Since water only flows from regions of higher to lower a_w, the organism must decrease a_w inside the cell by raising intracellular solute concentration. The solute which is used inside the cell for this adjustment must be nontoxic; such compounds are called compatible

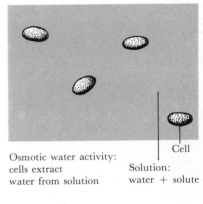

Osmotic water activity: cells extract water from solution

Cell

Solution: water + solute

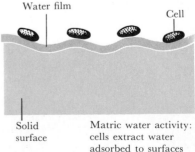

Water film

Cell

Solid surface

Matric water activity: cells extract water adsorbed to surfaces

FIGURE 8.12 Comparison of osmotic and matric water activity.

solutes. Three examples have been studied in detail: 1) halobacteria which live in habitats of 3.5 M Na^+, and use K^+ as a compatible solute, concentrating the latter from the environment until the interior of the cell is nearly 4M K^+; 2) some osmotolerant yeasts which can live in the presence of high sucrose concentrations and use polyalcohols (sorbitol and ribitol) as compatible solutes; 3) the halophilic alga *Dunaliella* which lives in extremely saline habitats and synthesizes the polyalcohol glycerol internally as a compatible solute. Note that the eucaryotes in these examples, the yeasts and the alga, synthesize their compatible solutes while the bacterium incorporates cations from the outside. There is a considerable energetic cost to this solute synthesis, but the organisms presumably benefit by being able to grow in environments from which almost all other organisms are excluded. We can also see that the ability to grow in high solute environments requires more than just osmotic adjustment. For example, halophilic bacteria reach osmotic and ionic balance in saline areas by incorporating another salt, while *Dunaliella* spp in similar saline areas apparently cannot tolerate high intracellular salt concentrations and have evolved the more elaborate mechanism of internal solute synthesis.

Ecological significance of water activity   Organisms living in soil or on surfaces exposed to the air are most commonly affected by matric water activity. Microbial activity in soil is markedly affected by its water status, although microorganisms vary considerably in ability to develop at low water activity, and some fungi can grow at water activities as low as 0.60 to 0.70 (Table 8.3).

We must distinguish between the effects of drying on microbial activity and those on viability. Although all microbial activity will cease at water activities around 0.60 to 0.65, many microorganisms are remarkably resistant to drying. Small cells are usually more resistant than are large cells, round cells are more resistant than rod-shaped cells, and cells with thick walls (for example, most Gram-positive microorganisms) are more resistant than those with thin walls. The spirochete *Treponema pallidum,* which causes syphilis, is a long cell with a thin wall and is so sensitive to drying that it dies almost instantly in air. On the other hand, *Mycobacterium tuberculosis,* which causes tuberculosis, is very resistant to drying because of its thick cell wall, which contains a heavy lipid coating. Bacterial spores, the sexual spores of algae and fungi, and the cysts of protozoa resist drying much more than do the vegetative cells. Resistance to drying prolongs the survival of microorganisms that are dispersed through the air. A dry cell is metabolically dormant; in the absence of heat or other external influences it may remain dormant for long periods of time, but revive quickly upon the introduction of moisture.

Ecology of saline environments   From an ecological viewpoint, osmotic effects mainly concern habitats with high concentrations of salts. Seawater contains about 3.5 percent sodium chloride (NaCl, a_w 0.980), plus small amounts of many other minerals and elements. Microorganisms found in the sea usually have a specific requirement for NaCl, but often at concentrations much below those of seawater; their growth is inhibited by higher or lower concentrations of salt

(Figure 8.13). Marine microorganisms require sodium ions (Na^+) for the stability of the cell membrane and in addition many of their enzymes require Na^+ for activity. This requirement is specific for Na^+: it cannot be replaced by a potassium ion (K^+) or another related ion.

Salt is often produced commercially by evaporating seawater in large basins, called "salt pans," where the NaCl concentration may reach 25 percent before crystallization occurs. Microorganisms that can grow in these brines are usually pink or red in color and are responsible for the brilliant coloration of salt pans (see Color Plate 1). They are mostly bacteria of the *Halobacterium* type, but the alga *Dunaliella salina* also lives under these conditions. Similar kinds of bacteria grow in food products preserved in salt; growth of red bacteria in salted fish leads to a pink discoloration.

Natural habitats containing high concentrations of salt are saline lakes such as the Great Salt Lake, Utah, and many smaller lakes of the Great Basin of western United States. Similar lakes are found in enclosed basins in other parts of the world; the best known are the Dead Sea and the Caspian Sea. Many saline lakes have an ionic composition similar to that of concentrated seawater, with NaCl predominating. Such lakes are called thalassohaline (*thalasso-* is a combining form meaning "sea"); the Great Salt Lake is in this category. Other saline lakes have an ionic composition quite different from that of concentrated seawater. The Dead Sea, for instance, has magnesium chloride ($MgCl_2$) as the predominating ionic constituent. These ionic differences influence markedly the numbers and kinds of organisms that can grow in these waters. At the high salt concentrations found in saline lakes, no higher plants can develop, although in addition to microorganisms a few types of animals are capable of living (brine shrimp, brine flies).

Microorganisms that require NaCl for growth are called **obligate halophiles,** whereas those that will grow in a NaCl solution but do not require it are called **facultative halophiles.** Among the obligate halophiles are organisms that show different degrees of NaCl requirement (Figure 8.13); those that grow only when the NaCl concentration approaches saturation are called **extreme halophiles.**

Halobacterium: salinity tolerance and light-generated ATP synthesis

The extreme halophile *Halobacterium* has been much studied. It is a pink-colored Gram-negative rod that is unable to form spores. Its specific requirement for Na^+ cannot be satisfied by replacement with the chemically related ion K^+. The ionic concentration inside the cells of this bacterium is similar to that outside, which means that the cells do not become dehydrated. However, the predominant ion inside is not Na^+ but K^+, and the cells concentrate K^+ in very high amounts from the medium. When the NaCl concentration of the medium is reduced, the rod-shaped cells become spherical; if the concentration is lowered still further the cells lyse. This is not a direct osmotic effect but is due to the fact that the cell wall of *Halobacterium* is stabilized by Na^+ ions: when there is insufficient Na^+ the wall breaks and the cell lyses. Many enzymes of *Halobacterium* require high salt concentrations; the internal enzymes specifically require K^+ for stability and activity. The extreme halophiles differ from other bacteria in other

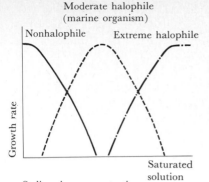

FIGURE 8.13 Effect of sodium ion concentration on growth of microorganisms of different salt tolerances.

ways. For instance, the cell wall of *Halobacterium* apparently varies from that of other bacteria in that it lacks muramic acid. The wall is largely composed of proteins which have an exceptionally high content of the acidic (negatively charged) amino acids, aspartic acid, and glutamic acid. The negative charges are shielded by the Na^+ ions; when these ions are diluted away, the negatively charged parts of the proteins actively repel each other, leading to loss of cell shape and eventual lysis, as described above. The ribosomes of *Halobacterium* also seem to be unusual, since they require high concentrations of K^+ for stability (ribosomes of other organisms have no K^+ requirement). The nutrition of *Halobacterium* is complex, and the organism requires many amino acids and vitamins for growth. Even under the best conditions, *Halobacterium* grows very slowly, showing a generation time of about 7 hours.

The plasma membrane of *Halobacterium* is of considerable interest because it is not constructed in the usual way (see Section 2.4). The phospholipids of *Halobacterium* do not contain fatty acids, as do those of other organisms, but instead contain long-chain isoprenoids, which are linked to glycerol not by ester linkage but by ether linkage. As will be discussed in chapter 19, one group of bacteria, the Archaebacteria, is recognized on the basis of a series of molecular properties including the biochemistry of the membrane lipids. All Archaebacteria so far discovered have ether-linked phospholipids similar to those of *Halobacterium.*

Halobacterium has the additional interesting property of showing a light-generated ATP synthesis that does not involve chlorophyll. As discussed in Chapter 6, one of the key aspects of photosynthesis is the synthesis of ATP by the development of a proton gradient across the photosynthetic membrane. The halobacterial membrane contains purple patches of *bacteriorhodopsin,* a carotenoid analogous to rhodopsin, the pigment of the animal eye, which functions in vision. Bacteriorhodopsin is able to absorb light and generate a proton gradient across the membrane, leading to ATP synthesis. This is the simplest type of photophosphorylation known and does not seem to be related to traditional photosynthesis, as it does not lead to autotrophic growth. Halobacteria are typical aerobic heterotrophs, and recent evidence suggests that one role of the rhodopsin-mediated ATP synthesis is that it enables the organism to synthesize ATP under anaerobic conditions, thus permitting it to survive in nature when oxygen availability is restricted. At the high salt concentrations at which halobacteria grow, oxygen solubility is very low, so that oxygen deficiencies can frequently develop. The discovery of the rhodopsin-mediated ATP synthesis in halobacteria has lead to new concepts about the utilization of light energy by living organisms and has also provided a useful model organism for studies on the manner by which proton gradients can develop and lead to ATP synthesis.

8.3

Hydrostatic Pressure

A long column of water exerts pressure on the bottom of the column because of the weight of water; this is called **hydrostatic pressure.** In nature, high hydrostatic pressure is found largely at the depths of the sea, and there is about a 10-atm increase in pressure for evéry 100 m increase in depth. The deepest part of the ocean is the Challenger Deep in the Mariana Trench, an area in the North Pacific Ocean east of the Philippines, which has a depth of 10,860 m, with a corresponding hydrostatic pressure of about 1100 atm. About 88 percent of the ocean is deeper than 1000 m, but only 1.2 percent is deeper than 6000 m. In terms of volume, about 75 percent of the entire mass of seawater is subjected to hydrostatic pressures between 100 and 600 atm, and about 0.1 percent to even higher pressures in the deep trenches.

Many microorganisms isolated from deep water tolerate high hydrostatic pressures to a variable degree, and we say they are **barotolerant.** Recent technological advances in sampling and culturing procedures have led to the isolation of truly **barophilic** bacteria, which grow better at higher than normal atmospheric pressure. These bacteria were obtained from deep-sea samples which were cultured for several months to allow the slow-growing bacteria to develop. Most bacteria isolated from shallow water or soil grow best at atmospheric pressure and are completely inhibited or killed by hydrostatic pressures of 200 to 600 atm. Hydrostatic pressure affects the activity of most enzymes, but the most sensitive sites are probably protein synthesis and membrane phenomena such as transport. When shallow water bacteria are grown at increased hydrostatic pressure, many of the cells grow without dividing and form long filaments, suggesting that cell-division processes are being affected.

Despite the detrimental effects of hydrostatic pressure, the bottoms of the deep oceans are not devoid of life. In fact, there is a surprising variety of animals present, including both fish and invertebrates. Many bacteria have been isolated from the bottom of the oceans that are unique in their ability to grow well at pressures corresponding to the deep-sea habitat. These bacteria are also psychrophiles. They grow well at 1 atm when the temperature is near 0°C, but are unable to grow at higher temperatures at 1 atm. High pressure counteracts the deleterious effects of temperature, since these organisms will grow at temperatures up to 20°C under high pressure.

8.4

Acidity and pH

The acidity or alkalinity of a solution is expressed by its pH value, the expression **pH** representing the negative logarithm of the hydrogen-ion concentration. The hydrogen-ion concentration of pure water is 10^{-7} M and hence pure water has a pH of 7 (Table 8.4). Those pH values that are less than 7 are acidic and those greater than 7 are alkaline. It should be remembered that pH is a logarithmic function; thus a solution that has a pH of 5 is 10 times as acidic as one with a pH of 6. In natural habitats pH values range widely; representative examples are given in Table 8.4

In culture media, it is desirable to maintain a relatively constant pH during microbial growth, and this is best accomplished by use of **buffers.** These are salts of weak acids or bases which take up or give off hydrogen ions as the hydrogen-ion concentration changes in the medium, and thus keep the hydrogen-ion concentration constant. Different buffers are suitable for different pH ranges. In the pH range

TABLE 8.4   The pH Scale

	Concentration of Hydrogen Ions (g/liter)	pH	Example
	10^{-0}	0	
	10^{-1}	1	Gastric fluids
Increasing acidity			Volcanic soils and waters
	10^{-2}	2	Lemon juice
			Acid mine drainage
	10^{-3}	3	Vinegar
			Rhubarb
			Peaches
	10^{-4}	4	Acid soil
			Tomatoes
	10^{-5}	5	Cheese
			Cabbage
	10^{-6}	6	Peas
			Corn, salmon
			Shrimp
Neutral	10^{-7}	7	Pure water
	10^{-8}	8	Seawater
	10^{-9}	9	Very alkaline natural soil
			Alkaline lakes
	10^{-10}	10	Soap solutions
Increasing alkalinity	10^{-11}	11	Household ammonia
			Concrete
	10^{-12}	12	Lime (saturated solution)
	10^{-13}	13	
	10^{-14}	14	

6 to 8, phosphate is an excellent buffer and is widely used in culture media. Under moderately acidic conditions, citrate is a good buffer, whereas at alkaline pH, borate and the amino acid glycine are buffers. There are hundreds of potential buffers, but it is always essential to be certain that the buffer has no toxic effect on the organism. For example, some bacteria are inhibited by relatively small concentrations of phosphate, a commonly used buffer. If a buffer cannot be used, pH control can be effected by adding small amounts of concentrated alkali or acid at intervals during growth to bring the pH back to its desired value. In large-scale culture operations, as in industry, automatic devices measure the pH and open a valve to let in alkali or acid as needed, thereby maintaining pH at a constant level.

Microbial growth and pH range   Each organism has a pH range within which growth is possible, and usually has a well-defined optimum pH. Most natural environments have pH values between 5 and 9, and organisms with optima in this range are most common. Only a few species can grow at pH values of less than 2 or greater than 10.

The lower pH limits for various groups of living organisms are given in Table 8.5. In general, this table shows that microorganisms are better adapted to life in acid environments than are higher organisms, although there is an interesting exception to this, the cyanobacteria. Members of this group are completely absent from habitats with pH less than 4, although a number of eucaryotic algae are able to thrive at pH values considerably below this. Indeed, even certain higher plants and mosses are more successful colonizers of acidic habitats than the cyanobacteria. The eucaryotic algae living at low pH are surprisingly diverse, and two of these algae, *Euglena mutabilis* and *Cyanidium caldarium*, are practically indicator species of acidic conditions. *Cyanidium caldarium* is a thermophile also, and hence is restricted to hot, acid springs and soils. *Euglena mutabilis* is much more widespread than *Cyanidium* because it does not require thermal conditions. It is found in coal, copper, and zinc mining operations throughout the world, living in the acid waters that drain these developments.

TABLE 8.5   Lower pH Limits for Different Groups of Organisms

Group	Approximate Lower pH Limit*	Examples of Species Found at Lower Limit
Animals:		
Fish	4	Carp
Insects	2	Ephydrid flies
Protozoa	2	Amoebae, heliozoans
Plants:		
Vascular plants	2.5–3	*Eleocharis sellowiana*
		Eleocharis acicularis
		Carex spp.
		Ericacean plants
		(heather, blueberries, cranberries, etc.)
Mosses	3	Sphagnum
Protists:		
Eucaryotic algae	1–2	*Euglena mutabilis*
		Chlamydomonas acidophila
		Chlorella spp.
	0	*Cyanidium caldarium*
Fungi	0	*Acontium velatum*
Bacteria:	0.8	*Thiobacillus thiooxidans*
		Sulfolobus acidocaldarius
	2–3	*Bacillus, Streptomyces*
Cyanobacteria	4	*Mastigocladus, Synechococcus*

*Lower pH limits are approximate and may vary depending on other environmental factors.

From Brock, T. D. 1978. Thermophilic microorganisms and life at high temperatures. Springer-Verlag, New York.

Organisms that live at low pH are called *acidophiles,* and both facultative and obligate acidophiles exist. Most fungi and yeasts grow fairly well at low pH, and a pH of 5 is often about optimal, but they also grow well even at pH values above 7; they are thus facultative acidophiles. The eucaryotic algae, which grow at pH values of 2 and below, are obligate acidophiles, completely unable to grow at neutral pH. There are also a number of obligately acidophilic bacteria, including several species of the genus *Thiobacillus,* and two bacteria which are also thermophilic, *Sulfolobus* and *Thermoplasma.* It is strange to consider that for obligate acidophiles a neutral pH is actually toxic. Probably the most critical factor for obligate acidophily is the plasma membrane. When the pH is raised to neutrality, the plasma membrane of obligate acidophilic bacteria actually dissolves, and the cells lyse, suggesting that high concentrations of H^+ are required for membrane stability.

Although microorganisms are found in habitats over a wide pH range, the pH within their cells is probably close to neutrality. In an acid environment, the organism can maintain a pH close to neutrality either by keeping H^+ ions from entering or by actively expelling H^+ ions as rapidly as they enter. The cell wall probably also plays some role in keeping hydrogen ions out. Neutrality is necessary because there are many acid and alkali labile components in the cell. Chlorophyll, DNA, and many proteins, for instance, are all destroyed by acid pH, and RNA and phospholipids are sensitive to alkaline pH. The optimum pH for intracellular enzymes is usually around neutrality, although enzymes in the periplasm (the layer between membrane and wall in Gram-negative bacteria) and extracellular enzymes may have pH optima for activity near the pH of the environment.

8.5

Oxygen

Oxygen (O_2) is an extremely interesting molecule because it is not only a vital substance for respiratory organisms, but is also able to form a variety of toxic derivatives, even in respiring organisms which require it. Air contains 20 percent O_2, and at one atmosphere air pressure there will be 0.2 atm O_2. One atmosphere of O_2 can be achieved by using pure O_2 instead of air, and even higher concentrations can be obtained if air or O_2 is placed under pressures greater than atmospheric (this is called hyperbaric oxygen). O_2 pressures greater than 0.2 atm are often toxic even to organisms that grow in air. Such high values will of course never be experienced in nature but can be induced in the laboratory and serve some experimental ends. Surprisingly, oxygen even at normal pressures is also frequently toxic to some organisms. O_2 pressures lower than 0.2 atm frequently occur in nature in partially aerated systems, and some organisms, called **microaerophilic**, require O_2 but actually grow better at O_2 pressures less than normal (see Table 8.6).

Certain terms are used to refer collectively to groups of organisms able to live under various reduction potentials and O_2 pressures, and these are outlined in Table 8.6. **Obligate aerobes** are found in aerobic environments and require O_2 usually because they are unable to generate enough energy for growth solely by fermentation. They

TABLE 8.6 Terms Used to Describe Redox and O_2 Relations of Microorganisms

Group	Environment		O_2 Effect
	Oxidizing	Reducing	
Aerobes:			
Obligate	Growth	No growth	Required
Facultative	Growth	Growth	Not required, but growth better with O_2
Microaerophilic	Growth, if level not too high	Growth, if level not too low	Required, but at levels lower than 0.2 atm
Anaerobes:			
Aerotolerant	Growth	Growth	Not required, and growth no better when O_2 present
Aerophobic (obligate anaerobes)	Death	Growth	Harmful

also often require O_2 for biosynthesis of sterols and unsaturated fatty acids (see Sections 5.5 and 5.6). **Facultative** organisms are able to obtain energy either by respiration or by fermentation and do not require O_2 for biosynthesis. We discussed the energy relations of these organisms in some detail in Chapter 4. Some facultative organisms can also use alternate electron acceptors such as nitrate when O_2 is absent. The **microaerophilic** organisms require O_2 but at pressures lower than 0.2 atm, probably because of direct O_2 toxicity. Anaerobes are those organisms unable to use O_2 as a terminal electron acceptor in the generation of energy, usually because they lack the terminal cytochromes that transfer electrons to O_2. They fall into two groups, aerotolerant and aerophobic. **Aerotolerant anaerobes** do not use O_2 but are not drastically harmed by it. They are able to grow in the presence or absence of O_2. **Aerophobic anaerobes,** more commonly called **obligate anaerobes** or strict anaerobes, do not use O_2 and are harmed by it, probably also because of direct O_2 toxicity.

Toxic forms of oxygen To understand oxygen toxicity, it is necessary to consider the chemistry of oxygen. Molecular oxygen, O_2, is unique among diatomic elements in that two of its outer orbital electrons are unpaired. It is partly because it has two unpaired electrons that oxygen has such a high reduction potential and is such a powerful oxidizing agent. However, the two electrons are in separate outer orbitals and have parallel spins (Table 8.7), whereas most other molecules have antiparallel spins. Since reactions between molecules occur much more readily if the spins of the outer electrons are the same, molecular oxygen is not immediately reactive in many situations, but needs activation.* The unactivated, ground state of oxygen, which

*The common occurrence of iron in respiratory systems may be due to the fact that ferrous iron does readily undergo redox reaction with O_2.

Form	Formula	Simplified Electronic Structure	Spin of Outer Electrons
Triplet oxygen (normal atmospheric form)	$^3O_2$	Ȯ—Ȯ	↑ ↑
Singlet oxygen	$^1O_2$	Ȯ—Ö	↑↓ ◯ or ↑ ↓
Superoxide free radical	O_2^-	Ö—Ȯ	↑↓ ↑
Peroxide	O_2^{2-}	Ö—Ö	↑↓ ↑↓

*Other oxygen species: Oxide, O^{2-}; hydroxyl free radical, OH·; water, H_2O.

is the most common, is called **triplet oxygen,** abbreviated $^3O_2$ (see Table 8.7).

Singlet oxygen, abbreviated $^1O_2$, is a higher energy form of oxygen. It is formed when the two outer electrons achieve antiparallel spins, either in the same orbital or in separate orbitals (Table 8.7). Singlet oxygen is extremely reactive and is one of the main forms of oxygen that is toxic to living organisms. Note that no additional electrons have been added to convert triplet to singlet oxygen, but energy has been added to change the spin of one of the electrons. Singlet oxygen is produced chemically in a variety of ways and is an atmospheric pollutant, being one of the components present in smog. Singlet oxygen may also be produced biochemically, either spontaneously or via specific enzyme systems. The most common biochemical means of singlet-oxygen production involves a reaction of triplet oxygen with visible light. This involves a dye molecule which acts as a mediator by absorbing light (see Section 8.6). Another important means of generating singlet oxygen is through the lactoperoxidase and myeloperoxidase enzyme systems. These systems are present in milk, saliva, and body cells (phagocytes) that eat and digest invading microorganisms. When a phagocyte ingests a microbial cell (see Figure 15.10), the peroxidase system is activated and singlet oxygen is generated. Due to the high toxicity of singlet oxygen, the phagocytes thus are able to kill invading organisms. In peroxidase systems, chloride ion is converted to hypochlorite (OCl^-), which reacts with hydrogen peroxide (H_2O_2, formation described below) in the following way: $H_2O_2 + OCl^- \rightarrow {}^1O_2 + OH^- + HCl$. The high reactivity of singlet oxygen means that if it is present in a biological system, a wide variety of uncontrolled and undesirable oxidation reactions can occur, leading to oxidative destruction of vital cell components such as phospholipids in the plasma membrane.

The reduction of O_2 to $2H_2O$, which occurs during respiration (see Section 4.7), requires addition of four electrons (Figure 8.14). This reduction usually occurs by single electron steps, and the first product formed in the reduction of O_2 is **superoxide** anion, O_2^- (see Table 8.7). Superoxide is probably formed transiently in small

$$O_2 + e^- \longrightarrow O_2^- \quad \text{Superoxide}$$
$$O_2^- + e^- + 2H^+ \longrightarrow H_2O_2 \quad \text{Hydrogen peroxide}$$
$$H_2O_2 + e^- + H^+ \longrightarrow H_2O + OH\cdot \quad \text{Hydroxyl radical}$$
$$OH\cdot + e^- + H^+ \longrightarrow H_2O \quad \text{Water}$$

$$\text{Overall:} \quad O_2 + 4e^- + 4H^+ \longrightarrow 2H_2O$$

FIGURE 8.14 Four-electron reduction of O_2 to water by stepwise addition of electrons. All of the intermediates formed are reactive and toxic to cells.

amounts during normal respiratory processes, and it is also produced by light through mediation of a dye and by one-electron transfers to oxygen. Flavins, flavoproteins, quinones, thiols, and iron-sulfur proteins all carry out one-electron reductions of oxygen to superoxide. Superoxide is highly reactive and can cause oxidative destruction of lipids and other biochemical components. It has the longest life of the various oxygen intermediates and may even pass from one cell to another. As we will see below, superoxide is probably important in explaining the oxygen sensitivity of obligate anaerobes.

The next product in the stepwise reduction of oxygen is peroxide, O_2^{2-} (Table 8.7). Peroxide anion is most familiar in the form of hydrogen peroxide, which is sufficiently stable so that it can be used as an item of chemical commerce. Peroxide is commonly formed biochemically during respiratory processes by a two-electron reduction of O_2, generally mediated by flavoproteins. Hydrogen peroxide is probably produced in small amounts by almost all organisms growing aerobically.

Hydroxyl free radical, $OH\cdot$, is the most reactive of the various oxygen intermediates. It is the most potent oxidizing agent known and is capable of attacking any of the organic substances present in cells. Hydroxyl radical is formed as a result of the action of ionizing radiation, and is probably one of the main agents in the killing of cells by X rays and gamma rays (see Section 8.6). Hydroxyl radical can also be produced chemically via a reaction (mediated by ferric iron) between superoxide and peroxide: $O_2^- + H_2O_2 \rightarrow OH^- + OH\cdot + O_2$. Thus, in any living system in which superoxide and peroxide are generated simultaneously, there is the possibility of the biochemical formation of hydroxyl radical.

Enzymes that act on oxygen derivatives With such an array of toxic oxygen derivatives, it is perhaps not surprising that organisms have developed enzymes that destroy certain oxygen products. The most common enzyme in this category is **catalase,** the activity of which is illustrated in Figure 8.15a. A simple means for detecting catalase activity in a microbial culture is illustrated in Figure 8.16. Another enzyme that acts on hydrogen peroxide is **peroxidase** (Figure 8.15b), which requires the presence of a reducing substance, usually NADH. Superoxide is destroyed by the enzyme **superoxide dismutase** (Figure 8.15c), which combines two molecules of superoxide to form one molecule of hydrogen peroxide and one molecule of oxygen. Superoxide dismutase and catalase working together can thus bring about the conversion of superoxide back to oxygen.

Anaerobic microorganisms Environments of low reduction potential (that is, anaerobic environments) include muds and other sediments

(*a*) **Catalase:**
$$H_2O_2 + H_2O_2 \longrightarrow 2H_2O + O_2$$
Hydrogen peroxide

(*b*) **Peroxidase:**
$$H_2O_2 + NADH + H^+ \longrightarrow 2H_2O + NAD^+$$

(*c*) **Superoxide dismutase:**
$$O_2^- + O_2^- + 2H^+ \longrightarrow H_2O_2 + O_2$$
Superoxide

FIGURE 8.15 Enzymes acting on toxic oxygen species. Catalases and peroxidases are generally porphyrin-containing proteins, although some flavoproteins may act in this manner. Superoxide dismutases are metal-containing proteins, the metal being either copper, zinc, manganese, or iron. The distribution of catalase and superoxide dismutase in bacteria with various oxygen tolerances is as follows:

	Catalase	Superoxide dismutase
Obligate and facultative aerobes	+	+
Aerotolerant anaerobes	−	+*
Obligate anaerobes	−	−

*Some of the lactic acid bacteria (*Lactobacillus plantarum, Streptococcus faecalis*) do not synthesize this enzyme, but instead use manganese (Mn^{2+}) ions to perform the necessary dismutation activity. (Y. Kono and I. Fridovich. 1981. J. Bacteriol. 145:442–451.)

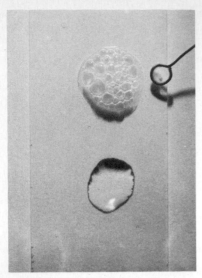

FIGURE 8.16 Method for testing a microbial culture for the presence of catalase. A heavy loopful of cells from an agar culture was mixed on a slide with a drop of 30 percent hydrogen peroxide. The immediate appearance of bubbles is indicative of the presence of catalase. The bubbles are O_2, produced by the reaction:

$$H_2O_2 + H_2O_2 \longrightarrow 2H_2O + O_2$$

of lakes, rivers, and oceans; bogs and marshes; waterlogged soils; canned foods; intestinal tracts of animals; the oral cavity of animals, especially around the teeth (see Section 14.4), certain sewage-treatment systems; deep underground areas such as oil pockets; and some underground waters. In most of these habitats, the low reduction potential is due to the activities of organisms, mainly bacteria, that consume oxygen during respiration. If no replacement oxygen is available, the habitat becomes anaerobic.

So far as is known, obligate anaerobiosis occurs in only two groups of microorganisms, the bacteria and the protozoa. The best known obligately anaerobic bacteria belong to the genus *Clostridium*, a group of Gram-positive spore-forming rods. Clostridia are widespread in soil, lake sediments, and intestinal tracts, and are often responsible for spoilage of canned foods. Other obligately anaerobic bacteria are found in the genera *Methanobacterium*, *Bacteroides*, *Fusobacterium*, and *Ruminococcus*. Even among obligate anaerobes, the sensitivity to oxygen varies; some organisms are able to tolerate traces of oxygen whereas others are not.

The reasons why obligate anaerobes are intolerant to O_2 are not completely understood, although a good correlation exists (Figure 8.15) between the presence of superoxide dismutase activity and the ability to grow in the presence of O_2. One idea is that obligate anaerobes form superoxide in various side reactions during growth; they cannot get rid of this toxic radical due to lack of superoxide dismutase activity and thus are killed. However, although it is well established that obligate anaerobes cannot grow in the presence of O_2, there has really been little work on the killing of these organisms by O_2. It is now known that certain methanogenic bacteria, usually considered to be among the most O_2-sensitive organisms, are not killed by O_2 but are only inhibited from growing. If a strong reducing agent is added to a culture which has been aerated, in order to scrub out all traces of the added O_2, then growth will resume.

Working with obligate anaerobes requires considerable care to avoid introduction of O_2 into the culture. Dissolved O_2 is driven out of the culture tube during heat sterilization, and can be prevented from returning by the use of an oxygen-impermeable stopper. A good reducing agent must be added to the medium to scrub out any traces of O_2 which may enter; suitable reducing agents are cysteine, sodium or ferrous sulfide, or titanium citrate. Since some reducing agents are themselves toxic, it is important to select an appropriate agent. If culture tubes must be opened in the air, then a stream of nitrogen or hydrogen should be directed down into the tube, to prevent entry of O_2. Anaerobic hoods are now widely used, permitting operations in an O_2-free atmosphere. For some obligate anaerobes, such as the clostridia, such extreme precautions are not necessary, and simple sealing of the liquid in the culture tubes with paraffin or mineral oil is sufficient, the O_2 having first been driven off by steaming (followed by slow cooling in an upright position). For incubation of petri plates, anaerobic jars can be used, the atmosphere in the jars containing O_2-free nitrogen or hydrogen. It is always much easier to grow obligate anaerobes in mixed culture than in pure culture, because in mixtures there are generally facultative organisms present that will consume traces of O_2 and maintain O_2-free conditions.

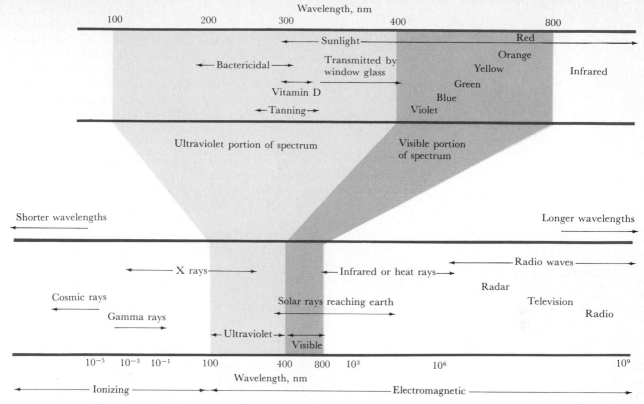

FIGURE 8.17 The electromagnetic spectrum.

8.6

Radiation

Radiation refers to those sources of energy that are transmitted from one place to another through air or outer space. They may consist either of particles or of electromagnetic waves. Particulate radiation consists of streams of atoms, electrons, or neutrons and will not be discussed here. **Electromagnetic waves** include radio waves, light, and X rays. The character of an electromagnetic wave is determined by its wavelength, as outlined in Figure 8.17. The longest-wavelength radiations are radio waves and have no detectable biological effects. Somewhat shorter are the **infrared rays,** which produce heat when absorbed, although, as we discussed in Chapter 6, infrared rays with wavelengths less than 1,000 nm may be used by photosynthetic bacteria as energy sources. The *visible* portion of the spectrum, which contains the radiations the human eye can see, ranges from about 380 to 760 nm. These wavelengths are the main energy sources used by algae for photosynthesis, as discussed in Chapter 6. Still shorter are the **ultraviolet radiations,** which range from 380 nm down to about 200 nm.* Ultraviolet radiations are damaging to living organisms,

*Strictly speaking, the word *light* refers only to the radiation seen by the human eye, so that the usages *ultraviolet light* and *infrared light* are incorrect. Proper usage is *ultraviolet radiation* and *infrared radiation.*

especially at the shorter wavelengths. The strongest source of infrared, visible, and ultraviolet radiations is the sun, although not all of the short ultraviolet and long infrared rays reach the earth (since they are absorbed in the atmosphere). The **ionizing radiations** are of still shorter wavelength and include X rays, mainly of artificial origin; gamma rays which are similar but are products of decay of radioactive materials; and cosmic rays, reaching earth from outer space. These radiations cause water and other substances to ionize and in general have harmful effects on organisms.

Ultraviolet radiation   Ultraviolet (UV) radiation is of considerable microbiological interest because of its frequent lethal action on microorganisms. Although the sun emits intense UV radiation of all wavelengths, only that of longer wavelengths penetrates the atmosphere of the earth and reaches its surface. Microorganisms that are wafted to high altitudes or dragged into space on the outside of rockets are quickly killed by the UV there. Shorter-wavelength UV radiation is produced in germicidal lamps for killing microorganisms.

The purine and pyrimidine bases of the nucleic acids absorb UV radiation strongly, and the absorption for DNA and RNA is at 260 nm. Proteins also absorb UV, but have a peak at 280 nm, where the aromatic amino acids (tryptophan, phenylalanine, tyrosine) absorb. It is now well established that killing of cells by UV radiation is due primarily to its action on DNA, so that UV radiation at 260 nm is most effective as a lethal agent. Although several effects are known, one well-established effect is the induction in DNA of thymine dimers, a state in which two adjacent thymine molecules are covalently joined, so that replication of the DNA cannot occur. The genetic effect of UV radiation on DNA are discussed in Chapter 9.

Many microorganisms have repair enzymes which correct damage induced in DNA by UV radiation. Damage in a given region of the DNA molecule is usually only on one of the two strands, and enzymes exist that will hydrolyze the phosphodiester bonds at each end of the damaged region and thus excise the altered material. Other enzymes then catalyze the insertion of new nucleotides that are complementary to the undamaged strand. One kind of repair enzyme works only when the damaged cell is activated by visible light in the blue region of the spectrum; recovery of UV radiation damage through action of these enzymes is called **photoreactivation** (Figure 8.18).

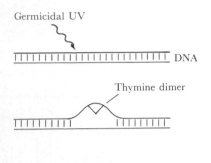

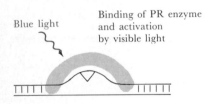

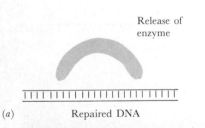

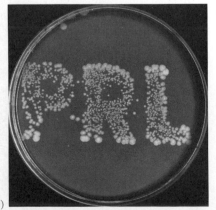

Germicidal UV

DNA

Thymine dimer

Blue light

Binding of PR enzyme and activation by visible light

Release of enzyme

Repaired DNA

(a)

(b)

FIGURE 8.18   The phenomenon of photoreactivation. (a) Mechanisms of ultraviolet damage and repair. Thymine dimers formed as a result of UV damage are repaired by a photoreactivation (PR) enzyme, which is activated by blue (visible) light. (b) Demonstration of photoreactivation of *Escherichia coli*. About 10^7 bacteria were spread over the surface of an agar plate and irradiated with a dose of UV sufficient to reduce survival to less than 1 in 10^7. The areas within the letters PRL were exposed to visible light for 20 minutes before the plate was incubated, resulting in recovery of viability. (Courtesy of B. A. Bridges.)

Other repair enzymes work in the absence of light (**dark reactivation**). Because of the existence of these repair enzymes, a given dose of ultraviolet radiation will cause death only in the event that the damage it induces is greater than that which can be repaired.

Although the atmosphere markedly reduces the amount of ultraviolet radiation from the sun, sufficient radiation reaches the surface of the earth down to wavelengths of about 300 nm to have some biological effects. The most effective wavelengths in this region for cell killing are around 340 nm. The mechanism by which ultraviolet radiation of these wavelengths kills cells is not known, but it is definitely not a direct effect on DNA. The germicidal nature of ultraviolet radiation that reaches the earth's surface may have some significant ecological effects, especially in habitats with direct exposure to the sun, such as shallow water bodies and surfaces of rocks and soil. Maximum UV radiation occurs, of course, at midday, when solar radiation is at its peak, and the UV effect is greatest on clear, dry days, when solar penetration is maximal.

Visible light   Visible light of sufficient intensity may cause cellular damage and death. Two mechanisms of cell killing by visible light have been recognized: one involves molecular oxygen and leads to the generation of singlet oxygen (see Section 8.5); the second functions independently of oxygen. Both mechanisms require the presence in the cell of photosensitizers, which are substances that absorb light. Virtually all organisms have light-absorbing components, such as the cytochromes, flavins, and chlorophylls. When such substances absorb light, they become activated and are raised to a higher energy state. They can sometimes return to the ground state by emitting light (fluorescence), but return to the ground state may also occur by the transfer of energy to another cellular component. Some transfers are beneficial, as we saw with the concept of reaction center chlorophyll in Section 6.2. However, many of these energy transfers can lead to cellular damage. In oxygen-independent damage, the sensitizer transfers energy to any of a variety of cellular components, leading to the formation of free radicals, which are highly reactive and can undergo destructive reactions. This oxygen-independent mechanism is of only minor consequence in comparison to the oxygen-dependent mechanism involving singlet oxygen. As we have seen in Section 8.5, singlet oxygen is highly reactive and is very lethal to living organisms.

How is visible light involved in the generation of singlet oxygen? When the photosensitizer (abbreviated S) is activated by light, the first activated state formed is the singlet excited species, $^1S^*$. This species has a very short lifetime, and the energy can be dissipated either through emission of light (fluorescence) or the sensitized dye can undergo a transition called intersystem crossing, in which it is converted to the activated triplet state:

$$^1S^* \longrightarrow {}^3S^* \qquad \textit{Intersystem crossing}$$

The activated triplet state of the sensitizer has a much longer lifetime than the singlet state and can now undergo direct reaction with ground-state oxygen (which, as we have seen, is normally in the triplet state) to form singlet oxygen:

$$^3S^* + {}^3O_2 \longrightarrow S + {}^1O_2^*$$

The singlet oxygen thus generated is very reactive and can cause a variety of lethal effects in the cell (see Section 8.5).

Evidence that a visible light effect involves O_2 can be obtained by measuring the effect both aerobically and anaerobically. As seen in Figure 8.19, killing by visible light only occurs under aerobic conditions.

Some microorganisms have special protective substances, which prevent singlet-oxygen attack. The most significant such substances are the **carotenoids** (see Figure 6.9), which are the most efficient of all compounds in quenching singlet oxygen. The reaction between carotenoid and singlet oxygen is not chemical (in the sense of an oxidation-reduction reaction) but physical, in the sense that it involves an interaction between the carotenoid and singlet oxygen to change the spin on the electron in the oxygen molecule. As a result, singlet oxygen is returned to the ground state, and the now activated carotenoid regenerates the ground state spontaneously. The dramatic effect of carotenoid in protecting against light-induced singlet oxygen action is illustrated in Figure 8.19, in which a mutant that lacks carotenoid is killed by visible light, whereas the carotenoid-containing wild type is protected. It was mentioned in Section 8.5 that singlet oxygen is generated by an enzymatic mechanism in phagocytic cells of the animal body, and is one of the agents involved in the process by which phagocytic cells kill invading microorganisms. It has been shown that carotenoid-containing bacteria are protected against killing by phagocytes, probably because the carotenoid neutralizes the action of singlet oxygen. This has been shown by the use of carotenoid-deficient mutants, but the same phenomenon has a clinical basis, in that pathogenic strains of *Staphylococcus* are almost always pigmented (and thus given the species designation *S. aureus*) whereas unpigmented strains (white in color and called *S. albus*) are generally nonpathogenic.

The relationship between pigmentation and photoprotection is also shown in the common observation that many airborne microorganisms are pigmented. In the air, visible light effects would be expected to be high, and a mechanism for protection against such effects should have selective advantage. Very few pathogenic bacteria survive long in bright sunlight, but *S. aureus* does; it is significant to note that this pathogen is often airborne and is also very abundant on the surface of human skin.

Ionizing radiation Ionizing radiation does not kill by directly affecting the cell constituents, but by indirectly inducing in the medium reactive chemical radicals (free radicals), of which the most important is the hydroxyl radical, $OH\cdot$ (see Section 8.5). Free radicals can react with and inactivate sensitive macromolecules in the cell. Ionizing radiation can act on all cellular constituents, but death usually results from effects on DNA. Dark reactivation of damage due to ionizing radiation occurs, but photoreactivation is absent. DNA is probably no more sensitive to ionizing radiation than other structures, but since each DNA molecule contains only one copy of most genes, inactivation of any critical gene can lead to death, whereas inactivation

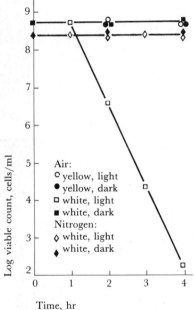

FIGURE 8.19 Effect of sunlight on the viability of white- and yellow-pigmented *Micrococcus luteus*. Note that death occurs in the white strain but not in the pigmented strain, from which it is inferred that the yellow pigment confers protection from sunlight. Killing does not occur anaerobically, but only in the presence of air. From this it is deduced that light-induced killing is a photooxidation reaction requiring oxygen. (From Mathews, M. M., and W. R. Sistrom. 1959. Nature 184:1892.)

of single protein molecules, which are virtually always present in multiple copies, will not cause death.

Bacteria vary markedly in sensitivity to ionizing radiation. At one end are sensitive organisms such as *Pseudomonas,* and at the other are resistant organisms such as *Micrococcus* and *Streptococcus.* One organism, *Micrococcus radiodurans,* is unusually resistant to ionizing radiation and is isolated frequently from irradiated foods and other products treated with ionizing radiation. The vegetative cells of the genera *Bacillus* and *Clostridium* are moderately radiation resistant, but their spores are exceptional in their resistance to ionizing radiation (Section 2.11). Probably the main reason that spores are radiation resistant is their low water content, which reduces the efficiency of ionization events. Vegetative cells are also more resistant to radiation when they are dry than when they are wet.

Another factor affecting radiation resistance is the presence of chemical protective agents. These are mainly organic sulfhydryl compounds that react with radiation-induced ionized substances and repair them, converting them back to normal substances. Such protective compounds may be present in high amounts in organisms that are unusually radiation resistant.

Naturally occurring ionizing radiations such as cosmic rays and emanations from natural radioactive materials probably never have any harmful effects on organisms, as their levels are too low.

8.7

Summary

We have seen in this chapter that a variety of environmental factors influence the growth and viability of microorganisms. The environmental factors we have considered are temperature, water and water activity, hydrostatic pressure, acidity and pH, oxygen (and its various derivatives), and visible and ionizing radiation. A striking generalization is that microorganisms vary widely in their ability to adapt to environmental extremes. Some microorganisms are rapidly killed under conditions where other organisms not only survive, but thrive. However, there do seem to be extreme limits above which life, in any form, is not possible. At these limits, physical-chemical forces become insurmountable.

An understanding of the influence of environmental factors is of great importance in developing means for controlling microbial growth in many practical situations. Sterilization of sensitive goods, preservation of foods and dairy products, prevention of deterioration, all require a detailed understanding of how the environment influences microbial development. Additionally, an understanding of the ecology of microorganisms can only be achieved when a clear understanding of environmental factors has been developed.

Supplementary Readings

Brock, T. D. 1967. Life at high temperatures. Science 158:1012–1019. **Brock, T. D.** 1969. Microbial growth under extreme conditions. Pages 15–41 *in* Microbial growth. 19th Symposium, Society for General Microbiology. Cambridge University Press, New York. **Brock, T. D.** 1970. High temperature systems. Annu. Rev. Ecol. Syst. 1:191–220. The ecology, evolution, and biochemistry of microorganisms living at high temperatures are discussed. **Brock, T. D.** 1978. Thermophilic microorganisms and life at high temperatures. Springer-Verlag, New York. A detailed, extensively illustrated review of microorganisms and thermal environments.

Fridovich, I. 1977. Oxygen is toxic! BioScience 27:462–466. A review of the chemistry of toxic oxygen products and the enzymes that destroy them.

Galápagos Biology Expedition Participants. 1979. Galápagos '79: initial findings of a biology quest. Oceanus 22:2–10. Description of newly discovered community on the ocean floor.

Horikoshi, K., and **T. Akiba.** 1982. Alkalophilic microorganisms. Springer-Verlag, New York. Systematic analysis of physiology of high pH bacteria.

Innis, W. E., and **J. L. Ingraham.** 1978. Microbial life at low temperatures: mechanisms and molecular aspects. *In* D. J. Kushner, ed. Microbial life in extreme environments. Academic Press, London.

Karl, D. M., H. W. Jannasch, and **C. O. Wirsen.** 1977. Deep-sea primary production at the Galápagos hydrothermal vents. Science 207:1345–1347. Experiments showing bacteria are active at great depths.

Kushner, D. J., ed. 1978. Microbial life in extreme environments. Adademic Press, London. Detailed reviews of growth parameters, with emphasis on effects of extremes.

Marquis, R. E., and **P. Matsumura.** 1978. Microbial life under pressure. *In* D. J. Kushner, ed. Microbial life in extreme environments. Academic Press, London. Excellent review of biochemistry and physiological effects of high pressure.

Mazur, P. 1970. Cryobiology: the freezing of biological systems. Science 168:939–949. An excellent review with emphasis on mechanisms of killing and survival.

Morris, J. G. 1975. The physiology of obligate anaerobiosis. Adv. Microb. Physiol. 12:169–246. A detailed review of obligate anaerobes, with a good section on oxygen chemistry and toxicity.

Mossel, D. A. A., and **M. Ingram.** 1955. The physiology of the microbial spoilage of foods. J. Appl. Bacteriol. 18:232–268. A clear and detailed discussion.

Nasim, A., and **A. P. James.** 1978. Life under conditions of high irradiation. *In* D. J. Kushner, ed. Microbial life in extreme environments. Academic Press, London. Thorough coverage of damaging effects of radiation as well as protective mechanisms.

Scott, W. J. 1957. Water relations of food spoilage microorganisms. Adv. Food Res. 7:83–127. Already a classic paper, this review of the effect of water activity on microorganisms emphasizes the quantitative and biophysical aspects.

Smith, D. W. 1982. Extreme natural environments. *In* R. G. Burns and J. H. Slater, eds. Experimental microbial ecology. Blackwell Scientific, Oxford. Review of major examples of extreme environments; current references.

Stoeckenius, W., and **R. A. Bogomolni.** 1982. Bacteriorhodopsin and related pigments of halobacteria. Annu. Rev. Biochem. 51:587–616. Mechanism of action of bacteriorhodopsin in gathering light energy for ATP generation.

Stumbo, C. R. 1965. Thermobacteriology in food processing. Academic Press, New York. Detailed discussion of thermal death time, and how the mathematical relationships can be applied to canning.

Yayanos, A. A., A. S. Dietz, and **R. Van Boxtel.** 1979. Isolation of a deep-sea barophilic bacterium and some of its growth characteristics. Science 205:808–810. First report of a truly barophilic bacterium.

Macromolecules and Molecular Genetics

We begin now a study of the genetics of microorganisms which will extend over the next four chapters. **Genetics** is the discipline which deals with the mechanisms by which traits are passed from one organism to another. The study of genetics is central to an understanding of the variability of organisms and the evolution of species. Genetics is also a major research tool in the attempts to understand the molecular mechanisms by which cells function. Further, genetics provides us with approaches to the introduction of new properties into organisms, and thus genetics has many industrial and medical applications.

A **gene** may be defined as an entity which specifies the structure of a single protein polypeptide. Genetic phenomena involve three types of macromolecules: deoxyribonucleic acid (DNA), the genetic material of the cell; ribonucleic acid (RNA), the intermediary or messenger, and proteins, the functional entities of the living cell. During growth, all three types of macromolecules are synthesized. DNA is **replicated,** leading to the synthesis of exact copies (Figure 9.1). The information in DNA is also **transcribed** into complementary sequences of nucleotide bases in RNA (Figure 9.1); this RNA, containing the information for the amino acid sequence of the protein, is called *messenger RNA* (mRNA). Messenger RNA is then **translated,** using the specific protein-synthesizing machinery of the ribosomes, and the translation product is protein. Because the steps from DNA to RNA to protein involved the transfer of information, these macromolecules are often called **informational macromolecules** to distinguish them from macromolecules such as polysaccharides and lipids which are large but not informational.

The properties of each protein or enzyme of the cell are determined ultimately by its amino acid sequence. There are 20 different amino acids commonly found in proteins, and often 100 or more amino acid residues occur per protein molecule. How is the cell able to ensure that for each separate enzyme, the proper amino acids are

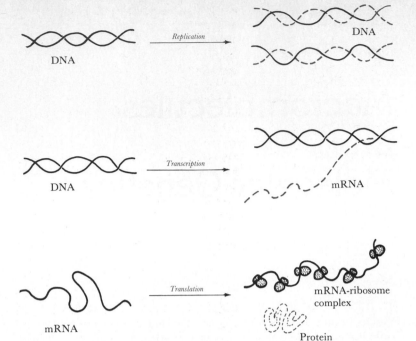

FIGURE 9.1 The three elements of macromolecular synthesis. The newly synthesized molecules are shown in color.

connected in the correct order? The amino acid sequence of each protein is specified by a gene, which is a portion of the DNA molecule, and it is the sequence of purine and pyrimidine bases within this gene that **codes** for the amino acid sequence of the protein.

Importance Microbial genetics is important for a number of reasons:

1. Gene function is at the basis of cell function, and basic research in microbial genetics is necessary to understand how microbes function.
2. Microbes provide relatively simple systems for studying genetic phenomena, and are thus useful tools in attempts to decipher the mechanisms underlying the genetics of all organisms.
3. Microbes are used for the isolation and duplication of specific genes from other organisms, a technique called **molecular cloning**. In molecular cloning, genes are manipulated and placed in a microbe where they can be induced to increase in number.
4. Microbes produce many substances of value in industry, such as antibiotics, and genetic manipulations can be used to increase yields and improve manufacturing processes. Also, genes of higher organisms that specify the production of particular substances, such as human insulin, can be transferred by molecular cloning into microorganisms, and the latter used for the production of these useful substances.
5. Many diseases are caused by microorganisms, and genetic traits underly these harmful activities. By understanding the genetics of disease-causing microbes, we can more readily control them and prevent their growth in the body. Viruses, although agents of dis-

ease, can be thought of as genetic elements, and understanding the genetics of viruses helps us to control virus disease.

Procaryotic and eucaryotic genetics Procaryotes have relatively simple genetic systems. Their chromosomes are single DNA molecules. Mechanisms for the transfer of genes from one procaryotic cell to another are also simple and easy to study. Eucaryotes, even the simplest eucaryotic microbes, have much more complex genetic systems. Their chromosomes are much more complex structurally, and are present in larger numbers. They have fairly complex mechanisms of sexual reproduction for bringing about genetic recombination between organisms. Scientists turning from procaryotic to eucaryotic genetics were surprised to learn that gene organization in eucaryotes had some unique features, and was not simply more complicated procaryotic genetics. Thus in this chapter we will be making contrasts between procaryotes and eucaryotes. We will develop first procaryotic genetics and will then highlight those features that are different in eucaryotes.

9.1

Overview

What transformations do informational macromolecules undergo during cell growth and division? It is the purpose of the present section to outline briefly some of the processes that will be discussed in this chapter.

A cell is an integrated system containing a large number of specific macromolecules. When a cell divides and forms two cells, all of these macromolecules are duplicated. The fidelity of duplication is very high, although occasional errors do occur. The molecular processes underlying cell growth can be divided into a number of stages which are described briefly here.

1. *Replication* The DNA molecule is a **double helix** of two long chains. During replication, DNA, containing the master *genetic code,* duplicates. The products of DNA replication are two molecules, each a double helix, the two strands thus becoming four strands.
2. *Genetic code* The specific sequence of amino acids in each protein is directed by a specific sequence of bases in DNA. It takes three bases to code for a single amino acid, and each triplet of bases is called a **codon.** There is a one-to-one correspondence between the base sequence of a gene and the amino acid sequence of a protein. A change in a single base changes the codon and can lead to a change in the amino acid in the protein. Such changes, which are frequently detrimental, are called **mutations.** Mutants are essential in genetic research, as they make possible the location of genes, through the construction of genetic crosses between related organisms.
3. *Transcription* DNA does not function directly in protein synthesis, but through an RNA intermediate. The transfer of the information of the genetic code to RNA is called **transcription,** and the RNA molecule carrying the information is called mes-

senger RNA (mRNA). Messenger RNA molecules frequently contain the instructions for making more than one protein. In most cases, at any particular location on the chromosome, only one strand of the DNA is transcribed, and the code of this strand is then contained in the mRNA.

4. *Translation*   The genetic code is translated into protein by means of the protein-synthesizing system. This system consists of **ribosomes, transfer RNA,** and a number of enzymes. The ribosomes are the structures to which messenger RNA attaches. Transfer RNA (tRNA) is the key link between codon and amino acid. There is one or more separate tRNA molecules corresponding to each amino acid, and the tRNA has a triplet of three bases, the **anticodon,** which is complementary to the codon of the messenger RNA. An enzyme brings about the attachment of the correct amino acid to the correct tRNA.

5. *Regulation*   Not all proteins are synthesized at equal rates. Complex systems of regulation exist which control the rates of synthesis of proteins. Some proteins, called **inducible,** are synthesized only when small molecules called **inducers** are present. Usually the inducer is a substrate of the enzyme, and induction thus ensures that the enzyme is only formed when it is needed. Another class of enzymes, called **repressible,** are synthesized only in the absence of specific small molecules, generally biosynthetic products. Enzyme repression and enzyme induction have as their basis the same underlying mechanisms, which is the regulation of mRNA synthesis. A specific protein called a **repressor protein** attaches to a region at the beginning of a gene and prevents the synthesis of mRNA. In the case of inducible enzymes, the repressor protein is inactivated when the inducer is present, whereas with repressible enzymes, the repressor protein is activated when the small molecule repressor is present.

6. *Eucaryotic gene structure and function*   Each eucaryotic chromosome consists of a single DNA molecule bound to proteins called **histones.** Eucaryotes generally have much more DNA per cell than procaryotes, and the DNA is present in a number of separate chromosomes. Transcription and translation are spatially separated in eucaryotes. Transcription occurs in the nucleus and the RNA molecules move to the cytoplasm for translation. The genes of eucaryotes are frequently split, with partial noncoding regions separating the coding regions. The coding sequences are called **exons,** and the intervening noncoding regions **introns.** Both intron and exon regions are transcribed into RNA, and the functional mRNA is subsequently formed by enzymatic removal of the noncoding regions.

Although the genetic processes of eucaryotes differ markedly from those of procaryotes, the genetics of eucaryotic organelles such as mitochondria and chloroplasts much more closely resembles that of procaryotes. From an evolutionary viewpoint, genetic and macromolecular studies provide strong support for the idea that mitochondria and chloroplasts of eucaryotes have arisen from procaryotic cells by a process of symbiosis (see Section 19.33).

DNA Structure and Synthesis

The problem of DNA replication is simply put: the nucleotide base sequence residing in each long DNA molecule must be precisely duplicated into two new molecules. The cell has solved this seemingly complex problem in an elegant fashion: the two strands of the DNA double helix are not identical but **complementary.** (A good example of two nonidentical but complementary objdcts are the left hand and the right hand.) The complementarity of DNA molecules arises because of the specific pairing of the purine and pyrimidine bases: adenine always pairs with thymine and guanine always pairs with cytosine (Figure 9.2). The structure conforms to a double helix (Figure 9.3), and replication occurs by unwinding of the existing helix and thc addition of new structure via the complementary pairing just mentioned (Figure 9.3).

The structure of a single DNA chain is shown in Figure 9.4. It can be seen that the backbone of the chain consists of alternating units of phosphate and the sugar *deoxyribose;* connected to each sugar is one of the nucleic acid *bases.* The numbering system for the positions of sugar and base is shown; the phosphate connecting two sugars spans from the 3' of one sugar to the 5' of the adjacent sugar. The phosphate linkage is a diester, since a single phosphate is connected by ester linkage to two separate sugars. At one end of the DNA molecule the sugar has a phosphate on the 5' hydroxyl, whereas at the

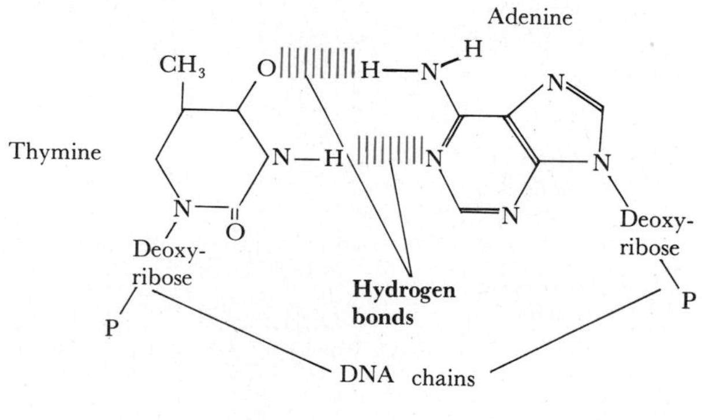

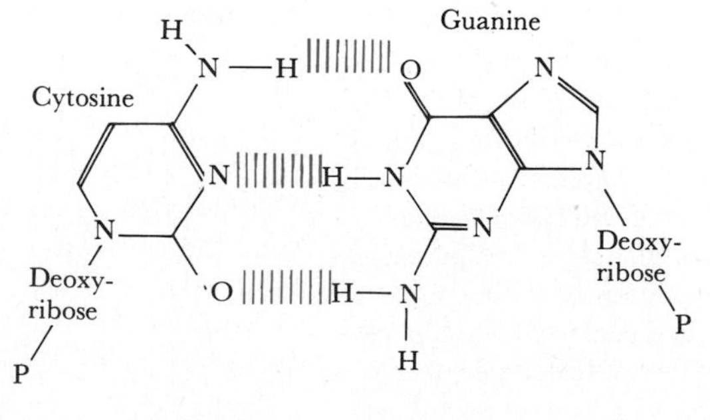

FIGURE 9.2 Specific pairing between adenine (A) and thymine (T) and between guanine (G) and cytosine (C), via hydrogen bonds. Because the G-C base pair involves three hydrogen bonds, it is stronger than the A-T base pair, which involves only two hydrogen bonds.

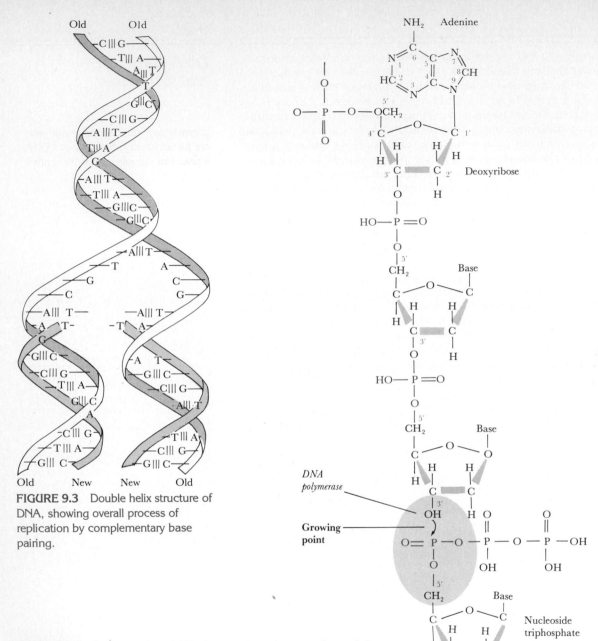

FIGURE 9.3 Double helix structure of DNA, showing overall process of replication by complementary base pairing.

Old   New   New   Old

Adenine

Deoxyribose

DNA polymerase

Growing point

Nucleoside triphosphate

FIGURE 9.4 Structure of the DNA chain and mechanism of growth by addition from a deoxyribonucleotide triphosphate at the 3′ end of the chain. The complete structure of one base, adenine, is shown to illustrate the numbering system of bases and of deoxyribose. Growth always proceeds from the 5′ to the 3′ end. The enzyme DNA polymerase catalyzes the addition reaction. The four deoxyribonucleotides that serve as precursors are deoxythymidine triphosphate (dTTP), deoxyadenosine triphosphate (dATP), deoxyguanosine triphosphate (dGTP), and deoxycytidine triphosphate (dCTP). The two terminal phosphates of the triphosphate are split off as pyrophosphate (PP_i).

other end the sugar has a free hydroxyl at the 3′ position. Replication of DNA proceeds by addition of a new base-sugar-phosphate unit at the free 3′ end; thus DNA synthesis always occurs 5′ → 3′. The precursor of the new unit added is a deoxyribonucleoside triphosphate (Figure 9.4).

DNA replication   DNA replication involves an opening up of the two-stranded double helix and insertion of new units by complementary base pairing (Figure 9.3). In the overall process, DNA synthesis occurs in both directions from a defined starting point, but because DNA replication occurs only at the 3′ hydroxyl end, and the two strands run in opposite directions, it is not possible for DNA replication to occur on both strands simply by successive addition of single nucleotides. Instead, DNA replication occurs by the addition of short fragments, which are subsequently joined (Figure 9.5). The enzyme **DNA polymerase** connects each nucleotide to the short fragment, and the fragments are subsequently joined by another enzyme, **DNA ligase,** thus forming the completed new strand.* In this sequence, the preexisting DNA acts as a template (preformed pattern) but not as a primer (point of origin). The strand opposite that being replicated acts as a template for ensuring that the proper nucleotide is inserted on the growing strand by complementary base pairing. In addition, the growing strand itself acts as a primer, to serve as a site at which the new nucleotide can be attached. DNA polymerase cannot act to synthesize DNA de novo, from a simple mixture of nucleotides, but only to copy preexisting DNA. In most cases, the initial priming function for DNA polymerase is met at the beginning of the chain by RNA or DNA fragments. When the double helix opens up at the beginning of replication (this opening takes place only at a specific site, the initiation point), a **primase** brings about the synthesis of a short strand of RNA or DNA complementary to the section of double-stranded DNA. Once the priming strand has been made, DNA polymerase brings about the addition of deoxyribonucleotides to the 3′ end of the priming strand. Subsequently, the priming strand is removed by a nuclease and the vacant region filled with complementary deoxyribonucleotides. Ultimately, the short fragments of DNA are joined, as described above.

DNA is circular in structure in procaryotic chromosomes, and in many viruses. Circular DNA molecules can become highly twisted, and such twisted circles are said to be **supercoiled.** The extent to which the DNA is supercoiled is determined by how many times the double strands are twisted about each other before they are connected at the ends. Enzymes called **topoisomerases** alter the twisting pattern of DNA. Supercoiling is an important factor in DNA replication and transcription. Unwinding is an essential feature of DNA replication and transcription, and because supercoiled DNA is under strain, it unwinds more easily than DNA that is not supercoiled. Thus, by regulating the degree of supercoiling, topoisomerases regulate the processes of replication and transcription. The topoisomerase that pro-

---

*There are actually several DNA polymerases, each having somewhat different functions. Some are more involved in repair of DNA than in the synthesis of new strands.

FIGURE 9.5   Replication of DNA always occurs at the 3′ end on both chains. Thus, the direction of DNA synthesis is always 5′→3′. Small DNA fragments are precursors of the long DNA chains. The individual nucleotides are inserted through the action of DNA polymerase, and the small fragments are joined through the action of DNA ligase. For the role of RNA as a primer, see text.

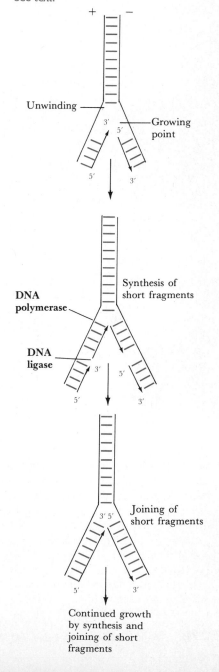

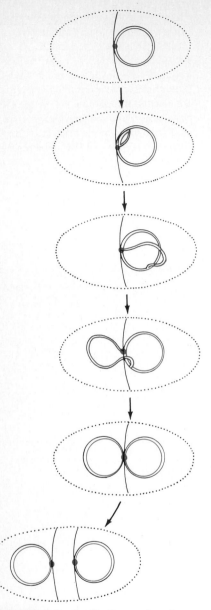

FIGURE 9.6 Replication and partitioning of DNA in a procaryote during cell division. Replication proceeds bidirectionally from a defined starting point.

motes DNA unwinding by introducing supercoiling is called **DNA gyrase,** and promotes parental strand separation at the replication fork.

DNA synthesis and cell division As we noted in Chapter 2, the DNA of procaryotes is contained in a single long molecule, which is arranged in the form of a circle. The replication of the DNA molecule begins at one point and moves bidirectionally at a constant rate around the DNA circle (Figure 9.6). The exact manner in which the DNA copies are partitioned into the two cells is not exactly understood, but one idea is that each DNA is attached at one point to the cell membrane and after replication one attachment point moves toward each of the developing cells. Cross-wall formation then occurs, followed by cell separation.

In eucaryotes, which have multiple chromosomes of considerably greater length, replication begins at a number of sites simultaneously and moves in both directions from each site. Clearly, the advantage of multiple initiation sites is that it permits replication of much larger amounts of DNA during reasonably short time spans.

Restriction and modification enzymes Organisms are occasionally faced with the problem of coping with foreign DNA, generally derived from viruses, that may derange cellular metabolism or initiate processes leading to death. Although a number of mechanisms for coping with foreign DNA exist, one of the most dramatic is that which results in its enzymatic destruction. The enzymes involved in the destruction of foreign DNA are remarkably specific in their action, an essential property if destruction of cellular DNA is to be avoided. This marvelous class of highly specific enzymes, called **restriction endonucleases,** combines with DNA only at sites with specific sequences of bases. Restriction enzymes have the unique property of making double-stranded breaks in DNA only at sequences which exhibit twofold symmetry around a given point. Thus one restriction endonuclease of *Escherichia coli* called EcoRI has the following recognition sequence:

$$5' \cdots \text{G--A--A--T--T--C} \cdots 3'$$
$$3' \cdots \text{C--T--T--A--A--G} \cdots 5'$$

The cleavage sites are indicated by arrows, and the axis of symmetry by a dashed line.

Nucleotide sequences with repetitious inversions, such as are recognized by restriction enzymes, are called **palindromes,** and have been found to be widespread in DNA. The palindromes recognized by restriction enzymes are relatively short, and probably are cleaved because the restriction enzymes are composed of identical subunits, one of which recognizes the sequence on a single chain. The significance of this specificity is that if the same sequence is found on each strand, such enzymes will always make double-stranded breaks, and such double-stranded breaks are not subject to correction by repair enzymes. This ensures that an invading nucleic acid will be destroyed.

Restriction enzymes are of great importance in DNA research, because they permit the formation of smaller fragments from large

TABLE 9.1   Recognition Sequences of a Few Restriction Endonucleases

Organism	Enzyme Designation	Recognition Sequence*
Escherichia coli	EcoRI	G↓AĀTTC
Escherichia coli	EcoRII	↓CĊAGG
Haemophilus influenzae	HindII	GTPy↓PuAĊ
Haemophilus influenzae	HindIII	Ā↓AGCTT
Haemophilus hemolyticus	HhaI	GĊG↓C
Bacillus subtilis	BsuRI	G↓ĊC
Brevibacterium albidum	BalI	TGG↓ĊCA
Thermus aquaticus	TaqI	T↓CGĀ

*Arrows indicate the sites of enzymatic attack. Asterisks indicate the site of methylation (modification). G = guanine; C = cytosine; A = adenine; T = thymine; Pu = any purine; Py = any pyrimidine. Only the 5′ → 3′ sequence is shown.

DNA molecules. Such fragments with defined termini, created as a result of the specificity of the restriction enzyme, are amenable to determination of nucleotide sequence, thus permitting the working out of the complete sequence of DNA molecules. A large collection of restriction enzymes has now been built up, which can be used in sequence determination. Recognition sequences for a few restriction enzymes are given in Table 9.1.

Another use of certain restriction enzymes is that they permit the conversion of DNA molecules into fragments which can be joined by DNA ligase. This enables laboratory researchers to create artificial genes, as will be discussed in Chapter 12.

An integral part of the cell's restriction mechanism is the **modification** of the specific sequences on its own DNA so that they are not attacked by its own restriction enzymes. Such modification generally involves methylation of specific bases within the recognition sequence so that the restriction nuclease can no longer act. Thus, for each restriction enzyme there must also be a modification enzyme, the two enzymes being closely associated. For example, the sequence recognized by the EcoRII restriction enzyme (also see Table 9.1) is:

$$C–C–A–G–G$$
$$G–G–T–C–C$$

and modification of this sequence results in methylation of two cytosines:

$$\begin{matrix} m & & & & \\ C & C & A & G & G \end{matrix}$$
$$\begin{matrix} C–C–A–G–G \\ G–G–T–C–C \\ m \end{matrix}$$

Note that a given nucleotide sequence can be a substrate for either a restriction enzyme or a modification enzyme but not both. This is because modification makes the sequence unreactive with restriction enzyme, and action of restriction enzyme destroys the recognition site of the modification enzyme.

TABLE 9.2 Enzymes Affecting DNA

Name	Action	Function in the Cell
Restriction endonuclease	Cuts DNA at specific base sequences	Destroys foreign DNA
DNA ligase	Links DNA molecules	Completion of replication process
DNA polymerase I	Attaches nucleotides to the growing DNA molecule	Fills gaps in DNA, primarily for DNA repair
DNA polymerase III	Attaches nucleotides to the growing DNA molecule	Replication of DNA
DNA gyrase	Increases the twisting pattern of DNA, promoting supercoiling	Strand separation during replication
DNase	Degrades DNA to nucleotides	General destruction of DNA
DNA methylase	Places methyl groups on DNA bases, thus inhibiting restriction endonuclease action	Modifies cellular DNA so that it is not affected by its own restriction endonuclease

We have now seen that a large number of enzymes are able to affect DNA structure and synthesis. These enzymes are summarized in Table 9.2.

Nucleic acid hybridization and homologies   Nucleic acid hybridization provides a method for detecting similarities in base sequence between different DNA molecules. If a double-stranded DNA preparation is heated, the two strands separate into single-stranded molecules that are complementary to each other. If the mixture is cooled slowly, the two complementary strands reassociate and the original double-stranded complex is reformed (Figure 9.7). Since reassociation requires the base sequences to be complementary, the presence of identical or nearly identical base sequences in two different organisms can be detected by measuring the degree to which their nucleic acids are able to interact specifically. Since the base sequence of RNA is complementary to that of one of the strands of the DNA, DNA-RNA hybridization can also be used to reveal similarities or differences among nucleic acids of organisms. Nucleic acid hybridization provides a powerful tool for studying the genetic relatedness of organisms. Hybridization between nucleic acids derived from different sources can be most conveniently studied if one of the nucleic acid preparations is radioactive and is broken into small pieces by mechanical shearing, while the nonradioactive preparation from the other source remains intact. After heating to separate the strands, the two preparations are mixed and cooled, and the degree to which the radioactive preparation associates with the nonradioactive preparation is measured. This association can be measured because double-stranded DNA and double-stranded DNA-RNA hybrids can be physically separated from single-stranded nucleic acids by adsorption and elution from cellulose nitrate filters.

Another way of measuring DNA relatedness, most suitable when the preparations are small and relatively similar, is to look for **heteroduplex** formation. A heteroduplex is a double-stranded DNA preparation in which one strand is from one source and the other strand from another, usually related, source. After heating, mixing, and cooling, the preparation is examined under the electron micro-

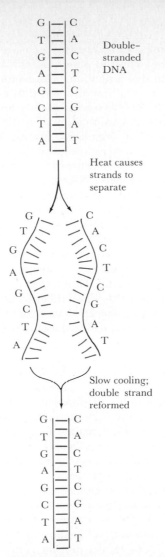

FIGURE 9.7 Denaturation of DNA by heat, leading to the formation of single-stranded molecules and the reformation of an intact double strand by slow cooling.

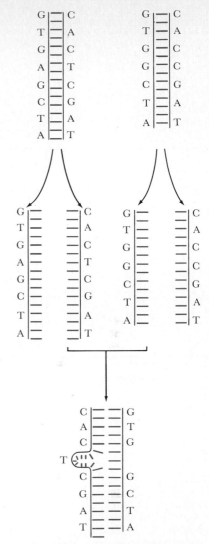

FIGURE 9.8 Heteroduplex formation between two DNA preparations, one of which has a small deletion. The single-stranded loop can be visualized in the electron microscope.

scope using a method that permits visualization of the DNA. If one of the strands differs from the other in a small region, there will be a region where the opposite strands will not be complementary, leading to the formation of a loop of unpaired bases (Figure 9.8). This technique is especially useful for looking for deletions of small regions of DNA and has been widely used in genetic studies on DNA viruses.

DNA base composition and sequences Techniques are now well developed for determining the precise base sequences of isolated pieces of DNA. Although the sequence of bases in a whole organism is much too complex to determine, sequences of bases in some viruses (see Chapter 10) and in many individual genes have been determined. Sequence determination has revolutionized research on molecular genetics. It is an important step in understanding the way in which proteins and nucleic acids interact, and is a vital procedure in the construction and manipulation of artificial genes.

For whole organisms, the proportions of the various bases in the total DNA can be assayed, and this is an important procedure in bacterial taxonomy. Because of the complementarity of the two strands, in double-stranded DNA the proportion of adenine always equals the proportion of thymine and that of guanine equals that of cytosine, but considerable variation in the frequency of adenine-thymine and guanine-cytosine base pairs occurs among various organisms, and these variations in base ratios are of value in taxonomic studies (see Section 18.2). By convention, the base composition of a DNA preparation is expressed as the mole percentage of guanine and cytosine $(G + C)$ of the total. Thus if the $G + C$ content is 40 percent, adenine and thymine $(A + T)$ would be 60 percent, since $(G + C) + (A + T) = 100$ percent. The DNA base compositions of different microorganisms vary widely, with values ranging from 22 to 74 percent $G + C$.

9.3

Genetic Elements

A genetic element is a particle or structure containing genetic material. A number of different kinds of genetic elements have been recognized. Although the main genetic element is the chromosome, other genetic elements are found and play important roles in gene function in both procaryotes and eucaryotes. Two key properties of genetic elements are (1) their ability for self-replication, and (2) their genetic coding properties.

Although single DNA molecules can be thought of as genetic elements, most genetic elements have a more complex structure than pure DNA. This structure is often associated with the condensation of DNA into compact bodies, a process necessary to fit the extremely long DNA molecules into the confines of a reasonable space. However, in all condensed genetic elements, the collinearity of the DNA molecule is preserved. When replication or transcription occurs, the condensed molecule must, at least temporarily, be unfolded.

Chromosome We have discussed the procaryotic nuclear body in Chapter 2 and the eucaryotic nucleus in Chapter 3. Each chromosome has at its base a single long DNA molecule. Because of the phosphate anions, DNA is very negatively charged, and the negative charges are mostly neutralized through combination with cations or positively charged proteins called **histones**. The single procaryotic chromosome, which is arranged in a circular fashion, contains most of the genetic information of the cell. In eucaryotes, more than one chromosome is present, the number constant within a species but varying widely between species. The overall structure of a eucaryotic chromosome can be visualized as a series of compact, highly folded units called **nucleosomes,** each containing about 200 DNA base pairs, separated by linkers of less extensively complexed DNA. Within each nucleosome, the DNA molecule is wound around a cluster of histone molecules organized in a precise and repeatable pattern. One function of the nucleosome structure is to permit packing of the long DNA molecules within the cell. For instance, if all of the DNA of the chromosomes of a human cell were stretched end to end, the length would be 180 cm, yet this DNA is packed into 46 chromosomes in a total length of 200 μm.

In procaryotes, there is no membrane separating the chromosomes from the cytoplasm, and there is the possibility of close and perhaps even direct association between the DNA and the protein-synthesizing machinery (ribosomes, transfer RNA, and so on). In eucaryotes, chromosomes are located inside the nucleus, and only at the time of division, when the nuclear membrane breaks down, are the chromosomes free in the cell. Because of this partitioning of chromosomes within the nucleus, transcription and translation are spatially separated. Transcription of DNA occurs within the nucleus and the messenger RNA molecules are transported out of the nucleus to the cytoplasm, where translation occurs. In procaryotes, transcription and translation can occur simultaneously, and it is even possible for translation to be initiated at one end of a messenger RNA molecule before transcription is complete at the other end. On the other hand, in eucaryotes, the initial RNA molecule, called the **primary transcript,** is extensively modified before it is finally translated, as we will discuss in Section 9.4.

The chromosomal organization in eucaryotes involves two features not found in procaryotes:

1. **Split genes.** Eucaryotic genes are discontinuous along the DNA, with noncoding sequences inserted between the sequences that actually code for protein. These noncoding intervening sequences are called **introns,** and the coding sequences are called **exons.** The number of introns per gene is variable, and ranges from none to over 50. During transcription, both introns and exons are copied, and the intron sequences are subsequently cut out and removed when the messenger RNA is processed into its final form in the cytoplasm.

2. **Repetitive sequences.** Eucaryotes generally contain much more DNA per genome than is needed to code for all of the proteins required for cell function. Eucaryotic DNA can be divided into three classes: **Single copy DNA** contains the coding sequences for the main proteins of the cell. **Moderately repetitive DNA,** found in a few to relatively large numbers of copies, codes for some major macromolecules of the cell: histones, immunoglobulins (involved in immune mechanisms, as discussed in Chapter 16), ribosomal RNA, transfer RNA. **Highly repetitive (satellite) DNA,** is found in a very large number of copies. In humans, about 20 to 30 percent of the DNA is found in repetitive sequences, and almost all eucaryotic DNAs studied have some repeated sequences. The function of highly repetitive DNA is unknown.

Nonchromosomal genetic elements Originally called *cytogenes,* genetic elements which are not part of the chromosome have been recognized for a long time. Some nonchromosomal genetic elements which we will discuss briefly here include viruses, plasmids, mitochondria, and chloroplasts.

Viruses are genetic elements, either DNA or RNA, which control their own replication and transfer from cell to cell. Viruses are of special interest because they are often (but not always) responsible for disease states, but our main emphasis in this text is on viruses as genetic elements. We discuss viruses in Chapter 10.

Plasmids are small, circular DNA molecules that exist and replicate separately from the chromosome. Plasmids differ from viruses in that they do not cause cellular damage (generally they are beneficial). Although plasmids have been recognized in only a few eucaryotes, they have been found in most procaryotic genera. Some plasmids are excellent genetic vectors, and find wide use in gene manipulation and genetic engineering, as outlined in Chapter 12.

Mitochondria and **chloroplasts** are nonchromosomal genetic elements found in eucaryotes. As we discussed in Chapters 3 and 4, mitochondria are the site of respiratory enzymes, and play a major role in energy generation in most eucaryotes. Chloroplasts are green, chlorophyll-containing structures which are the site of photosynthetic ATP formation. From a genetic viewpoint, mitochondria and chloroplasts can be viewed as independently replicating genetic elements. However, these organelles are much more complex than plasmids and viruses, since they contain not only DNA, but a complete machinery for protein synthesis, including their own ribosomes, transfer RNA, and all of the other components necessary for translation and formation of functional proteins. One intriguing feature of mitochondria and chloroplasts is that despite the fact that they contain many genes and a complete translation system, they are not independent from the chromosomes, since most proteins are coded not by the organelle DNA but by chromosomal DNA.

9.4

RNA Structure and Function

Ribonucleic acid (RNA) plays a number of important roles in the expression of genetic information in the cell. Three major types of RNA have been recognized: messenger RNA (mRNA), transfer RNA (tRNA), and ribosomal RNA (rRNA). There are three key differences between the chemistry of RNA and that of DNA: (1) RNA has the sugar *ribose* instead of *deoxyribose;* (2) RNA has the base *uracil* instead of the base *thymine;* and (3) except in certain viruses, RNA is not double stranded. A change from *deoxyribose* to *ribose* affects some of the chemical properties of a nucleic acid, and enzymes which affect DNA in general have no effect on RNA, and vice versa. The change from *thymine* to *uracil* does not affect base pairing, as the two nucleotide bases pair with adenine equally well.

RNA polymerase and the formation of messenger RNA The transcription of the genetic information from DNA to messenger RNA (mRNA) is carried out through the action of the enzyme RNA polymerase, which catalyzes the formation of phosphodiester bonds between ribonucleotides. RNA polymerase requires the presence of DNA, which acts as a template. The precursors of mRNA are the ribonucleoside triphosphates, ATP, GTP, UTP, and CTP, which are polymerized with the release of the two high-energy phosphate bonds.

In most cases, the DNA template for RNA polymerase is a double-stranded DNA molecule, but only one of the two strands is transcribed. (If both strands were copied, each gene would code for two proteins, but except for a few unusual cases one gene only codes for a single protein.)

The enzyme RNA polymerase is a complex protein consisting of several subunits, each of which has a specific function in the enzyme action. All bacterial RNA polymerases studied seem to be closely related in subunit structure. The enzyme from *Escherichia coli* has several subunits, designated β, β', α, and σ (sigma), with α appearing in two copies. The subunits interact to form the active enzyme, but the sigma factor is not as tightly bound as the others, and dissociates, leading to the formation of what is called the *core enzyme*. The core enzyme alone can catalyze the formation of mRNA, and the role of sigma is in the recognition of the appropriate site on the DNA for the initiation of RNA synthesis (Figure 9.9).

The site at which mRNA synthesis begins is not random. For each gene or group of related genes (called an **operon**) there is a spe-

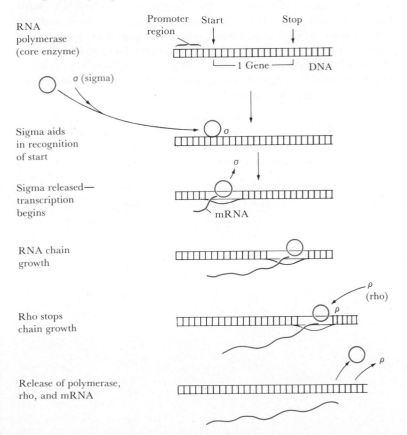

RNA polymerase (core enzyme)

σ (sigma)

Sigma aids in recognition of start

Sigma released— transcription begins

mRNA

RNA chain growth

Rho stops chain growth

ρ (rho)

Release of polymerase, rho, and mRNA

Promoter region   Start   Stop

1 Gene   DNA

FIGURE 9.9 Steps in messenger RNA synthesis. The start and stop sites are specific nucleotide sequences on the DNA. RNA polymerase moves down the DNA chain, causing temporary opening of the double helix and transcription of one of the DNA strands. Rho binds to the termination site and stops chain growth; termination can also occur at some sites without rho.

cific site, called the **promoter** region, at which the RNA polymerase first binds, and this specific binding requires the presence of sigma factor. The DNA sequence of the promoter region of a number of genes has been determined, and two regions have been recognized as important for promoter function. There is a small region 10 bases from the start of transcription (the −10 region, called the *Pribnow–Schaller box*) that contains the sequence TATG or TTAA or TATA or another closely related sequence. A second recognition site −35 bases from the start of transcription has also been identified. The variation in sequence at the promoter region is a function of the operon involved, rather than of the organism. Thus different promoters in *E. coli* show slightly different sequences at the Pribnow–Schaller box. Some promoter sequences are more effective than others in causing RNA polymerase to initiate transcription. The more effective promoters are called *strong promoters* and are of considerable value in genetic engineering, as will be discussed in Chapter 12.

The first base in the mRNA is almost always a purine, either adenine or guanine, most *E. coli* chains starting with guanine. The three phosphates of the initial nucleoside triphosphate remain intact, subsequent nucleotides being added on to the 3′-OH (hydroxyl group) of the ribose (Figure 9.10). Transcription occurs by selective opening of the DNA double helix at the transcription site, base pairing of the ribonucleotide with the proper base on the DNA, and enzymatic connection of the new base to the growing mRNA chain. The RNA polymerase moves progressively down the chain, and as it leaves a site the DNA double helix recloses (Figure 9.9). The growing mRNA chain remains attached to the transcription site, with the free end dissociated from the complex. Once transcription begins, sigma can dissociate; thus sigma is only involved in the formation of the initial polymerase-DNA complex.

FIGURE 9.10 Action of RNA polymerase in connecting nucleoside triphosphates to form messenger RNA. Note that the 5′ triphosphate of the initial nucleotide remains intact and that the second nucleotide is added to the 3′ hydroxyl. Thus, mRNA growth proceeds in the 5′ to 3′ direction.

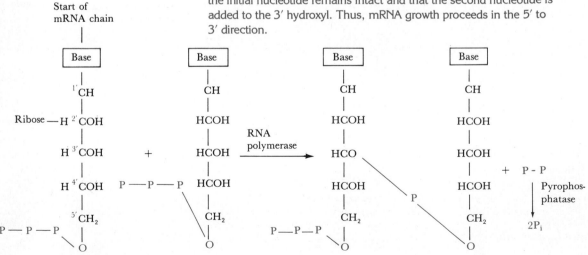

Termination of mRNA synthesis also occurs at specific sites on the DNA. Two different mechanisms of termination have been recognized. One involves a direct termination of growing RNA chains when the RNA polymerase reaches a region with a sequence that has a G-C-rich area followed by an A-T-rich area. The second type of termination involves another protein component called *rho,* which is able to recognize a specific DNA sequence that specifies termination and stops further action of RNA polymerase (Figure 9.9).

Specific inhibitors of RNA polymerase action   A number of antibiotics and synthetic chemicals have been shown to inhibit mRNA synthesis specifically. A group of antibiotics called the *rifamycins* inhibits by attacking the *beta* subunit of the enzyme. Rifamycin has marked specificity for procaryotes, but also inhibits RNA synthesis in chloroplasts and mitochondria of some eucaryotes. Rifamycin has been an especially useful tool in studying nucleic acid synthesis in virus-infected cells. A group of antibiotics called the *streptovaricins* is related to the rifamycins in structure and function. *Streptolydigin* is an antibiotic that also inhibits the beta subunit of RNA polymerase, but at a different site than rifamycin. Another chemical, *amanitin,* specifically inhibits mRNA synthesis in eucaryotes without affecting procaryotes.

   Actinomycin inhibits RNA synthesis by combining with DNA and blocking elongation. Actinomycin binds most strongly to DNA at guanine-cytosine base pairs, fitting into the groove on the double strand where RNA is synthesized.

Messenger RNA may code for more than one protein   In many cases a single mRNA molecule may have the information for several or many proteins. In bacteria, it is frequently the case that the genes coding for related enzymes occur together in a cluster. In these situations of polygenic mRNA, the RNA polymerase proceeds down the chain and transcribes the whole series of genes into a single molecule. Subsequently, when this polygenic mRNA participates in protein synthesis, the several proteins coded by the mRNA may be synthesized simultaneously.

Transfer RNA: structure and function   The translation of the RNA message into the amino acid sequence of protein is brought about through the action of transfer RNA (tRNA). Transfer RNA is an adaptor molecule having two specificities, one for a codon on mRNA, the other for an amino acid. The transfer RNA and its specific amino acid are brought together by means of an enzyme, the amino acid activating enzyme.

   The detailed structure of tRNA is now well understood. There are about 60 different types of tRNA in bacterial cells and 100 to 110 in mammalian cells. Transfer RNA molecules are short, single-stranded molecules, with lengths (among different tRNAs) of 73 to 93 nucleotides. When compared, it has been found that certain structural parts are constant for all tRNAs, and there are other parts that are variable. Transfer RNA molecules also contain some purine and pyrimidine bases differing slightly from the normal bases found in

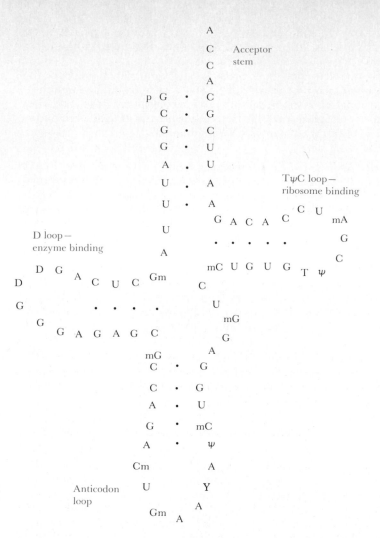

FIGURE 9.11 Structure of a transfer RNA, yeast phenylalanine tRNA. The conventional cloverleaf structure is shown; in actuality the molecule folds so that the D loop and TψC loops are close together and associate by hydrophobic interactions. The amino acid is attached to the ribose of the terminal A at the acceptor end. Abbreviations: A, adenine; C, cytosine; U, uracil; G, guanine; ψ, pseudouridine; D, dihydrouracil; m, methyl; p, phosphate (terminal); Y, a modified purine. (From Quigley, G. J., and A. Rich. 1976. Structural domains of transfer RNA molecules. Science 194:796–805.)

RNA in that they are chemically modified, often methylated. Some of these unusual bases are pseudouridine, inosine, dihydrouridine, ribothymidine, methyl guanosine, dimethyl guanosine, and methyl inosine. Although the molecular structure of tRNA is single stranded, there are extensive double-stranded regions within the molecule, as a result of folding back of the molecule on itself. The structure of tRNA is generally drawn in cloverleaf fashion, as shown in Figure 9.11. Since many of the unusual bases mentioned above cannot form conventional hydrogen-bonded base pairs, it is possible that their function in tRNA is to disrupt double helix formation, thus causing the formation of the "cloverleaf" loops.

Several parts of the tRNA molecule have specific functions in the translation process. One of the variable parts of the tRNA molecule contains the anticodon, the site recognizing the codon on the mRNA. The location of this **anticodon loop** is shown in Figure 9.11. There are just *three* nucleotides in the anticodon loop that are specifically involved in the recognition process. Another loop called the TψC loop (ψ is the symbol for pseudouridine) is believed to interact with the ribosome, whereas the D loop may be involved in binding to the activating enzyme (see below). As seen in Figure 9.11, there is also one region at the end of the chain where a sequence of three nucleotides projects from the rest of the molecule. The sequence of these three nucleotides is always the same, cytosine-cytosine-adenine (CCA), and it is to the ribose sugar of the terminal A that the amino acid is covalently attached, via an ester linkage. From this acceptor portion of the tRNA, the amino acid is transferred to the growing polypeptide chain on the ribosome by a mechanism that will be described in a subsequent section.

Another key element of the translation system is the enzymatic machinery activating the amino acid and placing it on the correct tRNA. The enzymes carrying out this process are called **aminoacyl-tRNA synthetases,** and they have the important function of recognizing *both* the amino acid *and* the specific tRNA for that amino acid. It should be emphasized at this point that the fidelity of this recognition process is crucial, for if the wrong amino acid is attached to the tRNA, it may be inserted in the improper place in the protein, leading to the synthesis of a faulty protein.

The specific chemical reaction between amino acid and tRNA catalyzed by the aminoacyl-tRNA synthetase involves *activation* of the amino acid by reaction with ATP:

$$\text{Amino acid} + \text{ATP} \rightleftharpoons \text{amino acid-AMP} + \text{P-P}$$

The amino acid-AMP intermediate formed normally remains bound to the enzyme until collision with the appropriate tRNA molecule, and the activated amino acid is then transferred to the tRNA:

$$\text{Amino acid-AMP} + \text{tRNA} \rightleftharpoons \text{amino acid-tRNA} + \text{AMP}$$

The pyrophosphate (P-P) formed in the first reaction is split by a pyrophosphatase, forming two molecules of inorganic phosphate. Since ATP is used and AMP is formed, two high-energy phosphate bonds are required for the activation of an amino acid. Once activation has occurred, the amino acid-tRNA leaves the synthetase and migrates through the cell to the ribosome-mRNA protein synthesizing machinery. The mechanism of protein synthesis is discussed in the next section.

RNA processing   In many cases, the primary RNA transcript is not used directly, but is converted into the active RNA by means of special enzymes. The conversion of a *precursor* RNA into a *mature* RNA is called **RNA processing,** and is found in both procaryotes and eucaryotes. For instance, in procaryotes, tRNA and rRNA are made initially as long precursor molecules which are then cut at several places to make the final mature RNAs. In eucaryotes, but apparently not in procaryotes, mRNA also undergoes extensive processing. As discussed in Section 9.3, the genes of eucaryotes are split, with noncoding intervening sequences, *introns,* separating the coding regions called *exons.* The primary RNA transcript must be extensively processed to remove the noncoding regions before the translation process can be initiated. One major feature of eucaryotic mRNA is that after transcription but before transfer to the cytoplasm, the mRNA molecules have added to them a long stretch of adenine nucleotides at one end, the *poly A tail,* and a methylated guanine nucleotide at the 5′ end, called the *cap.* Also, a number of the bases in mRNA are methylated after transcription and the 2′ hydroxyl group of the ribose is occasionally methylated.

We thus see that RNA synthesis is a complex and dynamic process that involves considerably more than the simple transcription of a DNA template.

9.5

Protein Synthesis

As we discussed in Chapter 4, it is the amino acid sequence that determines the structure of the final active protein. The key problem of protein synthesis is thus the placing of the proper amino acid at the proper place in the polypeptide chain. This is the role of the protein-synthesizing machinery of the cell.

Steps in protein synthesis Ribosomes are the site of protein synthesis. Each ribosome is constructed of two subunits, which in bacteria have sedimentation constants of 30S (Svedberg units) and 50S (Figure 9.12). Each large subunit consists of a number of individual proteins, 21 in the 30S ribosome and 34 in the 50S ribosome. The actual synthesis of a protein involves a complex cycle in which the various ribosomal components play specific roles. Here we shall present an overall outline of the process.

Although a continuous process, protein synthesis can be thought of as occurring in a number of discrete steps: **initiation, elongation, termination-release,** and **polypeptide folding.** These steps are out-

FIGURE 9.12 Subunit structure of a ribosome particle. The 50-S and 30-S particles exist free in the cell and come together to form a 70-S particle on the messenger RNA. The 50-S particle also contains a small 5-S RNA of unknown function.

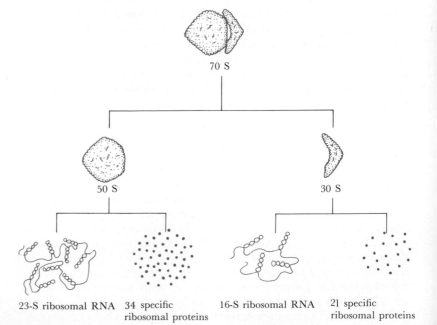

70 S

50 S   30 S

23-S ribosomal RNA   34 specific ribosomal proteins   16-S ribosomal RNA   21 specific ribosomal proteins

TABLE 9.3   Steps in Protein Synthesis

I. Initiation: formation of initial ribosome-tRNA-mRNA complex
   A. Combination of mRNA + 30S ribosome subunit + formylmethionine-tRNA (f-met-tRNA), forming initiation complex I. Guanosine triphosphate (GTP) and protein initiation factors are required. The mRNA location contains the codon AUG or GUG, which complements with the anticodon on f-met-tRNA.
   B. Addition of 50S ribosome subunit, forming initiation complex II. The f-met-tRNA is at site A of the ribosome.
   C. Translocation of f-met-tRNA to site P, forming initiation complex III. GTP is cleaved to GDP + P_i.

II. Elongation: formation of successive peptide bonds (compare with Figure 9.13)
   A. Addition of AA-tRNA at site A, the anticodon of the tRNA recognizing the codon on the RNA adjacent to the initiation codon AUG. GTP and protein elongation factors are required. GTP is cleaved to GDP + P.
   B. Peptide bond formation. Carboxyl group of f-met is transferred to amino group of second AA-tRNA. The tRNA which carried f-met is released. The enzyme peptidyl transferase which catalyzes this reaction is part of the 50S ribosome subunit.
   C. Translocation. The t-RNA carrying the f-met-AA dipeptide is shifted from site A to site P and the ribosome moves the length of one codon along the mRNA. Elongation factor is required and GTP is cleaved to GDP + P_i.
   D. Further elongation steps of addition, peptide bond formation, and translocation, leading to the synthesis of a polypeptide.

III. Termination-release: completion of polypeptide and release from ribosome
   A. Arrival of the growing chain at a termination codon, UAA, UAG, or UGA, which does not code for any amino acid. These are called nonsense codons.
   B. Release of polypeptide from tRNA and ribosome. Release factor is required. GTP is cleaved to GDP + P_i.

IV. Polypeptide folding and formation of functional protein
   A. Removal of formyl group from f-met. In some cases f-met is removed as a whole. Many proteins have methionine as the terminal amino acid.
   B. Folding of polypeptide to assume a secondary and tertiary conformation determined by its amino acid sequence.
   C. Association of polypeptide with other polypeptides to form the quarternary structure of the active enzyme.

---

lined in some detail in Table 9.3, and the elongation steps are diagrammed in Figure 9.13. In addition to mRNA, tRNA, and ribosomes, the process involves a number of protein factors designated initiation, elongation, and termination factors. Also guanosine triphosphate (GTP) provides energy for the process.

Several features of the process outlined in Table 9.3 should be noted. In procaryotes, initiation always begins with a special amino acyl-tRNA, which is formylmethionine-tRNA. Subsequently, the formyl group at the N-terminal end of the polypeptide is removed. In eucaryotes, initiation begins with methionine instead of formylmethionine.

Initiation always begins with a free 30S ribosome subunit, and an **initiation complex** forms consisting of 30S ribosome, mRNA, formylmethionine tRNA, and initiation factors. To this initiation complex a 50S subunit is added to make the active 70S ribosome. At the end of the process, the released ribosome breaks down again into 30S and 50S subunits. Just preceding the initiation codon is a se-

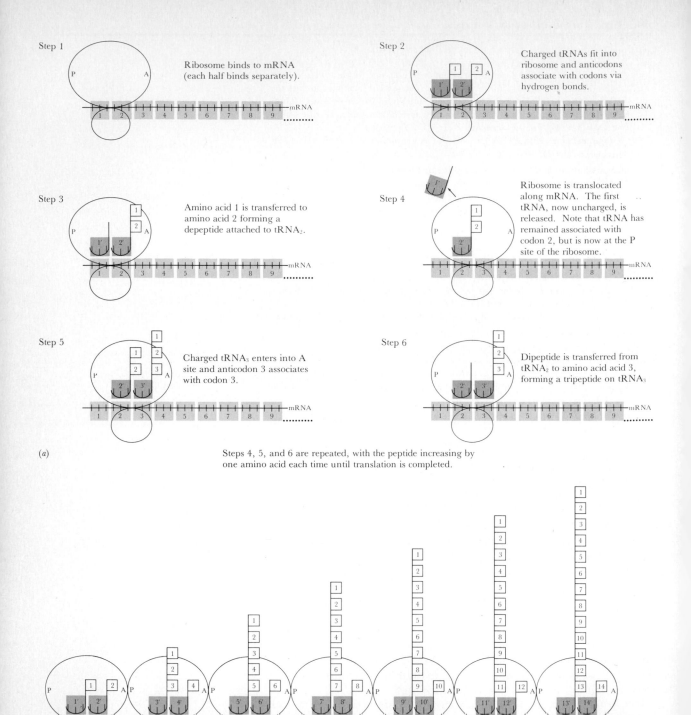

Step 1

Ribosome binds to mRNA
(each half binds separately).

Step 2

Charged tRNAs fit into
ribosome and anticodons
associate with codons via
hydrogen bonds.

Step 3

Amino acid 1 is transferred to
amino acid 2 forming a
depeptide attached to tRNA₂.

Step 4

Ribosome is translocated
along mRNA. The first
tRNA, now uncharged, is
released. Note that tRNA has
remained associated with
codon 2, but is now at the P
site of the ribosome.

Step 5

Charged tRNA₃ enters into A
site and anticodon 3 associates
with codon 3.

Step 6

Dipeptide is transferred from
tRNA₂ to amino acid acid 3,
forming a tripeptide on tRNA₃

(a)

Steps 4, 5, and 6 are repeated, with the peptide increasing by
one amino acid each time until translation is completed.

(b)

quence of from three to nine nucleotides (the Shine–Dalgarno sequence) which is involved in binding of the mRNA to the ribosome. This ribosome binding site is complementary to the 3' end of the 16S RNA of the ribosome, and it is thought that base pairing ensures effective formation of the ribosome-mRNA complex.

The mRNA is attached to the 30S subunit and the tRNA holding the growing polypeptide chain is attached to the 50S subunit (Figure 9.13). There are two sites on the 50S subunit, designated P and A. Site A, the **acceptor** site, is the site where the new AA-tRNA first attaches. Site P, the **peptide** site, is the site where the growing peptide is held by a tRNA. During peptide bond formation, the peptide moves to the tRNA at the A site as a new peptide bond is formed. Then after the now empty tRNA at the P site is removed, the tRNA holding the peptide moves (is translocated) from the A site back to the P site, thus opening up the A site for another AA-tRNA.

The termination of protein synthesis occurs when a codon is reached which does not specify an AA-tRNA. There are three codons of this type and they are called **nonsense** codons; they serve as the stopping points for protein synthesis.

Secretory proteins   Many proteins are used outside the cell and must somehow get from the site of synthesis on cytoplasmic ribosomes through the cell membrane. In procaryotes, periplasmic enzymes and extracellular enzymes are secretory proteins, and in eucaryotes various digestive enzymes and enzymes of the lysosome are in this category.

How is it possible for a cell selectively to transfer some proteins across a membrane, while leaving most proteins in place in the cytoplasm? This is explained by the **signal hypothesis,** which states that

FIGURE 9.13   (a) Schematic representation of translation. The P and A designations on the large ribosome subunit refer to distinct sites with separate functions: P for *peptide* bond formation and A for *attachment* of next charged tRNA. Vertical lines on mRNA denote individual ribonucleotides. Groups of three bases are referred to as *codons* and are shaded and numbered for emphasis. The vertical lines on the tRNA also represent individual ribonucleotides. Groups of three bases in a specific loop of the tRNA are referred to as *anticodons* and are also shaded and numbered for emphasis. Amino acids are shown as boxes and are numbered to correspond to the appropriate codon. (b) Part *a* was concerned with the translation of a single message by a single ribosome. However, the active cell often forms many copies of a given protein quite rapidly by repeated initiation of the translation process with additional ribosomes. The structure formed by the association of several ribosomes with one mRNA is termed a *polyribosome* or *polysome.* The lower diagram is a schematic representation of a polysome. Note that the length of the growing peptide chain associated with each ribosome is longer for ribosomes which are further along in translation. Polysomes and their associated protein chains are visible in electron micrographs. It should also be noted that proteins assume individual specific three-dimensional shapes as the result of secondary and tertiary interactions (see Chapter 4). These precise foldings occur while the peptide chain is still attached to the ribosome; the drawing of linear chains is a simplification.

secretory proteins are synthesized with an extra N-terminal peptide sequence, some 15 to 20 amino acids in length, which is called the **signal sequence.** In this signal sequence, hydrophobic amino acids predominate and may permit the enzyme to be threaded through the hydrophobic lipid membrane. In many cases, the ribosomes that synthesize secretory proteins are bound directly to the cell membrane, so that the protein is formed and passes through the membrane simultaneously. Once the protein has been secreted, the signal sequence is removed by a peptidase enzyme, an example of the process of **post-translational modification.**

The study of protein secretion has important practical implications for genetic engineering (see Chapter 12). If bacteria are genetically engineered to serve as agents for the production of foreign proteins, it is desirable to manipulate the signal sequence in order to arrange that the desired protein is excreted so that it can be readily isolated and purified.

Effect of antibiotics on protein synthesis A large number of antibiotics inhibit protein synthesis. Some of these antibiotics are medically useful, whereas others have proved ineffective in treating diseases but are still of interest as research tools. Many antibiotics act by affecting ribosome function in some way. One interesting aspect of antibiotic action, selectivity, was discussed in Section 7.6. Selectivity is especially seen in antibiotics acting on the ribosome, and antibiotics affecting ribosome function in procaryotes have no effect on cytoplasmic ribosomes in eucaryotes. A brief summary of the modes of action of certain antibiotics is given in Table 9.4. Note that antibiotics not only act at different stages of protein synthesis, but they usually affect only one of the two ribosomal subunits. Also, most antibiotics act against procaryotes, a few against eucaryotes, and

TABLE 9.4 Effects of Antibiotics on Protein Synthesis

	Streptomycin	Neomycin	Tetracycline	Puromycin	Chloramphenicol	Erythromycin	Cycloheximide	Spectinomycin
Function:								
Initiation	+						+	
Codon recognition	+	+	+					
Peptide formation				+	+	+		
Translocation							+	+
Termination			+					
Targets:								
Procaryotes	+	+	+	+	+	+		+
Eucaryotes				+			+	
Ribosome subunits:								
30S (40S)	+	+	+					+
50S (60S)				+	+	+	+	
Killing activity	+	+						

some act against both. Since mitochondria and chloroplasts have ribosomes of the procaryotic type, it is of interest that antibiotics inhibiting protein synthesis in procaryotes also generally inhibit protein synthesis in mitochondria and chloroplasts. This is one piece of evidence suggesting that mitochondria and chloroplasts might originally have been derived by intracellular infection of a eucaryotic cell by procaryotes (see endosymbiosis, Sections 3.3 and 19.33).

Regulation of Protein Synthesis: Induction and Repression

Not all enzymes are synthesized by the cell in the same amounts, some enzymes being present in far greater numbers of molecules than other enzymes. Clearly, the cell is able to regulate enzyme synthesis. Even more dramatic, the synthesis of many enzymes is greatly influenced by the environment in which the organism is growing, most particularly by the presence or absence of specific chemical substances. Often the enzymes catalyzing the synthesis of a specific product are not produced if this product is present in the medium. For example, the enzymes involved in the formation of the amino acid arginine are synthesized only when arginine is not present in the culture medium; external arginine inhibits or **represses** the synthesis of these enzymes. As can be seen in Figure 9.14, if arginine is added to a culture growing exponentially in a medium devoid of arginine, growth continues at the previous rate, but the formation of the enzymes involved in arginine synthesis stops. Note that this is a specific effect, as the syntheses of all other enzymes in the cell are found to continue at the same rates as previously.

Enzyme repression is a very widespread phenomenon in microorganisms—it occurs during synthesis of a wide variety of enzymes involved in the biosynthesis of amino acids, purines, and pyrimidines. In almost all cases it is the final product of a particular biosynthetic pathway that represses the enzymes of this pathway. In these cases repression is quite specific, and the process usually has no effect on the synthesis of enzymes other than those involved in a single biosynthetic pathway. The value to the organism of enzyme repression is obvious, since it effectively ensures that the organism does not waste energy synthesizing unneeded enzymes.

A phenomenon complementary to repression is **enzyme induction**, the synthesis of an enzyme only when a substrate is present. Figure 9.15 shows this process in the case of the enzyme β-galactosidase, which is involved in utilization of the sugar lactose. If lactose is absent from the medium the enzyme is not synthesized, but synthesis begins almost immediately after lactose is added. Enzymes involved in the breakdown of carbon and energy sources are often inducible. Again, one can see the value to the organism of such a mechanism, as it provides a means whereby the organism does not synthesize an enzyme until it is needed.

The substance that initiates enzyme induction is called an **inducer,** and a substance that represses enzyme production is called a **corepressor;** these substances, which are always small molecules, are often collectively called **effectors.** Not all inducers and corepressors are substrates of the enzymes involved. For example, analogs of these

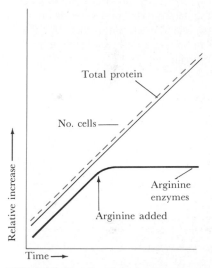

FIGURE 9.14 Repression of enzymes involved in arginine synthesis by addition of arginine to the medium. Note that the rate of total protein synthesis remains unchanged.

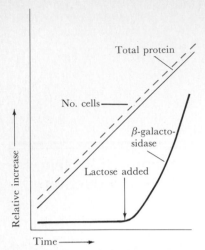

FIGURE 9.15 Induction of the enzyme β-galactosidase upon the addition of lactose to the medium. Note that the rate of total protein synthesis remains unchanged.

substances induce or repress although they are not substrates of the enzyme. Thiomethylgalactoside (TMG), for instance, is an inducer of β-galactosidase even though it cannot be hydrolyzed by the enzyme. In nature, however, inducers and corepressors are probably normal cell metabolites.

Mechanism of induction and repression   As we have just learned, the overall process of protein synthesis can be divided into two stages, transcription and translation. Although in principle enzyme induction and repression could be effected at either stage, in actuality control is exerted mainly at the stage of *transcription.* Thus when an enzyme inducer is added, it initiates enzyme synthesis by causing the formation of the mRNA that codes for the particular enzyme. When a substance (corepressor) that causes enzyme repression is added, it causes an inhibition of mRNA formation.

How can inducers and corepressors affect transcription in such a specific manner? They do this indirectly by combining with specific proteins, the **repressors,** which then in turn affect mRNA synthesis. In the case of a repressible enzyme, it is thought that the corepressor (for example, arginine) combines with a specific **repressor protein** that is present in the cell (Figure 9.16a). The repressor protein is probably an allosteric protein (see Section 5.12), its configuration being altered when the repressor combines with it. This altered repressor protein can then combine with a specific region of the DNA at the initial end of the gene, the **operator region,** where synthesis of mRNA is initiated. If this occurs, the synthesis of mRNA is blocked, and the protein or proteins specified by this mRNA cannot be synthesized. If the mRNA is polygenic, all of the proteins coded for by this mRNA will be repressed. A series of genes all regulated by one operator is called an **operon.**

For induction the situation described above is reversed. The specific repressor protein is thought to be active in the absence of the inducer, completely blocking the synthesis of mRNA, but when the inducer is added it combines with the repressor protein and inactivates it. Inhibition of mRNA synthesis being overcome, the enzyme can be made (Figure 9.16b). Induction and repression thus both have the same underlying mechanism, the inhibition of the synthesis of mRNA by the action of specific repressor proteins that are themselves under the control of specific small-molecule inducers and repressors.

It should be emphasized that not all enzymes of the cell are inducible or repressible. Those that are synthesized continuously are called **constitutive.** In fact, even inducible or repressible enzymes may become constitutive by genetic mutation. Two kinds of constitutive mutants are known: those that no longer produce the repressor protein, and those whose operator regions are altered so that they no longer respond to the action of the repressor protein. Although the net result of both of these mutations is the same, the two kinds of mutants can be distinguished by genetic tests.

Catabolite repression   Another type of enzyme repression is that called catabolite repression. In this phenomenon the syntheses of a variety of unrelated enzymes are inhibited when cells grow in a me-

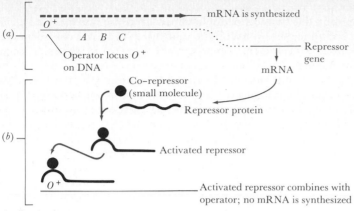

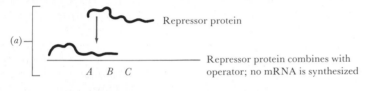

1. Repression
   (a) Unregulated; repressor is inactive and operon genes are transcribed
   (b) Repressed; repressor is active and operon genes are not transcribed

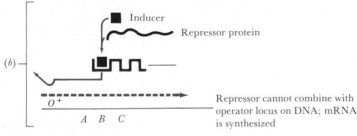

2. Induction
   (a) Unregulated; repressor is active and operon genes are not transcribed
   (b) Induced; repressor is inactive and operon genes are transcribed

FIGURE 9.16 Roles of repressor in enzyme repression and induction.

Although many eucaryotes do respond to repression, there is no good evidence for the kind of negative control so commonly found in procaryotes. However, positive control, which is rare in procaryotes, is common in eucaryotes. If operons exist in eucaryotes, they involve the control of only single enzymes, rather than the multi-enzyme control systems so commonly seen in procaryotes, and there is no evidence for messenger RNA molecules that contain the coding sequences for several enzymes. Thus, eucaryotes seem to have developed different kinds of control mechanisms than procaryotes, but the details of these mechanisms remain to be worked out.

Induction and Repression in Eucaryotes?

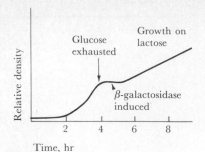

FIGURE 9.17 Diauxic growth on mixture of glucose and lactose. Glucose represses the synthesis of β-galactosidase. After glucose is exhausted, a lag occurs until β-galactosidase is synthesized, and then growth can resume on lactose.

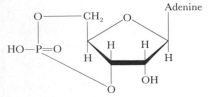

FIGURE 9.18 Cyclic adenosine monophosphate (cyclic AMP).

dium that contains an energy source such as glucose. Catabolite repression has been called the **glucose effect** because glucose was the first substance shown to initiate it, although in some organisms glucose does not cause enzyme repression. Catabolite repression occurs when the organism is offered an energy source that it can catabolize readily. One consequence of catabolite repression is that it can lead to so-called **diauxic** growth if two energy sources are present in the medium at the same time and if the enzyme needed for utilization of one of the energy sources is subject to catabolite repression. Growth first occurs on the one energy source, and there is then a temporary cessation before growth is resumed on the other energy source. This phenomenon is illustrated in Figure 9.17 for growth on a mixture of glucose and lactose. The enzyme β-galactosidase, which is responsible for utilization of lactose, is inducible, but its synthesis is also subject to catabolite repression. Thus, as long as glucose is present in the medium β-galactosidase is not synthesized; the organism grows only on the glucose and leaves the lactose untouched. When the glucose is exhausted, catabolite repression is abolished. After a lag β-galactosidase is synthesized, and growth on lactose can occur.

In recent years, one mechanism of catabolite repression has been determined, and it is now known to involve an additional type of control on transcription of the operon, at the level of the RNA polymerase. As discussed in Section 9.4, the RNA polymerase begins transcription by binding to the promoter site on the DNA. However, in the case of catabolite-repressible enzymes, binding of polymerase seems only to occur if another protein, called **catabolite activator protein** (CAP), has bound first.* Apparently an allosteric protein, CAP only binds if it has first bound a small-molecular-weight substance called *cyclic adenosine monophosphate* or **cyclic AMP**. Cyclic AMP (Figure 9.18) has been shown to be a key element in a variety of control systems, not only in bacteria but in higher organisms. Cyclic AMP is synthesized from ATP by an enzyme called adenylate cyclase, which is widespread in bacteria, and glucose either inhibits the synthesis of cyclic AMP or stimulates its breakdown. When glucose is present, the cyclic AMP level in the cell is low, and binding of RNA polymerase to the promoter does not occur. Thus catabolite repression is really a result of a deficiency of cyclic AMP, and can be overcome by adding this compound to the medium. The manner in which cyclic AMP is involved in catabolite repression is diagrammed in Figure 9.19.

Cyclic AMP has been shown to promote the synthesis of a number of enzymes in *E. coli,* mostly involved in catabolic processes. Cyclic AMP has a number of regulatory roles in eukaryotes that do not involve catabolite repression and is also an extracellular signal for the aggregation process in certain cellular slime molds.

Positive and negative control of protein synthesis Above, we have described two quite different mechanisms for the regulation of protein synthesis. The mechanism involving the *lac* repressor and operator is an example of **negative control**. Under normal conditions, the func-

*This protein is sometimes called *catabolite repression protein* (CRP).

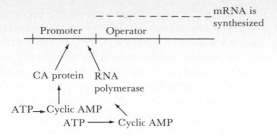

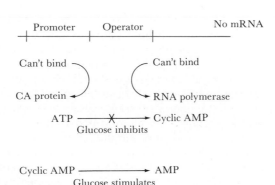

FIGURE 9.19 Action of cyclic AMP in promoting mRNA synthesis, and the mechanism of glucose (catabolite) repression.

tion of RNA polymerase is *blocked,* and this block can be overcome if an inducer is present. In the case of catabolite repression, a mechanism of **positive control** exists, since the catabolite activator protein normally promotes the binding of RNA polymerase, thus acting to *increase* mRNA synthesis, but this action can be overcome if cyclic AMP is deficient. There are a number of other positive control proteins which, by binding to specific promoters, bring about the synthesis of specific mRNAs. In negative control, if the regulatory protein is destroyed or inactivated, protein synthesis then occurs, whereas in positive control, if the regulatory protein is destroyed, protein synthesis does not occur.

Attenuation Another element of control, called **attenuation,** has been recognized in some operons controlling the biosynthesis of amino acids. The best studied case is that involving biosynthesis of the amino acid *tryptophan.* The tryptophan operon contains the structural genes for five proteins of the tryptophan biosynthetic pathway, plus regulatory sequences at the beginning of the operon (Figure 9.20). In addition to the promoter and operator regions, there is an additional sequence, called the **leader sequence,** within which is a region called the **attenuator,** which codes for a tryptophan-rich peptide. If tryptophan is plentiful in the cell, the leader peptide will be synthesized. On the other hand, if tryptophan is in short supply, the tryptophan-rich leader peptide will not be synthesized. The striking fact is that synthesis of the leader peptide results in termination

Excess tryptophan: termination

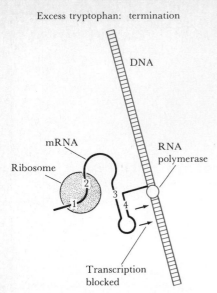

Tryptophan starved: no termination

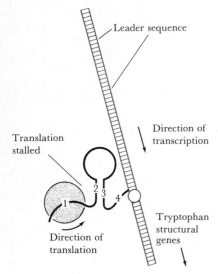

FIGURE 9.20 Model for attenuation in the *Escherichia coli* tryptophan operon. The leader peptide is coded by regions 1 and 2 of the mRNA. Two regions of the growing mRNA chain are able to form double-stranded loops, 2:3 and 3:4. Under conditions of excess tryptophan, the ribosome translates the complete leader peptide, so that region 2 cannot pair with region 3. Regions 3 and 4 then pair to form a loop which blocks RNA polymerase. If translation is stalled due to tryptophan starvation, loop formation via 2:3 pairing occurs, loop 3:4 does not form, and transcription proceeds past the leader sequence. (From Oxender *et al.* 1979, Proc. Nat. Acad. Sci. 76:5524.)

of transcription of the tryptophan structural genes, whereas if synthesis of the leader peptide is blocked by tryptophan deficiency, transcription of the tryptophan structural genes can occur.

How does *translation* of the leader peptide regulate *transcription* of the tryptophan genes downstream? Somehow or other, further action of RNA polymerase in transcription is blocked by translation of the leader peptide. One important hypothesis to explain this suggests that transcription and translation must be occurring virtually simultaneously. Thus, while transcription of downstream DNA sequences is still proceeding, translation of sequences already transcribed has begun. Apparently, as the mRNA is released from the DNA, the ribosome binds to it and translation begins. Attenuation occurs (RNA polymerase quits working on the mRNA) because a portion of the newly formed mRNA folds into a double-stranded loop which signals cessation of RNA polymerase action. There is a transcription site in the leader called the **transcription pause site,** at which RNA polymerase temporarily slows down if tryptophan is plentiful. In this case, translation of the leader peptide can occur and this slows down transcription sufficiently so that the double-stranded loop can form and signal RNA polymerase to cease functioning. If tryptophan is in short supply, however, translation of the leader peptide cannot occur and the RNA polymerase moves past the termination site and begins transcription of the tryptophan structural genes. Thus we see that there is a highly integrated system in which transcription and translation interact, with the rate of transcription being influenced by the rate of translation. The double-loop structure formed by mRNA is brought about because two stretches of nucleotide bases near each other are complementary (looping in a manner similar to that already described for tRNA; see Figure 9.11).

Thus, in the tryptophan biosynthetic pathway, two distinct mechanisms for the regulation of transcription exist, repression and attenuation. Repression is a mechanism that has large effects on the rate of enzyme synthesis, whereas attenuation brings about a finer control. Working together, these two mechanisms precisely regulate the synthesis of the biosynthetic enzymes, and hence the biosynthesis of tryptophan. Attenuation has also been shown to occur in the biosynthetic pathway for histidine, and for some other amino acids and essential metabolites as well.

The phenomenon of enzyme induction has had a long history in microbial physiology, and detailed studies of this phenomenon have played a major role in understanding macromolecular synthesis and its regulation.* In recent years, the concepts that have been developed from studies on enzyme induction in bacteria have been applied to eucaryotic cells and to an understanding of cancer. Emile Duclaux, an associate of Pasteur, first reported in 1899 that a fungus, *Aspergillus niger*, only produced the enzyme invertase (which hydrolyzes sucrose) when growing on a sucrose medium. Similar observations were made by other workers in the early 1900s. H. Karström of Finland first studied these phenomena systematically in the 1930s. He found that certain enzymes were always present, irrespective of the culture medium, whereas other enzymes were only formed when their substrates were present. Karström termed the enzymes formed all the time *constitutive,* and those formed only when substrate was present *adaptive.* In the 1940s, the French microbiologist Jacques Monod began a systematic study of bacterial growth and discovered the phenomenon he termed *diauxic growth* (Figure 9.17). From his study of diauxy, Monod turned to a more detailed study of the process of enzyme adaptation. For many years, the interpretation of enzyme adaptation had been complicated by the fact that the substrate of the enzyme seemed to be involved in the process. However, with the discovery of nonmetabolizable substances, such as thiomethylgalactoside (TMG), which brought about enzyme formation even though they were not substrates of the enzyme, it became clear that enzyme adaptation and enzyme function were two different things. Because the word *adaptation* has certain confusing connotations (among other things, adaptation does not distinguish between the selection of mutant strains from a culture and the synthesis of a new enzyme in a preexisting genotype), the term has been abandoned, and the word *induction* is used instead. Among the most critical experiments carried out by Monod and his colleagues was that which showed that enzyme induction resulted in the *synthesis of a new protein* and not the activation of some preexisting protein. This suggested that the enzyme inducer was somehow causing differential gene action.

While the phenomenon of enzyme induction had been known for a long time, enzyme repression, the specific inhibition of enzyme synthesis, was a relatively recent discovery, being reported simultaneously in 1953 by the laboratories of Monod, Edward A. Adelberg and H. Edwin Umbarger in Berkeley, and Donald D. Woods at Oxford, England. It soon became clear that enzyme induction and repression, although having opposite effects, were manifestations of a similar mechanism. Biochemical studies on enzyme synthesis provided considerable insight into the phenomena of induction and repression, but by themselves they probably never would have led to a final understanding of the picture. The introduction of the techniques of bacterial genetics was essential, and some of the aspects of this will be discussed in Chapters 11 and 12. Among the major contributors to this genetic approach was François Jacob, a student and colleague of Monod at the Pasteur Institute. Jacob isolated a large number of mutants in the lactose pathway and analyzed these genetically. The results led to the conclusion that induction and repression were under the control of specific proteins, called *repressors,* which were coded by regulatory genes. These regulatory genes were associated with, but were distinct from, the structural genes coding for the specific enzyme proteins. Finally, Walter Gilbert and Benno Müller-Hill at Harvard University developed a cell-free system of assaying for repressor protein, and carried out a biochemical isolation and purification process for

*Key reference for this section: Jacob, F., and J. Monod. 1961. Genetic regulatory mechanisms in the synthesis of proteins. J. Mol. Biol. 3:318–356.

the repressor. The proof that the repressor was indeed a protein not only confirmed the theory of Jacob and Monod, but provided the approaches necessary for studying the manner in which proteins are able to specifically interact with defined sequences on DNA molecules. Biochemical and genetic studies on RNA polymerase then clarified the manner in which the information in DNA is transcribed into RNA. The most significant result of this important fundamental research is that it shows that regulation of enzyme synthesis in bacteria occurs at the level of transcription rather than at the level of translation.

9.7

The Genetic Code

The genetic code was first elucidated in the bacterium *Escherichia coli* by use of a cell-free protein-synthesizing system and artificial RNA molecules made of repeating units of one or more nucleotides. With such a system, simple polypeptides composed of only single amino acids could be made. By determining the amino acid that was polymerized when each synthetic mRNA was used, the genetic code was determined. The genetic code determined in this way has been amply confirmed by other studies; it is presented in Table 9.5.

Characteristics of the code Perhaps the most interesting feature of the genetic code is that a single amino acid is frequently coded for by several different but related RNA triplets of bases. There are 64 possible combinations of four bases taken three at a time, and only 20 amino acids need be coded, but most amino acids are coded for by more than one codon. Recall that crucial to the translation of the genetic code are the tRNA molecules to which the amino acids are attached, and that an amino acid is attached to the growing polypep-

TABLE 9.5   The Genetic Code as Expressed by Triplet Sequences of Purine and Pyrimidine Bases in mRNA*

UUU	Phenylalanine	CUU	Leucine	GUU	Valine	AUU	Isoleucine
UUC	Phenylalanine	CUC	Leucine	GUC	Valine	AUC	Isoleucine
UUG	Leucine	CUG	Leucine	GUG (start)†	Valine	AUG (start)†	Methionine
UUA	Leucine	CUA	Leucine	GUA	Valine	AUA	Isoleucine
UCU	Serine	CCU	Proline	GCU	Alanine	ACU	Threonine
UCC	Serine	CCC	Proline	GCC	Alanine	ACC	Threonine
UCG	Serine	CCG	Proline	GCG	Alanine	ACG	Threonine
UCA	Serine	CCA	Proline	GCA	Alanine	ACA	Threonine
UGU	Cysteine	CGU	Arginine	GGU	Glycine	AGU	Serine
UGC	Cysteine	CGC	Arginine	GGC	Glycine	AGC	Serine
UGG	Tryptophan	CGG	Arginine	GGG	Glycine	AGG	Arginine
UGA	None (stop signal)	CGA	Arginine	GGA	Glycine	AGA	Arginine
UAU	Tyrosine	CAU	Histidine	GAU	Aspartic	AAU	Asparagine
UAC	Tyrosine	CAC	Histidine	GAC	Aspartic	AAC	Asparagine
UAG	None (stop signal)	CAG	Glutamine	GAG	Glutamic	AAG	Lysine
UAA	None (stop signal)	CAA	Glutamine	GAA	Glutamic	AAA	Lysine

*The codons in DNA are complementary to those given here. Thus U here is complementary to the A in DNA, C is complementary to G, G to C, and A to U.

†GUG and AUG, at the beginning of the mRNA, code for N-formylmethionine.

tide chain through the mediation of a molecule of tRNA. Although there may be more than one tRNA carrying a single amino acid, not every possible anticodon is present in a tRNA. To explain how each of the codons pairs with a tRNA anticodon, the **wobble** concept has been advanced. According to the wobble concept, pairing is not as critical for the third base in the triplet, so that imperfect pairing can occur. However, even with wobble, no single tRNA molecule can recognize more than three codons.

Another phenomenon of the code presented in Table 9.5 is that a few triplets do not correspond to any amino acid. These triplets are called **nonsense codons,** and they function as "punctuation" to signal the termination of translation of the gene coding for a specific protein. If no tRNA molecules correspond to these nonsense codons, no amino acid can be inserted, and the polypeptide is terminated and released from the ribosome.

In addition to stopping points, **starting** points are also needed. At least one mechanism for initiation of a polypeptide chain is known, involving the amino acid *N*-formylmethionine, as we discussed in Section 9.5. The importance of having a well-defined starting point is readily understood if we consider that with a triplet code it is absolutely essential that translation begin at the correct location, since if it does not, the whole **reading frame** would be shifted and an entirely different protein or none at all would be formed. The problem of reading-frame shift will be considered further when we discuss mutation.

Another problem in the translation of the genetic code is that error sometimes occurs. This means that a coding triplet of the mRNA may be "read" improperly and the wrong amino acid inserted. Amino acids whose codons differ by only a single base, for example, phenylalanine (UUU) and leucine (UUA), are most likely to be involved in errors because occasionally only two of the three bases are concerned with codon-anticodon recognition. In rare instances leucine may be added to the growing polypeptide instead of phenylalanine even when the codon is UUU. In the normal cell these rare errors probably occur in only a small number of all the protein molecules and hence have no detrimental effect. However, certain antibiotics which act on ribosomes, such as streptomycin, neomycin, and related substances (see Section 9.5), increase error to such an extent that many protein molecules in the cell are abnormal and the cell can no longer function properly. Other conditions that increase error are changes in cation concentration, pH, and temperatures that are not optimal for cell growth.

Overlapping genes   Although the evidence is strong that the nucleotide sequence specifying one product is separate and distinct from the sequence specifying another product, studies on the small bacterial virus φX174 have shown that this virus has insufficient genetic information in nonoverlapping genes to code for all of the proteins necessary for its reproduction, and that genetic economy is introduced by using the same piece of DNA for the coding of more than one product. This phenomenon of overlapping genes is illustrated in Section 10.5. It is a process made possible by the reading of the same nucleotide sequence in two different phases, beginning at different sites.

Mutants and Their Isolation

The basic biological material of the geneticist is a **strain** or **clone,** which is a population of genetically identical cells. A clone is a pure culture, and techniques such as those used for isolating pure cultures are directly applicable to isolating clones for genetic analysis.

A mutation is observed as a sudden inheritable change in the phenotype of an organism. It is common to refer to a strain isolated from nature as a **wild-type,** and strains isolated from the wild type via mutation as **mutants.** We can distinguish between two kinds of mutations, selectable and unselectable. An example of an unselectable mutation is that of loss of color in a pigmented bacterium (Figure 9.21a). White colonies have neither an advantage nor a disadvantage over the pigmented parent colonies when grown on agar plates (there may be a selective advantage for pigmented organisms in nature, however). This means that the only way we can detect such mutations is to examine large numbers of colonies and look for the "different" ones. A selectable mutation confers upon the mutant an advantage under certain environmental conditions, so that the progeny of the mutant cell are able to outgrow and replace the parent. An example of a selectable mutation is drug resistance: an antibiotic-resistant mutant can grow in the presence of antibiotic concentrations that inhibit the parent (Figure 9.21b). It is relatively easy to detect and isolate selectable mutants by choosing the appropriate environmental conditions.

Virtually any characteristic of a microorganism can be changed through mutation (Figure 9.21c). Nutritional mutants can be detected by the technique of **replica plating.** Using sterile velveteen cloth, an imprint of colonies from a master plate is made onto an agar plate lacking the nutrient. The colonies of the parental type will grow normally, whereas those of the mutant will not. Thus, the inability of a colony to grow on the replica plate (Figure 9.21d) will be a signal that it is a mutant. The colony on the master plate corresponding to the vacant spot on the replica plate (Figure 9.21e) can then be picked, purified, and characterized. A nutritional mutant that has a requirement for a growth factor is often called an **auxotroph;** the wild-type parent from which the auxotroph was derived is called a **prototroph.**

An ingenious method which is widely used to isolate mutants that require amino acids or other growth factors is the **penicillin-selection method** (Figure 9.22). Ordinarily, mutants that require growth factors are at a disadvantage in competition with the parent cells, so that there is no direct way of isolating them. However, penicillin kills only growing cells (see Section 2.5 and 5.11), so that if penicillin is added to a population growing in a medium lacking the growth factor required by the desired mutant, the parent cells will be killed, whereas the nongrowing mutant cells will be unaffected. Thus, after preliminary incubation in the absence of growth factor in penicillin-containing medium, the population is washed free of penicillin and transferred to plates containing the growth factor. Among the colonies that grow up (including some wild-type cells that have escaped penicillin killing) should be some representing growth factor mutants.

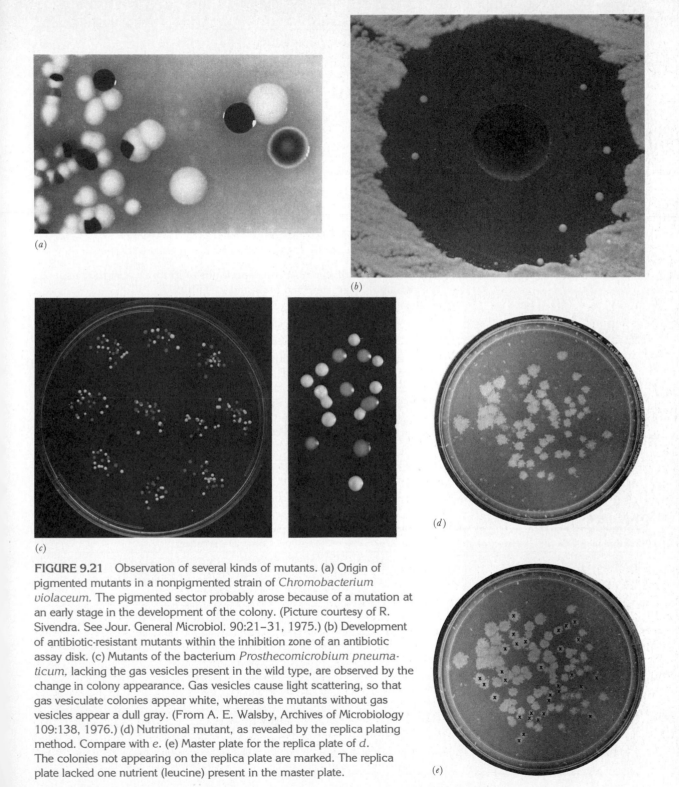

(a)

(b)

(c)

(d)

(e)

FIGURE 9.21 Observation of several kinds of mutants. (a) Origin of pigmented mutants in a nonpigmented strain of *Chromobacterium violaceum.* The pigmented sector probably arose because of a mutation at an early stage in the development of the colony. (Picture courtesy of R. Sivendra. See Jour. General Microbiol. 90:21–31, 1975.) (b) Development of antibiotic-resistant mutants within the inhibition zone of an antibiotic assay disk. (c) Mutants of the bacterium *Prosthecomicrobium pneuma-ticum,* lacking the gas vesicles present in the wild type, are observed by the change in colony appearance. Gas vesicles cause light scattering, so that gas vesiculate colonies appear white, whereas the mutants without gas vesicles appear a dull gray. (From A. E. Walsby, Archives of Microbiology 109:138, 1976.) (d) Nutritional mutant, as revealed by the replica plating method. Compare with *e.* (e) Master plate for the replica plate of *d.* The colonies not appearing on the replica plate are marked. The replica plate lacked one nutrient (leucine) present in the master plate.

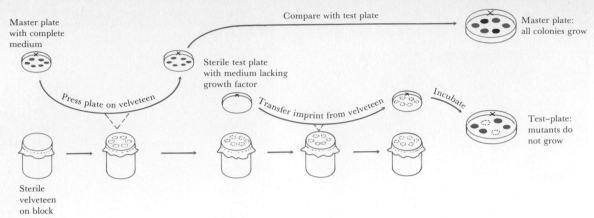

Master plate
with complete
medium

Compare with test plate

Master plate:
all colonies grow

Sterile test plate
with medium lacking
growth factor

Press plate on velveteen

Transfer imprint from velveteen

Incubate

Test–plate:
mutants do
not grow

Sterile
velveteen
on block

FIGURE 9.22 Replica-plating method for detecting nutritional mutants.

Some of the most common kinds of mutants and the means by which they are detected are listed in Table 9.6.

9.9

The Molecular Basis of Mutation

Although many mutations arise spontaneously, it is possible to increase the rate of mutation many times by treatment of cells with various **mutagenic** agents. An analysis of how mutagenic agents work provides many insights into the molecular mechanisms of mutation. Further, a study of the alterations in amino acid sequence in proteins as a result of mutation provides excellent confirmation for the details of the genetic code. All of the existing evidence indicates that mutations result from alterations in the purine-pyrimidine base sequences of DNA. Several kinds of changes in DNA can be visualized.

TABLE 9.6   Kinds of Mutants

Description	Nature of Change	Detection of Mutant
Nonmotile	Loss of flagella; nonfunctional flagella	Compact colonies instead of flat, spreading colonies
Noncapsulated	Loss or modification of surface capsule	Small, rough colonies instead of larger, smooth colonies
Rough colony	Loss or change in lipopolysaccharide outer layer	Granular, irregular colonies instead of smooth, glistening colonies
Nutritional	Loss of enzyme in biosynthetic pathway	Inability to grow on medium lacking the nutrient
Sugar fermentation	Loss of enzyme in degradative pathway	Do not produce color change on agar containing sugar and a pH indicator
Drug resistant	Impermeability to drug or drug target is altered or drug is detoxified	Growth on medium containing a growth-inhibitory concentration of the drug
Virus resistant	Loss of virus receptor	Growth in presence of large amounts of virus
Temperature sensitive	Alteration of any essential protein so that it is more heat sensitive	Inability to grow at a temperature normally supporting growth (e.g., 37°C) but still growing at a lower temperature (e.g., 25°C)
Cold sensitive	Alteration in an essential protein so that it is inactivated at low temperature	Inability to grow at a low temperature (e.g., 20°C) that normally supports growth

Point mutations **Point mutations** spring from changes of single bases; an adenine base may be replaced by guanine or thymine by cytosine, for instance (Figure 9.23). The consequences of a point mutation will depend greatly on where in the gene the mutation occurs. It might produce no change at all in the protein, an inconsequental amino acid substitution, a serious amino acid substitution, or no protein may be formed at all.

Let us consider the codon that specifies the amino acid glycine at a specific site in the protein tryptophan synthetase (Figure 9.24). Because of the multiplicity of codons for some amino acids, a phenomenon called **degeneracy,** three of the possible base changes have no effect on the amino acid; such base changes would of course be undetectable since they would not lead to an alteration in the protein. These are called **silent mutations.** Notice that these changes occur in the third base of the codon. As seen in Table 9.5, degeneracy is found most frequently in the third base, so that for many codons one-third of the mutations are not harmful. Changes in the first or second base of the triplet much more often lead to significant changes in the protein.

Not all mutations that cause amino acid substitution necessarily lead to nonfunctional proteins. The outcome depends greatly on where in the polypeptide chain the substitution has occurred, and on how it affects the folding and the catalytic activity of the protein.

Deletions and insertions **Deletions** are due to elimination of portions of the DNA of a gene (Figure 9.23). A deletion may be as simple as the

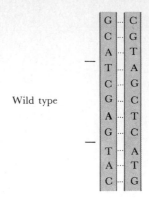

Wild type

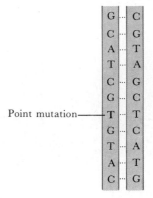

Point mutation

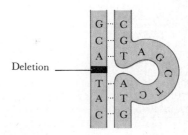

Deletion

FIGURE 9.23 DNA base changes from wild type, involving point mutation and deletion.

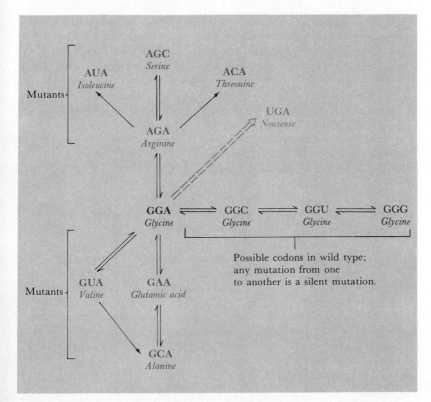

Possible codons in wild type; any mutation from one to another is a silent mutation.

FIGURE 9.24 Base changes affecting the occurrence, at a single site, of amino acids in the tryptophan synthetase of *E. coli.* For simplicity, the changes are shown in mRNA rather than in the DNA to which it is complementary.

removal of a single base, or it may involve hundreds of bases. Deletion of a large segment of the DNA results in complete loss of the ability to produce the protein. Such deletions cannot be restored through further mutations, but only through genetic recombination. Indeed, one way in which large deletions are distinguished from point mutations is that the latter are usually revertible through further mutations, whereas the former usually are not.

Insertions occur when new bases are added to the DNA of the gene. Insertions can involve only a single base or many bases. Generally, insertions do not occur by simple copy errors as do deletions, but arise as a result of mistakes that occur during genetic recombination (discussed in Chapter 11). Many insertion mutations are due to the insertion of specific identifiable DNA sequences 700 to 1400 base pairs in length called **insertion sequences** or **insertion elements.** The involvement of such insertion sequences in genetics and genetic engineering is discussed in later chapters.

Frame-shift mutations   Since the genetic code is read from one end in consecutive blocks of three bases, any deletion or insertion of a new base results in a **reading-frame shift,** and the translation of the gene is completely upset. Partial restoration of gene function can often be accomplished by insertion of another base near the one deleted. After correction, depending on the exact amino acids coded by the still faulty region and the region of the protein involved, the protein formed may have some enzymatic activity or be completely normal.

Polarity   As we have discussed, a cluster of genes may be transcribed together in a single mRNA. One consequence of the fact that translation occurs from one end is that nonsense mutations (see Section 9.7) near the beginning of translation (the 5′ end) will completely block the translation of all the successive genes, whereas nonsense mutations farther down will have fewer effects. This phenomenon is called **polarity.** Polarity gradients arise because when a ribosome reaches a chain-terminating nonsense codon, an incomplete polypeptide is released; the ribosome generally detaches from its mRNA template, so that the following genes are not translated. Because of polarity, nonsense mutations at the 3′ end will generally have no effect on genes transcribed at the 5′ end.

Back mutations or reversions   Many but not all mutations are revertible. A "revertant" is operationally defined as a strain in which the wild-type phenotype that was lost in the mutant is restored. Revertants can be of two types. In first-site revertants, the mutation that restores activity occurs at the site at which the original mutation occurred. In second-site revertants, the mutation occurs at some different site. Several mechanisms for second-site mutations are known: (1) a mutation somewhere else in the same gene can restore enzyme function, such as in a reading-frame shift mutation; (2) a mutation in another gene may restore the wild-type phenotype **(suppressor mutation);** and (3) the production of another enzyme may occur that can replace the mutant one by introducing a metabolic pathway different from that used by the mutant enzyme. In this last type no production of the original enzyme occurs although it does in the other types.

Suppressor mutations are new mutations that suppress the original one indirectly. One way in which a suppressor mutation can arise is through an alteration in a transfer RNA anticodon, so that it recognizes the wrong (but right!) mutant codon. In this way, the proper amino acid can be added to the polypeptide chain, at least some of the time. It is usually characteristic of suppressor mutations of this type that normal function is not completely restored; only some of the polypeptides formed on a single mRNA are normal. Therefore, revertants arising due to suppressor mutations often do not grow as rapidly as do the wild type. They can also be distinguished from true revertants by genetic mapping since the locus of the suppressor gene is in a different region of the genome from that of the original mutant gene.

9.10
Mechanisms of Mutagenesis

Many mutations appear to be *spontaneous,* occurring without any direct intervention of the investigator. Some may arise because of random errors that occur during the replication of DNA. The accuracy with which DNA is copied during replication is very high, some studies indicating the error to be less than once in 500,000 bases copied. Not all such errors will result in detectable mutations. The spontaneous mutation rate in bacteria is of the order of one in 10^8 to 10^9 DNA base pairs per generation, making the detection of spontaneous mutation fairly difficult.

It is now well established that a wide variety of chemicals can induce mutations. An overview of some of the major chemical mutagens and their modes of action are given in Table 9.7. Several classes of chemical mutagens occur. A variety of chemical mutagens are **base analogs,** resembling DNA purine and pyrimidine bases in structure, yet showing faulty pairing properties. When one of these base analogs is incorporated into DNA, replication may occur normally most of the time, but occasional copying errors occur, resulting in the incorporation of the wrong base into the copied strand. During subsequent segregation of this strand, the mutation is revealed.

A variety of chemicals react directly with DNA, causing changes in one or another base, which results in faulty pairing or other changes (Table 9.7). Such chemicals differ in their action from the base analogs in that the former are able to introduce direct changes even in nonreplicating DNA, whereas the base analogs are only incorporated during replication. One interesting group of chemicals, the acridines, are planar molecules which become inserted between two DNA base pairs, thereby pushing them apart. During replication, an extra base can then be inserted in acridine-containing DNA, lengthening the DNA by one base and thus shifting the reading frame.

Mutations that arise from the SOS response Many kinds of mutations arise as a result of faulty repair of damage induced in DNA by various agents. Conditions that cause damage to DNA include ultraviolet

TABLE 9.7 Chemical Mutagens and Their Modes of Action

Agent	Action	Result
Base analogs:		
5-Bromouracil	Incorporated like T; occasional faulty pairing with G	A–T pair $\longrightarrow$ G–C pair Occasionally G–C $\longrightarrow$ A–T
2-Aminopurine	Incorporated like A; faulty pairing with C	A–T $\longrightarrow$ G–C Occasionally G–C $\longrightarrow$ A–T
Chemicals reacting with DNA:		
Nitrous acid (HNO$_2$)	Deaminates A, C	A–T $\longrightarrow$ G–C G–C $\longrightarrow$ A–T
Hydroxylamine (NH$_2$OH)	Reacts with C	G–C $\longrightarrow$ A–T
Alkylating agents:		
Monofunctional (e.g., ethyl methane sulfonate)	Put methyl on G; faulty pairing with T	G–C $\longrightarrow$ A–T
Bifunctional (e.g., nitrogen mustards, mitomycin, nitrosoguanidine)	Cross-link DNA strands; faulty region excised by DNase	Both point mutations and deletions
Intercalative dyes (e.g., acridines, ethidium bromide)	Insert between two base pairs	Reading-frame shift
Radiation:		
Ultraviolet	Pyrimidine dimer formation	Repair may lead to error or deletion
Ionizing radiation (e.g., X rays)	Free-radical attack on DNA, breaking chain	Repair may lead to error or deletion

radiation, ionizing radiation, thymine starvation, alkylating agents (such as ethyl methane sulfonate and nitrosoguanidine), and certain antibiotics (nalidixic acid, mitomycin C). A complex cellular mechanism, called the **SOS regulatory system,** is activated as a result of DNA damage, bringing on a number of processes of DNA repair. In the SOS system, DNA repair occurs in the absence of template instruction, resulting in the creation of many errors, hence many mutations.

In the SOS regulatory system, DNA damage serves as a distress signal to the cell, resulting in the coordinate derepression (activation) of a number of cellular functions involved in DNA repair. The SOS system is normally repressed by the LexA protein, but LexA is inactivated by RecA, a protease that is activated as a result of DNA damage. Since the DNA repair mechanisms of the SOS system are inherently error-prone, many mutations arise. Thus, through the SOS regulatory system, DNA damage by various agents such as chemicals and radiation leads to mutagenesis.

The SOS system senses the presence in the cell of DNA damage and the repair mechanisms are activated. But once the DNA damage has been repaired, the SOS system is switched off, and further mutagenesis ceases. In addition to its effect on cellular mutagenesis, the SOS regulatory system plays a central role in the regulation of temperate virus replication, as will be discussed in Section 10.6.

It should be emphasized that not all DNA repair occurs in the absence of template instruction. Cells generally have a constitutive DNA repair system which does require template instruction and does lead to proper DNA repair. This system apparently works most of the time, but is not sufficient to repair certain, large amounts of damage done by some of the agents mentioned above.

The sensitivity by which mutations can be detected in bacterial populations is very high, because large populations of cells can be conveniently handled. Thus, even though mutations are rare events, they can be studied and quantified in bacteria. One striking observation that has been made is that many mutagenic chemicals are also carcinogenic, causing cancer in higher animals or humans. Although a discussion of cancer is beyond the scope of this text, in the present context the key point is that there seems to be a strong correlation between the mutagenicity of many chemicals and their carcinogenicity. This has led to the development of bacterial mutagenesis tests as screens for potential carcinogenicity of various compounds. The variety of chemicals, both natural and artificial, which the human population comes into contact with, through industrial exposure, is enormous. There is considerable need for simple tests to ascertain the safety of such compounds. There is good evidence that a large proportion of human cancers have environmental causes, most likely through the agency of various chemicals, making the detection of chemical carcinogens very urgent. It does not necessarily follow that because a compound is mutagenic, it will be carcinogenic. The correlation, however, is quite high, and the knowledge that a compound is mutagenic in a bacterial system serves as a warning of possible danger. Similarly, the fact that a compound is not mutagenic in a bacterial system does not mean that it is not carcinogenic, since the bacterial system cannot detect all compounds active in higher animals. The development of bacterial tests for carcinogenic screening has been carried out primarily by a group at the University of California in Berkeley, under the direction of Bruce Ames, and the mutagenicity test for carcinogens is sometimes called the *Ames test.*

The standard way to test chemicals for mutagenesis has been to measure the rate of back mutations in strains of bacteria that are auxotrophic for some nutrient. It is important, of course, that the original mutation be a point mutation, so that reversion can occur. When such an auxotrophic strain is spread on a medium lacking the required nutrient (for example, an amino acid or vitamin), no growth will occur, and even very large populations of cells can be spread on

Mutagenesis and Carcinogenesis: The Use of Bacterial Mutagenesis to Detect Potentially Carcinogenic Chemicals

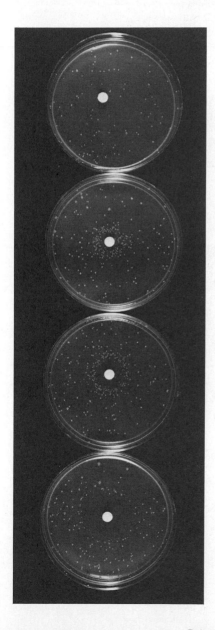

FIGURE 9.25 Testing of a carcinogen by the Ames test. *Salmonella typhimurium* strain TA100, a histidine-requiring mutant, was spread on plates of minimal medium. Disks containing known concentrations of aflatoxin B, a known carcinogen, were placed in the center of the plates, and the plates incubated at 37°C for 2 days. Colonies around the disks predominantly represent reversions induced by the carcinogen. Spontaneous revertants form a few colonies scattered over the plate. The aflatoxin was activated with an enzyme preparation from rat liver, with NADH and NADPH added. The dose-response series shown, from top to bottom, is: control, 0.05 μg aflatoxin, 0.1 μg aflatoxin, 1.0 μg aflatoxin. The highest concentration inhibits growth. Plates prepared by Dianne Chambliss.

the plate without formation of visible colonies. However, if back mutations have occurred, those cells will be able to form colonies. Thus if 10^8 cells are spread on the surface of a single plate, even as few as 10 to 20 back mutants can be detected by the 10 to 20 colonies they will form.

Although the simple testing of chemicals for mutagenesis in bacteria has been carried out for a long time, two elements have been introduced in the Ames test to make it a much more powerful test. The first of these is the use of strains of bacteria lacking DNA repair enzymes, so that any damage that might be induced in DNA is not corrected. The second important element in the Ames test is the use of liver enzyme preparations to convert the chemicals into their active mutagenic (and carcinogenic) forms. It has been well established that many potent carcinogens are not directly carcinogenic or mutagenic, but undergo chemical changes in the human body, which convert them into active substances. These changes take place primarily in the liver, where enzymes (mixed-function oxygenases) normally involved in detoxification cause formation of epoxides or other activated forms of the compounds, which are then highly reactive with DNA. In the Ames test, a preparation of enzymes from rat liver is first used to activate the compound. Next the activated complex is taken up on a filter-paper disk, which is placed in the center of a plate on which the proper bacterial strain has been overlayed. After overnight incubation, the mutagenicity of the compound can be detected by looking for a halo of back mutations in the area around the paper disk (Figure 9.25). It is always necessary, of course, to carry out this test with several different concentrations of the compound, because compounds vary in their mutagenic activity and are lethal at higher levels. A wide variety of chemicals has been subjected to the Ames test, and it has become one of the most useful prescreens to determine the potential carcinogenicity of a compound.

9.12

Summary

In this chapter we have presented a brief overview of some of the most basic and significant aspects of molecular genetics. Some of this material is summarized in Table 9.8.

The information for the synthesis of a specific protein is present in a portion of the DNA molecule, and the sequence of amino acids in the protein is coded by a sequence of purine and pyrimidine bases in the DNA. Three nucleotide bases code for a single amino acid, and the sequence of three bases is called a **codon.** One or several distinct codons exist for each amino acid. The information in the DNA code is first **transcribed** into RNA; the molecules of RNA carrying genetic information for protein synthesis are called **messenger RNAs.** Transcription is carried out through the action of the enzyme **RNA polymerase,** which catalyzes the formation of phosphodiester bonds between ribonucleotides, using DNA as a template. RNA polymerase is a complex protein with several subunits, and two additional components, *sigma* and *rho,* are involved in the control of the starting and stopping of mRNA synthesis. Messenger RNA synthesis starts at a specific site of the DNA called the **promoter,** and the specific binding

TABLE 9.8 Elements in the Regulation of Transcription, the Copying of Information from DNA into mRNA

A. Starting mRNA synthesis

Starting point: the *promoter* region, a specific DNA sequence where RNA polymerase binds. Attachment of enzyme to this region requires *sigma* factor. Regulation can occur by modification or inhibition of sigma.

B. Negative control of mRNA synthesis

Operator: site on DNA is upstream of structural gene, but downstream of promoter. Repressor binds here.

Repressor: protein which binds to operator and blocks mRNA synthesis.

Repression: specific inhibition of mRNA synthesis when the end product of pathway is present; mRNA for *first* enzyme in pathway is blocked.

    (a) *Corepressor:* small molecule (usually end product), which combines with repressor protein and activates it.

    (b) Corepressor + repressor = repression (inhibition of mRNA synthesis).

Induction: specific synthesis of mRNA when an inducer is present. Repressor protein normally prevents mRNA synthesis, and repressor is inactivated by inducer.

    (a) *Inducer:* small molecule which combines with repressor and activates it.

    (b) Inducer + repressor = induction (mRNA synthesis occurs).

C. Positive control of mRNA synthesis

Catabolite repression (glucose effect): overrides induction.

    (a) *CAP:* combines with promoter region and causes binding of RNA polymerase. Active only when cyclic AMP present.

    (b) *Cyclic AMP:* nucleotide derived from ATP. Glucose inhibits cyclic AMP formation.

    (c) In presence of glucose, cyclic AMP absent, CAP inactive, mRNA synthesis does not occur. Glucose thus prevents induction.

of RNA polymerase to the promoter region requires the presence of sigma. The RNA polymerase moves progressively down the chain and the growing RNA chain remains attached to the transcription site, with the free end dissociated from the complex. Termination of mRNA synthesis occurs at sites on the DNA, coded by specific stop signals.

The translation of the RNA message into the amino acid sequence of protein is brought about through the agency of **transfer RNA.** One part of the tRNA molecule contains the **anticodon,** a sequence of three nucleotides which are complementary to the triplet codon on the mRNA. The amino acid is attached to one end of the tRNA by a specific enzyme, **aminoacyl-tRNA synthetase,** which has the important function of recognizing both the amino acid and the specific tRNA. Protein synthesis occurs through an interaction between mRNA, amino acid-charged tRNA, and ribosome. The mRNA attaches to the 30-S subunit of the RNA, and the tRNA attaches to the 50-S subunit. Amino acid-charged tRNA molecules come in sequence, as specified by the genetic code, and peptide bond formation occurs between the growing protein chain and the amino acid attached to the tRNA. The now empty tRNA leaves the complex, and the next amino acid-charged tRNA arrives. Codons of the mRNA specify both a starting point, where the synthesis of a protein begins, and a stopping point, where the protein chain is terminated. The codons that signal a stopping point are called **nonsense** codons, since no tRNA anticodons correspond to them.

Not all proteins are synthesized by the cell in the same amounts, and synthesis of proteins is regulated. Although several mechanisms of regulation have been recognized, the most well known is the regulation by the presence of specific chemical substances present in the environment. Two phenomena are recognized: **enzyme induction,** in which a substance added to the environment of the cell brings about the synthesis of an enzyme which is not otherwise formed; and **enzyme repression,** in which the added substance brings about the inhibition of the synthesis of an enzyme. Although having opposite effects, induction and repression have similar underlying mechanisms, which cause selective inhibition or selective synthesis of mRNA. Messenger RNA synthesis in such systems is affected by the presence and activity of specific proteins called **repressors,** which combine with specific regions of the DNA at the initial end of the gene, the so-called **operator** regions. If the repressor binds to the operator, then mRNA synthesis cannot occur. In induction, the low-molecular-weight inducer molecules bind to the repressor and change its conformation so that it can no longer bind to the operator. In repression, the repressor protein does not normally bind to the operator, but if the low-molecular-weight substance (called the corepressor) binds to the repressor, the repressor is activated and now binds to the operator, and prevents enzyme synthesis.

Catabolite repression is another type of enzyme repression. Catabolite repression occurs when an organism is offered an energy source that it can catabolize readily, such as glucose; the synthesis of inducible enzymes involved with other, less easily used energy sources is repressed even though the inducer is present. Only after the readily catabolizable energy source is exhausted is catabolite repression abolished, and enzyme induction occurs. The mechanism of catabolite repression has been shown to involve two components, a catabolite activation protein called CAP, and a derivative of ATP called cyclic AMP. Cyclic AMP, a key element in a variety of control systems in organisms, is synthesized when glucose is absent, but cyclic AMP levels become low when glucose is present. Cyclic AMP binds with CAP and changes its configuration so that it can bind to the promoter region and facilitate the binding of RNA polymerase. In the absence of cyclic AMP, CAP cannot bind and mRNA synthesis does not occur. Thus, catabolite repression results in a shutting down of mRNA synthesis simultaneously for a variety of enzymes in the cell.

Supplementary Readings

Breathnach, R., and P. Chambon. 1981. Organization and expression of eucaryotic split genes coding for proteins. Annu. Rev. Biochem. 50:349–383. Includes a good discussion of RNA processing.

Dressler, D., and H. Potter. 1982. Molecular mechanisms in genetic recombination. Annu. Rev. Biochem. 51:727–761. A good discussion of the roles of the RecA protein.

Gellert, M. 1981. DNA topoisomerases. Annu. Rev. Biochem. 50:879–910.

Grivell, L. A. 1983. Mitochondrial DNA. Sci. Am. 248 (March):78–89. A fascinating presentation of the unique genetics of mitochondria.

Igo-Kemenes, T., W. Horz, and H. C. Zachau. 1982. Chromatin. Annu. Rev. Biochem. 51:89–121. Extensive discussion of the organization of the eucaryotic chromosome, including a discussion of the nucleosome and satellite DNA.

Jelinek, W. R., and **C. W. Schmid.** 1982. Repetitious sequences in eukaryotic DNA and their expression. Annu. Rev. Biochem. 51:813–844.

Kornberg, A. 1980. DNA replication. W. H. Freeman, San Francisco. 724 pp. An authoritative textbook on the biochemistry of DNA replication, by the Nobel laureate who discovered DNA polymerase.

Kreil, G. 1981. Transfer of proteins across membranes. Annu. Rev. Biochem. 50:317–348. Reviews the signal sequence hypothesis.

Lindahl, T. 1982. DNA repair enzymes. Annu. Rev. Biochem. 51:61–87.

Little, J. W., and **D. W. Mount.** 1982. The SOS regulatory system of *Escherichia coli*. Cell 29:11–22. An authoritative review of the mechanisms involved in the regulation of repair of damaged DNA.

Loeb, L. A., and **T. A. Kunkel.** 1982. Fidelity of DNA synthesis. Annu. Rev. Biochem. 52:429–457. Discusses the various proofreading mechanisms that reduce the error involved in DNA replication.

Meyer, D. I. 1982. The signal hypothesis—a working model. Trends Biochem. Sci. 7:320–321. A short review of the signal sequence hypothesis and secretion of proteins.

Singer, B., and **J. T. Kusmierek.** 1982. Chemical mutagenesis. Annu. Rev. Biochem. 51:655–693. What are chemical mutagens, and how do they work?

Stryer, L. 1981. Biochemistry, 2nd ed. W. H. Freeman, San Francisco. 949 pp. A well-illustrated textbook with good chapters on nucleic acid structure and function.

Wang, J. C. 1982. DNA topoisomerases. Sci. Am. 247 (July):94–109.

Yanofsky, C. 1981. Attenuation in the control of expression of bacterial operons. Nature (Lond.) 289:751–758. An authoritative review by the discoverer of attenuation.

Zimmerman, S. B. 1982. The three-dimensional structure of DNA. Annu. Rev. Biochem. 51:395–427. Illustrates the various shapes in which DNA can be found.

10

Viruses

A **virus** is an infectious agent that has a genome containing either DNA or RNA and that can alternate between two distinct states, intracellular and extracellular. In the extracellular or infectious state, viruses are submicroscopic particles containing nucleic acid surrounded by protein and occasionally containing other components. These **virus particles** or **virions** are metabolically inert and do not carry out respiratory or biosynthetic functions. The role of the virion is to carry the viral nucleic acid from the cell in which the virion has been produced to another cell where the viral nucleic acid can be introduced and the intracellular state initiated. In the latter phase replication occurs: more nucleic acid and other components of the virus are produced. Cells that viruses can infect and in which they can replicate are called **hosts.** The host performs most of the metabolic functions necessary for virus replication.

Viruses may be considered in two ways: as agents of disease and as agents of heredity. As agents of disease, viruses can enter cells and cause harmful changes in these cells, leading to disrupted function or death. As agents of heredity, viruses can enter cells and cause permanent, hereditable changes that usually are not harmful and may even be beneficial. In many cases, which role the virus plays depends on the host cell and on the environmental conditions.

Viruses vary widely in size, shape, chemical composition, range of organisms attacked, kinds of cell damage induced, and range of genetic capabilities. They are known to infect animals, plants, bacteria, and fungi. Bacterial viruses are often called **bacteriophages** (or **phage** for short). Practically all virions are of submicroscopic size, which means that the individual particle can rarely be seen with the light microscope, although they are fairly easy to visualize with the electron microscope.

Structure of the virion   An analysis of the chemical composition of virus particles shows that some contain RNA and others DNA, but never both. For both DNA and RNA viruses, types are known which have either single-stranded or double-stranded nucleic acid.

The structures of virus particles are exceedingly diverse. The nucleic acid is located in the center, surrounded by a protein coat called the **capsid;** the individual proteins that make up the capsid are called protein subunits, or **capsomeres.** In the simple viruses the protein coat is composed of a single kind of protein, whereas in the more complex viruses several kinds of capsomere proteins may be present. Shapes of several kinds of virus particles are illustrated in Figure 10.1. Many of the small viruses are geometrical structures

The Virus Particle or Virion

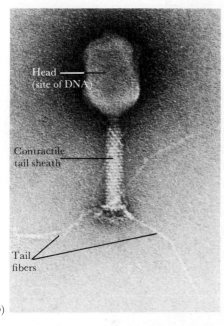

(b)

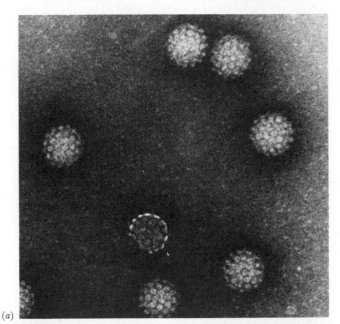

(a)

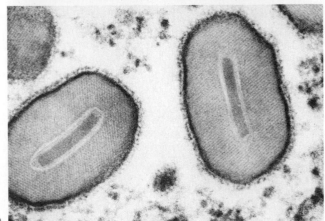

(c)

FIGURE 10.1   Electron micrographs of several types of virions. (a) Human wart virus, an icosahedral virus. Magnification, 179,500×. (From Noyes, W. F. 1964. Virology 23:65.) (b) Bacterial virus (bacteriophage) T4 of *Escherichia coli.* Note the complex structure. The tail components are involved in attachment of the virion to the host and injection of the nucleic acid. Magnification, 220,000×. (Courtesy of M. Wurtz, Biozentrum, Basel, Switzerland.) (c) Moth granulosis virus within infected insect cell. An enveloped virus: note the viral core within the outer envelope. Magnification, 95,000×. (From Arnott, H. J., and K. M. Smith. 1968. J. Ultrastructure Res. 21:251.)

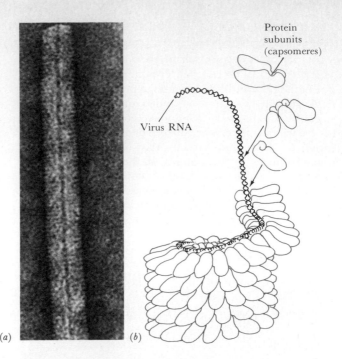

FIGURE 10.2 Structure and manner of assembly of tobacco mosaic virus. (a) Electron micrograph at high resolution of a portion of the virus particle. Magnification, 40,000×. (From Finch, J. T. 1964. J. Mol. Biol. 8:872.) (b) Assembly of the tobacco mosaic virion. The RNA assumes a helical configuration surrounded by the protein capsomeres. The center of the particle is hollow. (Adapted from Klug, A., and D. L. D. Caspar, 1960. Advan. Virus Res. 7:225.)

(a)

(b)

Protein subunits (capsomeres)

Virus RNA

called **icosahedrons,** with 20 triangular faces and 12 corners. Some virus particles are long rods, whose nucleic acid forms a central spiral surrounded by the protein capsomeres. The manner in which the capsomeres are arranged around the nucleic acid of tobacco mosaic virus (TMV) is shown in Figure 10.2. Some viruses have more complex structures, with membranous envelopes surrounding the central nucleic acid-protein core. The most complicated viruses in terms of structure are some of the bacterial viruses, which possess not only an icosahedral head structure but also a tail (Figure 10.1*b*). Sometimes, such as in T4 virus of *Escherichia coli,* the tail itself is a rather complicated structure. The range of sizes and shapes of bacterial viruses is shown in Table 10.1.

Quantification In any study of viruses it is necessary to count the virus particles accurately. This is most conveniently done by **plaque assay,** which is analogous to the colony count used for determining number of viable bacteria (Figure 10.3). It is assumed that each plaque has originated from one virus particle, just as each bacterial colony is derived from one bacterium. This procedure also permits the isolation of pure virus strains, since if a plaque has arisen from one virus particle or infected cell, all the virus particles in this plaque are genetically identical. Some of the particles from a single plaque can be picked and inoculated into a broth culture to make a culture that will establish a pure line. The development of this technique was as important for the advance of virology as was Koch's development of solid media (Section 1.5) for bacteriology.

Plaques may be obtained for animal viruses by using animal tissue-culture systems as a host. A monolayer of cultured animal cells

TABLE 10.1 Morphological Diversity and Nucleic Acid Character
of Bacterial Viruses

	Description	Examples* Escherichia coli	Other Bacteria
	Contractile tail; DNA double-stranded	T2, T4, T6	*Pseudomonas:* 12S, PB-1 *Bacillus:* SP50 *Myxococcus:* MX-1 *Salmonella:* 66t
	Long noncontractile tail; DNA double-stranded	T1, T5, lambda	*Pseudomonas:* PB-2 *Corynebacterium:* B *Streptomyces:* K1
	Short noncontractile tail; DNA double-stranded	T3, T7	*Pseudomonas:* 12B *Agrobacterium* R-1001 *Bacillus:* GA/1 *Salmonella:* P22
	No tail; large capsomeres; DNA single-stranded	ϕX174	*Salmonella:* ϕR
	No tail; small capsomeres; RNA single-stranded	f2, MS2, Qβ	*Pseudomonas:* 7S, PP7 *Caulobacter*
	Filamentous; no head; DNA single-stranded	fd, f1, M-13	*Pseudomonas*

1000 nm†

*A unique virus, not fitted into the scheme above, is ϕ6 of *Pseudomonas phaseolicola,* which has double-stranded RNA and a lipid envelope.

†The sizes of the drawings are to scale.

Based on Bradley, D. E. 1967. Bacteriol. Rev. 31:230–314.

is prepared in a petri plate or flat bottle, and the virus suspension overlayed. Plaques are revealed by zones of destruction of the animal cells (Figure 10.4).

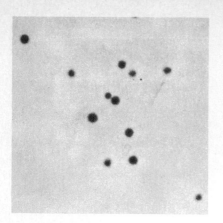

FIGURE 10.3 Quantification of bacterial virus by plaque assay. Portion of the surface of an agar plate containing a lawn of *Escherichia coli* cells infected with bacteriophage T4. A dilution of the suspension containing the virus material is mixed in a small amount of melted agar with the sensitive host bacteria, and the mixture poured on the surface of a nutrient agar plate. The host bacteria, which have been spread uniformly throughout the top agar layer, begin to grow, and after overnight incubation will form a lawn of confluent growth. Each virus particle that attaches to a cell and reproduces may cause cell lysis, and the virus particles released can spread to adjacent cells in the agar, infect them, be reproduced, and again lead to lysis and release. The size of the plaque formed depends on the virus, the host, and the conditions of culture. The plaques shown are about 1–2 mm in diameter.

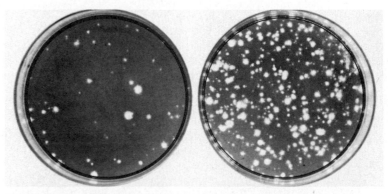

FIGURE 10.4 Quantification of an animal virus. Plaques of the tumor virus Rous sarcoma on a monolayer of mouse L cells. The cells have been stained to increase the contrast of the plaques.

A Bit of History

The word *virus* originally referred to any poisonous emanation, such as the venom of a snake, and later came to be used more specifically for the causative agent of any infectious disease.* Pasteur often referred to bacteria that caused infectious diseases as viruses. By the end of the nineteenth century, a large number of bacteria had been isolated and shown to be causal agents of specific infectious diseases, but there were some diseases for which a bacterial cause had not been shown. One of these was foot-and-mouth disease, a serious skin disease of animals. In 1898, Friedrich Loeffler and P. Frosch presented the first evidence that the cause of foot-and-mouth disease was an agent so small that it could pass through filters that could hold back all known bacteria. That the agent was not an ordinary toxin could be shown by the fact that it was active at very low dilution and could be transmitted in filtered material from animal to animal. Loeffler and Frosch concluded "that the activity of the filtrate is not due to the presence in it of a soluble

*Key references for this section: Brock, T. D. 1961. Milestones in microbiology. Reprinted 1975. American Society for Microbiology, Washington, D.C. Duckworth, D. H. 1976. Who discovered bacteriophage? Bacteriol. Rev. 40:793–802. Hughes, S. S. 1977. The virus: a history of the concept. Heinemann, London.

substance, but due to the presence of a causal agent capable of reproducing. This agent must then be obviously so small that the pores of a filter which will hold back the smallest bacterium will still allow it to pass. . . . If it is confirmed by further studies . . . that the action of the filtrate . . . is actually due to the presence of such a minute living being, this brings up the thought that the causal agents of a large number of other infectious diseases . . . which up to now have been sought in vain, may also belong to this smallest group of organisms."[†]

A year later, the Dutch microbiologist Martinus Beijerinck published his work on tobacco mosaic disease, a crippling leaf disease of tobaccos and tomatoes. In 1892, D. Ivanowsky of Russia had first shown that the causal agent of tobacco mosaic disease was filterable, but Beijerinck went much further and provided strong evidence that although the causal agent was filterable, it had many of the properties of a living organism. He called the agent a *Contagium vivum fluidum,* a living germ that is soluble. He postulated that the agent must be incorporated into the living protoplasm of the cell in order to reproduce, and that its reproduction must be brought about with the reproduction of the cell. This postulate comes very close to our current understanding of how viruses reproduce. Beijerinck also noted that there were other plant diseases for which causal agents had not been isolated, and these might also be caused by filterable agents. Soon a number of other filterable agents were shown to be the causes of both plant and animal diseases. Such agents came to be called filterable viruses, but as further work on these agents was carried out the word "filterable" was gradually dropped.

Today, the original meaning of "virus" has been forgotten, and the word is now used to refer to the kinds of agents discussed in this chapter. Bacterial viruses were first discovered by the British scientist F. W. Twort in 1915, and independently by the French scientist F. d'Herelle in 1917, who called them *bacteriophages* (from the combining form *phago* meaning "to eat"). In 1935, the American scientist Wendell Stanley published work describing the crystallization of tobacco mosaic virus, providing the first demonstration that some of the properties associated with living organisms can be found in agents which can be crystallized like chemicals. The crystals were shown to consist of only two components, protein and RNA. Stanley's work can well be said to mark the beginning of molecular biology, as it provided the first insight into the chemical nature of life itself.

10.2

General Features of Virus Replication

The basic problem of virus replication can be put simply; the virus must somehow induce a living host cell to synthesize all of the essential components needed to make more virus particles. These components must then be assembled in the proper order and the new virus particles must escape from the cell and infect other cells. The various phases of this replication process can be summarized in six steps (Figure 10.5): (1) attachment (adsorption) of virus particle to sensitive cell, (2) penetration into the cell by the virus or its nucleic acid, (3) replication of the virus nucleic acid, (4) production of protein capsomeres and other essential virus constituents, (5) assembly of nucleic acid and protein capsomeres into new virus particles, and (6) release of mature virus particles from the cell.

[†]Brock, Milestones, p. 152.

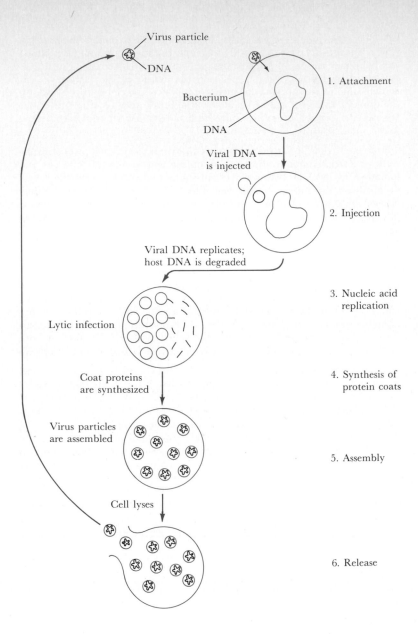

Virus particle

DNA

Bacterium

DNA

1. Attachment

Viral DNA
is injected

2. Injection

Viral DNA replicates;
host DNA is degraded

3. Nucleic acid
   replication

Lytic infection

Coat proteins
are synthesized

4. Synthesis of
   protein coats

Virus particles
are assembled

5. Assembly

Cell lyses

6. Release

FIGURE 10.5 The replication cycle
of a bacterial virus. The six general
stages of virus replication are indicated.

Attachment   There is a high specificity in the interaction between
virus and host. The most common basis for host specificity involves
the attachment process. Often a cell has specific surface components
that act as receptor sites. These are normal surface components of the
host, such as proteins, polysaccharides, or lipoprotein-polysaccharide
complexes, to which the virus particle attaches. In the absence of the
receptor site, the virus cannot adsorb, and hence cannot infect. If the
receptor site is altered, the host may become resistant to virus infection. However, mutants of the virus can also arise which are able to
adsorb to resistant hosts.

   In general, virus receptors carry out normal functions in the cell. For example, in bacteria some phage receptors are pili or flagella, others are cell-envelope components, and others are transport binding proteins. The receptor for influenza virus is a glycoprotein found on red cells and on cells of the mucous membrane of susceptible animals, whereas the receptor site of polio virus is a lipoprotein. However, many animal and plant viruses do not have specific attachment sites at all and the virus enters passively as a result of phagocytosis or some other endocytotic process.

Penetration   The means by which the virus penetrates into the cell depends on the nature of the host cell, especially on its surface structures. Cells with cell walls, such as bacteria, are infected in a different manner from animal cells, which lack a cell wall. The most complicated penetration mechanisms have been found in viruses that infect bacteria. The bacteriophage T4, which infects *E. coli,* can be used as an example.

The structure of the bacterial virus T4 is shown in Figure 10.1*b*. The particle has a **head,** within which the viral DNA is folded, and a long, fairly complex **tail,** at the end of which is a series of tail fibers. When a suspension of virus particles is added to a suspension of sensitive bacteria, the virus particles first attach to the cells by means of the tail fibers (Figure 10.6). These tail fibers then contract, and the core of the tail makes contact with the cell envelope of the bacterium. Changes then occur in the surface of the cell, probably through enzymatic action, resulting in the formation of a small hole. The tail sheath contracts, the DNA of the virus passes into the cell through

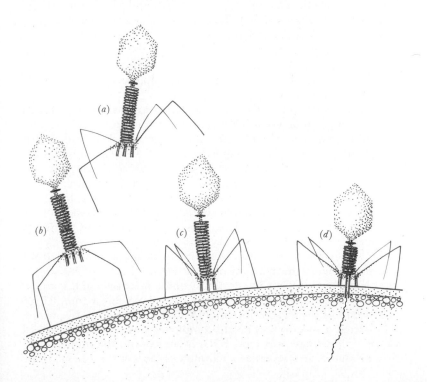

FIGURE 10.6   Attachment of T4 bacteriophage particle to the cell wall of *E. coli* and injection of DNA: (a) unattached particle; (b) attachment to the wall by the long tail fibers; (c) contact of cell wall by the tail pin; (d) contraction of the tail sheath and injection of the DNA. Compare with Figure 10.1*b*. (From Simon, L. D., and T. F. Anderson, 1967. Virology, 32:279.)

a hole at the tip of the tail, the protein coat remaining outside. The DNA of T4 has a total length of about 50 μm, whereas the dimensions of the head of the T4 particle are 0.095 μm by 0.065 μm. This means that the DNA must be highly folded and packed very tightly within the head.

With animal cells, the whole virus particle may penetrate the cell, being carried inside by endocytosis (phagocytosis or pinocytosis), an active cellular process.

10.3

Viral Nucleic Acid and Protein Synthesis

The outcome of lytic infection of the host by the virus is the synthesis of viral nucleic acid and viral protein coats. In effect, the virus takes over the biosynthetic machinery of the host and uses it for its own synthesis. The host supplies the energy-generating system, the ribosomes, amino-acid-activating enzymes, transfer RNA (with a few exceptions), and other soluble factors. However, although the host can supply most biosynthetic functions needed by the infecting virus, it does not necessarily supply all of them, and some new enzymes are synthesized using genetic information provided by the virus.

Viral mRNA synthesis   The synthesis of virus-specific enzymes and viral coat protein takes place using mRNA that has been synthesized after virus infection. Thus, the initial step in virus replication is the synthesis of viral mRNA. Exactly how the virus brings about new mRNA synthesis depends upon the type of virus, and especially whether its genetic material is RNA or DNA, and whether it is single-stranded or double-stranded. The essential features of mRNA synthesis were discussed in Chapter 9, and it was shown that mRNA represents a complementary copy of one of the two strands of the DNA double helix. In uninfected cells all mRNA is made on the DNA template, but viruses have evolved a surprising array of alternatives for directing mRNA synthesis.

These variations in mRNA synthesis are outlined in Figure 10.7. By convention, the mRNA is considered to be of plus ($+$) configuration. The complementarity of the viral nucleic acid is then indicated by a plus if it is the same as the mRNA and a minus if it is of opposite complementarity. As seen, if the virus has double-stranded DNA, then mRNA synthesis proceeds directly as in uninfected cells. The same is true if the virus has double-stranded RNA, this RNA serving as a template in a manner analogous to DNA. However, if the virus has a single-stranded DNA, then it is first converted to double-stranded DNA and the latter serves as the template for mRNA synthesis. There are three classes of viruses with single-stranded RNA, and they differ in the mechanism by which mRNA is synthesized. In one class the viral RNA has a minus complementarity and is copied directly. In the other two classes, the viral RNA has a complementarity identical to that of the mRNA so that before mRNA can be synthesized a strand of opposite complementarity must be formed. In the simplest case, the viral RNA is copied and the minus strand thus produced then serves as a template for mRNA synthesis. In the

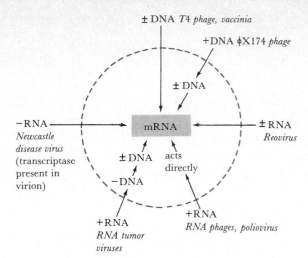

FIGURE 10.7 Formation of mRNA after infection of viruses of different types. The complementarity of the mRNA is considered as plus (+). The complementarities of the various virus nucleic acids are indicated as + if the same as mRNA, as − if opposite, or as ± if double stranded. The viruses are shown outside the circle and the events after injection are shown inside the circle. Examples are indicated next to the virus nucleic acids. Compare with Figure 10.8. (Adapted from Baltimore, D. 1971. Bacteriol. Rev. 35:235–241.)

case of reverse transcription (discussed below), a double-stranded DNA serves as the template for mRNA synthesis.

Replication schemes in viruses   Figure 10.8 outlines the variety of nucleic acid replication schemes found in viruses. The precise pattern followed depends on whether the virus contains RNA or DNA, and whether the nucleic acid is single-stranded or double-stranded. Nucleic acid replication with double-stranded DNA is similar to DNA replication in cells (Figure 10.8*a*). In some cases, a new virus-specific DNA polymerase is involved in this process. If the virion consists of single-stranded DNA, the first event is its conversion inside the cell into a double-stranded form, called the **replicative form,** which then replicates; the one strand of the DNA is discarded before the virion is assembled (Figure 10.8*b*).

In the case of RNA viruses, more variation exists. With double-stranded RNA viruses, such as reovirus, replication proceeds in a manner similar to double-stranded DNA (Figure 10.8*c*), although with ribonucleotides instead of deoxyribonucleotides. In the case of the single-stranded RNA viruses (by far the largest proportion of RNA viruses), two quite distinct mechanisms have been recognized. In the type where the RNA is of *plus* form, the single-stranded RNA enters the cell and serves as a messenger for the synthesis of a new enzyme, called **RNA replicase.** RNA replicase copies the viral RNA and makes a minus strand, which remains associated with the plus strand to make a double-stranded replicative form (Figure 10.8*d*). In the case where the RNA is of *minus* form, an enzyme (transcriptase) is packed in the virus particles and converts the minus strand to plus (that is, mRNA) upon infection

Another, rather surprising case of RNA virus replication, called **reverse transcription,** is found in certain tumor viruses. The viral RNA serves as a template for the synthesis of a *single-stranded* DNA of opposite complementarity; this DNA is then copied, resulting in the formation of double-stranded DNA (Figure 10.8*e*). The double-stranded DNA form may become integrated into the host genome,

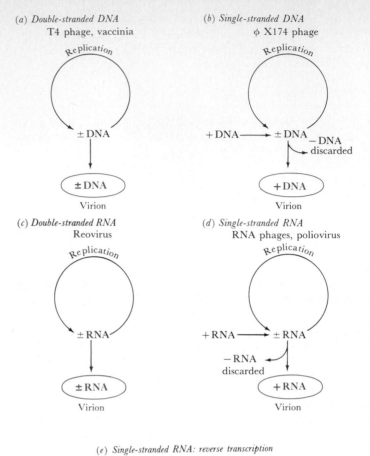

(a) *Double-stranded DNA*
T4 phage, vaccinia

Replication

±DNA

±DNA
Virion

(b) *Single-stranded DNA*
φ X174 phage

Replication

+DNA ⟶ ±DNA
−DNA discarded

+DNA
Virion

(c) *Double-stranded RNA*
Reovirus

Replication

±RNA

±RNA
Virion

(d) *Single-stranded RNA*
RNA phages, poliovirus

Replication

+RNA ⟶ ±RNA

−RNA discarded

+RNA
Virion

(e) *Single-stranded RNA: reverse transcription*

Replication

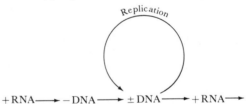

+RNA ⟶ −DNA ⟶ ±DNA ⟶ +RNA ⟶

FIGURE 10.8 Replication schemes of various viruses. Complementarity of the virion nucleic acid is given as + if single stranded or ± if double stranded.

where it is replicated indefinitely without the production of mature virions. It is after integration of virus-derived DNA into the host genome that the transformation of normal cells into tumor cells can occur. (Tumor cells are distinguished from normal cells by the fact that the former show uncontrolled growth in the animal. In tissue cultures tumor cells tend to grow in haphazard clumps, whereas normal cells grow as monolayers.) The transformed host cells may or may not produce virus, but in either case, the virus genome is present within the cell and is transmitted to both daughter cells at division. This DNA is also the source of information for the synthesis of new viral RNA, which may occur under certain conditions, thereby leading to the production of new virus particles (Figure 10.8e).

Alternative schemes in virus replication Exceptions to the schemes just discussed have been recognized. One of the most interesting exceptions concerns replication of **hepatitis B virus.** This virus consists of a partially single-stranded circular DNA (all of the minus DNA strand plus part of the plus strand) which replicates by way of an **RNA intermediate.** When the virus enters the cell, the minus-strand DNA enters the nucleus and is transcribed into plus strand RNA which is then transferred to the cytoplasm. In the cytoplasm, reverse transcriptase (see above) synthesizes viral DNA, which is subsequently packaged. This DNA virus is thus unique in its replication by way of an RNA intermediate, just the opposite of the RNA retrovirus discussed above.

Poliovirus is a well-studied RNA virus which has the interesting property of having a virus protein which is *covalently* linked to the 5′ terminus of the RNA. The protein seems to be necessary for assembly of the mature virus. The RNA with protein attached is encapsidated in a precursor capsid, and proteolytic cleavage leads to the formation of the capsid of the mature virus. The poliovirus RNA acts as an mRNA directly (Figure 10.7), and codes for a number of proteins, but these are synthesized as a large protein precursor which is cleaved (post-translational cleavage) by virus-specific protease.

Viroids An interesting group of infectious agents called **viroids** consists of small pieces of RNA uncomplexed with any protein coat. The best studied viroid is that causing the potato spindle tuber disease. The RNA of this agent has 359 nucleotides (about one-tenth the size of the smallest known virus), yet it is highly infectious. We have noted that an important role of the protein coat of a virus is to protect the nucleic acid from degradation by host nucleases. The potato spindle tuber viroid is resistant to nucleases. At least one reason for nuclease resistance is because its RNA is circular and mostly double-stranded.

Because of the small size of the viroid RNA, it can code for only one or two small proteins, but little is known about its manner of translation or replication. Also, the manner in which harm is brought about to the host is not known. Although most of the work has been done on viroids infectious for plants, a viroid has also been suggested to be the infectious agent of scrapie, a disease of sheep, and of certain "slow virus" diseases of humans.

10.4

Viral Genetics

As yet we have not approached the question of whether viruses are living organisms. This topic is discussed more fully in Section 10.8, and as we shall see, it is a question that is not easily answered. We do know, however, that viruses exhibit genetic phenomena similar to those of living organisms. Studies of viral genetics have played a significant role in understanding many aspects of genetics at the molecular level. In addition, knowledge of the basic phenomena of viral genetics has increased our understanding of processes involved in virus replication. Understanding these processes has also led to some practical developments, especially in the isolation of viruses which are of use in immunization procedures. Most of the detailed work on viral genetics has been carried out with bacteriophages, because of the convenience of working with these viruses. We mention here briefly some of the types of genetic phenomena of bacterial viruses, and we shall return to certain aspects of this discussion in Sections 10.5 and 10.6.

Mutations Much of our knowledge of viral reproduction and how it is regulated has depended on the isolation and characterization of virus mutants. Several kinds of mutants have been studied in viruses: host-range mutants, plaque-type mutants, temperature-sensitive mutants and nonsense mutants.

Host-range mutations are those that change the range of hosts that the virus can infect. Host resistance to phage infection can be due to an alteration in receptor sites on the surface of the host cell, so that the virus can no longer attach, and host-range mutations of the virus can then be recognized as virus strains able to attach to and infect these virus-resistant hosts. Other host-range mutants may involve changes in the viral and host enzymes involved in replication, or in the restriction and modification systems (Section 9.2).

Plaque-morphology mutations are recognized as changes in the characteristics of the plaques formed when a phage infects cells in the conventional agar-plate technique (Figure 10.9). Characteristics of the plaque, such as whether it is clear or turbid, and its size, are under genetic control. The underlying basis of plaque morphology lies in processes taking place during the virus replication cycle, such as the rate of replication and the rate of lysis. Under appropriate experimental conditions plaque morphology can be a highly reproducible characteristic of the virus. The advantage of plaque mutants for genetic studies is that they can be easily recognized on the agar plate, but a disadvantage is that there is no convenient way of selecting for them among the large background of normal particles.

Temperature-sensitive mutations are those which allow a virus to replicate at one temperature and not at another, due to a mutational alteration in a virus protein that renders the protein unstable at moderately high temperatures. For instance, temperature-sensitive mutants are known in which the phage will not be replicated in the host at 43°C but will at 25°C, although the host functions at both temperatures. Such mutations are called *conditionally lethal,* since the virus is unable to reproduce at the higher temperature, but replicates at the lower temperature.

Nonsense mutations change normal codons into nonsense codons (see Section 9.7). In viruses, nonsense mutations are recognized because hosts are available that contain suppressors able to read nonsense codons. The virus mutant will be able to grow in the host containing the suppressor, but not in the normal host.

Genetic recombination in viruses The availability of virus mutants makes possible the investigation of genetic recombination. If two virus particles infect the same cell, there is a possibility for genetic exchange between the two virus genomes during the replication pro-

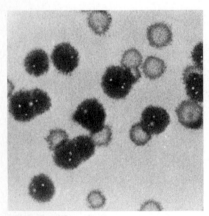

FIGURE 10.9 Plaque mutants of phage lambda in *Escherichia coli.* The turbid plaques are due to normal wild-type phage particles, and the clear plaques are due to mutant particles. A few host colonies from resistant mutants are seen within the clear plaques. The technique used here was similar to that described in Figure 10.3. The largest plaques are a few millimeters in diameter.

cess. If recombination does occur, the progeny of such a mixed infection should include not only the parental types, but recombinant types as well. With appropriate mutants, it is possible to recognize both the parental types and the recombinants and to study the events involved in the recombination process. Genetic recombination in viruses is an extremely complex process to analyze because recombination does not occur as a single discrete event during mixed infection, but may occur over and over again during the replication cycle. It has been calculated that the T-even bacteriophages (bacteriophages T2, T4, and T6) undergo, on the average, four or five rounds of recombination during a single infection cycle. By detailed and careful analysis of a wide variety of virus crosses, it has been possible to construct genetic maps of a number of bacterial viruses. Such maps have provided important information about the genetic structure of viruses.

Genetic recombination arises by exchange of homologous segments of DNA between viral genomes, most often during the replication process. The enzymes involved in recombination are DNA polymerases, endonucleases, and ligases, which also play a role in DNA repair processes (see Section 9.2).

Phenotypic mixing   During studies on genetic recombination between viruses, another phenomenon was discovered which superficially resembles recombination but has a quite different basis. Phenotypic mixing occurs when the DNA of one virus is incorporated inside the protein coat of a different virus. For phenotypic mixing to occur, the two viruses must be closely related, so that the protein coat is of proper construction of the packaging of either viral DNA. As an example of phenotypic mixing, in phage T2 of *E. coli* there is a gene called the *h* gene which controls host specificity through modification of the tail fibers of the phage. If a mixed infection is set up with two T2 phages, mutant T2*h* and wild-type T2*h*$^+$, tail fibers of *h* specificity may be incorporated into the particles containing DNA of *h*$^+$ specificity. Since it is the *h* function of the tail fibers that affects attachment, these mixed particles will show *h* specificity during the next round of infection, even though they contain *h*$^+$ DNA, but the particles resulting from this second round of infection will revert to the *h*$^+$ characteristic, because the DNA has been unchanged.

10.5

Regulation of DNA Virus Reproduction

Considerable information exists regarding how virus replication is regulated. The overall scheme of regulation involves many of the elements already discussed in Chapter 9. In the present section we discuss the regulation of reproduction of two well-studied DNA bacteriophages, T7 and T4. In the next section we shall discuss regulation in RNA viruses.

Virus infection obviously upsets the regulatory mechanisms of the host, since there is a marked overproduction of nucleic acid and protein in the infected cell. In some cases, virus infection causes a complete shutdown of host macromolecular synthesis while in other cases host synthesis proceeds concurrently with virus synthesis. In either case, the regulation of virus synthesis is under the control of the virus rather than the host. There are several elements of this control which are similar to the host regulatory mechanisms discussed in Chapter 9, but there are also some uniquely viral regulatory mechanisms.

The proteins synthesized as a result of virus infection can be grouped into two broad categories, the enzymes synthesized soon after infection, called the **early enzymes,** which are necessary for the replication of virus nucleic acid, and the proteins synthesized later, called the **late proteins,** which include the proteins of the virus coat. Both the time of appearance and the amount of these two groups of virus proteins are regulated. The early enzymes are synthesized in smaller amounts and the late proteins in much larger amounts.

Reproduction of double-stranded DNA phages Viral DNA contains genetic information for the production of a number of new enzymes. Viral enzyme formation involves the synthesis of new mRNA molecules, using the viral DNA as a template. The viral mRNA molecules associate with the ribosomes of the host, and new protein production occurs by way of the host's protein-synthesizing machinery. One function of the new enzymes synthesized after infection is to form the building blocks needed for viral DNA synthesis, some of which are virus-specific. Thus synthesis of viral DNA does not begin until these new enzymes have been formed. Once the building blocks are available, DNA synthesis proceeds and soon the cell contains a large amount of viral DNA.

During these events of virus replication, the metabolic machinery of the host continues to function, providing the energy and small molecules needed for the biosynthesis of viral enzymes and nucleic acid. In some cases, the protein-synthesizing machinery of the host is taken over completely by the virus, as a result of the inhibition of host mRNA synthesis. In phage T4, inhibition of host mRNA synthesis is accomplished by means of a phage-coded enzyme which modifies the host RNA polymerase so that it no longer recognizes host promoters, but now recognizes phage gene promoters.

Viral DNA also encodes the structural proteins of the virion. Soon after viral DNA is synthesized, head and tail proteins are formed and at this stage the cell contains large amounts of the protein subunits and DNA of the virion.

The protein coat of the virus particle forms spontaneously from the protein capsomers by a **self-assembly** process (see also Section 2.7), leading to the formation of headlike structures. Some of the phage proteins must be partially cleaved (by other phage-specific enzymes) before they can enter the self-assembly process. The DNA subsequently enters the head, but the manner in which this occurs is not known. In the case of T4, assembly of heads and tails occur on independent pathways. DNA is packaged into the assembled head and the tail and tail fibers are added subsequently (Figure 10.10). The release of the virions occurs upon breakdown of the cell wall through the action of enzymes whose synthesis is also directed by the virus DNA.

mRNA synthesis and regulation in bacteriophage T7 T7 is a relatively small double-stranded DNA phage which infects *E. coli.* The genes

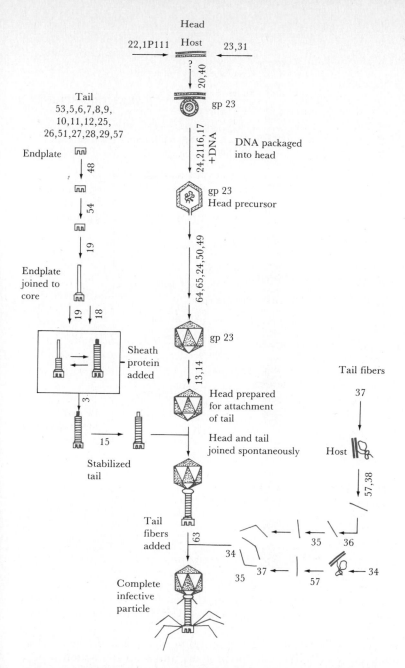

FIGURE 10.10 Steps in the assembly of a T4 bacterial virus. The numbers refer to mutants blocked at the various stages.

of the virus DNA that code for virus proteins are arranged along the DNA in a linear sequence that corresponds to the order in which the proteins appear after infection (Figure 10.11). Transcription begins at the end of the DNA that specifies the early proteins, using an RNA polymerase in the host, and continues until about 15 percent of the DNA is transcribed, at which point a stop signal in the DNA prevents further transcription.

The mRNA formed from these early genes is a single long mole-

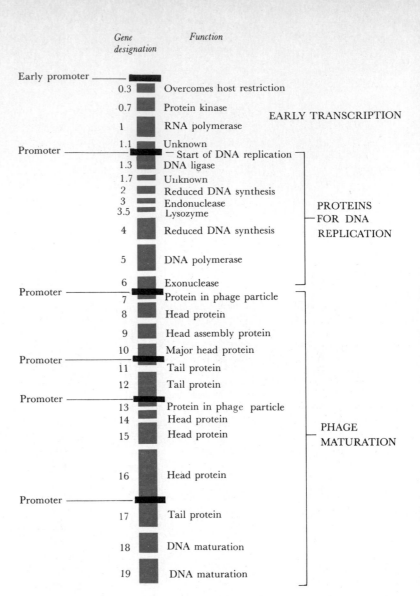

Gene designation Function

Early promoter — ▬

0.3 ▬ Overcomes host restriction

0.7 ▬ Protein kinase

1 ▬ RNA polymerase

EARLY TRANSCRIPTION

Promoter — 1.1 ▬ Unknown
— Start of DNA replication ⌐

1.3 ▬ DNA ligase

1.7 ▬ Unknown

2 ▬ Reduced DNA synthesis

3 ▬ Endonuclease

3.5 ▬ Lysozyme

4 ▬ Reduced DNA synthesis

5 ▬ DNA polymerase

PROTEINS
FOR DNA
REPLICATION

6 ▬ Exonuclease

Promoter — 7 ▬ Protein in phage particle

8 ▬ Head protein

9 ▬ Head assembly protein

10 ▬ Major head protein

Promoter — 11 ▬ Tail protein

12 ▬ Tail protein

Promoter — 13 ▬ Protein in phage particle

14 ▬ Head protein

15 ▬ Head protein

16 ▬ Head protein

Promoter — 17 ▬ Tail protein

PHAGE
MATURATION

18 ▬ DNA maturation

19 ▬ DNA maturation

FIGURE 10.11 Genetic map of phage T7, showing gene numbers, approximate sizes, and functions of the gene products. Transcription can be divided into two segments. Early transcription, which involves about 15 percent of the genome, uses host RNA polymerase. Transcription of the rest of the phage genes involves a phage RNA polymerase made from the early transcript. Several promoters have been identified. (The diagram is based on data in Studier, F. W. 1972. Science 176:367–376; and Studier, personal communication.)

cule, which is subsequently cleaved by a host enzyme into shorter chains. These latter chains then act as mRNAs in protein synthesis. One of these mRNA chains codes for the T7 RNA polymerase, a new enzyme necessary for transcription of the late T7 genes. Two other early mRNAs code for proteins that stop the action of host RNA polymerase, thus turning off transcription of the early genes as well as the transcription of host genes. The T7 RNA polymerase then begins transcription at the stop point of the early genes and proceeds to the end of the chromosome. The mRNA molecules for the last 80 percent of the viral DNA are then made, and translation of these mRNA molecules leads to synthesis of later proteins. It is thus seen that regulation of replication in T7 has both negative and positive control; negative, by means of the formation of proteins that stop host RNA polymerase and thus shut off transcription of the early T7 genes that are recognized by this enzyme, and positive, by means of the formation of a new RNA polymerase, which recognizes promoters of the late T7 genes.

mRNA synthesis and regulation in bacteriophage T4   In bacteriophage T4, the details of regulation of replication are more complex, but involve primarily positive control. T4 is a much larger phage than T7 and has many more genes and phage functions. In addition, the DNA of T4 contains an unusual base, 5-OH cytosine, which replaces the normal cytosine, and the OH (hydroxyl) groups of this base are glucosylated (glucosylation occurs after the DNA molecule is synthesized). Thus, enzymes for the synthesis of this unusual base and for its glucosylation must be formed after phage infection, as well as formation of an enzyme that breaks down the normal DNA precursor deoxycytidine triphosphate. In addition, T4 codes for a number of enzymes that have functions similar to those enzymes in DNA replication, but are formed in larger amounts, thus permitting faster synthesis of T4-specific DNA. In all, T4 codes for over 20 new proteins that are synthesized early after infection. It also codes for the synthesis of several new tRNAs, whose function is presumably to read more efficiently T4 mRNA. Overall, the T4 genetic map can be divided into three clusters, containing genes for early, middle, and late proteins. The early proteins are the enzymes involved in DNA replication. The middle proteins are also involved in DNA replication. For instance, a DNA unwinding protein (DNA gyrase) is formed which destabilizes the DNA double helix, forming short single-stranded regions at which DNA synthesis can be initiated. The late proteins are the head and tail proteins and the enzymes involved in liberating the mature phage particles from the cell. In T4, there is no evidence for a new phage-specific RNA polymerase, as in T7. The control of T4 mRNA synthesis involves the production of proteins that modify the specificity of the host RNA polymerase so that it recognizes different phage promoters. The early promoter, present at the beginning of the T4 genome, is read directly by the host RNA polymerase, and involves the function of host *sigma* factor. Host RNA polymerase moves down the chain until it reaches a stop signal. One of the early proteins blocks host sigma factor action. The early protein combines with the RNA polymerase core enzyme, and when this protein builds up, initiation of early phage genes is stopped. The RNA polymerase cores are now available to combine with new phage-specific activators, which control the transcription of the middle and late genes. The middle genes are transcribed along the same DNA strand as the early genes, but the late genes are transcribed along the opposite strand.

DNA replication   After the enzymes and precursors necessary for DNA synthesis have been formed, DNA synthesis itself can begin. Studies on the mechanism of synthesis of viral DNA have provided considerable insight into the molecular events surrounding DNA synthesis in general, and they also help us to understand the control of viral replication.

FIGURE 10.12 Role of repetitive ends in the replication of phage T7. The letters signify genes (arbitrary); the wavy lines represent newly synthesized DNA. Initiation of replication begins at an internal origin and proceeds in both directions (bidirectional), but growth toward the 5′ phosphate end is backwards since it cannot be initiated at the 5′ terminus. This leaves a single-stranded segment at the end of each double-stranded molecule. However, since the DNA is terminally repetitious, pairing of the unreplicated termini can occur, and when the ends are joined by DNA ligase, a completely replicated concatamer is formed. Continued replication can lead to very long concatamers, but action of a cutting enzyme ultimately converts these into pieces of appropriate length (compare with Figure 10.13).

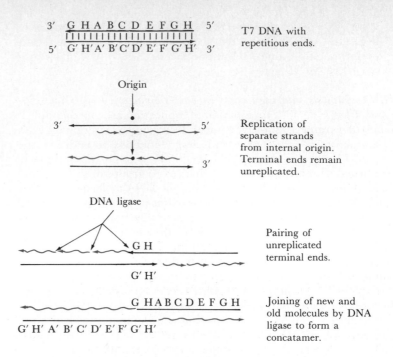

T7 DNA with repetitious ends.

Replication of separate strands from internal origin. Terminal ends remain unreplicated.

Pairing of unreplicated terminal ends.

Joining of new and old molecules by DNA ligase to form a concatamer.

The actual manner of viral DNA replication depends on the structure of the DNA of the virus. Many viral DNAs have repetitious sequences at the ends of the chains, which are necessary for replication. The DNA of T7 has identical sequences at both ends, as illustrated in Figure 10.12, and T7 is thus said to have ends with **repetitious sequences.** It should be recalled (Section 9.2) that replication of a DNA strand can proceed only from the 3′ hydroxyl end. In T7, replication begins at an internal location, and as seen in Figure 10.12, from this origin, replication from the 5′ phosphate toward the 3′ end hydroxyl of the new strand can be continuous, but toward the 5′ end it can only occur in fragments, which are ultimately joined by DNA ligase (see Figure 9.5). There is an unreplicated portion of the T7 DNA at the 5′ terminus of each strand. The two opposite single 3′ ends, being complementary, can pair, forming a DNA molecule twice as long as the original T7 DNA. The unreplicated portions of this end-to-end bimolecular structure are then completed through the action of DNA polymerase and DNA ligase, resulting in a linear bimolecule, called a **concatamer.** Continued replication can lead to concatamers of considerable length, but ultimately a cutting enzyme slices each concatamer at a specific site, resulting in the formation of virus-size linear molecules (Figure 10.13).

In T4, the cutting enzyme which forms virus-sized fragments does not recognize specific locations on the long molecule, but rather cuts off head-full packages of DNA irrespective of the sequence. Thus each virus DNA molecule not only contains repetitious ends, but the nucleotide sequences at the ends of different molecules are different, although each molecule contains the complete sequence of viral genes (Figure 10.14). Thus T4 is said to have **permuted sequences.** In

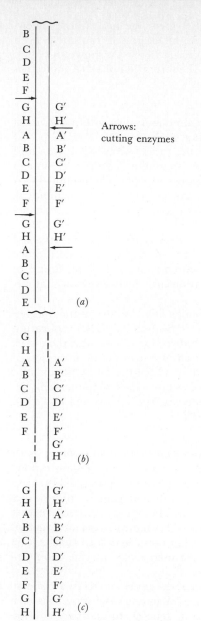

FIGURE 10.13 Production of mature viral DNA molecules from long T7 concatamers by action of cutting enzyme, an endonuclease. (a) The enzyme makes single-stranded cuts at specific sequences (arrows). (b) DNA polymerase completes the single-stranded ends. (c) The mature T7 molecule, with repetitious ends (compare with Figure 10.12).

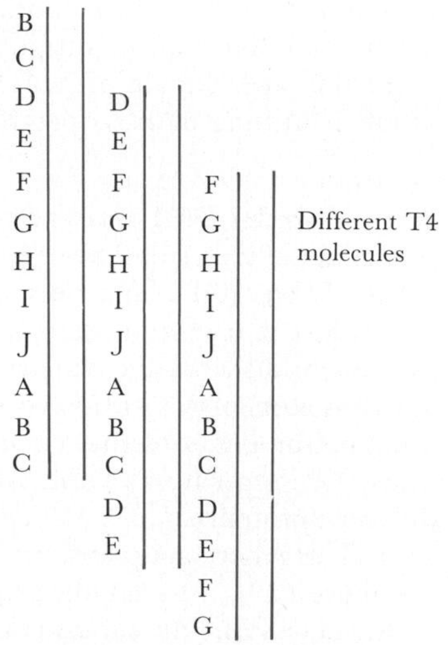

FIGURE 10.14 Role of permuted sequences in the replication of phage T4. The end of the T4 molecules are repetitious, but different molecules have ends located randomly.

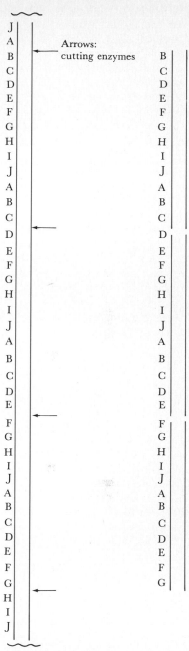

FIGURE 10.15 Generation in T4 of viral length molecules with permuted sequences by a cutting enzyme, which cuts off constant lengths of DNA irrespective of the sequence. Left, arrows, sites of enzyme attack. Right, molecules generated.

effect, such an arrangement of genes gives the appearance of circularity, and genetic maps of T4 can be expressed as a circle, even though the individual DNA molecules are linear. As shown in Figure 10.15, the cutting process results in the formation of DNA molecules with permuted sequences at the ends.

Circular DNA Viral DNA is often circular, either within the virion or during replication. Bacteriophages φX174, P1, P2, P22, and lambda have circular DNA, as do the tumor viruses polyoma and SV40. Bacterial chromosomal DNA is circular (Chapter 2), and in Chapter 11 we shall see that a number of nonchromosomal DNA elements of bacteria (plasmids) are also circular. Also, mitochondrial and chloroplast DNA is circular. Why are these diverse genetic elements circular?

One characteristic of circular DNA is that it is not attacked by exonucleases, which must begin at one end of the chain. Since DNA exonucleases are the most common sort of DNases found in cells, resistance to nuclease degradation may be one reason that circular DNA evolved.

A more general reason for circular DNA relates to the manner in which DNA replicates. As we saw in Section 9.2, DNA replication begins at one point, and proceeds bidirectionally around the circle. In a circular DNA, at the termination of one cycle the initiation of a second cycle can begin immediately. Once the DNA polymerase has become fixed to the origin, successive rounds of replication can proceed apace. Such a mechanism seems especially suited to a virus which is forming many copies of itself.

Single-stranded DNA viruses and the phenomenon of overlapping genes A number of very small bacterial viruses contain only single-stranded DNA (Table 10.1). The best studied of these viruses, φX174, is so small that it does not seem to have sufficient genetic information to code for all virus-specific proteins. Studies on the genetic structure and function of φX174 have led to the surprising conclusion that some of the genes of this virus overlap other genes, so that certain nucleotide sequences in the virus are used twice, for the coding of different proteins.

The genetic map and function of the genes of φX174 are shown in Figure 10.16. As seen, the sequences of genes *D* and *E* overlap each other, gene *E* being contained within gene *D*. In addition, the termination codon of gene *D* overlaps the initiation codon of gene *J* by one nucleotide. The reading of gene *E* must be in a different phase (starting point) from that of gene *D*. Obviously, any mutation in gene *E* will also lead to an alteration in the sequence of gene *D*, but whether a given mutation affects one or both proteins will depend on the exact nature of the alteration. There is also some evidence that genes *B* and *C* may overlap as well.

The existence of overlapping genes has been confirmed by the determination of the complete nucleotide sequence of φX174, the first virus for which the complete sequence was worked out.

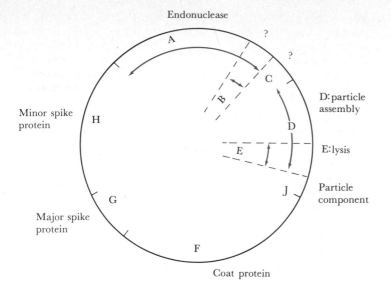

Endonuclease

Minor spike
protein

Major spike
protein

H

G

A

B

E

C

D

J

F

D: particle
assembly

E: lysis

Particle
component

Coat protein

FIGURE 10.16 Genetic map of the single-stranded DNA phage φX174, showing overlapping of genes A and B and genes D and E. In addition, the termination codon of gene D overlaps the initiation codon of gene J by one nucleotide. The functions identified for the various genes are given on the outside of the circle.

10.6

Temperate Bacterial Viruses: Lysogeny

Most of the bacterial viruses described above are called **virulent** viruses, since they usually kill the cells they infect. However, many other viruses, although also often able to kill cells, frequently have more subtle effects. Such viruses are called **temperate.** Their genetic material can become integrated into the host genome and is thus duplicated along with the host material at the time of cell division, being passed from one generation of bacteria to the next. Under certain conditions these bacteria can spontaneously produce virions of the temperate virus, which can be detected by their ability to infect a closely related strain of bacteria. Such bacteria, which appear uninfected but have the hereditary ability to produce phage, are called **lysogenic.** Temperate viruses are so far known only among the bacteria, and all are DNA viruses.

The ability of a bacterium to reproduce virus material without lysis of the cell may seem unusual since heretofore we have associated virus replication with eventual destruction of the cell. With most temperate phages, if the host simply makes a copy of the viral DNA, lysis does not occur; but if complete virion particles are produced, then the host cell lyses. In a lysogenic bacterial culture at any one time, a small fraction of the cells, 0.1 to 0.0001 percent, produce virus and lyse, while the majority of the cells do not produce virus and do not lyse. Although only rarely do cells of a lysogenic strain actually produce virus, every cell has the potentiality. Lysogeny can thus be considered a genetic trait of a bacterial strain.

The temperate virus does not exist in its mature, infectious state inside the cell, but rather in a latent form, called the **provirus** or **prophage** state, with the virus DNA inserted into the DNA of the host cell. In considering virulent viruses we learned that the DNA of the virulent virus contains information for the synthesis of a number of enzymes and other proteins essential to virus reproduction. The prophage of the temperate virus carries similar information, but in the lysogenic cell this information remains dormant because the expression of the virus genes is blocked through the action of a specific repressor coded for by the virus. When the repressor is inactivated, virus reproduction occurs, the cell lyses, and virus particles are released.

A lysogenic culture can be treated so that most or all of the cells produce virus and lyse. Such treatment, called **induction,** usually involves the use of agents such as ultraviolet radiation, nitrogen mustards, or X rays, known to damage DNA and activate the SOS system (see Section 9.10). However, not all prophages are inducible; in some temperate viruses, prophage expression occurs only by spontaneous events.

Although a lysogenic bacterium may be susceptible to infection by other viruses, it cannot be infected by virus particles of the type for which it is lysogenic. This immunity is conferred by the intracellular repression mechanism (see above) under the control of virus genes.

It is sometimes possible to eliminate the lysogenic virus (to "cure" the strain) by heavy irradiation or treatment with nitrogen mustards. Among the few survivors may be some cells that have been cured. Presumably the treatment causes the prophage to detach from the host chromosome but it does not replicate and is lost during subsequent cell growth. Such a cured strain is no longer immune to the virus and can serve as a suitable host for study of virus replication.

How is it possible to determine whether a strain is lysogenic? A sensitive host is necessary—that is, a strain closely related to the presumed lysogenic strain but not infected with the prophage. In practice, a large number of related strains are obtained, either from nature or from culture collections. The presumed lysogenic strain is cultured in a suitable medium, and irradiated in midlogarithmic phase to induce the prophage to replicate. After further incubation, the culture is filtered to remove live bacteria, and the filtrate is tested against the various test strains using the agar overlay technique described for use in assaying virus particles. If plaques are seen, it can be assumed that virus particles are present and that the strain is lysogenic. Sometimes a large number of strains must be tested to find a sensitive host. Most bacteria isolated from nature are lysogenic for one or more viruses.

Consequences of temperate virus infection   What happens when a temperate virus infects a nonlysogenic organism? The virus may inject its DNA and initiate a reproductive cycle similar to that described for virulent viruses (Figure 10.5), with the infected cell lysing and releasing more virus particles. Alternatively, when the virus injects its DNA, **lysogenization** may occur instead: the viral DNA becomes incorporated into the bacterial genetic material and the host bacterium is converted into a lysogenic bacterium (Figure 10.17). In lysogenization the infected cell thus becomes genetically changed. Sensitive cells can undergo either lysis or lysogenization; which of these occurs is often determined by the action of a complex repression

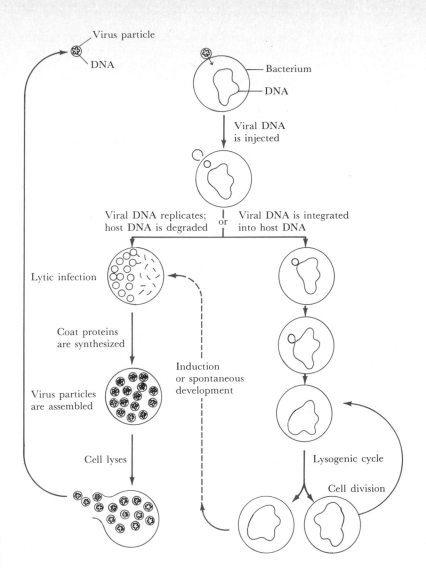

Virus particle

DNA

Bacterium

DNA

Viral DNA
is injected

Viral DNA replicates; | Viral DNA is integrated
host DNA is degraded or into host DNA

Lytic infection

Coat proteins
are synthesized

Induction
or spontaneous
development

Virus particles
are assembled

Cell lyses

Lysogenic cycle

Cell division

FIGURE 10.17 Consequences of infection by a temperate bacteriophage. The alternatives upon infection are integration of the virus DNA into the host DNA (lysogenization) or replication and release of mature virus (lysis). The lysogenic cell can also be induced to produce mature virus and lyse.

system, as will be described below. We thus see that the temperate virus can have a dual existence. Under one set of conditions, it is an independent entity able to control its own replication, but when its DNA is integrated into the host genetic material, replication is then under the control of the host.*

Regulation of lambda reproduction   One of the best studied temperate phages is lambda, which infects *E. coli,* and our knowledge of the molecular mechanisms involved in lysogenization and lytic processes in this phage is very advanced. Morphologically, lambda particles look like those of many other bacteriophages (Figure 10.18).

*There are also temperate viruses which maintain the prophage state, not by integration into the host genome, but as plasmids.

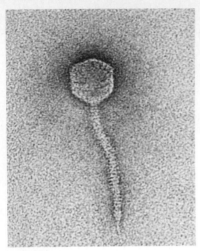

FIGURE 10.18 Electron micrograph by negative staining of lambda bacteriophage particle. Magnification, 220,000×. (Courtesy of M. Wurtz, Biozentrum, Basel, Switzerland.)

The lambda genome has two sets of genes, one controlling lytic growth, the other lysogenic integration. Upon infection, genes promoting both lytic growth and lysogenic integration are expressed. Which pathway succeeds is determined by the competing action of these early gene products and by the influence of host factors.

The genetic map of lambda is illustrated in Figure 10.19. Although lambda DNA is linear inside the virion, each molecule has cohesive ends (small single-stranded complementary segments at each end) and cyclizes upon release from the phage particle into the cell. The genetic map, although actually linear, can thus be oriented as a circle. The lambda map consists of several operons, each of which controls a set of related functions. Some of the phage genes are transcribed by RNA polymerase from one strand of the double helix, and others are transcribed from the other strand. Upon injection, transcription of the phage genes which code for the synthesis of the lambda repressor occurs, and if repressor builds up before lytic functions are expressed, lytic reproduction is blocked. The repressor protein, which has been purified and characterized, blocks the transcription of all later lambda genes, thus preventing expression of the genes involved in the lytic cycle.

Integration of the lambda DNA into the host chromosome occurs at a unique site on the *E. coli* genome. Integration occurs by

FIGURE 10.19 Genetic and molecular map of lambda. The two complementary DNA strands are indicated with 3′ and 5′ ends. The genes are designated by letters .att. attachment site for phage to host chromosome. DNA strand l is transcribed leftward (counterclockwise), with a 5′ G at its left cohesive end (m), and two operons are indicated as L1 and L2. DNA strand r is transcribed rightward (counterclockwise), with a 5′ A at its right cohesive end (m′); and two operons are indicated as R1 and R2. Genes of special interest: cI, repressor protein. O_R, operator for R1 transcription: O_L, operator for L2 transcription: cro, gene for second repressor, which depresses transcription of L1, L2, and R1: immunity region, where cI, cro, O_R, and O_L are located; N, positive regulator counteracting rho dependent termination. (Based on Szybalski, W. 1974. Bacteriophage lambda. Pages 309–322 in R. C. King, ed., Handbook of genetics, vol. 1, Plenum Press, N.Y.)

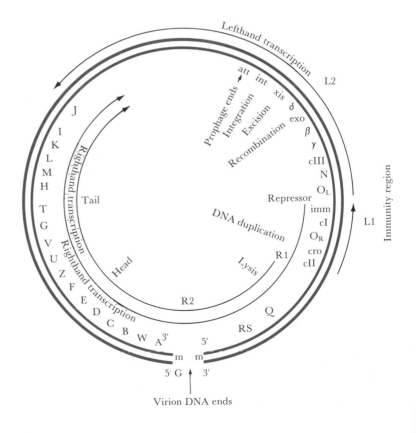

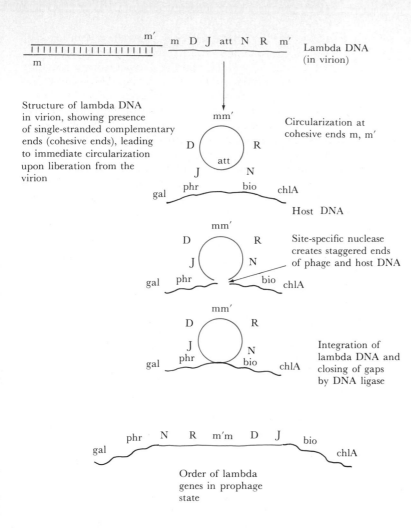

Structure of lambda DNA in virion, showing presence of single-stranded complementary ends (cohesive ends), leading to immediate circularization upon liberation from the virion

Lambda DNA (in virion)

Circularization at cohesive ends m, m′

Host DNA

Site-specific nuclease creates staggered ends of phage and host DNA

Integration of lambda DNA and closing of gaps by DNA ligase

Order of lambda genes in prophage state

FIGURE 10.20 Integration of lambda DNA into the host. Integration always occurs at a specific site on the host DNA, involving a specific attachment site (att) on the phage. Some of the host genes near the attachment site are given, and a few of the phage genes around the circular map. A site-specific enzyme (integrase) is involved, and specific pairing of the complementary ends results in integration of phage DNA. The linear arrangement of phage genes in the virion is

m-A-F-Z-J-b2-a•a′-int-N-imm-R-m′

where a•a′ are sites within the attachment site, att. After circularization at m, m′, the prophage opens between a and a′ in the att site. The order of host genes at the attachment site is

-gal-chlD-pgl-phr-b•b′-bio-uvrB-chlA-

where b•b′ are not genes but sites within the host attachment site. Upon insertion of lambda, the gene arrangement is:

-gal-chlD-pgl-phr-b•a′-int-N-imm-R-m′-m-A-F-Z-b2-a•b′-bio uvrB-chlA-

insertion of the virus DNA into the host genome (thus effectively lengthening the host genome by the length of the virus DNA). As illustrated in Figure 10.20, upon injection, the cohesive ends of the linear lambda molecule find each other and form a circle, and it is this circular DNA which becomes integrated into the host genome. To establish lysogeny, genes *cI* and *int* (Figure 10.19) must be expressed. The *cI* gene product is a protein which represses early transcription and thus shuts off transcription of all later genes. The integration process requires the product of the *int* gene, which is a site-specific topoisomerase catalyzing recombination of the phage and bacterial attachment sites (labeled *att* in Figures 10.19 and 10.20).

Once integration has occurred, the lambda repression system prevents the expression of the lambda genes except for gene *cI,* which codes for the lambda repressor. During host DNA replication, the integrated lambda DNA is replicated along with the rest of the host genome, and transmitted to progeny cells. Thus, in the integrated state, the lambda genome behaves as other host genes. The lambda repressor is present in each cell in about 10 to 20 copies, most of which are bound to two specific operators, one controlling transcription of the left genes (*L* strand, Figure 10.19) and the other controlling transcription of the right genes (*R* strand). In addition to blocking the transcription of all phage genes, the lambda repressor also has the interesting property of stimulating the expression of the *cI* gene itself. Thus both positive control (stimulation of its own synthesis) and negative control (repression of all other phage functions) keeps the lambda repression system in tight control.

Lytic growth of lambda lysogens occurs after the repressor is inactivated. Inactivation of repressor protein occurs under conditions leading to DNA damage, such as after sublethal treatment with ultraviolet radiation, or nitrogen mustards. In lytic growth, a different repressor protein called *cro* is made, which turns off the synthesis of lambda repressor. Both repressor and *cro* bind to the same region of the lambda DNA, the O_R region, but work in different ways. When *cro* is active, its own synthesis is promoted and transcription occurs in a right-handed direction from the promoter associated with O_R, leading to lytic growth. When *cI* is active, its own synthesis is promoted and transcription occurs leftward from the promoter associated with O_R, transcribing only the *cI* gene. Thus a "molecular switch" determines which of two sets of genes is turned on and therefore whether lytic growth or lysogeny occurs. Destruction of lambda repressor and activation of the *cro* gene occur as a result of the SOS response brought about by DNA damage (see Section 9.10).

Replication of lambda DNA occurs in two distinct fashions during different parts of the phage production cycle. Initially, liberation of the lambda DNA from the host results in replication of a circular DNA, but subsequently linear concatamers are formed, which replicate in a different way. Replication is initiated at a site close to gene *O* (Figure 10.19) and from there proceeds in opposite directions (bidirectional symmetrical replication), terminating when the two replication forks meet. In the second stage, generation of long linear concatamers occurs, and replication occurs in an asymmetric way by a mechanism called **rolling circle replication** (Figure 10.21). In this mechanism, replication proceeds in one direction only, and can result in very long chains of replicated DNA. This mechanism is efficient in permitting extensive, rapid, relatively uncontrolled DNA replication; thus it is of value in the later stages of the phage replication cycle, when large amounts of DNA are needed to form mature virions. The long concatamers formed are then cut into virus-sized lengths by a DNA cutting enzyme. In the case of lambda, the cutting enzyme makes staggered breaks at specific sites on the two strands, twelve nucleotides apart, which provide the cohesive ends involved in the cyclization process. Rolling circle replication occurs not only during replication of some phage DNAs, but also in cellular DNA replication when the number of copies of a gene must be increased. The involvement of rolling circle replication in bacterial mating is described in Figure 11.8.

Lambda is one of the agents of choice for use as a cloning vector for artificial construction of DNA hybrids with restriction enzymes. It has several features that make it an excellent system for genetic engineering. One feature of lambda that makes it of special use for

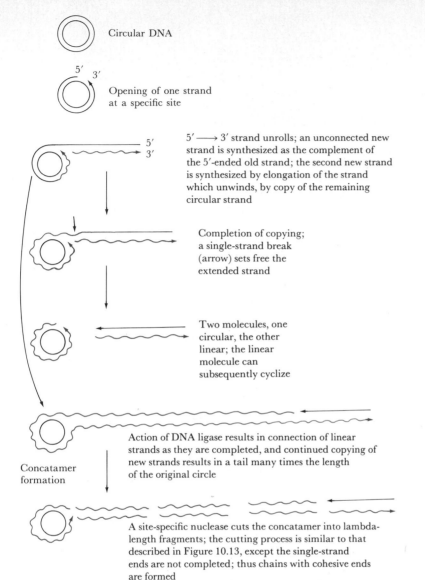

Circular DNA

Opening of one strand
at a specific site

$5' \longrightarrow 3'$ strand unrolls; an unconnected new
strand is synthesized as the complement of
the 5'-ended old strand; the second new strand
is synthesized by elongation of the strand
which unwinds, by copy of the remaining
circular strand

Completion of copying;
a single-strand break
(arrow) sets free the
extended strand

Two molecules, one
circular, the other
linear; the linear
molecule can
subsequently cyclize

Concatamer
formation

Action of DNA ligase results in connection of linear
strands as they are completed, and continued copying of
new strands results in a tail many times the length
of the original circle

A site-specific nuclease cuts the concatamer into lambda-
length fragments; the cutting process is similar to that
described in Figure 10.13, except the single-strand
ends are not completed; thus chains with cohesive ends
are formed

FIGURE 10.21 The rolling circle
pattern of DNA replication, seen during
the major replication phase in lambda.
Long concatamers formed are cut to
virus-sized lengths. Rolling circle
replication is also thought to occur
during bacterial mating.

cloning is that there is a long region of DNA, between genes *J* and
att (Figure 10.19), which does not seem to have any essential functions
for replication, and can be replaced with foreign DNA. We describe
the use of lambda as a cloning vector in Section 12.1.

10.7

Viruses of Vertebrates

Among vertebrates, viruses infecting warm-blooded animals are
probably best known, as they have been studied intensively in relation to infectious disease. Current knowledge permits classification
of these viruses in a number of separate groups. Such groups are
called *families,* in line with the similar usage in classifying organisms.

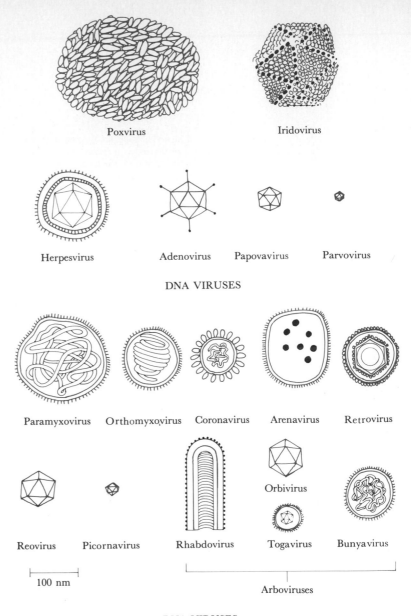

Poxvirus

Iridovirus

Herpesvirus

Adenovirus Papovavirus Parvovirus

DNA VIRUSES

Paramyxovirus Orthomyxovirus Coronavirus Arenavirus Retrovirus

Reovirus Picornavirus Rhabdovirus Togavirus Bunyavirus

Orbivirus

100 nm

Arboviruses

RNA VIRUSES

FIGURE 10.22 The shapes and relative sizes of vertebrate viruses of the major taxonomic groups. Bar = 100 nm. (From Fenner, F., et al. 1973. The biology of animal viruses, 2nd ed. Academic Press, N.Y.)

The diversity of size and morphology of viruses infecting vertebrates is illustrated in Figure 10.22, which presents drawings (to scale) of viruses of the various groups.

The vertebrate viruses having RNA as their genetic material are summarized in Table 10.2, and those having DNA as their genetic material are listed in Table 10.3. Aspects of infectious disease involving vertebrate viruses are discussed in Chapter 17.

TABLE 10.2   RNA Viruses Infecting Vertebrates*

Family	Nucleic Acid				Capsid				Examples
	Strands Single (S) or Double (D)	Number of Molecules	Mol Wt × 10⁶	mRNA†	Symmetry	Naked (N) or Enveloped (E)	Size (nm)	Site of Assembly	
Picornaviruses	S	1	2.6	Virus	Icosahedral	N	21–30	Cytoplasm	Polioviruses, Mengo
Togaviruses	S	1	4.0	Virus	Icosahedral	E	50	Cytoplasm	Rubella, hog cholera, yellow fever, some arboviruses
Rhabdoviruses	S	1	4.0	Transcriptase	Helical	E	70–80	Cytoplasm	Rabies
Paramyxoviruses	S	1	7.0	Transcriptase	Helical	E	125–300	Cytoplasm	Measles, mumps, Newcastle disease
Orthomyxoviruses	S	8	4.0	Transcriptase	Helical	E	80–120	Cytoplasm	Influenza virus
Retroviruses	S	2	10–12	Virus		E	110	Cytoplasm, reverse transcription	RNA tumor viruses
Reoviruses	D	10	15	Transcriptase	Icosahedral	N	75–80	Cytoplasm	Reovirus, orbiviruses

*Other RNA virus groups not yet fitted into this scheme: arenaviruses; coronaviruses, bunyaviruses.

†mRNA: virus serves mRNA except where presence of transcriptase is mentioned, in which case, the transcriptase (RNA polymerase) is a virus-specific enzyme present in the virus particle. See Figure 10.7.

TABLE 10.3   DNA Viruses Infecting Vertebrates

Family	Nucleic Acid			Capsid		Size (nm)	Site of Assembly	Examples
	Strands Single (S) or Double (D)	Linear (L) or Circular (C)	Mol Wt × 10⁶	Symmetry	Naked (N) or Enveloped (E)			
Papovaviruses	D	C	3–5	Icosahedral	N	45–55	Nucleus	Polyoma, papilloma
Adenoviruses	D	L	23	Icosahedral	N	70–80	Nucleus	Adenoviruses of mammals, birds
Herpesviruses	D	L	100	Icosahedral	E	180–200	Nucleus	Herpes simplex
Poxviruses	D	L	160	Complex	E	200–350	Cytoplasm, transcriptase in virus	Smallpox, vaccinia
Parvoviruses	S	L	1.8	Icosahedral	N	18–22	Nucleus	Canine parvovirus
Hepatitis B	S	C	1.8	—	E	42	Nucleus and cytoplasm	Hepatitis B

Chemical inhibition of animal viruses   Since viruses depend on their host cells for many functions of virus replication, it is difficult to inhibit virus multiplication without at the same time affecting the host cell itself. Because of this, the spectacular successes in the discovery of antibacterial and antifungal agents have not been followed by similar success in the search for specific antiviral agents. A few antiviral compounds are successful in controlling virus infections in laboratory situations (Table 10.4) and certain of these have been used in restricted clinical cases; but no substance has yet been found with more than limited practical use.

One interesting inhibitor listed in Table 10.4 is rifamycin, which is an inhibitor of RNA polymerase in procaryotes but not in

TABLE 10.4   Stages of Virus Replication at Which Chemical Inhibition of Virus Action is Known to Occur

Stage of Replication	Chemical	Virus
Free virus	Kethoxal	Influenza virus
Adsorption	None known	—
Entry of nucleic acid	Streptomycin	Certain bacteriophages
	Adamantanamine	Influenza virus
	Carbobenzoxypeptides	Measles
Nucleic acid replication	Benzimidazole, guanidine	Poliovirus
	5-Fluorodeoxyuridine (FUDR)	Herpesvirus
	5-Iododeoxyuridine (IUDR)	Herpesvirus
	Acyclovir	Herpesvirus
	Rifamycin	Vaccinia viris
Maturation (or late protein synthesis)	Isatin-thiosemicarbazone	Smallpox virus
Release	None known	—

eucaryotes (see Section 9.4). The RNA polymerase of vaccinia and other poxviruses is apparently inhibited by rifamycin, since this antibiotic specifically inhibits the replication of these viruses, although it has no effect on a wide range of other viruses affecting animal cells.

Interferon   Interferons are antiviral substances produced by many animal cells in response to infection by certain viruses. They are low-molecular-weight proteins that prevent viral multiplication. They were first discovered in the course of studies on virus interference, a phenomenon whereby infection with one virus interferes with subsequent infection with another virus, hence the name *interferon*. It has been found that interferons are formed in response not only to live virus but also to virus inactivated by radiation or to viral nucleic acid. Interferon is produced in larger amounts by cells infected with viruses of low virulence, whereas little is produced against highly virulent viruses. Apparently the virulent viruses inhibit cell protein synthesis before any interferon can be produced. Interferon is also induced by a variety of double-stranded RNA molecules, either natural or synthetic, and since double-stranded RNA does not exist in uninfected cells but exists as the replicative form in RNA-virus-infected cells, it has been suggested that double-stranded RNA serves as a signal of virus infection in the animal cell and brings into action the interferon-producing system.

Interferons are not virus specific but host specific; that is, an interferon produced by one type of animal (for example, chicken) in response to influenza virus will also inhibit multiplication of other viruses in the same species but will have little or no effect on the multiplication of influenza virus in other animal species. Interferon has little or no effect on uninfected cells; thus it seems to inhibit viral synthesis specifically. It acts by preventing RNA synthesis directed by virus, thus inhibiting synthesis of virus-specific proteins.

Interferons have been of interest as possible antiviral agents, and possibly also as anticancer agents. Their use as therapeutic agents was long hindered by the difficulty and expense of producing large quantities, but genetic engineering techniques (see Chapter 12) have now made possible the production of interferon on a commercial scale, and many clinical trials are under way.

10.8

Origin and Evolution of Viruses

Up to this point we have been content to discuss the phenomenology of viruses without touching on the question of whether they are living or nonliving. Let us say immediately that any answer to this question depends on how we define life itself. If we approach the question as a geneticist would, we might emphasize that viruses are undoubtedly genetic elements, capable of reproducing themselves and of controlling the synthesis of virus-specific proteins. Since self-replication is a property of living cells, the geneticist would probably consider the virus a living organism. The physiologist, concerned with the flow of materials and energy, might place most emphasis on the fact that virus replication does not occur without the agency of a cell, which provides the energy metabolism, the protein-synthesizing machinery, and the building blocks for the virus polymers; hence the physiologist would probably conclude that a virus is not living. Although both points of view are valid, neither alone is completely satisfactory. Viruses alternate between two states, the infectious, free-existing virion, and the noninfectious, replicating state of the intracellular genetic element. Neither state alone expresses the totality of the concept of a virus; only when we consider the two can we construct an acceptable definition of a virus.

Two theories for the origin of viruses have been advanced. Viruses may have arisen independently or they may have been derived from cells. The second mechanism is relatively easy to imagine when we consider the temperate DNA viruses. Perhaps a series of genetic changes took place in a region of host DNA and converted this region into virus DNA by making possible the synthesis of virus proteins, the replication of virus nucleic acid, and the assembly and release of mature virus. It is much more difficult to imagine how viruses might have arisen independently of cells; it seems more likely that the viruses that exist today, which have such intimate relations to cells, arose from cells by genetic modification.

Both mutation and genetic recombination are known in viruses, and these processes undoubtedly play a role in viral evolution since they allow the development of new virus types. We have already seen how a host may become resistant to virus infection through modification of its virus-receptor site. The virus in turn can then mutate to a form capable of attaching to the modified receptor site. Through interactions of this type, both virus and host become progressively transformed. Furthermore, viruses can undergo genetic recombination, so that genes from two different viruses can be brought together in the same virus. Viral genetics has played an important role in developing current concepts of the molecular basis of both mutation and genetic recombination.

Supplementary Readings

Birge, E. A. 1981. Bacterial and bacteriophage genetics. Springer-Verlag, New York. 359 pp. An elementary textbook which describes the principles of the genetics of bacterial viruses.

Bitton, G. 1980. Introduction to environmental virology. John Wiley, New York. 326 pp. Emphasis is on methods for assessing the virus content of environmental samples.

Davis, R. W., D. Botstein, and J. R. Roth. 1980. Advanced bacterial genetics: a manual for genetic engineering. Cold Spring Harbor Laboratory, Cold Spring Harbor, N. Y. 254 pp. Describes techniques for the use of bacterial viruses in genetic experiments.

Dulbecco, R. 1980. Virology. Harper & Row, Hagerstown, Md. pp. 854–1261. (Originally published as a section of Microbiology, 3rd ed. B. D. Davis, et al., eds.) An advanced textbook treatment of the basic concepts of virology. Good treatment of viruses of eucaryotic cells.

Fraenkel-Conrat, H., and P. C. Kimball. 1982. Virology. Prentice-Hall, Englewood Cliffs, N. J. 406 pp. A textbook for advanced college students on general virology.

Gerba, C. P., and S. M. Goyal, eds. 1982. Methods in environmental virology. Marcel Dekker, New York. Emphasis on methods for the recovery of viruses from water and other environmental samples.

Hahn, F. E., (ed.) 1980. Virus chemotherapy. S. Karger, Basel. 306 pp. Research-level book dealing with approaches for the development of chemical methods for the treatment of virus diseases.

Hughes, S. S. 1977. The virus: a history of the concept. Heinemann, London, 140 pp. A brief treatment of the history of virology, describing the research studies that led to the development of the concept of a virus.

Joklik, W. K. 1980. Principles of animal virology. Appleton-Century-Crofts, New York. 373 pp. A textbook treatment, primarily for medical students.

Luria, S. E., J. E., Darnell, Jr., D. Baltimore, and **A. Campbell.** 1978. General virology, 3rd ed. John Wiley, New York. 578 pp. The standard textbook on the principles of virology, readable, albeit a little dated on modern developments.

Matthews, R. E. F. 1981. Plant virology, 2nd ed. Academic Press, New York. 897 pp. An extensive textbook treatment of the viruses that infect plants.

McLean, D. M. 1980. Virology in health care. Williams & Wilkins, Baltimore. 286 pp. Brief textbook on the aspects of virology relevant to medical studies.

Ptashne, M., A. D. Johnson, and **C. O. Pabo.** 1982. A genetic switch in a bacterial virus. Sci. Am. 247:128–140. A fascinating account of the evidence for the role of two proteins in the regulation of lysis and lysogenization in lambda.

11

Bacterial Genetics

Now that we have introduced the main features of molecular genetics of cells and viruses, we can turn to a discussion of how genetic material is transferred from one organism to another. Gene transfer can occur in a number of different ways, and if it is accompanied by genetic recombination, it can lead to the formation of new organisms.

Genetic recombination is the process by which genetic elements contained in two separate genomes are brought together in one unit. Through this mechanism, new genotypes can arise even in the absence of mutation. Since the genetic elements brought together may enable the organism to carry out some new function, genetic recombination is a mechanism for adaptation to changing environments. Genetic recombination in eucaryotes is an ordered, regular process, which is generally part of the sexual cycle of the organism (see Chapter 3), but in procaryotes it is largely fragmentary.

Genetic recombination is an important tool in dissecting the genetic structure of an organism. It is also of major importance in the construction of new organisms for practical applications, a major activity in the field of genetic engineering. We present the basic principles of bacterial genetics in this chapter and then show in the next chapter how these principles apply to research in genetic engineering.

11.1

Kinds of Recombination

In procaryotes genetic recombination first involves the insertion into a recipient cell of a fragment of genetically different DNA derived from a donor cell; next the integration of this DNA fragment or its copy into the genome of the recipient cell must occur. We shall define briefly here the three means by which the DNA fragment is introduced into the recipient: (1) **transformation** is a process by which free DNA is inserted directly into a competent recipient cell;

(2) **transduction** involves the transfer of bacterial DNA from one bacterium to another within a temperate or defective virus particle; and (3) **conjugation** (mating) involves DNA transfer via actual cell-to-cell contact between the recipient and the donor cell. The last process most closely resembles sexual recombination in eucaryotes, but differs from it in several fundamental respects. These processes are contrasted in Figure 11.1.

Detection of recombination   Genetic recombination in procaryotes usually is a rare event, occurring in only a small percentage of the population. Because of its rarity, special techniques are usually necessary to detect its occurrence. One must usually use as recipients strains that possess selectable characterisics such as the inability to grow on a medium on which the recombinants can grow. Various kinds of selectable and nonselectable markers (such as drug resistance, nutritional requirements, and so on) were discussed in Section 9.8. The exceedingly great sensitivity of the selection process is shown by the fact that 10^8 or more bacterial cells can be spread on a single plate and, if proper selective conditions are used, no parental colonies will appear, whereas even a few recombinants can form colonies. The only requirement is that the reverse mutation rate for the characteristic selected must be low, since revertants will also form colonies. This problem can often be overcome by using double mutants, since it will be very unlikely that two back mutations will occur in the same cell. Much of the skill of the bacterial geneticist is exhibited in the choice of proper mutants and selective media for efficient detection of genetic recombination.

Transformation: incorporation of free DNA

Free DNA

Transduction: transfer of DNA via a phage particle

Phage

Conjugation: transfer of DNA via cell-to-cell contact

FIGURE 11.1   The three types of genetic recombination in bacteria.

11.2
Genetic Transformation

The discovery of genetic transformation in bacteria was one of the outstanding events in biology, as it led to experiments proving without a doubt that DNA is the genetic material (see "A Bit of History," page 376). This discovery became the keystone of molecular biology and modern genetics. Genetic transformation, as we have stated, involves uptake into the recipient bacterial cell of a free, fairly large fragment of bacterial DNA and its integration into the genome of the recipient.

A number of bacteria have been found to be transformable, including both Gram-negative and Gram-positive species. However, even within transformable genera, only certain strains, called *competent*, are transformable.

Since the DNA of procaryotes is present in the cell as a long single molecule, when the cell is gently lysed, the DNA pours out. Because of its extreme length (1100 to 1400 μm in *Escherichia coli*), the DNA molecule breaks easily; even after gentle extraction it fragments into 100 or more pieces (*E. coli* DNA of molecular weight 2.8×10^9 is converted into fragments of about 10^7 molecular weight). Since the DNA which corresponds to a single gene has a molecular weight of the order of 1×10^6 (corresponding to about 1000 nucleotides), each of the fragments of purified DNA will have about 50 genes. Any cell will usually incorporate only a few DNA fragments so that only a small proportion of the genes of one cell can be transferred to another by a single transformation event.

Competence   A cell that is able to take up a molecule of DNA and be transformed is said to be **competent**. Only certain strains are competent; the ability seems to be an inherited property of the organism. Further, competence is affected by the physiological state of the cells and the media in which they are grown, and it varies with the stages of the growth cycle. For instance, Figure 11.2 shows the proportion of competent cells present in a culture of *Streptococcus pneumoniae* at different stages of the growth cycle. There is a brief period, during middle exponential phase, when the competence of the population rises dramatically and then just as rapidly falls. Good evidence exists that during this brief period of competence the surfaces of the cells change so that DNA is now able to be bound to them. This surface change can also be brought about by an enzyme-like factor produced by the cells themselves, which appears in the culture medium at about the time competence appears; moreover, adding this factor to noncompetent cells of the same strain can induce them to convert into competent cells. At the same time that the peak of competence is reached, virtually all of the cells are able to take up DNA.

Uptake of DNA   At first, competent bacteria bind DNA reversibly; soon, however, the binding becomes irreversible. Competent cells bind much more DNA than do noncompetent cells—as much as 1000 times more. As we noted earlier, the sizes of the transforming fragments are much smaller than that of the whole genome; in *Streptococcus pneumoniae* each cell can incorporate only about 10 molecules of molecular weight 1×10^7. The DNA fragments in the mixture compete with each other for uptake, and if excess DNA that does not contain the genetic marker is added, a decrease in the number of transformants occurs. In preparations of transforming DNA, only about 1 out of 100 to 200 DNA fragments contains the marker being studied. Thus at high concentrations of DNA, the competition between DNA molecules results in saturation of the system so that even under the best conditions it is impossible to transform all of the cells in a population for a given genetic marker. The maximum frequency of transformation that has so far been obtained is about 10 percent of the population; actually the values usually obtained are between 0.1 and 1.0 percent. The minimum concentration of DNA yielding detectable transformants is about 0.00001 μg/ml (1×10^{-5} μg/ml), which is so low that it is undetectable chemically.

Integration of incorporated DNA   Soon after incorporation, one of the strands is broken down, while the other is integrated into the genome of the recipient (Figure 11.3). During replication of this hybrid DNA, one parental and one recombinant DNA molecule are formed. Upon segregation at cell division, the latter will be present in the transformed cell. The steps in the transformation process can be listed briefly: (1) pairing of incoming DNA with the homol-

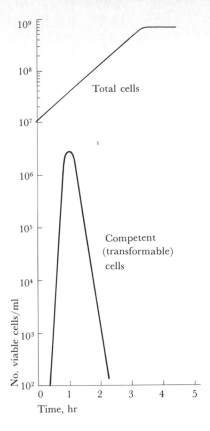

FIGURE 11.2 Appearance and disappearance of competence in *Streptococcus pneumoniae* (pneumococcus) during the exponential growth phase. Competence is plotted as the proportion of cells transformed for a single genetic character. In actuality, at the peak all of the cells are competent. (From Tomasz, A., and R. D. Hotchkiss. 1964. Proc. Nat. Acad. Sci. 51:480.)

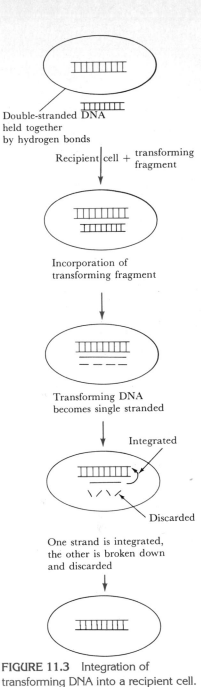

FIGURE 11.3 Integration of transforming DNA into a recipient cell.

ogous sequence of the recipient chromosome; (2) single-stranded nicking by a specific DNA cutting enzyme which opens up the DNA double helix; (3) insertion of the incoming strand; and (4) connection of the incoming strand by action of DNA ligase.

Increasing transformation efficiency Many organisms are transformed only poorly or not at all. Transformation has been found to be much more efficient in strains that are deficient in certain DNases, which would normally destroy incoming DNA. Determination of how to induce competence may involve considerable empirical study, with variation in culture medium, temperature, and other factors. The nature of the cell surface must be of importance in determining whether a cell can take up DNA. For genetic engineering techniques (see Chapter 12), it is essential that *Escherichia coli*, a Gram-negative bacterium, be transformed. It has been found that if *E. coli* is treated with high concentrations of calcium ions and then stored overnight in the cold, it becomes transformable at low efficiency. With proper procedures it is possible to select *E. coli* transformants. Why the calcium treatment works is not known, but this procedure also works with some other Gram-negative bacteria, and presumably with sufficient research most bacteria could be made transformable, at least at low efficiency.

Transduction

In transduction DNA is transferred from cell to cell through the agency of viruses. Genetic transfer of host genes by viruses can occur in two ways. The first, called **specialized transduction,** occurs only in some temperate viruses; a specific group of host genes is integrated directly into the virus genome—usually replacing some of the virus genes—and is transferred to the recipient during lysogenization. In the second, called **generalized transduction,** host genes deriving from virtually any portion of the host genome become a part of the mature virus particle in place of, or in addition to, the virus genome. The transducing virus particle in both cases is **defective** and not able to cause lysis of the host, probably because of a lack of some virus genes.

Transduction has been found to occur in a variety of bacteria. Not all temperate phages will transduce, and not all bacteria are transducible; but the phenomenon is sufficiently widespread for us to assume that it plays an important role in genetic transfer in nature.

Generalized transduction   In generalized transduction, virtually any genetic marker can be transferred from donor to recipient. An example of how transducing particles may be formed is given in Figure 11.4. When the population of sensitive bacteria is infected with a phage, the events of the phage lytic cycle may be initiated. In a lytic infection, the host DNA often breaks down into virus-sized pieces, and some of these pieces become incorporated inside virus particles. Upon lysis of the cell, these particles are released with the normal virus particles, so that the lysate contains a mixture of normal and transducing virus particles. When this lysate is used to infect a population of recipient cells, most of the cells become infected with normal virus particles. However, a small proportion of the population receives transducing particles, whose DNA can now undergo genetic recombination with the host DNA. Since only a small proportion of the particles in the lysate are of the defective transducing type and each particle contains only a small fragment of donor DNA, the probability of a defective phage particle containing a particular gene is quite low, and usually only about 1 cell in 10^6 to 10^8 is transduced for a given marker.

Phages that form transducing particles can be either temperate or virulent, the main requirement being that they do not cause complete degradation of the host DNA. The detection of transduction is most certain when the multiplicity of phage to host is low, so that a host cell is infected with only a single phage particle, since with multiple infection, the cell may be killed by the normal particles.

Specialized transduction   Generalized transduction is a rare genetic event, but in specialized transduction a very efficient transfer by phage of a specific set of host genes can be arranged. The example we shall use, which was the first to be discovered and is the best understood today, involves transduction of the galactose genes by the temperate phage lambda of *E. coli.*

As we discussed in Section 10.6, when a cell is lysogenized by lambda, the phage genome becomes integrated into the host DNA at

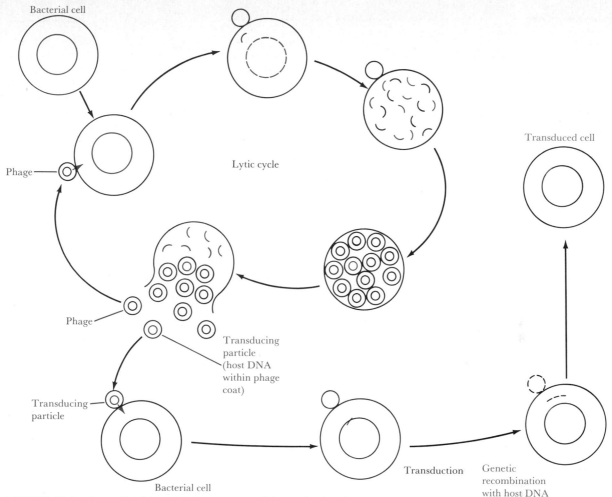

FIGURE 11.4 Generalized transduction: one possible mechanism by which virus (phage) particles containing host DNA can be formed.

a specific site. With lambda, this region is immediately adjacent to the cluster of host genes that control the enzymes involved in galactose utilization (Figure 10.20), and the DNA of lambda is inserted into the host DNA at that site. From then on, viral DNA replication is under host control. Upon induction, the viral DNA separates from the host DNA by a process that is the reverse of integration (Figure 11.5). Ordinarily when the lysogenic cell is induced, the phage DNA is excised as a unit. Under rare conditions, however, the phage genome is excised incorrectly. Some of the adjacent bacterial genes (the galactose cluster) are excised along with phage DNA. At the same time, some phage genes are left behind. One type of altered phage particle, called **lambda dg** (dg means "defective, galactose") is defective and does not make mature phage. However, if another phage called a **helper** is used together with lambda dg in a mixed

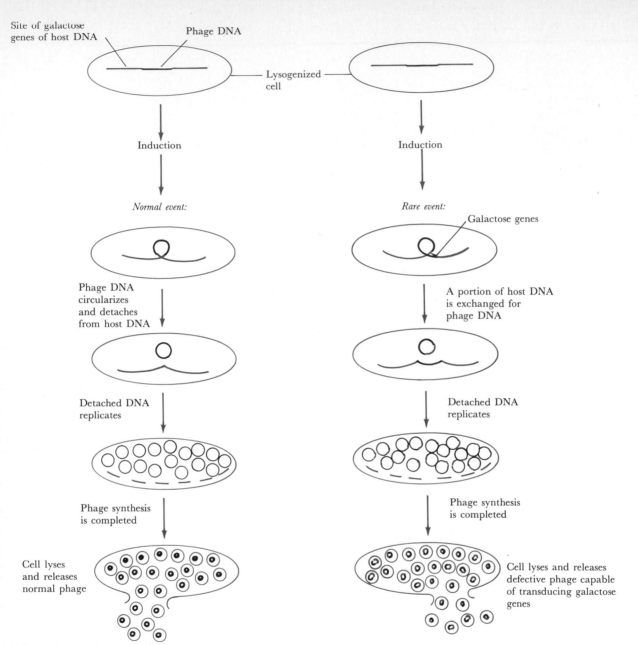

Site of galactose genes of host DNA **Phage DNA**

— Lysogenized cell —

Induction | *Induction*

Normal event: | *Rare event:*

Galactose genes

Phage DNA circularizes and detaches from host DNA | A portion of host DNA is exchanged for phage DNA

Detached DNA replicates | Detached DNA replicates

Phage synthesis is completed | Phage synthesis is completed

Cell lyses and releases normal phage | Cell lyses and releases defective phage capable of transducing galactose genes

FIGURE 11.5 The production of particles transducing the galactose genes in an *E. coli* cell lysogenic for lambda virus.

infection, then the defective phage can be replicated and can transduce the galactose genes. Thus the culture lysate obtained contains a few lambda dg particles mixed in with a large number of normal lambda particles.

If a galactose-negative culture is infected at high multiplicity with such a lysate, and *gal*+ transductants are selected, many are double lysogens, carrying both lambda and lambda dg. When such a double lysogen is induced, a lysate is produced containing about equal numbers of lambda and lambda dg. Such a lysate can transduce at high efficiency, although only for the restricted group of *gal* genes.

If the phage is to be viable, there is a maximum limit to the amount of phage DNA that can be replaced with host DNA, since the attachment region must be present and sufficient phage DNA must be retained in order to provide the information for production of the phage protein coat and for other phage proteins needed for lysis and lysogenization. However, if a helper phage is used together with the defective phage in a mixed infection, then even less information is needed in the defective phage for transduction. Only the attachment region and the replication origin are needed for production of a transducing particle, provided a helper is used.

One important distinction between specialized and generalized transduction is in how the transducing lysate can be formed. In specialized transduction this *must* occur by induction of a lysogenic cell, whereas in generalized transduction it can occur either in this way or by infection of a nonlysogenic cell by the temperate phage, with subsequent phage replication and cell lysis.

Although we have discussed specialized transduction only in the lambda-*gal* system, specialized transduction is also known for lambda *bio* and for phage ϕ80 of *E. coli*, as well as several others.

Phage conversion This is a phenomenon analogous in some ways to specialized transduction. When a normal temperate phage (that is, a nondefective one) lysogenizes a cell and its DNA is converted into the prophage state, the lysogenic cell is immune to further infection by the same type of phage. In certain cases other phenotypic alterations can be detected in the lysogenized cell, which seem to be unrelated to the phage immunity system. Such a genetic change, which is brought about through lysogenization by a normal temperate phage, is called **phage conversion.** Two cases have been especially well studied. One involves a change in structure of a polysaccharide on the cell surface of *Salmonella anatum* upon lysogenization with phage ε^{15}. The second involves the conversion of nontoxin-producing strains of *Corynebacterium diphtheriae* to toxin-producing, pathogenic strains, upon lysogenization with phage β. In these situations the information for production of these new materials is apparently an integral part of the phage genome and hence is automatically and exclusively transferred upon infection by the phage and lysogenization. Note that phage conversion is an exception to the general principle that phage genetic information is not expressed in the lysogenic state.

Lysogeny probably carries a strong selective value for the host cell, since it confers resistance to infection by viruses of the same type. Phage conversion seems also to be of considerable evolutionary significance, since it results in efficient genetic alteration of host cells. Many bacteria isolated from nature are lysogenic. It seems reasonable to conclude, therefore, that lysogeny is the normal state of affairs and may often be essential for survival of the host in nature.

11.4

Plasmids

We now introduce a third method of genetic transfer, **conjugation,** but before we do so we must introduce a new kind of genetic element called the **plasmid.** Plasmids are circular genetic elements that reproduce autonomously and have an extrachromosomal existence (Figure 11.6). Most plasmids can be eliminated from the cell by various treatments (called **curing**) without lethal effect on the cell, and many plasmids can be transmitted from cell to cell by means of the conjugation process, which is discussed below. Some plasmids also have the ability to become integrated into the chromosome, and under such conditions their replication comes under control of the chromosome. Subsequently, an integrated plasmid may be released and then resume independent replication. The sequence of independent replication, chromosome integration, and escape from integration are phenomena equivalent to those undergone by the temperate bacteriophage (Section 10.6). This similarity is actually more than a

FIGURE 11.6   Plasmid biology.

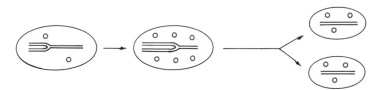

(a) Plasmid replication independent of chromosome

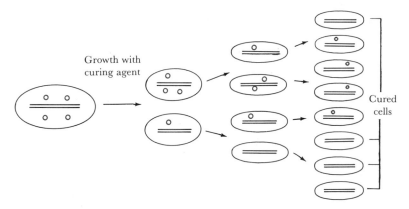

(b) Curing of plasmid

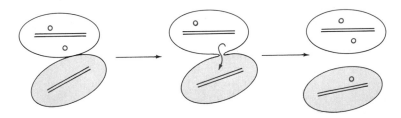

(c) Cell-to-cell transfer during conjugation

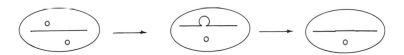

(d) Integration of plasmid into chromosome

mere analogy; by the definition given here temperate phages are also plasmids.

Many plasmids carry genes that control the production of toxins and provide resistance to antibiotics and other drugs. Many plasmids also carry genes that control the process of conjugation, such as genes which alter the cell surface to permit cell-to-cell contact and genes which bring about the transfer of DNA from one cell to another. Plasmids which govern their own transfer by cell-to-cell contact are called **conjugative,** but not all plasmids are conjugative. Transmissability by conjugation is controlled by a set of genes within the plasmid called the **sex factor.** The presence of a sex factor in a plasmid can have another important consequence, if the plasmid becomes integrated into the chromosome. In that case, the plasmid can mobilize the transfer of chromosomal DNA from one cell to another. Bacteria that transfer large amounts of chromosomal DNA during conjugation are called *Hfr* (high frequency of recombination). Although considerably rarer than bacteria transferring only plasmid DNA, Hfr bacteria are of considerable interest and importance, since the study of Hfr bacteria permits an analysis of the genetic organization of the whole bacterial chromosome.

Physical evidence for plasmids   Plasmids are generally less than one-twentieth the size of the chromosome. As usually isolated, the plasmid DNA is a double-stranded closed circle. When isolated, this circular double helix is twisted about itself in a form called a **supercoil** (Figure 11.7). A single break (called a "nick") in one of the two strands causes the supercoil to convert to an open circular form, and if breaks occur in both strands at the same place, a linear duplex structure will be formed. Most of the plasmid DNA isolated from cells is in the super-coiled configuration, which would be the most compact form within the cell. Isolation of plasmid DNA can generally be readily accomplished by making use of certain physical properties of supercoiled DNA molecules. Such molecules are more compact than nicked or linear duplexes and hence sediment more rapidly in an ultracentrifuge. Further, supercoiled molecules show a high sedimentation rate in liquids such as alkaline sucrose, which ordinarily separate the strands of linear duplexes. They also show less affinity for certain dyes such as ethidium bromide, which can become inserted into the DNA molecule. Since binding of ethidium bromide to DNA lowers its density, supercoiled plasmid DNA assumes a different position in the gradient than chromosomal DNA when placed in a cesium chloride density gradient. In this way, the plasmid DNA can be easily separated from chromosomal DNA in cell extracts which have been treated with saturating amounts of ethidium bromide. Plasmid DNA molecules of different sizes can also be readily separated by electrophoresis (movement induced by an electric current) on agarose gels (see Figure 12.7). This technique provides a good opportunity to analyze the action of endonucleases and other enzymes on plasmid DNA.

Plasmid DNA can be observed under the electron microscope, and one type of evidence for the existence and molecular size of a plasmid is observation of its structure and measurement of its length in electron micrographs. Plasmid DNA molecules vary widely in

Covalently closed circular duplex (supercoiled)

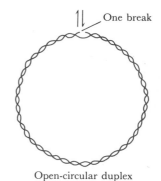

Open-circular duplex

Linear duplex

FIGURE 11.7 Supercoiled, open-circular, and linear duplex forms of DNA. Plasmid DNA is isolated from cells primarily in the supercoiled form.

molecular weight, from 3×10^6 to greater than 10^8. The larger plasmids are more often conjugative than the smaller ones.

Curing of plasmids One of the common features of plasmids is that they can be eliminated from host cells by various treatments. This process, termed **curing,** apparently results from inhibition of plasmid replication without parallel inhibition of chromosome replication, and as a result of cell division the plasmid is diluted out. Curing may occur spontaneously, but it is greatly increased by use of acridine dyes which become inserted into DNA, as well as by other treatments that affect DNA, such as ethidium bromide, thymine starvation (in thymine-deficient mutants), ultraviolet or ionizing radiation, and heavy metals. Growth at temperatures above the optimum may also result in elimination of the plasmid.

Replication of plasmids Although we have emphasized that plasmids are independently replicating genetic elements, replication of plasmids must be under some sort of cellular control, as the number of plasmid molecules per cell is determined to some extent by the cell and by environmental conditions. The manner of replication of plasmids is of interest not only in understanding the way in which plasmids maintain autonomy in the cell, but perhaps as a simplified model for how the much more complex chromosomal DNA is replicated. Some plasmids are present in the cell in only a few copies, perhaps one to three, and they probably replicate in a manner similar to that already described for the chromosome. This involves initiation of replication at a single point and bidirectional symmetrical replication around the circle. Because of the small size of the plasmid DNA, the whole replication process occurs very quickly, perhaps in one-tenth or less of the total time of the cell division cycle. There is some evidence that the enzymes involved in plasmid replication are normal cell enzymes, so that the genetic elements within the plasmid itself, which control its replication, may be concerned primarily with the control of the timing of the initiation process and with the apportionment of the replicated plasmids between daughter cells. There is evidence for an association of the plasmid DNA with the cell membrane during replication, and this association probably plays some role in the partitioning of the replicated plasmid DNAs between daughter cells at division, but membrane association need not occur at all times to ensure plasmid retention.

11.5

Conjugation and Plasmid: Chromosome Interactions

Bacterial conjugation or mating is a process of genetic transfer that involves cell-to-cell contact. The genetic material transferred may be a plasmid, or it may be a portion of the chromosome mobilized by a plasmid. In conjugation, one cell, the donor, transmits genetic information to another cell, the recipient.

Specific pairing between donor and recipient cells must occur. The donor cell, by virtue of its possession of a conjugative plasmid, possesses a surface structure, the **sex pilus** (see Figure 2.34), which is involved in pair formation and in the transfer of DNA. It is thought that the sex pili make possible specific contact between donor and recipient cells and then retract, pulling the two cells together so that a conjugation bridge can form through or on which DNA passes from one cell to another. Although the recipient cells lack sex pili, they must have some sort of recognition mechanism on their surface, as pair formation for conjugation generally occurs only between strains of bacteria that are closely related.

Conjugative plasmids possess the genetic information to code for sex pili and for some proteins needed for DNA transfer. In *E. coli,* one class of conjugative plasmids, called F factors (F for fertility), possesses the additional property of being able, occasionally, to become integrated into the chromosome and to mobilize the chromosome so that it can be transferred during cell-to-cell contact. When the F factor is integrated into the chromosome, large blocks of chromosomal genes can be transmitted, and genetic recombination between donor and recipient is then very extensive. As mentioned earlier, bacterial strains that possess a chromosome-integrated F factor and do show such extensive genetic recombination are called Hfr (for high frequency of recombination). When the F factor is not integrated into the chromosome, it behaves as a conjugative plasmid. Cells possessing an unintegrated F factor are called F^+, and strains which can act as recipients for F^+ are called F^-. F^- cells lack the F-factor plasmid; in general cells that contain a plasmid are very poor recipients for the same or closely related plasmids. We thus see that the presence of the F factor results in three distinct alterations in the properties of a cell: 1) ability to synthesize the F pilus; 2) mobilization of DNA for transfer to another cell; 3) alteration of surface receptors so that the cell is no longer able to behave as an F^- recipient.

Mechanism of DNA transfer during conjugation   DNA synthesis is necessary for DNA transfer to occur, and the evidence suggests that one of the DNA strands is derived from the donor cell and the other is newly synthesized in the recipient during the transfer process. A mechanism of DNA synthesis in certain bacteriophages, called **rolling circle replication,** was presented in Figure 10.21. This model best explains DNA transfer during conjugation, and a possible mechanism for this process is outlined in Figure 11.8. The whole series of events is probably triggered by cell-to-cell contact, at which time the plasmid DNA circle opens, and one parental and one new strand are transferred. According to this model, transfer can only occur if DNA synthesis can also occur, and this has been shown experimentally by use of chemicals which specifically inhibit DNA synthesis. The model also accounts for the fact that if the DNA of the donor is labeled, some labeled DNA is transferred to the recipient, but that only a single labeled strand is transferred. With the rolling circle mechanism, the donor cell also duplicates its plasmid at the time transfer occurs, so that at the end of the process both donor and recipient possess completely formed plasmids.

The high efficiency of the DNA transfer process is shown by the fact that under appropriate conditions, virtually every recipient cell which pairs can acquire a plasmid.

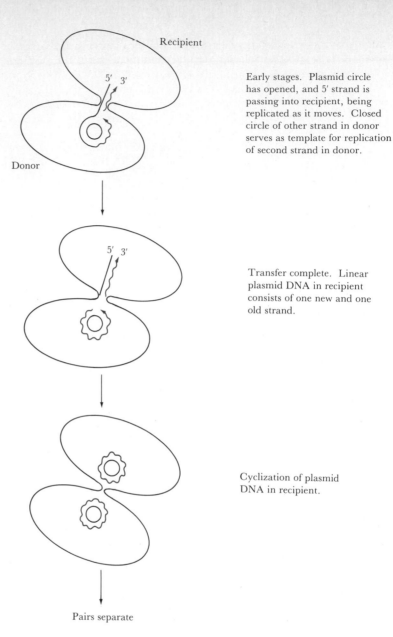

Early stages. Plasmid circle has opened, and 5' strand is passing into recipient, being replicated as it moves. Closed circle of other strand in donor serves as template for replication of second strand in donor.

Transfer complete. Linear plasmid DNA in recipient consists of one new and one old strand.

Cyclization of plasmid DNA in recipient.

Pairs separate

FIGURE 11.8 Rolling circle DNA replication during transfer of plasmid DNA from donor to recipient. Compare with Figure 10.21. Only the plasmid DNA is shown.

Infectiousness of conjugative plasmids If the plasmid genes can be expressed in the recipient, the recipient itself then soon becomes a donor and can then transfer the plasmid to other recipients. Conjugative plasmids can spread rapidly between populations, behaving in the manner of infectious agents (Figure 11.9). Thus plasmid-mediated drug resistance frequently has been called **infectious drug resistance.** The infectiousness of conjugative plasmids is of major ecological significance, since a few plasmid-positive cells introduced into an appropriate population of recipients, if they contain genes which confer a selective advantage, can convert the whole recipient population into a plasmid-bearing population in a short period of time. Under optimum conditions, the rate of spread of a conjugative plasmid through a population can be exponential, resembling a bacterial growth curve.

Formation and behavior of Hfr strains As noted above, an F factor can become integrated into the chromosome and mobilize the chromosome for conjugation. The process of integration of F factor into the chromosome is analogous to the integration of a temperate bacteriophage, such as lambda, as described in Figure 10.20. There are several specific sites in the chromosome at which F factors can be integrated, and these sites, called IS (for "insertion sequence") represent regions of homology between chromosome and F factor DNA. Enzymes are involved in the events leading to the insertion of F factor, such as a site-specific recombination enzyme similar to the *int* (integrase) enzyme of lambda. As seen in Figure 11.10, integration of F factor involves insertion in the chromosome at the specific site. In the particular Hfr shown, the integration site is between the chromosomal genes *pro* and *lac*. The site on the F factor (*origin*) at which transfer will initiate during conjugation is indicated by an arrow. At the time of specific cell pairing, the chromosome opens at the origin, and the host genes are inserted into the recipient beginning with the gene downstream from the origin (Figure 11.11).

The origin of Hfr strains is thus seen as a result of the integration of F factor into the chromosome. Since a number of distinct insertion sites are present, a number of distinct Hfr strains are possible. A given Hfr strain always donates genes in the same order, beginning with the same position, but Hfr strains of independent origin transfer genes in different sequences. During normal cell division, the DNA of the Hfr replicates normally, but at the time of pairing with an F^- cell, a DNA strand from the Hfr is inserted into the F^- cell, and replication occurs by the rolling circle process. The Hfr strain, after transfer, still remains Hfr, since it has retained a copy of the transferred genetic material.

Although Hfr strains transmit chromosomal genes at high frequency, they usually do not convert F^- cells to F^+ or Hfr, because the entire F factor is only rarely transferred. On the other hand, F^+ cells efficiently convert F^- to F^+ because of the infectious nature of the F plasmid.

At some insertion sites, the F factor is integrated with the origin in one direction, whereas at other sites the origin is in the opposite direction. The direction in which the F factor is inserted determines which of the chromosomal genes will be inserted first into the recipient. The manner in which a variety of Hfr strains can arise is illustrated in Figure 11.12. By use of various Hfr strains, it has been possible in *E. coli* to determine the arrangement and orientation of a large number of chromosomal genes, as will be described in Section 11.7.

Transfer of chromosomal genes to F Under rare conditions, integrated F factors may be excised from the chromosome, and the possibility exists for the incorporation at that time of chromosomal genes into the liberated F factor. Such F factors containing chromosomal genes are called F′ (F prime). These F factors differ from nor-

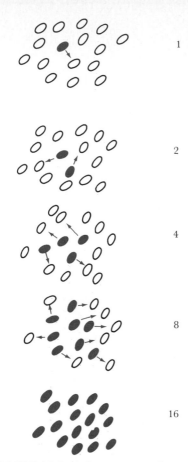

FIGURE 11.9 Infectiousness of a conjugative plasmid.

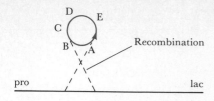

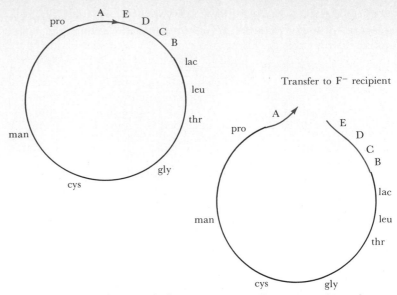

FIGURE 11.10 Integration of F factor into the chromosome with formation of an Hfr. The insertion of F factor occurs at a variety of specific sites, the one here being between chromosomal genes pro and lac. The letters on the F factor represent arbitrary genes. The arrow indicates the origin of transfer, with the arrow as the leading end. The site in F factor at which pairing with the chromosome occurs is between A and B. Compare with Figure 10.20.

FIGURE 11.11 Breakage of Hfr chromosome at the origin and transfer of DNA to the recipient. Replication occurs during transfer, as illustrated in Figure 11.8.

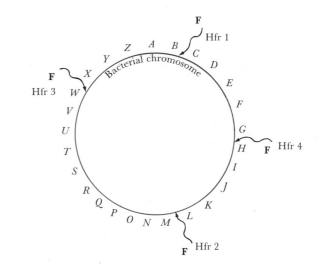

(a)

FIGURE 11.12 Manner of formation of different Hfr strains, which donate genes in different orders and from different origins. The bacterial chromosome is visualized as a circle (a) that can open at various locations, at which F particles become attached. The gene orders are shown in (b).

Hfr 1	←—	*CDE*			*XYZAB*	Gene *C* donated first; clockwise order
Hfr 2	←—	*LKJ*	*BAZYX*	*ONM*		Gene *L* donated first; counterclockwise order
Hfr 3	←—	*XYZAB*			*UVW*	Gene *X* donated first; clockwise order
Hfr 4	←—	*GFE*	*BAZYX*	*JIH*		Gene *G* donated first; counterclockwise order

(b)

mal F factors in that they contain identifiable chromosomal genes, and they transfer these genes at high frequency to recipients.

Oriented transfer of Hfr and the phenomenon of interrupted mating
The measurement of bacterial conjugation is usually done by use of parental strains having properties that can be selected against on agar plates. The recipient is usually resistant to an antibiotic and is auxotrophic for one or more nutritional characters. The donor is antibiotic sensitive but prototrophic for the nutritional characters. With proper agar media, only the recombinants can grow and the large background of nonrecombinants eliminated.

The oriented transfer of chromosomal genes from Hfr to F^- was most clearly shown by a procedure called **interrupted mating.** The mating pairs are not strongly joined and can be easily separated by agitation in a mixer or blender. If mixtures of Hfr and F^- cells are agitated at various times after mixing and the genetic recombinants scored, it is found that the longer the time between pairing and agitation, the more genes of the Hfr will appear in the F^- recombinant. It is also found that gene transfer always occurs in a specific order in a specific Hfr. As shown in Figure 11.13, genes present closer to the origin enter the F^- first and are always present in higher percentage of the recombinants than genes that enter late. In addition to showing that gene transfer from donor to recipient is a sequential process, experiments of this kind provide a method of determining the order of the genes on the bacterial DNA (genetic mapping). The arrangement of gene loci on the chromosome is called a **genetic map** (see Section 11.7).

Genetic recombination between Hfr genes and F^- genes in the F^- cell requires the presence of enzymes in the recipient cell. This has been shown by the isolation of mutants of F^- strains, which are unable to form recombinants when mated with Hfr. These mutants, called *rec*$^-$ (recombination-minus), are also unusually sensitive to ultraviolet radiation, and are deficient in enzymes involved in dark repair of DNA.

Insertion elements   We have stated that plasmids become integrated into the host chromosome at defined insertion sites. The sites at which plasmids become integrated contain specific nucleotide sequences, and these sequences function in a regular way to bring about integration and transfer of pieces of DNA from one chromosome to another. Such sites are called **insertion elements,** and are found not only in host and plasmid DNA, but also in phage DNA. In essence, an insertion element is a short segment of DNA, around 1000 nucleotides long, that can become integrated at a specific site. Genes attached to the insertion element may also become integrated. Required for integration is an enzyme, similar to the integrase enzyme (*int*) of lambda (see Section 10.6), which is able to recognize the insertion sequence. This enzyme can be coded for by either the host chromosome or by the plasmid or phage being integrated. In addition to serving as integration sites for plasmids, insertion elements have the ability to transpose themselves alone to different sites.

A number of insertion elements have been recognized, desig-

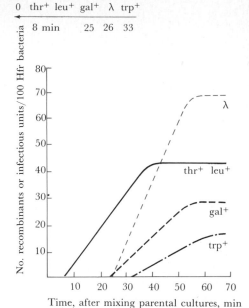

FIGURE 11.13   Rate of formation of recombinants containing different genes after mixing Hfr and F^- bacteria. The location of the genes along the Hfr chromosome is shown in the small diagram. Note that the genes closest to the origin are the first ones detected in the recombinants. The experiment is done by mixing Hfr and F^- cells under conditions in which essentially all Hfr cells find mates. The F^- recipient was streptomycin resistant but auxotrophic for the markers being scored. The Hfr donor was streptomycin sensitive. At various intervals, samples of the mixture are shaken violently to separate the mating pairs and plated on a selective medium in which only the recombinants can grow and form colonies (From Hayes, W. 1968. The genetics of bacteria and their viruses, John Wiley, N.Y.)

nated by numbers: IS1, IS2, IS3, and so on. IS1, the first to be discovered, is 800 nucleotides long, and IS2 is 1300 nucleotides long. Since there are many places that a single insertion element can be inserted, a large number of possibilities exist for movement of DNA from place to place.

Transposons are defined as genetic elements which can move from place to place. They contain an insertion element at each end and a series of genes between. Because they possess an insertion element at each end, transposons can readily move from place to place, carrying their genes with them. Several transposons have been identified which carry antibiotic resistance genes. Further, bacteriophage Mu, which is a temperate phage, has the interesting property of integrating at a large number of locations in the host chromosome. Mu can be viewed as a transposon containing viral genes.

The importance of insertion elements is that they not only bring about ready transfer of genetic information from place to place, but they permit rapid evolution of plasmids with a variety of genetic functions. In the next section, examples of plasmid diversity are presented.

Transposon mutagenesis   If the insertion site for a transposable element is within a bacterial gene, insertion of the transposon will result in loss of linear continuity of the gene, leading to mutation. Transposons thus provide a facile means of creating mutants throughout the chromosome.

The most convenient element for transposon mutagenesis is one containing an antibiotic resistance gene. Clones containing the transposon can then be selected by the isolation of antibiotic resistance colonies. If the antibiotic resistance clones are selected on rich medium where all auxotrophs can grow, they can be subsequently screened on minimal medium supplemented with various growth factors, to determine if a growth factor is required.

Transposons are also useful for incorporating an auxotrophic gene marker into a wild-type organism. Normally, auxotrophic recombinants cannot be isolated by positive selection, but if the auxotrophic marker to be introduced contains a transposable element with an antibiotic resistance marker, then one can select for antibiotic-resistant clones, a positive selection procedure, and automatically obtain clones which have incorporated the auxotrophic marker.

Two widely used transposons for mutagenesis are Tn5, which confers neomycin and kanamycin resistance, and Tn10, which contains a marker for tetracycline resistance.

11.6

Kinds of Plasmids and Their Biological Significance

It should be clear from our discussion of plasmids that these genetic elements are of great importance as tools for understanding a wide variety of genetic phenomena in procaryotes. Plasmids are also useful for understanding a number of biological and ecological phenomena, and it is now evident that many ideas of bacterial taxonomy and ecology will have to be modified to account for the presence and activity of plasmids.

F and I pili   At least two kinds of pili, called F and I pili, are known to be involved in cell-to-cell transfer of plasmids. Two classes of RNA phages are known to infect cells which carry transmissable plasmids. These phages can be used to demonstrate the presence on the cell of either F or I pili (see Figure 2.34). The two kinds of pili can also be distinguished immunologically. F pili are involved in the transfer of F factor and some antibiotic resistance plasmids. F pili are also present on Hfr cells. I pili are involved in the transfer of other antibiotic resistance plasmids, colicin-determining plasmids, and others.

Resistance-transfer factors   Among the most widespread and well-studied groups of plasmids are the resistance-transfer factors (R factors), which confer resistance to antibiotics and various other inhibitors of growth. R factors were first discovered in Japan in strains of enteric bacteria which had gained resistance to a number of antibiotics (multiple resistance), and have since been found in other parts of the world. The emergence of bacteria resistant to several antibiotics is of considerable medical significance, and was correlated with the increasing use of antibiotics for the treatment of infectious diseases. Soon after these resistant strains were isolated it was shown that they could transfer resistance to sensitive strains via cell-to-cell contact. This is probably the reason for the rapid rise of multiply resistant strains, since it would be unlikely for resistance to a number of antibiotics to develop simultaneously by mutation and selection. The infectious nature of the R factors permits rapid spread of the characteristic through populations.

Two classes of R factors are known, those which use the F pilus as agent of transfer and those which use the I pilus. A variety of antibiotic-resistance genes can be carried by an R factor; those most commonly observed carry resistance to four antibiotics (tetracycline, chloramphenicol, streptomycin, sulfonamides), but some may have fewer and others more resistance genes. R factors with genes for resistance to kanamycin, penicillin, and neomycin are known, while others confer resistance to inhibition by nickel, cobalt, or mercury. Many drug-resistant elements on R factors are transposable elements and can be used in transposon mutagenesis (see Section 11.5).

Genes for characteristics not related to antibiotic resistance may also be carried by R factors. Paramount among these is the production of F or I pili, but R factors also carry genes permitting their own replication and genes controlling production of proteins that prevent the introduction of other related plasmids. Thus the presence of one R factor inhibits the introduction of another of the same type.

Because R factors are able to undergo genetic recombination, genes from two R factors can be integrated into one. This of course is one means by which multiply drug-resistant organisms might have first arisen. The genes in an R factor are carried in a definite order, as are chromosomal genes, and this order can be mapped by genetic recombination. Many R factors and F factors use the same pilus for conjugation, and recombination can occur between F and R factors.

Biochemical mechanism of R-factor-mediated resistance   The isolation in the laboratory of mutants resistant to antibiotics generally results in the selection of mutants in chromosomal genes. On the other hand, the majority of drug-resistant bacteria isolated from patients contain the drug-resistant genes on R factors. The biochemical mechanism of R-factor resistance is different from that of chromosomal resistance. In most cases, antibiotic resistance mediated by chromosomal genes arises because of a modification of the target of antibiotic action (e.g., a ribosome). On the other hand, R-factor resistance is in most cases due to the presence in the R factor of genes coding for new enzymes which inactivate the drug (Figure 11.14). For instance, a number of antibiotics are known which have similar chemical structures, containing aminoglycoside units. Among the aminoglycoside antibiotics are streptomycin, neomycin, kanamycin, and spectinomycin. Strains carrying R factors conferring resistance to these antibiotics contain enzymes that chemically modify these aminoglycoside antibiotics, either by phosphorylation, acetylation, or adenylylation, the modified drug then lacking antibiotic activity. In the case of the penicillins, R-factor resistance is due to the formation of penicillinase (β-lactamase), which splits the β-lactam ring, thus destroying the molecule. Chloramphenicol resistance mediated by R factor arises because of the presence of an enzyme that acetylates the antibiotic. Thus, the fact that R factors can confer multiple antibiotic resistance does not imply that the mode of action of the R-factor genes is similar. The presence of multiple antibiotic resistance is due to the fact that a single R factor contains a variety of genes coding for different antibiotic inactivating enzymes.

Toxins and other virulence characteristics   We will discuss in Chapter 15 the physiological and genetic characteristics of microorganisms that enable them to colonize hosts and set up infections, which can lead to harm. In the present context, we merely note the two major characteristics involved in virulence: (1) the ability of microbes to attach to and colonize specific sites in the host; and (2) the formation of substances (toxins), which cause damage to the host. It has now been well established that at least in some kinds of pathogenicity due to *E. coli*, each of these virulence characteristics is carried on plasmids. Enteropathogenic *E. coli* is characterized by an ability to colonize the small intestine and to produce a toxin that causes symptoms of diarrhea. Colonization requires the presence of a cell surface protein called the K antigen, which confers on the cells the ability to attach to epithelial cells of the intestine. At least two toxins in enteropathogenic *E. coli* are known to be carried by plasmid: the hemolysin, which lyses red blood cells, and the enterotoxin, which induces extensive secretion of water and salts into the bowel. It is the enterotoxin which is responsible for the induction of diarrhea, as will be discussed in Chapter 15. Since other virulence factors of *E. coli* are carried by chromosomal genes, it is not clear why these specific factors are carried by plasmids, but the existence of such plasmids raises the question of how widespread plasmid-related virulence is in microorganisms.

In *Staphylococcus aureus,* a number of virulence-conferring properties are either known or suspected to be plasmid-linked; *S. aureus* is noteworthy for the variety of enzymes and other extracellular pro-

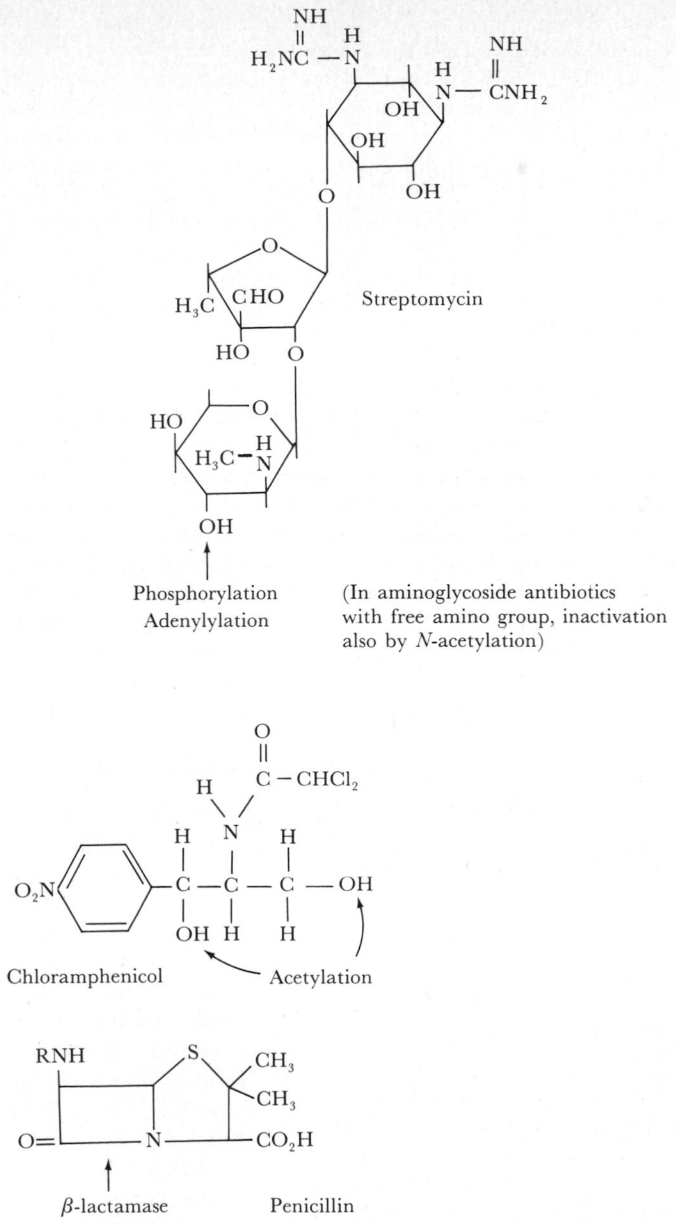

NH
‖
H_2NC — N
H

H
N — CNH_2
‖
NH

OH
OH

OH

O

O

H_3C CHO   Streptomycin
HO   O

HO   O

H_3C — N
H

OH

↑
Phosphorylation
Adenylylation

(In aminoglycoside antibiotics
with free amino group, inactivation
also by *N*-acetylation)

O
‖
C — $CHCl_2$

H   H
\   /
H   N   H
| | |
O_2N — C — C — C — OH
| | |
OH   H   H

Chloramphenicol   Acetylation

RNH   S CH_3
CH_3
O=   N CO_2H

↑
β-lactamase   Penicillin

teins it produces that are involved in its virulence, and the production
of coagulase, hemolysin, fibrinolysin, and enterotoxin is thought
to be plasmid-linked. In addition, the yellow pigment in *S. aureus,*
which is probably involved in its ability to resist the destructive action
of singlet oxygen in the phagocyte (see Section 8.6), may also be
plasmid-linked.

Presently, extensive research is underway to determine how
significant plasmid-mediated virulence is in the microbial world.
The implications of this research are widespread in medicine and
plant pathology.

Bacteriocins Most bacteria produce agents that inhibit or kill closely related species; these agents are called **bacteriocins** to distinguish them from the antibiotics, which have a wider spectrum of activity. Bacteriocins are a diverse group of substances, usually proteins and frequently of high molecular weight. The ability to produce some bacteriocins is inherited via plasmids. Those bacteriocins produced by *E. coli,* called *colicins,* have been best studied; the elements controlling their production are called Col factors. Some Col factors are conjugative, and they are found to fall in two groups, those transmitted by the F pilus and those transmitted by the I pilus. In general, Col factors are not transmitted as rapidly through populations as F or R factors.

Incompatibility of plasmids A phenomenon of considerable importance both in research on plasmids and in the evolution and ecology of plasmids is compatibility. When a plasmid is inserted into a cell which already carries another plasmid, a common observation is that the second plasmid may not be maintained and is lost during subsequent cell replication. The two plasmids are said to be **incompatible.** Plasmid incompatibility is controlled by genes in the plasmid, and a number of incompatibility groups have been recognized, the plasmids of one incompatibility group excluding each other but being able to coexist with plasmids from other groups. Evidence suggests that the plasmids of one incompatibility group have been derived from a common ancestral source.

Biological and taxonomic significance of plasmids A summary of plasmids which have been identified and organisms for which evidence of plasmids exist is given in Table 11.1. It is likely that virtually all bacterial groups possess plasmids. Techniques are now available for detecting the presence in an organism of plasmid-like DNA, and these techniques have been applied to a wide variety of bacteria. In some cases, virtually every strain tested has been shown to have plasmid-like DNA. Almost certainly, such DNA molecules control some genetic functions for the cell, although it is frequently difficult to relate the presence of the physical structure with a genetic function or functions. The only certain way of showing this is to transfer the plasmid to another strain, where the DNA can be distinguished by its physical characteristics, and to show that at the same time some genetic characteristic of the donor has also been transferred. Some plasmids are much more promiscuous than others, being able to be transferred to a variety of bacteria, some even quite unrelated.

As we have noted, many plasmids are not transmissible by cell-to-cell contact. In *Staphylococcus aureus* plasmid transfer never occurs by cell-to-cell transfer, and the demonstration of the existence of plasmids is much more difficult. Evidence for plasmids comes from curing experiments, and from the transfer of plasmids via transduction. On the other hand, cell-to-cell transfer of plasmids is extremely common in Gram-negative bacteria, not only within the enteric group, which has been so widely studied, but in other groups, such

TABLE 11.1 Types of Plasmids*

Type	Organisms
Conjugative plasmids	F-factor, *Escherichia coli;* pfdm, K, *Pseudomonas;* P, *Vibrio cholerae*
R plasmids:	
Wide variety of antibiotics	Enteric bacteria,
Resistance to mercury, cadmium, nickel, cobalt, zinc, arsenic	*Staphylococcus*
Bacteriocin and antibiotic production	Enteric bacteria; *Clostridium; Streptomyces*
Physiological functions:	
Lactose, sucrose, urea utilization, nitrogen fixation	Enteric bacteria
Degradation of octane, camphor, naphthalene, salicylate	*Pseudomonas*
Pigment production	*Erwinia, Staphylococcus*
Virulence plasmids:	
Enterotoxin, K antigen, endotoxin	*Escherichia coli*
Tumorigenic plasmid	*Agrobacterium tumefaciens*
Adherence to teeth (dextran)	*Streptococcus mutans*
Coagulase, hemolysin, fibrinolysin, enterotoxin	*Staphylococcus aureus*

*Plasmids have now been found in most bacterial genera.

as the pseudomonads. In fact, some pseudomonad plasmids are transferrable to a wide variety of other Gram-negative bacteria. Some pseudomonad plasmids have been shown to transfer the genetic information for biochemical pathways for the degradation of unusual organic compounds, such as camphor, octane, and naphthalene. These substances are highly insoluble, and the plasmid-coded enzymes convert them into more soluble substrates that are on the main line of biochemical catabolism, such as acetate, pyruvate, and isobutyrate, the catabolism of which is coded by chromosomal genes.

From a taxonomic point of view, the existence of plasmids that can be transmitted to quite unrelated Gram-negative bacteria raises the interesting point of the real significance of the complex taxonomic scheme for the bacteria (see Appendix 4). One possibility is that all Gram-negative bacteria are in fact fairly closely related, and that the differences which we observe and use in our taxonomies are really quite superficial. Since the genes that can be present on a plasmid are variable, and are really independent of the extent of conjugability of the plasmid, the possibility exists that all Gram-negative bacteria are really part of one large gene pool, connected via conjugative plasmids. However, the gene functions that have been shown to be carried by most plasmids are not a part of the main stream of cell function, but provide for characteristics such as antibiotic resistance, colicin production, and degradation of unusual substrates, and these are characteristics that cells, under many conditions, can do without. There is no evidence, for instance, for the presence on a plasmid of the genes for photosynthesis, such as the Calvin cycle, or the genes for the oxidation of inorganic energy sources. If such evidence were obtained, this would make taxonomists even more uncomfortable with the existing procaryotic scheme. At the moment, it is sufficient to point out that things may not be as clear cut as our taxonomic scheme suggests, and that further work on plasmids may lead to some major changes in taxonomic thinking.

Origin of plasmids   Although specific evidence for the origin of multiple drug resistant R factors is not available, a number of lines of circumstantial evidence suggest that plasmids with R-factor type character existed before the antibiotic era, and that the widespread clinical use of antibiotics provided selective conditions for the evolution of R factors with one or more antibiotic-resistance genes. Indeed, a strain of *E. coli* which was freeze-dried in 1946 was found to contain a plasmid with genes conferring resistance to tetracycline and streptomycin, even though neither of these antibiotics were used clinically until several years later. Also, strains carrying R-factor genes for resistance to a number of semisynthetic penicillins were shown to exist before the semisynthetic penicillins had been synthesized. Of perhaps even more ecological significance, R factors conferring antibiotic resistance have been detected in some nonpathogenic Gram-negative soil bacteria. In the soil, such resistance may confer selective advantage, since the antibiotic-producing organisms (*Streptomyces, Penicillium*) are also normal soil organisms (however, there is no real evidence that the common antibiotics are produced or act in the soil). Thus, it seems reasonable to conclude that R factors are not a recent phenomenon, but existed to a small extent in the natural bacterial population before the antibiotic era, and that widespread use of antibiotics provided selective conditions for the rapid spread of these R factors. The evolution of R factors is thus completely understandable, albeit frightening in the implications for the continued success of human medicine to combat rampant infections with the ever increasing use of larger and larger doses of antibiotics.

Engineered plasmids   The techniques of genetic engineering, discussed in the next chapter, have made possible the construction in the laboratory of a limitless number of new, artificial plasmids. Incorporation into artificial plasmids of genes from a wide variety of sources has made possible the transfer of genetic material across virtually any species barrier. It is even possible to synthesize completely new genes and introduce those into plasmids. Such artificial plasmids are useful tools in understanding plasmid structure and function, as well as for the more practical aims of genetic engineering that are discussed in the next chapter. In order to create an artificial plasmid, the main requirement is that the genetic material of the plasmid involved in replication and maintenance of the plasmid in the cell be connected to the new material of interest. Artificial plasmids are introduced into appropriate hosts via transformation and are selected for by means of antibiotic resistance characters. The ability to create artificial plasmids has greatly expanded the possibilities for plasmid research.

The procedure of interrupted mating (Figure 11.13) can be used to map the locations of various genes on the chromosome. Different Hfr strains are used, which initiate DNA transfer from different parts of the DNA (Figure 11.12). By using Hfr strains with origins in different sites, it is possible to map the whole bacterial gene complement. A circular reference map for *E. coli* strains K-12 is shown in Figure 11.15, which also indicates the leading transfer regions of a number of Hfr strains that have been used in genetic mapping. The map distances are given in minutes of transfer, with 100 minutes for the whole chromosome and with "zero time" arbitrarily set as that at which the first gene transfer (for the threonine gene) can be detected from strain Hfr-H, the first Hfr strain isolated.

Although interrupted mating experiments provide the best means of obtaining an overall picture of the bacterial genetic map, they are less convenient for mapping closely linked genes than genetic mapping by transduction. Bacteriophage P1 has been used extensively in *E. coli* to fill in the gaps in the genetic map, since it transduces fairly large segments of DNA, equivalent to about 2 minutes on the map. Transduction is especially useful for determining the order and

An Overview of the Bacterial Genetic Map

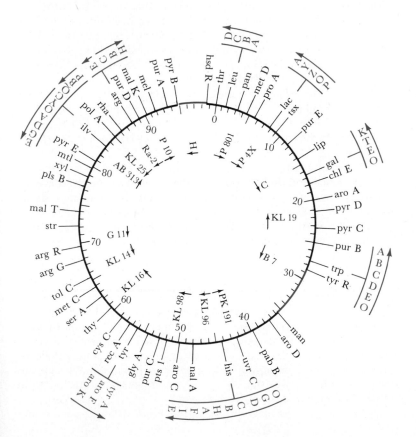

FIGURE 11.15 Circular reference map of *E. coli* K-12. The large numbers refer to map position in minutes, relative to the thr locus. From the complete linkage map, 52 loci were chosen on the basis of greatest accuracy of map location, utility in further mapping studies, and/or familiarity as long-standing landmarks of the *E. coli* K-12 genetic map. Inside the circle, the leading transfer regions of a number of Hfr strains are indicated. Arrows above operons indicate the direction of messenger RNA transcription for these genes. (From Bachmann, B. J., K. B. Low, and A. L. Taylor. 1976. Bacteriol. Revs. 40:116–167.)

location of genes that are closely linked, since interrupted mating experiments do not permit separation of genes that are very close together. By use of such transductional analysis, it has been determined that many genes that code for proteins involved in a single biochemical pathway are closely linked. Some of these gene clusters are indicated on the map (Figure 11.15).

Electron microscopic studies on heteroduplex DNA molecules (see Section 9.2) of known genetic composition make it possible to correlate genetic map distances with physical distances on the DNA. The total molecular length of the *E. coli* genome, 100 minutes on the genetic map, is equivalent to 4.1×10^6 base pairs, corresponding to a molecular weight of 2.7×10^9. Since the total length of DNA of an *E. coli* nuclear body is about 1100 to 1400 μm, 1 minute of transfer represents about 11 to 14 μm of DNA.

Gene clusters and operons Mapping of the genes that control the enzymes of a single biochemical pathway has shown that these genes are often clustered or closely linked. The gene clusters for several biochemical pathways on the *E. coli* chromosome are shown in Figure 11.15; letters are used to indicate the genes for the specific enzymes of the pathway (for instance, the galactose cluster at minute 17, with genes *K, T,* and *E*). It has also been found that all of the enzymes of a single gene cluster are often affected simultaneously by induction or repression. These related observations have been of considerable importance in the development of the **operon** concept, explaining not only how enzyme induction or repression can be brought about, but also how the synthesis of a series of related enzymes can be simultaneously controlled (Section 9.6).

The transcription of some operons proceeds in one direction along the chromosome, whereas with other operons transcription is in the opposite direction. The direction of mRNA transcription of the operons listed in Figure 11.15 is shown by arrows above the gene clusters.

Although many enzymes controlling specific biochemical pathways are linked and their synthesis is regulated by operons, this is not always the case. For some pathways, individual enzymes are at different locations on the chromosome, with the same operator present at each site. In such a situation, coordinate regulation may occur even though the enzymes do not map together.

11.8

Summary

In this chapter we have presented a brief overview of the field of bacterial genetics, concentrating on the general mechanisms of gene transfer and recombination. The details of bacterial genetics are so extensive that only the highlights can be presented, and it should be emphasized that a vast amount of experimentation by a large number of individuals was necessary to provide the information from which these generalizations were drawn.

Genetic transfer mechanisms in bacteria can be grouped under three broad headings, transformation, transduction, and conjugation. **Transformation** involves the transfer of free fragments of DNA. Only certain organisms are able to take up free DNA, and the process is probably of minor importance in nature. The discovery of genetic transformation in bacteria, however, was one of the outstanding events in biology, as it led to experiments proving beyond a doubt that DNA was the genetic material of living organisms.

Transduction involves transfer of DNA from one cell to another via a bacterial virus. The virus which carries the DNA is usually defective and does not initiate a productive infection. Two kinds of transduction are recognized: generalized and specialized. In generalized transduction, virtually any genetic marker of the donor can be transferred, the donor DNA fragment becoming incorporated into a virus particle by chance during the maturation process. Generalized transduction is an inefficient event, because of the improbability of incorporation of a specifically marked fragment into a defective phage, but it provides a general means for carrying out genetic recombination studies in bacteria. Specialized transduction involves the specific association of a donor DNA with a temperate phage, at the time that the temperate phage DNA is excised from the host chromosome. Only the host donor genes near the attachment site of the temperate phage can be transferred, but in appropriate situations the frequency of transduction can be very high.

Conjugation, also called mating, involves cell-to-cell contact between donor and recipient. Conjugation can involve either chromosomal genes or genes carried by plasmids. **Plasmids** are circular DNA molecules that are present in most bacteria and are able to replicate extrachromosomally. Not all plasmids are transmissable by conjugation, but conjugative plasmids are widely found in Gram-negative bacteria. For transmission to occur, the plasmid must contain genetic information for the formation of sex pili, and must also have other genes involved in opening up the circular DNA and injecting DNA into the recipient. Plasmids always carry an origin of replication, and frequently carry information for their transmissability, but many plasmids also carry genes which confer important properties on the cell carrying them. Many plasmids are known which carry genes for antibiotic resistance, called R factors. Plasmids carrying genes for bacteriocin production, antibiotic production, toxin formation, and others, also are known. In addition, some plasmids also have the unique ability to integrate into the chromosome, at specific attachment sites, and plasmids containing F factors (F for fertility) are able to integrate with and mobilize the chromosome for transfer by conjugation. Strains containing chromosome-integrated F factors, which transfer chromosomal genes at high frequency, are called Hfr strains (for high frequency of recombination).

By use of various genetic recombination processes, it is possible to develop a genetic map for the bacterial chromosome, in which all known genes can be located. The map of *E. coli* shows that genes with related functions are often clustered together, and evidence for such clustering provided the initial impetus for the development of the operon model for regulation of bacterial protein synthesis.

A Bit of History

Although genetic recombination in eucaryotes had been known for a long time, the discovery of genetic recombination in bacteria by transformation, transduction, and conjugation has been a relatively recent event. Of the three recombination processes, the discovery of transformation was the most significant, as it provided the first direct evidence that DNA is the genetic material. The first evidence of bacterial transformation was obtained by the British scientist Fred Griffith in the late 1920s. Griffith was working with *Streptococcus pneumoniae* (pneumococcus), a bacterium which owes its ability to invade the body in part due to the presence of a polysaccharide capsule. Mutants can be isolated which lack this capsule and are thus unable to cause infection; such mutants are called R strains, because their colonies appear rough on agar, in contrast to the smooth appearance of capsulated strains. A mouse infected with only a few cells of an S (smooth) strain will succumb in a day or two to pneumococcus infection, whereas even large numbers of R cells will not cause death when injected. Griffith showed that if heat-killed S cells were injected along with living R cells, a fatal infection ensued, and the bacteria isolated from the dead mouse were S types. A number of different polysaccharide capsules were known in different pneumococcus S strains, and it was possible to do this experiment with heat-killed S cells from a type different from that from which the R strain was derived. Since the isolated living S cells always had the capsule type of the heat-killed S cells, the R cells had been transformed into a new type, and the process had all the properties of a genetic event. The molecular explanation for the transformation of pneumococcus types was provided by Oswald T. Avery and his associates at Rockefeller Institute in New York, by a series of studies carried out during the 1930s, culminating in the now classic paper by Avery, McCarty, and MacLeod (*Journal of Experimental Medicine*) in 1944. Avery and his coworkers showed that under certain conditions the transformation process could be carried out in the test tube rather than the mouse, and that a cell-free extract of heat-killed cells could induce transformation. By a long series of painstaking biochemical experiments, the active fraction of cell-free extracts was purified and was shown to consist of DNA. The transforming activity of purified DNA preparations was very high, and only very small amounts of material were necessary. Subsequently, Rolin Hotchkiss, Harriet Ephrussi, and others at Rockefeller showed that transformation could occur in pneumococcus not only for capsular characteristics, but for other genetic characteristics of the organism, such as antibiotic resistance and sugar fermentation. In the 1950s, transformation was also shown to occur in *Haemophilus, Neisseria, Bacillus,* and a variety of other organisms. In 1953, James Watson and Francis Crick announced their model for the structure of DNA, providing a theoretical framework for how DNA could serve as the genetic material. Thus, two types of studies, the bacteriological and biochemical ones of Avery, and physical-chemical ones of Watson and Crick, solidified the concept of DNA as the genetic material. In the subsequent years, this work has led to the whole field of molecular genetics.

Although bacterial transformation resulted from an essentially accidental discovery, bacterial conjugation was initially shown to occur by Joshua Lederberg and E. L. Tatum in 1946, through experiments carefully designed to determine if a sexual process might occur in bacteria. Because it appeared that the process, if present, would be quite rare (no microscopical evidence for bacterial mating had ever been seen, although such evidence can easily be obtained in eucaryotes), Lederberg developed a method which involved the use of nutritional mutants of *Escherichia coli.* Fortunately, he isolated these mutants in strain K-12, one of the few strains now known to contain (in the wild-type state) the F factor. The principle was to mix two strains, one requiring biotin and methionine, the other requiring threonine and leucine, and plate the mixture on a minimal medium lacking all four growth factors. Neither parental type could grow on this medium, but any recombinants could, and when about 10^8 cells were plated, a small but significant number of colonies was obtained. Strains with two separate nutritional requirements were employed since it would be unlikely that spontaneous back mutation of both genes would occur in a single cell. Thus the only explanation for the phenomenon was some sort of genetic recombination. To show that the process required cell-to-cell contact, and hence could not be a type of transformation, it was shown that culture filtrates or extracts from either type would not transform the other. Also if the two cell types were separated by a sintered glass disk, permeable to macromolecules but not to cells, recombination did not occur. Although initially conjugation appeared to be a very rare event, by the early 1950s a strain of *E. coli* had been isolated by the Italian scientist L. L. Cavalli-Sforza, while he was working in Lederberg's laboratory, which showed a high frequency of recombination. The British physician William Hayes, who independently isolated an Hfr strain, then showed that genetic transfer during recombination between Hfr and F^- was a one-way event, with the Hfr serving as donor. The interrupted mating experiment and the demonstration of the circular genetic map of *E. coli* were then carried out by Elie Wollman and Francois Jacob, working with Jacques Monod at the Pasteur Institute in Paris. The distinction between Hfr and F^+ was made by Lederberg, who also showed that F^+ behaved in an infectious manner. Lederberg coined the term *plasmid* in the 1950s, to describe such apparently extrachromosomal genetic elements, although the term did not find wide usage until the 1970s, when infectious drug resistance became a major medical problem.

Bacterial transduction was discovered by the American scientist Norton Zinder when he was working at the University of Wisconsin as a graduate student with Lederberg on genetic recombination in *Salmonella typhimurium.* The original motivation for this work was to show that conjugation occurred in an organism other than *E. coli,* and the techniques involved isolation of mutants and quantification of recombination by observing colony growth on minimal medium. However, although evidence of recombination was obtained, it could be shown that cell-to-cell contact was not required. Although this suggested a type of transformation, the process was not affected by DNase, and the gene transfer agent behaved like a bacteriophage. The gene transfer agent could be purified by the same procedures used to purify virus particles, and transduction occurred only with recipient cells that had receptor sites for the virus in question. Further, transducing activity could be eliminated by treatment of a lysate with substances able to adsorb the virus, such as sensitive cells or antibodies. Thus, in all cases, transducing activity and virus activity behaved in similar ways. Zinder and Lederberg coined the word "transduction" to refer to any genetic recombinational process that was only fragmentary and did not involve cell-to-cell contact, intending in this way to encompass processes involving either free DNA (transformation) or phage, but subsequently the word transduction has been applied only to virus-mediated genetic transfer.

The relationship between lysogeny and bacterial genetics was initially discerned by the French scientist Andre Lwoff, at the Pasteur Institute, who showed by careful microculture experiments that the ability to produce bacteriophage was a hereditary characteristic of bacteria. The discovery of transduction and conjugation soon resulted in the merging of many lines of investigation, since Esther Lederberg showed that *E. coli* K-12 carried a virus, called lambda, which was shown by Morse and Lederberg to transduce at high frequency the *gal* (galactose) genes, and Jacob and Wollman showed that, as a prophage, lambda behaved as a chromosomal element during conjugation. Jacob and Wollman soon recognized the analogy between lambda and the F factor, and with the physical evidence of the presence of extrachromosomal circular DNA molecules in cells, the concept of the plasmid became established. Work on plasmid biology, which only burgeoned in the 1970s, is too recent for historical treatment; however, bacterial genetics will probably be as exciting in the future as it has been in the past.*

Supplementary Readings

Campbell, Allan. 1981. Evolutionary significance of accessory DNA elements in bacteria. Annu. Rev. Microbiol. 35:55–83. A speculative review that considers the evolutionary significance of transposons, plasmids, viruses, and insertion sequences.

Davis, R. W., D. Botstein, and **J. R. Roth.** 1980. Advanced bacterial genetics: a manual for genetic engineering. Cold Spring Harbor Laboratory, Cold Spring Harbor, N. Y. 254 pp. An advanced laboratory manual describing the techniques and procedures for the genetic manipulation of bacteria.

Elwell, L. P., and **P. L. Shipley.** 1980. Plasmid-mediated factors associated with virulence of bacteria to animals. Annu. Rev. Microbiol. 34:465–496. Evidence that many virulence properties of bacteria are mediated by plasmids.

Ganesan, A. T., S. Chang, and **J. A. Hoch,** eds. 1982. Molecular cloning and gene regulation in bacilli. Academic Press, New York. 359 pp. Papers given at an international conference on the genetics of bacteria of the genus *Bacillus.*

Glass, Robert E. 1982. Gene function: *E. coli* and its heritable elements. University of California Press, Berkeley, Calif. 487 pp. The genetics of this well-studied organism.

Hopwood, D. A. 1981. Genetic studies with bacterial protoplasts. Annu. Rev. Microbiol. 35:237–272. Fusion of bacterial protoplasts is a suitable method for carrying out genetic recombination studies. This review describes the procedures and results.

Lewin, B. 1983. Genes. John Wiley, New York. 715 pp. Textbook emphasizing molecular aspects of genetics.

Roberts, G. P., and **W. J. Brill.** 1981. Genetics and regulation of nitrogen fixation. Annu. Rev. Microbiol. 35:207–235. Detailed review of the genetics of this important bacterial process.

Smith, H. O., D. B. Danner, and **R. A. Deich.** Genetic transformation. Annu. Rev. Biochem. 50:41–68. This review covers not only transformation in bacteria, but in eucaryotes. Emphasis is on how genetic material is integrated into the recipient chromosome.

*Key reference for this section: Dubos, R. 1976. The professor, the institute, and DNA. Rockefeller University Press, New York. An account of Oswald T. Avery's research career, culminating in the discovery that DNA was the genetic material. Also provides brief historical background on other aspects of bacterial genetics.

Gene Manipulation and Genetic Engineering

12

The concepts of molecular genetics, described in the three preceding chapters, have made possible the development of sophisticated procedures for the isolation, manipulation, and expression of genetic material, a field called **genetic engineering.*** Genetic engineering has applications in both basic and applied research. In basic research, genetic engineering techniques are used to study the mechanisms of gene replication and expression in procaryotes, eucaryotes, and their viruses. Some of the most important basic discoveries of molecular genetics were made using genetic engineering techniques. For applied research, genetic engineering permits the development of microbial cultures capable of producing valuable products, such as human insulin, human growth hormone, interferon, vaccines, and industrial enzymes. The potential of genetic engineering for commercial application seems limitless.

It is the purpose of this chapter to discuss the general principles of genetic engineering and to show how these principles can be applied. This chapter builds on the material discussed in Chapters 9 through 11, and it may be desirable from time to time to refer back to these chapters to recall details of the processes of molecular genetics.

One important part of genetic engineering is the use of procedures which permit modification of microbial genetic systems so that the replication of a gene or the synthesis of a gene product (that is, a protein) is under the control of the investigator (Figure 12.1).

Gene cloning is one aspect of genetic engineering. Gene cloning is the isolation of a desired gene from one organism, its incorporation into a suitable vector by use of enzymes, the introduction of this vector into an appropriate host organism, and the production of large amounts of the desired gene. One purpose of gene cloning is to obtain large amounts of a given gene, perhaps for chemical determination

*The assistance of Dr. Julian Davies of Biogen S.A. in providing material for this chapter is gratefully acknowledged.

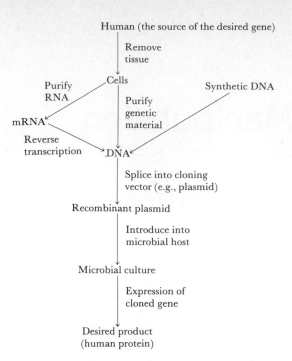

Human (the source of the desired gene)

Remove tissue

Cells

Purify RNA

Purify genetic material

Synthetic DNA

mRNA

Reverse transcription

DNA

Splice into cloning vector (e.g., plasmid)

Recombinant plasmid

Introduce into microbial host

Microbial culture

Expression of cloned gene

Desired product (human protein)

FIGURE 12.1 Various options for obtaining production of a human protein in a microbe.

of its base sequence, perhaps for use in another genetic engineering procedure. One common goal of gene cloning is to obtain a collection of bacterial strains, each of which contains a separate gene or cluster of genes from an organism of interest such as the human. Such a collection of strains, sometimes called a **clone library,** provides the investigator with material for study of gene organization. The term **recombinant DNA** is usually used to refer to molecules containing unrelated pieces of DNA. Recombinant DNA technology is at the basis of most studies in genetic engineering.

Frequently, one is interested in obtaining **expression** of a cloned gene. In gene expression, one is not interested in just producing large amounts of the DNA, but in producing large amounts of the product (for example, an enzyme) coded for by the cloned gene.

Uses of genetic engineering for commercial purposes Two quite distinct commercial uses of genetic engineering have developed: (1) increasing the production of existing microbial products, and (2) developing systems for the production in microbes of substances of animal or plant origin. The production of microbial substances can be improved by the use of genetic engineering techniques. For instance, increased yields of enzymes produced commercially can be obtained by engineering more effective regulation systems in the microorganisms used for the production of these enzymes. Increased yields of antibiotics used in the pharmaceutical industry can be brought about by gene manipulation. Although genetics has been used for years to increase gene copy number and hence to increase yields of desired products, the new techniques have greatly expanded the possibilities for improvement.

More dramatic have been the procedures discovered for making human product in microbes. For instance, it has been possible to introduce the gene for human insulin into the bacterium *Escherichia coli*, leading to the development of a commercial process for human insulin production. A whole new discipline, **biotechnology,** which deals with all phases of the production of useful products in microbial systems, is emerging. In the present chapter we consider the scientific principles underlying biotechnology.

Gene cloning is at the base of most genetic engineering procedures. Cloning involves four major steps: (1) *isolation and fragmentation* of the source DNA and incorporation of these fragments into a **cloning vector** using enzymes able to cut and rejoin DNA molecules (restriction endonuclease and ligase, respectively); (2) *incorporation* of the DNA (genetic transformation) into a recipient organism that is able to bring about replication of the cloning vector; (3) *detection* of transformed cells that contain the DNA, and isolation of a pure culture; and (4) *production* of large numbers of cells containing the cloned fragment, for isolation and study of the DNA of the clone.

Optionally, once obtained, the cloned DNA may itself be used for further gene cloning studies, perhaps in the development of new DNA constructs, perhaps for development of a more suitable system for *expression* of the gene product.*

A cloning system consists of two parts, a **cloning vector** which is able to incorporate the DNA and replicate it, and the **host** in which the vector will replicate.

Cloning vector   A cloning vector is a DNA molecule that can bring about the replication of foreign DNA fragments. All cloning vectors have the following properties in common: (1) they will replicate autonomously in a suitable host; (2) their DNA can be easily separated from the host DNA and purified; and (3) they contain regions of DNA that are not essential for their replication, which can thus be substituted with a fragment of source DNA. The ideal characteristics of the vector are: small DNA, able to enter the host and replicate to high copy number, stable in the host, able to express the inserted gene to high levels. The most useful cloning vectors are plasmids and viruses.

A number of cloning vectors have been developed, each of which is useful for different purposes. Among these cloning vectors are: plasmids, cosmids (phage-plasmid artificial hybrids), bacteriophage lambda, and bacteriophage M13. We discuss here plasmids and lambda as examples of two types of cloning vectors.

Plasmids   We have discussed plasmids in some detail in Section 11.4. Although plasmids in nature are generally transferred by cell-to-cell

---

*__Clone__ is frequently used to refer to an identical copy of a living organism. In genetic engineering, the word *clone* is being used in two different ways. A pure culture arising from a single cell is a clone of cells, but a foreign DNA molecule introduced into a culture is also called a clone. The DNA clone is something different than the culture clone, and the two uses of the word should not be confused.

contact, transfer in the laboratory can also be brought about by transformation procedures. Depending on the host-plasmid system, replication of the plasmid may be under tight cellular control, in which case only a few copies are made, or under relaxed cellular control, in which case a large number of copies are made. Achievement of large copy number is often important in gene cloning, and by proper selection of the host-plasmid system and manipulation of cellular macromolecule synthesis, plasmid copy numbers of several thousand per cell can be obtained.

An example of a suitable cloning plasmid is pBR322, which replicates in *Escherichia coli*. This plasmid, which was artificially constructed for use as a cloning vehicle, contains genes for ampicillin and tetracycline resistance and has a number of sites that can be attacked by specific restriction enzymes. Plasmid pBR322 is a relatively small plasmid, which makes it easier to handle (the DNA breaks less readily in the test tube) and leads to its replication to high copy number (large plasmids in general replicate only to low copy number). The complete base sequence of this 4362-nucleotide-long plasmid is known, and from this sequence and a knowledge of the specificity of restriction enzymes (see Table 9.1), a large number of restriction sites can be located. It is important that the plasmid has only a *single* recognition site for at least one restriction enzyme, so that treatment with that enzyme will open the plasmid but will not chop it into pieces. Plasmid pBR322 contains, among others, single restriction sites for the enzymes EcoRI, BamHI, and PstI (Figure 12.2*a*). The BamHI site is within the gene for tetracycline resistance, and the PstI site is within the gene for ampicillin resistance. If a piece of foreign DNA is inserted into one of these sites, the antibiotic resistance conferred by this site is lost, a phenomenon called **insertional inactivation** (Figure 12.2*b*). Insertional inactivation is used to detect the presence of foreign DNA within the plasmid. Thus, if pBR322 is digested with BamHI, linked with foreign DNA, and then transformed bacterial clones are isolated, those clones which are both ampicillin resistant and tetracycline resistant lack the foreign DNA (the plasmid incorporated into these cells represents DNA that had recyclized without picking up foreign DNA), whereas those cells still *resistant* to ampicillin but *sensitive* to tetracycline contain the plasmid with inserted foreign DNA. Since ampicillin resistance and tetracycline resistance can be determined independently on agar plates, isolation of bacteria containing the desired clones and elimination of cells not containing the plasmid can readily be accomplished.

Cloning in plasmids such as pBR322 is a versatile and fairly general procedure of wide use in genetic engineering. Plasmids are the best cloning vectors if the expression of the cloned gene is desired, as will be discussed in Section 12.2.

Bacteriophage lambda We have discussed temperate bacteriophages, and bacteriophage lambda in particular, in Section 10.6. Bacteriophage lambda is a useful cloning vector because its molecular genetics is well known, and because DNA can be efficiently packaged into phage particles which can be used to infect suitable host cells.

Lambda has a complex genetic map (Figure 10.19) and a large

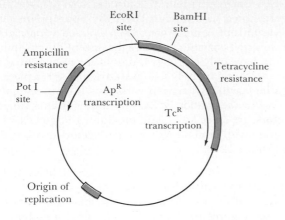

(a)

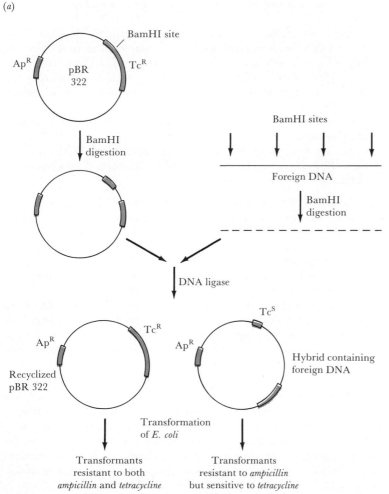

(b)

FIGURE 12.2 The use of a plasmid as a cloning vector. (a) Structure of plasmid pBR322, showing the origin of replication, antibiotic resistance genes, and locations of several restriction sites. (b) Formation of recombinant plasmids, showing how insertion of foreign DNA causes inactivation of tetracycline resistance, permitting easy isolation of transformants containing the cloned DNA fragment.

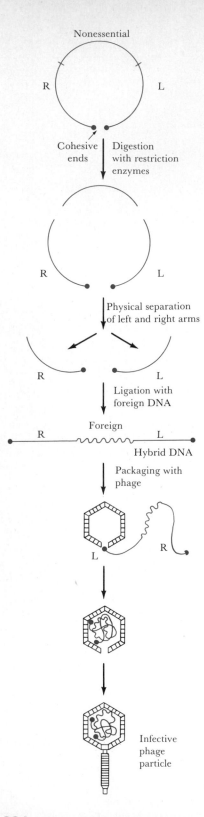

Nonessential

R   L

Cohesive | Digestion
ends | with restriction
         | enzymes

R   L

Physical separation
of left and right arms

R   L

Ligation with
foreign DNA

Foreign

R   L

Hybrid DNA

Packaging with
phage

L   R

Infective
phage
particle

number of gene functions. However, the central third of the lambda genome, between genes *J* and *N*, is not essential for lytic growth and can be substituted with foreign DNA. By appropriate genetic procedures, variants of lambda have been isolated which lack sites for certain restriction enzymes in the region essential for lambda reproduction and have only one or two sites in the nonessential region. In those variants that have only a single restriction site, a piece of foreign DNA can be *inserted,* whereas in variants with two sites, the foreign DNA can *replace* the lambda DNA. The latter variants, called **replacement vectors,** are especially useful for cloning large DNA fragments.

Cloning with lambda replacement vectors involves the following steps (Figure 12.3):

1. Isolation of the vector DNA from phage particles and digestion with the appropriate restriction enzyme.
2. Separation of the two fragments (the left- and right-hand fragments: see Figures 10.19 and 12.3) by centrifugation or electrophoresis.
3. Connection of the two lambda fragments to the foreign DNA using DNA ligase. Conditions are chosen so that molecules are formed of a length suitable for packaging into phage particles.
4. Packaging of the DNA by adding cell extracts containing the head and tail proteins and allowing the formation of viable phage particles.
5. Infection of *Escherichia coli* and isolation of phage clones by picking plaques on a host strain.
6. Checking recombinant phage for the presence of the desired foreign DNA sequence, by nucleic acid hybridization procedures or observation of genetic properties.

Selection of recombinants is less of a problem with lambda than with plasmids because (1) the efficiency of transfer of recombinant DNA into the cell by lambda is very high, and (2) lambda fragments which have not received new DNA are too small to be incorporated into phage particles.

Although lambda is a useful cloning vector, there are limits on how much DNA can be inserted. Viability of phage particles is low if the DNA is longer than 105 percent of normal lambda DNA, so that really large fragments (greater than 20,000 bases) cannot be efficiently cloned. Also, lambda is not as suitable as plasmids for use as an expression vector, since it replicates to high copy number only during the lytic cycle.

Hosts for cloning vectors The ideal characteristics of the host are: fast growing, capable of growth in cheap culture medium, not harmful or pathogenic, transformable by DNA, and stable in culture. The most useful hosts for cloning are microorganisms which grow well and for which a lot of genetic information is available, such as the bacteria *Escherichia coli* and *Bacillus subtilis* and the yeast *Saccharomyces cerevisiae.*

FIGURE 12.3 The use of bacteriophage lambda as a cloning vector. See text for details.

Although most molecular cloning has been done in *Escherichia coli,* there are disadvantages in using this host. *E. coli* presents dangers for large-scale production of products derived from cloned DNA, since it is found in the human intestinal tract and is potentially pathogenic. Also, even nonpathogenic strains produce endotoxins which could contaminate products, an especially bad situation with pharmaceutical injectibles. Finally, *E. coli* retains extracellular enzymes in the periplasmic space, making isolation and purification potentially difficult.

The Gram-positive organism *B. subtilis* can be used as a cloning vehicle. *B. subtilis* is not potentially pathogenic, does not produce endotoxin, and excretes proteins into the medium. Although the technology for cloning in *B. subtilis* is not nearly as well developed as for *E. coli,* plasmids suitable for cloning have been constructed. Transformation is a well-developed procedure in *B. subtilis,* and a number of bacteriophages that can be used as DNA vehicles are known. Some distinct disadvantages to *B. subtilis* as a cloning host exist, however. Plasmid instability is a real problem, and it is hard to maintain plasmid replication over many culture transfers. Also, foreign DNA is not well maintained in *B. subtilis* cells, so that the cloned DNA is often unexpectedly lost. Finally, few good regulation systems are known in *B. subtilis,* making its use as an expression vector (see later) difficult.

Cloning in *eucaryotic microorganisms* has some important uses, especially for understanding details of gene regulation in eucaryotic systems. The yeast *Saccharomyces cerevisiae* is the best known genetically, and is being extensively studied as a cloning host. Plasmids are known in yeast, and transformation using genetically engineered DNA can be accomplished. However, gene expression in eucaryotes is not well understood, making the development of high-yielding systems difficult. The ability to clone appropriate genetic material in yeast will advance our understanding of the complex transcription and translation systems of eucaryotes, and so should provide a better foundation for basic research.

For many purposes, gene cloning in *mammalian cells* would be desirable. Mammalian cell culture systems can be handled in some ways like microbial cultures, and find wide use in research on human genetics, cancer, infectious disease, and physiology. Vaccine production (see Chapter 16) is also done using mammalian cell cultures. The DNA virus SV40, a virus causing tumors in primates, has been developed as a cloning vector into human tissue culture lines. SV40 virus has double-stranded circular DNA and the entire nucleotide sequence is known. Derivatives of SV40 which do not induce cancer have been developed which permit cloning of mammalian genes, and expression of these genes has been obtained. SV40 or similar mammalian cloning vectors should prove very useful in understanding the events involved in gene expression in these complex organisms. One important advantage of mammalian cells as hosts for cloning vectors is that they already possess the complex RNA and protein processing systems that are involved in the production of gene products in higher organisms, so that these systems do not have to be engineered into the vector, as they need to be if production of the desired product is to be carried out in a procaryote. A disadvantage of mammalian cells as hosts is that they are expensive and difficult to produce under the large-scale conditions needed for a practical system.

   Plant-cell cultures are finding wide use in research in agriculture and plant physiology, and promise to find many practical uses in improvement of crop plants. One attractive feature of plant cell cultures is that under suitable conditions a single plant cell can be induced to reform a complete plant, from which seeds can be obtained. Thus the engineering of desirable traits into plant genes has great commercial potential. As an essential part of plant genetic engineering, gene cloning vectors are needed. One approach has been the use of plant DNA viruses. A number of plant viruses are known which contain DNA, and both single-stranded and double-stranded DNA viruses have been studied. Progress has been hampered by the poor knowledge of plant viruses and by the slow rate at which plant cell cultures grow. Another approach has been the use of the plant pathogenic bacterium *Agrobacterium tumefaciens*. This organism causes crown gall (Section 19.10), a disease resulting in the formation of tumors. The bacterium contains a large plasmid called the Ti plasmid that is responsible for virulence. Part of this plasmid becomes integrated into the plant DNA where its replication then comes under the control of the plant. The Ti plasmid can thus be used as a vehicle for introducing foreign genes into plant cells, and since it can be grown both in the bacterium and in the plant appears to have considerable promise as a cloning vector.

12.2

Expression Vectors

For practical developments it is essential that systems be available in which the cloned genes can be *expressed*. Organisms have complex regulatory systems, and many genes are not expressed all of the time. One of the major goals of genetic engineering is the development of vectors in which high levels of gene expression can occur. An **expression vector** is a vector which not only contains the desired gene but also contains the necessary regulatory sequences so that expression of the gene is kept under control of the genetic engineer. Some of the elements involved in gene expression are summarized in Figure 12.4.

Requirements of a good expression system Many factors influence the level of expression of a gene, and a vector must be constructed in which all these factors are under control. In addition, a host must be used in which the expression vector is most effective. We summarize the key requirements of a good expression system here.

1. *Number of copies* of the genes per cell. In general, more product is made if many copies of the gene are present. Vectors such as small plasmids (for example, pBR322) and temperate phage (for example, lambda) are valuable because they can replicate to large copy number.
2. *Strength of the transcriptional promoter.* The promoter region is the site at which binding of RNA polymerase first occurs (see Section 9.4), and native promoters in different genes vary considerably in RNA polymerase binding strength. For engineering a practical system,

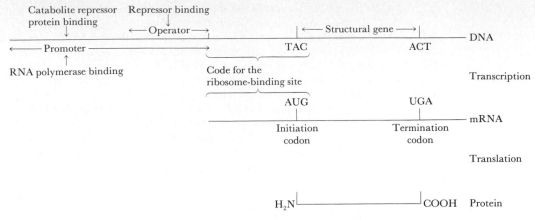

FIGURE 12.4 Factors affecting the expression of cloned genes in bacteria. Sequences and signals that must be appropriate for high levels of gene expression are indicated.

it is important to include a strong promoter in the expression vector. The DNA region around 10 and 35 nucleotides before the start of transcription (called the -10 and -35 *regions*) is especially important in the promoter. Many *E. coli* genes are controlled by relatively weak promoters, and eucaryotic promoters function poorly or not at all in *E. coli*. Strong *E. coli* promoters that have been used in the construction of expression vectors include *lac*uv 5 (which normally controls β-galactosidase), *trp* (which normally controls tryptophan synthetase), lambda P_L (which normally regulates lambda virus production), and *ompF* (which regulates production of outer membrane protein).

3. Presence of the bacterial *ribosome binding site*. The transcribed mRNA must bind firmly to the ribosome if translation is to begin, and an early part of the transcript contains the ribosome-binding site (Shine–Dalgarno sequence, Figure 12.4 and Section 9.5). Bacterial ribosome binding sites are different from eucaryotic ribosome binding sites and it is thus essential that the bacterial region be present in the cloned gene if high levels of gene expression are to be obtained. As part of the requirement for proper ribosome binding is the necessity for a proper distance between the ribosome binding site and the translation initiation codon. If these sites are too close or too far apart, the gene will be translated at low efficiency.

4. Avoidance of *catabolite repression* (see Section 9.6).

5. Proper *reading frame*. Because of the way the source DNA is being fused into the vector, three possible reading frames (see Section 9.7) could be obtained, only one of which is satisfactory. One approach to obtaining the correct reading frame is the use of three vectors, each having the restriction site into which new DNA will be inserted positioned such that the insert will be in a different reading frame. The gene is inserted into all three vectors and the one which gives proper expression is selected by testing.

6. *Codon usage.* There is more than one codon for most of the 20 amino acids (see Table 9.5), and some codons are used more frequently

than others. Codon usage is partly a function of the concentration of the appropriate tRNA in the cell. A codon frequently used in a mammalian cell may be used less frequently in the organism in which the gene is being cloned. Insertion of the appropriate codon would be difficult because it would have to be changed in all locations in the gene. One approach would be to engineer the host so that it used the required codon more frequently.

7. *Fate of the protein* after it is produced. Some proteins are susceptible to degradation by intracellular proteases and may be destroyed before they can be isolated. Excreted proteins must have the signal sequence attached (see Section 9.5) if they are to move through the cell membrane. Some eucaryotic proteins are toxic to the procaryotic host, and the host for the cloning vector may be killed before sufficient amount of the product is synthesized. Further engineering of either the host or the vector may be necessary to eliminate these problems.

The skill of the genetic engineer is thus essential in the construction of an appropriate vector which can be (1) efficiently incorporated into the proper host, (2) replicated to high copy number, (3) efficiently transcribed, and (4) efficiently translated. Many mammalian proteins are completely unexpressed when their genes are first cloned in *E. coli,* but with appropriate manipulation of the vector, expression can sometimes be achieved. The best example is the production on a commercial scale of human insulin in *E. coli,* as described in Section 12.4.

The role of regulatory switches in expression vectors   For large-scale production of a protein, it is very desirable that the synthesis of the protein be under the direct control of the experimenter. The ideal situation is to be able to grow up the culture containing the expression vector until a large population of cells is obtained, each containing a large copy number of the vector, and then turn on expression in all copies simultaneously by manipulation of a regulatory switch.

We have discussed regulatory controls of gene expression in Section 9.6. Recall the major importance of the repressor/operator system in regulating gene transcription. A strong repressor can completely block the synthesis of the proteins under its control by binding to the operator region. Repressor function can be turned off at the chosen time by adding an inducer, allowing the transcription of the proteins controlled by the operator.

For the repressor-operator system to work as a regulatory switch for the production of a foreign protein, it is desirable to retain in the expression vector a fragment of the structural gene and the operator controlled by the repressor, to which the source gene is fused. This permits proper arrangement of the sequence of genetic elements: promoter-operator-ribosome binding site-structural gene so that efficient transcription and translation can occur. The presence of a fragment of the normal protein can help render the foreign protein stable and capable of being excreted.

The construction of plasmid expression vectors containing the regulatory components of the *lac* operon provides one means of providing a suitable regulatory switch. As we have discussed in Section 9.6, the *lac* operon is switched on by inducers such as lactose or related β-galactosides. Phasing of cell growth and protein synthesis can thus be achieved by allowing growth to proceed in the absence of inducer until a suitable cell density is achieved, then adding inducer to bring about protein synthesis. Plasmids have been constructed containing the *lac* promoter, ribosome binding site, and operator. When the desired gene is inserted into such a system, expression can then be achieved by adding *lac* inducer. Production of high levels of mammalian proteins have been achieved in *E. coli* using such expression vectors (for instance: 10,000 to 15,000 molecules of beta globin per cell; 5000 to 10,000 molecules of human interferon per cell).

Using *lambda bacteriophage,* it is possible to integrate the engineered *lambda* genome into the *E. coli* chromosome, then induce replication of the lambda cloning vector by inactivating the lambda repressor. Under suitable conditions, replication of lambda will be followed by expression of the gene for the foreign protein. For large-scale production, a more suitable system for lambda induction is the use of a lambda repressor which is temperature sensitive. By raising the temperature of the culture to the proper value (usually 8 to 10°C higher than the growth temperature), the lambda repressor is inactivated and lambda replication begins. Once released from repression, other genes that have been inserted into the lambda genome will also be expressed.

Another regulatory system that has been used to construct expression vectors is the *tryptophan operon.* Although repressed by tryptophan, the *trp* operon can be induced by adding a tryptophan analog (such as β-indolacrylic acid) which brings about an apparent tryptophan deficiency.

12.3

The Cloning and Expression of Mammalian Genes in Bacteria

We have now laid out the principles behind the development of systems for obtaining the production of foreign genes in an organism such as *Escherichia coli.* How are these principles put together in the engineering of a desired expression system? In the human or some other complex organism, the sought-for gene will be buried within a large mass of DNA containing unwanted genes, as well as repetitive DNA sequences that have no coding function at all (see Section 9.3). Also, most eucaryotic genes are split, with noncoding introns interspersed among the coding exons. How can a gene be selected from this complex mixture? Although a number of approaches are available, this is the least well developed stage of genetic engineering. Not only skill, but also intuition and good luck play big roles in a successful outcome. A summary of approaches and procedures is given in Figure 12.5.

We start first with a consideration of the desired product and where it is produced in the human body. A human being is a highly differentiated organism, and many genes are expressed only in certain organs or tissues. The hormone *insulin,* for instance, is produced only in the pancreas. Of course, the insulin gene is found in all tissues and organs, but it is only expressed in this one organ.

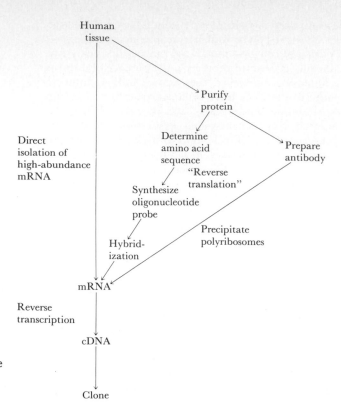

Human
tissue

Purify
protein

Direct
isolation of
high-abundance
mRNA

Determine
amino acid
sequence

Prepare
antibody

"Reverse
translation"

Synthesize
oligonucleotide
probe

Precipitate
polyribosomes

Hybrid-
ization

mRNA

Reverse
transcription

cDNA

Clone

FIGURE 12.5 Several routes to the isolation of mammalian genes for cloning in microorganisms.

Reaching the gene via messenger RNA One approach is to get to the gene through its mRNA. A major advantage of using mRNA is that the noncoding information present in the DNA (introns) has been removed (see Section 9.4). The isolated mRNA is used to make complementary DNA (cDNA), by means of reverse transcription (see Section 10.3). It is likely that a tissue expressing the gene will contain large amounts of the desired mRNA, although except in rare cases this will certainly not be the only mRNA produced. In a fortunate situation, where a single mRNA dominates a tissue type, extraction of mRNA from that tissue provides a useful starting point for gene cloning.

In a typical mammalian cell, about 80 to 85 percent of the RNA is ribosomal, 10 to 15 percent is transfer RNA and other low-molecular-weight RNAs, and 1 to 5 percent is messenger RNA. Although low in abundance, the mRNA in a eucaryote is identifiable because of the poly A tails found at the 3′ end (see Section 9.4). In maturing red blood cells, for instance, where virtually the only protein made is the globin portion of hemoglobin, from 50 to 90 percent of the poly A-containing cytoplasmic RNA consists of globin mRNA. By passing a poly A-rich RNA extract over a chromatographic column containing poly T fragments (linked to a cellulose support), most of the mRNA of the cell can be separated from the other cellular RNA by the specific pairing of A and T bases. Elution of the RNA from the column then gives a preparation greatly enriched in mRNA.

Synthesis of complementary DNA (cDNA) from mRNA   Once the RNA message has been isolated, it is necessary to convert the information into DNA. This is accomplished by use of the enzyme **reverse transcriptase,** which we have discussed earlier (Section 10.3). This remarkable enzyme, an essential component of retrovirus replication, copies information from RNA into DNA (Figure 12.6).

With appropriate enzymatic procedures, a single-stranded tail is attached to the cDNA which is complementary to a similar fragment in plasmid or other cloning vector. The plasmid DNA and cDNA are then linked by enzymatic ligation to form a linear double-stranded molecule which can be used to transform *E. coli.* Antibiotic-resistance genes on the plasmid are used to select for transformants, which are then tested to see if they contain the desired gene. Proof that the desired DNA has been cloned can be obtained by determining its nucleotide sequence. The code as revealed in the nucleotide sequence should correspond to the amino acid sequence of the protein, as determined from the genetic code (Table 9.5).

Hybridization of nucleic acids   For most of the rest of the procedures described, techniques for artificial hybridization of nucleic acids are used. By **hybridization** is meant the artificial construction of a double-stranded nucleic acid by complementary base pairing of two single-stranded nucleic acids. We discussed the formation of DNA : DNA hybrids by means of heating and slow cooling in Section 9.2. DNA : RNA hybrids can also be made. There must be a high degree of complementarity between two single-stranded nucleic acid molecules if they are to form a stable hybrid. In the most common use of hybridization in genetic engineering, one of the molecules is radioactive and formation of hybrids is detected by observing the formation of double-stranded molecules containing some of the radioactivity.

A common variation of this procedure is the use of hybridization to detect specific DNA or RNA fragments that have been separated by means of molecular size using the process of gel electrophoresis. **Gel electrophoresis** involves the use of a slab of agarose or polyacrylamide gel to which the nucleic acid preparation is affixed, and the movement of the molecules along the gel by means of an electric field. DNA fragments up to 50,000 base pairs can be separated by gel electrophoresis, provided that the fragments differ in length by at least 1 percent. Visualization of DNA fragments on gels can either be done by staining with ethidium bromide, a fluorescent compound that binds strongly to nucleic acids, or by autoradiography on X-ray films, if the fragments are radioactive (Figure 12.7).

One of the most common uses of hybridization is to detect DNA sequences that are complementary to mRNA molecules. Detection of DNA : RNA hybridization is usually done with membrane filters constructed of cellulose nitrate. RNA does not stick to these filters, but single-stranded DNA and DNA : RNA hybrids do. The single-stranded DNA probe is first immobilized on the filter and the radioactive RNA added. After appropriate incubation, the unhybridized RNA is washed out and the radioactivity still bound to the filter is measured.

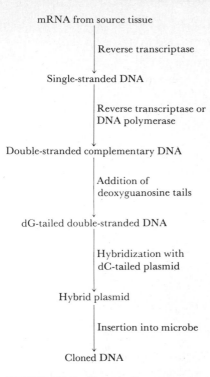

mRNA from source tissue

Reverse transcriptase

Single-stranded DNA

Reverse transcriptase or DNA polymerase

Double-stranded complementary DNA

Addition of deoxyguanosine tails

dG-tailed double-stranded DNA

Hybridization with dC-tailed plasmid

Hybrid plasmid

Insertion into microbe

Cloned DNA

FIGURE 12.6   Steps in the synthesis of complementary DNA (cDNA) from an isolated mRNA, using the enzyme reverse transcriptase.

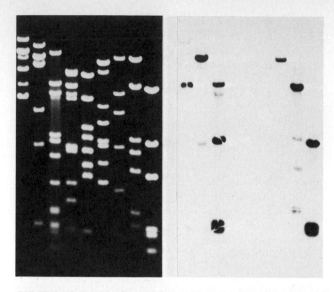

FIGURE 12.7 Gel electrophoresis and Southern blotting of fragments of DNA. Left panel: agarose gel electrophoresis of DNA molecules of a bacteriophage treated with restriction enzymes. The positions of the DNA fragments are revealed by the fluorescence of ethidium bromide treated gels. The various lines contain DNA fragments treated with different restriction enzymes. Right panel: Southern blotting is used to reveal fragments reacting with radioactively labeled RNA molecules. The labeled messenger RNA molecules were reacted with DNA that had been transferred from the gel to a cellulose nitrate filter by means of the Southern blotting technique. Note that only some of the DNA fragments contain sequences that react with the mRNAs. (Courtesy of Carl Marrs and Martha Howe.)

A widely used modification of this filter hybridization procedure is the *Southern transfer* or **Southern blot procedure** (named for the scientist E. M. Southern, who first developed it). In this procedure, DNA molecules are first separated by gel electrophoresis (see above), then transferred by blotting to a sheet of cellulose nitrate membrane filter. The pattern of DNA fragments on the gel is thus preserved on the membrane filter. After hybridization to radioactive RNA, autoradiography is done to detect which DNA fragments have hybridized with the RNA (Figure 12.7). This procedure permits the detection of DNA sequences that are complementary with mRNA molecules. This procedure is most commonly used after the DNA has been digested with restriction enzymes, thus permitting localization of genes or partial gene sequences on the DNA sequence.

Cloning of low-abundance messenger RNA   Frequently, the sought-for messenger RNA is present in a mixture with many other mRNAs. Hybridization procedures are then used to locate the mRNA (Figure 12.5).

One approach to cloning low-abundance mRNA is to make a **synthetic DNA** which is complementary to part of the mRNA and use this DNA as a *probe* to pull out by hybridization the mRNA of interest, using the Southern blot procedure. This requires that a partial or complete sequence of the protein be known. Then, from the genetic code, the nucleotide sequence of a section of the DNA is deduced, and this piece of DNA is synthesized. The best section of the DNA to synthesize is one which corresponds to a part of the protein rich in amino acids specified by only a single codon (methionine, AUG; tryptophan, UGG) or by two codons (phenylalanine: UUU, UUC; tyrosine: UAU, UAC; histidine: CAU, CAC) since this will increase the chances that the synthesized DNA will be complementary to the mRNA of interest. As shown in Table 9.5, several amino acids are coded for by four or more codons, only one of which would actually be used at the site in the mRNA of interest. Once the DNA probe has been made, it is fixed to an insoluble substrate and hybridized with the purified mixed RNA isolated from the tissue. The mRNA hybridized by this procedure is then transcribed into cDNA and cloned in the manner described earlier in this section.

Another approach involves separating from the tissue **ribosome complexes** that are in the process of synthesizing the desired protein, using an antibody specific for this protein. As we describe in Chapter 16, antibodies are proteins made in response to foreign substances which are able to combine with and specifically precipitate them. To use an antibody in this procedure, a sample of the protein of interest is first purified and injected into an animal to produce a specific antibody. In the ribosome complex, polypeptides of the protein of interest are still complexed with the protein-synthesizing machinery (which contains, among other things, the sought-for mRNA). When the antibody precipitates the protein in the ribosome complex, the mRNA also is precipitated. After isolation of the mRNA, it can be used to prepare cDNA as described earlier in this section.

Synthesis of the DNA   The direct approach to cloning a mammalian gene is to synthesize the DNA chemically and incorporate it into a suitable cloning vector. The techniques for synthesis of DNA molecules are now well developed, and it is possible to synthesize genes coding for proteins 100 to 200 amino acid residues in length (300 to 600 nucleotides). The approach here is one called **reverse translation** (see Figure 12.5). From the amino acid sequence of the desired protein, the appropriate genetic code is deduced, and the appropriate polynucleotide sequence is synthesized. (Note that *reverse translation* is not a cellular process but a mental exercise of the genetic engineer.) For use in cloning, the synthesized polynucleotide should contain suitable restriction enzyme sites at each end, so that ligation to a cloning vector can be accomplished. Also, for amino acids with multiple codons, the codons selected should be those used most frequently by the host. The use of the synthetic approach for production of human insulin by bacteria is described in Section 12.4.

We thus see that a number of procedures are available for obtaining a desired mRNA and producing cDNA suitable for cloning. By use of these procedures, a large number of mammalian genes have been cloned in *E. coli*, either for research purposes or for use in large-scale manufacturing processes. Once the desired gene is cloned, it is necessary to engineer it further to produce an expression vector, using one of the approaches described in Section 12.2.

Expression of mammalian genes in bacteria For the cloned mammalian gene to be expressed in a bacterium, it is essential that the mammalian gene be inserted adjacent to a strong *promoter* and that a bacterial *ribosome-binding site* be present. Also, it is essential that the *reading frame* be correct (see Section 12.2).

One method of providing a ribosome-binding site in the proper reading frame is to arrange for the mammalian DNA sequence to be expressed as part of a **fusion protein** which contains a short procaryotic sequence at the amino end and the desired eucaryotic sequence at the carboxyl end. Although less desirable for some purposes, fusion proteins are often more stable in bacteria than unmodified eucaryotic proteins, and the bacterial peptide portion can often be removed by chemical treatment after purification of the fused protein. One advantage of making a fusion protein is that the bacterial portion can contain the bacterial sequence coding for the signal peptide that enables transport of the protein across the cell membrane (see Section 9.5), making possible the development of a bacterial system which not only synthesizes the mammalian protein but actually excretes it. However, the main advantage of making a fusion protein is that the reading frame is ensured to be correct.

To obtain an expression vector that will bring about the synthesis of an **unfused protein,** it is essential that the eucaryotic gene be fused exactly to the initiation codon that is just downstream from the ribosome-binding site. In many mammalian proteins, the mature and active form lacks some of the amino acids present at the initiation site because they are cleaved after synthesis by a post-translational modification. Thus, to obtain synthesis of the active form of the protein directly, it is necessary to fuse to the initiation codon just the sequence coding for the final protein, and place this in proper position downstream of the ribosome-binding sequence. If high efficiency of translation is to be obtained, an intervening region of noncoding DNA of the proper length must be placed between the ribosome-binding sequence and the initiation codon. Several plasmids have been constructed which contain built-in promoters and suitable restriction enzyme sites, so that the proper tailoring of coding sequence to initiation site and ribosome-binding site can be achieved. Human proteins that have been expressed at high yield under the control of bacterial regulatory systems include human growth hormone, insulin, virus antigens, interferon, and somatostatin (a hypothalamic hormone).

12.4

Practical Results

A number of commercial products have been developed using genetic engineering techniques. Three main areas of interest for commercial development are:

1. *Microbial fermentations* A number of important products are made industrially using microorganisms, of which the antibiotics are the most significant. Genetic engineering procedures can be used to

manipulate the antibiotic-producing organism in order to obtain increased yields.

2. *Virus vaccines*   A vaccine is a material which can induce immunity to an infectious agent (see Chapter 16). Frequently, killed virus preparations are used as vaccines and there is always a potential danger to the patient if the virus has not been completely inactivated. Since the active ingredient in the killed virus vaccine is the protein coat, it would be desirable to produce the protein coat separately from the rest of the virus particle. By genetic engineering, viral coal protein genes can be cloned and expressed in bacteria, making possible the development of safe, convenient vaccines.

3. *Mammalian proteins*   A number of mammalian proteins are of great medical and commercial interest. Table 12.1 provides a list of some mammalian genes which have been expressed in bacteria, and their commercial or medical significance. In the case of human proteins, commercial production by direct isolation from tissues or fluids is complicated and expensive, or even impossible. By cloning the gene for a human protein in bacteria, its commercial production is possible.

In the rest of this section we discuss two particular applications of genetic engineering, the production of a vaccine for a virus, and the production of human insulin.

Virus vaccine   As we discussed in Chapter 10, a virus particle consists of a nucleic acid core surrounded by a coat containing one or more specific proteins. For vaccine purposes, it is the protein coat which is of interest and the gene for the virus protein is cloned (Figure 12.8).

Virus particles are generally produced by growing the virus in tissue culture. From a purified virus preparation, the nucleic acid of the virus is isolated. If a DNA virus is under study, then direct cloning of the virus DNA can be undertaken. With an RNA virus, a complementary DNA must first be made, as described in Section 12.2. The isolated nucleic acid is then fragmented with restriction enzymes and the fragments inserted into a suitable cloning vector, generally a plasmid, using DNA ligase. It is essential to find a restriction enzyme which does not cut within the gene of interest. Provision for proper reading frame and ribosome-binding site must be arranged, as outlined in Figure 12.4. The hybrid plasmids are then

TABLE 12.1   Some Mammalian Genes Expressed in Bacteria

Protein	Function
Interferon	Antiviral agent, anticancer (?)
Insulin	Treatment of diabetes
Serum albumin	Transfusion applications
Growth hormone	Growth defects
Urokinase	Blood-clotting disorders
Parathyroid hormone	Calcium regulation
Human viruses (Hepatitis B, cytomegalovirus, influenza)	Vaccines
Animal viruses (foot-and-mouth disease)	Vaccines

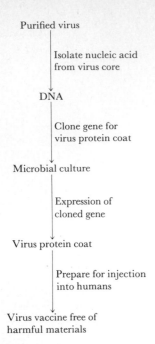

Purified virus

↓ Isolate nucleic acid
from virus core

DNA

↓ Clone gene for
virus protein coat

Microbial culture

↓ Expression of
cloned gene

Virus protein coat

↓ Prepare for injection
into humans

Virus vaccine free of
harmful materials

FIGURE 12.8 Steps in the preparation of a virus vaccine by genetic engineering.

inserted into bacteria by transformation and an antibiotic resistance character is used to select for plasmid-containing colonies. A large number of colonies are isolated and each one is tested for the production of the virus antigen. Once a clone of the virus protein gene has been obtained, further manipulation can be used to increase yield of the protein, and make it easier to purify. The ultimate goal is the development of a system which produces the antigenic virus protein in high yield. The purified protein is then used as a vaccine in humans without the dangers attendant on using killed virus particles.

Human insulin   One of the most dramatic examples of the value of genetic engineering is the production of human insulin. Insulin is a protein produced in the pancreas that is vital for the regulation of carbohydrate metabolism in the body. Diabetes, a disease characterized by insulin deficiency, afflicts millions of people. The standard treatment for diabetes is periodic injections of insulin, and because insulins of most mammals are similar in structure, it is possible to treat human diabetes by use of insulin isolated commercially from beef or pork pancreas. However, nonhuman insulin is not as effective as human insulin, and the isolation process is expensive and complex. Cloning of the human insulin gene in bacteria has hence been carried out.

Insulin in its active form consists of two polypeptides (A and B) connected by disulfide bridges (Figure 12.9a). These two polypeptides are coded by separate parts of a single insulin gene. The insulin gene codes for **preproinsulin,** a longer polypeptide containing a signal sequence (involved in excretion of the protein), the A and B polypeptides of the active insulin molecule, and a connecting polypeptide that is absent from mature insulin. **Proinsulin** (Figure 12.9a) is formed from preproinsulin and the conversion of proinsulin to insulin involves the enzymatic cleavage of the connecting polypeptide from the A and B chains.

Two approaches have been used to obtain production of human insulin in bacteria: (1) production of proinsulin and conversion to insulin by chemical cleavage, and (2) production in two separate bacterial cultures of the A and B chains, and joining of the two chains chemically to produce insulin. Because the insulin protein is fairly small, it was more convenient with either approach to synthesize the proper DNA sequence chemically rather than to attempt to isolate the insulin gene from human tissue. There are 63 bases coding for the A chain and 90 bases coding for the B chain (Figure 12.9b). In proinsulin there are an additional 105 bases for the peptide which connects the A and B chains. When the polynucleotides were synthesized, suitable restriction enzyme sites were placed at each end so that the polynucleotides could be ligated to plasmid pBR322. To obtain effective expression, the synthesized genes were fused to suitable *Escherichia coli* promoters, either *lac* or *trp,* in a manner such that a portion of the β-galactosidase or tryptophan synthetase protein was synthesized. An important advantage of making the fused protein is that the fusion product is much more stable in *E. coli* than insulin itself. The *trp* fusion in particular results in the formation of an insoluble protein that precipitates inside *E. coli,* thus preventing

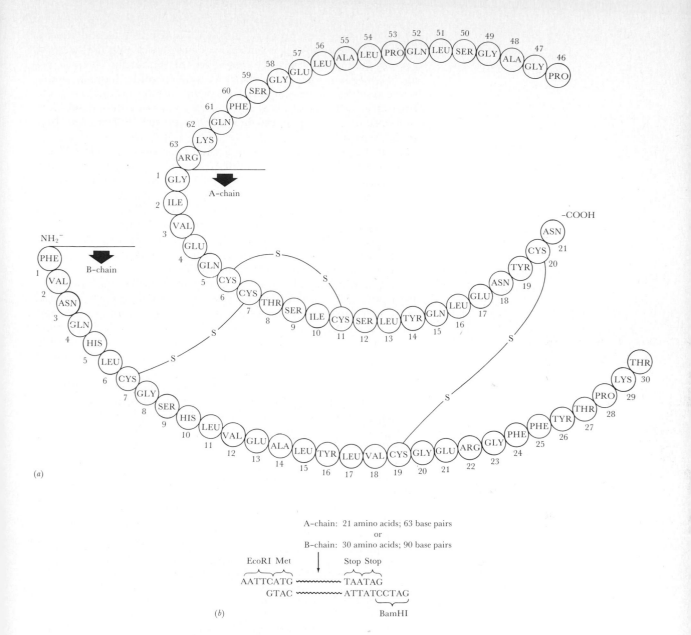

(a)

(b)

A–chain: 21 amino acids; 63 base pairs

or

B–chain: 30 amino acids; 90 base pairs

EcoRI Met   Stop Stop

AATTCATG ~~~~~ TAATAG
GTAC ~~~~~ ATTATCCTAG

BamHI

FIGURE 12.9 Genetic engineering for the production of human insulin in bacteria. (a) Structure of human proinsulin. (Based on Chance et al., 1981; see references) (b) Chemical synthesis of the insulin gene and suitable linkers, permitting cloning and expression. The synthesized fragments were linked via restriction sites EcoRI and BamHI to plasmid pBR322. The methionine coding sequence was inserted to permit chemical cleavage of the A and B chains from the fused protein made in the bacteria, since the reagent cyanogen bromide specifically cleaves at methionine residues and insulin does not contain methionine. Two stop signals were incorporated at the downstream (carboxyl) end of the coding sequence.

E. coli proteases from breaking it down. Finally, a nucleotide triplet coding for methionine was placed at the region joining the *trp* or *lac* gene to the insulin gene. The reason for this is that the chemical reagent *cyanogen bromide* specifically cleaves polypeptide chains at methionine residues, permitting recovery of the insulin product once the fused protein has been isolated from the bacteria. Insulin itself does not contain methionine and hence is unaffected by cyanogen bromide treatment.

If the proinsulin route is used, then the proinsulin isolated from the bacteria via cyanogen bromide treatment is converted to insulin by disulfide bond formation, followed by enzymatic removal of the connecting peptide of proinsulin. Proinsulin naturally folds so that the cysteine residues are opposite each other (Figure 12.9*a*), and chemical treatment then causes the formation of disulfide cross links. Once this has been accomplished, the connecting peptide can be removed by treatment with the proteases trypsin and carboxypeptidase B, which have no effect on insulin itself.

If insulin is produced by way of the separate A and B peptides, then each of the fusion proteins is isolated from a separate bacterial culture and the chains then released by cyanogen bromide cleavage. The cleaved chains are then connected by use of chemical treatment that results in disulfide bond formation. Under appropriate conditions, a yield of at least 60 percent of the theoretical yield can be obtained.

A summary of the procedures used is given in Figure 12.10. The final product, biosynthetic human insulin, is identical in all respects to insulin purified from the human pancreas, and can be marketed commercially. The principles discovered in the development of the insulin process should find wide use in genetic engineering.

FIGURE 12.10 Overall steps in engineering the production of insulin in bacteria.

1. *Synthesize* the DNA sequence coding for A and B chains, with suitable restriction enzyme sites. (The DNA sequences needed are determined by "reverse translation" of the amino acid sequences of the insulin molecule.)
2. *Ligate* each DNA sequence separately to a suitable expression vector. (Plasmid pBR322 containing the *trp* promoter and the bacterial ribosome-binding site.)
3. *Clone* each expression vector in a separate strain of *Escherichia coli.*
4. *Grow* the two bacterial cultures in large-scale equipment and *produce* large amounts of A and B chains, each fused to a *trp* protein.
5. *Purify* the fused A and B proteins.
6. *Cleave* the A and B proteins from the fused bacterial peptides by use of cyanogen bromide.
7. *Cross-link* the A and B peptides to make the final product: human insulin.

12.5

Principles at the Basis of Genetic Engineering

We have presented the fundamentals of genetic engineering and have shown how the approaches used have been derived from an understanding of basic concepts of molecular genetics. We now summarize the principles of genetic engineering by relating current knowledge back to the basic information presented in Chapters 9 through 11.

The following developments were essential for the perfection of genetic engineering:

1. *DNA chemistry:* development of procedures for isolation, sequencing, and synthesis of DNA.
2. *DNA enzymology:* discovery of restriction endonucleases, DNA ligases, and DNA polymerases.
3. *DNA replication:* understanding how DNA replication occurs, and the importance of DNA vectors capable of independent replication.
4. *Plasmids:* discovery of plasmids, and determination of the mechanisms by which plasmids replicate.
5. *Temperate bacteriophage:* understanding how replication and/or integration is controlled in the DNA of temperate bacteriophages.
6. *Transformation:* discovery of methods for getting free DNA into cells.
7. *RNA chemistry and enzymology:* understanding how to work with messenger RNA, how eucaryotic mRNA is constructed, and the importance of RNA processing in the formation of mature eucaryotic mRNA.
8. *Reverse transcription:* the discovery of the enzyme *reverse transcriptase* in retroviruses and its development as a means for transcribing information from mRNA back into DNA.
9. *Regulation:* understanding the factors involved in the regulation of transcription, including the discovery of promoter sites and operon control.
10. *Translation:* understanding the steps involved in translation, the importance of ribosome-binding sites on the mRNA, the role of the initiation codon, and the importance of a proper reading frame.
11. *Protein chemistry:* development of methods for isolation, purification, assay, and sequencing of proteins.
12. *Protein excretion and post-translational modification:* understanding how proteins are built with signal sequences that are removed during or after excretion. Discovery of other kinds of post-translational modification of proteins, such as the removal of polypeptides at the initiation end of the protein.
13. *The genetic code:* the discovery of the genetic code and the determination that it was the same in all organisms. The understanding of the importance of proper reading frame, and that certain codons were less frequently used in some organisms than in others.

Despite the tremendous advances in knowledge, and the high sophistication of molecular genetics, genetic engineering is as much an art as a science. The best way of achieving a practical system of gene expression is not always clear at the beginning of a study, and many possible approaches may be tried. Here are the key steps in the development of any process: (1) isolate or create (synthesize) the DNA of the desired gene; (2) link up this DNA to an expression vector; (3) get the linked DNA into a cell, and isolate clones; (4) get the gene expressed in the cloning system; and (5) isolate and purify the product.

12.6

Summary

We have seen in this chapter that techniques are now available for the manipulation of genetic material in highly sophisticated ways. DNA can be readily isolated, its base sequence determined, and fragments chemically synthesized. A variety of enzymes are available which permit cutting DNA molecules into small fragments, connecting fragments together, and creating DNA molecules having desired properties. **Gene cloning** is the process by which a gene from one organism is isolated, incorporated into a suitable vector, and introduced into a second organism where it can replicate and be produced in large amounts. Clones provide a plentiful supply of DNA containing the code for a single protein. A variety of **cloning vectors** are available that can be used to clone DNA molecules in microbial cell populations.

Under appropriate conditions, not only replication but also **expression** of the gene can be obtained. An **expression vector** is a DNA molecule, usually derived from a plasmid or temperate bacteriophage, which is able to replicate to high copy number in a suitable host and contains regulatory signals for efficient transcription and translation of the incorporated gene.

Obtaining suitable expression requires that all of the necessary regulatory signals of the procaryotic cell are present on the cloned fragment. A strong **promoter** is desirable, so that transcription will occur efficiently, and the cloned fragment must also contain the **ribosome-binding site** of the procaryote. The position of the **initiation codon** of the cloned fragment is also critical, since this will influence the **reading frame** of the translation system.

When obtaining the production of a eucaryotic gene in a procaryote, a number of problems arise. Because of the presence of introns in many eucaryotic genes, it is not possible to clone eucaryotic genes directly in procaryotes. Instead, messenger RNA of eucaryotes must be isolated and the information transcribed back into DNA by use of the enzyme **reverse transcriptase.** This provides a DNA molecule containing the desired gene but lacking the introns. Despite the universality of the genetic code, certain codons are used more frequently in some organisms than in others, and poor translation of a cloned gene may arise because of differences in codon usage.

A number of eucaryotic products have been successfully produced in bacteria. Human genes that have been expressed in bacteria include those for interferon, insulin, serum albumin, growth hormone, urokinase, globin, and parathyroid hormone. The proteins of a number of human and animal viruses have also been produced in bacteria, opening up the possibility of developing new and highly efficient vaccines for major virus diseases. The first human protein to be produced commercially in bacteria was insulin, an important substance for the treatment of diabetes.

Supplementary Readings

Anderson, W. F., and **E. G. Diacumakos.** 1981. Genetic engineering in mammalian cells. Sci. Am. 245(July):106–121. Discusses some of the approaches and promises of genetic engineering of mammalian cells, with emphasis on how this might be used to correct hereditary deficiencies.

Barton, K. A., and **W. J. Brill.** 1983. Prospects in plant genetic engineering. Science 219:671–676. Brief review of the possible uses of *Agrobacterium* and viruses in the genetic engineering of plants.

Chance, R. E., J. A. Hoffman, E. P. Kroeff, M. G. Johnson, E. W. Schirmer, and **W. W. Bromer.** 1981. The production of human insulin using recombinant DNA technology and a new chain combination procedure. American Peptide Symposium, Pierce Chemical Company, Rockford, Ill. pp. 721–728. Brief review of the procedures used for connecting the A and B peptides of insulin after separate production in bacteria.

Crea, R., A. Kraszewski, T. Hirose, and **K. Itakura.** 1978. Chemical synthesis of genes for human insulin. Proc. Natl. Acad. Sci. USA 75:5765–5769. Research paper describing the first synthesis of the polynucleotides coding for the A and B chains of insulin.

Davis, R. W., D. Botstein, and **J. R. Roth.** 1980. Advanced bacterial genetics: a manual for genetic engineering. Cold Spring Harbor Laboratory, Cold Spring Harbor, N. Y. 254 pp. Detailed descriptions of techniques and methods for bacterial genetics.

Goeddel, D. V., D. G. Kleid, F. Bolivar, H. L. Heyneker, D. G. Yansura, R. Crea, T. Hirose, A. Kraszewski, K. Itakura, and **A. D. Riggs.** 1979. Expression in *Escherichia coli* of chemically synthesized genes for human insulin. Proc. Natl. Acad. Sci. USA 76:106–110. First paper describing the techniques for obtaining expression of the human insulin gene in bacteria, via synthesis of a fusion protein.

Harris, T. J. R. 1983. Expression of eukaryotic genes in *E. coli.* Pages 128–155 *in* R. Williamson, ed. Genetic engineering, vol. 4. Academic Press, New York. Detailed review of the procedures that have been used for obtaining expression of eucaryotic genes in bacteria.

Hitzeman, R. A., D. W. Leung, L. J. Perry, W. J. Kohr, H. L. Levine, and **D. V. Goeddel.** 1983. Secretion of human interferons by yeast. Science 219:620–625. Procedures for cloning and expression of plasmids in yeast.

Johnson, I. S. 1983. Human insulin from recombinant DNA technology. Science 219:632–637. A brief discussion of legal and regulatory problems related to introduction of recombinant DNA technology for production of human insulin.

Kleid, D. G., D. Yansura, B. Small, D. Dowbenko, D. M. Moore, M. J. Grubman, P. D. McKercher, D. O. Morgan, B. H. Robertson, and **H. L. Bachrach.** 1981. Cloned viral protein vaccine for foot-and-mouth disease: responses in cattle and swine. Science 214: 1125–1129. The first report of a virus vaccine produced by recombinant DNA technology.

Maniatis, T., E. F. Fritsch, and **J. Sambrook.** 1982. Molecular cloning: a laboratory manual. Cold Spring Harbor Laboratory, Cold Spring Harbor, N. Y. 545 pp. A detailed manual describing many procedures for cloning and obtaining expression of cloned genes. Brief discussions of the principles behind each procedure.

Old, R. W., and **S. P. Primrose.** 1981. Principles of gene manipulation: an introduction to genetic engineering, 2nd ed. University of California Press, Berkeley, Calif. 214 pp. Brief textbook on genetic engineering procedures.

Ream, L. W., and **M. P. Gordon.** 1982. Crown gall disease and prospects for genetic manipulation of plants. Science 218:854–859. Brief review of the potentialities of using *Agrobacterium tumefaciens* as a vector for carrying genes into plant cells.

Schleif, R. F., and **P. C. Wensink.** 1981. Practical methods in molecular biology. Springer-Verlag, New York, 220 pp. Gives detailed procedures for doing many of the chemical and enzymatic techniques used in genetic engineering.

Vournakis, J. N., and **R. P. Elander.** 1983. Genetic manipulation of antibiotic-producing microorganisms. Science 219:703–709. Review of conventional genetic and genetic engineering procedures for increasing the yield of commercially valuable antibiotics.

Yelverton, E., S. Norton, J. F. Obijeski, and **D. V. Goeddel.** 1983. Rabies virus glycoprotein analogs: biosynthesis in *Escherichia coli.* Science 219:614–620. A genetic engineering approach to the production of rabies virus vaccine in bacteria.

Microbial Activities in Nature

In this chapter we consider some of the activities of microorganisms in the world at large, especially in soil and in aquatic habitats. Microorganisms play far more important roles in nature than their small sizes would suggest. To see properly the place of microorganisms in nature, we must first consider the concept of an **ecosystem.** Ecologists define an ecosystem as a total community of organisms, together with the physical and chemical environment in which they live. Each organism interacts with its physical and chemical environment and with the other organisms in the system, so that the ecosystem can be viewed as a kind of superorganism with the ability to respond to and modify its environment. A good example of an ecosystem is a lake. The sides of the lake define the boundaries of the ecosystem. Within these boundaries organisms live and carry out their activities, greatly modifying the characteristics of the lake as well as each other.

Energy enters an ecosystem mainly in the form of sunlight and is used by photosynthetic organisms in the synthesis of organic matter. Some of the energy contained in this organic matter is dissipated by the photosynthesizers themselves during respiration, and the rest is available to herbivores, which are animals that consume the photosynthesizers. Of the energy entering the herbivores, one portion is dissipated by them during respiration and the rest is used in synthesizing the organic matter of the herbivore bodies. Herbivores are themselves consumed by carnivorous animals and these carnivores are eaten by other carnivores, and so on. At each step in this chain of events, a portion of the energy is dissipated as heat. Any plants or animals that die, whether from natural causes, injury, or disease, are attacked by microorganisms and small animals, collectively called **decomposers.** These decomposers also utilize energy released by plants or animals in the form of excretory products. These reactions constitute a **food chain** or food web.

Although most of the energy fixed by photosynthesizers is ultimately dissipated as heat, the chemical elements that serve as nutrients usually are not lost from the ecosystem. For instance, carbon from CO_2 fixed by plants in photosynthesis is released during respiration by various organisms of the food chain and becomes available for further utilization by the plants. Nitrogen, sulfur, phosphorus, iron, and other elements taken up by plants are also released through the activity of the decomposers, and hence are made available for reassimilation by other plants. Thus, although energy flows through the ecosystem, chemical elements are carried through cycles within the system. In some parts of the cycle the element is oxidized, whereas in other parts it is reduced; for many elements a **biogeochemical cycle** can thus be defined, in which the element undergoes changes in oxidation state as it is acted upon by one organism after another. In addition to this redox (oxidation-reduction) cycle, it is also possible to define a **transport cycle,** which describes the movement of an element from one place to another on earth, as for instance from land to air or from air to water. Such a transport cycle may or may not also involve a redox cycle. For instance, when oxidation or reduction leads to conversion of a nonvolatile substance to one that is volatile, the latter can then be transported to the air, so that the transport cycle is coupled to the redox cycle. In other cases, oxidation or reduction does not lead to a change in state, and has no influence on transport.

In order to evaluate the roles of microorganisms in ecosystems, it is essential to understand their precise natural habitats, and how their activities in these habitats are measured. We thus open this chapter with a discussion of these questions.

13.1

Microorganisms in Nature

The natural habitats of microorganisms are exceedingly diverse. Any habitat suitable for the growth of higher organisms will also permit microbial growth, but in addition, there are many habitats unfavorable to higher organisms where microorganisms exist and even flourish. Because microorganisms are usually invisible, their existence in an environment is often unsuspected; yet microbial action is usually of considerable importance for the function of the ecosystem.

Microenvironments Because microbes are small, microbial environments are also small, and within a single soil crumb or upon a single root surface there may be a variety of distinct **microenvironments,** each suitable for the growth of certain kinds of organisms but not of others. Thus when we think of the existence of microbes in nature we must learn to "think small." Any study of the microenvironment requires the use of the microscope, and by carefully examining samples of the habitat under the microscope it is possible to observe where microbes are flourishing. Many habitats are opaque, and microscopic examination of such habitats cannot be done using ordinary transmitted light. Certain microscopic techniques are available for studying opaque environments (see Figures 13.7 and 13.8), but a simple procedure that permits use of ordinary microscopes is to immerse microscope slides in the environment (for example, soil or water).

The microscope slide serves as a surface upon which organisms can attach and grow, and after a period of time the slide is removed and examined, preferably with a phase-contrast microscope. The development of microbial colonies on glass slides (Figure 13.1) presumably is similar to the way the same organisms develop upon natural surfaces.

Surfaces Surfaces in general are of considerable importance as microbial habitats because nutrients from the environment adsorb to them; thus in the microenvironment of a surface the nutrient levels may be much higher than they are away from the surface. A chemical assay for a nutrient in a body of water (for example, lake, river, ocean), therefore, will not reveal the concentrations present on surfaces such as sand, silt, or rock within the body of water. On such surfaces microbial numbers and activity will usually be much higher than in the free water, due to these adsorption effects.

Nutrients Despite adsorption to surfaces, nutrient concentrations are probably much lower in nature than they are in the usual laboratory culture media. For instance, the level of organic matter in culture media is usually 1 to 10 g/liter, whereas in nature it can often be at levels of 1 to 10 mg/liter or less. Thus microorganisms in natural environments are often subjected to nutrient-poor conditions. Even where richer sources of nutrients are available, as for instance the leaf litter on the forest floor or an algal bloom in a lake, the nutrients are available only briefly, for they are rapidly colonized and consumed by microbes, which then are returned to a semistarvation diet. In the case of photosynthetic organisms, their energy source, light, is highly variable in amount from day to night and throughout the seasons of the year. Thus organisms in nature must be able to adapt to a feast-or-famine existence, probably one of the main reasons why highly sophisticated regulatory mechanisms have evolved.

Growth rates in nature Extended periods of exponential growth in nature are rare. Growth more often occurs in spurts, when substrate becomes available, followed by an extended stationary phase, after the substrate is used up. Even where extended growth periods do occur, they are rarely exponential, but bear more resemblance to growth in the chemostat. Measurements of growth rates have been made in nature and compared with those of the same organisms in laboratory culture in optimal cultural conditions. The generation time of *Escherichia coli* in the intestinal tract is about 12 hours (two doublings per day) whereas in culture it grows much faster, with a generation time of 20 to 30 minutes (48 doublings per day). The marine bacterium *Leucothrix mucor,* an epiphyte of seaweeds, has a generation time of about 11 hours in nature and about 2 hours in culture. Some soil bacteria grow very much slower in nature than in the laboratory, and generation times as long as 1200 hours have been estimated in grassland soil. If these estimates are accurate it would mean that the bacteria are dividing only a few times per year.

However, all of these estimates for natural growth rates are averages, and it is possible (even likely) that growth rates are much

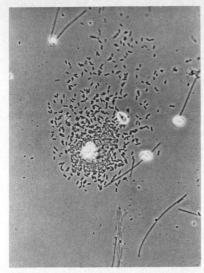

FIGURE 13.1 Bacterial microcolonies developing on a microscope slide immersed in a small river. The bright particles are mineral matter. Magnification, 310×.

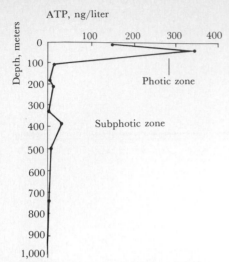

ATP, ng/liter

FIGURE 13.2 Distribution of ATP with depth in the ocean. Water samples were taken from the Pacific Ocean off the coast of California. Note the high ATP level in the photic zone, where light permits growth of photosynthetic organisms, and another small peak of ATP in the dark (subphotic) zone. This second peak is caused primarily by bacteria growing on organic matter that accumulated at this depth. (Data taken from Holm-Hansen, O., and C. R. Booth. 1966. Limnol. Oceanogr. 11:510–519.)

faster for shorter periods of time, followed then by long periods when the organisms are not growing and are essentially dormant. Dormancy is of course a common property of spores, but can also exist in many organisms even in the vegetative state.

Measurement of microbial activity in nature It is relatively easy to perform counts of the numbers of microorganisms in natural environments, but because of dormancy, counts do not indicate the activity of the organisms. Several procedures are available which do indicate microbial activity. One of the most widely used is measurement of respiration, as either O_2 uptake or CO_2 production. A sample of soil or water is incubated in a closed container under simulated natural conditions, and change in one of these gases is measured. Although adequate, this method is not very sensitive. Where low microbial numbers or low activity exist, measurement of ATP levels is now widely used. All organisms produce ATP as a result of energy metabolism, and under starvation the ATP level dips to a low value. In ecological studies a sample of water or soil is treated to extract the ATP from microbes, and the ATP level of the extract is measured. Sensitive methods are available for measuring ATP, the limit of detection being about 10^{-5} μg of ATP per liter of sample. One μg of ATP is equivalent to about 250 μg of carbon in living organisms. Since ATP is lost very rapidly from dead or dormant organisms, ATP measurements essentially provide a measure of living biomass. ATP measurements have been made most frequently in the oceans, where microbial numbers are low and where very sensitive methods are needed; some representative data are given in Figure 13.2.

For the measurement of a specific microbial process in a natural environment, radioactive isotopes are very useful. They provide extremely sensitive and specific means of measuring single processes. For instance, if photosynthesis is to be measured, the light-dependent uptake of radioactive $^{14}CO_2$ into microbial cells can be measured, or if sulfate reduction is of interest, the rate of conversion of $^{35}SO_4^{2-}$ to $H_2^{35}S$ can be studied. Isotope methods are extremely valuable to the microbial ecologist and are probably the most widely used means of evaluating the activity of microbes in nature. However, because there is always the possibility that some transformation of a labeled compound might be due to a strictly chemical rather than a microbial process, it is essential when using isotopes to show that the transformation is prevented by microbicidal agents or heat treatments that are known to block microbial action.

Significant population levels Microbial populations in natural environments vary widely, from values of a few hundred cells per milliliter or less to values as high as 10^9 to 10^{10} per milliliter. At what level is a microbial population likely to have a significant effect on an ecosystem? It is difficult to give any general figure, as it will depend on how active individual cells are and on what kinds of processes they carry out. For instance, an organism such as *Clostridium tetani* produces such a highly potent toxin that it can have a significant effect on its habitat (the human host) even when its population level is quite low. On the other hand, in ecosystems such as soil and water such highly specific effects of microbes are uncommon, and fairly large populations are probably necessary if any significant effect is to be observed. As a general rule, population levels probably have to be greater than 10^6 per milliliter or 10^6 per gram before any rapid effect of a microbial population on an ecosystem can be anticipated. However, if the system is relatively stable over a long period of time, lower population levels may have some significant effect.

13.2
Aquatic Habitats

Water evaporates from the surface of the earth, accumulates in the atmosphere in clouds, and falls back to the surface of the earth as precipitation in the form of rain, snow, hail, and so on. Water falling on land either runs off directly into rivers and lakes or percolates through the soil until it reaches a level, the **water table.** From the water table it slowly moves, following the contours of the land, ultimately to reach the surface as springs or seepages. Most of the water eventually finds its way to the sea. The interactions of this **hydrologic cycle** create a variety of aquatic habitats in which living organisms develop.

Typical aquatic environments are the oceans, estuaries, salt marshes, lakes, ponds, rivers, and springs. Aquatic environments differ considerably in chemical and physical properties, and it is not surprising that their species compositions also differ. The predominant photosynthetic organisms in most aquatic environments are microorganisms; in aerobic areas cyanobacteria and eucaryotic algae prevail, and in anaerobic areas photosynthetic bacteria are preponderant. Algae floating or suspended freely in the water are called **phytoplankton;** those attached to the bottom or sides are called **benthic algae.** Because these photosynthetic organisms utilize energy from light in the initial production of organic matter, they are called the **primary producers.** In the last analysis, the biological activity of an aquatic ecosystem is dependent on the rate of primary production by the photosynthetic organisms. The activities of these organisms are in turn affected by the physical environment (for example, temperature, pH, and light as was discussed in Chapter 8) and by the kinds and concentrations of nutrients available. Open oceans are very low in primary productivity, whereas inshore ocean areas are high, with some lakes and springs being highest of all. The open ocean is infertile because the inorganic nutrients needed for algal growth are present only in low concentrations. The more fertile inshore ocean areas, on the other hand, receive extensive nutrient enrichment from rivers. There are, however, some open ocean areas that are rather fertile; these are places where winds or currents cause an extensive upwelling of deep ocean water, bringing to the surface nutrients from the bottom of the sea. It is because of such upwellings that areas off the coasts of California and Peru are so productive. The amount of economically important crops such as fish or shellfish is determined ultimately by the rate of primary production; lakes and inshore ocean areas are high in primary production and thus are the richest sources of fish and shellfish.

Food chains The energy found as organic matter in primary producers reaches later stages of the food chain in several ways. Some of the organic matter is excreted in soluble form and serves as nutrient for the growth of heterotrophic microorganisms, primarily bacteria. The organic matter retained within the cells of the primary producers is a major source of food for small animals, primarily crustaceans, which are collectively called **zooplankton.** Some of the smallest zooplankton also consume bacteria as a major source of food. Zooplankton are themselves consumed by larger invertebrates, which in turn are devoured by fish. Thus a simplified food chain for an open water aquatic zone can be represented as:

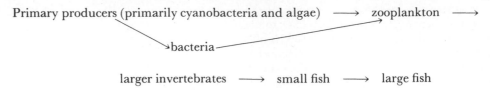

larger invertebrates $\longrightarrow$ small fish $\longrightarrow$ large fish

Since each step in the food chain involves a loss of energy, if one is interested in an end product such as fish it is obviously important to have a short food chain in order to utilize the maximum amount of energy in synthesizing the desired product.

In inland waters, especially rivers, surrounded as they are by land areas containing large plants, much of the organic matter is derived not from primary producers but from dead leaves, humic substances, and other forms of organic detritus originating on the surrounding land. These materials are acted upon primarily by bacteria and fungi and are converted at least partially into microbial protein. In such waters the food chain may begin not with primary producers but with these heterotrophic microorganisms.

13.3

Oxygen Relationships in Aquatic Environments

Although oxygen is now one of the most plentiful gases in the atmosphere, it has limited solubility in water, and in a large water mass its exchange with the atmosphere is slow. The photosynthetic production of oxygen occurs only in the surface layers of a lake or ocean, where light is available. Organic matter that is not consumed in these surface layers sinks to the depths and is decomposed by facultative microorganisms, using oxygen dissolved in the water. Once the oxygen is consumed, the deep layers become anaerobic; here strictly aerobic organisms such as higher plants and animals cannot grow, and the bottom layers have a species composition restricted to anaerobic bacteria and a few kinds of microaerophilic animals. In addition, there is a conversion from a respiratory to a fermentative metabolism, with important consequences for the carbon cycle.

Whether or not a body of water becomes depleted of oxygen depends on several factors. If organic matter is sparse, as it is in unproductive lakes or in the open ocean, there may be insufficient substrate available for heterotrophs to consume all the oxygen. Also important is how rapidly the water from the depths exchanges with surface water. Where strong currents or turbulence occurs, the water mass may be well mixed, and consequently oxygen may be transferred to

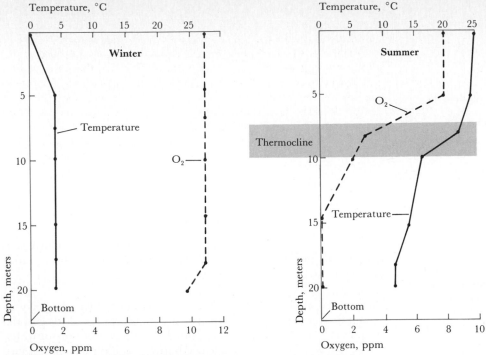

FIGURE 13.3 Development of anaerobic conditions in the depths of a temperate-climate lake as a result of summer stratification. Because the colder waters below the thermocline are denser than the surface waters, the two water masses do not mix. Data are for Lake Mendota, Wisconsin. (Taken from Birge, E. A., and C. Juday. 1911. Bull. Wis. Geol. Nat. Hist. Survey, No. 22, Sci. Ser. No. 7.)

the deeper layers. In many bodies of water in temperate climates, however, the water mass becomes stratified during the summer, with the warmer and less dense surface layers separated from the colder and denser bottom layers. After stratification sets in, usually in early summer, the bottom layers become anaerobic. In the late fall and early winter, the surface waters become colder and heavier than the bottom layers, and the water "turns over," leading to a reaeration of the bottom. Most lakes in temperate climates thus show an annual cycle in which the bottom layers of water pass from aerobic to anaerobic and back to aerobic (Figure 13.3).

Rivers The oxygen relations in a river are of particular interest, especially in regions where the river receives much organic matter in the form of sewage and industrial pollution. Even though the river may be well mixed because of rapid water flow and turbulence, the large amounts of added organic matter can lead to a marked oxygen deficit. This is illustrated in Figure 13.4. As the water moves away from the sewage outfall, organic matter is gradually consumed, and the oxygen content returns to normal. Oxygen depletion in a body of water is undesirable because most higher animals require O_2 and die under anaerobic conditions. Further, conversion to anaerobic condi-

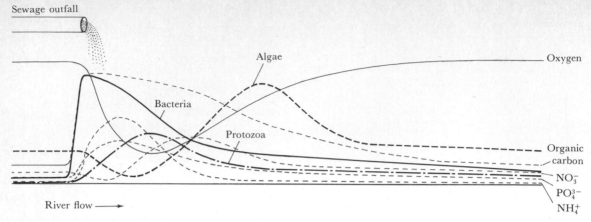

Sewage outfall

Algae

Oxygen

Bacteria

Protozoa

Organic
carbon

NO_3^-
PO_4^{3-}
NH_4^+

River flow ⟶

FIGURE 13.4 Chemical and biological changes in a river at various locations downstream from the source of a sewage outfall. Increased organic matter below the outfall leads to an increase in heterotrophic bacteria, a decrease in oxygen content, and an increase in NH_4^+. Farther downstream, NH_4^+ is oxidized to NO_3^- by nitrifying bacteria, and organic matter is oxidized by heterotrophs. Inorganic nutrients released from the decomposing organic matter make possible the growth of algae. Protozoa that feed on the bacteria are responsible for some of the decrease in bacterial numbers downstream.

tions results in the production by anaerobic bacteria of odoriferous compounds (for example, amines, H_2S, mercaptans), some of which are also toxic to higher organisms.

Biochemical oxygen demand Sanitary engineers term the oxygen-consuming property of a body of water its **biochemical oxygen demand** (BOD). The BOD is determined by taking a sample of water, aerating it well, placing it in a sealed bottle, incubating for a standard period of time (usually 5 days at $20°C$), and determining the residual oxygen in the water at the end of incubation. Although it is a crude method, a BOD determination gives some measure of the amount of organic material in the water that could be oxidized by microorganisms. As a river recovers from contamination with an organic pollutant, the drop in BOD is accompanied by a corresponding increase in dissolved oxygen (Figure 13.4).

We thus see that the oxygen and carbon cycles in a water body are greatly intertwined, and that heterotrophic microorganisms, mainly bacteria, play important roles in determining the biological nature and productivity of the body of water.

13.4

Terrestrial Environments

In the consideration of terrestrial environments, our attention is inevitably turned to the **soil** since it is here that many of the key processes occur that influence the functioning of the ecosystem. The process of soil development involves complex interactions between the parent material (rock, sand, glacial drift, and so on), topography, climate, and living organisms. Soils can be divided in two broad groups—**mineral soils** and **organic soils**—depending on whether they derive initially from the weathering of rock and other inorganic material or from sedimentation in bogs and marshes. Our discussion will concentrate on mineral soils, the predominant soil in most areas.

Soil formation Soils form as a result of combined physical, chemical, and biological processes. An examination of almost any exposed rock will reveal the presence of algae, lichens, or mosses. These organisms are able to remain dormant on the dry rock and then grow when moisture is present. They are photosynthetic and produce organic matter, which supports the growth of heterotrophic bacteria and fungi. The numbers of heterotrophs increase directly with the degree of plant cover of the rocks. Carbon dioxide produced during respiration by heterotrophs is converted into carbonic acid, which is an important agent in the dissolution of rocks, especially those composed of limestone. Many heterotrophs also excrete organic acids, which further promote the dissolution of rock into smaller particles. Freezing and thawing and other physical processes lead to the development of cracks in the rocks. In these crevices a raw soil forms, in which pioneering higher plants can develop. The plant roots penetrate farther into crevices and increase the fragmentation of the rock, and their excretions promote the development of a rhizosphere (the soil that surrounds plant roots) flora. When the plants die, their remains are added to the soil and serve as nutrients for an even more extensive microbial development. Minerals are further rendered soluble, and as water percolates it carries some of these chemical substances deeper. As weathering proceeds, the soil increases in depth, thus permitting the development of larger plants and trees. Soil animals become established and play an important role in keeping the upper layers of the soil mixed and aerated. Eventually the movement of materials downward results in the formation of layers, and a typical soil profile becomes outlined (Figure 13.5). The rate of development of a typical soil profile depends on climatic and other factors, but it is usually very slow, taking hundreds of years. The picture given here is a general one, and marked variation among climates and geography exists.

Soil as a microbial habitat The most extensive microbial growth takes place on the surfaces of soil particles (Figure 13.6). To examine soil particles directly for microorganisms, special reflected-light fluorescence microscopes are often used, the organisms in the soil being stained with a dye that fluoresces. In effect each microbial cell is its own light source, and its shape and position on the surface of the par-

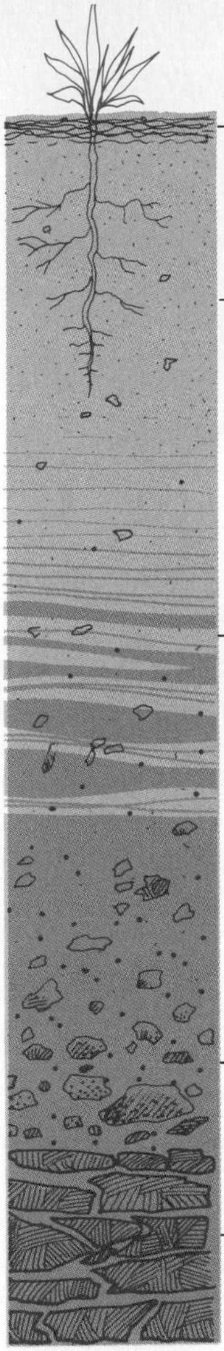

0 horizon
Layer of undecomposed plant materials
A horizon
Surface soil (high in organic matter, dark in color, is tilled for agriculture; plants and microorganisms grow here)
B horizon
Subsoil (minerals, humus, etc., leached from soil surface accumulate here; little organic matter)
C horizon
Soil base (develops directly from underlying bedrock)
Bedrock

FIGURE 13.5 Profile of a mature soil.

Microcolony
Quartz
Air
Organic matter
Quartz
Clay particle
Quartz
Water
Air
Clay particle

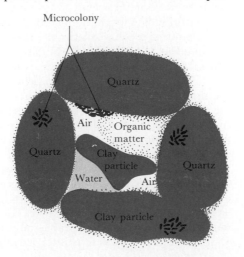

FIGURE 13.6 A soil aggregate composed of mineral and organic components, showing the localization of soil microorganisms. Very few microorganisms are found free in the soil solution; most of them occur as microcolonies attached to the soil particles.

Soil particles

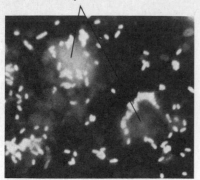

FIGURE 13.7 Bacterial microcolonies (*Rhizobium japonicum*) on the surface of soil particles, visualized by the fluorescent-antibody technique. Notice some areas with bacteria localized on soil particles. Magnification, 670×. (From Bohlool, B. B., and E. L. Schmidt, 1968. Science 162:1013 © 1968 by the American Association for the Advancement of Science.)

ticle can easily be seen. To observe a specific microorganism in a soil particle, **fluorescent-antibody staining** (see Chapter 16) can be used (Figure 13.7). Microorganisms can be visualized excellently on such opaque surfaces as soil by means of the **scanning electron microscope** (Figure 13.8), which provides much higher resolution and depth of field than does the light microscope.

One of the major factors affecting microbial activity in soil is the availability of water. Water is a highly variable component of the soil, its presence depending on rainfall, drainage, and plant cover. It is held in the soil in two ways, by adsorption onto surfaces or as free water existing in thin sheets or films between soil particles. Different soils vary greatly in their capacity to adsorb and retain water. The pore space is usually 30 to 50 percent of the total volume of a soil, and in a well-drained soil the volumes of the respective components are: soil particles, 50 percent; air, 10 percent; water, 40 percent.

Even when a soil lacks a free-water phase, each soil crumb is a surface upon which water may adsorb, and often the major part of the moisture in a soil is the portion that is adsorbed. As we discussed in Section 8.2, control of water potential in soil systems is mainly by matric (adsorption) effects rather than by osmotic effects. The adsorptive power of different soil constituents for water varies, sand having little affinity for water and clay binding water tightly. The water present in soils has a variety of materials dissolved in it; the whole mixture is called the soil solution. In well-drained soils air penetrates readily, and oxygen is rarely deficient even in the deepest portions. However, water and air compete for the space between the soil particles, and the former can drive out the latter. Thus in a waterlogged soil the only oxygen present is that dissolved in the water, and this is soon consumed by microorganisms. Such soils quickly become anaerobic, showing profound changes in their biological properties.

The nutrient status of a soil is the other major factor affecting microbial activity. The greatest microbial activity is in the organic-rich surface layers and in the regions adjacent to plant roots. In deeper layers of mineral soils, microbial numbers and activity are usually quite low; the organisms present may be dormant, and are most likely in the spore form. However, if a supply of organic material is introduced into such a location, microbial activity quickly develops. The few organisms present colonize the material and grow to large numbers in the localized area of the nutrient. They become dormant again when the nutrient supply has been exhausted.

The term **humus** is used to denote the organic fraction of the soil that is relatively resistant to decomposition, to distinguish it from the organic matter of the litter itself. Humus is not a homogeneous substance but is a complex mixture of materials. It is derived partly from the protoplasmic constituents of soil organisms which themselves have resisted decomposition, and partly from resistant plant material. If the plant cover is removed from a soil, as is done in agricultural practice, the humus slowly disappears. The rate of humus disappearance is greater in warmer than in cooler climates and is affected by rainfall and soil texture. Humus promotes the development of an agriculturally desirable soil texture. As we will see (Section 13.6), humic substances constitute one of the largest reservoirs of carbon on earth.

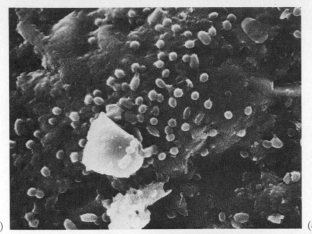

(a)

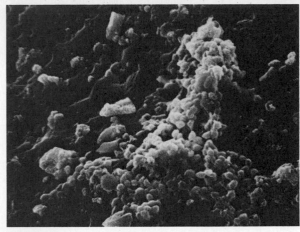

(b)

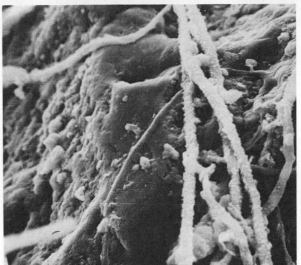

(c)

FIGURE 13.8 Visualization of microorganisms on the surface of soil particles by use of the scanning electron microscope: (a) bacteria (magnification, 1950×) (b) actinomycete spores (magnification, 1710×); (c) fungus hyphae (magnification, 990×). (Courtesy of T. R. G. Gray.)

13.5

Biogeochemical Cycles on a Global Scale

The redox cycles for various elements, catalyzed to a greater or lesser extent by microorganisms, are intimately related to the transfer of elements from place to place on earth. Despite the enormous size of the earth, a considerable amount of information is available about cycling of elements on a global scale, and an understanding of these events is important in predicting long-term consequences of human perturbations of these cycles. To a first approximation, the earth can be divided into a series of compartments: atmosphere, land surface, oceans, and earth's crust. Global cycling can then be expressed as the rate of movement of an element from one compartment to another (Figure 13.9). Over the short run, the major movement of an element from one place to another on earth occurs either as a gas in the atmosphere or as suspended or dissolved forms in water. Once a gas reaches the atmosphere, it can be carried with the general atmospheric circulation (winds, air currents, vertical circulation) around the earth.

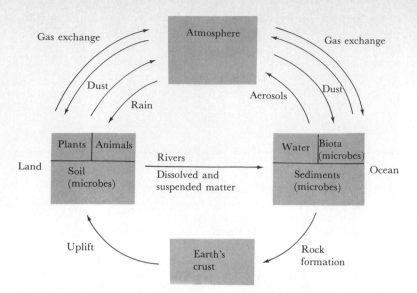

FIGURE 13.9 Global cycling of elements. Some of the major compartments are indicated.

This circulation within the atmosphere is relatively rapid, and microbial production of gases provides an important means of speeding up global cycling of the elements carbon, nitrogen, and sulfur. Movement in water is slower but still relatively quick. Elements are carried by streams and rivers to the sea, and there is a well-defined system of horizontal ocean currents, downwellings, and upwellings, which carry materials throughout the earth. Once an element sediments to the bottom and becomes incorporated into ocean sediments, its movement is much slower. Eventually it may become converted into rock (sedimentary rocks) and may then only reach the surface of the earth again when tectonic forces (that is, the forces of the earth's crustal structure) cause uplift and mountain building. As a result of rock weathering and soil erosion, elements are now being transferred from land to the oceans that had been deposited in oceanic sediments many millions of years ago.

Each box in Figure 13.9 can be considered to be a reservoir or compartment within which an element is stored for a definite period of time. The rate of movement from one compartment to another, called the *flux,* provides a quantitative expression of the intensity of the biogeochemical cycle. The length of time an element remains in a compartment is expressed as its **residence time** which is defined as the amount of an element in a compartment at any given time divided by the rate of addition (or substraction) of the element. Residence time is the reciprocal of the dilution rate, defined as the flow rate divided by the volume. One way of relating residence time to dilution rate is to consider the change in concentration of an element if the input is stopped, but the output of the element is continued. In one residence time, the concentration of the element would be reduced by one-half. Thus residence times are sometimes spoken of as half-lives ($t_{1/2}$). If we were to be concerned about the elimination of a pollutant as a result of dilution, knowing the residence time would make it possible for us to calculate the rate of removal. However, one residence time would only dilute the pollutant by one-half of what was remaining. Thus theoretically, by dilution alone, complete removal of the pollutant would never occur, the rate of decrease of concentration occurring at an ever decreasing rate. From a practical point of view, however, dilution would ultimately decrease the pollutant to an insignificant level. As a rough approximation, after six residence times, an element would be reduced to a level that could be considered insignificant.

13.6
The Carbon Cycle

The carbon cycle is central to a discussion of all other biogeochemical cycles, because other cycles are driven by energy derived from photosynthesis or the breakdown of organic materials. A brief overview of the redox cycle for carbon is given in Figure 13.10. This shows that three oxidation states of carbon are of main significance, CH_4 (methane), the most reduced; $(CH_2O)_n$, a general formula for protoplasm, which is approximately at the oxidation level of carbohydrate; and CO_2 (carbon dioxide), the most oxidized form of carbon. Another substance, carbon monoxide, CO, oxidation state $+2$, is a minor component of the carbon cycle.

To appreciate the significance of the various steps in the carbon cycle, it is necessary to consider the compartment sizes and residence times for the various components on a global scale. These are sum-

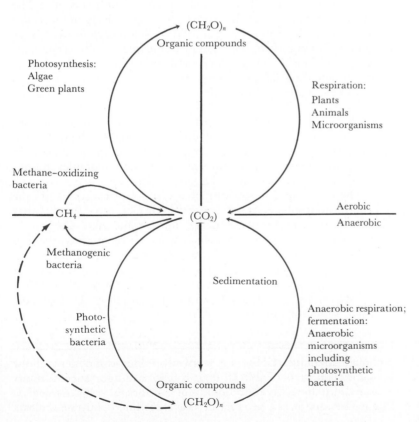

FIGURE 13.10 Redox cycle for carbon. Oxidation states of key components: CH_4, -4; $(CH_2O)_n$, 0 (approximate, for protoplasm); CO_2, $+4$.

TABLE 13.1 Carbon Reservoirs and Residence Times

Component	Reservoir (grams)	Residence time (years)	Main Removal Process
Atmosphere:			
CH_4 (1.6 ppm*)	$3-6 \times 10^{15}$	3.6	Photochemical oxidation in atmosphere
CO (0.1 ppm)	0.3×10^{15}	0.1	Photochemical oxidation in atmosphere
CO_2 (320 ppm)	670×10^{15}	4	Plant photosynthesis
Land:			
Living organic (mainly plants)	$500-800 \times 10^{15}$	16 (plants)	Death; grazing; predation
Dead organic	$700-1200 \times 10^{15}$	40	Microbial decomposition to CO_2
Oceans:			
Living organic	6.9×10^{15}	0.14	Death; grazing; predation
Dead organic	760×10^{15}	19	Microbial decomposition
Inorganic	$40,000 \times 10^{15}$	100,000	Formation of carbonate rocks; CO_2 exchange with the atmosphere
Sediments and rocks	$72,500,000 \times 10^{15}$	300×10^6	Weathering; fossil fuel burning

*Parts per million (μg gas/cm^3 of air).

Based on: Reiners, W. A. 1973. A summary of the world carbon cycle and recommendations for critical research. Pages 368–382 *in* G. M. Woodwell, and E. V. Pecan, eds. Carbon and the biosphere. Published by National Technical Information Service, Springfield, Va., as CONF-720510; and Garrels, R. M., F. T. Mackenzie, and C. Hunt. 1975. Chemical cycles and the global environment, William Kaufmann, Los Altos, Calif.

marized in Table 13.1. The largest carbon reservoir is that constituting the sediments and rocks of the earth's crust, but the residence time is so long that flux out of this compartment is relatively insignificant on a human scale. From the viewpoint of living organisms, a large amount of organic carbon is found in land plants. This represents the carbon of forests and grasslands and constitutes the major site of photosynthetic CO_2 fixation. However, far more carbon is present in dead organic material than in living organisms. This dead organic material is primarily humus and humus-related material in the soil. The nature of humus was discussed in Section 13.4. The humus-like materials of the soil are fairly stable, with a global residence time of about 40 years, although certain components decompose much more slowly than others. Some of the chemically more complex humus materials have residence times of hundreds of years.

The most rapid means of global transfer of carbon is via the CO_2 of the atmosphere. As seen, atmospheric CO_2 constitutes a fairly significant reservoir, but its residence time is quite short. Carbon dioxide is removed from the atmosphere primarily by photosynthesis of land plants and is returned to the atmosphere by respiration of animals and heterotrophic microorganisms. An analysis of the various processes suggests that the single most important contribution of CO_2 to the atmosphere is via microbial decomposition of dead organic material, including humus.

13.7

Methane and Methanogenesis

Although methane (CH_4) is a relatively minor component of the global carbon cycle (Table 13.1), it is of great importance in many localized situations. In addition, it is of considerable microbiological interest because it is a product of anaerobic microbial metabolism.

Methane production is carried out by a highly specialized group of organisms, the methanogenic bacteria, which are obligate anaerobes. Methane production in these organisms is a part of their energy metabolism. Most methanogenic bacteria (or **methanogens,** for short) use CO_2 as their terminal electron acceptor in anaerobic respiration, converting it to methane; the electron donor used in this process is generally hydrogen, H_2. The overall reaction of **methanogenesis** in this pathway is as follows:

$$4H_2 + CO_2 \longrightarrow CH_4 + 2H_2O \qquad \Delta G^{0'} = -32 \text{ kcal}$$

which shows that CO_2 reduction to methane is an eight-electron process (four H_2 molecules). A few other substrates can be converted to methane, including methanol, CH_3OH; formate, $HCOO^-$; methyl mercaptan, CH_3SH; acetate, CH_3COO^-; and methylamines (see Section 19.20).

Despite the exceedingly narrow range of substrates used by methanogenic bacteria, methane production is widespread during the decomposition of a wide variety of organic materials. Methane formation occurs during the decomposition of such diverse substrates as cellulose, starch and sugars, proteins and amino acids, fats and fatty acids, alcohols, benzoic acid, and a variety of other substances. Although at one time it was thought that methanogenic bacteria attacked some of these substrates directly, it is now well established that methane formation from these materials requires the participation of other anaerobic bacteria. These other bacteria ferment the substances to either acetate or hydrogen plus CO_2, which are then used by the methanogenic bacteria. Thus, methanogenesis from organic carbon is virtually always a process carried out by a mixture of bacteria, none of which can perform the complete process by themselves. An example of this interdependence is the process of methane formation from cellulose in the rumen which is outlined in Figure 14.15.

In most situations, the rate-limiting step in methane formation from organic materials is not the final reduction of CO_2 to CH_4, but the step involved in the breakdown of the complex organic material into fermentation products, such as acetate and H_2. Since CO_2 is very common in anaerobic environments, either in the form of carbonates or derived from microbial respiration or fermentation processes, the methanogenic bacteria are generally limited by the availability of H_2. As soon as any H_2 is formed by a fermentative microorganism, it is quickly consumed by a methanogen. The only situations in which H_2 ever accumulates in nature are when methanogenesis is inhibited in some way.

Despite the obligate anaerobiosis and specialized metabolism of methanogens, they are quite widespread on earth. Thus, although high levels of methanogenesis are generally only seen in anaerobic environments, such as swamps and marshes, the process also occurs in habitats that normally might be considered aerobic, such as forest and grassland soils. In such habitats, it is likely that methanogenesis is occurring in anaerobic pockets, for example, in the midst of soil crumbs. An overview of the rates of methanogenesis in different kinds of habitats is given in Table 13.2. It should be noted that biogenic production of methane by the methanogenic bacteria exceeds con-

TABLE 13.2   Sources of Atmospheric Methane

Source	Average Methane Production Rate ($g/km^2/year$)	Total Area (km^2)	Annual Production ($g/year$)
Rice paddies*	2.1×10^8	9.2×10^5	1.9×10^{14}
Swamps	2.1×10^8	2.6×10^6	5.4×10^{14}
Enteric fermentation, ruminants, other animals			4.5×10^{13}
Forests	0.9×10^4	4.4×10^7	4×10^{11}
Fields, grasslands, cultivated areas	4.4×10^5	3×10^7	1×10^{13}
Humid tropical areas	2.1×10^7	2.9×10^7	6.1×10^{14}
Total biogenic production rate			Approx. 1×10^{15}
Coal fields			2×10^{13}
Natural gas well output			5.2×10^{14}
Total, nonbiogenic production			5.4×10^{14}
Biogenic production rate, percentage of total: 65%			

*The data on methane production by paddies are probably too high, but more measurements are not available.

Modified from Ehhalt, D. H. 1973. Methane in the atmosphere. Pages 144–158 *in* G. M. Woodwell, and E. V. Pecan, eds. Carbon and the biosphere. Published by National Technical Information Service, Springfield, Va., as CONF-720510.

siderably the production rate from gas wells and other fossil fuel sources. An independent measure of the importance of biogenic processes in global methanogenesis can be obtained from measurement of the carbon 14 (^{14}C) content of atmospheric methane. Carbon 14 has a half-life of about 4000 years, so that any methane derived from fossil fuel deposits and other sedimentary rocks should be nonradioactive. Biogenic methane should have the same radioactivity as recently formed wood, since both are formed from CO_2 (which has some ^{14}C content because of production in the atmosphere from cosmic ray bombardment of ^{14}N). Atmospheric methane has 80 percent of the ^{14}C content of recent wood, suggesting that at least 80 percent of atmospheric methane is of a biogenic origin.

Sulfate inhibition of methanogenesis   Methanogenesis is much more common in freshwater and terrestrial environments than in the sea. The reason for this appears to be that marine waters and sediments contain rather high levels of sulfate, and sulfate-reducing bacteria effectively compete with the methanogenic population for available acetate and H_2, two major electron donors for sulfate-reducing bacteria (see Sections 13.10 and 19.16):

$$4H_2 + SO_4^{2-} \longrightarrow H_2S + 2H_2O + 2OH^- \qquad \Delta G^{0\prime} = -39.3 \, \text{kcal}$$
$$CH_3COO^- + SO_4^{2-} \longrightarrow 2HCO_3^- + HS^- \qquad \Delta G^{0\prime} = -11.3 \, \text{kcal}$$

The biochemical basis for the success of sulfate-reducing bacteria in scavenging H_2 appears to lie in the increased affinity sulfate-reducing bacteria have for H_2 as compared to typical methanogens. When H_2 levels get below 5 to 10 μM, as they often do in sulfate-rich environments, methanogens are no longer able to grow since their H_2 uptake systems are unable to function at such low H_2 concentrations. Sulfate reducers, on the other hand, can grow at these low partial pressures of H_2, effectively preventing H_2-mediated methanogenesis.

Sulfate reduction is also a significant process in fresh water, but because the sulfate concentration of fresh water is so low, sulfate is rapidly depleted at the surface of anaerobic sediments; thus, throughout the bulk of the sediment, methanogenesis is the major process consuming H_2.

A kinetic mechanism is also responsible for the ability of sulfate-reducers to effectively compete with methanogens for acetate. The affinity for acetate of some sulfate-reducers is over ten times that of methanogens, and this differential becomes even greater as acetate levels become limiting. In sulfate-rich environments, acetate levels are generally low; hence the majority of acetate consumed in these environments will be by sulfate-reducing, rather than methanogenic, bacteria.

As seen in Table 13.1, the residence time of methane in the atmosphere is short, 3.6 years. The main process consuming methane in the atmosphere is photochemical oxidation to CO_2. Some methane formed in soils and anaerobic sediments never reaches the atmosphere because it is oxidized by another group of bacteria, the methane-oxidizing bacteria (see Section 19.12). We discuss the role of methanogenesis in sewage treatment and waste disposal in Section 13.15.

13.8

The Nitrogen Cycle

The element nitrogen, N, a key constituent of protoplasm, exists in a number of oxidation states (Figure 13.11). Several of the key redox reactions of nitrogen are carried out in nature almost exclusively by microorganisms, so that microbial involvement in the nitrogen cycle is of great importance. Thermodynamically, nitrogen gas, N_2, is the most stable form of nitrogen, and it is to this form that nitrogen will revert under equilibrium conditions. This explains the fact that a major reservoir for nitrogen on earth is the atmosphere (Table 13.3).

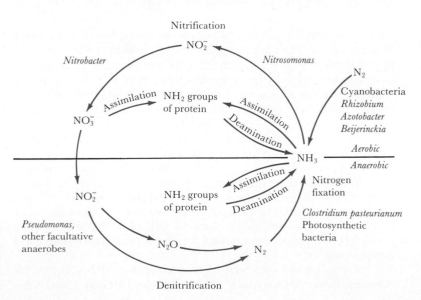

FIGURE 13.11 Redox cycle for nitrogen. Oxidation states of key compounds: Organic N (R—NH_2), -3; NH_3, -3; N_2, 0; N_2O, $+1$ (average per N); NO_2^-, $+3$; NO_3^-, $+5$.

TABLE 13.3 Nitrogen Reservoirs and Residence Times

Component	Reservoir (g)	Residence Time
Atmosphere:		
$NH_3 + NH_4^+$	0.003×10^{15}	Few days to few months
N_2	$3,800,000 \times 10^{15}$	44×10^6 years
N_2O	13×10^{15}	12–13 years
NO_3^-	0.0005×10^{15}	2–3 weeks
Organic nitrogen	0.001×10^{15}	Approx. 10 days
NO	0.03×10^{15}	Approx. 1 month
Land:		
Plant biomass	$11–14 \times 10^{15}$	16 years
Animal biomass	0.2×10^{15}	
Soil organic nitrogen	300×10^{15}	1–40 years
Soil inorganic nitrogen	16×10^{15}	Less than 1 year
Oceans:		
Plant biomass	0.3×10^{15}	0.14 year
Animal biomass	0.17×10^{15}	
Dead organic matter		
Dissolved	5.30×10^{15}	
Particulate	$3–24 \times 10^{15}$	
N_2 (dissolved)	$22,000 \times 10^{15}$	220,000 years
N_2O	0.2×10^{15}	2.5 years
NO_3^-	570×10^{15}	
NO_2^-	0.5×10^{15}	
NH_4^+	7×10^{15}	
Sediments and rocks:		
Rocks	$190,000,000 \times 10^{15}$	
Sediments	$400,000 \times 10^{15}$	400×10^6 years (organic nitrogen of sediments, fossil fuels)
Coal deposits	120×10^{15}	

Data from Söderlund, R., and B. H. Svensson. 1976. The global nitrogen cycle. In B. H. Svensson, and A. Söderlund, eds. Nitrogen, phosphorus and sulfur—global cycles. SCOPE Report 7, Ecol. Bull. (Stockholm) 22:23–74; and from Garrels, R. M., F. T. Mackenzie, and C. Hunt. 1975. Chemical cycles and the global environment. William Kaufmann, Los Altos, Calif.

This is in contrast to carbon, in which the atmosphere is a relatively minor reservoir. The high energy necessary to break the $N \equiv N$ bond of molecular nitrogen (Section 5.13) means that the utilization of N_2 is an energy-demanding process. Only a relatively small number of organisms are able to utilize N_2, in the process called **nitrogen fixation;** thus the recycling of nitrogen on earth involves to a great extent the more easily available forms, ammonia and nitrate. Globally, only about 3 percent of the net primary production of organic matter involves nitrogen derived from N_2. The remaining primary production uses nitrogen from "fixed" forms of nitrogen (nitrogen in combination with other elements). However, because N_2 constitutes by far the greatest reservoir of nitrogen available to living organisms, the ability to utilize N_2 is of great ecological importance. In many environments, productivity is limited by the short supply of combined nitrogen compounds, putting a premium on biological nitrogen fixation.

The global nitrogen cycle is given in Figure 13.12. Note that the chemical form in which transfer between compartments occurs is not

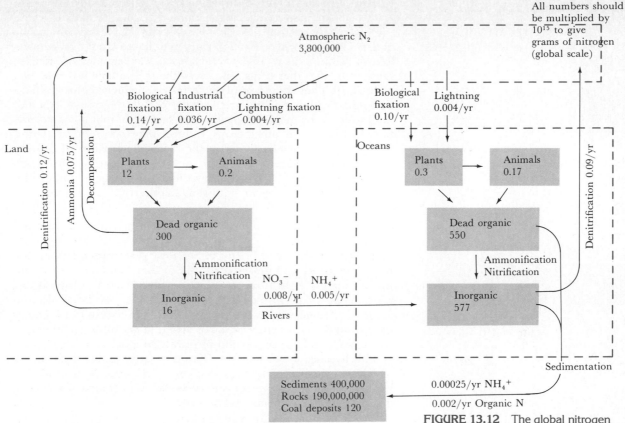

All numbers should be multiplied by 10^{15} to give grams of nitrogen (global scale)

Atmospheric N_2
3,800,000

Biological fixation 0.14/yr   Industrial fixation 0.036/yr   Combustion Lightning fixation 0.004/yr

Biological fixation 0.10/yr   Lightning 0.004/yr

Land

Denitrification 0.12/yr

Ammonia 0.075/yr

Decomposition

Oceans

Denitrification 0.09/yr

Plants 12   →   Animals 0.2

Plants 0.3   →   Animals 0.17

Dead organic 300

Dead organic 550

Ammonification Nitrification

Ammonification Nitrification

Inorganic 16

NO_3^- 0.008/yr NH_4^+ 0.005/yr

Rivers

Inorganic 577

Sedimentation

Sediments 400,000
Rocks 190,000,000
Coal deposits 120

0.00025/yr NH_4^+

0.002/yr Organic N

FIGURE 13.12 The global nitrogen cycle. Minor compartments and fluxes are not given. See also Table 13.3. Data from Söderlund, R., and B. H. Svensson. 1976. The global nitrogen cycle. In Svensson, B. H., and A. Söderlund, eds. Nitrogen, phosphorus and sulfur-global cycles. SCOPE report 7. Ecol. Bull. (Stockholm) 23:23–74 and Garrells, R. M., F. T. Mackenzie and C. Hunt. 1975. Chemical cycles and the global environment, William Kaufmann, Inc., Los Altos, Calif.

given. Transfer in and out of the atmosphere is to a great extent as N_2, with a smaller amount of transfer as nitrous oxide (laughing gas), N_2O, and as gaesous ammonia, NH_3. Transfer between terrestrial and aquatic compartments is primarily as organic nitrogen, ammonium ion, and nitrate ion.

Nitrogen fixation   We have discussed the biochemistry and microbiology of nitrogen fixation ($N_2 + 6H^+ + 6e^- \rightarrow 2NH_3$) in Section 5.13; and we shall discuss symbiotic nitrogen fixation by legumes in Section 14.10. Nitrogen fixation can also occur chemically in the atmosphere, to a small extent, via lightning discharges, and a certain amount of nitrogen fixation occurs in the industrial production of nitrogen fertilizers (labeled as industrial fixation in Figure 13.12). Some nitrogen fixation also occurs during artificial combustion processes, since air contains 78 percent N_2 by weight, and burning in air inevitably involves high-temperature combustion of some N_2 (to nitrogen oxides and ultimately to nitrate). However, as can be calculated from the fluxes given in Figure 13.12, about 85 percent of nitrogen fixation on earth is of biological origin. As can also be calculated from Figure 13.12, about 60 percent of biological nitrogen fixation occurs on land, and the other 40 percent in the oceans.

Estimations of the rate of nitrogen fixation in natural and agricultural ecosystems have been carried out extensively by use of the acetylene reduction technique (Section 5.13), but the data are more complete for terrestrial than aquatic systems. A breakdown can be made of global nitrogen fixation rates in different kinds of systems as follows (all in units of 10^{12} g/year): *agriculture:* legumes, 35; rice, 4; grasslands, 45; other crops, 5; *forests,* 40; *other habitats* (for example, deserts), 10. Thus about half of the terrestrial biological nitrogen fixation is in association with agricultural crops. In the case of legumes, the fixation is by means of symbiotic *Rhizobium* in root nodules (Section 14.10), but in other crops fixation may occur in free-living nitrogen-fixing bacteria living in the rhizosphere. Rates of fixation by free-living nitrogen fixers seem to be highly variable from one habitat to another; hence reliable estimates of nonleguminous biological nitrogen fixation are not yet available.

Denitrification   We have discussed the role of nitrate as an alternative electron acceptor in Section 4.11 and the enzymology of nitrate reduction was discussed briefly in Section 5.13. Assimilatory nitrate reduction, in which nitrate is reduced to the oxidation level of ammonia, for use as a nitrogen source for growth, and dissimilatory nitrate reduction, in which nitrate is used as an alternative electron acceptor in energy generation, are contrasted in Figure 13.13. Under most conditions, the end product of dissimilatory nitrate reduction is N_2 or N_2O, and the conversion of nitrate to gaseous nitrogen compounds is called **denitrification.** This process is the main means by which gaseous N_2 is formed biologically, and since N_2 is much less readily available to organisms than nitrate as a source of nitrogen, denitrification is a detrimental process.

The enzyme involved in the first step of nitrate reduction, nitrate reductase, is a molybdenum-containing enzyme. In general, assimilatory nitrate reductases are soluble proteins which are ammonia repressed, whereas dissimilatory nitrate reductases are membrane-bound proteins which are repressed by O_2 and synthesized under anaerobic (anoxic) conditions. Because O_2 inhibits the synthesis of dissimilatory nitrate reductase, the process of denitrification is strictly an anaerobic process, whereas assimilatory nitrate reduction can occur quite well under aerobic conditions. Assimilatory nitrate reduction occurs in all plants and most fungi, as well as in many bacteria, whereas dissimilatory nitrate reduction is restricted to bacteria, although a wide diversity of bacteria can carry out this process. In all cases, the first product of nitrate reduction is nitrite, NO_2^-, and another enzyme, nitrite reductase, is responsible for the next step. In the dissimilatory process, two routes are possible, one to ammonia and the other to N_2. The route to ammonia is carried out by a fairly large number of bacteria, but is of less significance geochemically because it does not result in the formation of a gaseous product. There are also some bacteria which do not reduce nitrate but do reduce nitrite to ammonia. This may be a detoxification mechanism, since nitrite can be fairly toxic under mildly acidic conditions (nitrous acid is an effective mutagen, Table 9.6). The pathway to nitrogen gas proceeds via two intermediate gaseous forms of nitrogen, nitric oxide (NO) and nitrous oxide (N_2O). Several organisms are known which produce

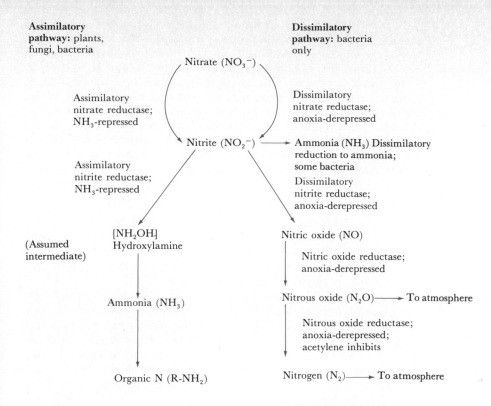

Assimilatory pathway: plants, fungi, bacteria	Dissimilatory pathway: bacteria only

Nitrate (NO_3^-)

Assimilatory nitrate reductase; NH_3-repressed

Dissimilatory nitrate reductase; anoxia-derepressed

Nitrite (NO_2^-) → Ammonia (NH_3) Dissimilatory reduction to ammonia; some bacteria

Assimilatory nitrite reductase; NH_3-repressed

Dissimilatory nitrite reductase; anoxia-derepressed

(Assumed intermediate)

[NH_2OH] Hydroxylamine

Nitric oxide (NO)

Nitric oxide reductase; anoxia-derepressed

Ammonia (NH_3)

Nitrous oxide (N_2O) → To atmosphere

Nitrous oxide reductase; anoxia-derepressed; acetylene inhibits

Organic N (R-NH_2)

Nitrogen (N_2) → To atmosphere

Representative genera of dissimilatory NH_3-formers:

Aeromonas
Bacillus (certain species)
Enterobacter
Erwinia
Flavobacterium
Micrococcus
Mycobacterium
Nocardia
Staphylococcus
Vibrio

Representative genera of N_2-formers:

Bacillus (certain species)
Hyphomicrobium
Moraxella
Pseudomonas
Spirillum
Thiobacillus (one species)

FIGURE 13.13 Comparison of various processes for the reduction of nitrate.

only N_2O during the denitrification process, while other organisms produce N_2 as the gaseous product. Because of their global significance, the formation of gaseous nitrogen compounds has been under considerable study.

Both nitrous oxide, N_2O, and nitric oxide, NO, are found in the atmosphere, although the former is quantitatively more important. Both of these gases are important in the regulation of the amount of ozone in the atmosphere. Since ozone in the upper atmosphere protects the surface of the earth from excessive ultraviolet radiation, destruction of ozone via reactions with nitrogen oxides could be of major global significance. It has been calculated that an *increase* of 1 percent in the emission rate of nitrous oxide will cause a 0.2 percent *decrease* in stratospheric ozone.

Although molecular nitrogen is the main product of denitrification under anaerobic conditions, nitrous oxide predominates under microaerobic conditions. The lower the pH of the habitat, the greater is the proportion of nitrous oxide formed, although at the same time the total rate of denitrification decreases. Under conditions of high nitrate, nitrous oxide is favored, whereas with a good supply of organic energy sources, complete denitrification to N_2 is favored. There are many complications in attempting to predict the ultimate rate at which nitrous oxides are released into the atmosphere, but it is almost certain that excessive fertilization can be a significant factor. Since even fertilization with ammonia leads to nitrate formation (see next section), switching from nitrate to ammonia as fertilizer may not lead to correction of the problem.

Ammonia fluxes and nitrification Ammonia is produced during the decomposition of organic nitrogen compounds (**ammonification**) and exists at neutral pH as ammonium ion (NH_4^+). Under anaerobic conditions, ammonia is stable, and it is in this form that nitrogen predominates in anaerobic sediments. In soils, much of the ammonia released by aerobic decomposition is rapidly recycled and converted into amino acids in plants. Because ammonia is volatile, some loss can occur from soils (especially highly alkaline soils) by vaporization, and major losses of ammonia to the atmosphere occur in areas of dense animal populations (for example, cattle feedlots). On a global basis, ammonia constitutes only about 15 percent of the nitrogen released to the atmosphere, the majority of the rest being in the form of N_2 or N_2O (from denitrification).

In aerobic environments ammonia can be oxidized to nitrogen oxides and nitrate, but ammonia is a rather stable compound and strong oxidizing agents or catalysts are usually needed for the chemical reaction. However, a specialized group of bacteria, the nitrifying bacteria, are excellent catalysts, oxidizing ammonia to nitrate in a process called **nitrification.** Nitrification proceeds in two steps, each catalyzed by a different group of bacteria:

$$NH_4^+ \longrightarrow NO_2^- \quad \textit{nitrosofying (nitrite-forming) bacteria}$$
$$NO_2^- \longrightarrow NO_3^- \quad \textit{nitrifying (nitrate-forming) bacteria}$$

It is surprising that a single organism does not catalyze the entire process, but despite the requirement for two organisms, nitrite rarely accumulates during nitrification, since both nitrosofying and nitrifying bacteria are generally present together in ecosystems. The characteristics of these interesting bacteria are discussed in Section 19.18.

Nitrification is strictly an aerobic process and occurs readily in well-drained soils at neutral pH; it is inhibited by anaerobic conditions or in highly acidic soils. If materials high in protein, such as manure or sewage, are added to soils, the rate of nitrification is increased. Although nitrate is readily assimilated by plants, it is very water soluble and is rapidly leached from soils receiving high rainfall. Consequently, nitrification is not necessarily beneficial in agricultural practice. Ammonia, on the other hand, is cationic and consequently is strongly adsorbed to negatively-charged clay minerals. Anhydrous ammonia is now used extensively as a nitrogen fertilizer, but some of the added ammonia is almost certainly converted to nitrate by nitrifying organisms. The significance of agricultural fertilization in nitrate formation can be appreciated by considering rates of nitrate discharge into rivers in agricultural as opposed to nonagricultural regions. In agricultural regions, it has been estimated that the rate of nitrate discharge is 152 mg of nitrogen per square meter per year, whereas in nonagricultural areas (both tropical and nontropical), the discharge rate has been estimated at about 24 mg of nitrogen per square meter per year. Because nitrate is readily soluble, it can leach into ground water. Excess nitrate can be detrimental because it is converted in the anaerobic intestines of humans and animals into nitrite. Nitrite can then combine with the hemoglobin of the blood to form methemoglobin, causing bodily harm and even death. Because of this, drinking water regulations require nitrate levels to be less than 10 mg/liter, although in some parts of the world (especially arid regions, where deep leaching of ground water occurs and heavy agricultural fertilization in irrigated cropland is practiced), concentrations of nitrate often exceed this standard. Another disadvantage of nitrate accumulation is that denitrification can occur in anaerobic pockets or under temporary conditions of waterlogging, resulting in formation of N_2 or nitrogen oxides, the latter of which can escape to the atmosphere and contribute to ozone depletion and nitrogen loss. In response to the nitrate problem, farmers using ammonia fertilizers have added certain chemicals known to inhibit the nitrification process. One of the most common inhibitors is a substituted pyridine called 2-chloro-6- (trichloromethyl) pyridine (commonly referred to as nitrapyrin or N-Serve). The addition of nitrification inhibitors has considerably reduced the conversion of ammonia to nitrate and has served to increase the efficiency of the fertilization process.

13.9

The Phosphorus Cycle, Algal Productivity, and Eutrophication

The driving force of the biogeochemical cycles is organic matter derived from photosynthetic activities. In aquatic systems, photosynthesis is primarily carried out by microorganisms (algae, cyanobacteria). A knowledge of the factors influencing algal productivity is thus of importance in understanding and predicting the rates of various steps in the biogeochemical cycles. In addition, excessive production of algal biomass, as a result of fertilization of aquatic habitats, has resulted in a number of serious practical problems. The word **eutrophication** is used to express the excessive fertilization of natural waters. A eutrophic lake is one with large amounts of algal nutrients, so that development of undesirably large amounts of algal biomass occurs. The opposite of a eutrophic lake is an **oligotrophic** lake, one in which algal nutrients are in low supply, and **mesotrophy** describes a condition between oligotrophy and eutrophy.

Formula for photosynthetic production of organic matter Analytical data on algal populations (for example, phytoplankton) and measurements of nutrient uptake during algal growth have permitted the construction of the following simple chemical equation:

$$106CO_2 + 16NO_3^- + PO_4^{3-} + 122H_2O + 18H^+$$

$$\text{(plus trace elements, solar energy)}$$

$$\longrightarrow \underset{\textit{Algal protoplasm}}{(C_{106}H_{263}O_{110}N_{16}P_1)} + 138O_2$$

This equation shows that there is a mole ratio of C : N : P of 106 : 16 : 1 in algal protoplasm. Although this equation is clearly an over-simplification, it has been shown to hold remarkably well for many aquatic systems, both marine and freshwater. This equation reflects in a simple way Liebig's law of the minimum, which states that plant growth is controlled by the availability of a single nutrient, the limit-ing nutrient. Any other nutrients in excess will be assimilated only in proportion to their needs for synthesis of protoplasm, the excess of such nonlimiting nutrients remaining unassimilated in the en-vironment.*

Phosphorus as a limiting nutrient A large number of data have been obtained which shows that phosphorus is more likely to be the limit-ing nutrient for algal productivity than any other element. (This does not mean that at times, in certain systems, other nutrients might not be limiting.) The evidence that phosphorus is the limiting algal nutrient comes from studies on changes in various algal nutrients in the water during development of extensive algal blooms (discussed below), fer-tilization experiments in whole lakes and in artificial lake models, and studies of the biogeochemical cycles of the various elements. Of the key elements of algal protoplasm, C, N, and P, only P does not have a volatile phase and does not undergo oxidation and reduction.† For C and N, as we have discussed earlier in this chapter, atmospheric sources exist, CO_2 and N_2, so that deficiencies of these elements in aquatic systems can be made up by assimilation from the atmosphere. Further, not only does phosphorus lack a volatile phase, but it also forms highly insoluble precipitates with Ca, Mg, and Fe, so that in many aquatic systems P is carried out of the water column into the sediment. Also, because of its insolubility, phosphorus does not move readily in groundwater, as do N (in the form of nitrate) and C (in the form of bicarbonate). Thus there are many reasons why phos-phorus is the most likely element to limit algal productivity in aquatic systems.

From the stoichiometry of algal production (see above), it can be concluded that algal uptake of N and P from water should be in a ratio of 16 N : 1 P. Some of the best evidence for this stoichiometry comes from analysis of nitrate and phosphate in the western Atlantic Ocean, as illustrated in Figure 13.14. It can be seen that there is a close correlation between concentrations of N and P in these waters and that the ratio is almost exactly 16 N : 1 P. The fact that the ratio is so close to the prediction is probably because productivity is originally

*This is an oversimplification; some organisms accumulate certain nutrients when they are in excess supply, using them when environmental deficiencies result, but these are only temporary perturbations of the system, and in the long run only limiting nutrients become completely depleted from the environment.

†Phosphate can be reduced to phosphite, hypophosphite, and phosphine, but the reduction potential of phosphate reduction is so low that the reaction rarely, if ever, occurs in natural environments.

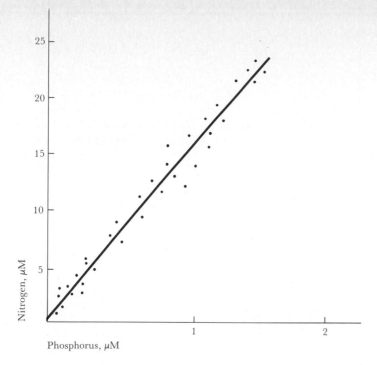

FIGURE 13.14 Correlations between soluble nitrogen and phosphorus in the western Atlantic Ocean. Analyses were performed on water samples collected in various locations and at various times of the year. The sizes of the values indicate the degree of algal activity. Note that the ratio of N:P is 16:1, reflecting the algal composition of these two elements (see text). (Data from Redfield A. C., B. H. Ketchum, and F. A. Richards. 1963. In Hill, M. N., ed. The sea, vol. 2. John Wiley, N.Y.)

controlled by P, and the balance of N is maintained by nitrogen fixation and denitrification. If there is more P than needed to make the 16 : 1 ratio, then nitrogen fixation by cyanobacteria will occur and increase the N value until this ratio is reached. In contrast, if there is excess N, then denitrification will occur, causing loss of N, and change the ratio to 16 : 1 again. This interpretation assumes that the processes involved are integrated over the whole water column and that sufficient time has been available for the development of appropriate equilibria; short-term fluctuations about this general equilibrium could well occur.

Correlation of phosphorus concentration with algal biomass in lakes
Eutrophication has been of greatest concern in lakes because of the vital importance of lakes to human welfare, for example, as sources of drinking water and recreational bases, and also because lakes are closely coupled to their surrounding basins. A large number of studies have been done relating algal biomass in lakes to phosphorus content. For such studies, algal biomass is conveniently quantified by measuring chlorophyll content, since chlorophyll is easy to measure chemically and is a substance unique to photosynthetic organisms. Figure 13.15 shows the relationship between total phosphorus and chlorophyll for a series of temperate lakes, and it can be seen that the correlation is exceedingly good. Data such as these provide further evidence for the importance of phosphorus in limiting algal growth in aquatic systems. To curb an important source of phosphorus input, many detergent manufacturers have reduced the phosphate content of their products and this has resulted in significant water quality improvements in some areas.

FIGURE 13.15 Relationship between phosphorus content of lakes and their algal biomass. Each dot represents a separate lake. The lakes are primarily in temperate climates, and total phosphorus (including both soluble and particulate P) is given during the period of maximum mixing of the lake water column in the spring of the year. Chlorophyll values are given for the summer period of maximum algal growth. (From Dillon, P. J. 1974. National Research Council of Canada Publication No. NRCC 13690, Ottawa, Canada.)

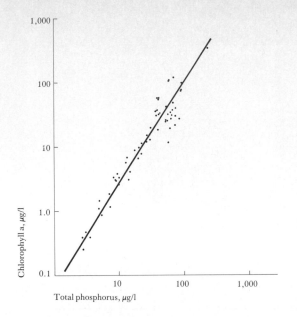

13.10

The Sulfur Cycle

Sulfur transformations are even more complex than those of nitrogen due to the variety of oxidation states of sulfur and the fact that some transformations occur at significant rates chemically as well as biologically. The redox cycle for sulfur and the involvement of some microorganisms in sulfur transformations are given in Figure 13.16. Although a number of oxidation states are possible, only three form significant amounts of sulfur in nature, -2 (sulfhydryl, $R\!-\!SH$ and

FIGURE 13.16 Redox cycle for sulfur. Oxidation states of key compounds: Organic S ($R\!-\!SH$), -2; Sulfide H_2S, HS^-, -2; Elemental sulfur S^0, 0; Thiosulfate $S\!-\!SO_3^{2-}$ ($S_2O_3^{2-}$), $+2$ (average); Tetrathionate $^{2-}O_3S\!-\!S\!-\!S\!-\!SO_3^{2-}$ ($S_4O_6^{2-}$), $+2$ (average); Sulfur dioxide SO_2, $+4$; Sulfite SO_3^{2-}, $+4$; Sulfur trioxide SO_3, $+6$; Sulfate SO_4^{2-}, $+6$.

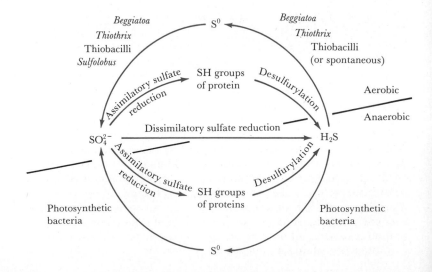

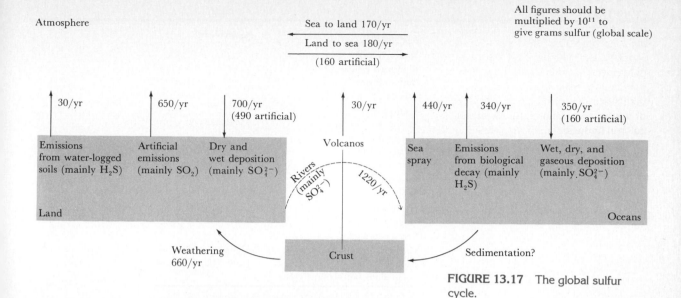

All figures should be multiplied by 10^{11} to give grams sulfur (global scale)

Atmosphere

Sea to land 170/yr

Land to sea 180/yr

(160 artificial)

| 30/yr | 650/yr | 700/yr (490 artificial) | 30/yr | 440/yr | 340/yr | 350/yr (160 artificial) |

Emissions from water-logged soils (mainly H_2S)

Artificial emissions (mainly SO_2)

Dry and wet deposition (mainly SO_4^{2-})

Volcanos

Sea spray

Emissions from biological decay (mainly H_2S)

Wet, dry, and gaseous deposition (mainly SO_4^{2-})

Land

Rivers (mainly SO_4^{2-})

1220/yr

Oceans

Weathering 660/yr

Crust

Sedimentation?

FIGURE 13.17 The global sulfur cycle.

sulfide, HS^-), 0 (elemental sulfur, S^0), and $+6$ (sulfate). Sulfur reservoirs on earth are given in Table 13.4, and it can be seen that in contrast to nitrogen, the atmosphere is quite insignificant as a sulfur reservoir. The bulk of the sulfur of the earth is found in sediments and rocks, in the form of sulfate minerals (primarily gypsum, $CaSO_4$) and sulfide minerals (primarily pyrite, FeS_2), although the oceans constitute the most significant reservoir of sulfur for the biosphere (in the form of inorganic sulfate). The global transport cycle for sulfur is given in Figure 13.17, and some of the components of this cycle are discussed below.

Hydrogen sulfide and sulfate reduction   A major volatile sulfur gas is hydrogen sulfide. This substance is formed primarily by the bacterial reduction of sulfate:

$$SO_4^{2-} + 8e^- + 8H^+ \rightleftharpoons H_2S + 2H_2O + 2OH^-$$

The form in which sulfide is present in an environment depends on pH due to the following equilibria:

$$\underset{\text{Low pH}}{H_2S} \rightleftharpoons \underset{\text{Neutral pH}}{HS^-} \rightleftharpoons \underset{\text{High pH}}{S^{2-}}$$

At high pH, the dominant form is sulfide, S^{2-}. At neutral pH, HS^- predominates, and at acid pH, H_2S is the major species. For simplicity, the reactions will be written with HS^-, the major form at neutral pH, unless a gaseous product is involved, in which case H_2S is written. HS^- and S^{2-} are very water soluble, but H_2S is not and readily volatilizes. Even at neutral pH, some volatilization of H_2S from HS^- can occur, because there is an equilibrium between HS^- and H_2S, and as volatilization occurs, the reaction is pulled toward H_2S.

A wide variety of organisms can use sulfate as a sulfur source and carry out assimilatory sulfate reduction, converting the HS^- formed to organic sulfur, $R-SH$. HS^- is ultimately formed from the decomposition of this organic sulfur by putrefaction and desulfuryla-

TABLE 13.4 Sulfur Reservoirs on Earth

Component	Reservoir (g)
Atmosphere:	
Hydrogen sulfide	9.6×10^{11}
Sulfur dioxide	6.4×10^{11}
Sulfate	16×10^{11}
Land:	
Living organic (mainly plants)	$25,000–40,000 \times 10^{11}$
Dead organic	$35,000–60,000 \times 10^{11}$
Inorganic	Unknown
Oceans:	
Living organic	0.0000035×10^{11}
Dead organic	0.00038×10^{11}
Inorganic (sulfate)	$13,760,000,000 \times 10^{11}$
Sediments and rocks:	
Calcium sulfate (gypsum)	$63,000,000,000 \times 10^{11}$
Metal sulfides predominantly pyrite FeS_2)	$47,000,000,000 \times 10^{11}$

Sulfur in living and dead organic matter was calculated from the carbon reservoirs in Table 13.1, assuming that the ratio of C : S is 200 : 1. Rest of data from Garrels, R. M., F. T. Mackenzie, and C. Hunt. 1975. Chemical cycles and the global environment. William Kaufmann, Los Altos, Calif.

tion (Figure 13.16), and this is a significant source of HS^- in fresh water. In the marine environment, because of the vast amount of sulfate present (Table 13.4), dissimilatory sulfate reduction is the main source of HS^-. Dissimilatory sulfate reduction, in which sulfate serves as an electron acceptor, is carried out by a variety of bacteria, collectively referred to as the sulfate-reducing bacteria (Sections 4.12, 19.16), which are all obligate anaerobes. It should be emphasized that the sulfate anion is very stable chemically, and its reduction does not occur spontaneously in nature at normal environmental conditions.

The reduction of sulfate to sulfide proceeds through a number of intermediate oxidation states, of which only sulfite, SO_3^{2-}, has been positively identified. As discussed in Section 5.14, the initial step in sulfate reduction involves the activation of the sulfate anion with ATP, leading to the formation of adenosyl phosphosulfate (APS), and it is this activation step which is the key biological reaction in sulfate reduction. Once activated, reduction to sulfite occurs readily with reduced NAD. Sulfite itself is much more readily reduced than sulfate. This reaction can occur chemically, and a variety of bacteria also carry it out. Other intermediate oxidation states of sulfur: thiosulfate and tetrathionate, may be intermediates in dissimilatory sulfate reduction, although details are not known, but these substances, in addition, can be reduced by a variety of bacteria. Thus, once the initial activation and reduction of sulfate to sulfite occurs, the rest of the reduction process occurs readily. Some of the environmental conditions for sulfate-reducing bacteria and electron donors involved in sulfate reduction are given in Table 13.5. As discussed in Section 13.7, there is a competition in nature between methanogenic and sulfate-

TABLE 13.5 Characteristics of Sulfate Reduction

Organisms: *Desulfovibrio, Desulfotomaculum, Desulfomonas, Desulfobulbus, Desulfobacter, Desulfococcus, Desulfosarcina, Desulfonema*

Environmental limits:

Redox	$+350$ to -500 mV O_2, NO_3^-, or Fe^{3+} absent
pH	4.2 to 10.4
Pressure	1 to 1000 atm (deep oceans)
Temperature	0 to 100°C
Salinity	Fresh water to 5 percent NaCl

Electron donors for sulfate reduction:*

Generally used	Restricted use (some organisms)
H_2	Formate
Lactate	Choline
Pyruvate	Ethanol
	Fumarate
	Malate
	Propionate
	Acetate
	Carbon monoxide
	Butyrate
	Long-chain fatty acids
	Benzoate

*Some electron donors (e.g., pyruvate, fumarate) are also used fermentatively by sulfate-reducing bacteria. Others may be used fermentatively with H_2 production if a methanogenic bacterium is present to remove the H_2 as it is formed.

reducing bacteria for available electron donors, especially H_2 and acetate, and as long as sulfate is present the sulfate-reducing bacteria are favored. Because of the necessity of organic electron donors (or molecular H_2, which is itself derived from the fermentation of organic compounds), sulfate reduction only occurs extensively where significant amounts of organic matter are present. Thus there is a relationship between primary production of organic carbon and intensity of sulfate reduction. For example, a reaction carried out by many sulfate-reducing bacteria is the reduction of sulfate with lactate as electron donor:

$$2\,\text{Lactate} + SO_4^{2-} \longrightarrow 2\,\text{Acetate} + 2CO_2 + HS^- + H_2O + OH^-$$

where the ratio of carbon oxidized to sulfate reduced is 2 : 1. Thus 2 moles of organic carbon are needed to reduce 1 mole of sulfate to HS^-. In many marine sediments, the rate of sulfate reduction is carbon limited, and the rate can be greatly increased by addition of organic matter. This is of considerable importance for marine pollution, since disposal of sewage, sewage sludge and garbage in the sea can lead to marked increases in organic matter in the sediments. Since HS^- is a toxic substance to many organisms, formation of HS^- by sulfate reduction is potentially detrimental.

A major factor in determining the fate and significance in the environment of HS^- is the availability of iron and heavy metals. Most metals form highly insoluble sulfides, and metal sulfide formation will occur spontaneously if the metal is available in the environment. Iron is fairly common in all sediments, so that sulfate reduction almost always results in the formation of iron sulfide, FeS, which is black and gives anaerobic sediment its characteristic dark color. If heavy metals (for example, Cu, Hg) are present, they are also precipitated. Thus only after all of the iron has reacted with sulfide will the formation of free sulfide occur. A number of economically important mineral deposits have been formed as a result of bacterial sulfate reduction. Biogenic sulfide mineral deposits are characteristically found in thin layers (stratiform) in association with other sedimentary rocks (for example, shales). They must be distinguished from sulfide mineral deposits of volcanic and hydrothermal origin, which are not stratiform and are generally associated with volcanic activity in the immediate area. Typical biogenic stratiform sulfide ore deposits are those of Mount Isa (Australia), Northern Rhodesia and Katanga (Africa), and the Kupferschiefer deposits in Germany.

Isotope fractionation during sulfate reduction   The role of sulfate-reducing bacteria in the formation of economic sulfur deposits is known from studies of the fractionation of stable (that is, nonradioactive) isotopes of sulfur. Sulfur has two major stable isotopes, ^{32}S, the normal isotope, which comprises 95 percent of the total sulfate of seawater, and ^{34}S, which comprises about 4 percent. When sulfate-reducing bacteria convert sulfate to HS^-, they show a slight preference for the lighter ^{32}S isotope, a process referred to as **isotope fractionation**. Hence the HS^- formed has less ^{34}S than did the sulfate (Table 13.6). Nonbiogenic sulfide (as, for instance, from volcanic deposits) does not show this bias toward the light isotope (Figure 13.18). Also, oxidation of sulfide to elemental sulfur or sulfate, either aerobically or anaerobically, shows a slight preference for the lighter isotope (Table 13.6). Thus measurement of $^{34}S : ^{32}S$ ratios of various sulfur and sulfide deposits can give a clue as to their microbial origin.

FIGURE 13.18   Summary of the isotope geochemistry of sulfur, indicating the range of values for ^{34}S and ^{32}S in various sulfur-containing substances. The arrows indicate the mean values. For definition of $\delta^{34}S$ see the legend to Table 13.6. Note that sulfide and sulfur of biogenic origin tend to be light in ^{34}S. (From Holser, W. T., and I. R. Kaplan. 1966. Isotope geochemistry of sedimentary sulfates. Chemical Geology 1:93–135.)

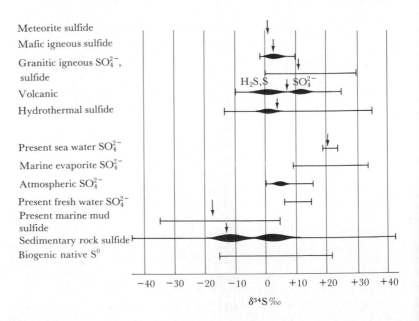

TABLE 13.6 Selective Utilization of ^{32}S over ^{34}S (Isotope Fractionation) in Certain Parts of the Microbiological Sulfur Cycle

Process	Starting Substance	End Product	Isotope Fractionation*
Dissimilatory sulfate reduction (*Desulfovibrio*)	SO_4^{2-}	HS^-	−46.0
Dissimilatory sulfite reduction (*Desulfovibrio*)	SO_3^{2-}	HS^-	−14.3
Assimilatory sulfate reduction (*E. coli*)	SO_4^{2-}	Organic S	−2.8
Putrefaction and desulfurylation (*Proteus*)	Organic S	HS^-	−5.1
Lithotrophic sulfide oxidation (*Thiobacillus*)	HS^-	S^0	−2.5
	HS^-	SO_4^{2-}	−18.0
Phototrophic sulfide oxidation (*Chromatium*)	HS^-	S^0	−10.0
	HS^-	SO_4^{2-}	0

*Isotope fractionation is expressed as:

$$\delta S^{34}\,^0/_{00} = \frac{S^{34}/S^{32}\ sample - S^{34}/S^{32}\ standard}{S^{34}/S^{32}\ standard} \times 1000$$

(The standard is an iron sulfide mineral from the Canyon Diablo Meteorite.)

Based on Goldhaber, M. B., and I. R. Kaplan. 1974. The sulfur cycle. Pages 569–655 *in* E. D. Goldberg, ed. The sea, vol. 5. John Wiley, New York.

Sulfide and elemental sulfur oxidation Sulfide (HS^-) rapidly oxidizes spontaneously at neutral pH, if O_2 or ferric iron is present. When HS^- reacts with O_2, oxidation generally does not proceed all the way to sulfate spontaneously, but stops either at elemental sulfur or thiosulfate. The precise product formed depends on the ratio of sulfide to oxygen, the initial concentration of sulfide, and the presence or absence of metals as catalysts. The sulfur-oxidizing bacteria are also able to catalyze the oxidation of sulfide, but because of the rapid spontaneous reaction, bacterial oxidation of sulfide only occurs in narrow zones or regions where H_2S rising from anaerobic areas meets O_2 descending from aerobic areas. If light is available, anaerobic oxidation of HS^- can also occur, catalyzed by the photosynthetic bacteria (see Sections 6.4, 6.5, and 19.1), but this only occurs in restricted areas, usually in lakes, where sufficient light can penetrate to anaerobic zones.

Elemental sulfur, S^0, is chemically stable in most environments in the presence of oxygen, but is readily oxidized by sulfur-oxidizing bacteria. Although a number of sulfur-oxidizing bacteria are known, members of the genus *Thiobacillus* (Section 19.19) are most commonly involved in elemental sulfur oxidation. Elemental sulfur is very insoluble, and the bacteria that oxidize it attach firmly to the sulfur crystals (Figure 6.17*b*). Oxidation of elemental sulfur results in the formation of sulfate and hydrogen ions, and sulfur oxidation characteristically results in a lowering of the pH. Elemental sulfur is sometimes added to alkaline soils to effect a lowering of the pH, reliance being placed on the ubiquitous thiobacilli to carry out the acidification process.

We will discuss the interrelationships between iron and sulfide oxidation in the next section, and this will be followed by a discussion of iron-sulfur interactions in mining practices, a process which leads to the formation of sulfuric acid, a key compound of acid mine drainage.

13.11

Iron and Manganese Transformations

Iron is one of the most abundant elements in the earth's crust, but is a relatively minor component in aquatic systems because of its relative insolubility in water. Iron exists in two oxidation states, ferrous ($+$II) and ferric ($+$III).* The form in which iron is found in nature is greatly influenced by pH and oxygen. Because of the high electrode potential of the Fe^{3+}/Fe^{2+} couple, 0.76 mV, the only electron acceptor able to oxidize ferrous iron is oxygen, O_2. At neutral pH, ferrous iron oxidizes spontaneously in air to ferric iron, which forms highly insoluble precipitates of ferric hydroxide and ferric oxides. Thus, at neutral pH, the only way that iron is maintained in solution is by chelation with organic materials (see below). Although ferrous iron is also fairly insoluble in water, it is considerably more soluble than ferric iron, so that reduction of ferric to ferrous iron results in some solubilization of iron.

Bacterial iron reduction and oxidation The bacterial reduction of ferric iron to the ferrous state is a major means by which iron is solubilized in nature. As we noted in Section 4.11, a number of organisms can use ferric iron as an electron acceptor. Many of these organisms also reduce nitrate, and since they are facultative anaerobes, they can also use O_2. The overall reaction of ferric iron reduction can be represented as follows (using H_2, a common electron donor for many organisms):

$$Fe(OH)_3 + \tfrac{1}{2}H_2 \xrightarrow{\text{bacteria}} Fe^{2+} + 2OH^- + H_2O$$

In addition to the bacterially catalyzed reduction, if hydrogen sulfide is present, as it is in many anaerobic environments, ferric iron is also reduced chemically to FeS (ferrous sulfide). Thus, there are complex interactions in many environments between the iron and sulfur cycles.

Ferric iron reduction is very common in waterlogged soils, bogs, and anaerobic lake sediments. In many waterlogged soils, the ferrous iron so formed is leached out of the soil, leaving the soil gray and mottled, a condition called **gleying**. (The faintly brownish or reddish color of most normal surface soils is due to the presence of oxidized iron.) Movement of iron-rich groundwater from anaerobic bogs or waterlogged soils can result in the transport of considerable amounts of iron. Once this iron-laden water reaches aerobic regions, the ferrous iron is quickly oxidized spontaneously and ferric compounds precipitate, leading to the formation of a brown deposit (Color Plate 3c). Such brown deposits frequently are a serious problem in the pipes used to carry industrial or drinking water from deep wells. Iron de-

---

*Elemental iron, Fe^0, does not exist significantly in nature but is a widespread artificial product. It is subject to biological corrosion processes.

posits are very common at the edges of bogs, and many of the great iron-ore beds of the world are bog-iron deposits. A local manifestation of this phenomenon is the iron spring or iron seep, in which an extensive flow of iron-rich water results in the movement of ferrous iron considerable distance downstream before oxidation and precipitation occur. The overall reaction of ferrous iron oxidation can be represented as follows:

$$Fe^{2+} + \tfrac{1}{4}O_2 + H^+ \longrightarrow Fe^{3+} + \tfrac{1}{2}H_2O$$
$$Fe^{3+} + 3OH^- \longrightarrow Fe(OH)_3 \text{ precipitates}$$
$$\text{Summation: } Fe^{2+} + \tfrac{1}{4}O_2 + 2OH^- + \tfrac{1}{2}H_2O \longrightarrow Fe(OH)_3$$

Note that although the initial oxidation of ferrous iron consumes hydrogen ions and thus leads to a rise in pH, the hydrolysis of Fe^{3+} and formation of $Fe(OH)_3$ consumes hydroxyl ions and leads to an acidification of the medium. This is one way in which iron oxidation leads to the formation of acidic conditions in the environment.

Although ferric iron forms very insoluble hydroxides, some ferric iron can be kept in solution in natural waters by chelation with organic materials. We described the iron chelation compounds involved in iron transport in Section 5.16; these plus a variety of other iron chelation compounds exist in natural waters and serve to keep some ferric iron in solution. If an organism is present that can oxidize the organic chelator, then the iron present will precipitate. This is probably a major mechanism of iron precipitation in many neutral pH environments. In addition, at neutral pH, organisms such as *Gallionella* and *Leptothrix* contribute to the oxidation of ferrous iron, but it is unlikely that the ferrous to ferric iron conversion serves as an energy source for these bacteria as it does for acidophilic thiobacilli (see below).

Ferrous iron oxidation at acid pH   At acid pH values, spontaneous oxidation of ferrous iron to the ferric state is low. However, the acidophilic lithotrophic organism *Thiobacillus ferrooxidans* is able to catalyze the oxidation (Figure 13.19). In the fixation of CO_2, *T. ferrooxidans* oxidizes ferrous iron as its primary energy-generating process. Because very little energy is generated in the oxidation of ferrous to ferric iron (see Section 6.10), these bacteria must oxidize large amounts of iron in order to grow, and consequently even a small number of cells can be responsible for precipitating a large amount of iron. This iron-oxidizing bacterium, which is a strict acidophile, is very common in acid mine drainages and in acid springs, and is probably responsible for most of the iron precipitated at acid pH values.

Thiobacillus ferrooxidans lives in environments in which sulfuric acid is the dominant acid and large amounts of sulfate are present. Under these conditions, ferric iron does not precipitate as the hydroxide, but as a complex sulfate mineral called *jarosite*, which has the approximate formula of: $MFe_3(SO_4)_2(OH)_6$ where M = K^+, NH_4^+, or H^+. In culture medium, M is usually K^+ or NH_4^+, but in nature, where these cations are low, M is generally H^+. Jarosite is a yellowish or brownish precipitate and is responsible for one of the manifestations of acid mine drainage, an unsightly yellow stain, called "yellow boy" by U.S. miners (see Color Plate 3*b*).

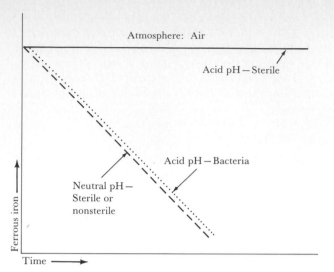

Atmosphere: Air

Acid pH — Sterile

Acid pH — Bacteria

Neutral pH —
Sterile or
nonsterile

Ferrous iron

Time

FIGURE 13.19 Oxidation of ferrous iron as a function of pH and the presence of the bacterium *Thiobacillus ferrooxidans*.

Pyrite oxidation One of the most common forms of iron and sulfur in nature is pyrite, which has the overall formula FeS_2. Pyrite is formed from the reaction of sulfur with ferrous sulfide (FeS) to form a highly insoluble crystalline structure, and pyrite is very common in bituminous coals and in many ore bodies. The bacterial oxidation of pyrite is of great significance in the development of acidic conditions in mines and mine drainages. Additionally, oxidation of pyrite by bacteria is of considerable importance in the process called microbial leaching of ores (see below). The oxidation of pyrite is a combination of spontaneous and bacterially catalyzed reactions. Two electron acceptors for this process can function: molecular oxygen (O_2) and ferric ions (Fe^{3+}). However, ferric ions are only present when the solution is acidic, at pH below about 2.5. At pH values above 2.5, ferric ion reacts with water to form the insoluble ferric hydroxide. When pyrite is first exposed, as in a mining operation, a slow spontaneous reaction with molecular oxygen occurs as shown in the following reaction:

$$FeS_2 + 3\frac{1}{2}O_2 + H_2O \longrightarrow Fe^{2+} + 2SO_4^{2-} + 2H^+$$

This reaction, called the *initiator reaction,* leads to the development of acidic conditions; once acidic conditions develop, the ferrous iron which is formed by the reaction is relatively stable in the presence of oxygen. However, *T. ferrooxidans* catalyzes under acid conditions the oxidation of ferrous to ferric ions. The ferric ions formed under these conditions, being soluble, can readily react with more pyrite to oxidize the pyrite to ferrous ions plus sulfate ions:

$$FeS_2 + 14Fe^{3+} + 8H_2O \longrightarrow 15Fe^{2+} + 2SO_4^{2-} + 16H^+$$

The ferrous ions formed are again oxidized to ferric ions by the bacteria, and these ferric ions again react with more pyrite. Thus there is a progressive, rapidly increasing rate at which pyrite is oxidized, called the *propagation cycle:*

Under natural conditions some of the ferrous iron generated by the

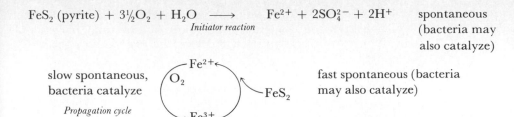

$$FeS_2 \text{ (pyrite)} + 3\tfrac{1}{2}O_2 + H_2O \longrightarrow Fe^{2+} + 2SO_4^{2-} + 2H^+$$

Initiator reaction

spontaneous
(bacteria may
also catalyze)

slow spontaneous,
bacteria catalyze

Propagation cycle

fast spontaneous (bacteria
may also catalyze)

bacteria leaches away, being carried by groundwater into surrounding streams. Since the pH of many of these acid mine drainages is quite low, the ferrous iron is stable in the absence of bacteria; but because oxygen is present in the aerated drainage, bacterial oxidation of the ferrous iron takes place. Since this is occurring in the absence of pyrite, the ferrous iron is oxidized to the ferric state, and an insoluble ferric precipitate is formed, as described in the preceding section.

Manganese   Manganese (Mn) is a transition element similar to iron. Although manganese can form more oxidation states than can iron ($+II$, $+III$, $+IV$, and $+VI$), the two elements show similar transformations. The divalent and tetravalent oxidation states of manganese are most common and are the ones that will be mentioned here. Divalent Mn is water soluble and predominates aerobically at pH values less than 5.5 or at higher pH values anaerobically. At pH values above 8 in air, Mn(II) is oxidized spontaneously to the water-insoluble Mn(IV), which forms manganese oxide, MnO_2. In some parts of the ocean and in some freshwater lakes, precipitation of manganese as insoluble MnO_2 occurs in the form of characteristic **manganese nodules.** These nodules are of potential economic importance for mining operations, because they often contain significant amounts of copper, cobalt, and other economically useful metals. The manganese deposited in these nodules has been derived from anaerobic sediments beneath them. Manganese is oxidized and precipitated when it diffuses into the aerobic zone. Although freshwater nodules develop fairly rapidly, the rate of development of the marine nodules is exceedingly slow, of the order of 1 mm thickness per million years.

The manganese redox changes can occur spontaneously as a result of changes in pH or O_2 availability, and since these latter changes are brought about by microbial action, manganese transformations can give the appearance of being carried out by specific manganese organisms. However, there is some evidence that certain bacteria are able to catalyze manganese oxidation specifically.

13.12

Mining Microbiology

Bacterial oxidation of sulfide minerals is the major factor in the formation of **acid mine drainage,** a common environmental problem in coal mining regions. The same bacterium, *Thiobacillus ferrooxidans,* also carries out a beneficial oxidation of sulfide minerals in the process called **microbial leaching,** which plays a major role in the concentration of copper from low-grade copper ores. In both cases, bacterial attack on pyrite is also involved, following the steps outlined in Section 13.11.

Acid mine drainage Not all coal seams contain iron sulfide; thus acid mine drainage does not occur in all coal mining regions. Where acid mine drainage does occur, however, it is often a very serious problem. Mixing of acidic mine waters with natural waters in rivers and lakes causes a serious degradation in the quality of the natural water, since both the acid and the dissolved metals are toxic to aquatic life. In addition, such polluted waters are unsuitable for human consumption and industrial use (see Color Plate 3*b*).

We have outlined in Section 13.11 the steps in the bacterial oxidation of pyrite, and we noted that certain of the steps occurred spontaneously, but that the rate-limiting step, the oxidation of ferrous to ferric iron, occurred at acid pH only in the presence of the bacterium. The breakdown of pyrite leads ultimately to the formation of sulfuric acid and ferrous iron, and pH values can be as low as pH 2. The acid formed attacks other minerals in the rock associated with the coal and pyrite, causing breakdown of the whole rock fabric. A major rock-forming element, aluminum, is only soluble at low pH, and often large amounts of aluminum are brought into solution. The typical composition of an acid mine water is as follows: pH, 2 to 4.5; Fe^{2+}, 500 to 10,000 mg/liter; Al^{3+}, 100 to 2,000 mg/liter; SO_4^{2-}, 1000 to 20,000 mg/liter; with small amounts of N, P, and trace elements.

The requirement for O_2 in the oxidation of ferrous to ferric iron helps to explain how acid mine drainage develops. As long as the coal is unmined, oxidation of pyrite cannot occur, since neither air nor the bacteria can reach it. When the coal seam is exposed, it quickly becomes contaminated with *T. ferrooxidans,* and O_2 is introduced, making oxidation of pyrite possible. The acid formed can then leach into the surrounding streams.

Since acid mine drainage would not develop in the absence of bacterial activity, it might seem logical to prevent its occurrence by using chemicals or other agents that kill or inhibit the growth of bacteria. A number of such agents are available, including antibiotics, antiseptics, and organic acids. None of these agents have yet proved of practical value in controlling acid mine drainage. They may be too expensive to use in the enormous quantities necessary to have significant impact, or it may be difficult or impossible to deliver them at appropriate concentrations to the active sites of acid production. To date, the only effective way of controlling acid mine drainage is to seal or cover acid-bearing material, and to use mining practices that keep air and bacteria from the significant sites. The acid produced from coal-mining operations can be neutralized by use of lime. However, after a time the lime particles become coated with a layer of ferric hydroxide, and their neutralizing power is thus reduced.

Microbial leaching Sulfide forms highly insoluble minerals with many metals, and many ores used as sources of these metals are sulfides. If the concentration of metal in the ore is low, it may not be

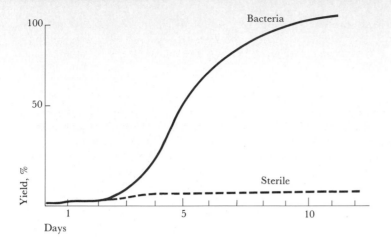

TABLE 13.7 Sulfide Minerals Listed in Order of Increasing Resistance to Oxidation

Pyrrhotite	FeS
Chalcocite	Cu_2S
Covellite	CuS
Tetrahedrite	$3Cu_2S \cdot Sb_2S_3$
Bornite	Cu_5FeS_4
Galena	PbS
Arsenopyrite	FeAsS
Sphalerite	ZnS
Pyrite	FeS_2
Enargite	$3Cu_2S \cdot As_2S_5$
Marcasite	FeS_2
Chalcopyrite	$CuFeS_2$
Molybdenite	MoS_2

economically feasible to concentrate the mineral by conventional chemical means. Under these conditions, **microbial leaching** is frequently practiced. Microbial leaching is especially useful for copper ores, because copper sulfate, formed during the oxidation of the copper sulfide ores, is very water soluble. We have noted that sulfide itself, HS^-, oxidizes spontaneously in air. Most metal sulfides will also oxidize spontaneously, but the rate is very much slower than that of free sulfide. Bacteria such as *T. ferrooxidans* are able to catalyze a much faster rate of oxidation of the sulfide minerals, thus aiding in solubilization of the metal. The relative rate of oxidation of a copper mineral in the presence and absence of bacteria is illustrated in Figure 13.20. The susceptibility to oxidation varies among minerals, as seen in Table 13.7, and those minerals which are most readily oxidized are most amenable to microbial leaching.

In the general microbial leaching process, low-grade ore is dumped in a large pile (the leach dump), and a dilute sulfuric acid solution (pH around 2) is percolated down through the pile (see Color Plate 3*a*). The liquid coming out of the bottom of the pile, rich in the mineral, is collected and transported to a precipitation plant where the metal is reprecipitated and purified. The liquid is then pumped back to the top of the pile and the cycle is repeated. As needed, more acid is added to maintain the low pH.

There are several mechanisms by which the bacteria can catalyze oxidation of the sulfide minerals. To illustrate these, examples will be used of the oxidation of two copper minerals, chalcocite, Cu_2S, in which copper has a valence of $+1$, and covellite, CuS, in which copper has a valence of $+2$. As illustrated in Figure 13.21*a*, *T. ferrooxidans* is able to oxidize monovalent copper in chalcocite to divalent copper, thus removing some of the copper in the soluble form, Cu^{2+}, and forming the mineral covellite. Note that in this reaction, there is no change in the valence of sulfide, the bacteria apparently utilizing the reaction Cu^+ to Cu^{2+} as a source of energy. This is analogous to the oxidation by the same bacterium of Fe^{2+} to Fe^{3+}.

A second mechanism, and probably the most important in most mining operations, involves an indirect oxidation of the copper ore with ferric ions that were formed by the bacterial oxidation of ferrous

(a) Direct oxidation of monovalent copper in chalcocite (Cu_2S) by *Thiobacillus ferrooxidans:*

$$2Cu_2S + O_2 + 4H^+ \longrightarrow 2CuS + 2Cu^{2+} + 2H_2O$$

Chalcocite   *Covellite*

$$(4Cu^+ \longrightarrow 4Cu^{2+} + 4e^-)$$

No change in valence of S. Electrons from Cu^+ oxidation probably used for energy generation by the bacteria.

(b) Indirect oxidation of covellite (also other ores) with ferric iron (Fe^{3+}) and regeneration of ferric iron from ferrous by bacterial oxidation:

$$CuS + 8Fe^{3+} + 4H_2O \longrightarrow Cu^{2+} + 8Fe^{2+} + SO_4^{2-} + 8H^+$$

$\uparrow$
chemical

$$8Fe^{2+} + 2O_2 + 8H^+ \longrightarrow 8Fe^{3+} + 4H_2O$$

$\uparrow$
bacterial

Overall: $CuS + 2O_2 \longrightarrow Cu^{2+} + SO_4^{2-}$

(c) Recovery of elemental copper from copper ions by reaction with scrap iron:

$$Cu^{2+} + Fe^0 \longrightarrow Fe^{2+} + Cu^0$$

(d) Indirect oxidation of uranium ore with ferric iron:

$$UO_2 \text{ (insoluble)} + 2 Fe^{3+} + SO_4^{2-} \longrightarrow UO_2SO_4 \text{ (soluble)} + 2 Fe^{2+}$$
(U^{4+}) (U^{6+})

Fe^{2+} is reoxidized by *T. ferrooxidans* in the same manner as in *b*.

FIGURE 13.21 Reactions involved in the microbial leaching of copper and uranium minerals.

ions (Figure 13.21*b*). In almost any ore, pyrite is present, and the oxidation of this pyrite (see Section 13.11) leads to the formation of ferric iron. Ferric iron is a very good oxidant for sulfide minerals, as has already been illustrated in the process of pyrite oxidation itself, and reaction of the copper sulfide with ferric iron results in the solubilization of the copper and the formation of ferrous iron. In the presence of O_2, at the acid pH values involved, *T. ferrooxidans* reoxidizes the ferrous iron back to the ferric form, so that it can oxidize more copper sulfide. Thus the process is kept going indirectly by the action of the bacterium on iron.

Another source of iron in leaching operations is at the precipitation plant used in the recovery of the soluble copper from the leaching solution. Scrap iron, Fe^0, is used to recover copper from the leach liquid by the reaction shown in Figure 13.21*c*, and this results in the formation of considerable Fe^{2+}. In most leaching operations, the Fe^{2+}-rich liquid remaining after the copper is removed is conducted to an oxidation pond, where *T. ferrooxidans* proliferates and forms Fe^{3+}. Acid is added at the pond to keep the pH low, thus keeping the Fe^{3+} in solution, and this ferric-rich liquid is then pumped to the top of the pile and the Fe^{3+} is available to oxidize more sulfide mineral.

Because of the huge dimensions of copper leach dumps, penetration of oxygen from air is poor, and the interior of these piles usually becomes anaerobic. Although most of the reactions written in Figure 13.21 require molecular O_2, it is also known that *T. ferrooxidans* can use Fe^{3+} as an electron acceptor in the absence of O_2, and thus catalyze the oxidation reactions described in Figure 13.21 anaerobically. Because of the large amounts of Fe^{3+} added to the leach solution from scrap iron, the process can continue to occur under anaerobic conditions.

Next to copper, the most important microbial process is **uranium leaching.** Uranium exists in many ores in the tetravalent form, an insoluble oxide, whereas the hexavalent form, containing the uranyl ion (UO_2^{2+}) is soluble. It is possible to oxidize tetravalent to hexavalent uranium with Fe^{3+}, with the reaction shown in Figure 13.21d. The Fe^{2+} so formed is then reoxidized to the Fe^{3+} by *T. ferrooxidans.* Most uranium ores contain associated pyrite, which serves as a source of Fe^{3+}, or it is possible to add ferric iron to trigger the process. Once the initial reaction has occurred, the iron can continue to cycle between the oxidized and reduced form, as the uranium is oxidized and solubilized. The soluble uranyl ion formed in the process is recovered from the leach solution by means of organic solvent extraction processes, which do not have a bacterial involvement. Uranium ores can be leached in large piles such as those described for copper, but it is also possible to carry out the leaching process without removing the ore from the ground, a procedure called **in situ leaching.** The leach solution is percolated down into the mine, the oxidation reaction occurs in the ore body, and the uranium-rich solution is then pumped up and the metal recovered. A similar in situ process has also been carried out on a relatively small scale for copper, but has not yet been widely adopted. In situ leaching has the advantage that it does not involve widespread disturbance of the landscape, as does conventional mineral mining.

13.13
Petroleum Microbiology

Microbial decomposition of petroleum and petroleum products is of considerable economic and environmental importance. Since petroleum is a rich source of organic matter and the hydrocarbons within it are readily attacked aerobically by a variety of microorganisms, it is not surprising that when petroleum is brought into contact with air and moisture, it is subject to microbial attack.

Hydrocarbon-oxidizing bacteria and fungi are the main agents responsible for decomposition of oil and oil products. A wide variety of bacteria, several molds and yeasts, and certain cyanobacteria and green algae have been shown to be able to oxidize hydrocarbons. Bacteria and yeasts, however, appear to be the prevalent hydrocarbon degraders in aquatic ecosystems. Small scale oil pollution of aquatic and terrestrial ecosystems from human as well as natural activities is very common, and hence it is not surprising that a diverse microbial population exists capable of using hydrocarbons as an electron donor. Hydrocarbon-oxidizing microorganisms develop rapidly within oil films and slicks. However, significant aliphatic hydrocarbon oxidation occurs only in the presence of O_2; if the oil gets carried into anaerobic bottom sediments, it will not be decomposed and may remain in place for many years (natural oil deposits in anaerobic environments are millions of years old). Even in aerobic environments, hydrocarbon-oxidizing microbes can act only if other environmental conditions, such as temperature, pH, and nutrients, are adequate. Because oil is

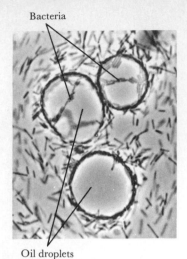

Bacteria

Oil droplets

FIGURE 13.22 Hydrocarbon-oxidizing bacteria in association with oil droplets. The bacteria are concentrated in large numbers at the oil-water interface but are not within the droplet. Magnification, 610×.

insoluble in water and is less dense, it floats to the surface and forms slicks. Hydrocarbon-oxidizing bacteria are able to attach to insoluble oil droplets, and can often be seen there in large numbers (Figure 13.22). The action of these bacteria eventually leads to decomposition of the oil and dispersal of the slick.

Massive spills of oil are not common, but when they do occur, such as the spill from the supertanker *Amoco Cadiz* in March 1978, huge amounts (in this case, greater than 190,000 metric tons) of oil can be released into the environment, causing obvious environmental problems. Microbiological studies begun a few months after the *Cadiz* spill occurred indicated that hydrocarbon-degrading bacteria had increased by a factor of nearly 10^5 over control sites which were not contaminated with oil. Although volatile hydrocarbons evaporated rather quickly, the oil-oxidizing microbial population removed up to 80 percent of the nonvolatile aliphatic and aromatic components within 7 months of the spill. Certain oil fractions, such as those rich in branched alkanes and polycyclic hydrocarbons, were removed at much slower rates, presumably due to the refractory nature of these compounds to microbial attack. Anaerobic rates of oil biodegradation were extremely slow. A further understanding of the ecological factors affecting hydrocarbon-oxidizing microorganisms would greatly aid in the prediction of the fate of oil spills in natural environments.

13.14

Trace Elements Cycles: Mercury

Trace elements are those elements that are present in low concentrations in rocks, waters, and atmosphere. Some trace elements (for example, cobalt, copper, zinc, nickel, molybdenum) are nutrients (Section 5.15), but a number of trace elements are toxic to organisms. Of these toxic elements, several are sufficiently volatile so that they exhibit significant atmospheric transport, and hence are of some environmental concern. These include mercury, lead, arsenic, cadmium, and selenium. Many of these trace elements undergo redox reactions catalyzed by microorganisms, and several of these elements are also converted into organic form (alkylated) via microbial action. Because of environmental concern and significant microbial involvement, we discuss the biogeochemistry of the element mercury.

Mercury   Although mercury is present in quite low concentrations in most natural environments, it is a widely used industrial product, and is the active component of many pesticides which have been introduced into the environment. Because of its unusual ability to be concentrated in living tissues and its high toxicity, mercury is of considerable environmental importance.

In nature, mercury exists in three oxidation states, Hg^0, Hg^+, and Hg^{2+}. Monovalent mercury generally exists as the dimer, $^+Hg\text{-}Hg^+$. It undergoes the following chemical disproportionation:

$$^+Hg\text{-}Hg^+ \rightleftharpoons Hg^{2+} + Hg^0$$

The main mercury ore is the sulfide, HgS, called *cinnabar*. The solubility of HgS is extremely low, so that in anaerobic environments mercury will be present primarily in this form, but upon aeration, oxidation of HgS can occur, probably primarily by thiobacilli, leading to the formation of mercury ion, Hg^{2+}. The soluble Hg^{2+} is quite toxic, but many bacteria have a detoxification mechanism for converting Hg^{2+} into elemental mercury, Hg^0. An NADP-linked Hg^{2+} reductase (that is frequently coded for by a plasmid) catalyzes the following reaction:

$$Hg^{2+} + NADPH + H^+ \rightleftharpoons Hg^0 + 2H^+ + NADP^+$$

Some bacteria employ a second detoxification mechanism for Hg^{2+}, converting it to methylmercury and dimethylmercury. This methylation involves a vitamin B_{12} coenzyme and can be represented as follows:

$$Hg^0 + CH_3 - B_{12} \longrightarrow CH_3 - Hg^0 \quad \textit{methylmercury}$$
$$CH_3 - Hg^+ + CH_3 - B_{12} \longrightarrow CH_3 - Hg - CH_3$$
$$\textit{dimethylmercury}$$

Both methylmercury and dimethylmercury are lipophilic and tend to be concentrated in cellular lipids. Methylmercury is about 100 times more toxic than Hg^+ or Hg^{2+}, and is concentrated to a considerable extent in fish, where it acts as a potent neurotoxin, eventually causing death. In addition, dimethylmercury is volatile and can be transported to the atmosphere. The rate of formation of dimethylmercury is considerably slower than that of methylmercury, and the steady-state concentrations of the two methylated forms will depend on the rate of synthesis and rate of removal. Some microorganisms possess a detoxification mechanism for methylmercury, reducing it to Hg^0 plus methane. A summary of these transformations is given in Figure 13.23.

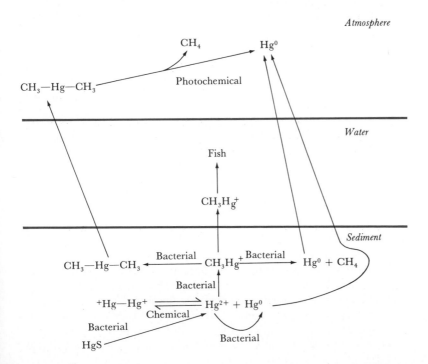

FIGURE 13.23 Bacterial activities influencing the global cycle of mercury.

A variety of plasmids isolated from both Gram-positive and Gram-negative bacteria have been found to code for heavy-metal resistance. Certain antibiotic resistance plasmids also have genes for resistance to mercury and arsenic. Other plasmids code only for heavy-metal resistances. A large plasmid isolated from *Staphylococcus aureus* has been found to code for resistance to mercury, cadmium, arsenate, and arsenite. The mechanism of resistance to any specific metal varies. For example, arsenate and cadmium resistances are due to the action of enzymes that facilitate the immediate pumping out of any arsenate or cadmium ions incorporated, while resistance to mercury involves mechanisms of detoxifying the heavy metal.

13.15

Sewage and Wastewater Treatment

Wastewater treatments are processes in which microorganisms play crucial roles, and illustrate well some of the principles of biogeochemistry discussed above. Wastewaters are materials derived from domestic sewage or industrial processes, which for reasons of public health and for recreational, economic, and aesthetic considerations cannot be disposed of merely by discarding them untreated into convenient lakes or streams. Rather, the undesirable materials in the water must first be either removed or rendered harmless. Inorganic materials such as clay, silt, and other debris are removed by mechanical and chemical methods, and microbes participate only casually or not at all. If the material to be removed is organic in nature, however, treatment usually involves the activities of microorganisms, which oxidize and convert the organic matter to CO_2. Wastewater treatment usually also results in the destruction of pathogenic microorganisms, thus preventing these organisms from getting into rivers or other supply sources. Water treatment can be carried out by a variety of processes, which may be separated broadly into two classes, anaerobic and aerobic.

Anaerobic treatment processes   Anaerobic sewage treatment involves a complex series of digestive and fermentative reactions in which the organic materials are converted into CO_2 and methane gas (CH_4), and the latter can be removed and burned as a source of energy. Since both end products, CO_2 and CH_4, are volatile, the liquid effluent is greatly decreased in organic substances. The efficiency of a treatment process is expressed in terms of the percent decrease of the initial biochemical oxygen demand (BOD); the efficiency of a well-operated plant can be 90 percent or greater, depending on the nature of the organic waste.

Anaerobic decomposition is usually employed for the treatment of materials that have much insoluble organic matter, such as fiber and cellulose, or for concentrated industrial wastes. The process occurs in four stages: (1) initial digestion of the macromolecular materials by extracellular polysaccharidases, proteases, and lipases to soluble materials; (2) conversion of the soluble materials to organic acids and alcohols by acid-producing fermentative organisms; (3) fermentation of the organic acids and alcohols to acetate, CO_2, and H_2; and

(a)

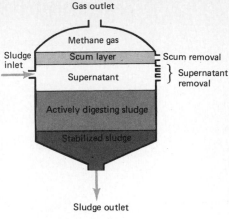

(b)

FIGURE 13.24   (a) Anaerobic sludge
digestor. Only the top of the tank is
shown; the remainder is underground.
(b) Inner workings of an anaerobic
sludge digestor.

(4) the conversion of H_2 plus CO_2 and acetate to CH_4 by methanogenic bacteria.

An anaerobic decomposition process is operated semicontin-
uously in large enclosed tanks called **sludge digestors,** into which the
untreated material is introduced and from which the treated material
is removed at intervals (Figure 13.24). The detention time in the tank
is on the order of 2 weeks to a month. The solid residue consisting of
indigestible material and bacterial cells is allowed to settle and is removed periodically and dried for subsequent burning or burial.

Aerobic treatment processes   There are several kinds of aerobic decomposition processes used in sewage treatment. A **trickling filter** is
basically a bed of crushed rocks, about 6 ft thick, on top of which the
liquid containing organic matter is sprayed (Figure 13.25). The liquid slowly trickles through the bed, the organic matter adsorbs to the
rocks, and microbial growth takes place. As the waste passes downward, air is passed upward. The microbial film that develops is at first
aerobic, but as it increases in thickness, the diffusion of oxygen to the
inner portion of the film is impeded, and this portion becomes anaerobic. The bulk of the organic matter is removed in the aerobic
zone, where it is assimilated by slime-forming bacteria, filamentous
bacteria, and filamentous fungi. These organisms are then eaten by
protozoa, which are very active in the film, and the protozoa are in
turn eaten by multicellular animals, so that there is a miniature food
chain in the film. This food chain is of great importance in the efficient removal of organic matter because at each step in the food chain
a portion of the organic matter is converted to CO_2 by respiratory processes so that eventually the organic matter is completely eliminated.

Oxidation of organic matter in the film leads to deamination of
organic nitrogen compounds and the release of NH_3, which is then
converted to nitrate by the lithotrophic nitrifying bacteria. Similarly,
H_2S is produced by decomposition of sulfur-containing organic compounds and is converted by lithotrophic sulfur oxidizers to SO_4^{2-}. Inorganic phosphate is also formed in the film by the hydrolysis of
nucleic acids and other phosphorus-containing compounds.

FIGURE 13.25 A trickling filter installation. The bed of crushed rock is being sprayed with liquid sewage.

The complete mineralization of organic matter to carbonate, nitrate, sulfate, and phosphate requires that the trickling filter be operated in proper biological balance. But under many conditions, overloading of the filter may occur, due to inadequate sewage treatment facilities, so that the effluent may contain the reduced forms, of which ammonia and sulfide are the most critical. It is a common experience that the heterotrophic organisms involved in the initial oxidation of the organic matter grow more rapidly than the lithotrophic organisms involved in oxidation of the inorganic substances. Although sulfide oxidizes spontaneously in air (see Section 13.10) and thus does not present a problem, ammonia is chemically stable and is oxidized at a significant rate in nature only by the nitrifying bacteria (Section 13.8). Thus overloaded sewage treatment facilities often have effluents high in ammonia, and since ammonia can be toxic to fish and other animals, this is an unsatisfactory situation. Also, ammonia in the effluent will contribute to the BOD of the receiving stream, because of the activities of nitrifying bacteria.

Another aerobic treatment system is the **activated sludge** process. Here, the wastewater to be treated is mixed and aerated in a large tank (Figure 13.26). Some municipalities force pure oxygen through their activated sludge tanks to increase the rate of biodegradation. Slime-forming bacteria (primarily a species called *Zoogloea ramigera*) grow and form flocs (so-called **zoogloeas**), and these flocs form the substratum to which protozoa and other animals attach. Occasionally, filamentous bacteria and fungi also are present. The basic process of oxidation is similar to that in a trickling filter. The effluent containing the flocs is pumped into a holding tank or clarifier, where the flocs settle. Some of the floc material is then returned to the aerator to

(a)

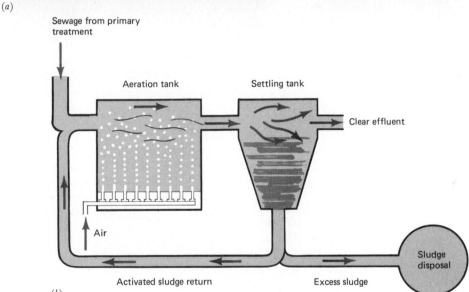

(b)

FIGURE 13.26 Activated sludge process. (a) Aeration tank of an activated sludge installation. (b) Inner workings of an activated sludge installation.

serve as inoculum, while the rest is sent to the anaerobic digestor. The residence time in an activated sludge tank is generally 5 to 10 hours, too short for complete oxidation of organic matter. The main process occurring during this short time is *adsorption* of soluble organic matter to the floc, and incorporation of some of the soluble materials into microbial cell material. The BOD of the liquid is thus considerably reduced by this process (75 to 90 percent), but the overall BOD (liquid plus solids) is only slightly reduced, because most of the adsorbed organic matter still resides in the floc. The main process of BOD destruction thus occurs in the anaerobic digestor to which the floc is transferred. It is possible to run an activated sludge tank so that most of the BOD reduction occurs by aerobic processes in the tank, but the required residence time is so long that the system becomes unwieldy (on the large scale required in urban sewage treatment). Because of the short residence time in the activated sludge tank, very little nitrification occurs, and the soluble effluent is high in ammonia, presenting the same problem already discussed with the trickling filter. Ammonia removal in both activated sludge and trickling filter effluent can be carried out by chemical treatments (for example, absorption on activated carbon, oxidation with Cl_2), or by bacterial nitrification in a second tank. The latter process is still experimental but does have considerable potential for converting the toxic ammonia into the nontoxic nitrate.

Another problem with sewage plant effluents is that they are high in phosphate, an algal nutrient that we have shown is of importance in the control of primary productivity in aquatic systems (Section 13.9). Because of this, sewage system effluents are a major cause of eutrophication, and considerable effort has been expended to find ways to remove phosphate before discharge. So far, these processes (generally categorized under the heading of tertiary treatment) have not been widely adopted.

13.16

Pesticide Biodegradation

The use of **herbicides** and **pesticides** has increased dramatically in the past generation. These compounds are of a wide variety of chemical types, such as phenoxyalkyl carboxylic acids, substituted ureas, nitrophenols, chlorinated organic acids, phenylcarbamates, and others. Some of these substances are suitable as carbon sources and electron donors for certain soil microorganisms, whereas others are not. If a substance can be attacked by microorganisms, it will eventually disappear from the soil. Such degradation in the soil is usually desirable, since toxic accumulations of the compound are avoided. However, even closely related compounds may differ remarkably in their degradability, as is shown for 2,4-dichlorophenoxyacetic acid (2,4-D) and 2,4,5-trichlorophenoxyacetic acid (2,4,5-T) in Figure 13.27. The relative persistence rates of a number of herbicides is shown in Table 13.8. However, these figures are only approximate since a variety of environmental factors, such as temperature, pH, aeration, and organic-matter content of the soil, influence decomposition. Some of the chlorinated insecticides are so indestructible that they have persisted for over 10 years. Disappearance of a pesticide from an ecosystem does not necessarily mean that it was degraded by microorganisms, since pesticide loss can also occur by volatilization, leaching, or spontaneous chemical breakdown.

The organisms that are able to metabolize pesticides and herbicides are fairly diverse, including genera of both bacteria and fungi. Some pesticides serve well as carbon and energy sources and are oxidized completely to CO_2. However, other compounds are much more recalcitrant, and are attacked only slightly or not at all, although they may often be degraded either partially or totally provided some other

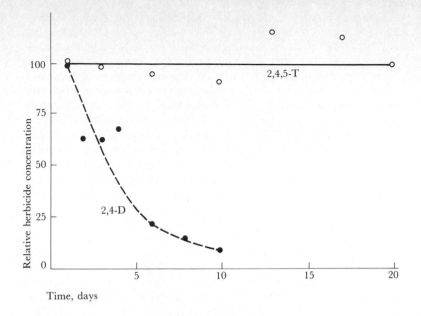

FIGURE 13.27 Rate of microbial decomposition in a soil suspension of two herbicides, 2,4-dichlorophenoxy-acetic acid (2,4-D) and 2,4,5-trichloro-phenoxyacetic acid (2,4,5-T). (From Whiteside, J. S., and M. Alexander. 1960. Weeds 8:204–213.)

organic material is present as primary energy source, a phenomenon called **co-metabolism.** Where the breakdown is only partial, the microbial degradation product of a pesticide may sometimes be even more toxic than the original compound.

The existence of organisms able to metabolize foreign chemical molecules such as the artificial pesticides is of considerable evolution-

TABLE 13.8 Persistence of Herbicides and Insecticides in Soils

Substance	Time for 75–100% Disappearance
Chlorinated insecticides:	
DDT [1,1,1-trichloro-2,2-bis(p-chlorophenyl)ethane]	4 years
Aldrin	3 years
Chlordane	5 years
Heptachlor	2 years
Lindane (hexachloro-cyclohexane)	3 years
Organophosphate insecticides:	
Diazinon	12 weeks
Malathion	1 week
Parathion	1 week
Herbicides:	
2,4-D (2,4-dichloro-phenoxyacetic acid)	4 weeks
2,4,5-T (2,4,5-trichloro-phenoxyacetic acid)	20 weeks
Dalapin	8 weeks
Atrazine	40 weeks
Simazine	48 weeks
Propazine	1.5 years

Data from Kearney, P. C., J. R. Plimmer, and C. S. Helling. 1969. Encycl. Chem. Technol. 18:515.

ary interest, since these compounds are completely new to the earth in the past 50 years. Observations on the rapidity with which organisms capable of metabolizing new compounds arise could give us some idea of the rates of microbial evolution. In Section 11.6 we discussed the evolution in the past decades of plasmids conferring resistance to antibiotics. There is some evidence that evolution of pesticide-degrading strains has also been due to genetic changes in plasmids.

Assessing biodegradability   One of the important research studies carried out before a new compound is introduced into commerce is to determine its relative biodegradability in nature. A variety of procedures can be used, but the most generally useful is a procedure called the **die-away test.** The compound of interest is added to natural soil, water, or sewage, and the concentration assayed at various times during incubation of the materials under natural conditions. The rate of disappearance of the compound in such systems will give a good indication of the potential biodegradability in natural (that is, unadapted) ecosystems. However, because of the phenomenon of genetic and physiological adaptation, it cannot be concluded that because a compound does not decompose under these conditions that it is nonbiodegradable. It is thus important to attempt to adapt various natural systems to biodegradation, by incubating for periods of time and then using these materials as inoculum for fresh samples into which the compound is introduced. Such adaptation experiments should be extended over at least a year, preferably longer, in order to obtain a firm picture of the potential biodegradability of a compound.

Supplementary Readings

Atlas, R. M. 1981. Microbial degradation of petroleum hydrocarbons: an environmental perspective. Microbiol. Rev. 45:180–209. A review article that stresses the taxonomy of hydrocarbon utilizers and the effect of the environment on hydrocarbon utilization.

Atlas, R. M., and R. Bartha. 1981. Microbial ecology. Fundamentals and applications. Addison-Wesley, Reading, Massachusetts. An intermediate-level textbook of microbial ecology.

Brierley, C. L. 1982. Microbiological mining. Sci. Am. 247:44–53. A description of the current major uses of bacteria in mining with emphasis on copper and uranium leaching.

Brock, T. D. 1966. Principles of microbial ecology. Prentice-Hall, Englewood Cliffs, N.J. Fairly elementary textbook emphasizing fundamental principles (out of print, but available in many libraries).

Cosgrove, D. J. 1977. Microbial transformations in the phosphorous cycle. Pages 95–134 *in* M. Alexander, ed. Advances in Microbial ecology, vol. 1. Plenum Press, New York. A consideration of the microbial aspects of the phosphorus cycle.

Delwiche, C. C. (ed.) 1981. Denitrification, nitrification and atmospheric nitrous oxide. John Wiley, New York. A series of articles by experts dealing with the microbial nitrogen cycle and the environmental implications of nitrous oxide.

Garrels, R. M., F. T. Mackenzie, and C. Hunt. 1975. Chemical cycles and the global environment. William Kaufmann, Los Altos, Calif. An elementary treatment of geochemical cycles, with emphasis on human influences upon these cycles. Quite useful presentation of basic principles and good quantitative diagrams of the key cycles.

Goldhaber, M. B., and **I. R. Kaplan.** 1974. The sulfur cycle. Pages 569–655 *in* E. D. Goldberg, ed. The sea, vol. 5. John Wiley, New York. A very detailed and extremely useful review on the marine sulfur cycle, with an extensive exposition of the microbiological and biochemical aspects.

Gundlach, E. R., P. D. Boehm, M. Marchard, R. M. Atlas, D. M. Ward, and **D. A. Wolfe.** 1983. The fate of *Amoco Cadiz* oil. Science 221:122–129.

Hutchinson, G. E. 1957, 1967. A treatise on limnology, vols. 1 and 2. John Wiley, New York. The standard reference work on limnology, highly detailed and useful mainly as a source of references for various topics. Volume 1 covers chemical and physical limnology, including mineral cycles, and Volume 2 is mainly on zooplankton and phytoplankton.

Lundgren, D. G., and **M. Silver.** 1980. Ore leaching by bacteria. Annu. Rev. Microbial. 34:263–283. A short review covering the theory and practical aspects of microbial leaching.

Revelle, R. 1982. Carbon dioxide and world climate. Sci. Am. 247:35–43. A recent assessment of the effects of rising CO_2 levels on a global scale.

Rheinheimer, G. 1980. Aquatic microbiology, 2nd edition, John Wiley & Sons, New York. A brief, but excellent treatment of the ecology of aquatic microorganisms and their biogeochemical activities.

Stumm, W., and **J. J. Morgan.** 1981. Aquatic chemistry. 2nd ed. John Wiley, New York. An extremely valuable basic textbook, with good discussion of redox relations of the major elements.

Stumm, W., and **E. Stumm-Zollinger.** 1972. The role of phosphorus in eutrophication. Pages 11–42 *in* R. Mitchell, ed. Water pollution microbiology. John Wiley, New York. An excellent discussion of the concepts of phosphorus limitation and phosphorus loading in aquatic systems.

Summers, A. O., and **S. Silver.** 1978. Microbial transformations of metals. Annu. Rev. Microbiol. 32:637–672. Emphasis on microbial transformations of mercuric compounds.

14

Microbial Symbiosis

Microbial species rarely exist alone in nature. When two or more kinds of organisms are present in a limited space, possibilities exist for interactions that can be beneficial or harmful to one or more of them. When two organisms live together, the relationship is called **symbiotic.*** When both organisms are benefited by the association, the relationship is called **mutualistic,** and when one organism is benefited and the other is harmed, the relationship is **parasitic.** Occasionally, one of the organisms is benefited and the other is unaffected; such a relationship is termed **commensalistic.** Finally, if two populations living together have no effect on each other, the relationship is called **neutralistic.**

Beneficial relationships in the microbial world vary widely, but they can be roughly placed in two classes: relationships of microorganisms to higher organisms (plants and animals), and relationships between two or more microorganisms. In this chapter we shall discuss both types of relationships but will concentrate on microbe-plant and microbe-animal symbioses, because most is known about these. Among the microbe-microbe relationships, the most common and interesting is that found in the lichens, and our discussion will open with the lichen symbiosis.

14.1

Lichens

The word **lichen** refers to any regular association of an alga or cyanobacterium with a fungus in which a functional relationship between the two partners exists. This relationship usually leads to the forma-

*Anton de Bary, the nineteenth-century botanist, first defined "symbiosis" to mean any relationship between two organisms, whether or not both organisms were benefited. Thus symbiosis could mean either a beneficial or a harmful relationship. Subsequently, the word has been used to refer only to beneficial relationships, but the proper word for the latter is "mutualism." In this chapter we use the word "symbiosis" in its original, and accurate, meaning.

tion of a plantlike structure of definite shape and morphology called a **thallus,** which can usually be seen with the naked eye. Lichens are widespread in nature, and are often found growing on bare rocks, on tree trunks, on house roofs, and on the surfaces of bare soils (Figure 14.1). They are in many cases the predominant photosynthetic organisms of extreme environments, such as deserts and polar regions, where other plants are not able to grow.

The lichen thallus usually consists of a tight association of many fungal hyphae, within the matrix of which the algal or cyanobacterial cells are embedded (Figure 14.2). The shape of the lichen thallus is determined primarily by the fungal partner, of which a variety of types are found in lichens. The diversity of algal types in lichens is smaller: many different kinds of lichen have green algae of the genera *Trentepohlia* and *Trebouxia.* Some lichens have cyanobacteria, and these usually fix nitrogen. Although both the alga and the fungus can be cultivated separately in the laboratory, in nature they almost always are found living together. The photosynthetic alga produces organic materials from carbon dioxide, water, and other substances in the air, water, and rock; some of these organic compounds are then used as nutrients by the fungus. In many cases the fungus produces special absorptive hyphae, **haustoria,** which penetrate into the algal cells (Figure 14.3). Even in the absence of haustoria, the intimate contact between algal cells and fungal hyphae promotes nutrient exchange.

In 1867, Simon Schwendener, a Swiss botanist, first suggested that lichens were a dual system, composed of a fungus and an alga. Up to that time, lichens had been considered a specific group of plants, and the study of lichens had been carried out by specialists who called themselves *lichenologists.* Schwendener's hypothesis elicited bitter opposition from the lichenologists, who felt their discipline was threatened, but eventually his idea became accepted. Interestingly, lichenology still exists as a discipline, separate from mycology and phycology, primarily because of the unique and complex relationships of the two partners in the lichen thallus.

A Bit of History

Direct evidence of movement of organic materials from alga to fungus has been obtained by radioisotope techniques. If the lichen is allowed to fix $^{14}CO_2$ in the light, it is found that the radioactivity appears first in the algal layer, but then some of the radioactivity passes to the fungal layer. In some cases, as much as 20 to 40 percent of the total carbon fixed by the alga in short-term experiments moves to the fungus, although in other cases it is much lower. The chemical composition of the carbon compound moving from alga to fungus depends on the type of photosynthetic organism present in the lichen. In those cases where the photosynthetic organism is a cyanobacterium, the compound moving is glucose, whereas with green algae the compound is ribitol, erythritol, or sorbitol, all sugar alcohols. After the carbon compound enters the fungus it is converted into another sugar alcohol, usually arabitol or mannitol, in which form

FIGURE 14.1 A lichen in its natural environment. Foliose lichen, *Letharia vulpina*, growing on a dead lodgepole pine branch.

FIGURE 14.2 Photomicrograph of a vertical section through a lichen thallus, *Physcia* sp., showing the location of the algal layer. Note also the tangle of fungal hyphae in the medula. Magnification, 220×.

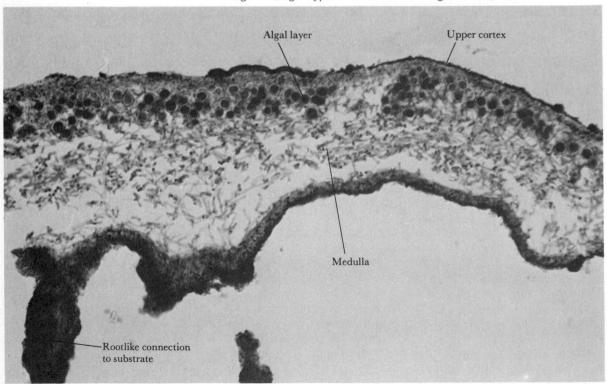

Algal layer

Upper cortex

Medulla

Rootlike connection
to substrate

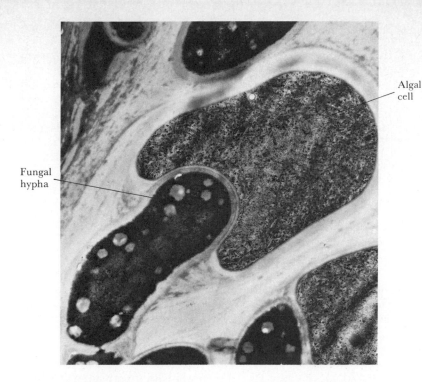

Algal cell

Fungal hypha

FIGURE 14.3 Electron micrograph of a photosynthetic cell being invaded by a fungal hypha in the lichen *Gonohymenia mesopotamica*. The photosynthetic organism is a cyanobacterium of the genus *Gloeocapsa*. Magnification, 9360×. (From Paran, N., Y. Ben-Shaul, and M. Galun. 1971. Arch. Mikrobiol. 76:103–113.)

the carbon is stored within the fungus and utilized as an energy source.

The fungus thus clearly benefits from the association, but how does the alga or cyanobacterium benefit? The fungus provides a firm substratum within which the alga can grow protected from the danger of erosion by rain or wind. In addition, the fungus is probably the agent by which the inorganic nutrients needed by the alga are absorbed from the surface on which the lichen is growing. Another role of the fungus may be to protect the alga from desiccation and lowered water potential. As we discussed in Section 8.2, fungi exhibit the most resistance among the microorganisms to lowered water potential. Within the lichen thallus the alga may be living in a micro-environment of higher water potential than the area around the thallus. In this way the alga or cyanobacterium may live in habitats too dry for it to colonize otherwise. Indeed, lichens often live in habitats exposed to direct sunlight, where they are subject to intense variations in moisture.

Lichens, however, are extremely sensitive to air pollutants and quickly disappear from urbanized areas (Figure 14.4). It is thought that this great sensitivity to pollutants is due, in part, to the fact that lichens absorb and concentrate elements from rainwater and air and possess no means of excreting them; thus lethal concentrations of toxic pollutants are gradually reached. However, two common pollutants to which lichens are unusually sensitive, ozone (O_3) and sulfur dioxide (SO_2), are not found concentrated in lichen thalli. Rather, these two gases are highly reactive and probably affect lichens when exposed directly to the lichen thallus in the air, rather than via water.

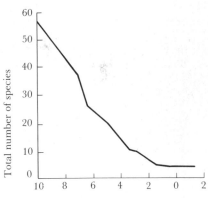

FIGURE 14.4 The decrease in diversity of lichen flora as the city center is approached. Data for Newcastle-upon-Tyne, an industrial city in northern England. Lichens can be used as biological indicators of long-term trends in air pollution. (From Gilbert, O. L. 1965. In Ecology and the Industrial Society. John Wiley & Sons, Inc., New York.)

With normal algae or cyanobacteria, living in an aquatic environment, O_3 and SO_2 would become diluted upon dissolving in water. Thus it may be the aerial growth habit of lichens that makes them unusually sensitive to these two air pollutants.

Many lichens produce characteristic pigments of striking color and unique chemical structure; many of these pigments have antibacterial properties and conceivably function in nature as a defense against bacterial attack. For a long time it was thought that the production of these pigments was a unique expression of the lichen association, but it is now known that the fungus alone is sometimes able to produce the lichen pigment in culture. Some of these lichen pigments can be used as dyes for clothing and other products and have been widely used by primitive societies for this purpose.

The growth rates of lichens are unusually low. We do not think of lichen growth in terms of hours or days, but of years. Indeed, some lichens grow so slowly that they have been used by geologists to estimate the age of glacial deposits.

If a lichen is removed from its natural habitat and placed under favorable laboratory conditions (for example, with organic substrates or high humidity), the symbiosis breaks down; the fungus may proliferate and destroy the alga, or the alga may overgrow the fungus. It is possible to cultivate algae and fungi separately in the laboratory, but it is difficult to resynthesize the lichen from these isolated components. Resynthesis never occurs in laboratory media in which the isolated components are able to grow well, but takes place only where starvation conditions are in effect, so that the fungus and alga are "forced" to reunite. We can therefore look upon the lichen thallus in nature as an ecological response of two organisms to extremely unfavorable environments.

14.2

Microbial Interactions with Higher Organisms

In the rest of this chapter we shall consider how microorganisms may live on or within higher organisms and how such relationships may be mutually beneficial. Animal and plant bodies provide favorable environments for the growth of many microorganisms. They are rich in organic nutrients and growth factors required by heterotrophs, they provide relatively constant conditions of pH and osmotic pressure, and warm-blooded animals have highly constant temperatures. However, an animal or plant body should not be considered as one uniform microbial environment throughout. Each region or organ differs chemically and physically from other regions and thus provides a selective environment where certain kinds of microbes are favored over others. In the higher animal, for instance, the skin, respiratory tract, gastrointestinal tract, and so on, each provide a wide variety of microenvironments in which different microorganisms can grow selectively. Further, higher animals and plants possess a variety of defense mechanisms that act in concert to prevent or inhibit microbial invasion and growth. The microorganisms that ultimately colonize successfully are those that have developed ways of circumventing these defense mechanisms. Interestingly, some animals and plants have developed mechanisms to encourage the growth of beneficial microorganisms.

Actually, it is often difficult to determine whether a relationship between a microorganism and a higher organism is beneficial, harmful, or neutral. This is because the outcome of an interaction may be influenced by external factors, so that under certain conditions the relationship may be beneficial, whereas under other conditions it may be neutral or harmful.

Our discussion here will emphasize warm-blooded animals, especially mammals, as it is for this group that we have the most information. Microorganisms are almost always found in those regions of the body exposed to the outside world, such as the skin, oral cavity, respiratory tract, intestinal tract, and genitourinary tract. They are not found normally in the organs and blood and lymph systems of the body; if microbes are found in any of these latter areas in significant quantities, it is usually indicative of a disease state (see Chapter 15).

14.3

Normal Flora of the Skin

Figure 14.5 indicates diagrammatically the anatomy of the skin and suggests some regions in which bacteria may live. The skin surface itself is not a favorable place for microbial growth, as it is subject to periodic drying. Only in certain areas of the body, such as the scalp, face, ear, underarm regions, genitourinary and anal regions, and palms and interdigital spaces of the toes, are moisture conditions on the surface sufficiently high to support resident microbial populations; in these regions characteristic surface microbial floras do exist. Most skin microorganisms are associated directly or indirectly with the sweat glands, of which there are several kinds. The **eccrine glands** are not associated with hair follicles and are rather unevenly distributed over the body, with denser concentrations on the palms, finger pads, and soles of the feet. They are the main glands responsible for the perspiration associated with body cooling. Eccrine glands seem to be relatively devoid of microorganisms, perhaps because of the extensive flow of fluid, since when the flow of an eccrine gland is blocked, bacterial invasion and multiplication do occur. The **apocrine glands** are more restricted in their distribution, being confined mainly to the underarm and genital regions, the nipples, and the umbilicus. They are inactive in childhood and become fully functional only at puberty. Bacterial populations on the surface of the skin in these warm, humid places are relatively high, in contrast to the situation on the smooth surface skin. Underarm odor develops as a result of bacterial activity on the secretions of the apocrines; aseptically collected apocrine secretion is odorless but develops odor upon inoculation with certain bacteria isolated from the skin. Each hair follicle is associated with a **sebaceous gland** which secretes a lubricant fluid. Hair follicles provide an attractive habitat for microorganisms; a variety of aerobic and anaerobic bacteria, yeasts, and filamentous fungi inhabit these regions, mostly within the area just below the surface of the skin. The secretions of the skin glands are rich in microbial nutrients. Urea, amino acids, salts, lactic acid, and lipids are present in considerable amounts. The pH of human secretions is almost always acidic, the usual range being between pH 4 and 6.

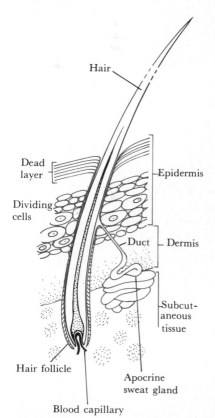

FIGURE 14.5 Anatomy of the human skin. Microbes are associated primarily with the sweat ducts and the hair follicles.

The microorganisms of the normal flora of the skin can be characterized as either **residents** or **transients.** Residents are organisms that are able to multiply, not merely survive, on the skin. The skin as an external organ is continually being inoculated with transients, virtually all of which are unable to multiply and usually die. The normal flora of the skin consists primarily of bacteria restricted to a few groups. These include several species of *Staphylococcus* and a variety of both aerobic and anaerobic corynebacteria. Of the latter, *Propionibacterium acnes* is ordinarily a harmless resident but can incite or contribute to the condition known as acne. Gram-negative bacteria are almost always minor constituents of the normal flora, even though such intestinal organisms as *Escherichia coli* are being continually inoculated onto the surface of the skin by fecal contamination. It is thought that the lack of success of Gram-negative bacteria is due to their inability to compete with Gram-positive organisms that are better adapted to the skin; if the latter are eliminated by antibiotic treatment, the Gram-negative bacteria can flourish. The death of many microorganisms inoculated onto the surface of the skin is thought to result primarily from two factors: the skin's low moisture content and organic acid content. Those organisms that survive and grow are able to resist these adverse conditions.

14.4

Normal Flora of the Oral Cavity: Dental Plaque

The oral cavity, despite its apparent simplicity, is one of the more complex and heterogeneous microbial habitats in the body. This cavity includes the teeth and tongue and the central space that they fill. As a first approximation, the teeth can be viewed only as the surface upon which saliva and materials derived from the food adsorb, rather than as a direct source of microbial nutrients. Although saliva is the most pervasive source of microbial nutrients in the oral cavity, in point of fact it is not an especially good microbial culture medium. Saliva contains about 0.5 percent dissolved solids, about half of which are inorganic (mostly chloride, bicarbonate, phosphate, sodium, calcium, potassium, and trace elements); the predominant organic constitutents of saliva are proteins, such as salivary enzymes, mucoproteins, and some serum proteins. Small amounts of carbohydrates, urea, ammonia, amino acids, and vitamins are also present. A number of antibacterial substances have been identified in saliva, of which the most important are the enzymes lysozyme and lactoperoxidase. Lactoperoxidase, an enzyme present in both milk and saliva, kills bacteria by a reaction involving chloride ions and H_2O_2, in which singlet oxygen is probably generated (see Sections 8.5 and 15.7). The pH of saliva is controlled primarily by a bicarbonate buffering system and varies between 5.7 and 7.0, with a mean near 6.7. The composition of saliva varies from individual to individual, and even within the same individual variations due to physiological and emotional factors are seen.

The teeth and dental plaque   In the infant, the lack of teeth is probably of considerable importance in determining the nature of the microbial flora. Bacteria found in the mouth during the first year of life are predominantly aerotolerant anaerobes such as streptococci and lactobacilli, but a variety of other bacteria, including some aerobes, can occur in small numbers. When the teeth appear, there is a pronounced shift in the balance of the microflora from aerobes towards anaerobes, and a variety of bacteria specifically adapted for growth on surfaces and in crevices of the teeth develop. A film forms on the surface of the teeth, called **dental plaque,** consisting mainly of bacterial cells surrounded by a polysaccharide matrix. Dental plaque can be readily observed by staining with dyes such as basic fuchsin or erythrosin; substances that stain the plaque are called disclosing agents. If effective tooth brushing is practiced, dental plaque is present only in crevices which are protected from the brush, but plaque rapidly accumulates if brushing is stopped (Figure 14.6).

Dental plaque consists mainly of filamentous bacteria closely packed and extending out perpendicular to the surface of the tooth, embedded in an amorphous matrix. These filamentous organisms are usually classified as *Leptotrichia buccalis* (or *Fusobacterium fusiforme*). They are obligate anaerobes on initial isolation but after subculturing become microaerophilic; they ferment carbohydrates to lactic acid. Associated with these predominant filamentous organisms are streptococci, diphtheroids, Gram-negative cocci, and others. The anaerobic nature of the flora may seem surprising, considering that the mouth has good accessibility to oxygen. It is likely that anaerobiosis develops through the action of facultative bacteria growing aerobically upon organic materials on the tooth, since the dense matrix of the plaque greatly decreases the diffusion of oxygen onto the tooth surface. The microbial populations of the dental plaque are thus seen to exist in a microenvironment partly of their own making, and are probably able to maintain themselves in the face of wide variations in the macroenvironment of the oral cavity.

The bacterial colonization of smooth tooth surfaces occurs as a result of firm attachment of single bacterial cells, followed by growth in the form of microcolonies. Beginning with a freshly cleaned tooth surface, the first event is the formation of a thin organic film (Figure 14.7a), as a result of the attachment of acidic glycoproteins from the saliva. This film then provides a firmer attachment site for the colonization and growth of bacterial microcolonies (Figure 14.7b). The colonization of this glycoprotein film is highly specific, and only a few species of *Streptococcus* (primarily *S. sanguis* and *S. mitis*) are involved. As a result of extensive growth of these organisms, a thick bacterial zone is formed (Figure 14.8a and b). Subsequently, the plaque is colonized by other organisms, and in mature plaque filamentous types may actually predominate.

The role of the oral flora in tooth decay (dental caries) has now been well established through studies on germ-free animals, although the precise mechanisms of the process are still under study. The smooth surfaces of the teeth that are exposed to frequent cleaning by the tongue, cheek, saliva, or toothbrush or to the abrasive action of food mastication are relatively resistant to dental caries. The tooth surfaces in crevices, where food particles can be retained, are the sites where tooth decay predominates. The shape of the teeth is an im-

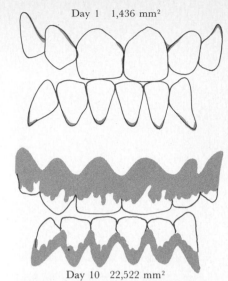

Day 1   1,436 mm²

Day 10   22,522 mm²

FIGURE 14.6 Distribution of dental plaque, as revealed by use of a disclosing agent, on brushed and unbrushed teeth. The numbers give the total area of dental plaque. (Redrawn from Lang, N. P., E. Ostergaard, and H. Löe. 1972. J. Periodont. Res. 7:59–67.)

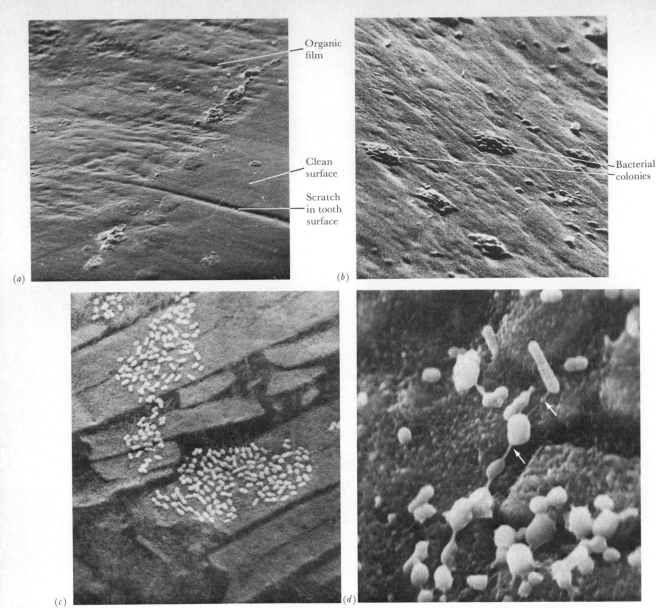

Organic
film

Clean
surface

Scratch
in tooth
surface

Bacterial
colonies

(a)

(b)

(c)

(d)

FIGURE 14.7 Early stages in the formation of dental plaque, as seen in the scanning electron microscope. (a) The formation of an organic film on a cleaned tooth surface. A piece of tape protected one-half of a cleaned tooth. After 1 hour, the tape was removed and the scanning electron micrograph made. Note how the organic film masks the scratch in the enamel surface. Magnification, 170×. (b) The formation of microcolonies on a cleaned tooth surface, 24 hours after cleaning. Individual organisms could be seen even a few hours after cleaning. Magnification, 1520×. (Parts *a* and *b* courtesy of Unilever, Ltd. from Caxton, C.A. 1973. Scanning electron microscope study of the formation of dental plaque. Caries Res. 7:102–119.) (c) Bacterial colonies growing on a model tooth inserted into the mouth for 6 hours. (d) Higher magnification of a preparation such as (c). Note the diverse morphology and the glue-like material (arrows) holding the organisms together. (Parts *c* and *d* from Lie, T. 1977. Early dental plaque morphogenesis. Jour. Periodontal Res. 12:73–89.)

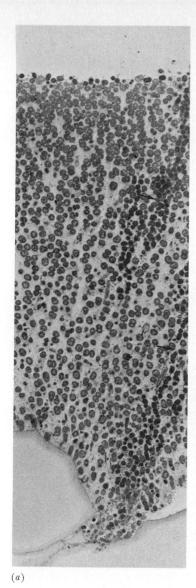

(a)

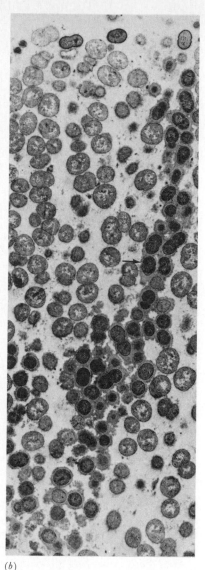

(b)

FIGURE 14.8 Electron micrographs of thin sections of dental plaque. Bottom is the base of the plaque; top is the portion exposed to oral cavity. (a) Low-power electron micrograph. Organisms are predominantly streptococci. The species *S. sanguis* has been labeled by an antibody-microchemical technique, and these cells appear darker than the rest. They are seen as two distinct lines (arrows). Magnification, 2200×. (b) Higher-power electron micrograph showing the region with *S. sanguis* cells (dark; arrow). Magnification, 9000×. (From Lai, C., Listgarten, M. A., and Rosan, B. 1975. Immunoelectron microscopic identification and localization of *Streptococcus sanguis* with peroxidase labeled antibody: localization of *Streptococcus sanguis* in intact plaque. Infection and Immunity 11:200–210.)

portant factor in the degree to which such crevices develop; dogs are highly resistant to tooth decay because the shape of their teeth does not favor retention of food. Diets high in sugars are especially cariogenic because lactic acid bacteria ferment the sugars to lactic acid, which causes decalcification of the hard dental tissue of the tooth. Once the breakdown of the hard tissue has begun, proteolysis of the matrix of the tooth enamel occurs, through the action of proteolytic enzymes released by bacteria. Microorganisms penetrate further into the decomposing matrix, but the later stages of the process may be exceedingly slow and are often highly complex.

Two organisms that have been implicated in dental caries are *S. sanguis* and *S. mutans,* both lactic acid-producing bacteria. As noted

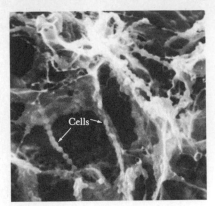

FIGURE 14.9 Scanning electron micrograph of the cariogenic bacterium *Streptococcus mutans.* The sticky dextran material can be seen as masses of filamentous particles. Magnification, 13,920×. Courtesy of I. L. Shechmeister and J. Bozzolla.

above, *S. sanguis* is able to colonize smooth tooth surfaces because of its specific affinity for salivary glycoproteins, and this organism is probably the primary organism involved in tooth decay. *S. mutans* is found predominantly in crevices and small fissures, and its ability to attach seems to be related to its ability to produce a dextran polysaccharide, which is strongly adhesive (Figure 14.9). *S. mutans* dextran is only produced when sucrose is present, by means of the enzyme dextransucrase (see Section 5.3 and Figure 5.11). Sucrose is a common sugar in the human diet, and its ability to act as a substrate for dextransucrase may be one reason that sucrose is highly cariogenic.

Susceptibility to tooth decay varies greatly among individuals and is affected by inherent traits in the individual as well as by diet and other extraneous factors. The structure of the calcified tissue plays an important role. Incorporation of fluoride into the calcium phosphate crystal matrix makes the latter more resistant to decalcification by acid, hence the use of fluorides in drinking water or dentifrices to aid in controlling tooth decay. Although tooth decay is an infectious disease, we tend to place it in a different category from other infectious diseases. However, microorganisms in the mouth can also cause infections that are more typically disease states, such as periodontal disease, gingivitis, infections of the tooth pulp, and so on.

14.5

Normal Flora of the Gastrointestinal Tract

FIGURE 14.10 Section through a Gram-stained preparation of the stomach wall of a 14-day-old mouse, showing extensive development of lactic acid bacteria (*Lactobacillus* sp.) in association with the epithelial layer. To avoid introducing changes in the bacterial flora during sample preparation, the tissue removed from the stomach was immediately frozen and was sectioned using a freezing microtome. Magnification, 670×. (Courtesy of Dwayne C. Savage.)

The human gastrointestinal tract, the site of food digestion, consists of the stomach, small intestine, and large intestine. The acidity of the stomach fluids is high, about pH 2. The stomach can thus be viewed as a microbiological barrier against penetration of foreign bacteria into the intestinal tract. Although the bacterial count of the stomach contents is generally low, the walls of the stomach are often heavily colonized with bacteria. These are primarily acid-tolerant lactobacilli and streptococci and can be seen in large numbers in histological sections of the stomach epithelium (Figure 14.10). These bacteria appear very early after birth of an animal, being well established by the first week. In humans, under abnormal conditions such as cancer of the stomach that produce higher pH values, a characteristic microbial flora consisting of yeasts and bacteria (genera *Sarcina* and *Lactobacillus*) may develop.

The **small intestine** is separated into two parts, the duodenum and the ileum. The former, being adjacent to the stomach, is fairly acidic and resembles the stomach in its microbial flora, although it may lack heavy populations on the epithelium. From the duodenum to the ileum the pH gradually becomes more alkaline and bacterial numbers increase. In the lower ileum, bacteria are found in the intestinal cavity (the lumen), mixed with digestive material.

In the **large intestine,** bacteria are present in enormous numbers, so much so that this region can be viewed as a specialized fermentation vessel. Many bacteria live within the lumen itself, probably using as nutrients some products of the digestion of food. Facultative anaerobes, for example *Escherichia,* are present but not abundantly so with respect to other bacteria; total counts of facultative anaerobes are generally less than 10^7/per gram of intestinal contents. The activities of facultative anaerobes consume any oxygen present, making the environment of the large intestine strictly anaerobic and favorable for the profuse growth of obligate anaerobes. The proportion of anaerobes to aerobes is often grossly underestimated, owing to the difficulties of cultivating the anaerobes. In recent years better anaerobic techniques have been developed, and organisms previously uncultured are now being isolated. Many of these bacteria are long, thin, Gram-negative rods, with tapering ends (called *fusiform*) and can be seen with the scanning electron microscope attached end-on to small indentations in the intestinal wall (Figure 14.11). Other obligate anaerobes include species of *Clostridium* and *Bacteroides*. The total number of obligate anaerobes is enormous. Counts of 10^{10} to 10^{11} cells per gram of intestinal contents are not uncommon, with various species of *Bacteroides* accounting for the majority of intestinal obligate anaerobes. In addition, *Streptococcus faecalis* is almost always present in significant numbers.

The intestinal flora of the newborn becomes established early. In breast-fed human infants the flora is often fairly simple, consisting largely of *Bifidobacterium* spp. (formerly called *Lactobacillus bifidus*). In bottle-fed infants the flora is usually more complex. The flora is conditioned partly by the fact that the infant's main source of food is milk, which is high in the sugar lactose. The reason that the

FIGURE 14.11 Scanning electron micrographs of the microbial community on the surface of the columnar epithelium in the mouse ileum. (a) An overview at low magnification. Note the long, filamentous organisms lying on the surface. Magnification, 600×. (b) Higher magnification, showing several filaments attached at a single depression. Note that the attachment is at the end of the filaments only. Magnification, 2800×. (From Savage, D. C., and R. V. H. Blumershine. 1974. Surface-surface associations in microbial communities populating epithelial habitats in the murine gastrointestinal ecosystem: scanning electron microscopy. Infection and Immunity 10:240–250.)

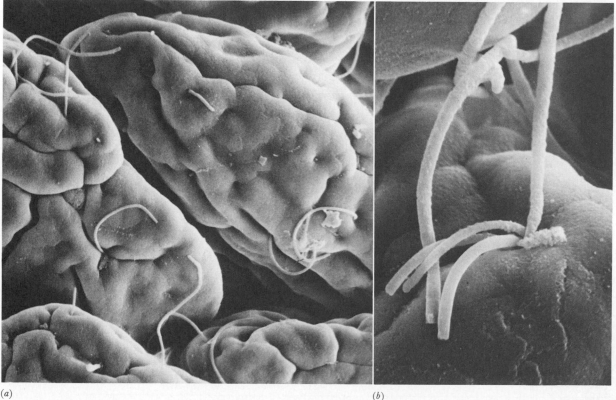

(a) (b)

flora of breast-fed infants differs from that of bottle-fed ones is not completely understood, but it is known that human milk contains a disaccharide amino sugar that is required as a growth factor by *Bifidobacterium*. As the infant ages and its diet changes, the composition of the intestinal flora also changes, ultimately approaching that of the adult.

The intestinal flora has profound influence on the animal, carrying out a wide variety of metabolic reactions (Table 14.1). Not all microorganisms carry out these reactions, and changes in the intestinal flora due to diet or disease may thus affect the animal. Of special note in Table 14.1 are the roles of the intestinal flora in modifying compounds of the bile, the bile acids. Bile acids are steroids produced in the liver and excreted into the intestine via the gall bladder. Their role is to promote emulsification of fats in the diet so that these can be effectively digested. Intestinal microbes cause a variety of transformations of these bile acids so that the materials excreted in the feces are quite different from bile acids. Also of interest is the role of the microbial flora in the production of gastrointestinal gas or flatus. A normal human expels about 400 to 650 ml of flatus per day, composed of about 40 percent CO_2, small amounts of CH_4 and H_2 from microbial fermentation, and about 50 percent N_2, derived from the air. Other products of microbial fermentation are the odor-producing substances listed in Table 14.1. Composition of the microbial flora as well as diet influences the amount of gas and the amount of odoriferous materials present. Further discussion of the role of the gastrointestinal flora in the animal will be given in Section 14.7 in the discussion of germ-free animals.

One rather dramatic instance of a beneficial role of a component of the intestinal flora has been the discovery of significant numbers of nitrogen-fixing bacteria of the species *Klebsiella pneumoniae* in the intestines of New Guineans subsisting on a diet in which 80 to 90 percent of their calories are derived from sweet potatoes, a food virtually devoid of protein nitrogen. This bacterium fixes N_2 anaerobically (Section 5.13) and is a constant member of the intestinal flora of these people. Apparently sufficient nitrogen is fixed and passes through the intestinal wall into the bloodstream so that these individuals can subsist on a diet that would be unsuitable otherwise.

During the passage of food through the gastrointestinal tract, water is withdrawn from the digested material, and it gradually becomes more concentrated and is converted into feces. Bacteria, chiefly dead ones, make up about one-third of the weight of fecal

TABLE 14.1 Biochemical and Metabolic Contributions of the Intestinal Microflora

Vitamin synthesis	Product: thiamin, riboflavin, pyridoxine, B_{12}, K
Gas production	Product: CO_2 CH_4, H_2 (and N_2 from air)
Odor production	Product: H_2S, NH_3, amines, indole, skatole, butyric acid
Organic acid production	Product: acetic, propionic, butyric acids
Nitrogen fixation	Agent: *Klebsiella pneumoniae* (in humans on high-carbohydrate diet)
Glycosidase reactions	Enzyme: β-glucuronidase, β-galactosidase, β-glucosidase, α-glucosidase, α-galactosidase
Sterol metabolism	Process: esterification, dehydroxylation, oxidation, reduction, inversion

matter. Organisms living in the lumen of the large intestine are being continuously displaced downward by the flow of material, and if bacterial numbers are to be maintained, those bacteria that are lost must be replaced by new growth. Thus, the large intestine resembles in some ways a chemostat. The time needed for passage of material through the complete gastrointestinal tract is about 24 hours in humans; the growth rate of bacteria in the lumen is one to two doublings per day.

When an antibiotic is given orally it may inhibit the growth of the microorganisms present; continued movement of the intestinal contents then leads to loss of the preexisting bacteria and the virtual sterilization of the intestinal tract. In the absence of the normal flora the environmental conditions of the large intestine change, and there may become established exotic microorganisms such as antibiotic-resistant *Staphylococcus, Proteus,* or the yeast *Candida albicans,* which usually do not grow in the intestinal tract because they cannot compete with the normal flora. Occasionally, establishment of these exotic organisms can lead to a harmful alteration in digestive function. The normal flora eventually becomes reestablished, but often only after considerable time.

14.6
Normal Flora of Other Body Regions

In the **upper respiratory tract** (throat, nasal passages, and nasopharynx) microorganisms live primarily in areas bathed with the secretions of the mucous membranes. Bacteria enter the upper respiratory tract from the air in large numbers during breathing, but most of these are trapped in the nasal passages and expelled again with the nasal secretions. The resident organisms most commonly found are staphylococci, streptococci, diphtheroid bacilli, and Gram-negative cocci. Each person generally has a characteristic flora, which may remain constant over extended periods of time. Potentially harmful bacteria, such as *Staphylococcus aureus, Streptococcus pneumoniae, Streptococcus pyogenes,* and *Corynebacterium diphtheriae* are often part of the normal flora of the nasopharynx of healthy individuals.

The **lower respiratory tract** (trachea, bronchi, and lungs) is essentially sterile, in spite of the large numbers of organisms potentially able to reach this region during breathing. Dust particles, which are fairly large, are filtered out in the upper respiratory tract. As the air passes into the lower respiratory tract, its rate of movement decreases markedly, and organisms settle onto the walls of the passages. These walls are lined with ciliated epithelium, and the cilia, beating upwards, push bacteria and other particulate matter toward the upper respiratory tract where they are then expelled in the saliva and nasal secretions. Only droplet nuclei smaller than 10 μm in diameter are able to reach the lungs (see Figure 17.9).

The female **urethra** is usually sterile, while that of the male contains Gram-positive cocci and diphtheroids in the lower one-third. The vagina of the adult female generally is weakly acidic and contains significant amounts of the polysaccharide glycogen. A *Lactobacillus,* sometimes called *Döderlein's bacillus,* which ferments glycogen and produces acid, usually is present in the vagina and may be responsible for the acidity. Other organisms, such as yeasts, strepto-cocci, and *E. coli,* are also present. Before puberty, the female vagina is alkaline and does not produce glycogen, Döderlein's bacillus is absent, and the flora consists predominantly of staphylococci, strep-tococci, diphtheroids, and *E. coli.* After menopause, glycogen dis-appears and the flora again resembles that found before puberty.

14.7

Germ-Free Animals

For a number of kinds of studies on the role of the normal flora, **germ-free animals** are useful. Colonies of germ-free animals are now widely established and are finding uses in many research studies. Establishing a germ-free animal colony is sometimes simplified by the fact that mammals frequently are microbially sterile until birth, so that if the fetus is removed aseptically just before the time of expected birth, germ-free offspring can often be obtained. These newborns must then be placed in germ-free isolators (Figure 14.12), and all air, water, food or other objects entering the isolators must be sterile. Within the isolators the infants must then be hand-fed until they have developed to the stage where they can feed themselves. Once established, a germ-free colony can be maintained by continued mating between germ-free males and females. With birds, establish-ment of germ-free colonies is easier, as the inside of the egg is usually sterile and the newly hatched chick is able to feed itself immediately. Germ-free colonies of mice, rats, guinea pigs, rabbits, hamsters, monkeys, and chickens have been established and similarly, germ-free individual lambs and pigs have been kept for considerable periods of time. Raising large animals in the germ-free condition is obviously much more difficult than the raising of small animals, since large

FIGURE 14.12 A germ-free isolator. (Courtesy of the Germfree Laboratories, Inc., Miami, Fla.)

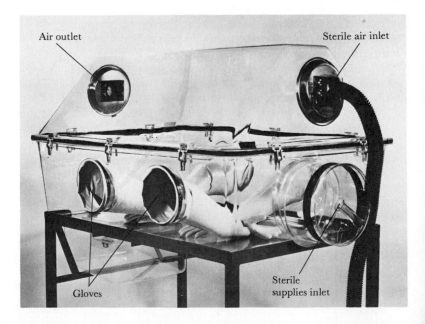

isolators are expensive and difficult to keep sterile. Germ-free mice are now available commercially from laboratory-animal supply houses.

To ascertain that such animals are really germ-free, microbiological culture studies using a variety of media must be performed, and body fluids and tissues of the animal should never yield microbes. "Germ-free" is a negative term that implies a complete absence of microorganisms; yet all that can be determined is that with the methods used no microorganisms can be detected. This does not mean that, if other methods were employed, microorganisms might not be found in a presumed germ-free animal.*

Germ-free individuals differ from normal animals in several important respects. Structures such as the lymphatic system, the antibody-forming system, and the reticuloendothelial system, which participate in defense against bacterial invaders (see Section 15.7), are poorly developed in germ-free animals. Also some organs that would normally have natural populations of bacteria are often reduced in size or capacity in the germ-free animal. However, in the germ-free guinea pig, rat, and rabbit the cecum is greatly enlarged (Figure 14.13). In the normal rodent the cecum is the part of the intestinal tract that has the largest bacterial population; in the germ-free state the cecum might lack the stimulus to evacuate caused by the bacteria and hence continuously fill up. Furthermore, the whole intestinal wall of the germ-free animal is thin and unresponsive to mechanical stimuli. The conclusion seems inescapable that bacteria are necessary for the normal development of the intestine.

Germ-free animals also differ in nutritional needs from conventional ones. For instance, vitamin K, which usually is not required by conventional animals, is required by germ-free animals. *Escherichia coli* synthesizes vitamin K; when this organism is established in the intestinal tract of germ-free animals, the vitamin K deficiency disappears, thus showing that *E. coli* probably is responsible for synthesizing this vitamin in the conventional animal.

Germ-free animals are much more susceptible to bacterial infections than are conventional animals. Organisms like *Bacillus subtilis* and *Micrococcus luteus,* which are harmless to conventional animals, are harmful to germ-free forms. It is difficult to infect conventional animals with *Vibrio cholerae* (the causal agent of cholera) and *Shigella dysenteriae* (the cause of bacterial dysentery); yet these organisms can readily be established in germ-free animals. It is likely that the normal flora of the conventional animal has a competitive advantage in the intestinal tract, preventing establishment of exotic organisms; with the normal flora gone, the foreign organisms can become established easily. Germ-free animals are also much more susceptible to infection by intestinal worms. On the other hand, they are resistant to infection by the intestinal amoeba that causes amoebic dysentery (*Entamoeba histolytica*). This is probably because the amoeba uses intestinal bacteria as a primary source of food and cannot grow in their absence. Also, germ-free animals do not show tooth decay (dental caries) even if they are fed a diet high in sucrose, which is very conducive to tooth decay in conventional animals. However,

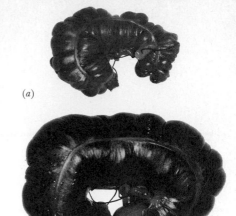

(a)

(b)

FIGURE 14.13 Comparison of the cecum size of a normal (a) and a germ-free (b) rodent. (Courtesy of Medical Audio-Visual Branch, Walter Reed Army Institute of Research, Washington, D. C.)

*Frequently the term *gnotobiotic* is used instead of *germ-free.* A gnotobiotic system is one in which the composition of the microbial flora is known.

if germ-free animals are inoculated with a pure culture of certain *Streptococcus* strains, especially *S. mutans,* isolated from the teeth, tooth decay does occur on a high sucrose diet.

Many of the characteristics of germ-free animals can be mimicked in conventional animals if they are administered antibiotics orally that cause the elimination of the intestinal flora. Such animals often are more susceptible to infections than are conventional animals and show vitamin deficiencies. We can for these reasons conclude that the conventional animal does derive considerable benefit from its normal flora, even though the normal flora causes some harmful effects (for example, tooth decay).

14.8

Rumen Symbiosis

Ruminants are herbivorous mammals that possess a special organ, the **rumen,** within which the digestion of cellulose and other plant polysaccharides occurs through the activity of special microbial populations. Some of the most important domestic animals, the cow, sheep, and goat, are ruminants. Since the human food economy depends to a great extent on these animals, rumen microbiology is of considerable economic significance.

Rumen fermentation The bulk of the organic matter in terrestrial plants is present in insoluble polysaccharides, of which cellulose is the most important. Mammals, and indeed almost all animals, lack the enzymes necessary to digest cellulose, but all mammals that subsist primarily on grasses and leafy plants can metabolize cellulose by making use of microorganisms as digestive agents. Unique features of the rumen as a site of cellulose digestion are its relatively large size (100 to 150 liters in a cow, 6 liters in a sheep) and its position in the alimentary tract as the organ where ingested food goes first. The high constant temperature (39°C) and the anaerobic nature of the rumen are also important factors. The rumen operates in a more or less continuous fashion, and in some ways can be considered analogous to a chemostat.

The relationship of the rumen to the other parts of the ruminant digestive system is shown in Figure 14.14. Food enters the rumen mixed with saliva and is churned in a rotary motion during which the microbial fermentation occurs. The food mass then passes gradually into the reticulum, where it is formed into small portions called cuds, which are regurgitated into the mouth where they are chewed again. The now finely divided solids, well mixed with saliva, are swallowed again, but this time the material passes down a different route, ending in the abomasum, an organ more like a true stomach. Here true digestion begins and continues into the small and large intestine.

Food entering the rumen is mixed with the resident microbial populations and remains there on the average about 9 hours. During this time, cellulolytic bacteria and protozoa hydrolyze cellulose to the disaccharide cellobiose and the monosaccharide glucose. These sugars then undergo a microbial fermentation with the production of

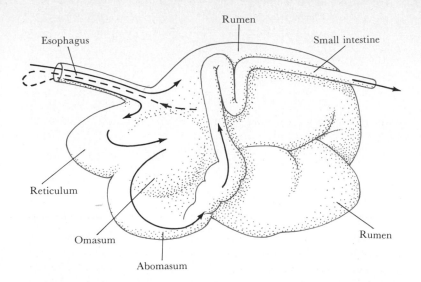

Esophagus

Rumen

Small intestine

Reticulum

Omasum

Abomasum

Rumen

volatile fatty acids, primarily acetic, propionic, and butyric, and the gases carbon dioxide and methane (Figure 14.15). The fatty acids pass through the rumen wall into the bloodstream and are oxidized by the animal as its main source of energy. In addition to their digestive functions, rumen microorganisms synthesize amino acids and vitamins that are the main source for the animal of these essential nutrients. The rumen contents after fermentation consist of enormous numbers of microbial cells (10^{10} to 10^{11} bacteria per milliliter of rumen fluid) plus partially digested plant materials, which pass through the gastrointestinal tract of the animal, where they undergo digestive processes similar to those of other animals. Microbial cells formed in the rumen and digested in the gastrointestinal tract are the main source of proteins and vitamins for the animal. Since many of the microbes of the rumen are able to grow on urea as a sole nitrogen source, it is often supplied in cattle feed in order to promote microbial protein synthesis. The bulk of this protein will end up in the animal itself. A ruminant is thus nutritionally superior to a nonruminant when subsisting on foods that are deficient in protein, such as grasses.

The rumen microorganisms The biochemical reactions occurring in the rumen are complex and involve a wide variety of microorganisms. Properties of some of the important rumen bacteria are outlined in Table 14.2. All of these organisms are obligate anaerobes and must be cultured under strictly anaerobic conditions. A number of rumen bacteria require as growth factors certain branched-chain acids (for example, isovaleric and isobutyric) that are present in the rumen fluid.

As seen in Table 14.2, different rumen bacteria carry out specific functions. A variety of types hydrolyze cellulose to sugars and then ferment these to acids. If the animal is fed a diet high in starch (grain, for instance), then starch-digesting bacteria are common, although on a low starch diet these organisms are usually in a minority.

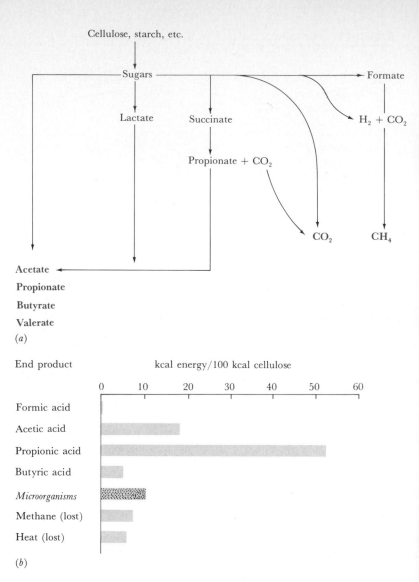

Cellulose, starch, etc.

Sugars → Formate

Lactate   Succinate $H_2 + CO_2$

Propionate + CO_2

CO_2 CH_4

Acetate ←

Propionate

Butyrate

Valerate

(a)

End product   kcal energy/100 kcal cellulose

Formic acid

Acetic acid

Propionic acid

Butyric acid

Microorganisms

Methane (lost)

Heat (lost)

(b)

FIGURE 14.15 (a) Biochemical reactions in the rumen. The end products are shown in boldface type. (b) Energy balance of the fermentation of cellulose in the rumen. The bars show the relative proportions of the energy of the cellulose that ends up in various products.

If an animal is fed on legume pasture, which is high in pectin, then the pectin-digesting bacterium *Lachnospira multiparus* is a common member of the rumen flora. Some of the fermentation products are themselves used by other rumen bacteria. Thus succinate is converted to propionate and CO_2 (Figure 14.15), and lactate is fermented to acetic and other acids by *Selenomonas* and *Peptostreptococcus* (Table 14.2). A number of rumen bacteria produce ethanol as a fermentation product when grown in pure culture, yet ethanol rarely accumulates in the rumen because it is fermented to H_2 and CO_2. Also, a variety of bacteria produce H_2 during growth in pure culture, yet H_2 never accumulates in the rumen because it is quickly used with CO_2 by *Methanobrevibacter ruminantium* in the production of methane. Another

TABLE 14.2 Characteristics of Some Rumen Bacteria

Organism	Gram Stain	Shape	Motility	Fermentation Products
Cellulose decomposers:				
Bacteroides succinogenes	Neg.	Rod	−	Succinate, acetate, formate
Butyrivibrio fibrisolvens	Neg.	Curved rod	+	Acetate, formate, lactate, butyrate, H_2, CO_2
Ruminococcus albus	Pos.	Coccus	−	Acetate, formate, H_2, CO_2
Clostridium lochheadii	Pos.	Rod (spores)	+	Acetate, formate, butyrate, H_2, CO_2
Starch decomposers:				
Bacteroides amylophilus	Neg.	Rod	−	Formate, acetate, succinate
Bacteroides ruminicola	Neg.	Rod	−	Formate, acetate, succinate
Selenomonas ruminantium	Neg.	Curved rod	+	Acetate, propionate, lactate
Succinomonas amylolytica	Neg.	Oval	+	Acetate, propionate, succinate
Streptococcus bovis	Pos.	Coccus	−	Lactate
Lactate decomposers:				
Selenomonas lactilytica	Neg.	Curved rod	+	Acetate, succinate
Peptostreptococcus elsdenii	Pos.	Coccus	−	Acetate, propionate, butyrate, valerate, H_2, CO_2
Pectin decomposer:				
Lachnospira multiparus	Pos.	Curved rod	+	Acetate, formate, lactate, H_2, CO_2
Methane producer:				
Methanobrevibacter ruminantium	Pos.	Rod	−	CH_4 (from H_2 + CO_2 or formate)

source of H_2 and CO_2 for the methane-producing bacterium is formate (Figure 14.15).

In addition to bacteria, the rumen has a characteristic protozoal fauna, composed almost exclusively of ciliates. Many of these protozoans are obligate anaerobes, which is a rare property among eucaryotes. Although these are not essential for rumen fermentation, the protozoa definitely contribute to the process. They are able to ferment sugars, starch, and cellulose with the production of the same organic acids that are formed by the bacteria. Protozoal population densities are usually larger in animals on a good ration than in those fed a poor diet, and there is some reason to believe that protozoal counts could be used as indicators of the well-being of the animal. It has been well established that the protozoa ingest rumen bacteria; animals without protozoa usually have higher bacterial populations, which suggests that the protozoa may control bacterial density to at least some extent.

Studies of rumen development in calves (calves are born with an undeveloped rumen) have shown that microorganisms are intimately involved in the formation of the rumen. Regurgitated food containing the rumen microflora is transferred from the mother to the calf in the first few weeks of life, and this inoculum initiates production of volatile fatty acids which trigger rumen development. Calves fed mixtures of volatile fatty acids develop functional rumens in the absence of the rumen microflora, showing that these low-molecular-weight compounds do induce rumen growth. The mechanism of fatty acid induction of rumen development is thought to involve a stimulation of blood flow through the rumen wall which presumably serves to deliver the nutrients required for proper rumen development.

14.9

Normal Flora of Plants

As microbial habitats, plant bodies are clearly vastly different from those of animals. Compared with warm-blooded animals, plants vary greatly in temperature, both diurnally and throughout the year. The communication system throughout the plant is only poorly developed, so that transfer of microorganisms from one part to another is inefficient. The aboveground parts of the plant, especially the leaves and stems, are subjected to frequent drying and for this reason have developed waxy coatings that retain moisture and, incidentally, probably also keep out microorganisms. The roots, on the other hand, have an environment whose moisture is less variable, and also the nutrient concentrations are higher. For this reason, the roots of plants are a main area of microbial action.

As nutrient sources, plants are high in carbohydrates but, except for the seeds, are relatively deficient in proteins. Vitamin B_{12} is not formed by plants. Many plants produce toxic chemicals that hinder or prevent microbial growth; examples are phenols produced by onions and many cereal and root plants, hydrogen cyanide produced by flax, and allyl sulfides produced by onions. Woody plants are usually resistant to microbial invasion; only if the bark is penetrated by insects or damaged by other means can microorganisms reach the interior of a tree trunk. Even then, movement through the trunk may be slow or absent.

The **rhizosphere** is the region immediately outside the root; it is a zone where microbial activity is usually high. The bacterial count is almost always higher in the rhizosphere than it is in regions of the soil devoid of roots, often many times higher. This is because roots excrete significant amounts of sugars, amino acids, and vitamins, which promote such an extensive growth of bacteria and fungi that these organisms often form microcolonies on the root surface. There is some evidence that the rhizosphere microorganisms benefit the plant by promoting the absorption of nutrients, but this is probably only of minor importance.

The **phyllosphere** (also called the **phylloplane**) is the surface of the plant leaf, and under conditions of high humidity, such as develop in wet forests in tropical and temperate zones, the microbial flora of leaves may be quite high. In tropical rain forests, the phyllosphere bacteria often produce gums or slimes, which presumably help the bacterial cells to stick to leaves under the onslaught of heavy tropical rains. Many of the bacteria on leaves are nitrogen-fixing forms, and nitrogen fixation presumably aids these organisms in growing with the predominantly carbohydrate nutrients provided by leaves. Some of the nitrogen fixed by these bacteria may be absorbed by the plant, although there is little evidence that this occurs on a large scale.

14.10

Symbiotic Nitrogen Fixation

The legume-Rhizobium symbiosis   One of the most interesting and important symbiotic relationships is that between leguminous plants and bacteria of the genus *Rhizobium*. **Legumes** are a large group that includes such economically important plants as soybean, clover, alfalfa, string beans, and peas, and are defined as any plant which bears its seeds in pods. *Rhizobium* organisms are Gram-negative motile rods. Infection of the roots of one of these legumes with the appropriate strain of *Rhizobium* leads to the formation of **root nodules** (Figure 14.16), which are able to convert gaseous nitrogen into combined nitrogen, a process called nitrogen fixation (see Section 5.13). Nitrogen fixation by the legume-*Rhizobium* symbiosis is of considerable agricultural importance, as it leads to very significant increases in combined nitrogen in the soil. Since nitrogen deficiencies often occur in unfertilized bare soils, legumes are at a selective advantage under such conditions and can grow well in areas where other plants cannot (Figure 14.17).

Under normal conditions, neither legume nor *Rhizobium* alone is able to fix nitrogen; yet the interaction between the two leads to the development of nitrogen-fixing ability. In culture, the *Rhizobium* is able to fix N_2 alone when grown under strictly controlled micro-aerophilic conditions. Apparently *Rhizobium* needs some O_2 to generate energy for N_2 fixation, yet its nitrogenase (like those of other nitrogen-fixing organisms) is inactivated by O_2. In the nodule, precise O_2 levels are controlled by the O_2-binding protein **leghemoglobin.** This is a red, hemoglobin-like protein which is always found in healthy, N_2-fixing nodules. Neither plant nor *Rhizobium* alone synthesizes leghemoglobin, but formation is induced somehow through the symbiotic interaction of these two organisms.

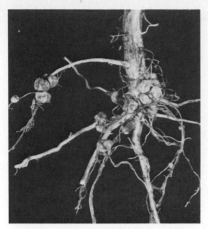

FIGURE 14.16 Soybean root nodules.

FIGURE 14.17 Unnodulated (left) and nodulated alfalfa plants growing in nitrogen-poor soil.

A Bit of History

It had been known for centuries that legumes had root nodules, but they were considered to be insect galls or some other pathological manifestation. The beneficial importance of root nodules for the nitrogen nutrition of leguminous plants was demonstrated in 1886–1888, by the German agricultural scientist Hermann Hellriegel. He showed that legumes could grow readily in soils that were free of nitrogen fertilizer, while nonlegumes could not. He concluded that legumes were obtaining their needed nitrogen from the atmosphere. The probable importance of bacteria in the formation of nodules was shown by B. Frank in 1879, when he proved that legume seeds allowed to grow in sterilized soil did not become nodulated. The British scientist H. M. Ward then showed in 1887, that crushed nodules could be used to inoculate leguminous seeds and induce nodulation even in sterilized soil. The next year the great Dutch microbiologist Martinus Beijerinck developed methods for the isolation and culture of the root nodule bacteria, and showed that his pure cultures, when inoculated onto legumes, induced the formation of root nodules. Although Beijerinck made many attempts, he could not demonstrate that his pure cultures would use atmospheric nitrogen, and he concluded that the ability to fix N_2 only developed after the bacteria had become associated with the plant in the root nodule. Only in very recent years has it been possible to show that under microaerophilic conditions, *Rhizobium* will fix N_2 in culture. An understanding of root nodule bacteria was one of the first major developments in agricultural microbiology, and raised considerable interest in other possible roles of microorganisms in soil fertility and plant growth. This led to the establishment of departments of agricultural bacteriology in most colleges of agriculture, and it was in such departments that the field of general bacteriology (meaning, primarily, nonmedical bacteriology) flourished.

About 90 percent of all leguminous species are capable of becoming nodulated. There is a marked specificity between species of legume and strains of *Rhizobium*. A single *Rhizobium* strain may be able to infect certain species of legumes but not others. A group of *Rhizobium* strains able to infect a group of related legumes is called a **cross-inoculation group.** Even if a *Rhizobium* strain is able to infect a certain legume, it is not always able to bring about the production of nitrogen-fixing nodules. If the strain is **ineffective** the nodules formed will be small, greenish-white, and incapable of fixing nitrogen; if the strain is **effective,** on the other hand, the nodule will be large, reddish, and nitrogen fixing. Effectiveness is determined by genes in the bacterium that can be lost by mutation or gained by genetic transformation.

Stages in nodule formation   The stages in the infection and development of root nodules are now fairly well understood (Figure 14.18). The roots of leguminous plants secrete a variety of organic materials, which stimulate the growth of a rhizosphere microflora. This stimulation is not restricted to the rhizobia but involves a variety of rhizosphere bacteria. If there are rhizobia in the soil, they grow in the rhizosphere and build up to high population densities. Infection of the root occurs by way of the root hairs.

Specificity in the legume–*Rhizobium* symbiosis depends initially upon a surface macromolecule of the root hair surface, which interacts with capsular polysaccharide or perhaps outer membrane lipopolysaccharide of the *Rhizobium* cell. This process has been studied in most detail in the white clover–*R. trifolii* symbiosis. The cells of *R. trifolii* have a fairly extensive capsule (see Figure 2.35), and it has been shown that there is a specific binding of the bacterial cells to the root hairs of white clover. *R. trifolii* does not bind strongly to root hairs of other legumes, and other species of *Rhizobium*, not infective for white

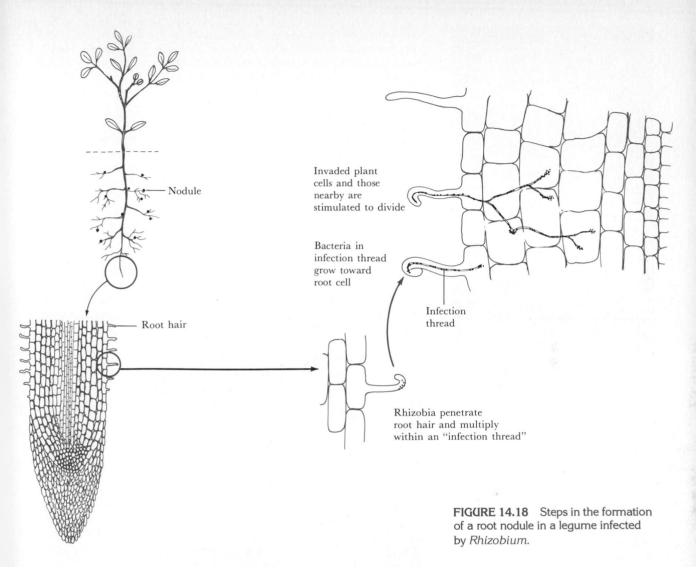

FIGURE 14.18 Steps in the formation of a root nodule in a legume infected by *Rhizobium.*

Labels within figure:
- Nodule
- Root hair
- Invaded plant cells and those nearby are stimulated to divide
- Bacteria in infection thread grow toward root cell
- Infection thread
- Rhizobia penetrate root hair and multiply within an "infection thread"

clover, do not bind strongly to white clover root hairs. A substance can be extracted and purified from white clover roots that specifically agglutinates *R. trifolii* cells, and this substance, a **lectin*** called trifolin, is apparently the agent that binds *R. trifolii* cells to the root hair. Binding involves a polysaccharide portion of the *R. trifolii* outer layer, which contains the sugar 2-deoxyglucose. If 2-deoxyglucose is added to roots, the binding of *R. trifolii* is specifically blocked. The interaction of the polysaccharide capsular material from *R. trifolii* has been studied by converting it into a fluorescent derivative, and then observing fluorescence of root hairs when treated with this material. As seen in Figure 14.19, the root hairs bind this material more strongly than do the cells beneath the root hairs, and the binding seems to be especially strong near the root hair tips. It is known from other work

---

*Lectins are plant proteins with a high affinity for specific sugar residues.

that the initial penetration of the *Rhizobium* into the root hair is via the root hair tip. After entry into the root hair, the bacteria proliferate, and form the so-called **infection thread,** which spreads down the root hair (Figure 14.20 and Color plate IV).

A root cell adjacent to the root hair then becomes infected. If this cell is a normal diploid cell, it usually is destroyed by the infection, undergoing necrosis and degeneration; if it is a tetraploid cell, however, it can become the forerunner of a nodule. There are always in the root a small number of tetraploid cells of spontaneous origin, and if one of these cells becomes infected, it is stimulated to divide. Progressive divisions of such infected cells leads to the production of the tumorlike nodule (Figure 14.21*a*). In culture, rhizobia produce substances called **cytokinins,** which cause tetraploid cells to divide, and it is likely that production of cytokinins also occurs in the infected cells.

The bacteria multiply rapidly within the tetraploid cells and become surrounded singly or in small groups by portions of the host cell membrane (Figure 14.21*b*). The bacteria then are sometimes transformed into swollen, misshapen, and branched forms called **bacteroids.** Only after the formation of bacteroids does nitrogen fixation begin. Eventually the nodule deteriorates, releasing bacteria

FIGURE 14.19 Specific attachment to clover root hairs of the capsular lipopolysaccharide from *Rhizobium trifolii.* The LPS extracted from the bacteria was rendered fluorescent by conjugation with fluorescein isothiocyanate, and after treating roots with the preparation, they were viewed under a fluorescence microscope. The root hair tips are especially heavily stained. Magnification, 95×. (From Dazzo, F. B., and W. J. Brill. 1977. Receptor site on clover and alfalfa roots for *Rhizobium.* Appl. Environ. Microbiol. 33:132–136.)

FIGURE 14.20 An infection thread formed by *Rhizobium trifollii* on white clover (*Trifolium repens*). A number of bacteria can still be seen attached to the root hair. The infection thread consists of a cellulose deposit down which the bacteria move to the cortical cells of the root. Magnification, 880×. (Courtesy of Frank Dazzo.)

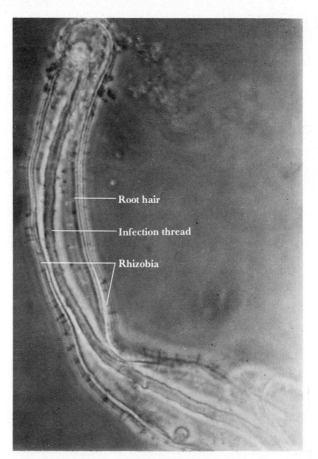

Root hair

Infection thread

Rhizobia

into the soil. The bacteroid forms are incapable of division, but there are always a small number of dormant rod-shaped cells present. These now proliferate, using as nutrients some of the products of the deteriorating nodule, and can initiate the infection in other roots or maintain a free-living existence in the soil.

Biochemistry of nitrogen fixation in nodules   As discussed in Section 5.13, nitrogen fixation involves the activity of the enzyme nitrogenase, a large two-component protein containing iron and molybdenum. Nitrogenase in nodules has characteristics similar to the enzyme of free-living N_2-fixing bacteria, including O_2 sensitivity and ability to reduce acetylene as well as N_2. The nitrogenase is found within the bacteroids themselves and is undoubtedly synthesized by the bacteroids, as pure cultures of *Rhizobium* synthesize nitrogenase when grown microaerophilically.

Organic compounds (probably organic acids such as succinate) synthesized by the plant leaves during photosynthesis are translocated to the bacteroids to serve as the electron donor for ATP production and for the reduction of N_2. Although the precise reductant used in the N_2-fixing process is not known, the biochemical steps are probably similar to those in free-living N_2-fixation systems. ATP is required for bacteroid N_2 fixation, as it is in other systems. The first stable product of N_2 fixation by bacteroids is NH_3; this is converted into amino acids, and the latter are in turn transferred from the bacteroid to the plant root cells and then to the rest of the plant.

Genetics of nodule formation   The important O_2-binding protein in the root nodule, leghemoglobin, is genetically coded for in part by both the plant and the bacterium. The globin (protein) portion of leghemoglobin is coded for by plant DNA while heme synthesis is genetically directed by the bacterium. In addition to leghemoglobin, however, there are thought to be 18 to 20 additional polypeptides that play crucial roles in the overall legume *Rhizobium*–symbiosis. Nitrogen fixation itself is clearly a process genetically directed by the bacterium (since several species of *Rhizobium* have been shown to fix N_2 in culture) while lectin synthesis is clearly a genetic property of the plant. Between the point of correct plant-bacterium recognition and the formation of an effective nodule, however, several genetically directed steps are likely to occur. There is good evidence that a number of nodulation-specific proteins are coded for by a large plasmid in the rhizobial cell. Such a plasmid was first found in *R. leguminosarum*, but similar-size plasmids have now been identified in a variety of *Rhizobium* species. Proof that these plasmids dictate host specificity is shown by experiments in which the plasmid has been conjugatively transferred from one species of *Rhizobium* to another. When the plasmid is transferred from *R. leguminosarum* (whose host is pea) to *R. trifolii* (whose host is clover) for example, the latter species become capable of effectively nodulating pea.

In addition to the nodulation process, the initial plant-bacterium recognition event may also be plasmid-mediated. Factors expressed by the bacterium that are critical for binding to the correct host lectin are coded for by a plasmid, but this may not be the same plasmid as that involved in the later stages of nodulation. Finally,

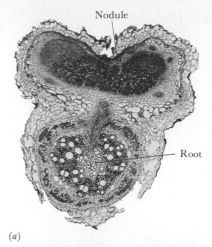

(a)

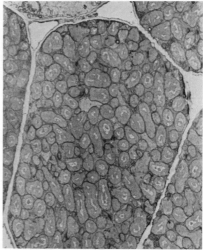

(b)

FIGURE 14.21   (a) Cross section through a legume root nodule. The darkly stained region contains plant cells filled with bacteria. Magnification, 110×. (b) Electron. micrograph of a thin section through a single bacteria-filled cell of a subterranean clover nodule. Magnification, 2680×. (From Dart, P. J., and F. V. Mercer. 1964. Arch. Mikrobiol. 49:209–235.)

the genes for hydrogenase, an enzyme whose role in *Rhizobium* is to take up H_2 for use as a reductant in nitrogen fixation, is also a plasmid coded protein.* Hydrogenase genes constitute part of the nodulation plasmid. Although there is little evidence that nitrogenase genes are plasmid encoded in other nitrogen-fixing bacteria, certain nitrogenase structural genes reside on a plasmid in *R. leguminosarum* and perhaps in other *Rhizobium* species as well. It is clear that plasmid DNA is an important source of genetic information for symbiotic nitrogen fixation.

Practical aspects of the legume-Rhizobium symbiosis When a crop is planted in an area where it has not been previously raised or where it has not been grown for some time, the appropriate *Rhizobium* strain will probably be lacking in the soil. To ensure nodulation, the seed is usually inoculated with the correct strain before sowing. Whether or not seed inoculation is necessary depends on when the crop was planted last and on the nature of the soil. In some soils *Rhizobium* strains can survive for many years after the specific legume host has been planted, whereas in other soils, *Rhizobium* dies out in a few years. In nitrogen-deficient soils legumes benefit greatly from being nodulated (see Section 13.8), and even in relatively rich soils some benefit is probably obtained although the effect may not be very dramatic. The amount of nitrogen fixed annually by living organisms is about 10^8 tons for the whole world, most of which is from symbiotic sources.

It is important to emphasize that despite appearances, nitrogen fixation is not a free lunch for the plant. A large amount of energy is required to fix molecular nitrogen; as discussed in Section 5.13, six electrons are required per molecule, in addition to considerable amounts of ATP. This energy must come from somewhere, and the source in symbiotic nitrogen fixation is, of course, the plant. The energy which the plant puts into the nitrogen fixation process is not available for other plant processes and must be considered on the negative side of the energy balance. Indeed, crop yields per acre of nitrogen-fixing plants, such as soybean, are generally lower than crop yields for non-nitrogen-fixing plants such as corn and wheat. It is not certain that these lower yields are due to the energy expenditure for nitrogen fixation, but such considerations do put the interpretation of the benefits of nitrogen fixation in a different light. In the final analysis, the benefits from nitrogen fixation must be considered in the context of whether it is more efficient to produce fixed nitrogen directly within the plant root, using energy from the sun, or to produce it chemically in fertilizer factories, using fossil fuel energy, in which case the plant must assimilate the combined nitrogen through its roots from the soil. There are some theoretical reasons to believe that the latter process is less efficient, and some of the nitrogen added as fertilizer may be lost via soil leaching. Before a final conclusion on this question can be drawn, a detailed energy balance of the whole plant system must be determined.

---

*Certain species of *Rhizobium* are also capable of using H_2 as an electron donor for lithotrophic growth with O_2 as electron acceptor and CO_2 as sole carbon source.

Nonlegume nitrogen-fixing symbioses In addition to the legume–*Rhizobium* relationship, nitrogen-fixing symbioses occur in a variety of nonlegumes, involving microorganisms other than rhizobia. Nitrogen-fixing cyanobacteria form symbioses with a variety of plants. The water fern *Azolla* contains N_2-fixing cyanobacteria of the genus *Anabaena* within small pores of its fronds. *Azolla* has been used for centuries to enrich rice paddies with fixed nitrogen. Before planting rice, the farmer allows the surface of the rice paddy to become densely covered with the fern. As the rice plants grow they crowd out the *Azolla–Anabaena* mixture, leading to death of the fern and release of fixed nitrogen which is assimilated by the rice plants. By repeating this process each growing season, the farmer can obtain high yields of rice without the need for nitrogenous fertilizers. The alder tree (genus *Alnus*) has nitrogen-fixing root nodules which may lack leghemoglobin but harbor a filamentous, streptomycete-like organism which has now been cultured and named *Frankia*. *Frankia* grows very slowly but has been shown to fix nitrogen in culture and is presumably the nitrogen-fixing agent in the alder nodules. Alder is a characteristic pioneer tree, able to colonize bare soils in nutrient poor sites, and this is likely due to its ability to enter into a symbiotic nitrogen-fixing relationship.

Azospirillum lipoferum is a N_2-fixing bacterium that lives in a rather casual association with roots of tropical grasses. It is found in the rhizosphere, where it grows on products excreted from the roots. It also has the ability to grow around the roots of cultivated grasses, such as corn (*Zea mays*), and inoculation of corn with *A. lipoferum* may lead to small increases in growth yield of the plant. It is likely that many casual relationships of this type exist between bacteria and plants, although only in the case of the nodule is easy recognition of a relationship possible.

The agronomic potential of legume and nonlegume nitrogen-fixing symbioses may be far greater than the agricultural applications made to date. Intensive research efforts are now under way to refine and improve known symbioses such as that of *Rhizobium* and the legume, and to discover new relationships which may have practical value for increased protein and fiber production on a global scale.

14.11

Mycorrhizae

Mycorrhiza literally means "root fungus" and refers to the symbiotic association that exists between plant roots and fungi. Probably the roots of the majority of terrestrial plants are mycorrhizal. There are two general classes of mycorrhiza: **ectomycorrhiza,** in which the fungal hyphae form an extensive sheath around the outside of the root with only little penetration of hyphae into the root tissue itself, and **endomycorrhiza,** in which the fungal mycelium is embedded in the root tissue. The present discussion will deal only with ectomycorrhizae.

Ectomycorrhizae are found mainly in forest trees, especially conifers, beeches, and oaks, and are most highly developed in temperate forests. In a forest, almost every root of every tree will be mycorrhizal. The root system of a mycorrhizal tree is composed of

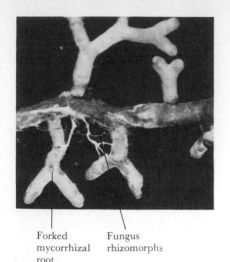

Forked   Fungus
mycorrhizal   rhizomorphs
root

FIGURE 14.22 Typical mycorrhizal root of a pine. *Pinus rigida,* with rhizomorphs of the fungus *Thelophora terrestris.* (From Schramm, J. R. 1966. Trans. Am. Phil. Soc. 56[Pt. 1].)

both long and short roots. The short roots, which are characteristically dichotomously branched (Figure 14.22), show the typical fungal sheath, whereas the long roots are usually uninfected. The fungi that participate in the mycorrhizal association are all Basidiomycetes, and most of them form typical mushroom fruiting bodies. Most mycorrhizal fungi do not attack cellulose and leaf litter, as do many other mushrooms, but instead use simple carbohydrates for growth and usually have one or more vitamin requirements; they obtain their nutrients from root secretions. The mycorrhizal fungi are never found in nature except in association with roots and hence can be considered obligate symbionts. These fungi produce plant growth substances that induce morphological alterations in the roots, causing characteristically short dichotomously branched mycorrhizal roots to be formed. Despite the close relationship between fungus and root, there is little species specificity involved. For instance, a single species of pine can form mycorrhizae with over 40 species of fungi.

The beneficial effect on the plant of the mycorrhizal fungus is best observed in poor soils, where trees that are mycorrhizal will thrive, but nonmycorrhizal ones will not. If trees are planted in prairie soils, which ordinarily lack a suitable fungal inoculum, trees that were artificially inoculated at the time of planting grow much more rapidly than uninoculated trees (Figure 14.23). However, in nutrient-rich conditions, the mycorrhizal plant does not grow any better than

FIGURE 14.23 Six-month-old seedlings of Monterey Pine (*Pinus radiata*) growing in prairie soil: left, nonmycorrhizal; right, mycorrhizal. (Courtesy of S. A. Wilde.)

an uninfected control, and growth may even be slightly retarded. There is no evidence that nitrogen fixation occurs in mycorrhizal plants, but it is well established that the mycorrhizal plant is able to absorb nutrients from its environment more efficiently than does a nonmycorrhizal one. This improved nutrient absorption is probably due to the greater surface area provided by the fungal mycelium.

The relationship between the fungus hyphae on the tree roots and the mushroom fruiting bodies seen on the surface of the soil has been well established in some cases. Fungus hyphae growing out from the surface of the root frequently form multistranded structures called **rhizomorphs,** which eventually approach the soil surface, differentiate, and form fruiting-body structures. The nutrients for the growth of the fruiting bodies come from the plant roots since rhizomorphs that have been isolated from the tree roots by cutting never form fruiting bodies.

14.12
Summary

We have seen in this chapter that microorganisms can enter into a variety of beneficial relationships with higher organisms. Some of these relationships are rather casual, whereas others are highly specific. Among all groups of microorganisms—bacteria, algae, fungi, and protozoa—are species that are symbionts, and they have developed relationships with many kinds of plants and animals. We have a fairly clear understanding of the underlying physiological mechanisms for a few symbioses of economic significance, such as the legume-*Rhizobium* or the rumen symbioses.

In symbioses between microbes and higher organisms the enormous difference in size between the two members should be emphasized. The protoplasmic mass of a microbe is almost always vanishingly small compared to the mass of the multicellular creature with which it interacts. In spite of its size, the symbiotic microbe is able to affect its partner by producing such substances as enzymes, vitamins, or other growth factors active at quite high dilutions; and as we shall see in the next chapter, the pathogenic microbe also produces potent substances, many of which harm the host. The study of the interactions of microbes with higher organisms thus provides us with some dramatic insights into the biochemical mechanisms that microbes have evolved to ensure their evolutionary success.

Supplementary Reading

Bauer, W. D. 1981. Infection of legumes by rhizobia. Annu. Rev. Plant Physiol. 32:407–449. A detailed review of the processes of plant-bacterium recognition, infection thread formation, and regulation of infectivity.

Bowden, G. H. W., D. C. Ellwood, and **I. R. Hamilton.** 1979. Microbial ecology of the oral cavity. Pages 135–217 *in* M. Alexander, ed. Advances in microbial ecology, vol. 3. Plenum Press, New York. A detailed review of microbial interactions in the oral cavity and the microbial contribution to dental caries.

Brill, W. J. 1979. Nitrogen fixation: basic to applied. Am. Sci. 67:458–466. An overview of the process of nitrogen fixation with discussion of the biochemistry, genetics, and practical aspects of the *Rhizobium*–legume symbiosis.

Brown, D. H., D. L. Hawksworth, and **R. H. Bailey,** eds. 1976. Lichenology: progress and problems. Academic Press, New York. This book gives a good picture of the current status of lichen research, and deals with all aspects of lichens: structure, taxonomy, chemistry, physiology, ecology, and effects of pollutants. Each chapter is written by an expert.

Gordon, H. A., and **L. Pesti.** 1971. The gnotobiotic animal as a tool in the study of host microbial relationships. Bacteriol. Rev. 35:390–429. How the germ-free animal can be used in the study of host-parasite relationships.

Marples, M. J. 1969. Life on the human skin. Sci. Am. 220:108–115. The skin is discussed as a microbial habitat.

Oomen, H. A. P. C. 1970. Interrelationship of the human intestinal flora and protein utilization. Proc. Nutr. Soc. 29:197–206. Reviews the work suggesting that sweet-potato-eating New Guineans fix N_2 via their intestinal flora.

Roberts, G. P., and **W. J. Brill.** 1981. Genetics and regulation of nitrogen fixation. Annu. Rev. Microbiol. 35:207–235. A review focusing on the genetics of N_2 fixation in *Klebsiella,* but deals briefly with genetics of the *Rhizobium* system and certain free-living N_2 fixers.

Savage, D. C. 1977. Microbial ecology of the gastrointestinal tract. Annu. Rev. Microbiol. 31:107–133. This review makes a clear-cut distinction between the indigenous and transient microorganisms of the intestine, and shows the importance of the former in animal function.

Wolin, M. J. 1979. The rumen fermentation: a model for microbial interactions in anaerobic ecosystems. Pages 49–77 *in* M. Alexander, ed. Advances in microbial ecology, vol. 3. Plenum Press, New York. An excellent treatment of the rumen microbial ecosystem.

Host-Parasite Relationships

The ability to cause infectious disease is one of the most dramatic properties of microorganisms. The understanding of the physiological and biochemical basis of infectious disease has led to therapeutic and preventive measures that have had far-reaching influence on medicine and human affairs. The present chapter discusses the nature of infectious disease and its control.

A **parasite** is an organism that lives on or in and causes damage to another living organism, called the **host.** The relationship between host and parasite is dynamic, since each modifies the activities and functions of the other. The outcome of the host-parasite relationship depends on the **pathogenicity** of the parasite, that is, on its ability to inflict damage, and on the resistance or susceptibility of the host. The term **virulence** is used to indicate the *degree* of pathogenicity of the parasite and is usually expressed as the LD_{50}: the dose or cell number that will damage or kill 50 percent of the inoculated animals within a given time period (LD stands for *lethal dose*). Neither the virulence of the parasite nor the resistance of the host are constant factors, however, each varying under the influence of external factors or as a result of the host-parasite relationship itself.

In all cases there is some specificity in a host-parasite relationship in that a given parasite can grow only in a restricted variety of hosts. Often this specificity is such that the parasite is restricted to a single host species, but the specificity may be much broader.

Infection is not synonymous with **disease** because infection does not always lead to injury of the host, even if the pathogen is potentially virulent. In a diseased state the host is harmed in some way, whereas infection refers to any situation in which a microorganism is established and growing in a host, whether or not the host is harmed.

483

Entry of the Pathogen into the Host

A pathogen must first gain access to host tissues and multiply before damage can be done. In most cases this requires that the organism penetrate the skin, mucous membranes, or intestinal epithelium, surfaces which normally act as microbial barriers. Passage through the skin into the subcutaneous layers almost always occurs through wounds; only in rare instances is there any evidence that pathogens can penetrate through the unbroken skin.

FIGURE 15.1 Adherence of the pathogen *Vibrio cholerae* to the cells of the ileal epithelium of the adult rabbit. (a) Scanning electron micrograph, showing flagellated bacteria attached to the epithelium one hour after infection. The microvilli tips appear as a regular array of closely packed particles. Magnification, 25,200×. (b) Transmission electron micrograph of a thin section of *V. cholerae* adhering to the brush border of the rabbit villus. Magnification, 56,700×. (From Nelson, E. T., J. D. Clements, and R. A. Finkelstein. 1976. Infection and Immunity 14:527–547.)

Specific adherence Most microbial infections begin on the mucous membranes of the respiratory, alimentary, or genitourinary tracts. There is considerable evidence that bacteria or viruses able to initiate infection are able to adhere specifically to epithelial cells (Figure 15.1). The evidence for specificity is of several types. First, an infecting microbe does not adhere to all epithelial cells equally, but shows selectivity by adhering to cells in the particular region of the body where it normally gains entrance. For example, *Neisseria gonorrhoeae,* the causative agent of the veneral disease gonorrhea, adheres much more strongly to urogenital epithelia than to other tissues. Second, adherence is strain specific: a bacterial strain that normally infects humans will adhere more strongly to the appropriate human epithelial cells than to similar cells in another animal (for example, the rat), whereas

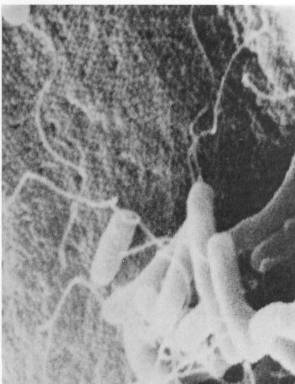

(a)

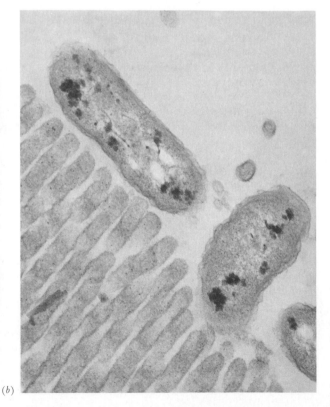

(b)

a strain which specifically colonizes the rat will adhere more firmly to rat cells than to human cells. Thus there is a host-strain specificity for adherence, reminiscent of the host-strain specificity so well worked out in the legume-*Rhizobium* symbiosis (see Section 14.10). Many bacteria possess specific surface macromolecules that bind to complementary receptor molecules on the surfaces of certain animal cells, thus promoting specific and firm adherence. Certain of these macromolecules are polysaccharide in nature and form a sticky meshwork of fibers called the bacterial **glycocalyx** (see Section 2.9 and Figure 15.2). The glycocalyx is important not only in attaching bacterial cells to eucaryotic cell surfaces, but also in adherence between bacterial cells (Figure 15.2). In addition, fimbriae (see Section 2.9) may be important in the attachment process, since it now appears that fimbriae of *N. gonorrhoeae* play a key role in the attachment of this organism to urogenital epithelium.

Evidence of the specific interaction between mucosal epithelium and pathogen comes from studies on diarrhea caused by *Escherichia coli*. Most strains of *E. coli* are nonpathogenic and are part of the normal flora of the *large* intestine. A few strains (only a handful of the 160 different *E. coli* serotypes) are enteropathogenic, possessing the ability to colonize the *small* intestine and initiate diarrhea. Such strains possess a specific surface structure, the K antigen, an acidic polysaccharide capsule, which is involved in specific attachment to intestinal mucosa. As we discussed in Section 11.6, the ability to produce the K antigen is controlled by a conjugative plasmid. We shall see in Section 15.5 that colonization of the small intestine alone is not sufficient to initiate symptoms of diarrhea; another factor, the enterotoxin, must also be formed. Thus two kinds of *E. coli* can be recognized: pathogenic strains, which are able to adhere to the mucosal surface of the small intestine and cause disease symptoms; and "normal" *E. coli*, which are unable to adhere to the small intestine or produce enterotoxin, which grow in the large intestine (cecum and colon), and which enter into a long-lasting symbiotic relationship with the mammalian host.

Penetration In diseases such as whooping cough, caused by *Bordetella pertussis*, diphtheria, caused by *Corynebacterium diphtheriae*, and cholera, caused by *Vibrio cholerae*, the pathogen can remain localized at the mucosal surface and initiate damage by liberating toxins, as described later. However, in most cases the pathogen penetrates the epithelium and either grows in the submucosa, or spreads to other parts of the body where growth is initiated. Access to the interior of the body may occur in those areas where lymph glands are near the surface, such as the nasopharyngeal region, the tonsils, and the lymphoid follicles of the intestine. Many times small breaks or lesions in the mucous membrane in one of these regions will permit an initial entry. Motility may be of some value to an invader, although many pathogens are nonmotile. Among factors inhibiting establishment of the pathogen is the normal microbial flora itself, with which the pathogen must compete for nutrients and living space. If the normal flora is altered or eliminated by antimicrobial therapy, successful colonization by a pathogen may be easier to accomplish.

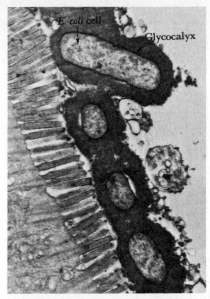

FIGURE 15.2 Enteropathogenic *E. coli* in fatal model infection in the newborn calf. The bacterial cells are attached to the brush border of the villus via an extensive glycocalyx. Magnification, 11,700×. Courtesy of J. W. Costerton.

Growth in Vivo

The initial inoculum is almost never enough to cause damage; a pathogen must grow within host tissues in order to produce a disease condition. If the pathogen is to grow it must find in the host appropriate nutrients and environmental conditions. Temperature, pH, and reduction potential are environmental factors that affect pathogen growth, but of most importance is the availability of microbial nutrients in host tissues. Although at first thought it might seem that a vertebrate animal would be a nutritional paradise for microbes, not all conceivable nutrients are in plentiful supply, and there is probably considerable selectivity in determining what kinds of organisms can grow. Soluble nutrients such as sugars, amino acids, and organic acids may often be in short supply and organisms able to utilize high-molecular-weight components may be favored. Not all vitamins and other growth factors are necessarily in adequate supply in all tissues at all times. *Brucella abortus,* for example, can grow slowly in most tissues of infected cattle, but grows rapidly only in the placenta, where it causes bovine abortion. This specificity is due to the elevated concentration of erythritol found in the placenta, which greatly stimulates growth of *B. abortus.* Trace elements may also be in short supply and influence establishment of the pathogen. In the latter category, the most evidence exists for the influence of iron on microbial growth. A specific protein called transferrin, present in animals, binds iron tightly and transfers it through the body. Such is the affinity of this protein for iron that microbial iron deficiency may be common; indeed, administration of a soluble iron salt to an infected animal may greatly increase the virulence of some pathogens. As we noted in Section 5.16, many bacteria produce iron-chelating compounds which help them to obtain iron from the environment.

Growth in body fluids   Body fluids such as blood, lymph, and urine are able to support the growth of some microbes, but there is considerable selectivity between host and microbe. Blood contains a wide variety of substances harmful to microbes including lysozyme, complement (see Section 15.7), and a wide variety of antibodies which inhibit microbial growth, and it is also relatively low in low-molecular-weight organic compounds that might support the growth of most pathogens. Lymph is similar in composition to the fluid portion of the blood. Urine has been known to support the growth of microbes since the time of Pasteur, but it is also quite selective in its action, partly because of its pH. The pH of urine is generally somewhat acidic although a few individuals may have urine with pH as high as 7. That acidity alone is an inhibitory factor for microbial growth can be shown experimentally by adjusting the pH to various values; if the pH of urine is adjusted to 6 or higher it readily supports the growth of *E. coli,* a common cause of urinary infection. Other enteric bacteria would probably be expected to show a similar response. Interestingly, the pH of females' urine is generally higher than that of males, and the urinal pH of pregnant females is sufficiently high so that it would almost invariably be optimal for the growth of

E. coli (urinary tract infections are very common in pregnant women). One way of controlling a urinary tract infection is to administer the sulfur-containing amino acid methionine, which is metabolized to sulfuric acid, leading to an acidification of the urine.

Localization in the body   After initial entry, the organism often remains localized and multiplies, producing a small **focus of infection,** such as the boil, carbuncle, or pimple that commonly arises from *Staphylococcus* infections of the skin. Alternatively, the organisms may pass through the lymphatic vessels and be deposited in lymph nodes. If an organism reaches the blood it will be distributed to distant parts of the body, usually concentrating in the liver or spleen. Spread of the pathogen through the blood and lymph systems can result in a generalized (systemic) infection of the body, with the organism growing in a variety of tissues. If extensive bacterial growth in tissues occurs, some of the organisms are usually shed into the bloodstream in large numbers, a condition called **bacteremia.** However, generalized infection of this type is rare; a more common situation is for the pathogen to localize in a specific organ.

Enzymes involved in invasion   Streptococci, staphylococci, pneumococci, and certain clostridia produce **hyaluronidase,** an enzyme that promotes spreading of organisms in tissues by breaking down hyaluronic acid, a polysaccharide that functions in the body as a tissue cement. Production of this enzyme may therefore enable these organisms to spread from an initial focus. Streptococci and staphylococci also produce a vast array of proteases, nucleases, and lipases which serve to depolymerize host proteins, nucleic acids, and fats, respectively. Clostridia that cause gas gangrene produce **collagenase,** which breaks down the collagen network supporting the tissues; the resulting dissolution of tissue is a factor in enabling these organisms to spread through the body. Fibrin clots are often formed by the host in a region of microbial invasion and serve to wall off the organism and prevent its spread through the body. Some organisms are able to produce fibrinolytic enzymes to dissolve these clots, and make further invasion possible. One such fibrinolytic substance, produced by streptococci, is known as **streptokinase.** On the other hand, some organisms produce enzymes that actually promote fibrin clotting, which causes localization of the organism rather than its spread. The best studied fibrin-clotting enzyme is **coagulase,** produced by many staphylococci, which causes the fibrin material to be deposited on the cocci and may offer them protection from phagocytosis. Production of coagulase probably accounts for localization of many staphylococcal infections as boils and pimples.

15.3

Exotoxins

The ways in which pathogens bring about damage to the host are diverse. Only rarely are symptoms due to the presence of large numbers of microorganisms per se. Although a large mass of cells can block vessels or heart valves or clog the air passages of the lungs, in

TABLE 15.1   Exotoxins Produced by Certain Bacteria Pathogenic for Humans

Organism	Disease	Toxin	Action
Clostridium botulinum	Botulism	Neurotoxin	Flaccid paralysis
C. tetani	Tetanus	Neurotoxin	Spastic paralysis
C. perfringens	Gas gangrene, food poisoning	α-Toxin	Hemolysis (lecithinase)
		β-Toxin	Hemolysis
		γ-Toxin	Hemolysis
		δ-Toxin	Hemolysis
		θ-Toxin	Hemolysis (cardiotoxin)
		κ-Toxin	Collagenase
		λ-Toxin	Protease
		Enterotoxin	Alters permeability of intestinal epithelium
Corynebacterium diphtheriae	Diphtheria	Diphtheria toxin	Inhibits protein synthesis
Staphylococcus aureus	Pyogenic infections (boils, etc.), respiratory infections, food poisoning	α-Toxin	Hemolysis
		Leukocidin	Destroys leukocytes
		β-Toxin	Hemolysis
		γ-Toxin	Kills cells
		δ-Toxin	Hemolysis, leukolysis
		Enterotoxin	Induces vomiting, diarrhea
Streptococcus pyogenes	Pyogenic infections, tonsillitis, scarlet fever	Streptolysin O	Hemolysis
		Streptolysin S	Hemolysis
		Erythrogenic toxin	Causes scarlet fever rash
Vibrio cholerae	Cholera	Enterotoxin	Induces fluid loss from intestinal cells
Escherichia coli (enteropathogenic strains only)	Gastroenteritis	Enterotoxin	Induces fluid loss from intestinal cells
Shigella dysenteriae	Bacterial dysentery	Neurotoxin	Paralysis, hemorrhage
Yersinia pestis	Plague	Plague toxin	Kills cells
Bordetella pertussis	Whooping cough	Whooping cough toxin	Kills cells
Pseudomonas aeruginosa	Various *P. aeruginosa* infections	Dermonecrotic toxin	Kills cells

most cases more specific factors are involved. Many pathogens produce **toxins** that are responsible for all or much of the host damage.

Toxins released extracellularly as the organism grows are called **exotoxins.** These toxins may travel from a focus of infection to distant parts of the body and hence cause damage in regions far removed from the site of microbial growth. Table 15.1 provides a summary of the properties and actions of some of the best known exotoxins.

Diphtheria toxin   The toxin produced by *Corynebacterium diphtheriae*, the causal agent of diphtheria, was the first exotoxin to be discovered. It is a protein of molecular weight 65,000, which differs markedly in its action on different animal species; rats and mice are relatively resistant, whereas humans, rabbits, guinea pigs, and birds are susceptible. Diphtheria toxin is very potent in its action, with only a single molecule being required to cause the death of a single cell. It binds irreversibly to the cell, and within a few hours the cell loses its ability to synthesize protein because the toxin blocks transfer of an amino acid from a tRNA to the growing peptide chain. The toxin specifically inactivates one of the elongation factors (elongation fac-

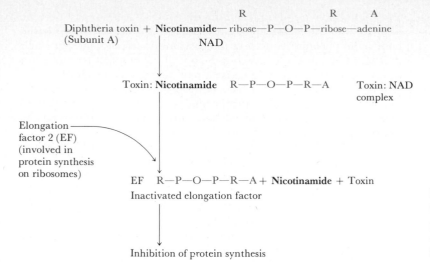

$$
\underset{\text{(Subunit A)}}{\text{Diphtheria toxin}} + \overset{R}{\underset{\text{NAD}}{\textbf{Nicotinamide}}} \text{— ribose—P—O—P—ribose—adenine}
$$

$$\downarrow$$

$$
\text{Toxin: } \textbf{Nicotinamide} \quad \text{R—P—O—P—R—A} \qquad \underset{\text{complex}}{\text{Toxin: NAD}}
$$

Elongation
factor 2 (EF)
(involved in
protein synthesis
on ribosomes)

$$\downarrow$$

$$
\text{EF} \quad \text{R—P—O—P—R—A} + \textbf{Nicotinamide} + \text{Toxin}
$$
Inactivated elongation factor

$$\downarrow$$

Inhibition of protein synthesis

FIGURE 15.3 Catalysis by diphtheria toxin of attachment of adenyl diphosphate ribose portion of NAD to elongation factor 2, leading to inhibition of protein synthesis.

tor 2) involved in growth of the polypeptide chain (see Figure 9.15) by catalyzing the attachment of the adenosine diphosphate ribose moiety of NAD to the elongation protein (Figure 15.3). The toxin is a single polypeptide chain whose function is separable into two parts: fragment A containing the enzymatic activity for inactivating elongation factor 2, and fragment B being responsible for the specific association of the toxin with the sensitive cell cytoplasm.

Diphtheria toxin is formed only by strains of *C. diphtheriae* that are lysogenized by phage β, and its production is coded for by genetic information present in the phage genome, as shown by the fact that mutants of phage β can be isolated which cause the production of toxin molecules with altered proteins. Nontoxigenic and hence nonpathogenic strains of *C. diphtheriae* can be converted to pathogenic strains by infection with the phage (the process of phage conversion, see Section 11.3).

A nongenetic factor of significance for toxin production is the concentration of iron present in the environment in which the bacteria are growing. In media containing sufficient iron for optimal growth no toxin is produced, whereas when the iron concentrations is reduced to suboptimal levels, toxin production occurs. Some evidence exists that the role of iron is to bind to and activate a regulatory protein in *C. diphtheriae,* and that the iron-binding protein combines with a control region of the DNA of β phage and prevents expression of the diphtheria toxin gene. When iron is absent, the regulatory protein does not act, and toxin synthesis can occur. Mutants in the regulatory region are known which synthesize toxin even in the presence of high iron concentrations.

Exotoxin A of *Pseudomonas aeruginosa* has been shown to have an action quite similar to diphtheria toxin, also transferring the ADP-ribosyl portion of NAD to elongation factor 2.

Tetanus and botulinum toxins These toxins are produced by two species of obligately anaerobic bacteria, *Clostridium tetani* and *C.*

botulinum, which are normal soil organisms that occasionally become involved in disease situations in animals. *Clostridium tetani* grows in the body in deep wound punctures that become anaerobic, and although it does not invade the body from the initial site of infection, the toxin it produces can spread and cause a neurological disturbance, which results in death. *Clostridium botulinum* rarely if ever grows directly in the body, but it does grow and produce toxin in improperly canned foods. Ingestion of the food without proper heat treatment results in neurological disease and death.

The tetanus toxin is a protein of molecular weight 67,000. Upon entry into the central nervous system this toxin becomes fixed to nerve synapses, binding specifically to one of the lipids. To understand how it induces its characteristic spastic paralysis, it is essential to discuss briefly how normal nerve function in muscle contraction operates. As seen in Figure 15.4, in normal muscle contraction, nerve impulses from the brain travel through the spinal cord and initiate a sequence of events at the motor end plate (the structure forming a junction between a muscle fiber and its motor nerve), which results in muscle contraction. When one set of muscles contracts, the opposing set of muscles becomes stretched. A stretch-sensitive receptor in the oppos-

FIGURE 15.4 Contrasting actions of tetanus and botulinum toxins. See text for details.

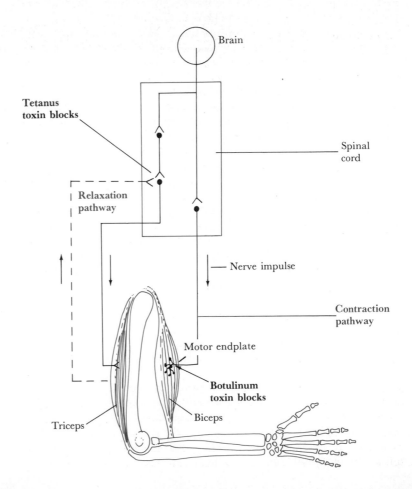

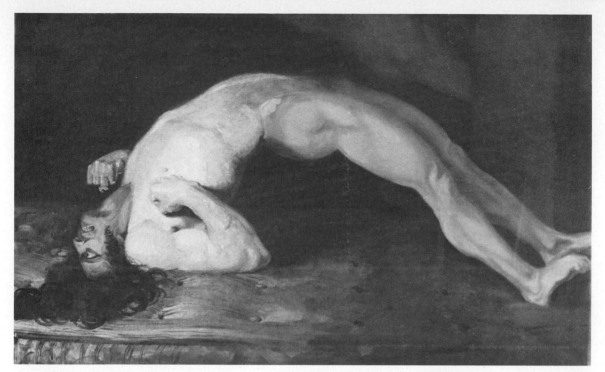

FIGURE 15.5 A soldier dying from tetanus. Painting by Charles Bell in the Royal College of Surgeons, Edinburgh. (Courtesy of the Royal College of Surgeons of Edinburgh.)

ing muscles would cause neurons enervating these muscles to fire and oppose the stretching, but the firing is inhibited by impulses from an inhibitory nerve. Thus the opposing set of muscles relaxes. Tetanus toxin binds specifically to these inhibitory motoneurons and blocks the inhibitory action, resulting in contraction of both sets of muscles. This leads to a spastic paralysis, as the two types of muscles attempt to oppose each other. Since the flexor muscles are usually dominant over the extensor muscles, the symptoms in advanced tetanus are generalized flexor muscle contractions (see Figure 15.5). If the muscles of the mouth are involved, the prolonged spasm restricts the mouth's movement, resulting in the condition known as **lockjaw.** If the respiratory muscles are involved, death may be due to asphyxiation.

Botulinum toxin is an even more potent neurotoxin, perhaps the most poisonous substance known. One milligram of pure botulinum toxin is enough to kill more than 1 million guinea pigs. At least six distinct botulinum toxins have been described and the syntheses of at least two of these are coded for by genes residing on lysogenic bacteriophages specific for *C. botulinum.* The major toxin is a protein of about 150,000 molecular weight which readily forms complexes with nontoxic botulinum proteins to give an active species of almost 1,000,000 molecular weight. Toxicity occurs due to binding of the toxin to presynaptic terminal membranes at the nerve-muscle junction, blocking the release of acetylcholine. Since transmission of the nerve impulse to the muscle is by means of acetylcholine action, muscle contraction is inhibited, causing a flaccid paralysis. The fatality rate from botulism poisoning can approach 100 percent, but can be significantly reduced by quick administration of antibody against the toxin (antitoxin, see Section 16.3) and by use of the artificial respirator to prevent breathing failure. Death is usually due to respiratory or cardiac failure.

Hemolysins   Various pathogens produce proteins that are able to act on the animal cell membrane, inducing cell lysis and hence cell death. The action of these toxins is most easily detected with red blood cells, hence they are often called **hemolysins;** in probably all cases, however, they also work on cells other than erythrocytes. The production of such toxins is most readily demonstrated by streaking the organism on a blood agar plate. During growth of the colonies, some of the hemolysin is released and lyses the surrounding red blood cells, typically clearing a zone (Figure 15.6). Various degrees of hemolysis can occur, from partial to total cell lysis, and these stages have clinical significance, especially in the diagnosis of streptococcal diseases (see Section 19.23). Some hemolysins have been shown to be enzymes that attack the phospholipid of the host cell membrane. Because the phospholipid lecithin (phosphatidylcholine) is often used as a substrate, these enzymes are called **lecithinases** or **phospholipases.** There are several sites in the phospholipid molecule that can be hydrolyzed (Figure 5.13*b*). Since the cell membranes of all organisms, both procaryotes and eucaryotes, contain phospholipids, hemolysins that are phospholipases sometimes destroy bacterial as well as animal cell membranes. Some hemolysins are not phospholipases, however. Streptolysin O, a hemolysin produced by streptococci, affects the sterols of the host cell membrane, and its action is neutralized by addition of cholesterol or other sterols. **Leukocidins** are lytic agents capable of lysing white blood cells and hence serve to decrease host resistance.

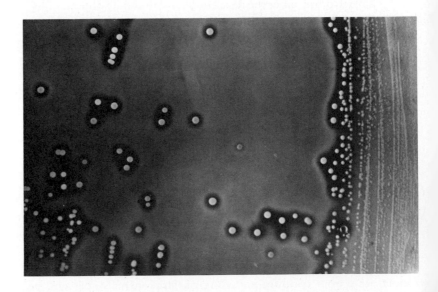

FIGURE 15.6 Zones of hemolysis around colonies of *Streptococcus pyogenes* growing on a blood agar plate.

Gram-negative bacteria produce lipopolysaccharides as part of the outer layer of their cell walls (Figures 2.18 and 2.19), which under many conditions are toxic. These are called **endotoxins** because they are generally cell bound and are released in large amounts only when cells lyse. Unfortunately, the word *endotoxin* suggests that the material is within the cell, rather than part of the outer layer. In fact, endotoxin is sometimes even liberated from the surface of the intact cell, making the distinction between an exotoxin and an endotoxin even less clear. For the present discussion, the word *endotoxin* should be equated with lipopolysaccharide toxin, and we shall confine its use to those outer layers of Gram-negative bacteria that are toxic to animals.* The major differences between exotoxins and endotoxins are listed in Table 15.2.

Endotoxins

TABLE 15.2 Basic Properties of Exotoxins and Endotoxins

Property	Exotoxins	Endotoxins
Chemical composition	Proteins, excreted by certain Gram-positive or Gram-negative bacteria	Lipopolysaccharide/lipoprotein; released upon cell lysis as part of the outer membrane of Gram-negative bacteria
Toxicity	Highly toxic, often fatal	Weakly toxic, rarely fatal
Immunogenicity	Highly immunogenic; stimulate the production of neutralizing antibody (antitoxin)	Poorly immunogenic; immune response not sufficient to neutralize toxin
Toxoid potential	Treatment of toxin with formaldehyde will destroy toxicity, but treated toxin (toxoid) remains immunogenic	None
Fever potential	Do not produce fever in host	Often produce fever in host

Endotoxin structure and function When injected into an animal, endotoxins cause a variety of physiological effects. Fever is an almost universal symptom of endotoxemia, because endotoxin stimulates host cells to release proteins called *pyrogens,* which affect the temperature-controlling center of the brain. In addition, however, the animal may develop diarrhea, experience a rapid decrease in lymphocyte, leukocyte, and platelet numbers, and enter into a generalized inflammatory state. Large doses of endotoxin can cause death, primarily through hemorrhagic shock and tissue necrosis. In general, the toxicity of endotoxins is much lower than that of exotoxins. For instance, in the mouse the LD_{50} (see page 483) is 200 to 400 μg per mouse for an average endotoxin preparation, whereas the LD_{50} for botulinum toxin is about 0.000025 μg per mouse.

The overall structure of lipopolysaccharide is diagrammed in Figure 2.18. Lipopolysaccharide contains a **core polysaccharide,** which in *Salmonella* is the same for many species, consisting of ketodeoxyoctonate, seven-carbon sugars (heptoses), glucose, galactose, and

*Even the lipopolysaccharides of some Gram-negative photosynthetic bacteria show endotoxin activity, although these organisms, as far as is known, never infect animals.

N-acetylglucosamine. Connected to the core is the *O-polysaccharide,* variable from strain to strain, which usually contains galactose, glucose, rhamnose, and mannose, and generally contains one or more unusual dideoxy sugars such as abequose, colitose, paratose, or tyvelose. These sugars are connected in four- or five-sugar sequences (often branched) which then repeat to form the long O-polysaccharide. The structure of the lipid is least well known. The lipid, usually referred to as **lipid A,** is not a normal glycerol lipid, but instead the fatty acids are connected by ester linkage to *N*-acetylglucosamine. Fatty acids frequently found in the lipid include β-hydroxymyristic, lauric, myristic, and palmitic acids.

Most of the evidence suggests that it is the lipid portion of the lipopolysaccharide that is responsible for toxicity, and that the polysaccharide acts mainly to render the lipid water soluble. It should be emphasized, however, that not all lipopolysaccharides of Gram-negative bacteria are endotoxins. Endotoxins have been studied primarily in the genera *Escherichia, Shigella,* and especially *Salmonella.*

Limulus assay for endotoxin An endotoxin assay of unusual sensitivity has been developed using lysates of amoebocytes from the horseshoe crab, *Limulus polyphemus.* Although the mechanism of this assay is not known, it has been shown that endotoxin specifically causes the gelation or precipitation of the lysate, and the degree of reaction can be measured by a spectrophotometer, an instrument that compares turbidities of various samples. A measurable reaction can be obtained with as little as 10 to 20 picograms/ml of lipopolysaccharide (a picogram is 10^{-12} g or 10^{-6} μg). Apparently the active component of the *Limulus* extract reacts with the lipid component of lipopolysaccharide. The *Limulus* assay has been used to detect the presence of minute quantities of endotoxin in serum, cerebrospinal fluid, drinking water, and fluids used for injection. The test is so exquisitely sensitive that considerable care must be taken to avoid contamination of the equipment, solutions, and reagents used, since Gram-negative bacteria abound in the laboratory and clinical environment, for example, as contaminants in the distilled water. In clinical work, detection of endotoxin by the *Limulus* assay in serum or cerebrospinal fluid can be taken as evidence of Gram-negative infection of these body fluids.

15.5

Enterotoxins

Enterotoxins act on the small intestine, generally causing massive secretion of fluid into the intestinal lumen, leading to the symptoms of diarrhea. The word *enterotoxin* should not be confused with *endotoxin;* actually enterotoxins are exotoxins whose site of action is the intestine. The action of enterotoxins has been most commonly studied using animal models, of which the most precise is the ligated ileal loop. For most work, rabbits have been used, but pigs have also been studied. The procedure is done under anaesthesia, an incision being made and segments of the ileum being tied off with sutures. Experimental materials, such as cultures, culture filtrates, or purified

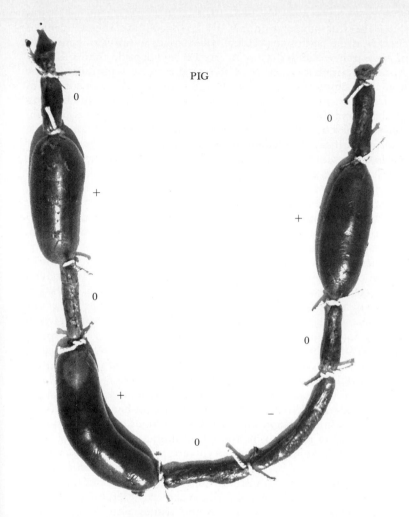

PIG

0

0

+

+

0

0

+

−

0

FIGURE 15.7 Action of *Escherichia coli* enterotoxin in the ligated ileum of the pig. The various segments were isolated by sutures, and inoculations made. Inoculated with enterotoxin-producing culture: + segments. Inoculated with culture not producing enterotoxin: − segments. Uninoculated segment: 0 segments. Enterotoxin action can be seen visually, but for precise quantification, the amount of fluid accumulating in each segment can be determined by removal with a syringe. (From Smith, H. W., and S. Halls. 1967. Jour. Pathology and Bacteriology 93:499–529.)

toxins, are injected into one or more of the segments (called loops), and control segments are either uninjected or are injected with sterile saline. After 18 to 24 hours, the animal is sacrificed and the ileal loops removed. Visual observation readily indicates the segments in which positive enterotoxin action has occurred (Figure 15.7), but for more precise evaluation, the fluid from each segment is removed and its volume determined. Because of the expense and time involved in the ileal loop assay, a number of tissue culture assays have been developed, but it is always essential to check these simpler models with the ileal loop model, since the latter duplicates the clinical action of the enterotoxin. Enterotoxins are produced by a variety of bacteria, including the food-poisoning organisms *Staphylococcus aureus, Clostridium perfringens,* and *Bacillus cereus,* and the intestinal pathogens *Vibrio cholerae, Escherichia coli,* and *Salmonella enteritidis.* The *S. aureus* and *E. coli* enterotoxins are plasmid encoded. It is likely that the plasmid which codes for the enterotoxin of *E. coli* also codes for synthesis of the specific surface antigens that are essential for attachment of enteropathogenic *E. coli* to intestinal epithelial cells (see Section 15.1).

The enterotoxin produced by *V. cholerae,* the causal agent of cholera, is the best understood. Cholera toxin is a protein consisting of three polypeptides, the A_1, A_2, and B polypeptide chains. Chains A_1 and A_2 are connected together covalently by a disulfide bridge to make a dimer called subunit A, and this is loosely associated with a variable number of B chains. The B subunit contains the binding site by which the cholera toxin combines specifically with a ganglioside (complex glycolipid) in the epithelial cell membrane, but the B subunit itself does not cause alteration in membrane permeability. Rather, the toxic action of cholera toxin is found in the A_1 chain, which activates the cellular enzyme adenyl cyclase, causing the conversion of ATP to **cyclic AMP.** As we have discussed in Section 9.6, cyclic AMP is a specific mediator of a variety of regulatory systems in cells. In mammals, cyclic AMP is involved in the action of a variety of hormones, as well as in synaptic transmission in the nervous system, and in inflammatory and immune reactions of tissues, including allergies. Although the A_1 subunit is responsible for activation of adenyl cyclase, A_1 itself must be activated by an enzymatic activity of the cell, which requires NAD and ATP. In the action of cholera enterotoxin, the increased cyclic AMP levels bring about the active secretion of chloride and bicarbonate ions from the mucosal cells into the intestinal lumen. This change in ionic balance leads to the secretion into the lumen of large amounts of water (Figure 15.8). In the acute phase of cholera, the rate of secretion of water into the small intestine is greater than reabsorption of water by the large intestine, so that massive fluid loss occurs. Cholera victims generally die from extreme dehydration, and the best treatment for the disease is the oral administration of electrolyte solutions containing solutes (Figure 15.8) to make up for the loss of fluid and ions. It is of some interest that at the molecular level cholera enterotoxin has a mode of action (formation of cyclic AMP) identical to some normal mammalian hormones, and it has been suggested that cholera toxin may represent an ancestral hormone. Since cholera enterotoxin activates adenyl cyclase in a variety of cells and tissues, pathological manifestations of cholera toxin thus seem to be related more to the specific site at which it binds, the epithelial cells of the small intestine, than to its activation of adenyl cyclase. Indeed, it has been shown that purified B subunits devoid of adenyl cyclase activity can actually prevent the action of cholera enterotoxin, if they are administered first, because they bind to the specific cholera receptors on the mucosal cells and prevent the binding of the complete toxin. There is some evidence that the enterotoxins produced by *Escherichia* and *Salmonella* have similar modes of action to the cholera toxin, and it is of interest that antibody against cholera enterotoxin also inactivates these other enterotoxins. Also, the active component of *Escherichia* enterotoxin is also activated by a cellular enzyme system requiring ATP and NAD. As discussed in Section 11.6, *Escherichia* enterotoxin is controlled by a conjugative plasmid, but the enterotoxin gene of *Vibrio cholerae* is chromosomal, although transmissible by conjugation. However, the enterotoxins produced by the food-poisoning bacteria (*S. aureus, C. perfringens, B. cereus*) may be quite different in their modes of action, since their action is at least partly systemic and is not limited to alterations in intestinal permeability alone.

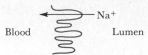

Blood   Lumen

1. Normal ion movement, Na$^+$ from lumen
    to blood, no net Cl$^-$ movement

Blood   Lumen

Epithelium

2. Bacterial colonization of
    the small intestine and
    production of cholera toxin

Cholera
toxin
↓
Adenyl cyclase

ATP ⟶ Cyclic AMP

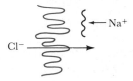

3. Activation of epithelial
    adenylcyclase by
    cholera toxin

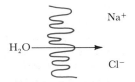

Cl$^-$

4. Na$^+$ movement blocked,
    net Cl$^-$ movement to lumen

Na$^+$

H$_2$O

Cl$^-$

5. Osmotic balance upset,
    massive water movement to
    lumen = diarrhea

Oral solution for cholera therapy (per liter):
Glucose, 20g; NaCl, 4.2; NaHCO$_3$,
4.0; KCl, 1.8

FIGURE 15.8 Action of cholera
enterotoxin.

15.6

Virulence
and Attenuation

Virulence is a quantitative term that refers to the relative ability of a parasite to cause disease, that is, its degree of pathogenicity, (see page 483). Virulence is determined by the **invasiveness** of the organism and by its **toxigenicity.** Both are quantitative properties and may vary over a wide range from very high to very low. An organism that is only weakly invasive may still be virulent if it is highly toxigenic. A good example of this is the organism *Clostridium tetani.* The cells of this organism rarely leave the wound where they were first deposited; yet they are able to bring about death of the host because they produce the potent tetanus exotoxin, which can move to distant parts of the body and initiate paralysis. On the other hand, a weakly toxigenic organism may still be able to produce disease if it is highly invasive. *Streptococcus pneumoniae* is not known to produce any potent toxin, but is able to cause extensive damage and even death because it is highly invasive, being able to grow in the lung tissues in enormous numbers and initiate responses in the host that lead to disturbance of the functions of the lung. These two organisms exemplify the extremes of invasiveness and toxigenicity; most pathogens fall somewhere between these two extremes.

It is often observed that, when pathogens are kept in laboratory culture and not passed through animals for long periods, their virulence is decreased or even completely lost. Such organisms are said to be **attenuated.** Attenuation probably occurs because nonvirulent mutants grow faster and, through successive transfers to fresh media, such mutants are selectively favored. Attenuation often occurs more readily when culture conditions are not optimal for the species. If an attenuated culture is reinoculated into an animal, virulent organisms are sometimes reisolated, but in many cases loss of virulence is permanent. Attenuated strains find frequent use in the production of vaccines (see Section 16.8).

15.7

Interaction of Pathogens with Phagocytic Cells

The host is anything but passive to attack by invading microorganisms. Two mechanisms of host defense against infection exist, which generally can be classified as **cellular** and **humoral** (see Chapter 16). Humoral defenses include those agents that are soluble (or at least, noncellular). Such agents include antibodies, a large class of specific protein molecules, which will be discussed in the next chapter, and several enzymes and chemical substances, which act as nonspecific antimicrobial agents. The best known enzyme involved in defense is **lysozyme,** which is found in tears, nasal secretions, saliva, mucus, and tissue fluids. As discussed in Section 2.5, lysozyme hydrolyzes the peptidoglycan layer of bacteria, causing lysis and death. Cellular defense mechanisms involve the activity of **phagocytes** (literally, "cells that eat"), which are able to ingest and destroy invading microbes. Phagocytes are found both in tissues and in body fluids, such as blood and lymph. In this section we shall concentrate our discussion on cellular defenses.

Blood and its components   Many of the substances and cells involved in defense are found in the blood. Additionally, changes in blood constituents and properties are sensitive reflections of disease states, and because blood is a readily available material for clinical analysis, many analytical procedures involve sampling of blood. Blood consists of cellular and noncellular components. The most numerous cells in the blood are **red blood cells (erythrocytes),** which are non-nucleated cells that function to carry oxygen from the lungs to the tissues. The white blood cells, or leukocytes, include a variety of phagocytic cells, as well as cells involved in antibody production. Red blood cells outnumber leukocytes by roughly a factor of one thousand. Platelets are small cell-like constituents that lack a nucleus and play an important role in stopping the leakage of blood from a damaged blood vessel. Platelets clump together to form a temporary plug in a damaged vessel until a permanent clot forms through the action of various clotting agents, some of which are released from the platelets themselves. When the cells and platelets are removed from blood, the remaining fluid is called plasma. An important component of plasma is fibrinogen, a clotting agent, which undergoes a complex set of reactions during the formation of the fibrin clot. Clotting can be prevented by addition of anticoagulants, such as potassium oxalate, potassium citrate, heparin, or sodium polyanetholsulfonate. Plasma is stable only when such an anticoagulant is added. When plasma is allowed to clot, the fluid components left behind, called serum, consist of all of the proteins and other dissolved materials of the plasma except fibrin. Since serum contains antibodies, it is widely used in immunological investigations (see Chapter 16).

Phagocytes   Some of the leukocytes found in whole blood are phagocytes, and phagocytes are also found in various tissues and fluids of the body. Phagocytes are usually actively motile by amoeboid action. Attracted to microbes by chemotactic phenomena, the phagocytes engulf the microbes and kill and digest them. One of two types of phagocytes are granulocytes (Figure 15.9a), also called polymorphonuclear leukocytes (sometimes abbreviated PMN), which are small, actively motile cells containing many distinctly staining membranous granules called lysosomes. These granules contain several bactericidal substances and enzymes, such as hydrogen peroxide, lysozyme, proteases, phosphatases, nucleases, and lipases. Granulocytes are short-lived cells that are found predominantly in the bloodstream and bone marrow and appear in large numbers during the acute phase of an infection. They can move rapidly, up to 40 μm/min, and are attracted chemotactically to bacteria and cellular components by immune mechanisms described in Chapter 16. Because they appear in the blood in large numbers during acute infection, they can serve as indicators of infection. Granulocyte production is severely retarded by ionizing radiation, leading to severe bacteremia from bacteria of the normal intestinal flora, such as *E. coli.*

Macrophages (also called monocytes) are phagocytic cells that can readily be distinguished from granulocytes by their nuclear morphology (Figure 15.9b). They contain few lysosomes and hence do not appear granular when stained. Macrophages are of two types: wandering cells, which are found free in the bloodstream, lymph vessels (where they are called monocytes), and fixed phagocytes, or histiocytes, which are found embedded in various tissues of the body and have only a limited mobility.

The total system of phagocytic cells is often called the reticuloendothelial system (RE system); it consists of the large macrophages of the loose connective tissue, the lymphatic tissue, bone marrow, lung, blood, spleen, and liver that are involved in the uptake and

Granulocyte

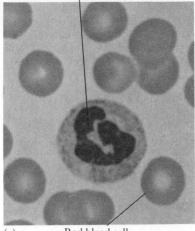

(a)   Red blood cell

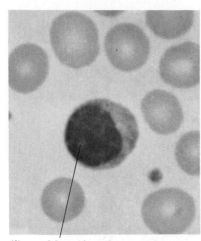

(b)   Macrophage (monocyte)

FIGURE 15.9   Two major phagocytic cell types. (a) Granulocyte. (b) Macrophage. Magnifications, 1500×.

removal of particulate matter from lymph and blood. The RE system can be seen experimentally by injecting intravenously into an animal an insoluble dye suspension; the dye particles are phagocytized by the various cells of the RE system, and since the dye is indigestible, it remains in place in the cells and can be seen upon autopsy. Macrophages are long-lived cells that play roles in both the acute and chronic phases of infection and they also play a role in antibody formation (see Section 16.5). The efficiency of the phagocytic system in clearing foreign particles and microbial cells from the bloodstream and lymph system is high. For instance, 80 to 90 percent of the particles entering the liver are removed.

Phagocytosis As noted, phagocytes are attracted chemotactically to invading microbes. Some aspects of the chemotactic process involve the action of the complement system and lymphocytes, and will be discussed in Section 16.4. As a result of chemotactic attraction, large numbers of phagocytes are often seen around foci of infection, and such phagocytic accumulations are often the first indication of the presence of infection. Phagocytes work best when they can trap a microbial cell upon a surface, such as a vessel wall, a fibrin clot, or even particulate macromolecules. After adhering to the cell, an invagination of the phagocyte's plasma membrane envelopes the foreign cell, and the entire complex is pinched off and enters the cytoplasm as a **phagocytic vacuole.** The microbial cell is then released in a region of the phagocyte containing granules, and the latter are disrupted and release enzymes, which digest and destroy the invader (Figure 15.10).

During the process of phagocytosis, the metabolism of the granulocyte converts from aerobic pathways to anaerobic glycolysis. Glycolysis results in the formation of lactic acid and a consequent drop in pH; this lowered pH is partly responsible for the death of the microbe, and it also provides an environment in which the hydro-

FIGURE 15.10 Phagocytosis: engulfment and digestion of a *Bacillus megaterium* cell by a human phagocyte, observed by phase-contrast microscopy. Magnification, 1250×. (From Hirsch, J. G. 1962. J. Exp. Med. 116:827.)

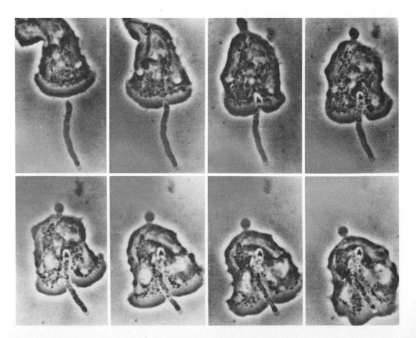

lytic enzymes, all of which have acid pH optima, can act more effectively. The initial act of phagocytosis conditions a cell so that it is more efficient in subsequent phagocytic action—a cell that has recently phagocytized can take up bacteria about 10 times better than a cell that has not.

Role of singlet oxygen and halogenation in phagocytic killing   An important microbicidal activity in the granulocyte is provided by biochemical systems that lead to the production of **singlet oxygen ($^1O_2$).** As we discussed in Section 8.5, singlet oxygen is very toxic to cells. There are two ways in which singlet oxygen is generated in phagocytes. Subsequent to engulfment, there is a marked increase in activity of an NADP oxidase system, which carries out a single-electron reduction of O_2 to superoxide. At the acid pH of the phagocyte, superoxide is unstable and disproportionates spontaneously to singlet oxygen and hydrogen peroxide (Figure 15.11). Another enzyme system, the myeloperoxidase system, generates singlet oxygen from hydrogen peroxide, using halide ion (chloride, iodide) as cofactor. Chloride ion is oxidized to hypochlorous acid (HOCl), using hydrogen peroxide as oxidant, and hypochlorous acid reacts chemically with further peroxide to generate singlet oxygen (Figure 15.11). Hypochlorous acid can also kill directly by catalyzing the lethal halogenation of bacteria and viruses.

Direct evidence for the generation of singlet oxygen in phagocytes comes from measurements of chemoluminescence, since singlet oxygen can decay to the ground state with the emission of light, and subsequent to phagocytosis, phagocytes show light emission at the wavelength at which singlet oxygen chemoluminesces. As we discussed in Section 8.6, carotenoids quench singlet oxygen, and bacteria containing carotenoids are much more resistant to killing within phagocytes than those that do not contain carotenoids. It is of interest that *Staphylococcus aureus,* a pathogen which commonly causes infections where large numbers of phagocytes consequently develop, contains yellow carotenoids (hence the name *aureus*), and this bacterium is quite resistant to phagocytic killing.

Leukocidins   Some pathogens produce proteins called **leukocidins,** which destroy phagocytes. In such cases the pathogen is not killed when ingested but instead kills the phagocyte and is released alive. Destroyed phagocytes make up much of the material of pus, and organisms that produce leukocidins are therefore usually **pyogenic** (pus-forming) and bring about characteristic abscesses. One leukocidin produced by certain strains of *Staphylococcus aureus* binds to the cytoplasmic membrane of phagocytic cells and increases permeability. This protein also stimulates membrane fusions between the cytoplasmic membrane and the membranes which surround lytic granules; such fusions effectively transport the contents of the granules outside the phagocytic cell and away from the invading pathogen. Streptococci and staphylococci are the most common leukocidin producers.

Capsules and phagocytosis   An important mechanism for defense against phagocytosis is the bacterial capsule. Capsulated bacteria are

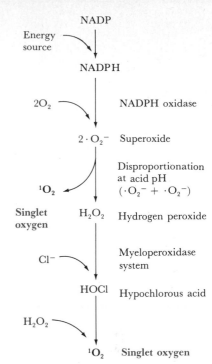

FIGURE 15.11   Generation of singlet oxygen in the phagocyte.

frequently highly resistant to phagocytosis, apparently because the capsule prevents in some way the adherence of the phagocyte to the bacterial cell. The clearest case of the importance of a capsule in permitting invasion is that of *Streptococcus pneumoniae*. If only a few cells of a capsulated strain of this species are injected into a mouse, an infection is initiated that leads to death within a few days. On the other hand, noncapsulated mutants of capsulated strains are completely avirulent, and injection of even large numbers of bacteria usually causes no disease. If a noncapsulated strain is transformed genetically with DNA from a capsulated strain (see Section 11.2), both capsulation and virulence are restored at the same time. Furthermore, enzymatic removal of the capsule renders the organism noninvasive. Surface components other than capsules can also inhibit phagocytosis. Pathogenic streptococci produce on both the cell surface and fimbriae a specific protein called the *M protein*, which apparently alters the surface properties of the cell in such a way that the phagocyte cannot act.

15.8

Intracellular Growth

One group of organisms, the **intracellular parasites,** are readily phagocytized but are not killed, nor do they kill the phagocyte. Instead, they can remain alive for long periods of time and even reproduce within the phagocyte. In most situations the pathogen enters the host cell in a phagocytic vacuole called a **phagosome,** and the pathogen grows within this structure until the phagosome bursts. Intracellular parasites are a diverse group. Some, such as *Mycobacterium tuberculosis, Salmonella typhi* (cause of typhoid fever), and *Brucella melitensis* and *Brucella suis* (causes of undulant fever), are facultative intracellular parasites, and can live either intracellularly or extracellularly. In acute infections they multiply in the extracellular body fluids, but in chronic conditions they may live only intracellularly. When growing intracellularly the organism is protected from immune mechanisms of the host and is less susceptible to drug therapy. Some important intracellular parasites are unable to grow outside of living cells and are called obligate intracellular parasites. Included in this category are the viruses, the chlamydias, the rickettsias, and some protozoa, such as the one that causes malaria (see Section 17.2).

The intracellular environment differs markedly from the extracellular environment in physical and chemical characteristics (Table 15.3), being much richer in organic nutrients and cofactors; the advantages to a parasite of living intracellularly are thus partly nutritional. In addition to the normal microbial nutrients, the intracellular environment also provides ATP, which is completely absent extracellularly; some of the obligate intracellular parasites are unable to generate their own ATP from organic compounds and require preformed ATP provided by the host (see Section 19.31).

Some intracellular pathogens (for example, *Listeria monocytogenes, Rickettsia rickettsii*) grow within the cell nucleus as well as the cytoplasm. The intracellular growth of *Rickettsia rickettsii* has been studied in some detail. The organism is able to penetrate the plasma membrane readily and initiate growth (Figure 15.12), but progeny are not

TABLE 15.3   Physical and Chemical Differences Between the Intracellular and Extracellular Mammalian Environment

Character	Intracellular	Extracellular (Plasma)
Ca^{2+}	Very low	0.0025 M
Mg^{2+}	0.02 M	0.0015 M
Na^+	0.01 M	0.15 M
K^+	0.15 M	0.005 M
Cl^-	0.003 M	0.1 M
$HPO_4{}^{2-}$	0.05 M	0.002 M
$HCO_3{}^-$	0.01 M	0.03 M
Overall ionic strength	High	About 75% of intracellular ionic strength
Organic nutrients: cofactors, acids, and high-energy compounds	High levels	Relatively low levels

Data courtesy of R. R. Brubaker.

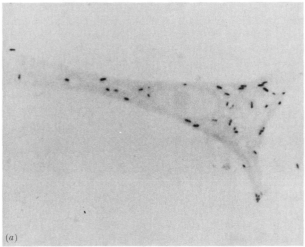
(a)

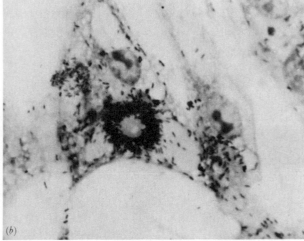
(b)

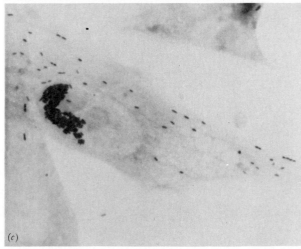
(c)

FIGURE 15.12   Intracellular growth of *Rickettsia rickettsii* in chicken embryo cells. The preparations were stained and examined by the light microscope. (a) Early stage of growth, 28 hours after infection. Note the generally dispersed cytoplasmic distribution. (b) Extensive intracytoplasmic colonies after 120 hours incubation. The dark mass represents extensive development of rickettsiae in a ring or doughnut radially arranged around a cytoplasmic vacuole. (c) Intranuclear growth. Note the compact intranuclear mass and the dispersed cytoplasmic distribution. 45 hours after incubation. Not all cells showed intranuclear growth. All magnifications, 550×. (From Wisseman, C. L., E. A. Edlinger, A. D. Waddell, and M. R. Jones. 1976. Infection cycle of *Rickettsia rickettsii* in chicken embryo and L-929 cells in culture. Infection and Immunity 14:1052–1064.)

restricted to the intracellular environment, and readily pass out of the plasma membrane without causing damage to the host cell. Thus, despite sustained growth of the intracellular parasite, massive accumulation within the host cell does not always occur. The cells released from the cytoplasm readily infect adjacent cells, thus resulting in a rapidly spreading infection. However, if growth occurs in the nucleus, the progeny do not leave, so that sustained growth results in an extensive bacterial colonization of the nucleus (Figure 15.12c). Thus the rickcttsial traffic is bidirectional across the plasma membrane and dominantly monodirectional across the nuclear membrane. One significant aspect of these observations is that release of rickettsial cells from the host can occur without apparent damage to the host cell.

15.9

Inflammation

The tissues of the animal react to infection and to mechanical injury by an **inflammatory response,** the characteristic symptoms of which are redness, swelling, heat, and pain. The initial effect of the foreign stimulus is to cause local dilation of blood vessels and an increase in capillary permeability. This results in an increase flow of blood and passage of fluid out of the circulatory system into the tissues, causing swelling (edema) (Figure 15.13). Phagocytes also pass through the capillary walls to the inflamed area. Initially granulocytes appear, followed later by macrophages (Figure 15.13). Within the inflamed area, a fibrin clot is usually formed, which in many instances will localize the invading microbe. Pathogens that produce fibrinolytic enzymes may be able to escape and continue to invade the body. One of the factors involved in initiating the inflammatory response is the chemical **histamine,** which is released by damaged cells and acts on the capillaries to increase their permeability. The use of antihistamine drugs in controlling inflammation is based on their ability to counteract some of the effects of histamine. However, although anti-

FIGURE 15.13 Sequence of events in a typical inflammatory response.

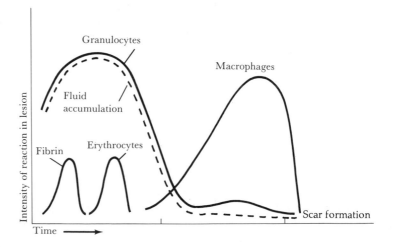

histamines may reduce the painful aspects of inflammation, they also counteract some of the beneficial effects so that their use is not an unmixed blessing.

Inflammation is one of the most important and ubiquitous aspects of host defense against invading microorganisms and is present in a small way virtually continuously. However, inflammation is also an important aspect of microbial pathogenesis since the inflammatory response elicited by an invading microorganism can result in considerable host damage. **Fever,** for example, is a frequent inflammatory response caused by the release from phagocytic cells of a low-molecular-weight protein called **endogenous pyrogen.** This protein affects the thermoregulatory control centers in the hypothalamus and results in an increase of the normal 37°C (98.6°F) human body temperature. Slight temperature increases benefit the host by accelerating phagocytic and antibody responses, while strong fevers of 40°C (104°F) or greater may benefit the pathogen if host tissues are further damaged. Pathogens vary in the degree to which they induce inflammation; in some cases a major reaction by the host to only a minor microbial invasion may occur, resulting in serious damage to the host.

15.10

Clinical Stages of Infectious Disease

In terms of clinical symptoms the course of disease can be conveniently divided into stages:

1. *Infection,* when the organism becomes lodged in the host.
2. *Incubation period,* the time between infection and the appearance of disease symptoms. Some diseases have short incubation periods; others, longer ones. The incubation period for a given disease is determined by inoculum size, virulence of pathogen, resistance of host, and distance of site of entrance from focus of infection. Average incubation periods for some well-known diseases are given in Figure 15.14.
3. *Prodromal period,* a short period sometimes following incubation in which the first symptoms such as headache and feeling of illness appear.
4. *Acute period,* when the disease is at its height, with overt symptoms such as fever and chills.
5. *Decline period,* during which disease symptoms are subsiding, the temperature falls, usually following a period of intense sweating, and a feeling of well-being develops. The decline may be rapid (within 1 day), in which case it is said to occur by *crisis,* or it may be slower, extending over several days, in which case it is said to be by *lysis.*
6. *Convalescent period,* during which the patient regains strength and returns to normal.

During the later stages of the infection cycle, the immune mechanisms of the host become increasingly important, and in most cases recovery from the disease requires the action of these immune mechanisms.

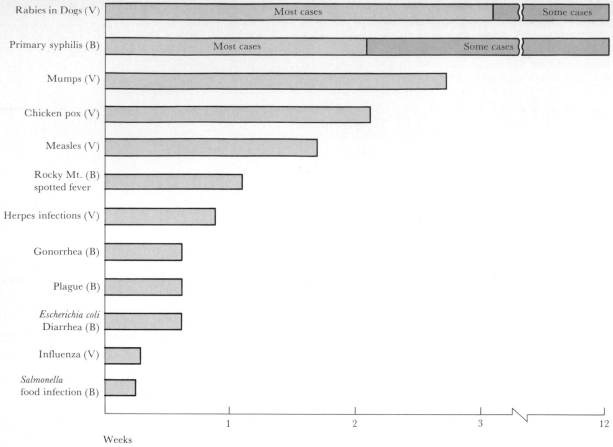

Rabies in Dogs (V)
Primary syphilis (B)
Mumps (V)
Chicken pox (V)
Measles (V)
Rocky Mt. (B) spotted fever
Herpes infections (V)
Gonorrhea (B)
Plague (B)
Escherichia coli Diarrhea (B)
Influenza (V)
Salmonella food infection (B)

Most cases   Some cases
Most cases   Some cases

Weeks   1   2   3   12

FIGURE 15.14 Average incubation period for some well-known diseases. The causative agent of each disease listed is either a bacterium (B) or a virus (V). Adapted from Smith, A. L., 1980. Microbiology and Pathology. C V Mosby Co., St. Louis.

15.11

Clinical Microbiology

The most important activity of the microbiologist in medicine is to isolate and identify the causal agents of infectious disease. This major area of microbiology is called **clinical microbiology.** In recent years this field has greatly expanded because of the increasing awareness of the importance of identification of the pathogen for proper treatment of the disease. The days when the physician can substitute antibiotics for diagnosis are over; the microbiologist is a major force in ensuring proper diagnosis and proper choice of antibiotics for therapy.

The physician, on the basis of careful examination of the patient, may decide that an infectious disease is present. Samples of infected tissues or fluids are then collected for microbiological analysis. Depending on the kind of infection, materials collected may include blood, urine, feces, sputum, cerebrospinal fluid, or pus. A sterile swab may be passed across a suspected infected area. Small pieces of living tissue may be aseptically removed (biopsy). The sample must be carefully taken under aseptic conditions so that contamination is avoided. Once taken, the sample is analyzed as soon as possible. If it cannot be analyzed immediately, it is usually refrigerated to slow down deterioration.

A Bit of History

The history of the discovery of the microbial role in infectious disease has been described in some detail in Chapter 1. Once the concept of specific microbial disease agents had been clarified, and the procedures for culture of microorganisms had been developed, it was a relatively simple procedure to isolate a large number of microbial pathogens. The two decades after the enunciation of Koch's postulates were indeed fruitful for medical microbiology. The rapid development of this field is indicated by Table 15.4, which lists the main pathogens isolated during the years of 1877 to 1906.

TABLE 15.4 The Discoverers of the Main Bacterial Pathogens

Year	Disease	Organism	Discoverer
1877	Anthrax	*Bacillus anthracis*	Koch, R.
1878	Suppuration	*Staphylococcus*	Koch, R.
1879	Gonorrhea	*Neisseria gonorrhoeae*	Neisser, A. L. S.
1880	Typhoid fever	*Salmonella typhi*	Eberth, C. J.
1881	Suppuration	*Streptococcus*	Ogston, A.
1882	Tuberculosis	*Mycobacterium tuberculosis*	Koch, R.
1883	Cholera	*Vibrio cholerae*	Koch, R.
1883	Diphtheria	*Corynebacterium diphtheriae*	Klebs, T. A. E.
1884	Tetanus	*Clostridium tetani*	Nicolaier, A.
1885	Diarrhea	*Escherichia coli*	Escherich, T.
1886	Pneumonia	*Streptococcus pneumoniae*	Fraenkel, A.
1887	Meningitis	*Neisseria meningitidis*	Weichselbaum, A.
1888	Food poisoning	*Salmonella enteritidis*	Gaertner, A. A. H.
1892	Gas gangrene	*Clostridium perfringens*	Welch, W. H.
1894	Plague	*Yersinia pestis*	Kitasato, S., Yersin, A. J. E. (independently)
1896	Botulism	*Clostridium botulinum*	van Ermengem, E. M. P.
1898	Dysentery	*Shigella dysenteriae*	Shiga, K.
1900	Paratyphoid	*Salmonella paratyphi*	Schottmüller, H.
1903	Syphilis	*Treponema pallidum*	Schaudinn, F. R., and Hoffmann, E.
1906	Whooping cough	*Bordetella pertussis*	Bordet, J., and Gengou, O.

Blood cultures Bacteremia means the presence of bacteria in the blood (see Section 15.2). Because of the rapidity with which bacteria are cleared from the bloodstream (Section 15.7), bacteremia is uncommon in healthy individuals, and presence of bacteria in the blood is generally indicative of infection. The classic type of blood infection is **septicemia,** resulting from a virulent organism entering the blood from a focus of infection, multiplying, and traveling to various body tissues to initiate new infections. Septicemia is indicated by the presence of severe systemic symptoms, usually with fever and chills, followed by prostration. However, there are a number of other disease conditions in which bacteremia may be present, and transient bacteremia (usually without any symptoms) can occur as a result of minor traumatic events, such as tooth extractions, catheterization of the urinary tract, examination of the lower bowel (proctosigmoidoscopy), and liver biopsy. However, in these cases the bacteremia is short-lived. In many disease situations, culture of the blood provides the only immediate way of isolating and identifying the causal agent, and diagnosis may thus depend on careful and proper blood culture.

The standard blood culture procedure is to swab the collection site (usually the middle portion of the arm just opposite the elbow) with 70 to 95 percent ethanol followed by 2 percent iodine, aseptically remove 10 ml of blood from a vein, and inject it into a blood culture bottle containing an anticoagulant and an all-purpose culture medium. Replicate cultures are set up with one bottle being incubated aerobically and one anaerobically, and the bottles are examined at intervals for signs of visible growth (turbidity). Subcultures are made on blood agar plates after 24 to 48 hours, and if no growth is obtained after 5 to 7 days, the blood culture is scored as negative.

Because a certain amount of skin contamination is probably unavoidable during initial drawing of the blood, even under optimal conditions a contamination rate of 2 to 3 percent can be expected. Because of this, identification of positive blood cultures is essential, and contamination may be indicated if certain organisms commonly found on the skin are isolated, each as *Staphylococcus epidermidis,* coryneform bacteria, or propionibacteria, although even these organisms can occasionally cause infection of the wall of the heart (subacute bacterial endocarditis). Thus considerable microbiological and clinical experience is necessary in interpreting blood cultures.

Urine cultures Urinary infections are very common, and because the causal agents are often identical or similar to bacteria of the normal flora (for example, *Escherichia coli*), considerable care must be taken in bacteriological analysis. Since urine supports extensive bacterial growth under many conditions (see Section 15.2), fairly high cell numbers are often found in urinary infection. In most cases, the infection occurs as a result of an organism ascending the urethra from the outside.

Significant urinary infection generally results in bacterial counts of 10^5 or more organisms per milliliter, whereas in the absence of infection, contamination of the urine from the external genitalia (almost unavoidable to some extent) results in less than 10^3 organisms per milliliter. The most common organisms are members of the enteric bacteria, with *E. coli* accounting for about 90 percent of the cases. Other organisms include *Klebsiella, Enterobacter, Proteus, Pseudomonas,* and *Streptococcus faecalis. Neisseria gonorrhoeae,* the causal agent of gonorrhea, does not grow in the urine itself, but in the urethral epithelium, and must be diagnosed by different methods (see below). Because of the relatively high number of organisms found in a significant urinary infection, direct microscopic examination of the urine is of considerable value and is recommended as part of the basic procedure. A small drop of urine is allowed to dry on a mi-

TABLE 15.5 Colony Characteristics of Frequently Isolated Gram-Negative Rods Cultured on Various Clinically Useful Agars*

Organism	Agar Media			
	EMB	MC	SS	BS
Escherichia coli	Dark center with greenish metallic sheen	Red or pink	Red to pink	Mostly inhibited
Enterobacter	Similar to *E. coli,* but colonies are larger	Red or pink	White or beige	Mucoid colonies with silver sheen
Klebsiella	Large mucoid, brownish	Pink	Red to pink	Mostly inhibited
Proteus	Translucent, colorless	Transparent, colorless	Black center, clear periphery	Green
Pseudomonas	Translucent, colorless to gold	Transparent, colorless	Mostly inhibited	No growth
Salmonella	Translucent, colorless to gold	Translucent, colorless	Opaque	Black to dark green
Shigella	Translucent, colorless to gold	Transparent, colorless	Opaque	Brown or inhibited

*BS, Bismuth Sulfite agar, EMB, eosin methylene blue agar; MC, MacConkey agar; SS, *Salmonella-Shigella* agar.
Modified from Lennette, E. H., A. Balows, W. H. Hausler, and J. P. Truant, eds. 1980. Manual of clinical microbiology, 3rd ed. American Society for Microbiology, Washington, D.C.

croscope slide and a Gram stain performed. If more than 10^5 organisms per milliliter are present, there is usually one or more bacteria per microscope field (100 × objective). Cultural examination of the specimen must be by quantitative means, because contamination of the specimen by small numbers of bacteria from the external parts of the body is almost unavoidable. Ideally two media should be used in the initial culture, blood agar as an unselective general medium, and a medium selective for enteric bacteria, such as MacConkey or EMB agar. Either of these latter media permit the initial differentiation of lactose fermenters from nonfermenters, and the growth of Gram-positive organisms such as *Staphylococcus* (a common skin contaminant) is inhibited. Organisms isolated can be identified, and antibiotic susceptibility tests performed. Experienced clinical microbiologists may make a tentative identification of an isolate by observing the color and morphology of colonies of the suspected pathogen grown on various selective media as described in Table 15.5. Such an identification must be followed up with more detailed analyses, of course, but clinical microbiologists will use this information in conjunction with more detailed test results in order to make a positive identification.

Laboratory diagnosis of gonorrhea   Although rarely fatal, gonorrhea is one of the most common infectious diseases, and laboratory procedures are central to its diagnosis. The causal agent, *Neisseria gonorrhoeae* (referred to clinically as the *gonococcus*) colonizes mucosal surfaces of the urethra, uterine cervix, anal canal, throat, and conjunctiva. The organism is quite sensitive to drying, and because of this it is transmitted almost exclusively by direct person-to-person contact, usually by sexual intercourse. The major goal of public health measures to control gonorrhea involves identification of asymptomatic carriers, and this requires microbiological analysis. Because the organism is a Gram-negative coccus, and similar organisms are not very common in the normal flora of the urogenital

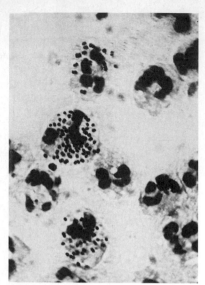

FIGURE 15.15 Photomicrograph of *Neisseria gonorrhoeae* within human granulocytes. Magnification, 1200×. (Courtesy of Theodor Rosebury.)

tract, direct microscopy of Gram-stained material is of value. There are also several culture media that are selective for *N. gonorrhoeae.* Because the organism is delicate, drying and contact with antiseptics must be avoided, and cultural procedures must be initiated as soon as possible after specimen collection.

For direct microscopy, smears are prepared from urethral swabs and conventional Gram stains made. During visualization under the oil-immersion lens, attention should be paid to polymorphonuclear leukocytes (granulocytes), since in acute gonorrhea such cells will frequently contain groups of Gram-negative, kidney-shaped diplococci (Figure 15.15), and there will be few if any other types of organisms present.

Cultural procedures have a higher degree of sensitivity than microscopy. Most media for the culture of *N. gonorrhoeae* contain heated blood or hemoglobin (referred to as *chocolate agar* because of its deep brown appearance), the heating causing the formation of a precipitated material, which is quite effective in absorbing toxic products present in the medium. After streaking, the plates must be incubated in a humid environment in an atmosphere containing 3 to 7 percent CO_2 (CO_2 is required for growth of gonococci). The plates are examined after 24 and 48 hours, and portions of colonies should be immediately tested by the oxidase test, since all *Neisseria* are oxidase-positive (Section 19.17). Oxidase-positive Gram-negative diplococci growing on the antibiotic-containing chocolate agar can be presumed to be gonococci if the inoculum was derived from genitourinary sources, but definite identification requires determination of carbohydrate utilization patterns. *Neisseria gonorrhoeae* degrades glucose with acid production, but does not produce acid from maltose, sucrose, or lactose, and this pattern is considered a confirmation of identification. Except in special situations, all isolates must be subjected to the carbohydrate degradation tests to confirm identification. A second primary isolation medium, called *Thayer-Martin agar,* also is used for isolation of *N. gonorrhoeae.* This medium incorporates the antibiotics vancomycin, nystatin, and colistin, to which most clinical isolates of *N. gonorrhoeae* are naturally resistant.

For many years, all isolates of *N. gonorrhoeae* were found to be sensitive to penicillin or ampicillin (a semisynthetic penicillin), and sensitivity testing was not necessary. Penicillin-resistant gonococci (see Section 17.11) have now been isolated from patients in various parts of the world, and resistance has been linked to the production of a plasmid-encoded penicillinase. Simple and rapid laboratory procedures are now available for screening colonies from direct isolation plates for penicillinase production, and such procedures are now recommended for detecting penicillin-resistant strains. To cure infections due to penicillin-resistant gonococci, the antibiotic spectinomycin can be used, as cross-resistance between spectinomycin and penicillin does not occur.

Culture of anaerobes Obligately anaerobic bacteria are common causes of infection and will be completely missed in clinical diagnosis unless special precautions are taken for their isolation and culture.

We have discussed anaerobes in general in Section 8.5, and we noted that many anaerobes are extremely susceptible to oxygen. Because of this, specimen collection, handling, and processing require special attention if it is thought that an obligate anaerobe is involved. There are several habitats in the body (for example, the intestinal tract, the teeth) that are generally anaerobic, and in which obligately anaerobic bacteria can be found as part of the normal flora. However, other parts of the body can become anaerobic as a result of tissue injury or trauma, which results in reduction of blood supply to the injured site, and such anaerobic sites can then become available for colonization by obligate anaerobes. In general, the pathogenic anaerobic bacteria are part of the normal flora and are opportunistic pathogens, although two important pathogenic anaerobes, *Clostridium tetani* (casual agent of tetanus) and *C. perfringens* (causal agent of gas gangrene), both sporeformers, are predominantly soil organisms.

With anaerobic culture, the microbiologist is presented not only with the usual problems of obtaining and maintaining an uncontaminated specimen, but ensuring that at no time does the specimen come in contact with air. If only exudate can be collected, the best procedure is to remove the specimen directly from the infected tissue by aspiration with a syringe and needle, care being taken to be certain that no air is drawn into the syringe. It is preferable to obtain a sample of infected tissue rather than just exudate, since viability will be maintained much longer in tissue. Although swabs can be used to collect material, this is not generally recommended, because it is difficult to avoid aeration; also the number of organisms obtained on the swab is much lower than those obtained in tissue or exudate. The collected sample must be immediately placed in a tube containing oxygen-free gas, preferably containing a small amount of dilute salt solution containing the redox indicator resazurin. This dye is colorless when reduced, and becomes pink when oxidized, thus any oxygen contamination of the specimen will be quickly indicated. If a proper anaerobic transport tube is not available, the syringe itself can be used to transport the specimen, the needle being inserted into a sterile rubber stopper so that no air is drawn into the syringe.

For anaerobic incubation, agar plates are placed in a sealed jar which is made anaerobic by either replacing the atmosphere in the jar with an oxygen-free gas mixture (a mixture of N_2 and CO_2 is frequently employed), or by adding some compound to the enclosed vessel which removes O_2 from the atmosphere. Alternative means of isolating anaerobes include the use of prereduced culture media or anaerobic hoods. The latter are large gas impermeable bags filled with an oxygen-free gas such as nitrogen or hydrogen and which are fitted with an airlock for inserting and removing cultures.

In general, media for anaerobes do not differ greatly from those used for aerobes, except that they are generally richer in organic constituents, and contain reducing agents (usually cysteine or thioglycolate) and a redox indicator such as resazurin. Once positive cultures have been obtained, it is essential that they be characterized and identified, to be certain that the isolate is not a member of the normal flora.

Chemotherapy

Chemical agents able to cure infectious diseases are generally called **chemotherapeutic agents.** The hallmark of a chemotherapeutic agent is *selective toxicity,* that is, toxicity to the pathogen but not to the host. We have discussed antimicrobial action in some detail in Chapter 7, and the modes of actions of many antibiotics in Chapter 9. Here we discuss the principles involved in the use of antimicrobial agents in the therapy of infectious disease. The action of antimicrobial agents in the test tube (*in vitro*) may be quite different from the action of such agents in the animal (*in vivo*). An agent highly effective in vitro may be completely ineffective in vivo, for the animal body is not a neutral environment for chemical agents, and many agents are modified so that they are no longer biologically active.

Drug distribution and metabolism in the body When a drug is administered it becomes distributed through various compartments of the body, as outlined in Figure 15.16. After initial absorption into the blood, a major portion of the drug may be found bound to plasma proteins. For some agents, as much as 90 percent of the drug may be found bound to blood proteins. This binding is reversible, but the bound drug is not microbially active. Because of protein binding, the minimum inhibitory concentration of antibiotic when tested in plasma or serum may be three- to four-fold higher than when tested in culture medium. However, the bound drug is not inactivated and can be looked upon as a reservoir, to be released when concentration of the free drug is lowered. In the tissues, the drug may be metabolized, and the metabolites are generally less active than the administered drug. The most common organ in which drug metabolism occurs is the liver. From the viewpoint of the animal body, the drug is a foreign agent, and metabolic systems for natural detoxification function to detoxify many drugs. Detoxification enzymes are generally less well developed in infants, so that the drugs may be much more toxic (per unit weight) in infants than in older individuals.

FIGURE 15.16 Movement of drug through various compartments of the body.

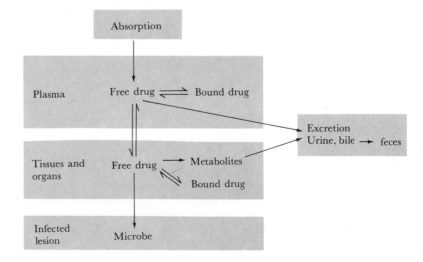

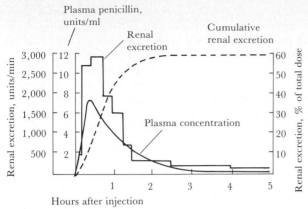

FIGURE 15.17 Kinetics of penicillin absorption and excretion in a normal adult human. One intramuscular injection of 300,000 units of aqueous penicillin G was given at zero time. Samples of blood and urine were assayed periodically for antibiotic concentration. The results given are average values found in humans with good renal function. Note that the peak plasma level is reached within 15 to 30 minutes, and that about 60 percent of the dose was excreted in the urine within 5 hours. (From Goodman, L. S. and A. Gillman, 1975. The pharmacological basis of therapeutics, 5th ed. Macmillan, N.Y.)

Excretion of the drug and its metabolites generally occurs rapidly. Two main routes of excretion exist: renal (kidney) excretion to the urine, and hepatic (liver) excretion to the bile, which is excreted with the feces. Minor routes of excretion are through sweat, saliva, and milk (lactating animals).

The characteristic pattern of drug absorption and excretion is illustrated by data for penicillin in Figure 15.17. As seen, after injection there is a rapid increase in concentration of antibiotic in the blood, followed by a gradual fall in concentration as the drug is excreted. The antibiotic concentration in the blood reaches a peak within 15 to 30 minutes, and quickly thereafter large amounts of antibiotic appear in the urine. Within 5 hours about 60 percent of the injected dose has been eliminated in the urine. Data of these types indicate that if bacteriostatic or bactericidal concentrations of antibiotic are to be maintained in the body, periodic doses must be given.

Toxicity Virtually all drugs have some **toxicity** (that is, cause some harm) to the host, and a knowledge of host toxicity is vital for the intelligent use of chemotherapeutic agents. Two broad classes of toxicity are recognized: acute and chronic. Acute toxicity is expressed by pathological manifestations observed within a few hours after administration of a single dose. Generally, acute toxicity occurs as a result of drug overdosage.

Chronic toxicity is expressed by gradual changes, which take place during continuous administration of a drug over a long period of time. A variety of toxic manifestations require considerable time to develop, and if the course of treatment with the agent must be extended, chronic toxicity must be taken into consideration. For example, extended treatment with the antibiotic streptomycin in humans is thought to cause inner ear problems which can result in deafness. For treatment of infectious disease, very long courses of treatment are the exception rather than the rule, although for tuberculosis (see Section 19.27) and certain other chronic infectious diseases, therapy may continue over many months or years. Also, some antimicrobial agents are used continuously as prophylactic agents (to prevent future attacks of the disease) in certain high-risk patients, and under these conditions, chronic toxicity becomes an important consideration. However, some manifestations of chronic toxicity may be exhibited in even as short a time as 1 to 2 weeks.

Testing of antimicrobial agents in vivo   Once the acute and chronic toxicity of a new antimicrobial agent has been determined, in vivo tests in experimental animals are initiated. Animal models can be developed for many infectious diseases, and such models are then used to evaluate chemotherapeutic agents. For economy and ease of operation, the mouse is the preferred experimental animal. Groups of mice can be infected with a pathogen, and the effect of drug administration observed. The in vivo testing of antibiotics in experimental animals is an important and highly interesting procedure, and much can be learned about drug action from careful observation of such studies. The effect of the drug can be measured either by simply counting the number of infected animals that do not die after various periods of time, or by actually measuring the growth of the pathogen in the animal, and determining if the drug inhibits growth. The precise experimental design will depend upon the pathogen and upon the aim of the experiment. It should be emphasized, however, that animal models do not closely mimic most human infections, since animal models make use of highly virulent strains, and large doses of pathogens are often used, so that the course of the infection is quick and severe. Also, the animals used are of undefined immunological background, but have probably never come in contact with the pathogen before, so that antibodies to the pathogen are not present. The value of these animal models is that they provide precise, reproducible systems for evaluating quantitatively the action of the antibiotic.

Antibiotic resistance   Mutations to drug resistance occur in pathogens, and in the presence of the drug the mutant form has a selective advantage and may replace the parent type. Resistance can develop to virtually all chemotherapeutic agents and is known to occur in vivo as well as in vitro (Figure 15.18). The resistant strain may be just as virulent as the parent, may not be controllable by other chemotherapeutic agents, and may be passed on to other individuals. Resistance can be minimized if drugs are used only for serious diseases and are given in sufficiently high doses so that the population level is reduced before mutants have a chance to appear. Resistance can also be minimized by combining two unrelated chemotherapeutic agents since

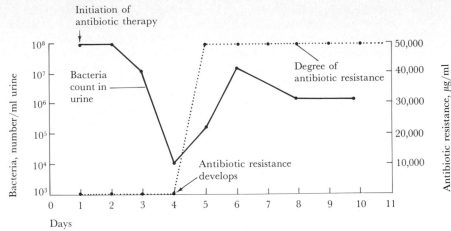

FIGURE 15.18 Development of antibiotic-resistant mutant in a patient undergoing antibiotic therapy. The patient, suffering from chronic pyelonephritis caused by a Gram-negative bacterium, was treated with streptomycin by intramuscular injection. Antibiotic sensitivity is expressed as the minimum inhibitory concentration of the antibiotic in micrograms per milliliter. (Data from Finland, M., et al. 1946. J. Am. Med. Assoc. 132:16–21.

it is likely that a mutant resistant to one will still be sensitive to the other. With the increasing prevalence of resistance transfer factors in pathogenic bacteria (see Section 11.6), however, multiple antibiotic therapy is proving less attractive as a clinically useful strategem. Some organisms, such as the streptococci, do not seem to develop drug resistance readily in vivo, wheres the staphylococci are notorious for developing resistance to chemotherapeutic agents. The rise of strains resistant to antibiotics because of the presence of conjugative plasmids (resistance transfer factors) is illustrated in Figure 15.19.

Clinical uses of antibiotics Although there are presently available a wide variety of relatively nontoxic antimicrobial agents, only a restricted number actually are in use in chemotherapy. Many agents that are effective against organisms in vitro have no effect in an infected host. There are several possible explanations for this: (1) the drug might be destroyed, inactivated, bound to body proteins, or too rapidly excreted; (2) the drug might remain at the injection site or might not penetrate as far as the site of infection; or (3) the parasite in vivo might be different from the parasite in vitro, perhaps showing different physiological properties or possibly growing intracellularly, where it is protected from the action of the drug.

In medical practice the decision whether or not to use a chemotherapeutic agent is complex, and the answers to several questions should be taken into consideration: (1) Is the organism sensitive to the agent as shown by antibiotic-sensitivity tests? (2) Are the symptoms due to the organism itself or to a toxin? If to the latter, then attack on the organism alone will not suffice to effect a cure. (3) Is it possible for the drug to reach the site of infection? Many skin infections are hard to treat because the drug will not penetrate to the infected site; again,

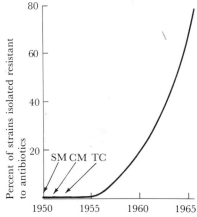

FIGURE 15.19 Rise in antibiotic-resistant *Shigella* in Japan following introduction of antibiotic therapy. Arrows indicate the years when the three antibiotics were introduced. SM, streptomycin; CM, chloramphenicol; TC, tetracycline. (Adapted from Mitsuhashi, S., ed. 1971. Transferable drug resistance factor R. University Park Press, Baltimore.)

many drugs will not enter the spinal fluid and therefore cannot be used in meningitis. (4) Is the patient allergic to the drug? (5) Will the chemotherapeutic agent cause adverse side effects or interfere with the host's defense mechanisms?

The physician's decision to treat an infection with a certain antibiotic is usually reached after consideration of all the questions posed above. In general, the physician will use the antibiotic that is most effective at a relatively low concentration. To guide such decisions a wealth of data exists on the antibiotic susceptibility of routinely encountered pathogens. Although in vitro and in vivo antibiotic susceptibilities may differ as already pointed out, data such as those in Table 15.6 are useful to the physician in choosing the best antibiotic for a specific bacterial infection. Fortunately, many potentially serious pathogens (for example, *Streptococcus pyogenes*) are highly susceptible to a number of different antibiotics and this allows the physician considerable latitude in the course of treatment. Most *Pseudomonas aeruginosa* infections, on the other hand, are very difficult to treat with the majority of common antibiotics, and the physician's choice is limited to a rather restricted group of drugs.

It should be emphasized that in very few infections is the drug alone responsible for a cure. The specific and nonspecific host defenses discussed above and in the next chapter must also function to bring the treatment to a successful conclusion. This is especially true in the case of drugs that are bacteriostatic rather than bactericidal. A bacteriostatic agent will only prevent further growth of the pathogen; it remains for the host defenses to eliminate the organisms already present. With antibiotics such as penicillin, which kill only growing cells, an occasional problem is the presence of **persisters**. These are organisms that are sensitive to the antibiotic, but exist alive in some nongrowing state in an infected region of the body. Since they are nongrowing, they will not be killed by penicillin. At a later time, after drug therapy has been stopped, they may begin to grow.

TABLE 15.6   Minimum Inhibitory Concentrations of Various Antibiotics
for Several Common Pathogens*

Organism	Antibiotic						
	Ampi-cillin	Erythro-mycin	Chloram-phenicol	Tetra-cycline	Kana-mycin	Genta-mycin	Tobra-mycin
Staphylococcus aureus	0.1	0.2	8.0	0.5	1.0	<0.2	1.0
Streptococcus pneumoniae	<0.06	<0.06	4.0	0.2	>128	8.0	16
Streptococcus pyogenes	0.02	0.06	2.0	0.2	16	2.0	32
Neisseria gonorrhoeae	0.06	0.1	0.5	0.2	4.0	1.0	—
Legionella pneumophila	0.2	0.1	0.5	8.0	2.0	0.06	0.1
Escherichia coli	8.0	>128	8.0	4.0	4.0	1.0	1.0
Pseudomonas aeruginosa	>256	>128	>128	32	>64	4.0	2.0

*All values are expressed in micrograms of drug per milliliter required to inhibit growth in laboratory culture (that is, in vitro assays). The values listed represent the average results obtained with susceptible strains of a given species. Mode of action: ampicillin interferes with peptidoglycan synthesis; all others listed interfere in various ways with bacterial protein synthesis.

Data from Lennette, E. H., A. Balows, W. J. Hausler, Jr., and J. P. Truant, eds. 1980. Manual of clinical microbiology, 3rd ed. American Society for Microbiology, Washington, D.C.

In addition to helping to cure diseases actively in progress, drugs may be given to prevent future infections in individuals who are unusually susceptible, a procedure called **chemoprophylaxis.** For example, penicillin is used to prevent streptococcal sore throats in rheumatic fever patients, since these streptococcal infections often lead to a recurrence of rheumatic fever symptoms.

15.13

Summary

In this chapter we have examined some of the processes by which pathogenic microorganisms interact with mammalian hosts. The emphasis has been on general principles rather than specific disease states. We have seen that **infection** implies the growth of the parasite in the host. Whether or not harm will occur as a result of infection will depend upon properties of both parasite and host. The initial stage of infection generally involves the **attachment** of the parasite to epithelial cells of the mucous membranes in respiratory, gastrointestinal, or genitourinary tracts. Attachment of parasite to epithelium has a major element of specificity, apparently arising from interactions between surface macromolecules of the parasite with surface macromolecules of host cells.

Before significant harm occurs, some **growth** of the parasite in the host is necessary, as the number of cells in the inoculum will almost never be large enough to cause damage directly. There is considerable selectivity in determining the extent to which a parasite can grow in the host. Certain nutrients, such as sugars, amino acids, and organic acids are often in short supply, and organisms able to utilize high-molecular-weight components are favored. Not all vitamins and growth factors required by parasites are necessarily available in adequate amounts, and trace elements such as iron are frequently deficient. Body fluids such as blood and urine are unfavorable for the growth of many organisms, due to nutritional inadequacies, improper pH, or the presence of host defenses.

Damage to the host is commonly brought about through the action of enzymes or toxins. **Exotoxins** are protein molecules released from pathogens; they may travel considerable distance from a focus of infection and hence may cause damage far removed from the site of microbial growth. A considerable amount of information is available about the chemistry and mode of action of exotoxins. **Endotoxins** are components of the outer membrane of Gram-negative bacteria, the lipopolysaccharide (LPS) layer. They are released in large amounts upon cell lysis, but are also released in small amounts from living cells. The toxic component of LPS is the lipid portion, the remainder of the LPS serving primarily to render the lipid water soluble. **Enterotoxins** are exotoxins that act on the small intestine, generally causing massive secretion of fluid into the intestinal lumen, leading to the symptoms of diarrhea.

Virulence is a quantitative expression of the ability of a parasite to cause disease. Virulence is genetically determined, and the virulence of a pure culture can be lost as a result of continued passage in culture media. **Attenuation** is the gradual loss of virulence of a culture and is probably due to selection of nonvirulent mutants. Attenuated cultures may be of use in immunization procedures.

The host has a variety of mechanisms for counteracting the invading microorganism. In the next chapter we shall discuss specific immunological mechanisms; in the present chapter we discussed the role of **phagocytic cells** in host resistance. The host has a variety of phagocytic cells, which have the ability to ingest invading microbes (or other foreign materials) and digest them. Some phagocytes are highly mobile, and move into regions where infection foci have been set up. Other phagocytes are fixed, serving as agents for the clearing of pathogens from tissues. The efficiency of the phagocytic system for clearing foreign particles and microbial cells from the bloodstream and lymph system is high.

Phagocytes are attracted chemotactically to invading foreigners, and work best when they can trap a microbial cell upon a surface. Upon ingestion, a microbial cell is killed by one of a variety of microbicidal mechanisms present in the phagocyte, including various hydrolytic enzymes, low pH, halogenation reactions, and production of a toxic form of oxygen, singlet oxygen. Some pathogens are able to resist phagocytic killing by producing agents called **leukocidins,** which destroy phagocytes; the ingested organism is then released unharmed. In other cases, a pathogen can avoid phagocytosis completely, through possession of surface structures such as capsules, which somehow prevent phagocytic engulfment. One group of pathogens are phagocytized and are neither killed nor kill, but actually grow within the phagocyte. These are the **intracellular parasites,** which somehow have become adapted to an intracellular existence. Some intracellular pathogens (for example, rickettsias, chlamydias and the viruses) are obligate, being unable to grow outside the phagocyte, whereas other intracellular pathogens (for example, *Salmonella, Brucella, Mycobacterium tuberculosis*) can grow either within or without phagocytes.

As a result of the host-parasite interaction, **inflammation** often develops. Inflammation represents a response of the host to an external stimulus and is characterized by the development of symptoms of redness, swelling, heat, and pain. The primary cause of inflammation is an increase in capillary permeability, which results in an increased flow of blood to the damaged site. Inflammation is one of the most important and ubiquitous aspects of host defense against invading microorganisms. However, inflammation is also an important aspect of microbial pathogenesis, since the inflammatory response elicited by an invading microorganism can result in considerable host damage.

The present chapter is the first of three dealing with the broad subject of infectious disease. In Chapter 16, we discuss immunological mechanisms, so vital to the development of specific resistance in the host to infection. In Chapter 17, the manner in which pathogens move through populations and cause epidemics will be discussed. In that chapter we shall introduce a variety of public health methods for controlling infectious disease. It is commonly agreed that the most important way of controlling disease is by prevention rather than cure, and this topic is the main theme of Chapter 17.

Bizzini, B. 1979. Tetanus toxin. Microbiol. Rev. 43:224–240. A short review highlighting the mode of action of tetanus toxin.

Boyd, R. F., and **J. J. Marr.** 1980. Medical microbiology. Little, Brown, Boston. An elementary textbook of medical microbiology.

Costerton, J. W., R. T. Irvin, and **K.-J. Cheng.** 1981. The bacterial glycocalyx in nature and disease. Annu. Rev. Microbiol. 35:299–324. A well-illustrated review covering the mechanisms of bacterial attachment and the role of external polysaccharides in the attachment process.

Davis, B. D., R. Dulbecco, H. N. Eisen, and **H. S. Ginsberg.** 1980. Microbiology, 3rd ed. Harper & Row, Hagerstown, Md. A brief treatment of host-parasite relationships can be found in part 4 of this standard medical school textbook. A good reference source for information on diseases caused by specific groups of organisms.

Goodman, L., and **A. Gillman.** 1975. The pharmacological basis of therapeutics, 5th ed. Macmillan, New York. An excellent discussion of the use of antibiotics and other chemotherapeutic agents in the treatment of infectious disease can be found in a series of chapters in this standard medical textbook.

Jawetz, E., J. L. Melnick, and **E. A. Adelberg.** 1982. Review of medical microbiology, 15th ed. Lange Medical, Los Altos, Calif. A brief but very up-to-date treatment of infectious diseases.

Lennette, E. H., A. Balows, W. J. Hausler, Jr., and **J. P. Truant.** 1980. Manual of clinical microbiology, 3rd ed. American Society for Microbiology, Washington, D.C. The standard reference work of clinical microbiological procedures.

Mims, C.A. 1982. The pathogenesis of infectious disease, 2nd ed. Grune & Stratton, New York. A nicely illustrated elementary treatment of host-parasite relationships. Useful as a brief overview of principles.

Montie, T. C., S. Kadis, and **S. J. Ajl,** eds. 1970. Microbial toxins, vol. 3, Bacterial protein toxins. Academic Press, New York. Covers the hemolysins, leukocidins, enterotoxins, and some lesser known exotoxins.

Simons, K., H. Garoff, and **A. Helenius.** 1982. How an animal virus gets into and out of its host cell. Sci. Amer. 246:58–66. A well-illustrated description of the uptake and release of enveloped animal viruses using Semliki forest virus as a model system.

Smith, A. L. 1980. Microbiology and pathology, 12th ed. C. V. Mosby, St. Louis, Mo. A textbook suitable for a course in infectious diseases. Well illustrated; emphasizes the pathology of the disease process.

Smith, H. 1977. Microbial surfaces in relation to pathogenicity. Bacteriol. Rev. 41:475–500. Discusses the surface components of microorganisms (includes bacteria, fungi, and viruses), which are involved in adherence to host surfaces and in invasion and virulence.

Sugiyama, H. 1980. *Clostridium botulinum* neurotoxin. Microbiol. Rev. 44:419–448. A detailed review describing the various forms of botulism and the structure and biological properties of the botulinum toxins.

Weiss, E. 1982. The biology of Rickettsiae. Annu. Rev. Microbiol. 36:345–370. A concise but useful review of the physiology, ecology, pathogenesis, and immunology of the major rickettsial representatives.

16

Immunology and Immunity

Higher animals possess a highly sophisticated mechanism, the **immunological response,** for developing resistance to specific microorganisms. The immunological response occurs because the body has a general system for the neutralization of foreign macromolecules and microbial cells. This system is based on two properties: (1) the presence of a variety of cells that specialize in destroying foreign substances; and (2) the formation of specific proteins called **antibodies.** The former is called **cell-mediated immunity** and the latter is called **humoral immunity.*** Since an invading microorganism contains a variety of macromolecules foreign to the host, antibodies and specific cell types develop against it. We shall discuss first the nature and formation of antibodies, and then show how they confer specific resistance to infection. Later, we will discuss the concepts of cell-mediated immunity.

Several features of the humoral response will be listed here and then discussed in some detail below.

1. Antibodies are formed against a variety of foreign macromolecules, but ordinarily not against macromolecules of the animal's own tissues; thus the animal is able to distinguish between its own and foreign macromolecules.
2. In virtually every case, antibody against a foreign macromolecule is formed only if the animal is challenged with the foreign substance.
3. Many but not all foreign macromolecules elicit the immunological response; those that do are called **antigens.**
4. There is a high specificity in the interaction between antibody and antigen; antibodies will react against closely related antigens but usually with reduced efficiency.

*The term humoral comes from the fact that antibodies reside in the body fluids (humors), specifically in the blood serum.

5. Antigen-antibody reactions are manifested in a wide variety of ways, depending on the nature of the antigen, the antibody, and the environment in which the reaction takes place.
6. Not all antigen-antibody reactions are beneficial; some, such as those involved in allergy and autoimmune reactions are harmful.
7. The high specificity of antigen-antibody reactions makes them useful in many research and diagnostic procedures.
8. The immunological response can be made the basis of specific immunization procedures for the prevention and control of specific diseases.

The distinction between the words "immune" and "immunological" should be clear. **Immune** refers to the ability of the animal to resist infectious disease, whereas **immunological** refers to the processes of antibody formation and antigen-antibody reaction, whether or not these lead to immunity to a particular disease. The word *immunological* was first used in the context of immunity because antibodies were first discovered through studies on the development of immunity; but as further study revealed that antibodies were also involved in other situations, the meaning of the word has been broadened to cover all aspects of the immune response.

16.1
Antigens

Antigens are defined as substances that under appropriate conditions induce formation of specific antibodies; antigens then react specifically with these antibodies. Antigens are distinguished from **haptens** (see below) by the fact that although the latter react specifically with antibodies they do not induce their formation. Thus, the decision as to whether or not a substance is an antigen must be operational; the material must be injected into an animal that is then shown to form antibodies against it. However, the antibody response is not always an automatic reaction whenever an antigen is injected. It depends very much on the species (or even strain) of the animal, its previous history, the route of injection, the frequency of injection, the concentration of antigen injected, and the presence or absence of various nonspecific antibody-stimulating agents administered with the antigen, which are known as **adjuvants.**

An enormous variety of macromolecules can act as antigens under appropriate conditions. These include virtually all proteins and lipoproteins, many polysaccharides, some nucleic acids, and certain of the teichoic acids. One important requirement is that the molecules must be of fairly high molecular weight, usually greater than 10,000. However, the antibody is directed not against the antigenic macromolecule as a whole, but only against restricted portions of the molecule that are called its **antigenic determinants.** Chemically, antigenic determinants include sugars, amino acid side chains, organic acids and bases, hydrocarbons, and aromatic groups. Antibodies are formed most readily to determinants that project from the foreign molecule or to terminal residues of a polymer chain. In general, the specificity of antibodies is comparable to that of enzymes, which are able to distinguish between closely related substrates. For instance, antibodies can distinguish between the sugars glucose and galactose, which differ only in the position of the hydroxyl group on carbon 3. However, specificity is not absolute, and an antibody will react at least to some extent with determinants related to the one that induced its formation. The antigen which induced the antibody is called the **homologous antigen** and others, if any, that react with the antibody are called **heterologous antigens.** As we shall see below, antibodies also are not uniform substances but exist as a spectrum of molecules having varying degrees of affinity for the antigenic determinant. The chemical substances such as sugars and amino acids, which function as antigenic determinants, do not themselves induce antibody formation, but will, if added, combine with antibody molecules. These low-molecular-weight substances that combine with antibody molecules but do not induce antibody formation are called **haptens.** By combining with antibody and blocking the binding site, haptens have the ability to inhibit antibody-antigen reactions.

16.2

Antibodies

Antibodies are found predominantly in the serum fraction of the blood, although they may also be found in other body fluids, as well as in milk. Serum is the fluid portion of the blood that is left when the blood cells and the materials responsible for clotting (fibrin, platelets, and various cofactors, see Section 15.7) are removed. Serum containing antibody is often called **antiserum.** When serum proteins are separated by movement in an electric field (electrophoresis), four predominant fractions are seen: serum albumin and alpha, beta, and gamma globulins. Antibody activity occurs predominantly in the gamma globulin fraction, which is composed of many distinct proteins. Those gamma globulin molecules with antibody properties are called **immunoglobulins** (abbreviated Ig) and can be separated into five major classes on the basis of their physical, chemical, and immunological properties: **IgG, IgA, IgM, IgD,** and **IgE** (Table 16.1). Immunoglobulin class IgG has been further resolved into four immunologically distinct subclasses called IgG_1, IgG_2, IgG_3, and IgG_4. The basis for this separation will be discussed below. Antibody molecules specific for a given antigenic determinant are found in each of the several classes, even in a single immunized individual. Upon initial immunization, it is IgM (the so-called **macroglobulin**), a protein with a molecular weight of about 900,000, that first appears; IgG appears later. In most individuals about 80 percent of the immunoglobulins are IgG proteins, and these have therefore been studied most extensively.

Immunoglobulin structure An understanding of the structure and function of antibodies has been greatly advanced by the study of immunoglobulins of the class IgG. Immunoglobulin G has a molecular weight of about 150,000 and is composed of four polypeptide chains (Figure 16.1a). Both intrachain and interchain disulfide (S—S) bridges are present (Figure 16.1a) The two light (short) chains are identical in amino acid sequence, as are the two heavy (longer) chains. The molecule as a whole is thus symmetrical (Figure 16.1a–c). Each

TABLE 16.1 Properties of the Immunoglobulins

Class Designation	Molecular Weight	Structure	Proportion of Total Antibody (Percent)	Antigen Binding Sites	Properties	Distribution
IgA	160,000 (monomer) 320,000 (dimer)	Secretory component	10–13	2 (monomer) 4 (dimer)	Secretory antibody	Secretions: extra-cellular and blood fluids; exists as a mono-mer in serum and as a dimer in secretions
IgG	150,000		80	2	Major circulating antibody; four subclasses exist: IgG_1, IgG_2, IgG_3, IgG_4; binds comple-ment	Extracellular fluid; blood and lymph; crosses placenta
IgM	900,000 (pentamer)		5–10	10	First antibody to appear after immunization; binds comple-ment	Blood and lymph only
IgD	185,000		1–3	2	Minor circulating antibody; heat labile	Blood and lymph; lymphocyte surfaces
IgE	200,000	Mast cell binding fragment	0.002–0.05	2	Involved in aller-gic reactions: contains mast cell binding fragment	Blood and lymph only

light chain consists of about 220 amino acids and each heavy chain consists of about 440 amino acids.

When an IgG molecule is treated with a reducing agent and the proteolytic enzyme papain under carefully controlled conditions, it breaks into several fragments. The two fragments containing the complete light chain plus the amino terminal half of the heavy chain are the portions that combine with antigen and are called *Fab* fragments (*a*ntigen *b*inding *f*ragment), whereas the fragment containing the carboxyl terminal half of both heavy chains, called *Fc*, does not combine with antigen, although it does contribute to the specificity of antigen-antibody reactions (Figure 16.1*c*). Therefore, each antibody molecule of the IgG class contains *two antigen combining sites.* This bivalency is of considerable importance in understanding the manner in which some antigen-antibody reactions occur (see below). Within *Fab*, the antigen binding site is found in a small region of the amino terminal portion of both the heavy and the light chains (Figure 16.1*c*). Immunoglobulins also contain small amounts of carbohydrate, mainly hexose and hexosamine, which are attached to portions of the heavy chain (Figure 16.1*a*); however, the carbohydrate is not involved in the antigen binding site.

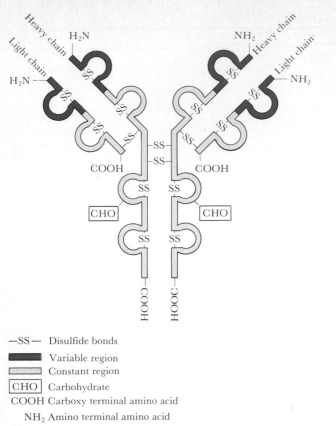

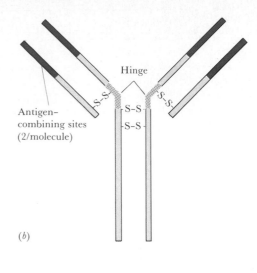

—SS— Disulfide bonds

▰ Variable region

▱ Constant region

CHO Carbohydrate

COOH Carboxy terminal amino acid

NH₂ Amino terminal amino acid

(a)

(b)

(c)

FIGURE 16.1 Structure of immunoglobulin G (IgG). (a) Structure showing disulfide linkages within and between chains. (b) Alternative structural diagram which deletes the intrachain disulfide bonds to simplify the diagram. (c) Effect of papain treatment on immunoglobulin structure.

Present in serum in larger amounts than immunoglobulins of other classes, IgG found in normal serum represents a complex mixture of antibody molecules directed against a number of different antigenic determinants. To effectively study the chemistry of immunoglobulins, it is necessary to have large amounts of a single kind. Certain antibody-producing cell types that have become cancerous, referred to as **myeloma tumors,** usually produce homogenous antibody, and humans suffering from myelomas produce huge amounts of antibody of a single antigenic specificity (monoclonal antibody, see Section 16.5). Amino acid sequence studies of these monospecific immunoglobulins have greatly advanced the understanding of humoral immunity.

Light chains of IgG   The sequence of amino acids in a major portion of the light chains of immunoglobulins of the class IgG may be identical, even in IgG's directed against different antigenic determinants. This is because the amino acid sequence in the carboxy terminal half of the light chain constitutes one of two specific sequences, referred to as the λ sequence or the κ sequence. One IgG molecule will have either two λ chains or two κ chains but never one of each. It is thus possible that two IgG molecules directed against different antigens will have light chains containing 50 percent or more amino acid sequence homology. The sequence in the amino terminal domain (antigen binding site) of their light chains will, however, be different, since they react with different antigens (Figure 16.1a).

Heavy chains of IgG   Analogous to the situation that exists in the light chain, all immunoglobulins of the class IgG have a portion of their heavy chain (the carboxy terminal region) in which the amino acid sequence is identical (see Figure 16.1a) from one IgG molecule to another, as well as a region in the amino terminal end (antigen binding site) where considerable amino acid sequence variation occurs from one IgG to the next.* An antibody molecule is thus a multisubunit protein consisting of four polypeptide chains each of which contains a region of amino acid sequence variability (**variable region domain**) and a region of constant sequence (**constant region domain**) for all immunoglobulin molecules of a given class. The great specificity of a given antibody molecule for a particular antigen lies in the unique three-dimensional structure of the antigen binding site dictated by the amino acid sequence in the variable regions of the heavy and light chains.

Other classes of immunoglobulins   How do immunoglobulins of the other classes differ from IgG? κ or λ sequences are found in the light chains of immunoglobulins of all classes, but distinct sequence differences exist in the heavy chain regions. The heavy-chain *constant* regions of a given immunoglobulin molecule can have five distinctly

---

*This statement must be qualified in the case of immunoglobulins of the class IgG because four subclasses, IgG_1, IgG_2, IgG_3, and IgG_4, have been identified in the human and the mouse, and these four subclasses differ slightly from one another in amino acid sequence in the "constant" region domain of their heavy chains.

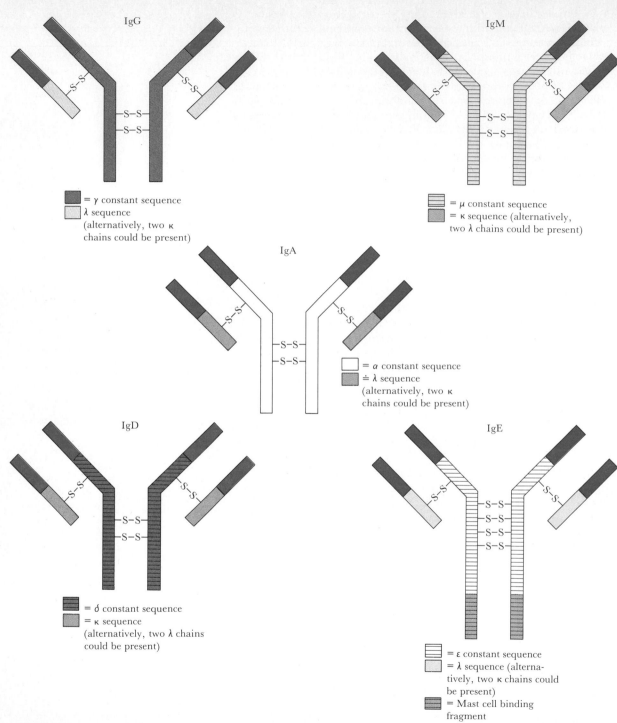

FIGURE 16.2 Major amino acid sequence differences in the five classes of immunoglobulins. Note that major differences exist in the constant regions of the heavy chains where the amino acid sequence, although constant within a given class, varies considerably from one class to another.

different amino acid sequences called γ, α, μ, δ, or ζ, and these sequences constitute the carboxy terminal three-fourths of the heavy chains of immunoglobulins of the class IgG, IgA, IgM, IgD, or IgE, respectively (Figure 16.2). Each antibody of the class IgM, for example, will contain a stretch of amino acids in its heavy chain which constitutes the μ sequence. If two immunoglobulins of different classes react with the same antigenic determinant, their entire light chains and the variable regions of their heavy chains would be identical, but their class-determining sequences, specific to their heavy chains, would be different. It is not unusual in a typical immune response to observe the production of antibodies of two different classes to the same antigenic determinant.

Immunoglobulin A (IgA) is of interest because it is present in body secretions. IgA is the dominant antibody in all fluids bathing organs and systems in contact with the outside world: saliva, tears, breast milk and colostrum, gastrointestinal secretions, and mucus secretions of the respiratory and genitourinary tracts. IgA is also present in serum, but the IgA of secretions has an altered molecular structure, having attached to it a protein component high in carbohydrate called the *secretory piece* (see Table 16.1), which may be involved in the ability of secretory IgA to pass into the fluid secretions. Because of its locations, IgA probably provides a first line of immunologic attack against bacterial invaders, which as we have seen, first become established on tissue surfaces.

Immunoglobulin E (IgE) is found in serum in extremely small amounts (in an average human about 1 of every 50,000 serum immunoglobulin molecules is IgE). Immediate-type hypersensitivities (allergies; see Section 16.6) are mediated by IgE, following the binding of these antibodies to certain cell types which respond by releasing chemicals that elicit allergic reactions. The molecular weight of an IgE molecule is significantly higher than that of other immunoglobulins (Table 16.1) because it contains additional polypeptides (of about 100 to 110 amino acids in length) attached to the carboxy-terminal portions of the heavy chains. This fragment is thought to function in binding IgE to mast cell surfaces (see Section 16.6), which is an important prerequisite for certain allergic reactions.

Immunoglobulin D (IgD) is also present in low concentrations (but 10 times higher than that of IgE), and its function in the overall immune response is unclear. Recent experiments have shown that IgD is abundant on the surfaces of antibody-producing cells (lymphocytes, see Section 16.5) and IgD may play a role in binding antigen as a signal to the lymphocyte to begin antibody production.

Immunoglobulin M (IgM) is usually found as an aggregate of five immunoglobulin molecules and accounts for 5 to 10 percent of the total serum immunoglobulins. IgM is the first class of immunoglobulin made in a typical immune response to a bacterial infection, but immunoglobulins of this class are generally of low affinity. The latter problem is compensated, however, by the high valency of the pentameric molecule; 10 binding sites are available for interaction with antigen (Table 16.1). IgM plays little if any role in the secondary immune response (see Figure 16.13), which is due instead to the rapid production of IgG.

Antigen-Antibody Reactions

Antigen-antibody reactions are most easily studied in vitro using preparations of antigens and antisera. The study of antigen-antibody reactions in vitro is called **serology,** and is especially important in clinical diagnostic microbiology. A variety of different kinds of serological reactions can be observed (Table 16.2), depending on the natures of the antigen and antibody and on the conditions chosen for reaction. Serology has many ramifications, only a few of which will be discussed here.

Neutralization   **Neutralization** of microbial toxins by specific antibody can occur when toxin molecules and antibody molecules directed against the toxin combine in such a way that the active portion of the toxin that is responsible for cell damage is blocked (Figure 16.3). Neutralization reactions of this type are known for a wide variety of exotoxins, including most of those listed in Table 15.1. Reactions of similar nature also occur between viruses and their specific antibodies. For instance, antibodies directed against the protein coats of viruses may prevent the adsorption of the viruses to host cells. Neutralizing antibodies are also known for a variety of enzymes; these act by combining with the enzyme at or near the active site and prevent formation of the enzyme-substrate complex. Neutralizing antibody requires only a single antigen-combining site for action and thus two antigen molecules may be bound by one antibody molecule. An antiserum containing neutralizing antibody against a toxin is sometimes referred to as an **antitoxin** (see Section 16.8).

Precipitation   Since an antibody has two combining sites (that is, is bivalent), it is possible for each site to combine with a separate antigen molecule, and if the antigen also has more than one combining site, a **precipitate** may develop consisting of aggregates of antibody and antigen molecules (Figure 16.4). Precipitation reactions for a wide variety of soluble polysaccharide and protein antigens are known. Because they are easily observed in vitro, precipitation reactions are very useful serological tests, especially in the quantitative measurement of antibody concentrations. Precipitation occurs maximally only when there are optimal proportions of the two reacting

TABLE 16.2   Types of Antigen-Antibody Reactions

Location of Antigen	Accessory Factors	Reaction Observed
Soluble antigen	None	Precipitation
On cell or inert particle	None	Agglutination
Flagellum	None	Immobilization or agglutination
On bacterial cell	Complement	Lysis
On bacterial cell	Complement	Killing
On erythrocyte	Complement	Hemolysis
Toxin	None	Neutralization
Virus	None	Neutralization
On bacterial cell	Phagocyte, complement	Phagocytosis (opsonization)

substances since, when either reactant occurs in excess, the formation of large antigen-antibody aggregates is not possible (Figure 16.4).

Precipitation can also be inhibited by the hapten corresponding to the antigenic determinant. In fact, hapten inhibition is an extremely useful way of studying the specificity of serological reactions since the inhibiting power of a series of related haptens can be compared quantitatively. Such studies show that although antigen-antibody reactions have a high degree of specificity, this specificity is not absolute, and antibody will almost always react with heterologous antigens. Also, antisera almost always contain mixtures of antibodies with varying specificities, and such heterogeneity must be taken into consideration in any serological study. Cross reactions can often be eliminated or minimized by allowing the antiserum to react first with heterologous antigens and then removing the precipitate that forms. This will remove antibodies specific for the heterologous antigen and will leave behind in the supernatant those antibodies which do not react with heterologous antigens but react with the homologous antigen.

Precipitation reactions carried out in agar gels have proved to be of considerable utility in the study of the specificity of antigen-antibody reactions. Both antigen and antibody diffuse out from separate wells cut in the agar gel and precipitation bands form in the region where antibody and antigen meet in equivalent proportions (Figure 16.5). The shapes of the precipitation bands and the distance from the wells are characteristic for the reacting substances, and it is possible to determine whether two antigens reacting against antibodies in an antiserum are identical or not by observing the bands formed when the two antigens are placed in adjacent wells near the antiserum well.

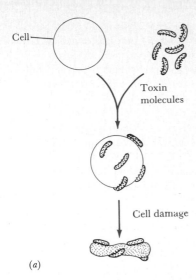

(a)

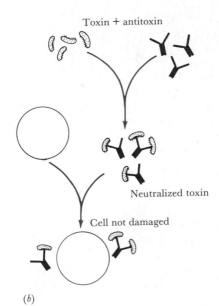

(b)

FIGURE 16.3 (a) Toxin action and (b) mechanism of neutralization of a toxin by a specific antibody (antitoxin).

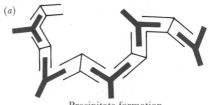

(a)

Precipitate formation at equivalence between antibody and antigen

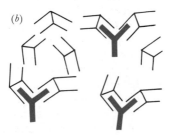

(b)

Soluble antibody–antigen complexes (no precipitation) when antigen is in excess of antibody (b) or when antibody is in excess of antigen (c).

(c)

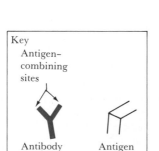

Key
Antigen–combining sites

Antibody   Antigen

FIGURE 16.4 Formation of precipitate between bivalent antibody and antigen at equivalence, and formation of soluble antibody-antigen complexes when antigen or antibody is in excess.

FIGURE 16.5 Formation of precipitation bands in agar gels due to reactions between antibody and soluble antigens. Wells labeled S contain anti-*Proteus mirabilis* antiserum. Wells labeled A, B, and C contain soluble extracts of *P. mirabilis* cells. In *b,* well C contained a foreign antigen that did not react with the antiserum. (From Weibull, C., W. D. Bickel, W. T. Haskins, K. C. Milner, and E. Ribi. 1967. J. Bacteriol. 93:1143–1159.)

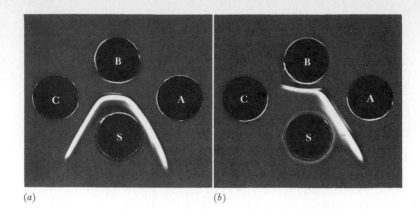

(*a*)   (*b*)

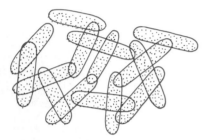

Antiserum against
somatic antigen

(*a*)

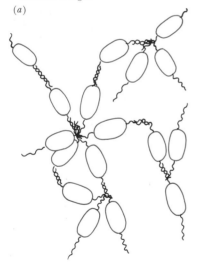

Antiserum against
flagellar antigen

(*b*)

FIGURE 16.6 Appearance of agglutinated bacterial cells when antibody is directed against (a) somatic, or O, antigen, and (b) flagellar, or H, antigen.

Agglutination   If the antigen is not in solution but is present on the *surface* of a cell or other particle, an antigen-antibody reaction can lead to clumping of the particles, called **agglutination.** When foreign cells are injected into an animal, antibodies are formed against a wide variety of their macromolecular constituents, including those in the cytoplasm as well as on the surface of the cells. Only the surface antigens are involved in agglutination, however, since only these are exposed in the intact cell. Considerable attention has been focused on the chemical nature of the surface macromolecules that might be involved in agglutination of bacterial cells. Antibodies against both the flagella (so-called H antigens) and the lipopolysaccharide layer (somatic or O antigens) will agglutinate cells, but the nature of the clump differs; that of flagellar agglutination is much looser and more flocculent (Figure 16.6).

Soluble antigens can be adsorbed to or coupled chemically to cells or other particulate structures such as latex beads or colloidal clay, and they can then be detected by agglutination reactions, the cell or particle serving only as an inert carrier. This greatly increases the ability to detect the presence of antibodies against soluble antigens since, as noted above, agglutination is much more sensitive than precipitation. Agglutination of red blood cells can also occur, a phenomenon called **hemagglutination.** Antibodies can react to antigens of red cells, or other antigens can be adsorbed to red cells and hemagglutination observed when antibody specific to the adsorbed antigen is used.

Immobilization   Flagellar antibodies can induce flagellar paralysis, causing **immobilization** of a flagellated organism. Immobilization occurs at antibody concentrations considerably lower than those required for flagellar agglutination, and for this reason it is a sensitive measure of antigen-antibody reaction. It also is one of the few immunological reactions that can be detected with single cells.

Fluorescent antibodies   Antibody molecules can be made fluorescent by attaching them chemically to fluorescent organic compounds such as rhodamine B or fluorescein isothiocyanate (Figure 16.7). This does not alter the specificity of the antibody significantly, but makes it possible to detect the antibody adsorbed to cells or tissues by use of the fluorescence microscope (Figure 16.8). **Fluorescent antibodies** have been of considerable utility in diagnostic microbiology since they permit the study of immunological reactions on single cells. The fluorescent-antibody technique is very useful in microbial ecology as one of the few methods for identifying microbial cells directly in the natural environments in which they live. Two distinct procedures, the **direct** and the **indirect** staining methods, are used (Figure 16.9). In the direct method, the antibody against the organism is itself fluorescent. In the indirect method, the presence of an antibody on the surface of the cell is detected by the use of another fluorescent antibody directed against the specific antibody itself. This is possible because immunoglobulins, as all proteins, are antigenic, and the immunoglobulins of one animal species can induce antibody formation in those of another. For this reason, fluorescent goat anti-rabbit immunoglobulin can be used to detect the presence of rabbit immunoglobulin adsorbed to cells. One of the advantages of the indirect staining method is that it eliminates the need to make the fluorescent antibody for each organism of interest.

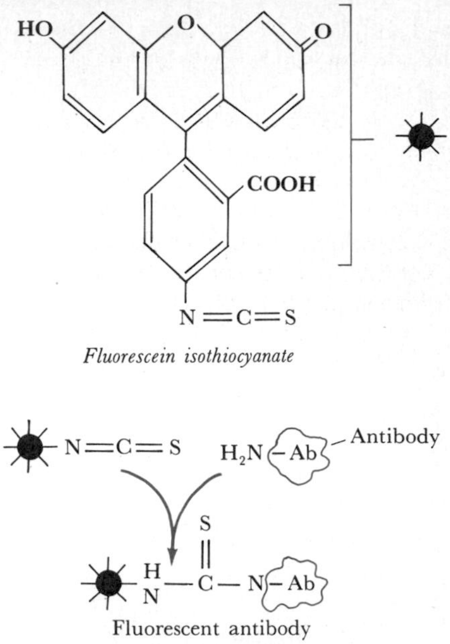

Fluorescein isothiocyanate

FIGURE 16.7 Preparation of fluorescent antibody by coupling the fluorescent dye fluorescein isothiocyanate to the antibody protein.

FIGURE 16.8 Fluorescent antibody reactions. (a) *Rhizobium japonicum*. Magnification, 1000×. (Courtesy of B. B. Bohlool and E. L. Schmidt.) (b) Actinomycete filaments. Magnification, 850×. (From Bohlool, B. B., and E. L. Schmidt, 1970. Soil Science 110: 229–236.)

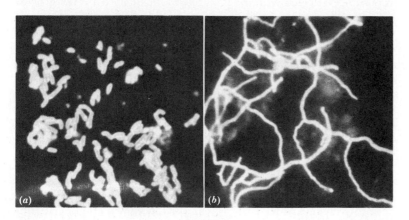

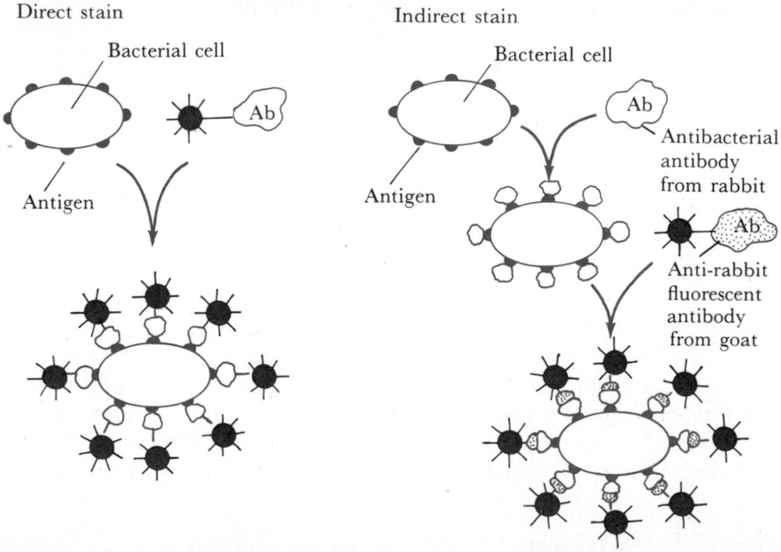

Direct stain

Indirect stain

FIGURE 16.9 Direct and indirect methods of using fluorescent antibody to demonstrate the presence of antigen on a bacterial cell.

531

Radioimmunoassay Various proteins, many of which serve as hormones or other specific mediators, are routinely assayed from the serum of individuals by a procedure known as **radioimmunoassay (RIA)**. The principle of RIA is based on competition for antibody between a known amount of radioactively labeled antigen and an unknown amount of the same antigen taken from a serum sample (Figure 16.10a). For example, using RIA it is possible to detect serum proteins such as insulin found in small amounts in normal human serum. For the RIA insulin assay, purified insulin, labeled with radioactive iodine (^{125}I) is mixed with a patient's serum containing an unknown amount of insulin and the mixture is reacted with anti-insulin antibodies. If the serum levels of insulin are sufficiently high (that is, the patient is not suffering from diabetes), the nonradioactive insulin will successfully compete with the radioactively labeled insulin, leading to the formation of precipitates which are not very radioactive. If serum insulin levels are low, however, more of the radioactive insulin will exist in the precipitates. Comparison of the test precipitates with the results of a calibration curve prepared with known concentrations of insulin allows the precise quantitation of insulin levels in the patient's blood. Other antigens of biochemical importance detected with the RIA technique include human growth hormone, glucagon, vasopressin, testosterone, and many other low-molecular-weight serum proteins.

ELISA Technique Because of the health hazards and expense of radioactive assays and the short shelf life of radioimmunoassay reagents, a less expensive and equally sensitive immunological assay method has been developed. Called **ELISA** (*e*nzyme-*l*inked *i*mmuno*s*orbent *a*ssay), this method makes use of antibodies to which enzymes have been covalently bound (Figure 16.10b). The linkage between enzyme and antibody is made such that the enzyme's catalytic properties and the antibody's immunologic properties are retained. Typical linked enzymes include alkaline phosphatase and β-galactosidase, both of which catalyze reactions whose products can be sensitively measured. In a typical assay, a known amount of antigen is added to the enzyme-labeled antibody preparation, and precipitation allowed to occur. After separation of unreacted antibody from the precipitated complexes, the enzyme activity of the complexes is measured. The procedure is repeated using patient's serum as a source of antigen. If the antigen is present in high concentration, a large amount of antigen–antibody complex will form, resulting in high levels of enzyme activity in the complexes (Figure 16.10b). Conversely, if the antigen is present in low concentrations, fewer antigen–antibody complexes will form, resulting in lower enzyme activity. A variation of the ELISA technique, employing an enzyme-labeled antigen, can be set up for measuring the concentration of the corresponding antibody in a patient's serum.

Since the ELISA technique requires very little in the way of expensive equipment and is just as sensitive as RIA, it is widely used in clinical laboratory situations. The ELISA technique has been used to assay hormones, quantitate antibodies to rubella virus (German measles), and to detect certain drugs in serum, and the method has

Radioimmunoassay (RIA)

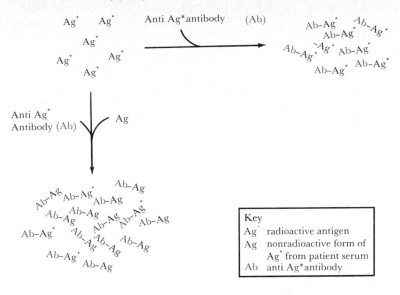

Key
Ag˙ radioactive antigen
Ag nonradioactive form of
 Ag˙ from patient serum
Ab anti Ag*antibody

ELISA

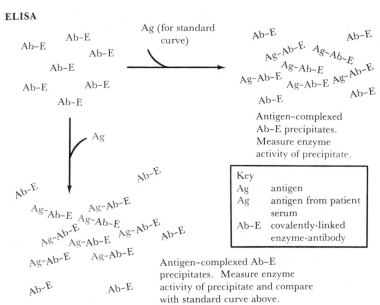

Antigen–complexed
Ab-E precipitates.
Measure enzyme
activity of precipitate.

Key
Ag antigen
Ag antigen from patient
 serum
Ab-E covalently-linked
 enzyme-antibody

Antigen–complexed Ab-E
precipitates. Measure enzyme
activity of precipitate and compare
with standard curve above.

FIGURE 16.10 Basic principles of radioimmunoassay and the enzyme-linked immunosorbent assay (ELISA) technique. (a) Radioimmunoassay. In this example RIA is being used to measure the concentration of a specific antigen in a patient's serum; the higher the concentration of antigen in the patient's serum the fewer radioactive antigens are bound. (b) ELISA. In this example ELISA also is being used to measure the concentration of a specific antigen in a patient's serum; the enzyme activity in the Ag·Ab-E complexes prepared with a patient's serum as source of Ag is compared with a standard curve prepared with known amounts of Ag.

shown promise for the rapid and sensitive detection of antigens of the causative agents of syphilis, brucellosis, salmonellosis, and cholera.

16.4

Complement and Complement Fixation

In addition to the antigen-antibody reactions described above, other types of reactions exist that are of great importance in immunity. Before these reactions are discussed, however, it will be necessary to introduce another group of serum substances, called **complement**. Complement acts in concert with specific antibody to bring about several kinds of antigen-antibody reactions that would not otherwise occur. These reactions are included in Table 16.2.

Complement is not an antibody but is a series of enzymes found in blood serum that attack bacterial cells or other foreign cells, causing lysis or leakage of cellular constituents as a result of damage to the cell membrane. These enzymes, found even in unimmunized individuals, are normally inactive, but become active when an antibody-antigen reaction occurs. One of the main functions of antibody is to recognize invading cells and activate the complement system for attack. There is considerable economy in an arrangement such as this, since a wide variety of antibodies, each specific for a single antigen, can call into action the complement enzymatic machinery; thus the body does not need separate enzymes to attack each kind of invading agent.

Certain components of complement are very heat labile, being destroyed by heating at $55°C$ for 30 minutes; antibody, on the other hand, is heat stable. The usual procedure for studying complement-requiring antigen-antibody reactions is to destroy the complement activity of the antiserum by heating, and then add some fresh serum from an unimmunized animal as a source of complement. In this way one can set up reaction mixtures with defined amounts of complement.

Some reactions in which complement participates include (1) bacterial lysis, especially in Gram-negative bacteria, when specific antibody combines with antigen on bacterial cells in the presence of complement; (2) microbial killing, even in the absence of lysis; and (3) phagocytosis, which may not occur during infection if the invading microorganism possesses a capsule or other surface structure that prevents the phagocyte from acting. When specific antibody combines with the cell in the presence of complement, the cell is changed in such a way that phagocytosis can occur. (This process in which antibody plus complement renders a cell susceptible to phagocytosis is sometimes called **opsonization.**)

Activation of the complement system   Complement is a system of 11 proteins, designated C1, C2, C3, and so on. Activation of complement occurs only with antibodies of the IgG and IgM classes; when such antibodies combine with their respective antigens they are altered in such a way that the first component of complement, C1 (which is really a complex of three subunits called C1q, C1r, and C1s), combines with the antibody-antigen aggregate (Figure 16.11*a*.1) Protein C4 then combines with the C1 complex and is converted by the enzy-

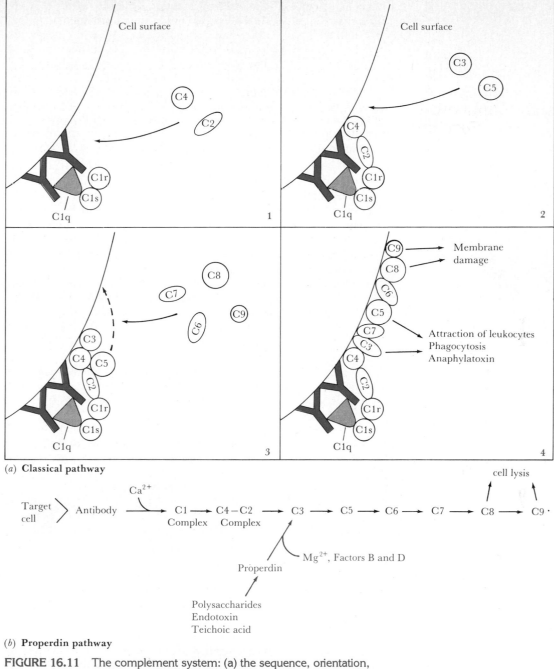

(a) Classical pathway

(b) Properdin pathway

FIGURE 16.11 The complement system: (a) the sequence, orientation, and activity of the various components. Panel 1, binding of the antibody recognition unit C1q, and other C1 proteins; panel 2, the C4-C2 complex; panel 3, the C4-C2-C3-C5 complex, after activation the C-5 unit travels to an adjacent membrane site; panel 4, binding of C6, C7, C8, and C9, resulting in membrane damage; (b) the relation of the classical to the properdin pathways.

Holes

FIGURE 16.12 Electron micrograph of a negatively stained preparation of *Salmonella paratyphi,* showing holes created in the cell envelope as a result of reaction between cell-envelope antigens, specific antibody, and complement. Magnification, 255,500×. (Courtesy of E. Munn.)

matic activity of C1 to an active fragment. This activated C4 binds C2, and the C2 molecule is then converted into an activated complex, C4-C2, which possesses enzymatic activity. The C4-C2 complex binds directly to the cell membrane adjacent to the C1-antibody binding site (Figure 16.11a.2). The complement protein C3 is the substrate for the C4-C2 complex, and when C3 becomes activated, C5 can combine with it (Figure 16.11a.3). Following C5 binding, C6 and C7 bind and the C5-C6-C7 complex binds to the cell membrane at a new site. C8 binds to the C5-C6-C7 complex and this event leads to the formation of a small hole in the membrane (Figure 16.11a.4). The binding of C9 enlarges the hole causing the loss of internal ions and a large influx of water. The sequence and some of the resulting reactions are summarized in Figure 16.11. As seen, reactions at the C3 level result in chemotactic attraction of phagocytes to invading agents, and to phagocytosis (opsonization). Reaction at C5 also leads to leukocyte attraction. The terminal series of reactions from C5 through C9 results in cell lysis and death. Lysis itself results from a destruction of the integrity of the cell membrane, leading to the formation of holes through which cytoplasm can leak (Figure 16.12).

The complement system is thus seen to act in cascade fashion, the activation of one component resulting in the activation of the next. In summary, the steps are (1) binding of antibody to antigen (initiation); (2) recognition (C1); (3) C4-C2 binding to an adjacent membrane site; (4) activation of C3; (5) formation of the C5-C6-C7 complex, causing attraction of leukocytes; and (6) formation of the C8-C9 complex causing cell lysis. This pathway is called the *classical* pathway, to distinguish it from the *properdin* pathway discussed below.

A variety of complement reactions The complement system is involved in other immunological reactions besides those leading to the destruction of invading organisms. The reactions of importance here include certain aspects of the inflammatory response and the production of proteins called **anaphylatoxins** which produce symptoms resembling those of allergic responses (discussed below). Anaphylatoxins consist of fragments of C3 and C5 and cause release of histamines from tissue mast cells. Histamine increases the permeability of capillaries (Section 15.9), enabling leukocytes and fluid to escape into the tissue cells. One serious type of allergic reaction, **anaphylactic shock** (see Section 16.6), may also involve components of the complement system, although most such allergic responses involve the activity of immunoglobulins of the class IgE.

In addition to the activation of the complement system beginning with C1, which was described above, there is another means of activating C3 which bypasses the C1 complex and C2 and C4. This is the **properdin** system (Figure 16.11b). Properdin is a protein complex in normal blood serum (it may also be formed as a result of antigenic stimulation) that combines with bacterial polysaccharides or certain aggregated immunoglobulins to activate C3 and C5 directly, leading to opsonization and the production of inflammatory reactions. The importance of the properdin system is that it leads to defense against invading organisms without the production of specific antibodies, thus it plays a role in innate (nonspecific) immunity.

Another important reaction in which complement participates is hemolysis, the lysing of red blood cells. If an antiserum against erythrocytes is mixed with a suspension of erythrocytes and some normal serum added as a source of complement, lysis occurs within 30 minutes on incubation at 37°C. Erythrocyte hemolysis is often used to test for the presence of complement in unknown sera or to measure complement fixation (see below).

Complement is necessary in the bactericidal and lytic actions of antibodies against many Gram-negative bacteria. (Interestingly, Gram-positive bacteria are not killed by specific antibody, in either the presence or absence of complement, although Gram-positive bacteria are opsonized.) Probably death (including that from lysis) involves antibodies against antigens on the surface of the cell; complement perhaps brings about an actual change in the cell surface, possibly by an enzymelike reaction, after antibodies have prepared the way. No cytocidal or lytic effect is seen when cells and complement are mixed alone, whereas if cells have adsorbed antibody first, death or lysis occurs rapidly after complement is added.

As noted, complement is necessary for opsonization, the promotion of phagocytosis by antibodies against capsules. This does not require the whole complement pathway, but only the reactions through C3. Opsonization occurs because the binding of C3 causes the cells to adhere to phagocytes. Phagocytes have C3 receptors on their surfaces and it is presumably these receptor molecules which are involved in adherence of the C3-coated bacterial cell to the phagocyte. A similar immune adherence occurs as a result of C3 binding to antibody-antigen aggregates (for example, precipitates formed from soluble antigens).

Complement fixation   An important property of the complement system is that the components are enzymatically altered during reaction, so that they will no longer react in a new sequence of reactions. Complement thus appears to be used up during antibody-antigen reactions. This is called **complement fixation** and occurs whenever an IgG or IgM antibody reacts with antigen in the presence of complement, even if complement is not required in the reaction. Complement fixation is measured by assaying the concentration of complement after an antibody-antigen reaction has occurred. After the initial reaction to permit complement fixation, an indicator system is added consisting of sheep red blood cells and antibody to the red blood cells. In the absence of complement, lysis of the red cells will not take place, but if complement is present, the normal series of reactions leading to cell lysis occurs. Thus, if complement has *not* been fixed, lysis occurs, but if complement *has* been fixed, lysis *will not* occur. Sheep cells are used since their lysis is readily observed by eye. Appropriate controls must be set up to be sure that nothing in the system is inactivating complement nonspecificially. By measuring complement fixation, one has a means of determining the occurrence of an antigen-antibody reaction even if less sensitive immunological assays such as precipitation or agglutination have not produced a visible reaction.

The Mechanism of Antibody Formation

How is it possible for higher animals to produce such a large variety of specific proteins, the antibodies, in response to invasion by foreign macromolecules? In considering the mechanism of antibody formation, we must first explore what happens to the antigen. The main sites of antigen localization in the body are the lymph nodes, the spleen, and the liver. It has been well established that antibodies are formed in both the spleen and lymph nodes; the liver seems not to be involved. If the antigen is injected intravenously, the spleen is the site of greatest antibody formation, whereas subcutaneous, intradermal, and intraperitoneal injections lead to antibody formation in lymph nodes. Fragments of lymph node or spleen from immunized animals can continue to produce antibody when placed in tissue culture or when injected into other, nonimmunized, animals.

Following the first injection of an antigen that an animal is seeing for the first time, there is a lapse of time (latent period) before any antibody appears in the circulation, followed by a gradual increase in **titer** (that is, concentration) and then a slow fall. This reaction to a single injection is called a **primary response** (Figure 16.13). When a second injection is made some days or weeks later, the titer rises rapidly to a maximum 10 to 100 times above the level achieved after the primary injection. This secondary injection is sometimes called a **booster** dose, and results in a secondary antibody response with high antibody titers. The titer slowly drops again, but later injections can bring it back up. The secondary response is the basis for the vaccination procedure known as a "booster shot," (for example, the yearly rabies shot given to domestic animals) to maintain high levels of circulating antibody specific for a certain antigen.

Cellular aspects It is now certain that the cells responsible for immunoglobulin production are a class of white blood cells referred to as **lymphocytes.** These cells are dispersed throughout the body and are one of the most prevalent mammalian cell types (the average

FIGURE 16.13 Primary and secondary antibody responses.

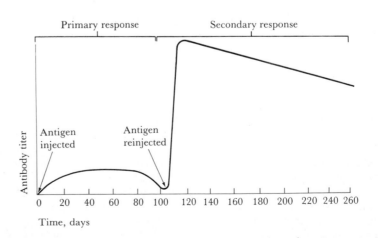

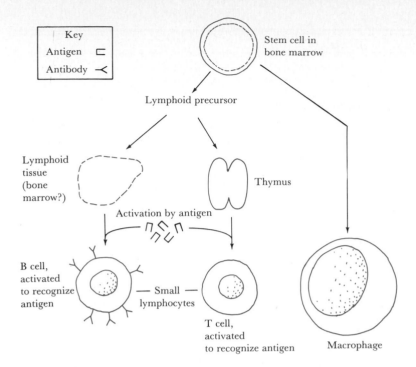

Key
Antigen ⊏
Antibody ≺

Stem cell in
bone marrow

Lymphoid precursor

Lymphoid
tissue
(bone
marrow?)

Thymus

Activation by antigen

B cell,
activated
to recognize
antigen

— Small —
lymphocytes

T cell,
activated
to recognize antigen

Macrophage

FIGURE 16.14 Formation of B and T cells and macrophages from stem cells in the bone marrow.

adult human has about 10^{12} lymphocytes). Two types of lymphocytes, B-lymphocytes (**B cells**) and T-lymphocytes (**T cells**) are involved in immune responses.

Immunoglobulins are made only by lymphocytes of the B type, while T cells play a variety of alternative roles in the overall immune response. Both B and T cells are derived from stem cells in the bone marrow, and their subsequent differentiation is determined by the organ within which they become established (Figure 16.14). Several subsets of T-lymphocytes are known (Table 16.3) but they are all differentiated in the thymus (hence the designation "T"). **T-helper (T_H) cells** stimulate B cells to produce high levels of immunoglobulin; in most cases little if any antibody is made without T_H cell interaction. **T-suppressor (T_S) cells** perform a regulatory function in their ability to suppress further immunoglobulin production by B cells once the antigen has been removed. Cytotoxic or **T-killer (T_K) cells** secrete toxic substances that can kill foreign cells, and together with **T-delayed type hypersensitivity (T_{DTH}) cells**, play primary roles in the cellular immune responses (Table 16.3 and Section 16.6). In birds, B cells differentiate in an organ called the *bursa of Fabricius* (hence the designation "B"). A similar organ has not been found in mammals, and it is thought that B cells differentiate in lymph nodes, in lymphoid structures such as the appendix or tonsils, or possibly in the bone marrow itself. In addition to B and T cells, **macrophages** (a type of phagocytic cell, see Section 15.7) also play a role in the overall formation of antibody.

How do B cells, T cells, and macrophages act together to produce a humoral response? Macrophages act nonspecifically, phagocytizing antigens so that antigenic materials become attached to

TABLE 16.3 Comparison of B and T Cells

T Cells	B Cells
Origin: bone marrow	Origin: bone marrow
Maturation: thymus	Maturation: lymphoid tissue or bone marrow; bursa in birds
Long-lived: months to years	Short-lived: days to weeks
Mobile	Relatively immobile (stationary)
No complement receptors	Have complement receptors
No immunoglobulins on surface	Immunoglobulins on surface
Restricted antigenic specificity	Restricted antigenic specificity
Proliferate upon antigenic stimulation	Proliferate upon antigenic stimulation into plasma cells
No immunoglobulin synthesis	and memory cells
Produce cell-mediated immunity (T_{DTH} and T_K cells)	Synthesize immunoglobulin (antibody)
Show delayed hypersensitivity (T_{DTH} cells)	
Produce lymphokines (T_{DTH} cells)	
Participate in transplantation immunity (T_K cells)	
Help in antibody production by B cells (T_H cells)	
React against intracellular bacteria, viruses, and parasites (cellular immunity) (T_K cells)	
Perform as killer T-lymphocytes in cell-mediated immunity (T_K cells)	
Modulate degree of humoral response (T_S cells)	

macrophage cell surfaces where they can interact with other cell types (Figure 16.15). Macrophages are very sticky and attach well to surfaces; their stickiness probably also promotes the attachment of B and T cells. In this fashion, macrophages are thought to "present" the antigen to B and T lymphocytes, thus initiating the process of antibody production. Both B and T cells contain specific antigen receptors on their cell surfaces. The receptors on B cell surfaces consist of the antibody molecules that the cells are genetically programmed to produce.* The receptors on T cells appear to consist of only a small portion of an antibody molecule, probably including only the variable region of the heavy-chain portion of the molecule.

Once an antigen has been presented to the correct B cell– T cell pair, it is bound to the immunoglobulin receptors of the B cell and this event stimulates growth and division of the B cell to form a clone of B cells, each capable of producing identical antibody molecules. Further differentiation of the expanded B-cell population then occurs, resulting in the formation of large antibody-secreting cells called **plasma cells** and an activated form of B cell called a **memory cell** (Figure 16.15). Plasma cells are relatively short-lived (less than 1 week) but excrete large amounts of antibody during this period. Memory cells, on the other hand, are very long-lived cells, and upon second exposure to antigen, these cells are quickly transformed into plasma cells. This accounts for the rapid and more abundant antibody response observed upon antigenic stimulation the second time (secondary response; see Figure 16.13). T-helper cells stimulate the initial B cell $\rightarrow$ plasma cell + memory cell conversion by releasing a

*One B lymphocyte may produce immunoglobulins of more than one class but each will react with only one antigenic determinant (that is, their variable region domains will be identical).

(a) **Macrophage: antigen interaction**

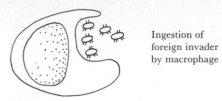

Ingestion of
foreign invader
by macrophage

(b) **Macrophage: T cell: B cell complex**

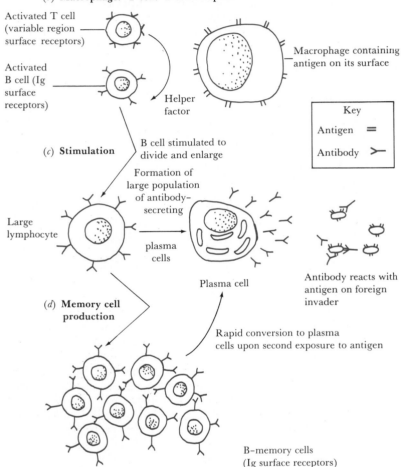

Activated T cell
(variable region
surface receptors)

Activated
B cell (Ig
surface
receptors)

Helper
factor

Macrophage containing
antigen on its surface

Key

Antigen =

Antibody >—

(c) **Stimulation**

B cell stimulated to
divide and enlarge

Formation of
large population
of antibody-
secreting

Large
lymphocyte

plasma
cells

Plasma cell

Antibody reacts with
antigen on foreign
invader

(d) **Memory cell
production**

Rapid conversion to plasma
cells upon second exposure to antigen

B–memory cells
(Ig surface receptors)

FIGURE 16.15 Interaction between
B and T cells and macrophages to
produce a line of antibody-secreting
plasma cells and a population of
B-memory cells.

soluble "helper factor" which functions both to accelerate the differentiation process and to intensify the antibody response (Figure
16.15). Certain antigens can stimulate low-level antibody production
in the absence of T-cell interactions (these are the so-called T-inde-
pendent antigens), but the immunoglobulins produced (usually of
the class IgM) are of low antigenic affinity, and the B cells which
produce them do not have immunogenic memory.

The discussion above describes the general principles of antibody production. Each antigen (strictly speaking, each antigenic
determinant) will catalyze the expansion of a different B-cell line,
the B-cell type in each case being genetically controlled to produce

antibodies that react specifically with that antigen. In this fashion it is thought that the normal animal can respond to at least several million distinct antigens by expanding a specific B-cell clone in response to stimulation by antigen.

Genetics of antibody production   How are antibody molecules coded for at the DNA level? If one B-lymphocyte produces one type of immunoglobulin, it should only require one gene for the two identical light chains and one gene for the two identical heavy chains. But the genetic information present in an *immature* B lymphocyte as it is formed in the stem cells of the bone marrow is no different from that of a neighboring B cell, yet the latter B-lymphocyte might eventually make an immunoglobulin molecule which reacts with a completely different antigen (and therefore must contain different amino acid sequences in its light and heavy chains). How can this be accounted for in genetic terms?

Recent research has shown that the variable regions of immunoglobulin heavy and light chains are more complex than that shown in Figure 16.1. Although variation in amino acid sequences is readily apparent in the variable region of different immunoglobulins, amino acid variability is especially extreme in several so-called **hypervariable regions** (Figure 16.16). It is at these hypervariable sites that combination with antigen actually occurs. Each variable region in the light and heavy chains is thought to have three hypervariable regions with a portion of the third region on the heavy chain being coded for by a distinct gene called the *D* (for "diversity") *gene,* with the first two hypervariable sites being coded for by the variable region gene itself (Figure 16.16). In addition, at the site of joining between the diversity region and the constant region DNA, there is a stretch of nucleotides about 40 bases in length called the *J* (for "joiner") region that is coded for by a distinct gene (*J gene*). Finally, the class-defining constant region of the immunoglobulin molecule is coded for by its own gene, the *C* (for "constant") *gene.* Light chains are coded for by their own variable region genes, joining region genes, and constant region genes, but apparently do not contain diversity regions as in the heavy chains (Figure 16.16).

All the genetic information required to make antibodies against a virtually unlimited pool of antigens is thought to exist in each lymphocyte as it is formed in the stem cells of the bone marrow. Each *immature* B-lymphocyte is thought to contain about 150 light-chain variable region genes and five distinct joining sequences, while approximately 80 variable region genes, six joining sequences, and about 50 diversity region genes are thought to exist for the heavy chains (Figure 16.17). These genes are not located adjacent to one another, but are separated by noncoding sequences (introns) typical of gene arrangements in eucaryotes (see Section 9.3). During maturation of lymphocytes destined to become antibody-producing (that is, B) cells, genetic recombination occurs, resulting in the construction of an **active heavy gene** and an **active light gene** that are transcribed and translated to make the heavy and light chains of the immunoglobulin molecule, respectively (Figure 16.17). When gene rearrange-

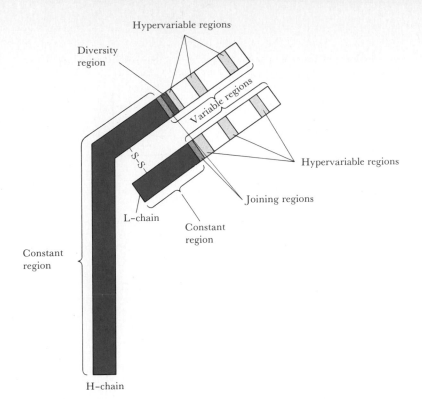

Diversity region

Hypervariable regions

Variable regions

S—S

L-chain

Constant region

Hypervariable regions

Joining regions

Constant region

Constant region

H-chain

FIGURE 16.16 Detailed structure of variable regions of immunoglobulin light and heavy chains. Only one-half of a typical immunoglobulin molecule is shown.

ments occur, the variable and constant encoding sequences not employed in the final gene product are somehow eliminated, resulting in a B cell genetically capable of producing antibody to only a single antigenic determinant. The final light and heavy gene structure of any given B cell is strictly a matter of chance rearrangement, and all possible gene combinations appear equally probable. The number of possible gene combinations is such that over 1 *billion* different antibody molecules theoretically can be made. One of the truly fascinating discoveries in biology has been the realization that this enormous number of antibodies can be genetically coded for by a relatively small amount of DNA, as a result of genetic recombination.

The discovery of gene rearrangements in lymphocytic cells has had a marked impact on biology. Biologists have always assumed that every cell of a multicellular organism is genetically identical (barring any mutation, of course) and that the differences between cell types were a function of differential gene expression. This clearly is not the case with B lymphocytes, as it is likely that several million (if not billion) genetically distinct lymphocytes exist in the mammalian body. In addition, the molecular biological dogma that states that one gene codes for the production of one polypeptide has also had to be abandoned in the case of antibody molecules. For example, although an immunoglobulin heavy chain consists of a single polypeptide, no fewer than four distinct genes are required to code for this molecule. For light chains at least three genes are required. It is not

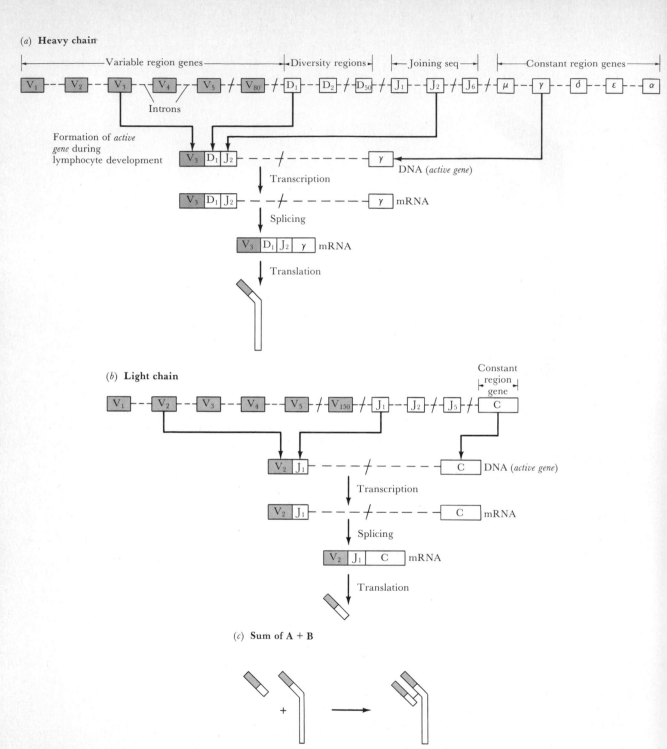

FIGURE 16.17 Gene arrangement for immature lymphocyte heavy and light chain regions, and the mechanism of active gene formation. (a) Heavy chain; (b) light chain; (c) formation of one-half an antibody molecule.

known whether lymphocytes are the only eucaryotic cell line in which gene rearrangements are commonplace, but it is hoped that an understanding of the mechanisms which direct these genetic shifts may help our understanding of the complex changes involved in cellular differentiation in animals.

Clonal selection   The selection of certain B-lymphocytes for expansion and subsequent antibody production is to some degree a function of the antigenic history of the animal. From the diverse pool of B-lymphocytes discussed above, specific B cells are stimulated by antigen to expand and form a clone of B cells, all of identical genetic makeup (Figure 16.18). This idea, known as the **clonal selection hypothesis,** was proposed long before the genetics of antibody production was known. Clonal selection predicts that many B-cell types will remain "silent" within the animal's body due to lack of exposure to the correct antigen. Presumably, however, this large pool of antibody diversity remains available for future selective expansion when reactive antigens are encountered.

FIGURE 16.18   Expansion of a particular antibody-producing cell line after stimulation with antigen—the clonal selection hypothesis.

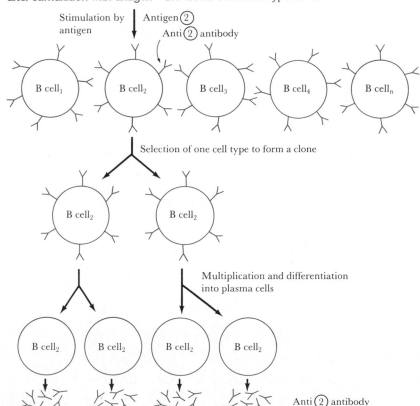

Monoclonal antibodies   A typical immune response results in the production of a broad spectrum of antibodies of varying affinities for the antigen. The antibodies directed toward a particular determinant will represent but a small proportion of the total antibody pool. If one could isolate a clone of B cells responsible for making a particular antibody of interest, however, a source of monospecific or **monoclonal antibody** would be available.

Techniques are now available for isolating and growing single B lymphocytes for indefinite periods of time. This procedure, called the **hybridoma technique,** combines the unlimited division properties of a cancer cell with the monoclonal antibody production of a single B-lymphocyte. In practice, myeloma cells (see Section 16.2) are fused with a pool of B cells removed from a mouse previously immunized with the antigen of interest; agents such as polyethylene glycol are added to the mixture to promote fusion (Figure 16.19). Since myeloma cells are themselves B-lymphocytes, a variant myeloma cell line that has lost the ability to make immunoglobulin is routinely employed in hybridoma selections. After fusing, the heterokaryotic hybrid cell undergoes a nuclear fusion and can then be injected into mice, where it grows as a myeloma tumor secreting monoclonal antibody (Figure 16.19). The immunoglobulin molecules produced by a hybridoma are characteristic of the normal B cell to which the malignant cell was fused. The resulting hybridoma population is screened

FIGURE 16.19 Monoclonal antibody production using the hybridoma technique.

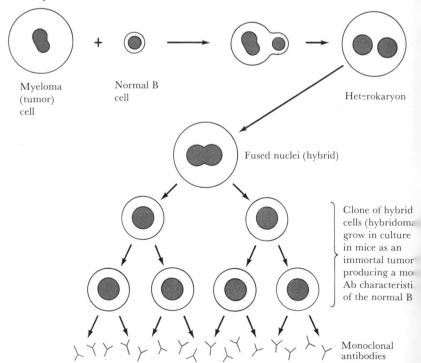

Myeloma (tumor) cell

Normal B cell

Heterokaryon

Fused nuclei (hybrid)

Clone of hybrid cells (hybridoma) grow in culture in mice as an immortal tumor producing a mo Ab characteristi of the normal B

Monoclonal antibodies

using radioimmunoassay and other techniques to identify the clone producing the monoclonal antibody of interest.

Monoclonal antibodies can be used as medical tools for the identification or destruction of specific antigens. For example, it is now known that subtle antigenic differences exist between the surfaces of malignant cells and normal cells. Monoclonal antibodies directed against cancer-related determinants might serve as effective weapons against cancer cells without harming normal cells. Monoclonal antibodies have also been described that can distinguish between human transplantation antigens, thus greatly simplifying the process of tissue matching for transplantation purposes. Monoclonal antibodies also appear promising as tools for use in routine diagnostics (for example, blood grouping analyses) and in the treatment of a variety of diseases.

16.6

Cellular Immunity, Hypersensitivities, and Autoimmunity

T cells also can act in the absence of B cells to eliminate invading microorganisms and other foreign material by releasing large molecules called **lymphokines.** These molecules are of several types. The lymphokine **chemotactic factor** attracts macrophages which ingest the invading particle, **migration inhibition factor** prevents the migration of macrophages away from the antigen, and **macrophage activating factor** stimulates macrophages to become more effective killers of microbial cells (Figure 16.20). Other effects of lymphokines

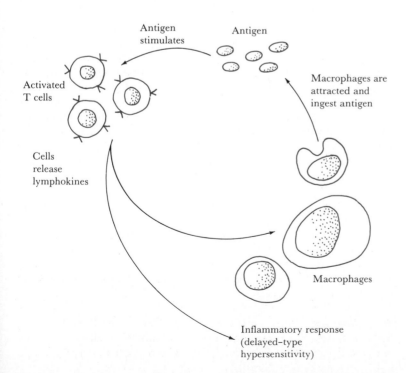

FIGURE 16.20 Cell-mediated immunity. Antigen stimulation of previously activated T cells induces the release of substances called lymphokines, which act on macrophages and other cells such as leukocytes and basophils.

resemble those previously described as inflammatory responses (Section 15.9). Because conventional antibody is not involved in these reactions, they are generally classified under the category of cell-mediated immune reactions. Cellular immunity differs from antibody-mediated immunity in that the immune response cannot be transferred from animal to animal by antibodies or serum containing antibodies, but can be transferred only by lymphocytes removed from the blood. The lymphocytes that function in transfer of cellular immunity are T cells, which have first been activated by previous antigenic stimulation. Most of the lymphocytes circulating in the blood are T cells; B cells are not generally circulating but are localized in lymphoid tissues.

The responses in cellular immunity are much slower than the responses in antibody-mediated immunity. Antibody action is a rather rapid phenomenon, quickly observable after antigenic injection in an immune animal, whereas cellular immunity develops rather slowly, and is for this reason sometimes called **delayed-type immunity.**

Skin reactions in cellular immunity The best example of a cellular immune response is the development of immunity to the causal agent of tuberculosis, *Mycobacterium tuberculosis.* This cellular immune response was first discovered by Robert Koch during his classical work on tuberculosis and has been widely studied. Antigens derived from the bacterium, when injected subcutaneously into an animal previously immunized with the same antigen, elicit a characteristic skin reaction which develops only after a period of 24 to 48 hours. (In contrast, skin reactions to antibody-mediated responses, as seen in conventional allergic reactions, develop almost immediately after antigen injection.) In the region of the injected antigen, T cells become stimulated by the antigen and release lymphokines, which attract large numbers of macrophages. The macrophages are responsible for the ingestion and digestion of the invading antigen. The characteristic skin reaction seen at the site of injection is a result of an inflammatory response arising as a result of the release of lymphokines by the activated T cells. This skin response serves as the basis for the **tuberculin test** for determining prior infection with *M. tuberculosis.*

A number of microbial infections elicit cellular immune reactions. In addition to tuberculosis, these include leprosy, brucellosis, psittacosis (all caused by bacteria), mumps (caused by a virus), and coccidioidomycosis, histoplasmosis, and blastomycosis (caused by fungi). In all of these cellular immune reactions, characteristic skin reactions are elicited upon injection of antigens derived from the pathogens, and these skin reactions can be used in diagnosis of prior exposure to the pathogen. Immunity to tumors, rejection of transplants, and some drug allergies also involve cell-mediated immune responses.

Macrophage activation A key property of activated T cells is that they can cause changes in macrophages so that the macrophages are able to kill intracellular bacteria that would normally multiply. As we have noted (Section 15.8), some bacteria are able to survive and multiply within macrophages, whereas most bacteria taken into macrophages are killed and digested. Bacteria multiplying within macrophages include *M. tuberculosis, M. leprae* (causal agent of leprosy), *Listeria monocytogenes* (causal agent of listeriosis) and various *Brucella* species (causal agents of undulant fever and infectious abortion). Animals given a moderate dose of *M. tuberculosis* are able to overcome the infection and become immune, because of the development of a T-cell-mediated immune system. Surprisingly, such immunized animals are also immune to infection by an unrelated organism such as *Listeria,* and it can be shown that macrophages in the immunized animal have been somehow changed so that they more readily kill the secondary invader. The macrophages have become activated in some way. An example of the difference in activity of normal and activated macrophages is illustrated in Figure 16.21. This nonspecific immunity to intracellular bacterial infection can be induced by various delayed-type hypersensitivity reactions (see below). Evidence suggests that when activated T cells come into contact with antigen, the T cells release soluble substances that modify some macrophages so that they are more able to kill ingested microorganisms. The macrophages that are altered by interaction with activated T cells are probably cells that are in an early stage of development. These unusual "killer" macrophages can be recognized because they contain large numbers of hydrolytic granules, suggesting that they have a more highly developed system for killing foreign cells. Certain of these granules contain hydrogen peroxide (H_2O_2), and when this compound is released it causes oxidative killing of foreign cells.

It is of some interest that macrophages not only kill foreign pathogens, but are also involved in the destruction of foreign mammalian cells. This shows up in the development of transplantation immunity, and is the major problem in the transplanation of tissues from one person to another. Tumor cells contain some specific antigens not seen on normal cells, and tumors can function in a manner

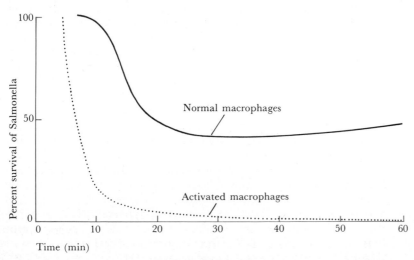

FIGURE 16.21 Experiment illustrating the activation of macrophages in cell-mediated immunity. The experiment measures the rate at which opsonized *Salmonella typhimurium* is killed by normal and activated macrophages. (From MacKaness, G. B. 1971. Cell-mediated immunity to infection. Pages 45–54 in R. A. Good and D. W. Fisher, eds. Immunobiology. Sinauer Assoc., Stamford, Conn.)

like self-inflicted transplants. There is some evidence that tumor cells are recognized as foreign and are destroyed by macrophages and cytotoxic T-lymphocytes (see below). As evidence of this, some immunity to cancer can be brought about by immunization with materials that induce macrophage and cytotoxic T-cell activity, such as *M. tuberculosis* cells. A strain of *M. tuberculosis* called BCG (bacillus of Calmette-Guerin) is nonpathogenic and has been used in attempts to induce immunity to cancer. Although clinical trials have been inconclusive, there is some evidence that repeated injections of strain BCG can lengthen the lifespan of cancer patients, with treatment in the early stages of the malignancy being much more effective than treatment in the later stages.

Cytotoxic T-lymphocytes In addition to macrophages, a class of T-lymphocytes called **cytotoxic T cells** (T-killer cells; see Section 16.5) also are involved in the destruction of foreign cells. Any cell carrying a foreign antigen, such as those introduced by incompatible tissue grafts, can be lysed by T_K cells. Contact between a T_K cell and the target cell is required for lysis, although the exact mechanism of lysis is not known. Unlike the situation in which antibody and complement work together to form holes in the cell membrane of the target cells (see Section 16.4), cells attacked by cytotoxic T cells disintegrate shortly after coming in contact with the T_K cells. Interestingly, in contrast to the nonspecific nature of activated macrophage killing, T_K cells are only effective at lysing the specific target cells containing the foreign antigen; killing of other foreign cells occurs at much lower frequencies.

Cytotoxic T cells have been implicated in the immune response to a variety of viral infections. This is due to the fact that cells infected with certain viruses undergo antigenic changes on their surfaces. This initiates the activity of a population of T_K cells which lyse the virus-infected cells. In addition, a major set of cell-surface antigens called the **histocompatibility antigens** seem to be the target of T_K cells involved in tissue graft rejection. If tissue grafts are made between two different species, or even between two strains of the same species with insufficiently similar histocompatibility antigens, cytotoxic T cells respond by killing the foreign tissue, resulting in rejection of the graft. In humans, tissue cross-matching to ensure that the major histocompatibility antigens are identical in donor and recipient is now a routine clinical procedure and is required for successful skin grafting or organ transplants.

Hypersensitivity Reactions Although immunological responses are often beneficial in protecting against infectious disease, some responses are harmful enough to cause severe symptoms or even death. These harmful responses are called *hypersensitivity reactions* and two broad classes of hypersensitivities have been identified. **Immediate-type hypersensitivity (anaphylaxis)** is an allergic response which occurs shortly (within a few minutes) after exposure to the antigen and is antibody mediated. **Delayed-type hypersensitivity** involves the activities of a special subset of T-cells called T-delayed type hypersensitivity (T_{DTH}) cells, with symptoms appearing several hours to a few days after exposure to the allergen.

Delayed-type hypersensitivity reactions occur against invasion by certain microorganisms and as a result of skin contact with certain sensitizing chemicals. The latter phenomenon, known as **contact dermatitis,** is responsible for the majority of common allergic skin disorders in humans, including those to poison ivy, cosmetics, and certain drugs or chemicals. Delayed-type hypersensitivities are mediated by T cells of the T_{DTH} and cytotoxic (T_K) subclasses. Shortly after exposure to the allergen the skin feels itchy at the site of contact, and within several hours reddening and swelling appear, indicative of a general inflammatory response. Macrophages, white blood cells, and cytotoxic T cells are attracted to the site of contact by lymphokines secreted by T_{DTH} cells, and localized tissue destruction occurs due to the lytic and phagocytic activities of these cells.

From 10 to 20 percent of the human population suffer from immediate-type hypersensitivities, involving allergic (anaphylactic) reactions to specific allergens such as pollens, animal dander, and a variety of other agents (Table 16.4). In a typical anaphylactic reaction an allergen will elicit (upon first exposure) the production of immunoglobulins of the class IgE. Instead of circulating like immunoglobulins of the class IgG or IgM, IgE molecules tend to become attached via their cell binding fragment (see Figure 16.2) to the surfaces of mast cells and basophils. **Mast cells** are nonmotile connective tissue cells found adjacent to capillaries throughout the body, and **basophils** are motile white blood cells (leukocytes) that make up about 1 percent of the total leukocyte population. Upon subsequent exposure to antigen, the IgE molecules attached to these cells bind the antigen which triggers the release from mast cells and basophils of several allergic mediators. There is a requirement that an antigen bridge at least two IgE molecules on the cell to initiate release of these active substances. The primary chemical mediators are **histamine** and **serotonin** (both are modified amino acids), but other mediators have been characterized and most are small peptides. The release of histamine and serotonin causes dilation of blood vessels and contraction of smooth muscle which initiate the typical symptoms of the immediate-type allergic response. These symptoms include, among others, difficulty in breathing, flushed skin, copious mucus production, and itchy, watery eyes. In general, the symptoms are relatively short-lived, but once initially sensitized by an allergen an individual can respond repeatedly upon subsequent exposure to the antigen. Depending on the individual (a genetic propensity to allergy exists) the magnitude of the anaphylactic reactions may vary from mild symptoms (or none), to such severe symptoms that the individual goes into a state of **anaphylactic shock.** In humans, the latter is characterized by severe respiratory distress, capillary dilation (causing a sharp drop in blood pressure), and flushing and itching. If severe cases of anaphylactic shock are not treated immediately with large doses of adrenalin to counter smooth muscle contraction and promote breathing, death can occur.

Autoimmunity   When proteins from a foreign source are injected into a human they are recognized as foreign by the immune system and an antibody response occurs. Similarly, if human proteins are injected into another mammalian species, an antibody response against the

TABLE 16.4   Common Immediate-Type Hypersensitivities (Allergies)

Hay fever (pollen and fungal spores)
Asthma (some forms)
Hives
Arthus reaction (swelling at site of injection of soluble antigen)
Penicillin and other drug allergies
Certain food allergies
Animal dander
Mites in house dust
Serum sickness syndrome (fever, swollen lymph nodes, and rash due to injection of large amounts of foreign serum)

human proteins occurs. That is, human proteins serve as antigens in other species and elicit a typical immune response. Why is it then that humans and other animals are not constantly making antibodies against their own self constituents?

The clonal selection hypothesis (Section 16.5) assumes that clones of B and T cells destined to cooperate and make antibody against self constituents are somehow eliminated or inactivated during embryological development of the animal. This, indeed, seems to be the case for most such lymphocyte clones, and the immunological basis for this selective destruction appears to reside in the class of T-lymphocytes called T-suppressor (T_s) cells. Besides regulating the extent of antibody production (see Section 16.5), T_s cells are thought to catalyze the selective elimination or inactivation of B- and T-cell clones which make antibodies or other substances against self constituents. Nevertheless, sensitive immunological assays have occasionally detected the presence of low numbers of lymphocytes that do react with various self constituents in apparently healthy individuals. In some cases, however, the activities of these lymphocytes remain unchecked, leading to one or more of a series of immunological disorders referred to as **autoimmune diseases.**

TABLE 16.5   Some Autoimmune Diseases of Humans

Disease	Organ or Area Affected	Mechanism
Juvenile diabetes	Pancreas	Autoantibodies against surface and cytoplasmic antigens of islets of Langerhans
Myasthenia gravis	Skeletal muscle	Autoantibodies against acetylcholine receptors on skeletal muscle
Goodpastures' syndrome	Kidney	Autoantibodies against basement membrane of kidney glomeruli
Rheumatoid arthritis	Cartilage	Autoantibodies against self IgG antibodies, leading to cartilage breakdown
Hashimoto's disease	Thyroid	Autoantibodies to thyroid surface antigens
Male infertility (some cases)	Sperm cells	Autoantibodies agglutinate host sperm cells
Pernicious anemia	Intrinsic factor	Autoantibodies prevent absorption of vitamin B_{12}
Systemic lupus erythematosis	DNA, cardiolipin, nucleoprotein, blood clotting factors	Massive autoantibody response to various cellular constituents
Addison's disease	Adrenal glands	Autoantibodies to adrenal cell antigens
Allergic encephalomyelitis	Brain	Cell-mediated response against brain tissue
Multiple sclerosis	Brain	Cell-mediated and autoantibody response against central nervous system

A number of diseases have now been recognized as manifestations of the autoimmune response (Table 16.5). Depending on the specific disorder, the problem may be due to the production of harmful antibodies or, more rarely, to a cellular immune response to self constituents. **Juvenile diabetes** (insulin-dependent diabetes) and **myasthenia gravis,** for example, are autoimmune diseases that result from the production of antibodies that react with insulin-producing cells in the pancreas or antigenic determinants on skeletal muscle, respectively. **Allergic encephalomyelitis** and **multiple sclerosis,** on the other hand, appear to be due to cellular autoimmune responses involving activated T-lymphocytes which attack brain tissue. In the case of multiple sclerosis, autoantibodies to myelin proteins (the substances which surround nerve fibers) also can be demonstrated. As more becomes known about control of the immune response it is likely that certain other disorders will be recognized as autoimmune diseases. Hopefully, however, a better understanding of the immune response also will yield methods of controlling or eliminating autoimmune diseases.

16.7
Antibodies and Immunity

Although many aspects of immunological reactions do not concern immunity to infectious disease, the major role of antibodies is in protecting the animal from the consequences of infection. The importance of antibodies in disease resistance is shown most dramatically in individuals with the inherited disorder **agammaglobulinemia,** in whom antibodies are not produced because their antibody-forming cells are defective. Such individuals are unusually sensitive to bacterial diseases, and in the days before antibiotic therapy, few of them survived infancy. Antibodies defend against infectious diseases in a variety of ways, but it should be emphasized that the only direct function of antibodies is to *combine* with antigens; all other phenomena are secondary consequences of this primary reaction. In diseases whose symptoms spring from the action of an exotoxin, neutralization of the toxin by antibody (antitoxin) confers resistance to the disease, even though the antibody is not directed against the pathogen itself. Once the disease symptoms are eliminated, other immune mechanisms can be brought into play to eliminate the pathogen.

Opsonization is an important function of antibody since it makes possible the phagocytosis of pathogens that would otherwise not be engulfed. Agglutination, although probably not lethal, is important in immunity, as the agglutinated clumps of cells are readily filtered out in the reticuloendothelial system (fixed and circulating phagocytic cells) thus promoting phagocytosis. Although the bacteriolytic action of antibody can be readily demonstrated in the test tube, there is no evidence that it plays an important role in immunity.

16.8
Immunization for Infectious Disease

The induction of specific immunity to infectious diseases provided one of the first real triumphs of the scientific method in medicine, and was one of the outstanding contributions of microbiology to the treatment and prevention of infectious diseases. An animal or human being may be brought into a state of immunity to a disease in either of two distinct ways. (1) The individual may be given injections of an antigen that is known to induce formation of antibodies, which will confer a type of immunity known as **active immunity** since the individual in question produced the antibodies itself. (2) The individual may receive injections of an antiserum that was derived from another individual, who had previously formed antibodies against the antigen in question. The second type is called **passive immunity** since the individual receiving the antibodies played no active part in the antibody-producing process.

An important distinction between active and passive immunity is that in *active* immunity the immunized individual is fundamentally changed, since it is able to continue to make the antibody in question, and it will exhibit a secondary or booster response if it later receives another injection of the antigen in question. Active immunity often may remain throughout life. A *passively* immunized individual will never have more antibodies than it received in the initial injection, and these antibodies will gradually disappear from the body; moreover, a later inoculation with the antigen will not elicit a booster response. Active immunity is usually used as a *prophylactic* measure, to protect a person against future attack by a pathogen. Passive immunity is usually *therapeutic,* designed to cure a person who is presently suffering from the disease. For example, tetanus toxoid (see the following section) actively immunizes an individual against future encounters with *Clostridium tetani* exotoxin, while tetanus antiserum (antitoxin; see below) is administered to passively immunize an individual suspected of coming in contact with *C. tetani* exotoxin via growth of the organism following a penetrating wound.

Vaccination   The material used in inducing active immunity, the antigen or mixture of antigens, is known as a **vaccine.** (The word derives from Jenner's vaccination process using cowpox virus; *vacca* is Latin for "cow".) However, to induce active immunity to toxin-caused diseases, it is clearly not desirable to inject the toxin itself; to overcome this problem, many exotoxins can be modified chemically so that they retain their antigenicity but are no longer toxic. Such a modified exotoxin is called a **toxoid.** One of the common ways of converting toxin to toxoid is by treating it with formaldehyde, which blocks some of the free amino groups of the toxin. Toxoids are usually not such efficient antigens as the original exotoxin but have the advantage that they can be given safely and in higher doses. When immunization against whole microorganisms is necessary, such as for endotoxin-producing organisms, the microorganism in question may first be killed by agents such as formaldehyde, phenol, or heat, and the dead cells then injected. Endotoxin-caused diseases for which vaccines are made routinely are whooping cough and typhoid. Formaldehyde treatment is also used to inactivate viruses in preparing some vaccines, such as the Salk polio vaccine.

Immunization with live cells or virus is usually more effective than with dead or inactivated material. Often it is possible to isolate a mutant strain of a pathogen which has lost its virulence but which

TABLE 16.6   Available Vaccines for Infectious Diseases in Humans

Disease	Type of Vaccine Used
Bacterial diseases:	
Diphtheria	Toxoid
Tetanus	Toxoid
Pertussis	Killed bacteria (*Bordetella pertussis*)
Typhoid fever	Killed bacteria (*Salmonella typhi*)
Paratyphoid fever	Killed bacteria (*Salmonella paratyphi*)
Cholera	Killed cells or cell extract (*Vibrio cholerae*)
Plague	Killed cells or cell extract (*Yersinia pestis*)
Tuberculosis	Attenuated strain of *Mycobacterium tuberculosis* (BCG)
Meningitis	Purified polysaccharide from *Neisseria meningitidis*
Bacterial pneumonia	Purified polysaccharide from *Streptococcus pneumoniae*
Typhus fever	Killed bacteria (*Rickettsia prowazekii*)
Viral diseases:	
Smallpox	Attenuated strain (Vaccinia)
Yellow fever	Attenuated strain
Measles	Attenuated strain
Mumps	Attenuated strain
Rubella	Attenuated strain
Polio	Attenuated strain (Sabin) or inactivated virus (Salk)
Influenza	Inactivated virus
Rabies	Inactivated virus (human) or attenuated virus (dogs and other animals)

still retains the immunizing antigens; strains of this type are called **attenuated** strains (see Section 15.6).

A summary of vaccines available for use in humans is given in Table 16.6.

Vaccination practices   Infants possess antibodies derived from their mothers and hence are relatively immune to infectious disease during the first six months of life. It is desirable to immunize infants for key infectious diseases as soon as possible, so that their own active immunity can replace the passive immunity received from the mother. However, infants have a rather poorly developed ability to form antibodies, so that immunization is not begun until a few months after birth. As discussed in Section 16.5, a single injection of antigen does not lead to a high antibody titer; it is desirable therefore to use a series of injections, so that a high titer of antibody is developed.

The importance of immunization procedures in controlling infectious diseases is well established. Upon introduction of a specific immunization procedure into a population, the incidence of the disease often drops markedly (Figure 16.22). The degree of immunity obtained by vaccination varies greatly, depending on the individual and on the quality and quantity of the vaccine. However, rarely is life-long immunity achieved by means of a single injection, or even a series of injections, and the population of antibody-producing cells induced by immunization will gradually disappear from the body. One way in which antigenic stimulation occurs even in the absence of

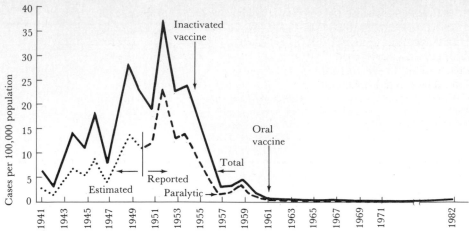

FIGURE 16.22 Case reports in the United States for polio and the consequences of the introduction of polio vaccine. (Courtesy of the Center for Disease Control, Atlanta, Ga.)

immunization is by periodic subclinical infection. A natural infection will result in a rapid booster response, leading to both a further increase in activated antibody-producing cells and to production of antibody, which will attack the invading pathogen. It is not known how long immunity will last in the complete absence of antigenic stimulation, but probably the immune period varies considerably from person to person.

Immunization procedures are not only beneficial to the individual, but are effective public health procedures, since disease spreads poorly through a population in which a large proportion of the individuals are immune (see Section 17.5).

Passive immunity   The material used in inducing passive immunity—the serum containing antibodies—is known as a **serum**, an **antiserum**, or an **antitoxin** (the last applies to a serum containing antibodies directed against a toxin). Antisera are obtained either from large immunized animals, such as the horse, or from human beings who have high antibody titers (that is, who are **hyperimmune**). The antiserum or antitoxin is standardized to contain a known amount of antibody, using some internationally agreed upon arbitrary unit of antibody titer; a sufficient number of units of antiserum must be inoculated to neutralize any antigen that might be present in the body. Sometimes the gamma-globulin fraction of pooled human serum is used as a source of antibodies. This contains a wide variety of antibodies that normal people have formed through the years by artificial or natural exposure to various antigens; it is used when hyperimmune antisera are not available.

The most common use of passive immunization is in the prevention of infectious hepatitis (due to hepatitis A virus, see Section 17.9). Pooled human gamma globulin contains fairly high titers of antibody against hepatitis A virus, due to the fact that infection with hepatitis A virus is widespread in the population. Travelers to areas where incidence of infectious hepatitis is high, such as North and tropical Africa, the Middle East, Asia, and parts of South America, may be given prophylactic doses of pooled human gamma globulin. A single dose of 0.02 ml/kg body weight should protect for up to 2 months, but for more prolonged exposures doses at repeated intervals should be given. Pooled human gamma globulin may also be of value in the therapy of infectious hepatitis, if given early in the incubation period.

A Bit of History

The discovery of immunological phenomena has been at once a major scientific and a major practical achievement. Some of the most significant advances in public health and clinical medicine have come about via immunological manipulations, and at the same time, the study of the mechanism of antibody formation has provided important insights into cellular and genetic phenomena in the animal body.

Immunology arose out of the development of procedures for immunization against infectious diseases, and two people are most closely identified with the early history, the British physician Edward Jenner and the French scientist Louis Pasteur. In 1798, Jenner first published his work showing that inoculation with material from cowpox lesions could induce immunity to smallpox infection. This was the first example of the use of an attenuated living agent to bring about immunity; the cowpox virus was sufficiently closely related to the smallpox virus to induce immunity, yet sufficiently low in virulence in humans so that frank infection did not occur. Jenner's procedure came to be known as vaccination (from *vacca*, the Latin for cow). It was almost 20 years after Pasteur began his career that he started to work on infectious disease. After some initial work on anthrax (Koch had just proved that anthrax was caused by a bacterium; see Section 1.5), Pasteur turned to fowl cholera, a bacterial disease which at that time was affecting 10 percent of the chickens in France. He isolated the causal agent and was able to maintain its virulence by periodic passage through chickens, but found that if the organism was cultured for a while in the laboratory it lost its virulence. Being aware of Jenner's work, he concluded that the nonvirulent cultures were attenuated and had the same relationship to the virulent cultures that cowpox virus had to smallpox virus. He was able to show that such attenuated cultures could be used to vaccinate chickens and confer immunity to infection with the most virulent cultures. Pasteur realized that he had derived a general principle for immunization, and he quickly turned to another infectious disease where immunization could be assumed to have more practical benefits, anthrax. He developed an attenuated culture to anthrax which could be used as a vaccine, and he used this attenuated culture in a dramatic public experiment in 1881 at Pouilly-le-fort. On 5 May 1881, he vaccinated 24 sheep, 1 horse, and 6 cows, and on 17 May they were revaccinated. Then on 31 May all the vaccinated animals plus 29 unvaccinated animals (24 sheep, 1 horse, 4 cows) were inoculated with a virulent culture. Boldly, Pasteur invited a large number of government officials, scientists, veterinarians, and journalists (including a newspaper reporter from the London *Times*) to view the results. The visitors arrived on 2 June, 2 days after the inoculation with virulent organisms. The vaccinated animals appeared completely healthy, whereas 20 of the unvaccinated sheep and the horse had died. Two other unvaccinated sheep died before the eyes of the spectators, and the last died before the end of the day. The experiment was thus a complete success and demonstrated to the world the importance of this procedure.

Pasteur's final work was with rabies virus, and he developed inactivated strains which could not only be used to induce immunity, but could also be used prophylactically to protect individuals who had succumbed to bites from rabid animals. In 1881, Pasteur and his co-workers found that the rabies virus could be grown in rabbits and that the spinal cord and brain tissue of inoculated rabbits served as an abundant source of active virus. A colleague of Pasteur's, Emile Roux, initiated experiments on the survival of rabies virus in dried rabbit spinal cords. During experiments to test the infectivity of dried rabies virus preparations, Pasteur observed that spinal cords dried for 14 days contained no active rabies virus. Shorter drying periods resulted in infective preparations, and Pasteur noted that virulence of the rabies virus was inversely proportional to the length of time the spinal cord had been dried. In additional inoculation tests, Pasteur found that rabbits inoculated with the inactive 14 day-old spinal cord preparation and given daily injections of spinal cord material starting with the 13 day-old preparation through to the 3 day-old preparation (which was highly infectious) did not develop rabies, but had now become immune. In a dramatic experiment, Pasteur then demonstrated the medical relevance of his findings by successfully inoculating and saving the life of a young boy, Joseph Meister, who had been badly bitten by a rabid dog.

By drying the rabies virus Pasteur had reduced its virulence, presumably through the inactivation of some key viral component(s). The viral particles retained their immunogenicity, however, and elicited in the animal an immune response capable of inactivating infectious rabies virus. Although Pasteur's inactivated virus would not be termed an attenuated strain in the usual sense of the word, rabies vaccine preparations used to vaccinate domestic animals today are attenuated virus suspensions obtained from successive passage of the virus through chick embryo tissue cultures. Humans bitten by a suspected rabid animal, however, are given a series of 14 to 21 injections of rabies virus preparations of increasing virulence in a fashion not unlike the early experiments of Pasteur. A recently approved vaccine, grown in human cell culture, has reduced the number of injections necessary to five or six.

The discovery of a mechanism for immunity via antibody formation was first made by Emil von Behring and Shibasaburo Kitasato in 1890, as a result of studies on the action of tetanus and diphtheria toxins. These men showed that as a result of immunization, substances appeared in the blood that could neutralize exotoxins, and that the neutralization was a specific phenomenon. Thus, animals immunized to diphtheria toxin contained a substance which neutralized this toxin but not tetanus toxin, and vice versa. Kitasato and von Behring showed that the substance was present in either blood or serum, was fairly stable, and could be used to prevent disease symptoms if injected into another animal. This discovery immediately opened up the possibility of specific therapy for diseases through the injection of immune serum, as well as the possibility of prevention of disease through the induction of specific antibody production in a potential host. Von Behring went on to develop this practical side with diphtheria antitoxin, and Paul Ehrlich (whom we have already mentioned, page 232, as one of the founders of chemotherapy) went on to develop the theoretical side of immunology, attempting to come to grips with the fascinating question of antibody specificity. Although it was many years before immunology as a discipline was placed on a firm footing (indeed, if there is a golden age of immunology, it is the present), these discoveries around the turn of the twentieth century were to have major impact on therapy and prevention of infectious disease. As an indication of the importance of immunology, von Behring was awarded the first Nobel prize in 1901, and Paul Ehrlich was awarded the Nobel prize (jointly with Elie Metchnikoff, the discoverer of phagocytosis) in 1908.

In the diagnosis of an infectious disease, isolation of the pathogen is not always possible, and an alternative is to measure antibody titer to a suspected pathogen. The principle here is that if an individual is infected with the suspected pathogen, the antibody titer to that pathogen should be elevated. Antibody titer can be measured by agglutination, precipitation, or immobilization methods, or by complement fixation, depending upon the situation. The general procedure is to set up a series of dilutions of the serum (usually twofold dilutions: $1:2, 1:4, 1:8, 1:16, 1:32$, and so on) and to determine the highest dilution at which the serological reaction occurs.

It should be emphasized that a single measure of antibody titer does not indicate active infection. Many antibodies remain at high titer for long times after infection, so that in order to establish that an acute illness is due to a particular agent, it is essential to show a *rise* in antibody titer in successive samples of serum from the same patient. Frequently, the antibody will be low during the acute stage of the infection and rise during convalescence (Figure 16.23). Such a rise in antibody titer is the best indication that the illness was due to the suspected agent. Measurement of such a rise in antibody titer is also useful in diagnosis of infectious diseases of a rather chronic nature, such as typhoid fever and brucellosis. In some cases, however, the mere presence of antibody may be sufficient to indicate infection.

Immunology in the Diagnosis of Infectious Disease

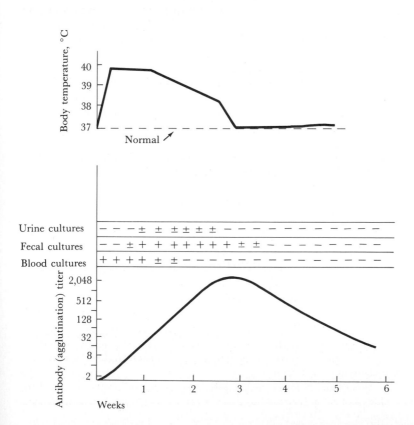

FIGURE 16.23 The course of infection in a typical untreated typhoid fever patient. Measurement of body temperature provides a measure of the course of clinical symptoms. The antibody titer was measured by determining the highest dilution (two-fold series) causing agglutination of a test strain of *Salmonella typhi*. Presence of viable bacteria in blood, feces, and urine was determined from periodic cultures. Note that the pathogen clears from the blood as the antibody titer rises, and clearance from feces and urine requires longer time. Body temperature gradually drops to normal as the antibody titer rises. The data given do not represent a single patient, but are a composite of the picture seen in large numbers of patients.

TABLE 16.7 Some Clinical Immunological Procedures for Diagnosis of Infectious Disease

Pathogen or Disease	Antigen	Serological Procedure*
Streptococcus (group A)	Streptolysin O (exotoxin)	Neutralization of hemolysis
	DNase (extracellular protein)	Neutralization of enzyme
Neisseria meningitidis	Capsular polysaccharide	Passive hemagglutination (*N. meningitidis* polysaccharide adsorbed to red cells)
	N. meningitidis cells	Indirect fluorescent antibody (see Figure 16.9)
Salmonella	O or H antigen	Agglutination (Widal test)
Vibrio cholerae	O antigen	Agglutination
		Bactericidal (in presence of complement)
Brucella	Cell wall antigen	Agglutination
Corynebacterium diphtheriae	Toxin	Skin test (Schick test)
Mycobacterium tuberculosis	Tuberculin (partially purified bacterial proteins, PPD)	Skin test (Tuberculin test)
Syphilis (*Treponema pallidum*)	Cardiolipin-lecithin-cholesterol	Flocculation [Veneral Disease Research Laboratory (VDRL) test]
	Motile *T. pallidum* cells	Immobilization [*Treponema pallidum* immobilization (TPI) test]
	Killed *T. pallidum* cells	Indirect fluorescent antibody (FTA) (see Figure 16.9)
Rickettsial diseases (Q fever, typhus, Rocky Mountain Spotted fever)	Killed rickettsial cells	Complement fixation or cell agglutination tests
Influenza virus	Influenza virus suspensions	Complement fixation test
	Nasopharynx cells containing influenza virus	Immunofluorescence

*Except for the skin tests, the serum of the patient is assayed for antibody against the specific antigen by the methods shown.

This is the case for a pathogen which is quite rare in a population, so that the presence of antibody is sufficient to indicate that the individual has experienced an infection. In these cases, some information can be gotten even if only a single serum sample can be obtained.

Another point to emphasize is that not all infections result in formation of systemic antibody. If a pathogen is extremely localized in its action, there may be little induction of an immunological response and no rise in antibody titer, even if the pathogen is proliferating profusely at its site of infection. It should be also obvious that presence of antibody in the serum may have been due to vaccination. In fact, measurement of rise in antibody titer during vaccination is one of the best ways of indicating that the vaccine being used is effective.

Some of the most common immunological diagnostic procedures are outlined in Table 16.7.

16.10

Summary

In this chapter we have learned about the specific mechanisms for the formation of **antibodies** against foreign macromolecules, and how antibody formation is important in the development of resistance to infectious disease. We have also learned that some immune responses are not associated with formation of antibodies, but with changes in cellular properties that can be generally included under the category of cell-mediated immunity.

A variety of antigen-antibody reactions are known, depending on the nature of the antigen and antibody and on the conditions for reaction. **Neutralization** of microbial toxins can occur when a specific antibody combines with the toxin molecule. **Precipitation** occurs when a bivalent antibody combines in appropriate molecular proportions with a bivalent antigen. If the antigen is present on the surface of cells, reaction of these cells with bivalent antibody can result in the formation of cell clumps, a phenomenon called **agglutination.** If the antibody is directed to flagellar antigens, **immobilization** of motile cells may occur. Antibodies can be rendered fluorescent by conjugation with fluorescent dye, and the attachment of such **fluorescent antibodies** to antigens on cells can be detected by use of the fluorescence microscope. Antibodies also can render microbial cells susceptible to phagocytosis, a phenomenon called **opsonization. Radioimmunoassay** and the **ELISA technique** are extremely sensitive methods of detecting specific antigens or antibodies and have considerable clinical significance.

An important complex in serum is **complement,** a series of enzymes in normal blood serum that are activated as a result of the occurrence of an antigen-antibody reaction. Complement is involved in cell killing and lysis, in the chemotactic attraction of phagocytes, and in the opsonization reaction. During antibody-antigen reactions involving complement, the complement components are changed so that they are no longer active, complement thus appearing to be used up during the reaction. This phenomenon is called **complement fixation,** and a measurement of complement fixation is a very sensitive way of indicating that an antigen-antibody reaction has occurred.

Five major classes of immunoglobulin molecules are known: **IgG, IgM, IgA, IgD,** and **IgE.** Each immunoglobulin molecule contains two short (light) and two long (heavy) polypeptide chains. Each class of immunoglobulin is defined by a sequence of amino acids found in the carboxy terminus of the heavy chain called the **constant regions.** The antigen-binding sites **(variable regions),** on the other hand, contain highly diverse amino acid sequences (even for immunoglobulin's of a single class), because a specific sequence defines a unique three-dimensional antigen binding site. The mechanism by which cells of the animal body are induced to form specific antibodies is now fairly well understood. The cells involved in antibody formation are **lymphocytes,** and two kinds are recognized, called B and T cells. The B cells interact with T cells and **macrophages** to generate a **humoral** (antibody) response. After contacting antigen, B cells differentiate into **plasma cells,** which are the main antibody-producing cells of the body, or into **memory cells,** which are very long-lived cells that differentiate into plasma cells upon subsequent encounter with antigen. A special subset of T cells **(T-helper cells)** serve to help the B cells produce antibody, but other subsets of T-lymphocytes play a prime role in cell-mediated immunity. Each B-lymphocyte is genetically programmed to produce antibody to one antigenic determinant only. This is the result of random genetic rearrangements that produce one active heavy and one active light chain gene per B cell. In the absence of antigenic stimulation, a given B-cell line may persist only at a low population level, but when an antigen complementary to the antibody produced by that cell line is introduced, **clonal selec-**

tion occurs, producing a large population of a specific B cell type. Recent cell fusion experiments joining a single B-lymphocyte to a malignant lymphoid cell have resulted in the production of monospecific or **monoclonal antibodies.**

In **cell-mediated immunity,** circulating antibody is not involved, but T cells become differentiated, and attack invading microorganisms and other foreign material by releasing large molecules called **lymphokines.** These molecules have several effects, including the attraction of macrophages, which ingest the invading particle. Another class of T cells known as **cytotoxic T cells (T-killer cells)** are also capable of destroying foreign cells, primarily by cell lysis. Cell-mediated immunity cannot be transferred from animal to animal by serum, but can be transferred only by T cells. The responses in cellular immunity are much slower than the responses in antibody-mediated immunity, and for this reason are sometimes called delayed-type immunity. Delayed-type immunity is the major factor in development of resistance to tuberculosis, leprosy, brucellosis, and to some viral and fungal diseases.

In all of these cellular immune reactions, characteristic skin reactions called **hypersensitivity reactions** are elicited upon injection of antigens derived from the pathogens, and these skin reactions can be used in diagnosis of prior exposure to the pathogen. The best example of this type of diagnostic procedure is the tuberculin test. **Immediate-type hypersensitivity (allergy)** is mediated by antibodies of the class IgE and not by cellular immune reactions. **Autoimmunity** is a severe malfunctioning of the immune response wherein self constituents, normally not recognized as antigens, serve to generate a humoral or cellular immune response. From an understanding of immune responses, a variety of procedures have been developed for the immunization of individuals for infectious disease. **Vaccination** is a procedure in which an individual is treated with a killed or attenuated organism, or an antigen derived from an organism, to induce antibody formation. Vaccination is one of the most important measures for the control of infectious diseases and has major public health significance, as will be discussed in the next chapter.

Supplementary Readings

Bier, O. G., W. Dias da Silva, D. Götze, and I. Mota. 1981. Fundamentals of immunology. Springer-Verlag. New York. A detailed treatment of the fundamental principles of immunology. Well illustrated.

Davis, B. D., R. Dulbecco, H. N. Eisen, and H. S. Ginsberg. 1980. Microbiology, 3rd ed. Harper & Row, Hagerstown, Md. Section 3 (Chapters 15–23) of this text provides a thorough treatment of immunology, with special emphasis on the molecular and cellular aspects of antibody formation.

Langer, W. L. 1976. Immunization against smallpox before Jenner. Sci. Am. 234:112–117. An interesting discussion of the early history of vaccination.

Leder, P. 1982. The genetics of antibody diversity. Sci. Am. 246:102–115. A concise review of the fascinating events involved in the rearrangement of genes in B-lymphocytes.

Roitt, I. 1980. Essential Immunology, 4th ed. Blackwell Scientific, Oxford. Probably the best short treatment of immunology, with primary emphasis on general phenomena of antibody formation and activity. Only a brief treatment of infectious diseases.

Rose, N. R. 1981. Autoimmune diseases. Sci. Am. 244:80–103. Emphasis on the immunology of autoimmune diseases rather than on the clinical syndromes.

Sites, D. P., J. D. Stobo, H. H. Fudenberg, and **J. V. Wells,** eds. 1982. Basic and clinical immunology, 4th ed. Lange Medical Publications, Los Altos, California. An advanced text/laboratory manual of immunological techniques useful in the clinical laboratory.

Smith, H. R., and **A. D. Steinberg.** 1983. Autoimmunity—A perspective. Annu. Rev. Immunol. 1:175–210. An overview of autoimmune diseases.

Yelton, D. E. and **M. D. Scharff.** 1980. Monoclonal antibodies. Am. Sci. 68: 510–516. A brief description of the methods involved in the production of hybridomas.

17

Epidemiology and Environmental Microbiology

In the two preceding chapters we have considered infectious diseases as they occur in isolated individuals. However, animals and humans do not live isolated but in populations; and when we consider infectious diseases in populations, some new factors arise. The study of infectious disease in populations is part of the field of **epidemiology.**

To continue existing in nature the pathogen must be able to grow and reproduce. For this reason, an important aspect of the epidemiology of any disease is a consideration of how the pathogen maintains itself in nature. In most cases the pathogen cannot grow outside the host, and if the host dies, the pathogen will also die. Pathogens that kill the host before they are transmitted to a new host would thus become extinct. This raises the question of why pathogens occasionally kill their hosts. Actually, a well-adapted parasite lives in harmony with its host, taking only what it needs for existence, and it causes only a minimum of harm. Serious host damage most often occurs when new races of pathogens arise for which the host has not developed resistance, or when the resistance of the host changes because of nutritional or other environmental factors. Pathogens are selective forces in the evolution of the host, just as hosts are selective forces in the evolution of pathogens. When equilibrium between host and pathogen exists, both are able to live more or less harmoniously together.

The dispersal and survival of pathogens in the environment are major aspects of the study of **environmental microbiology.** The environmental microbiologist is concerned with developing methods for tracing the spread of pathogens. By use of these methods, the microbiologist can assess the microbial safety of various agents by determining which microbes are passed from person to person: for example, air, water, or food. A major part of this chapter will deal with assessing the microbiological safety of the human environment.

Infection can be recognized in a population in one of three ways:

1. Obvious clinical symptoms of a disease known to be caused by a specific pathogenic agent are detected. Persons suffering from the disease present themselves to physicians or are found in health surveys and are diagnosed. However, as we have emphasized in previous chapters, infection and disease are not synonymous; not all cases of infection lead to frank disease symptoms.
2. Immunological surveys of populations are carried out to look for evidence of (a) circulating antibody or (b) hypersensitivity reactions to specific pathogens. Tuberculin testing (Table 16.7) is a good example of this approach. As noted earlier, however, immunological reactivity to a pathogen or one of its products does not indicate active infection, but only that at some time in the past the individual was either infected or immunized for that pathogen.
3. Routine surveys using microbiological methods (culture, direct microscopy) are performed to detect the presence of a pathogen in the population. Although this is theoretically the best way of demonstrating the presence of infection, it is complicated, expensive, and subject to considerable uncertainty, since clinical microbiological procedures may not always be suitable for detection of pathogens when pathogen numbers are low.

Variation in severity of disease   As detection of clinical symptoms is the main way in which data on infection of populations are obtained, it is important to note that diseases vary markedly in the severity of their manifestations. The characteristics of three quite different types of infectious disease are illustrated in Figure 17.1. The distribution of manifestations in this figure is given in several categories: inapparent infections, mild infections, moderate infections, severe (but nonfatal) infections, and fatal infections. In most cases, only when moderate, severe, or fatal infections exist will the presence of the infection be realized.

Tuberculosis is an example of an infectious disease in which most infections are inapparent, so that the extent of the disease in populations is not directly realized. In unvaccinated populations, measles is an infection in which frank symptoms develop in most infected individuals, and the presence of the disease in the population is easily recognized. Rabies, on the other hand, is such a severe disease that any infection is likely to be realized. Note that in this figure, we are not concerned about the frequency of infection in the population. Rabies, despite its severity, is very uncommon in human populations.

Mortality and morbidity   In practice, the prevalence of disease is determined by obtaining statistics of illness and death. From these data a picture of the public health in a population can be obtained. The population under consideration could range in size from the

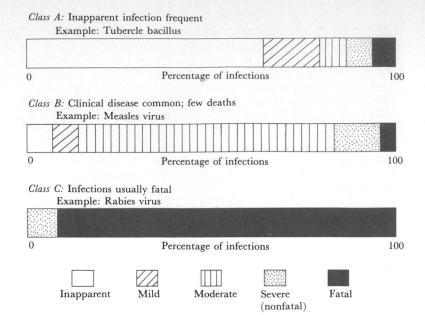

Class A: Inapparent infection frequent
     Example: Tubercle bacillus

0   Percentage of infections   100

Class B: Clinical disease common; few deaths
     Example: Measles virus

0   Percentage of infections   100

Class C: Infections usually fatal
     Example: Rabies virus

0   Percentage of infections   100

Inapparent   Mild   Moderate   Severe   Fatal
                                  (nonfatal)

FIGURE 17.1 Distribution of clinical severity for three classes of infection (not to scale.) (From Mausner, J. S., and A. K. Bahn. 1974. Epidemiology, W. B. Saunders Co., Philadelphia.)

total global population of humans down to the population of a localized region of a country or district. Public health varies from region to region, as well as with time; thus a picture of public health at a given moment provides only an instantaneous picture of the situation. By continuing to examine health statistics over many years, it is possible to assess the value of various public health protective measures in influencing the incidence of disease.

Mortality expresses the incidence of *death* in the population. Infectious diseases were the major causes of death in 1900, whereas currently they are of much less significance, and diseases such as heart disease and cancer are of greater importance. However, the current situation could rapidly change if a breakdown in public health measures were to occur.

Morbidity refers to the incidence of *disease* in populations and includes both fatal and nonfatal diseases. Clearly, morbidity statistics will more precisely define the health of the population than mortality statistics, since many diseases that affect health in important ways have only a low mortality. The major causes of illness are quite different from the major causes of death. Major illnesses are acute respiratory diseases (the common cold, for instance) or acute digestive system conditions, which are generally due to infectious causes.

17.2

Reservoirs

Reservoirs are sites in which viable infectious agents remain alive and from which infection of individuals may occur. Reservoirs may be either animate or inanimate. Some pathogens are primarily saprophytic (living on dead matter) and only incidentally infect and cause disease. Examples are *Clostridium tetani* (the causal agent of tetanus) and *C. perfringens* (one of the causal agents of gas gangrene), whose normal habitats are the soil. Infection by these bacteria is not essential for their continued existence and can be considered only an accidental event; if all susceptible hosts died, these organisms would still be able to survive in nature.

More critical, from the viewpoint of epidemiology, are pathogens whose only reservoirs are living organisms. In these cases, the reservoir is an essential component of the cycle by which the infectious agent maintains itself in nature. Some infections occur only in humans, so that maintenance of the cycle involves person to person transmittal. This type of pathogen cycle is the most common for the following diseases: most viral and bacterial respiratory diseases, venereal diseases, staphylococcal and streptococcal infections, diphtheria, typhoid fever, and mumps.

A large number of infectious diseases that occur in humans also occur in animals. Diseases which occur primarily in animals but are occasionally transmitted to people are called **zoonoses.** Because public health measures for animal populations are much less highly developed, infection rate for these diseases will be much higher in animals, and animal to animal transmission will be the rule. However, occasionally transmission will be from animal to human. It would be very unlikely, in such diseases, for transmission to also occur from person to person. Thus the maintenance of the pathogen in nature depends on animal to animal transfer. Diseases in this category include bovine tuberculosis (cows), brucellosis (cows, pigs, goats), anthrax (cows, sheep), leptospirosis (rodents), rabies (dogs, bats, foxes, wild rodents), and tularemia (wild rodents, rabbits). It should be obvious that control of the disease in the human population in no way eliminates these diseases as public health problems. Indeed, more effective human control can generally be achieved through elimination of the disease in the animal reservoir. Marked success has been achieved in the control of two of the diseases that occur primarily in domestic animals, bovine tuberculosis and brucellosis. Pasteurization of milk was also of considerable importance in the prevention of the spread of bovine tuberculosis to humans, since milk was the main vehicle of transmission.

Certain infectious diseases have more complex cycles, involving an obligate transfer from animal to human to animal. These are organisms with a complex life cycle and are either metazoans (for example, tapeworms) or protozoa (for example, malaria, see Section 17.12). In such cases, control of the disease in the population can be either through control in humans or in the alternate animal host.

Carriers A **carrier** is an infected individual not showing obvious signs of clinical disease. Carriers are potential sources of infection for others, and are thus of considerable significance in understanding the spread of disease. Carriers may be individuals in the incubation period of the disease, in which case the carrier state precedes the development of actual symptoms. Carriers of this sort are prime sources of infectious agents for respiratory infections, since they are not yet aware of their infection and so will not be taking any precautions against infecting others. Such persons can be considered **acute carriers,** because the carrier state will only last for a short while. More significant from the public health standpoint are chronic carriers, who may remain infected for long periods of time. **Chronic carriers** may be either individuals who had a clinical disease and recovered, or they may have had an infection that remained inapparent throughout.

Carriers can be identified by routine surveys of populations, using cultural, radiological (chest X ray), or immunological techniques. In general, carriers are only sought among groups of individuals who may be sources of infection for the public at large, such as food handlers in restaurants, groceries, or processing plants. Two diseases in which carriers have been of most significance are typhoid fever and tuberculosis, and routine surveys of food handlers for inapparent cases of these diseases are sometimes made. (Foodborne diseases will be considered in Section 17.10.)

A Bit of History

The classic example of a chronic carrier was the woman known as "Typhoid Mary," a cook in New York City and Long Island in the early part of this century. Typhoid Mary (her real name was Mary Mallon) was employed in a number of households and institutions, and as a cook she was in a central position to infect large numbers of people. Eventually she was tracked down after an extensive epidemiological investigation of a number of typhoid outbreaks revealed that she was the likely source of contamination. When her feces was examined bacteriologically, it was found that she had practically a pure culture of the typhoid bacterium, *Salmonella typhi.* She remained a carrier for many years, probably because her gallbladder was infected, and organisms were continuously being excreted from there into her intestine. Public health authorities offered to remove her gallbladder but she refused the operation, and to prevent her from continuing to serve as a source of infection she was imprisoned. After almost 3 years in prison, she was released on the pledge that she would not cook or handle food for others and that she was to report to the health department every three months. She promptly disappeared, changed her name, and cooked in hotels, restaurants, and sanitariums, leaving behind a wake of typhoid fever. After 5 years she was captured as a result of the investigation of an epidemic at a New York hospital. She was again arrested and imprisoned and remained in prison for 23 years. She died in prison in 1938, 32 years after epidemiologists had first discovered she was a chronic typhoid carrier.

17.3

Mechanisms of Transmission

A key job of the epidemiologist is to gather information relating to the extent and distribution of disease throughout a population. Epidemiologists are not only interested in the **prevalence** of a disease, that is, the total number of persons infected at any one time, but they also must follow the **incidence** of a disease by correlating geographical, seasonal, and age-group distribution of a disease with possible modes of transmission. A disease limited to a restricted geographical location, for example, may suggest a particular vector, as will be described later in this chapter for bubonic plague and malaria (see Section 17.12). A marked seasonality to a disease is often indicative of certain modes of transmission, such as in the case of chickenpox or measles, where the number of cases jumps sharply when children enter school and come in close contact (see Figures 17.4 and 17.10). Finally, the age-group distribution of a disease can be an important epidemiological statistic, and frequently suggests or eliminates particular routes of transmission.

Different pathogens have different modes of transmission, which are usually related to the habitats of the organisms in the body. For instance, respiratory pathogens are generally airborne, whereas intestinal pathogens are spread by food or water. If the pathogen is to survive in nature it must undergo transmission from one host to another. Thus pathogens generally have evolved features or mechanisms that permit or ensure transmittal. Transmission involves three stages: (1) escape from the host, (2) travel, and (3) entry into a new host. We give here a brief overview of transmission mechanisms, and several of these will be discussed in detail in subsequent sections.

Direct transmission   Some pathogens are so sensitive to environmental influences that they are unable to survive for significant periods of time away from the host. Such pathogens are transmitted from host to host by direct contact. The best examples of pathogens transmitted in this way are those causing venereal disease, such as *Treponema pallidum* (syphilis) and *Neisseria gonorrhoeae* (gonorrhea). These agents are extremely sensitive to drying and do not survive away from the body even for a few moments. Intimate person to person contact, such as by kissing or sexual intercourse, provides a direct means for the transmission of such pathogens. However, it should be obvious that such intimate transfer can only occur if the viable pathogen is present on the transmitting person at the body site that comes in direct contact with that of the recipient. Thus the pathogens causing venereal diseases live in genitalia, mouth, or anus, since these are the only sites involved in intimate person to person contact.

Direct contact is also involved in the transmittal of skin pathogens, such as staphylococci (boils and pimples) or fungi (ringworm). Because these pathogens are relatively resistant to environmental influences such as drying, intimate person to person contact is not the only means of transmission, as it is with the venereal diseases. It could also be considered that many respiratory pathogens are transmitted by direct means, since they are spread by droplets emitted as a result of sneezing or coughing, and many of these droplets do not remain airborne for long. Transmission, therefore, almost always requires close, although not necessarily intimate, person to person contact.

Indirect transmission   Indirect transmission can occur by either living or inanimate means. Living agents transmitting pathogens are called **vectors;** they are generally arthropods (for example, insects, mites, or fleas) or vertebrates (for example, dogs, rodents).

Arthropod vectors may not actually be themselves infected with the agent, but merely carriers of the agent from one host to another. Large numbers of arthropods obtain nourishment by biting, and if the pathogen is present in the blood, the arthropod vector will receive some of the pathogen and may transmit it when biting another individual. In some cases, the pathogen actually replicates in the arthropod, which is then considered an alternate host, and such replication leads to a buildup of the inoculum, increasing the probability that a subsequent bite will lead to infection.

Inanimate agents involved in transmission are generally called **vehicles.** Vehicles can be objects such as bedding, toys, books, or surgical instruments, which come in contact with individuals only casually (such materials are sometimes collectively called **fomites**). More significant vehicles are food and water, which are actively consumed in fairly large amounts. Food- and waterborne diseases are primarily intestinal diseases, the pathogen leaving the body in fecal material, contaminating food or water via improper sanitary procedures, and then entering the intestinal tract of the recipient during ingestion. Because food- and waterborne diseases are some of those which are most amenable to control by public health measures, we shall discuss them in some detail later in this chapter.

17.4

Epidemics

A disease is said to be **epidemic** when it occurs in an unusually high number of individuals in a community at the same time (a **pandemic** is a worldwide epidemic). In contrast, an **endemic** disease is one which is constantly present in a population. In an endemic disease, the pathogen may not be highly virulent, or the majority of the individuals may be immune, so that the disease incidence is low. However, as long as the endemic situation lasts, there will remain a few individuals who may serve as reservoirs of infection.

Common-source and propagated epidemics Two major types of epidemics can be distinguished: common source and propagated (person to person). These two types are contrasted in Figure 17.2. A **common-source epidemic** arises as the result of infection (or intoxication) of a large number of people from a contaminated source, such as food or water. Usually such contamination occurs because of a malfunction in some aspect of the distribution system providing food or water to the population. The shape of the epidemic curve for a common-source outbreak is characterized by a sharp rise to a peak, since a large number of individuals succumb within a relatively brief period of time. The common-source outbreak also declines rapidly, although the decline is less abrupt than the rise. Cases continue to be reported for a period of time approximately equal to the duration of one incubation period of the disease.

In a **propagated** or **person to person epidemic,** the curve of disease incidence shows a relatively slow, progressive rise (Figure 17.2) and a gradual decline. Cases continue to be reported over a period of time equivalent to several incubation periods of the disease. The

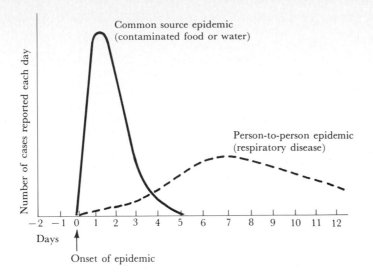

FIGURE 17.2 The shape of the epidemic curve helps to distinguish the likely origin. In a common-source outbreak, such as from contaminated food or water, the curve is characterized by a sharp rise to a peak, with a rapid decline, which is less abrupt than the rise. Cases continue to be reported for a period approximately equal to the duration of one incubation period of the disease. In a person-to-person epidemic, the curve is characterized by a relatively slow, progressive rise, and the cases will continue to be reported over a period equivalent to several incubation periods of the disease.

epidemic may have been initiated by the introduction of a single infected individual into a susceptible population, and this individual has infected one or a few people in the population. Then the pathogen replicated in these individuals, reached a communicable stage, and was transferred to others, where it replicated and again became communicable.

17.5

Quantitative Aspects of Propagated Epidemics and the Concept of Herd Immunity

To understand the kinetics of propagated epidemics, consider the diagram in Figure 17.3.* The number of individuals infected each day, and the cumulative numbers of infected and recovered individuals (possibly immune) are graphed in this figure also. The first graph presents the kind of data that would actually be observed in an epidemic, since all that is directly known by control personnel are the number of individuals who are sick at any time. If the cumulative number of individuals who have shown illness is graphed, however, the curve markedly resembles a bacterial growth curve, with an exponential increase followed by a stationary phase. If the number of individuals who have recovered from the disease is graphed, the duration of illness is given by the horizontal distance between the two lines (Figure 17.3).

Mathematical expression of propagated epidemics and the occurrence of a threshold   It is useful in attempting to understand the severity and probability of occurrence of epidemics to express epidemics mathematically. The intensity of an epidemic (N') can be expressed quantitatively by the ratio of the total number of individuals involved in the epidemic, N, divided by the total number of susceptibles present in

*Figure 17.3 strikingly resembles Figure 11.9, which describes the transfer of a conjugative plasmid. As we noted in connection with Figure 11.9, the transfer of a conjugative plasmid resembles an infection process.

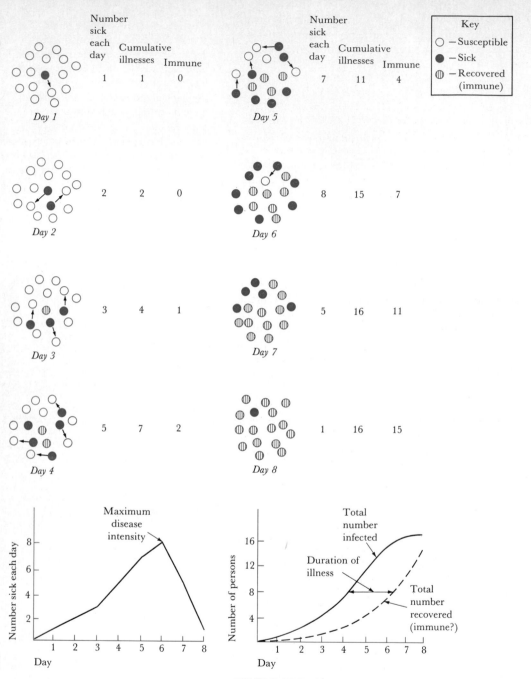

	Number sick each day	Cumulative illnesses	Immune
Day 1	1	1	0
Day 2	2	2	0
Day 3	3	4	1
Day 4	5	7	2
Day 5	7	11	4
Day 6	8	15	7
Day 7	5	16	11
Day 8	1	16	15

Key
○ — Susceptible
● — Sick
◍ — Recovered (immune)

FIGURE 17.3 Kinetics of spread of an imaginary propagated epidemic. It is assumed that the epidemic was initiated by the arrival of a single infected individual. The incubation period is assumed to be 1 day, and recovery occurs in 2 days. The number of susceptibles is assumed to be the total population at day 1 (15 persons). The number of infected and recovered are graphed in the lower part of the figure.

the population (s). Thus $N' = N/s$. Let D represent the average duration of infection, and I is a constant, which may be termed the *infectivity* of the disease (that is, the effectiveness with which the causal agent is transmitted from one person to another); the triple product sID will then be related to the intensity of the epidemic. In general, a relationship exists between the triple product sID and N' such that there is a threshold value of sID below which no epidemic at all would occur. A population may be described as subcritical to a particular infection when the triple product sID lies below a value of one. This low value can occur either because the number of susceptibles is too low, or because the infectivity of the agent is too low, or because the length of the period of communicability (D) is too short. However, the birth of new susceptibles and the loss of immunity by previously infected individuals will lead to a gradual increase in the number of susceptibles. Once the population exceeds a critical value, the probability of an epidemic increases markedly, reaching a situation where virtually all individuals can become infected.

Herd immunity   An analysis of herd immunity is of great importance in understanding the role of immunity in the development of epidemics. Because a threshold exists (see above), it can be concluded that not all individuals in a population must be immune in order to prevent infection of the population. **Herd immunity** is a concept used to explain the resistance of a group to invasion and spread of an infectious agent, due to the immunity of a high proportion of the members of the group. If the proportion of immune individuals is sufficiently high so that the triple product sID (see above) is below a threshold, then the whole population should be protected. Obviously, the fraction of resistant individuals necessary to prevent an epidemic is higher for a highly infectious agent or one with a long period of infectivity, and lower for a mildly infectious agent or one with a brief period of infectivity.

    The proportion of the population that must be immune to prevent infection of the rest of the population can be estimated from data on poliovirus immunization in the United States to be no higher than 70 percent. Thus there is essentially no polio disease in the United States, yet only 70 percent of the population age thirteen or younger have been immunized. Clearly, these immunized individuals are protecting the rest of the population. (One of the consequences of herd immunity is that the decreased incidence of the disease leads to a reduction in motivation of many individuals to become immunized; the proportion of immunized individuals decreases with time, leading again to an increase in the likelihood of disease.) For a highly infectious disease such as smallpox, the proportion of immunes necessary to confer herd immunity has been estimated to be 90 to 95 percent. A value of about 70 percent has been estimated for diphtheria, but further study of several small diphtheria outbreaks has shown that in densely settled areas a much higher proportion may have to be immunized to prevent development of an epidemic. Apparently in dense populations, person-to-person transmission can occur even if the agent is not highly infectious. In the case of diphtheria an additional

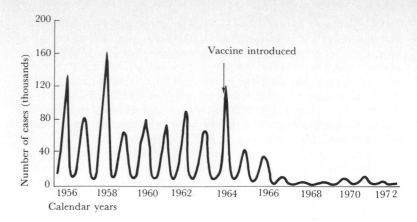

FIGURE 17.4 Cycles of disease, as illustrated by measles. Reported cases of measles by 4-week periods in the United States. The sharp decline in incidence after introduction of immunization is also seen. (From White, J. J. 1974. Recent advances in public health. Amer. Jour. Public Health 64: 939–944.)

complication arises because immunized persons can harbor the pathogen (inapparent infection) and thus act as chronic carriers, serving as potential sources of infection.

Cycles of disease The concepts of propagated epidemics and herd immunity can also explain why certain diseases occur in cycles. A good example of a cyclical disease is measles, which, in the days before the measles vaccine, occurred in a high proportion of school children. Since the measles virus is transmitted by the respiratory route, its infectivity is high in crowded situations such as schools. Upon entry into school at age five, most children would be susceptible, so that upon the introduction of measles virus into the school, an explosive propagated epidemic would result. Virtually every individual would become infected and develop immunity, and as the immune population built up, the epidemic died down. Measles showed an annual cycle (Figure 17.4), probably because a new group of nonimmune children arrived each year; the phasing of the epidemic would be related to the time of the year at which school began after the summer vacation. The consequences of vaccination for the development of measles epidemics are also illustrated in Figure 17.4.

17.6

Evolution of Host and Parasite

Our discussion of populations of hosts and parasites leads us to a consideration of the effect of each on the evolution of the other (usually referred to as **coevolution**). Parasites are elements in the evolution of host populations, just as hosts are elements in the evolution of parasites. The colonization of a susceptible, unimmunized host by a parasite may first lead to an explosive infection and an epidemic. As the host population develops resistance, however, the spread of the parasite is checked, and eventually a balance is reached in which host and parasite are in equilibrium. A subsequent genetic change in the parasite could lead to the formation of a more virulent form, which would then initiate another explosive epidemic until the host again responded, and another balance was reached.

There are numerous examples of the consequences of the introduction of virulent pathogens into susceptible human populations. Such introductions were widespread during the period 1400 to 1900 when colonization of remote parts of the world was being carried out by Europeans. The conquest of Mexico by the Spaniards, for instance, was due as much to the introduction of smallpox as to military prowess, and measles and tuberculosis had disastrous effects on Polynesian populations of the South Pacific when these islands were colonized in the nineteenth century. With time, these affected populations became resistant and achieved a sort of balance with the newly introduced pathogens, but not without a vast amount of disruption of their societies. However, such situations in humans are not amenable to research study, thus the details of evolution of host and parasite cannot be determined. Effective research can be done in wild-animal populations, and the best studied example is one in which a virus was introduced for purposes of population control, myxoma virus in the wild rabbits of Australia.

The wild rabbit was introduced into Australia from Europe in 1859, and quickly spread until it was overrunning large parts of the continent. Myxoma virus was discovered in South American rabbits, which are of a different species. In South America the virus and its hosts are apparently in equilibrium, as the virus causes only minor symptoms. However, this same virus was found to be extremely virulent to the European rabbit and almost always caused a fatal infection. The virus is spread from rabbit to rabbit by mosquitoes and other biting insects, and hence is capable of rapid spread in areas where appropriate insect vectors are present. The virus was introduced into Australian rabbits in 1950, with the aim of controlling the rabbit population. Within several months, the virus was well established in the population and spread over an area in Australia as large as all of Western Europe. The disease showed a marked seasonal pattern, rising to a peak in the summer when the mosquito vectors were present and declining in the winter. Research on the epidemiology of myxoma virus was initiated as a model of a virus-induced epidemic by Australian workers, under the direction of Professor Frank Fenner. Isolations of the virus from wild rabbits were carried out, and the virus strains isolated were characterized for virulence with laboratory rabbits. At the same time, baby rabbits were removed from their dens before infection could occur and reared in the laboratory. Then these rabbits were challenged with standard virulent strains of myxoma virus to determine their susceptibility. The changes in virulence of the virus and in susceptibility of the rabbit are illustrated by Figure 17.5.

During the early years of the epidemic, as high as 99 percent of the infected rabbits were dying. However by 1957, both the virus and the rabbit population had changed. Rabbit mortality had dropped to about 90 percent, and the virus isolated was of decreased virulence, although better able to survive effectively in the wild than was the original highly virulent virus. Improved survival in the wild was probably due to the fact that the virus did not kill its host as fast, or did not kill such a high percentage of its host, so that a reservoir was present which could serve for subsequent infections. If the virus did not cause a fatal infection in all cases, however, it could still be of significance in population control, because sick rabbits (which might subsequently recover) were easier prey to predators, such as foxes, feral cats, and eagles.

Perhaps more interesting was the change in resistance of the rabbit. In parts of Australia where the virus had been first introduced, the remaining rabbit population had been subjected to selective pressure by the virus for seven years. Because of the rapidity with which rabbits multiply, a number of generations of rabbits developed, and the surviving population had withstood at least five years of intensive selective pressure, surviving epidemics in which over 90 percent of their cohorts had been killed. As seen in Figure 17.5, within five years, the resistance of the rabbits had increased dramatically. It should be emphasized that this resistance is innate and is not due to immunological responses, for the rabbits tested had been removed from their mothers at birth and had never been in contact with the virus. Their resistance was probably inherited, due to some genetic change in the animal, which made it less likely to become infected.

As a result of introduction of myxoma virus, the rabbit population was greatly controlled, but the genetic changes in virus and host served to prevent a complete eradication of the rabbit from Australia. It has been estimated that with time, an equilibrium between rabbit and myxoma virus will be reached with the rabbit population at about 20 percent of that present before the introduction of myxoma virus. The virus was thus a major factor in population control, but could clearly never be responsible for the complete elimination of

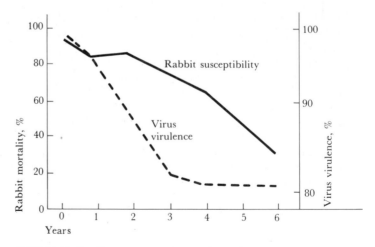

FIGURE 17.5 Changes in virulence of myxoma virus and in susceptibility of the Australian rabbit during the years after the virus was first introduced into Australia in 1950. Virus virulence is given as the average mortality in a standard laboratory breed of rabbits, for virus recovered from the field during each year. Rabbit susceptibility was determined by removing young rabbits from their dens and challenging with a virus strain of moderately high virulence, which killed 90 to 95 percent of normal laboratory rabbits. (Data from Burnet, M., and D. O. White. 1972. Natural history of infectious disease, 4th ed. Cambridge University Press, N.Y.)

rabbits in an area as large as Australia. In a smaller area, such as an island, the introduction of the virus might have resulted in eradication of the rabbit, because the initial virulence would have reduced the rabbit population to such a low level that successful breeding might not have occurred, and new migration to the island would have been improbable.

From the present point of view, the significance of the Australian experiment is that it reveals how quickly an equilibrium is reached between host and parasite. The manner in which the malaria parasite has affected biochemical evolution in humans will be discussed in Section 17.12.

17.7
Public Health Measures for the Control of Epidemics

An understanding of the epidemiology of an infectious disease generally makes it possible to develop methods for the control of the disease. **Public health** refers to the health of the population as a whole, and to the activities of public health authorities in the control of epidemics. It should be emphasized that the incidence of many infectious diseases has dropped dramatically over the past 100 years not because of any specific control methods, but because of general increases in the well-being of the population. Better nutrition, less crowded living quarters, and lighter work loads have probably done as much as public health measures to control diseases such as tuberculosis. However, diseases such as typhoid fever, diphtheria, brucellosis, and poliomyelitis owe their low incidence to active and specific public health measures. Finally, one infectious disease, gonorrhea, is much more prevalent today than it was 50 years ago. This increased incidence can be linked directly to changes in sexual behavior of the population. From an understanding of sources of infection and mechanisms of transmission, specific measures can be devised for control of diseases such as gonorrhea.

Controls directed against the reservoir   If the disease occurs primarily in domestic animals, then infection of humans can be prevented if the disease is eliminated from the infected animal population. Immunization procedures or slaughter of infected animals may be used to wipe out the disease in animals. These procedures have been quite effective in eliminating brucellosis and bovine tuberculosis from humans. Not incidentally, the health of the domestic animal population is also increased, with likely economic benefits to the farmer. However, a program for eradication of a disease in a domestic animal is expensive, and the government must compensate farmers for lost animals. In the long run, the money spent will be returned many times over in reduced medical costs. When the reservoir is a wild animal (for example, tularemia, plague, see Section 17.12) then eradication is much more difficult. Rabies is a disease that occurs both in wild and domestic animals, but is transmitted to domestic animals primarily by wild animals. Thus control of rabies can be achieved by immunization of domestic animals, although this will never lead to eradication of the disease completely. Indeed, statistics show that the ma-

TABLE 17.1   Reported Cases of Rabies in Animals,
United States 1981

Domestic Animals	Cases	Wild Animals	Cases
Cattle	460	Skunks	4480
Cats	283	Bats	858
Dogs	200	Foxes	481
Other domestic*	117	Racoons	196
		Other wild†	43
Total:	1060	Total:	6058
Total domestic and wild:	7118		
Percent domestic: 15%			
Percent wild: 85%			

*Horses, 88; sheep, 19; goats, 6; swine, 4; mule, 1.

†Arctic foxes, 15; woodchucks, 5; coyotes and badgers, 4 each; wolves
and muskrats, 3 each; bobcats, 2; ringtail civet, opossum, rabbit, mouse,
ferret, squirrel, 1 each.

From Morbidity and Mortality Weekly Report, Annual Summary, 1981;
issued October, 1982. Centers for Disease Control. Atlanta, Ga.

jority of rabies cases are in wild rather than domestic animals (Table
17.1). In England, where wild animal populations are fairly low,
rabies has been virtually eradicated by strict laws and quarantine
requirements for newly introduced animals. When humans are the
reservoir (for example, gonorrhea), then control and eradication are
much more difficult, especially if there are asymptomatic carriers.

Controls directed against transmission of the pathogen   If the organism
is transmitted via food or water, then public health procedures can be
instituted to prevent either contamination of these vehicles or to destroy the pathogen in the vehicle. Water purification methods (see
Section 17.9) have been responsible for dramatic reductions in the
incidence of typhoid fever, and the pasteurization of milk has helped
in the control of bovine tuberculosis in humans. Food protection laws
have been devised that greatly decrease the probability of transmission of a number of enteric pathogens to humans. Transmission of
respiratory pathogens is much more difficult to prevent. Attempts at
chemical disinfection of air have been unsuccessful. In Japan, many
individuals wear face masks when they have upper respiratory infections to prevent transmission to others, but such methods, although
effective, are voluntary, and would be difficult to institute as public
health measures.

Immunization procedures   Immunization has been the prime means
by which smallpox, diphtheria, tetanus, pertussis (whooping cough),
and poliomyelitis have been eliminated. As we have discussed in Section 17.5, 100 percent immunization is not necessary in order to control the disease in a population, although the percentage needed to
ensure control varies with the disease and with the condition of the
population (for example, crowding).

The application of public health measures can lead to a classic confrontation between the rights of the private individual and the welfare of the population. Many diseases could be readily controlled if known control procedures were universally applied, but marked resistance exists to universal application of public health measures, such as immunization, which directly affect the individual. Over the past several decades, the proportion of children vaccinated for diphtheria, tetanus, pertussis, and polio has been slowly decreasing, apparently because the public has become less fearful of contracting these diseases due to their very low incidence in the population. However, since none of these diseases has been eradicated from the United States (indeed, the reservoir of tetanus is the soil, so that it will never be eradicated), and with a decrease in proportion of individuals immunized, the phenomenon of herd immunity (see Section 17.5) may be overcome, and diseases such as diphtheria, pertussis, and polio could reappear in epidemic form.

Quarantine Quarantine involves the limitation of the freedom of movement of individuals with active infections to prevent spread of disease to other members of the population. The time limit of quarantine is the longest period of communicability of the disease. Quarantine must be done in such a manner that effective contact of the infected individual with those not exposed is prevented. Quarantine is not as severe a measure as strict isolation, which is used for unusually infectious diseases in hospital situations.

At one time, quarantine was required for a number of infectious diseases of childhood, such as measles, chickenpox, and mumps, and residences in which quarantined children were housed had placards affixed to the outside. Such measures were found to have little public health significance in the control of the spread of these diseases, and quarantine is no longer required, although it is still advisable as much as possible to prevent contact of infected children with other, possibly susceptible, children.

Currently, quarantine is required only for smallpox, a disease that has essentially been eradicated from all countries. (Quarantine is, however, strongly recommended for other diseases such as plaque.) Introduction of an individual infected with smallpox into many areas could currently have disastrous effects, as immunization for smallpox is no longer practiced. Thus it would be essential to quarantine such an infected individual for the period of communicability.

A summary of the control measures implemented to prevent epidemics is shown in Figure 17.6.

17.8

Respiratory Infections and Airborne Transmission

Air usually is not a suitable medium for the growth of microorganisms; those organisms found in it derive from the soil, water, plants, animals, people, or other sources. In outdoor air, soil organisms predominate. Microbial numbers indoors are considerably higher than those outdoors, and the organisms are mostly those commonly found in the human respiratory tract.

1. Eliminating the reservoir

Reservoir

Insects
Poor sanitation
Contaminated food
or water

Vectors

Susceptibles

2. Eliminating the vectors

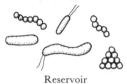

Reservoir

Insects
Poor sanitation
Contaminated food
or water

Vectors

Susceptibles

3 . Immunizing the susceptibles

Reservoir

Insects
Poor sanitation
Contaminated food
or water

Vectors

Susceptibles

4. Quarantine

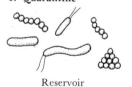

Reservoir

Insects
Poor sanitation
Contaminated food
or water

Vectors

Isolation
of diseased
individuals

FIGURE 17.6 Summary of control measures for the prevention of epidemics.

Air dispersal Dispersal can be divided into several stages: takeoff, flight, and landing, analogous to those of airplane travel. At each stage, specific microbial adaptations may be found.

Outdoors the source of virtually all airborne bacteria is the soil. Windblown dust carries with it significant microbial populations, which can be carried long distances. Indoors, the main source of airborne microbes is the human respiratory tract. Most of these organisms survive only poorly in air, so that effective transmittal to a suitable habitat (another human) occurs only over short distances. However, certain human pathogens or commensals (*Staphylococcus*, *Streptococcus*) are able to survive under dry conditions fairly well and so may remain alive in dust for long periods of time. Gram-positive bacteria are in general more resistant to drying than Gram-negative bacteria, and it may be for this reason that these bacteria are more likely to be involved in air dispersal. Spore-forming bacteria are also resistant to drying, but are generally not present in humans in the spore form. Other airborne microbes found to be derived from soil are also Gram-positive (*Micrococcus*). The reason Gram-positive bacteria are more resistant to drying than Gram-negative ones probably relates to the stability to drought effected by the thicker and more rigid Gram-positive cell wall.

An enormous number of droplets of moisture are expelled during sneezing (Figure 17.7), and a considerable number are expelled during coughing or even merely talking. Each infectious droplet has a size of about 10 μm and contains one or two bacteria. The speed of the droplet movement is about 100 m/s (over 200 miles/h) in a sneeze and about 16 to 48 m/s during coughing or loud talking. The number of bacteria in a single sneeze varies from 10,000 to 100,000. Because of the small size of the droplets, the moisture evaporates quickly in the air, leaving behind a nucleus of organic matter and mucus, to which the bacterial cells are attached.

Movement of microbes through the air has some resemblance to movement of gases, but there are considerable complications in this comparison, because often the sizes of microbes are large enough that gravitational forces enter the picture. Thus although movement is not strictly a diffusion process, the word *diffusion* is often used, since it permits a simple expression of a rather complex phenomenon. In still air, movement would be primarily downward, but air is of course never still, so that movement can be horizontal, or even upward. Figure 17.8 illustrates the manner in which a cloud of microbes would move and spread out in a horizontal direction. Note how the cloud initially remains intact but gradually becomes dissipated as it moves downwind.

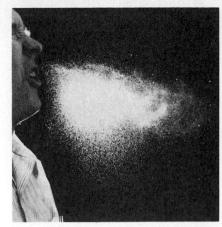

FIGURE 17.7 High-speed photograph of an unstifled sneeze.

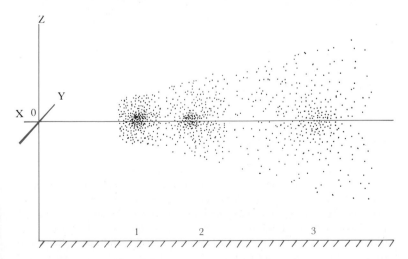

FIGURE 17.8 Diffusion of a microbial cloud during horizontal travel in the wind. The source of liberation is at 0. Three positions of movement downwind are shown; note the gradual dispersal of the cloud. (From Gregory, P. H. 1973. The microbiology of the atmosphere, 2nd ed. Blackie, Glasgow.)

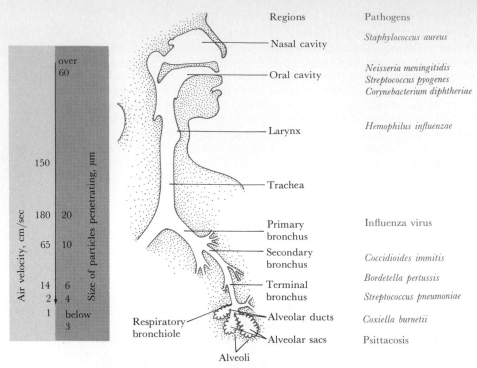

Air velocity, cm/sec	Size of particles penetrating, μm	Regions	Pathogens

(diagram labels)

Air velocity, cm/sec — Size of particles penetrating, μm

over 60

150

180 — 20

65 — 10

14 — 6
2 — 4
1 — below 3

Regions:
- Nasal cavity
- Oral cavity
- Larynx
- Trachea
- Primary bronchus
- Secondary bronchus
- Terminal bronchus
- Respiratory bronchiole
- Alveolar ducts
- Alveolar sacs
- Alveoli

Pathogens:
- *Staphylococcus aureus*
- *Neisseria meningitidis*
- *Streptococcus pyogenes*
- *Corynebacterium diphtheriae*
- *Hemophilus influenzae*
- Influenza virus
- *Coccidioides immitis*
- *Bordetella pertussis*
- *Streptococcus pneumoniae*
- *Coxiella burnetii*
- Psittacosis

FIGURE 17.9 Characteristics of the respiratory system of humans and locations at which various organisms generally initiate infections. (Modified from Mitchell, R. I. 1960. Am. Rev. Resp. Dis. 82:630. Austwick, P. K. C. 1966. In Madelin, M. F., ed. The fungus spore. Butterworth, London. Dimmick, R. L., A. B. Akers, R. J. Heckly, and H. Wolochow. 1969. An introduction to aerobiology. Wiley-Interscience, N.Y.)

Respiratory infection The average human breathes several million cubic feet of air in a lifetime, much of it containing microbe-laden dust, which is a potential source of inoculum for upper-respiratory infections caused by streptococci and staphylococci resistant to drying. The speed at which air moves through the respiratory tract varies, and in the lower respiratory tract the rate is quite slow. As the air slows down, particles in it stop moving and settle, the larger particles first and the smaller ones later, and as seen in Figure 17.9, in the tiny bronchioles only particles below 3 μm will be present. As also shown in Figure 17.9, different organisms reach different levels in the tract, thus accounting for the differences in the kinds of infections that occur in the upper and lower respiratory tracts.

Current status of respiratory diseases Because of their nature, respiratory diseases cannot be controlled by the sorts of public health measures so effective for food and waterborne diseases (see Sections 17.9 and 17.10), but must be controlled through immunization programs. Because of the availability of effective vaccines, such respiratory diseases as diphtheria, measles, and pertussis (whooping cough) are virtually absent in the United States. Chickenpox is a viral disease for which no available control methods are effective. The virus, varicella-zoster virus, causes an acute generalized infection of a relatively mild nature, with eruptions on the skin. The disease occurs primarily in children and is rarely fatal. In chickenpox epidemics, the virus is transmitted from person to person by direct contact, droplets, or by airborne spread of secretions from the respiratory tract, as well as by indirect transfer via fomites. The monthly incidence of chickenpox is illustrated in Figure 17.10 and the marked winter-spring seasonality is characteristic of a disease transmitted primarily by the respiratory route.

Influenza Influenza is caused by an RNA virus of the myxovirus group. Human influenza virus exists in nature only in humans. It is transmitted from person to person through the air, primarily in droplets expelled during coughing and sneezing. The virus infects the mucous membranes of the upper respiratory tract and occasionally invades the lungs. Localized symptoms occur at the site of infection. Systemic symptoms include an abrupt fever of 38 to 40°C for 3 to 7 days, chills, fatigue, headache, and general aching. Recovery is usually spontaneous and rapid. Most of the serious consequences of influenza infection do not occur because of the viral infection but because bacterial invaders may be able to set up severe infections in persons whose resistance has been lowered. Especially in infants and elderly people, influenza is often followed by bacterial pneumonia; death, if it occurs, is usually due to the bacterial infection.

Influenza often occurs in pandemics. Early pandemics, of which the one in 1918 is the most famous, occurred before knowledge was sufficiently advanced to make careful analysis possible, but the 1957 pandemic of the so-called Asiatic flu provided an opportunity for a careful study of how a worldwide epidemic develops. The epidemic began with the development of a mutant virus strain of marked virulence and differing from all previous strains in antigenicity. Since immunity to this strain was not present in the population, the virus was able to advance rapidly throughout the world. It first appeared in the interior of China in late February 1957 and by early April had been brought to Hong Kong by refugees. It spread from Hong Kong along air and naval routes and was apparently transferred to San Diego, California, by naval ships. An outbreak occurred in Newport, Rhode Island, on a naval vessel in May. Other outbreaks occurred in various parts of the United States. Peak incidence occurred in the last two weeks of October, during which time 22 million new bed cases developed. From this period on there was a progressive decline.

Influenza virus is unusual in its ability to undergo **antigenic shift** in its protein coat. Immunity in humans is dependent largely on the presence of secretory antibody (IgA) in the respiratory secretions (see Section 16.2). Once a virus has passed through the population, a majority of the people will be immune to that strain, and it will be impossible for a strain of similar antigenic type to cause an epidemic for about three years. However, antigenic shift of the virus can be either minor or major. A minor antigenic shift can result in some increase in the infectivity of the virus, but a major shift can result in a highly infective virus, able to infect virtually everyone in the population. This is apparently what happened in the pandemics of 1918 and 1957. There is some suggestion that in 1918 (an unusually serious

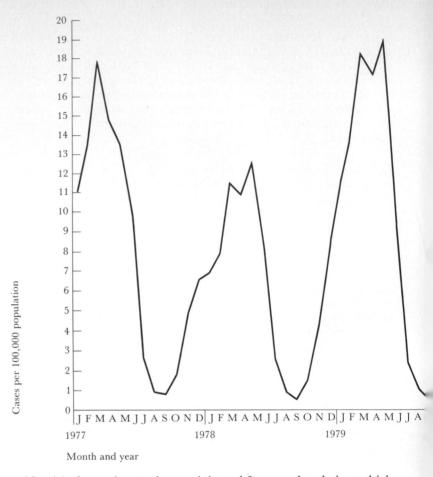

FIGURE 17.10 Reported cases of chicken pox by month in the United States, 1977–1980. Note the marked winter–early spring seasonality typical of a respiratory disease. (From Morbidity and Mortality Weekly Report, Annual Summary, 1981; issued October, 1982.)

Cases per 100,000 population

Month and year

epidemic), the strain may have originated from a related virus which infects swine (swine flu), and the 1957 strain may have arisen from a similar animal reservoir (perhaps a wild animal) somewhere in Asia. Although a vaccine can be prepared to any strain, the large number of strains and the phenomenon of antigenic shift make it difficult to prevent influenza epidemics.

Legionnaire's disease During a convention of the American Legion in the summer of 1976, an outbreak of pneumonia occurred, resulting in 182 cases and 29 deaths. The causative agent remained unknown for several months, but eventually was isolated by a team of microbiologists from the Centers for Disease Control in Atlanta, Georgia. The organism turned out to be a previously unrecognized bacterium and was given the name *Legionella pneumophila*.

Legionella is a thin Gram-negative rod with complex nutritional requirements, including an unusually high iron requirement, and is immunologically distinct from any known pathogen associated with respiratory infections. *Legionella* can be detected by immunofluorescence techniques (see Section 16.3) and can be isolated from many terrestrial and aquatic habitats. The major outbreak mentioned above affected almost exclusively a group of conventioneers housed in a certain Philadelphia hotel. Subsequent epidemiological studies showed

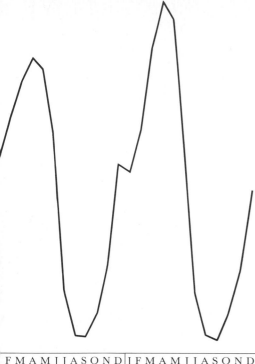

that *Legionella* was present in the cooling towers of the huge air conditioning units required to cool the hotel, and that infectious *Legionella* aerosols were being spread via the air ducts of the hotel's central air conditioning system. Curiously, however, although apparently spread via an airborne route, no evidence of direct person-to-person transmission of *Legionella* was obtained. Consistent with these findings is the fact that cases of Legionnaire's disease peak in the late summer months (Figure 17.11). This is in contrast to an airborne disease such as chickenpox, which is highly contagious and peaks in the winter months (Figure 17.10) when people are more frequently indoors and in close contact.

    Legionella infections may be totally asymptomatic and occasionally result in only mild symptoms such as headache and fever. In the majority of cases where pneumonia develops, patients are frequently elderly individals whose resistance has been previously compromised. In addition, certain serotypes of *Legionella* (four are known) are more strongly associated with the pneumonic form of the illness than others. Prior to the onset of pneumonia, intestinal disorders are common, followed by high fever, chills, and muscle aches. These symptoms precede the dry cough and chest and abdominal pain typical of Legionnaire's pneumonia. *Legionella pneumophila* is sensitive to the antibiotics rifampicin and erythromycin, and intravenous administration of erythromycin is the treatment of choice in most instances.

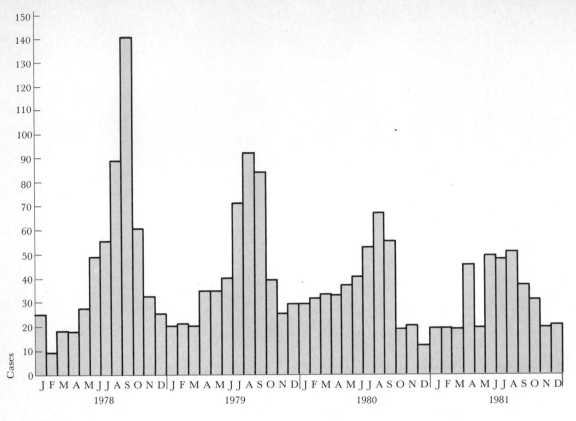

FIGURE 17.11   Seasonal distribution of cases of Legionnaire's Disease, 1978–1980. Note the tendency for cases to peak in mid-summer. (From *Morbidity and Mortality Weekly Report*, Annual Summary, 1981, issued October, 1982.)

17.9

Waterborne Disease and Water Purification

Pathogenic organisms transmitted by water   Organisms pathogenic to humans that are transmitted by water include bacteria, viruses, and protozoa (Table 17.2). Organisms transmitted by water usually grow in the intestinal tract and leave the body in the feces. Fecal pollution of water supplies may then occur, and if the water is not properly treated, the pathogens enter a new host when the water is consumed. Because water is consumed in large quantities, it may be infectious even if it contains only a small number of pathogenic organisms. The pathogens lodge in the intestine, grow, and cause infection and disease.

Probably the most important pathogenic *bacteria* transmitted by the water route are *Salmonella typhi,* the organism causing typhoid fever, and *Vibrio cholerae,* the organism causing cholera. Although the causal agent of typhoid fever may also be transmitted by contaminated food and by direct contact from infected people, the most com-

TABLE 17.2 Waterborne Disease Outbreaks in the United States, 1946–1980*

Disease	Causal Agent	Outbreaks[†]	Cases
Bacterial:			
Typhoid fever	*Salmonella typhi*	58	836
Shigellosis	*Shigella* species	59	13,261
Salmonellosis	*Salmonella paratyphi* etc.	22	17.914
Gastroenteritis	*Escherichia coli*	5	1188
	Campylobacter species	2	3800
Leptospirosis	*Leptospira* species	1	9
Tularemia	*Francisella tularensis*	2	6
Viral:			
Infectious hepatitis	Hepatitis A virus	71	2342
Poliomyelitis	Polio virus	1	16
	Norwalk agent	10	3147
Protozoal:			
Dysentery	*Entamoeba histolytica*	5	75
Giardiasis	*Giardia lamblia*	47	19,883
Unknown etiology:			
Gastroenteritis		366	87,439

*One waterborne disease still important in some parts of the world, but not seen in the United States, is cholera, caused by the bacterium *Vibrio cholerae*.

[†]A waterborne outbreak is defined as an incident in which two or more persons experience similar illness after consumption of water, and epidemiologic evidence implicates the water as the source of illness. Only water used for drinking is included.

Data from Craun, G. F., and L. J. McCabe, 1973. Review of the causes of waterborne-disease outbreaks. Jour. American Water Works Association 65:74–84; also data from Centers for Disease Control, Atlanta, Ga. Foodborne and waterborne disease outbreaks, 1976 through 1980.

mon and serious means of transmission is the water route. Typhoid fever has been virtually eliminated in many parts of the world, primarily as a result of the development of effective water treatment methods. However, typhoid fever does still occur occasionally, usually in the summer months, when swimmers are active in polluted water supplies. A breakdown in water purification methods, contamination of water during floods, earthquakes, and other disasters, or cross contamination of water pipes from leaking sewer lines occasionally results in epidemics of typhoid fever. Water that has been contaminated can be rendered safe for drinking by boiling for 5 to 10 minutes or by adding chlorine, as discussed later. Usually, contaminated water is safe to use for laundry or other domestic cleaning purposes, provided care is taken that it is not swallowed.

The causal agent of cholera is transmitted only by the water route. At one time cholera was common in Europe and North America, but the disease has virtually been eliminated from these areas by effective water purification. The disease is still common in Asia, however, and travelers from the West are advised to be vaccinated for cholera. Both *V. cholerae* and *S. typhi* are eliminated from sewage during proper sewage treatment and hence do not enter water courses receiving treated sewage effluent. More frequent than typhoid, but less serious a disease, is salmonellosis caused by species of *Salmonella* other than *S. typhi.* As seen in Table 17.2, the largest number of cases of waterborne bacterial disease in the United States during the last several decades have been due to salmonellosis.

Bacteria are effectively eliminated from water during the water purification process (see below), so that they should never be present in properly treated drinking water. Most outbreaks of waterborne disease in the United States are due to breakdowns in treatment systems, or as a result of post-contamination in the pipelines. This latter problem can be controlled by maintaining a concentration of free chlorine in the pipelines.

Viruses transmitted by the water route include poliovirus and other viruses of the enterovirus group, as well as the virus causing infectious hepatitis. Poliovirus has several modes of transmittal, and transmission by water may be of serious concern in some areas. This is one of the pathogens more commonly encountered during the summer months in polluted swimming areas.

Infectious hepatitis is caused by a virus (hepatitis A virus), which resembles the enteroviruses, but hepatitis A virus has not been successfully cultured so that its characteristics are poorly known. Although one of the most serious waterborne viral disease agents at present (see Table 17.2), hepatitis is also transmitted in foods, and probably most of the infectious hepatitis cases arise from foodborne rather than waterborne means.

Because viruses are acellular, they are more stable in the environment and are not as easily killed as bacteria. However, both poliovirus and infectious hepatitis virus are eliminated from water by proper treatment practices, and the maintenance of 0.6 ppm free chlorine in a water supply will generally ensure its safety.

Indicator organisms and the coliform test   Even water which looks clear and pure may be sufficiently contaminated with pathogenic microorganisms to be a health hazard. Some means are necessary to ensure that drinking water is safe. One of the main tasks of water microbiology is the development of laboratory methods which can be used to detect the microbiological contaminants that may be present in drinking water. It is not usually practical to examine drinking water directly for the various pathogenic organisms that may be present. As stated earlier, a wide variety of organisms may be present, including bacteria, viruses, and protozoa. To check each drinking water supply for each of these agents would be a difficult and time-consuming job. In practice, **indicator organisms** are used instead. These are organisms associated with the intestinal tract, whose presence in water indicate that the water has received contamination of an intestinal origin. The most widely used indicator is the **coliform group** of organisms. This group is defined in water bacteriology as all the aerobic and facultatively anaerobic, Gram-negative, nonspore-forming, rod-shaped bacteria that ferment lactose with gas formation within 48 hours at 35°C. This is an operational rather than a taxonomic definition, and the coliform group includes a variety of organisms, mostly of intestinal origin. In practice, the coliform organisms are almost always members of the enteric bacterial group (see Section 19.13). The coliform group includes the organism *Escherichia coli,* a common intestinal organism, plus the organism *Klebsiella pneumoniae,* a less common intestinal organism. The definition also currently includes organisms of the species *Enterobacter aerogenes,* not generally associated with the intestine.

The coliform group of organisms are suitable as indicators because they are common inhabitants of the intestinal tract, both of humans and warmblooded animals, and are generally present in the intestinal tract in large numbers. When excreted into the water environment, the coliform organisms eventually die, but they do not die at any faster rate than the pathogenic bacteria *Salmonella* and *Shigella,* and both the coliforms and the pathogens behave similarly during water purification processes. Thus it is likely that if coliforms are found in drinking water, the water has received fecal contamination and may be unsafe. There are very few organisms in nature that meet the definition of the coliform group that are not associated with the intestinal tract. (One exception is an organism called *Aeromonas,* which is common in some source waters and may appear in a coliform analysis. *Aeromonas* can be distinguished from the intestinal coliforms by means of a special diagnostic test, the oxidase test, see Table 18.2.) It should be emphasized that the coliform group includes organisms derived not only from humans but from other warm-blooded animals. Since many of the pathogens (for example, *Salmonella, Leptospira*) found in warm-blooded animals will also infect humans, an indicator of both human and animal pollution is desirable.

There are two types of procedures that are used for the coliform test. These are the **most-probable-number** (**MPN,** see Section 7.2) procedure and the **membrane filter** (**MF**) procedure. The MPN procedure employs liquid culture medium in test tubes, the samples of drinking water being added to the tubes of media. In the MF procedure, the sample of drinking water is passed through a sterile membrane filter, which removes the bacteria, and the filter is then placed on a culture medium for incubation.

There are three stages in the MPN procedure: **presumptive, confirmed,** and **completed** (Figure 17.12). In the *presumptive* test, a sample of water and dilutions from it are inoculated into tubes of a lactose broth medium, and the tubes are incubated for 24 to 48 hours at 35°C. If gas is produced, the test is considered a positive presumptive test, and the further stages are followed. For the *confirmed* test, samples are streaked from the positive tubes at the highest dilutions onto plates containing a special indicator agar, usually eosin-methylene blue (EMB)-lactose agar. Because coliforms produce acid from lactose, they form colonies of very dark color and with a metallic sheen, due to accumulation under acid conditions of the eosin and methylene blue dyes in the colonies (see Table 15.5). The presence of such colonies is considered a positive confirmed test. For the *completed* test, typical colonies are picked and inoculated into lactose broth. If gas is produced upon incubation, further tests are run to ensure that the organism so isolated is a typical Gram-negative non-sporulating rod and has certain biochemical properties characteristic of coliforms. If these observations are satisfactory, it is a positive completed test.

In well-regulated water systems, coliforms will always be nega-

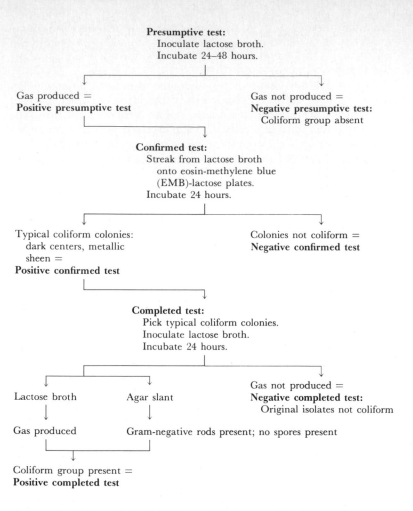

Presumptive test:
  Inoculate lactose broth.
  Incubate 24–48 hours.

Gas produced =
Positive presumptive test

Gas not produced =
Negative presumptive test:
  Coliform group absent

Confirmed test:
  Streak from lactose broth
    onto eosin-methylene blue
    (EMB)-lactose plates.
  Incubate 24 hours.

Typical coliform colonies:
  dark centers, metallic
  sheen =
Positive confirmed test

Colonies not coliform =
Negative confirmed test

Completed test:
  Pick typical coliform colonies.
  Inoculate lactose broth.
  Incubate 24 hours.

Lactose broth   Agar slant

Gas not produced =
Negative completed test:
  Original isolates not coliform

Gas produced   Gram-negative rods present; no spores present

Coliform group present =
Positive completed test

FIGURE 17.12 Standard method of water analysis to detect the presence of coliforms.

tive, and so the confirmed and completed tests will not have be to run. The determination of coliform numbers in the MPN procedure follows the principles described in Section 7.2 and in Appendix 2. For drinking water, five tubes are inoculated, each with 10 ml of water. The number of positive tubes is counted, and from the MPN table (Table A2.2), the MPN is determined. Confirmed and completed tests are then performed on positive tubes. For grossly polluted waters, all tubes may be positive if 10-ml samples are used. To obtain a count on such a water, tubes should also be set up using 1-ml and 0.1-ml water samples. The MPN is then determined using the inoculum size in which there are some negative and some positive tubes.

When using the membrane filter method with drinking water, at least 100 ml of water should be filtered, although in clean water systems, even larger volumes could be filtered. After filtration of a known volume of water, the filter is placed on one of several special culture media, which are selective and which will indicate the presence of coliforms. The coliform colonies are counted and from this value the number of coliforms in the original water sample can be determined. The membrane filter method permits the determination of coliform numbers in 1 day instead of the 3 to 4 days otherwise required, since presumptive, confirmed, and completed tests are combined into one. For some applications, this newer membrane filter technique is replacing the older MPN method, although it does not work well on water samples with large amounts of suspended matter, silt, algae, and bacteria, since these materials interfere with both the filtration and the development of colonies.

A Bit of History

The importance of drinking water as a vehicle for the spread of cholera was first shown in 1855 by British physician John Snow, who at that time was even without any knowledge of the bacterial causation of the disease. Snow's study is one of the great classics of epidemiology and serves as a model for how a careful study can lead to clear and meaningful conclusions.

In London, the water supplies to different parts of the city were from different sources and were transmitted in different ways. In a large area south of the Thames River, across the river from Westminster Abbey and Parliament Building, the water was supplied to houses by two competing private water companies, the Southwark and Vauxhall Company and the Lambeth Company. It was the water of the former company that was the major vehicle for the transmission of cholera. When Snow began to suspect the water supply of the Southwark and Vauxhall Company, he made a careful survey of the residence of every death in this district and determined which company supplied the water to that residence. In some parts of the area served by these two companies, each had a monopoly, but in a fairly large area the two companies competed directly, each having run independent water pipes along the various streets. Houses had the option of connecting with either supply, and the distribution of houses between the two companies was random. The clear-cut results of Snow's survey were completely convincing, even to those skeptical about the importance of polluted water in the transmission of cholera: in the first 7 weeks of the epidemic, there were 315 deaths per 10,000 houses supplied by the Southwark and Vauxhall Company, and only 37 per 10,000 houses supplied by the Lambeth Company. In the rest of London, there were 59 deaths per 10,000 houses, showing that those supplied by the Lambeth Company had fewer deaths than the general population. In the districts where each company had exclusive rights, it could of course be argued that it was not the water, but some other factor (soil, air, general layout of houses and so on), which might have been responsible for the differences in disease incidence, but in the districts where the two companies competed, all of these other factors were the same, yet the incidence was high for those supplied with Southwark and Vauxhall water and low for those supplied with Lambeth water. Snow attempted to relate these differences in disease incidence to the sources of the waters used by the two companies. As he knew that the excrement and evacuations from cholera patients were highly infectious, he considered that sewage contamination of the water supply might exist. In those days, sewage treatment did not exist and raw sewage was dumped directly into the Thames River. The Southwark and Vauxhall Company obtained its water supply from the Thames right in the heart of London, where much opportunity for sewage contamination could occur, while the Lambeth Company obtained its water from a point on the river considerably above the city, and hence was relatively free of pollution. It is almost certain that it was this difference in source which accounted for the difference in disease incidence. In Snow's words: "As there is no difference whatever, either in the houses or the people receiving the supply of the two Water Companies, or in any of the physical conditions with which they are surrounded, it is obvious that no experiment could have been devised which would more thoroughly test the effect of water supply on the progress of cholera than this. . . . The experiment, too, was on the grandest scale. No fewer than three hundred thousand people of both sexes, of every age and occupation, and of every rank and station, from gentlefolk down to the very poor, were divided into two groups without their choice, and, in most cases, without their knowledge; one group being supplied with water containing the sewage of London, and, amongst it, whatever might have come from cholera patients, the other group having water quite free from such impurity."

Water purification It is a rare instance when available water is of such a clarity and purity that no treatment is necessary before use. Water treatment is carried out both to make the water safe microbiologically and to improve its utility for domestic and industrial purposes. Treatments are performed to remove pathogenic and potentially pathogenic microorganisms and also to decrease turbidity, eliminate taste and odor, reduce or eliminate nuisance chemicals such as iron or manganese, and soften the water to make it more useful for the laundry.

The kind of treatment that water is given before use must depend on the quality of water supply. A typical treatment installation for a large city using a source water of poor quality is shown in Figure 17.13.

Water is first pumped to **sedimentation basins,** where sand, gravel, and other large particles settle out. A sedimentation basin should be used only if the water supply is highly turbid, since it has the disadvantages that algal growth may occur in the basin, adding odors and flavors, and that pollution of the water by surface runoff may occur. Bacteria may grow in the bottom mud and add further problems.

Most water supplies are subjected to **coagulation.** Chemicals containing aluminum and iron are added, which under proper control of pH form a flocculent, insoluble precipitate that traps organisms, absorbs organic matter and sediment, and carries them out of the water. After the chemicals are added in a mixing basin, the water containing the coagulated material is transferred to a settling basin where it remains for about 6 hours, during which time the coagulum separates out. Around 80 percent of the turbid material, color, and bacteria are removed by this treatment.

After coagulation, the clarified water is usually **filtered** to remove the remaining suspended particles and microbes. Filters can be of the slow or rapid sand type. **Slow sand filters** are suitable for small installations such as resorts or rural places. The water is simply allowed to pass through a layer of sand 2 to 4 ft deep. Eventually the top of the sand filter will become clogged and the top layer must be removed and replaced with fresh sand. **Rapid sand filters** are used in large installations. The rate of water flow is kept high by maintaining a controlled height of water over the filter. When the filter becomes clogged, it is clarified by backwashing, which involves pumping water up through the filter from the bottom. From 98 to 99.5 percent

Ohio River

River pumping station

Coagulation basins

Pumping station
Chemical building

Underground clear-water reservoir

Filter buildings

Sedimentation basins

Softening basins

Chlorination

FIGURE 17.13 Aerial view of water treatment plant of Louisville, Ky. The arrows indicate direction of flow of water through the plant. (Courtesy of Billy Davis and the Courier-Journal and Louisville Times.)

of the total bacteria in raw water can be removed by proper settling and filtration.

Chlorination is the most common method of ensuring microbiological safety in a water supply. In sufficient doses it causes the death of most microorganisms within 30 minutes. In addition, since most taste- and odor-producing compounds are organic, chlorine treatment reduces or eliminates them. It also oxidizes soluble iron and manganese compounds, forming precipitates which can be removed. Chlorine can be added to water either from a concentrated solution of sodium or calcium hypochlorite or as a gas from pressure tanks. The latter method is used most commonly in large water treatment plants, as it is most amenable to automatic control.

When chlorine reacts with organic materials it is used up. Therefore, if a water supply is high in organic materials, sufficient chlorine must be added so that there is a residual amount left to react with the microorganisms after all reactions with organic materials have occurred. The water plant operator must perform chlorine analyses on the treated water to determine the residual level of chlorine. A chlorine residual of about 0.2 to 0.6 ppm is an average level suitable for most water supplies.

After final treatment the water is usually pumped to **storage tanks,** from which it flows by gravity to the consumer. Covered storage tanks are essential to ensure that algal growth does not take place (no light is available) and that dirt, insects, or birds do not enter.

Drinking water standards   Drinking water standards in the United States first were instituted in 1914, when standards were prescribed by the federal government for the drinking water provided by common carriers (for example, railroads). Revised standards for common carriers were prescribed in 1925, 1943, and 1962. Some states established drinking water standards that were similar to the federal standard for common carriers. In 1974, the United States Congress passed the Safe Drinking Water Act, which provided a framework for the development by the Environmental Protection Agency of drinking water standards for the whole country. There seems to be general agreement that the measurement of coliforms provides the greatest assurance of the microbiological safety of drinking water. The maximum coliform levels permitted by the regulations are given in reference to the type of method used, membrane filter (MF) or the most-probable-number (MPN).

When the MF technique is used, 100-ml samples must be filtered, and the number of coliform bacteria shall not exceed any of the following: (1) 1 per 100 ml as the arithmetic mean of all samples examined per month; (2) 4 per 100 ml in more than one sample when less than 20 are examined per month; or (3) 4 per 100 ml in more than 5 percent of the samples when 20 or more are examined per month.

When the MPN technique is used, the sample size used can vary, but the intent of the regulations is that the coliform density should not exceed that specified by the MF procedure. In most water supply operations, the MPN procedure is carried out using five tubes, each of which is inoculated with a 10-ml sample. After incubation the number of tubes giving positive reactions is noted, and from Table A2.3, the coliform density can be calculated. In good water practice, all of the tubes inoculated should be negative, indicating a coliform density less than 2.2 per 100 ml.

Public health significance of drinking water purification   Today the incidence of waterborne disease is so low that it is difficult to appreciate the significance of treatment practices and drinking water standards. Most intestinal infection today is not due to transmission by the water route, but via food (see Section 17.10). It was not always so. At the beginning of the twentieth century, effective water treatment practices did not exist, and there were no bacteriological methods for evaluating the health significance of polluted drinking water. The first coliform counting procedures were introduced about 1905. Up until then, water purification, if practiced at all, was primarily for aesthetic purposes, to remove turbidity. Actually, turbidity removal by filtration provides for a significant decrease in the microbial load of a water, so that filtration did play a part in providing safer drinking water. But filtration alone was of only partial value, since many organisms passed through the filters. It was the discovery of the efficacy of chlorine as a water disinfectant in about 1910, which was to have major impact. Chlorine is so effective and so inexpensive that its use spread widely, and it is almost certain that the practice of chlorination was of major significance in reducing the incidence of waterborne disease. However, the effectiveness of chlorination would not have been realized, and the necessary doses could not have been determined, if the standard methods for assessing the coliform content of drinking water had not been developed. Thus engineering and microbiology moved forward together.

17.10

Foodborne Diseases and Epidemics

Food can be the vehicle for some of the same enteric pathogens that are transmitted by water, but in addition, there are certain diseases which are primarily foodborne. There are also certain foodborne illnesses of microbial origin that are intoxications rather than infections, the toxin having been produced by the organism during growth in improperly stored food, and ingestion of this toxin-containing food results in illness. Table 17.3 gives a breakdown of foodborne illness in the United States. As seen, the majority of the cases are bacterial or viral, although parasites (metazoan animals) and chemicals are responsible for a significant number of cases.

    One difference between food and water is that pathogens rarely or never grow in water, rather, they die off, but in food, active growth of a pathogen may occur (for example, due to improper storage) leading to marked increases in the microbial load. We have discussed some aspects of food storage and preservation (heat sterilization, pasteurization) in Section 8.1.

Food poisonings and toxins   Botulism is the most severe type of food poisoning; it is usually fatal, and occurs following the consumption of food containing the exotoxin produced by the anaerobic bacterium *Clostridium botulinum*. This bacterium normally lives in soil or water, but its spores may contaminate raw foods before harvest or slaughter. If the foods are properly processed so that the *C. botulinum* spores are killed, no problem arises; but if viable spores are present, they may initiate growth and even a small amount of the resultant neurotoxin can render the food poisonous. We discussed the nature and action of botulinum toxin in Section 15.3 (see also Figure 15.4). The toxin itself is destroyed by heat (80°C for 10 minutes) so that a properly cooked food should be harmless, even if it did originally contain the toxin. Most cases of botulism occur as a result of eating foods that are not cooked after processing. Canned vegetables and beans are often used without cooking in making cold salads. Similarly, smoked fish and meat and most of the vacuum-packed sliced meats are often eaten directly, without heating. If these products contain the botulinum toxin, then ingestion of even a small amount will result in this severe and highly dangerous type of food poisoning.

    Staphylococcal food poisoning is the most common type of

TABLE 17.3 Foodborne Disease Outbreaks in the United States, 1973–79

	Number of outbreaks*
Bacterial:	
Arizona hinshawii	2
Bacillus cereus	13
Brucella sp.	3
Clostridium botulinum	107
Clostridium perfringens	81
Escherichia cloacae	1
Escherichia coli	1
Salmonella	264
Shigella	36
Staphylococcus	216
Streptococcus	7
Vibrio cholerae	4
Vibrio parahemolyticus	9
Yersinia enterocolitica	1
Other bacteria	1
Viral	32
Parasites (metazoans)	89
Chemicals	266

*An outbreak is defined as an incident in which two or more persons experience a similar illness, usually gastrointestinal, after ingestion of a common food, and epidemiologic analysis implicates the food as the source of the illness. However, one case of botulism or one chemical poisoning constitutes an outbreak.

Data from Foodborne disease surveillance, Annual Summary 1979, issued April 1981, Centers for Disease Control, Atlanta, Ga.

FIGURE 17.14 Seasonal distribution of *Salmonella* infection, as revealed by isolations from humans in the *Salmonella* Surveillance Program of the Centers for Disease Control. There is a marked rise in isolations in mid- to late summer. Compare with the data in Figure 17.10. (From Morbidity and Mortality Weekly Report, Annual Summary, 1981; issued October, 1982.)

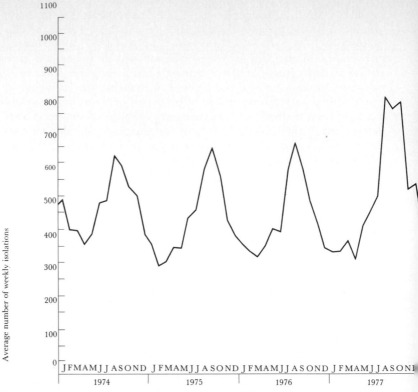

food poisoning and is caused by varieties of *Staphylococcus aureus.* This organism produces an enterotoxin (see Section 15.5) that is released into the surrounding medium or food; if food containing the toxin is ingested, severe reactions are observed within a few hours, including nausea with vomiting and diarrhea. The kinds of foods most commonly involved in this type of food poisoning are custard- and cream-filled baked goods, poultry, meat and meat products, gravies, egg and meat salads, puddings, and creamy salad dressings. If such foods are kept refrigerated after preparation, they remain relatively safe, as the *Staphylococcus* is unable to grow at low temperatures. In many cases, however, foods of this type are kept warm for a period of hours after preparation, such as in warm kitchens or outdoors at summer picnics. Under these conditions, *Staphylococcus,* which might have entered the food from a food handler during preparation, grows and produces enterotoxin. Many of the foods involved in staphylococcal food poisoning are not cooked again before eating, but even if they are, this toxin is relatively stable to heat and may remain active. Staphylococcal food poisoning can be prevented by careful sanitation methods so that the food does not become inoculated, by storage of the food at low temperatures to prevent staphylococcal growth, and by the discarding of foods stored for any period of hours at warm temperatures.

Although sometimes called a food poisoning, gastrointestinal disease due to foodborne *Salmonella* is more aptly called **Salmonella food infection,** because symptoms arise only after the pathogen grows in the intestine (hence symptoms can be experienced as late as several

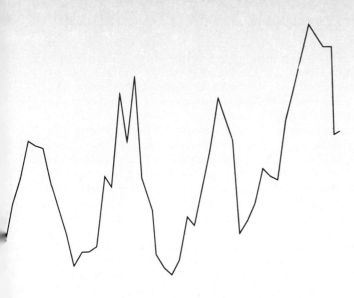

days after eating a contaminated food). Virtually all species of *Salmonella* are pathogenic for humans: one, *S. typhi,* causes the serious human disease typhoid fever, while a small number of other species cause foodborne gastroenteritis. The ultimate sources of the foodborne salmonellas are humans and warm-blooded animals. The organism reaches food by contamination from food handlers; or in the case of foods such as eggs or meat, the animal which produced the food may be the source of contamination. The foods most commonly involved are meats and meat products (such as meat pies, sausage, cured meats), poultry, eggs, and milk and milk products. If the food is properly cooked, the organism will be killed and no problem will arise, but many of these products are eaten uncooked or partially cooked. *Salmonella* strains causing foodborne gastroenteritis are often traced to products made with uncooked eggs such as custards, cream cakes, meringues, pies, and eggnog. Previously cooked foods that have been warmed and held without refrigeration or canned foods held for awhile after opening often support the growth of *Salmonella* if they have become contaminated by an infected food handler. *Salmonella* infection is more common in summer than in winter, probably because environmental conditions are more favorable for growth of salmonellae in foods. The seasonal distribution of *Salmonella* infections in humans is shown by the frequency of isolation in the *Salmonella* Surveillance Program, illustrated in Figure 17.14.

Assessing microbial content of foods   All fresh foods will have some viable microbes present. The purpose of assay methods is to detect evidence of abnormal microbial growth in foods or to detect the presence of specific organisms of public health concern, such as *Salmonella, Staphylococcus,* or *Clostridium botulinum.* With nonliquid foods, preliminary treatment is usually required to suspend in a fluid medium those microorganisms embedded or entrapped within the food. The most suitable method for treatment is high-speed blending. Examination of the food should be done as soon after sampling as possible, and if examination cannot begin within 1 hour of sampling the food should be refrigerated. A frozen food should be thawed in its original container in a refrigerator and examined as soon as possible after thawing is complete. For coliform counts, a most-probable-number method such as described in Section 17.9 should be performed, using a medium selective for this bacterium. For *Salmonella* several selective media are available, and tests for its presence are most commonly done on animal food products, such as raw meat, poultry, eggs, and powdered milk, since *Salmonella* from the animal may contaminate the food. For staphylococcal counts, a medium high in salt (either sodium chloride or lithium chloride at a final concentration of 7.5%) is used, since of the organisms present in foods, staphylococci are the only common ones resistant to salt. Since *Staphylococcus aureus* is responsible for one of the most common types of food poisoning, staphylococcal counts are of considerable importance.

Traveler's diarrhea   Traveler's diarrhea, sometimes called "turista" or "Montezuma's revenge," is an extremely common enteric infection in North American travelers to Mexico and other Central and South American countries. The primary causal agent is enteropathogenic *Escherichia coli* (EEC), although *Salmonella* and *Shigella* are sometimes implicated. The *E. coli* is an enterotoxin-producing strain, and as noted in Section 11.6 the gene for enterotoxin is often plasmid-linked. The K anitgen, necessary for successful colonization of the small intestine, is also generally plasmid-linked (Section 15.1).

Several studies have been done on groups of U.S. citizens traveling in Mexico. Such studies have shown that the infection rate is often quite high, greater than 50 percent, and that the prime vehicles are foods, such as uncooked vegetables (for example, lettuce in salads) and water. The impressively high infection rate in travelers has been shown to be due to the fact that the local population has a marked immunity to the infecting strains, due undoubtedly to the fact that they have lived with the agent for a long period of time. Secretory antibodies present in the bowel may prevent successful colonization of the pathogen in local residents, but when the organism colonizes the intestine of a non-immune person, it finds a hospitable environment. Also, stomach acidity, so often a barrier to intestinal infection, may not be able to act if only small amounts of liquid are consumed (as, for instance, the melting ice of a cocktail), since small amounts of liquid induce rapid emptying of the stomach and hence pass through so quickly that stomach acidity may have no effect on an enteric pathogen present in the liquid.

The sexually transmitted (venereal) diseases **gonorrhea** and **syphilis,** both of bacterial etiology, used to be the only such diseases mentioned in microbiology texts. Improved isolation techniques, however, have shown that a host of distinct **venereal diseases** exist and that the causative agents can be either bacterial, viral, or protozoan. Table 17.4 summarizes the major venereal diseases commonly encountered in medical practice today. Although infections caused by *Chlamydia trachomatis* are currently the most prevalent venereal diseases in the United States, we will begin our discussion with a consideration of the classic venereal diseases, gonorrhea and syphilis.

Control of the venereal diseases syphilis and gonorrhea presents an unusually difficult public health problem. Despite the advent of modern therapeutic drugs, gonorrhea has undergone a resurgence since the early 1960s so that currently it can be considered of epidemic importance (Figure 17.15). It seems very likely that gonorrhea is rampant because of the introduction of birth-control pills, reducing fear of pregnancy and thus eliminating one of the greatest checks against sexual promiscuity. To understand why gonorrhea, but not syphilis, has become epidemic (even pandemic), it is necessary to consider briefly the host-parasite relationships of these two infections.

Venereal Diseases

TABLE 17.4 Summary of Some Sexually Transmitted Diseases and Treatment Guidelines

Disease	Causative Organisms*	Recommended Treatment†
Gonorrhea	*Neisseria gonorrhoeae* (B)	Penicillin (tetracycline or spectinomycin for penicillin resistant strains)
Syphilis	*Treponema pallidum* (B)	Penicillin
Chlamydia trachomatis infections	*Chlamydia trachomatis* (B)	Tetracycline
Nongonococcal urethritis	*Chlamydia trachomatis* (B) or *Ureaplasma urealyticum* (B)	Tetracycline
Lymphogranuloma venereum	*C. trachomatis* (B)	Tetracycline
Chancroid	*Haemophilus ducreyi* (B)	Erythromycin
Genital herpes	Herpes Simplex Type 2 (V)	No known cure, symptoms can be controlled with topical application of acyclovir (see Figure 17.17)
Anogenital warts (venereal warts)	Papilloma virus (certain strains) (V)	10–25% podophyllin in tincture of benzoin (alternatively, warts can be removed surgically)
Trichomoniasis	*Trichomonas vaginalis* (P)	Metronidazole
Acquired immune deficiency syndrome (AIDS)	Unknown, possibly viral	None currently available

*B, bacterium; V, virus; P, protozoan.

†U.S. Department of Health and Human Services, Public Health Service Recommendations.
From: Sexually transmitted diseases treatment guidelines. Morbidity and Mortality Weekly Report, Suppl., 31(25), August 20, 1982.

FIGURE 17.15 Venereal diseases: reported cases of gonorrhea and syphilis per 100,000 population, United States. After 1940, military cases are excluded. (From Morbidity and Mortality Weekly Report, Annual Summary, 1981; issued October, 1982. Centers for Disease Control, Atlanta, Ga.)

Gonorrhea   Gonorrhea is a venereal disease that apparently occurs naturally only in human beings; the chimpanzee is the only experimental animal that can be infected with *N. gonorrhoeae* and develop the clinical symptoms of gonorrhea, although the disease does not occur naturally in these animals. Gonorrhea is one of the most widespread human diseases, and in spite of the availability of excellent drugs it is still a common disease even in countries where the cost of drugs is no economic problem. As opposed to syphilis, gonococcal infections rarely result in serious complications or death. The disease symptoms are quite different in the male and female. In the female the symptoms are usually a mild vaginitis that is difficult to distinguish from vaginal infections caused by other organisms, and the infection may easily go unnoticed; in the male, however, the organism causes a painful infection of the urethral canal, and the disease is often given the colloquial name of "strain" or "clap." In addition to gonorrhea, the organism also causes eye infections in the newborn and adult. The organism is killed quite rapidly by drying, sunlight, and ultraviolet light; this extreme sensitivity probably explains in part the venereal nature of the disease, the organism being transmitted from person to person only by intimate direct contact. Infants may become infected during birth to diseased mothers, showing the infection in the eyes. Prophylactic treatment of the eyes of all newborns with silver nitrate or an ointment containing penicillin is mandatory in most states and has helped to control the disease in infants.

We discussed the clinical microbiology of the causal agent, *Neisseria gonorrhoeae,* in Section 15.11, and the general bacteriology of the genus *Neisseria* is discussed in Section 19.17.

The organism enters the body by way of the mucous membranes of the genitourinary tract, customarily being transmitted from a member of the opposite sex. Treatment of the infection with penicillin has been successful in the past, with a single injection usually resulting in elimination of the organism and complete cure. Strains of *N. gonorrhoeae* resistant to penicillin have recently begun appearing, and as discussed in Section 15.11, resistance is due to a plasmid-encoded penicillinase. The incidence of **penicillinase-producing *N. gonorrhoeae*** (or PPNG in the epidemiologist's shorthand) has increased dramatically since the discovery of these strains in the mid-1970s (Figure 17.16). Fortunately, however, the majority of PPNG strains respond to alternative antibiotic therapy, with spectinomycin being the drug of choice in most cases. Where penicillin therapy is effective, it generally results in a more rapid cure in males than in females.

Despite the ease with which gonorrhea can be cured, the incidence of gonococcus infection remains relatively high. The reasons for this are threefold: (1) acquired immunity does not exist, hence repeated reinfection is possible (whether this is due to lack of local immunity or to the fact that at least 16 distinct serotypes of *N. gonorrhoeae* have been isolated is not known); (2) the widespread use of oral contraceptives. The latter cause a mimicking of the pregnant state, which results, among other things, in a lack of glycogen production in the vagina and a raising of the vaginal pH. Lactic acid bacteria, normally found in the adult vagina (see Section 14.6) fail

FIGURE 17.16 Reported penicillinase-producing *Neisseria gonorrhoeae* (PPNG) cases by quarter, April 1976–March 1981. (From Morbidity and Mortality Weekly Report, Annual Summary, 1981; issued October, 1982.)

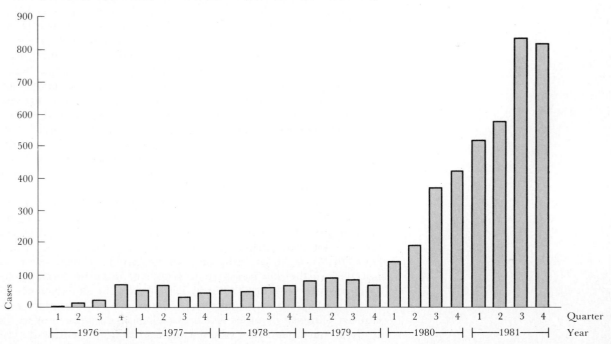

to develop under such circumstances, and this allows *N. gonorrhoeae* transmitted from an infected partner to colonize more easily than in an acidic vagina containing lactobacilli; and (3) symptoms in the female are such that the disease may go unrecognized, and an infected female can serve as a reservoir for the infection of many males. The disease could be controlled if the sexual contacts of infected persons were quickly identified and treated, but it is often difficult to obtain the necessary information and even more difficult to arrange treatment. The incidence of gonorrhea correlates closely with the promiscuity of the society; elimination of this disease is extremely difficult because, despite notable advances in drug therapy, gonorrhea remains a social rather than a medical problem.

Syphilis The venereal disease syphilis is potentially much more serious than gonorrhea, but because of differences in pathobiology, it has been better controlled (Figure 17.15). Syphilis is caused by a spirochete, *Treponema pallidum*, an obligate anaerobe that has been very difficult to cultivate. We mentioned the clinical immunology and diagnostic methods for syphilis in Table 16.7 and the biology of the spirochetes is discussed in Section 19.6.

The disease in humans exhibits variable symptoms. The organism does not pass through unbroken skin, and initial infection most probably takes place through tiny breaks in the epidermal layer. In the male, initial infection is usually on the penis, whereas in the female it is most often in the vagina, cervix, or perineal region. In about 10 percent of the cases infection is extragenital, usually in the oral region. During pregnancy, the organism can be transmitted from an infected woman to the fetus; the disease acquired in this way by an infant is called **congenital syphilis.** The organism multiplies at the initial site of entry and a characteristic *primary* lesion known as a **chancre** is formed within 2 weeks to 2 months. Dark-field microscopy of the exudate from syphilitic chancres often reveals the actively motile spirochetes. In most cases the chancre heals spontaneously and the organisms disappear from the site. Some, however, spread from the initial site to various parts of the body, such as the mucous membranes, the eyes, joints, bones, or central nervous system, and extensive multiplication occurs. A hypersensitive reaction to the treponeme takes place, which is revealed by the development of a generalized skin rash; this rash is the key symptom of the *secondary* stage of the disease. At this stage the patient's condition may be highly infectious, but eventually the organisms disappear from secondary lesions and infectiousness ceases. The subsequent course of the disease in the absence of treatment is highly variable. About one-fourth of the patients undergo a spontaneous cure and another one-fourth do not exhibit any further symptoms, although the infection may persist. In about half of the patients the disease enters the *tertiary* stage, with symptoms ranging from relatively mild infections of the skin and bone to serious or fatal infections of the cardiovascular system or central nervous system. Involvement of the nervous system is the most serious phase of the illness, since generalized paralysis or other severe neurological damage may result. In the tertiary stage only very few organisms are present, and most of the symptoms probably result from hypersensitivity reactions to the spirochetes.

Early methods of therapy involved administration of arsenic, mercury, or bismuth compounds, all of which were fairly toxic and relatively ineffective. Use of these drugs has now been completely superseded by highly effective penicillin therapy, and the early stages of the disease can usually be controlled by a series of injections over a period of 1 to 2 weeks. In the secondary and tertiary stages treatment must extend for longer periods of time. It is interesting that the incidence of syphilis has remained low during the current epidemic of gonorrhea. Almost certainly, this is due to the fact that syphilis shows similar symptoms in male and female, so that both sexes obtain treatment. The inapparent infections in females, so common in gonorrhea, do not occur with syphilis.

Chlamydial infections   Although not reportable diseases as is the case for gonorrhea or syphilis, it now appears that a host of venereal diseases can be ascribed to the obligate intracellular parasite *Chlamydia trachomatis,* with the total incidence of such diseases far outnumbering that of gonorrhea or syphilis. *C. trachomatis* also causes a serious disease of the eye called trachoma (see Section 19.31), but the strains responsible for venereal infections consist of a group of *C. trachomatis* serotypes distinct from those causing trachoma. Chlamydial infections cause urethritis in males, and urethritis, cervicitis, and pelvic inflammation in females. In many cases the symptoms are so mild that inapparent infections, in both the male and female, undoubtedly account for the elevated incidence of chlamydial infections. **Lymphogranuloma venereum,** also caused by *C. trachomatis,* can cause severe damage to regional lymph nodes, and is considered to be one of the most serious sexually transmitted chlamydial syndromes.

Genital Herpes   Herpes simplex virus infections are responsible for a number of blisterlike conditions in humans. **Herpesvirus type 1** (HV1) is generally associated with cold sores and fever blisters in and around the mouth and lips. The incubation period of HV1 infections is short (3 to 5 days) and the lesions heal without treatment in 2 to 3 weeks. Relapses of HV1 infections are relatively common and it is thought that the virus is spread primarily via the respiratory route.* **Herpesvirus type 2** (HV2) on the other hand, is associated primarily with the anogenital region, where it causes painful blisters on the penis of males or the cervix, vulva, or vagina of females. HV2 infections are transmitted by direct sexual contact, and the disease is most easily transmitted during the active blister stage rather than during periods of inapparent (presumably latent) infection.

Genital herpes infections are incurable at the present time, although a limited number of drugs have been found successful in controlling the infectious blister stages. The guanine analog **acyclovir** (Figure 17.17) is particularly effective in limiting the shed

*Latent herpes infections are apparently quite common with the virus persisting in low numbers in nerve tissue. Recurrent acute herpes infections may thus be due to a periodic triggering of virus activity.

FIGURE 17.17 Structure of guanine and the guanine analog acyclovir. Acyclovir has been used therapeutically to control genital herpes (HV2) blisters. See text for mode of action.

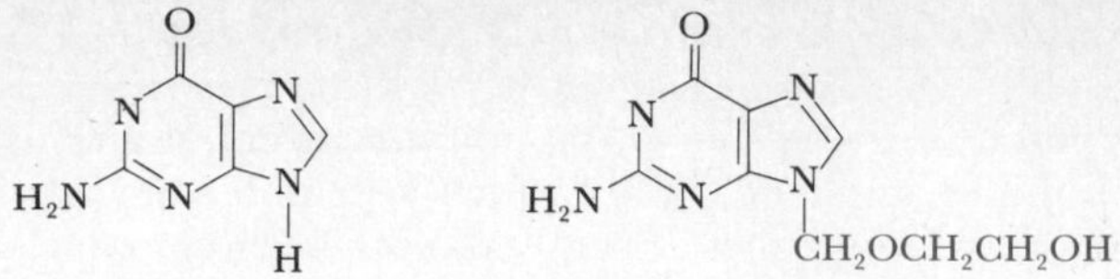

Guanine

Acyclovir

of active virus from blisters and promoting the healing of blisterous lesions. Acyclovir acts by interfering with herpesvirus DNA polymerase, hence inhibiting viral DNA replication.

The overall implications of genital herpes infections are not yet understood, but epidemiological studies have shown a significant correlation between genital herpes infections and cervical cancer in females. In addition, herpesvirus type 2 can be transmitted to the newborn at birth by contact with herpetic lesions in the birth canal. The disease in the newborn varies from latent infections with no apparent damage to systemic disease which can result in death. Severely affected infants who survive may suffer permanent brain damage. To avoid infection, delivery by caesarean section is advised for pregnant woment showing genital herpes lesions.

17.12

Malaria and Plague: Examples of Insect Transmitted Diseases

Malaria is a disease caused by a protozoan, a member of the Sporozoa group. The malaria parasite is one of the most important human pathogens and has played an extremely significant role in the development and spread of human culture; indeed, it has even affected the human evolutionary process. Four species infect humans, of which the most widespread is *Plasmodium vivax*. This parasite carries out part of its life cycle in humans, and part in the mosquito; the mosquito is the vector by which the parasite spreads from person to person. Only mosquitoes of the genus *Anopheles* are involved, and since these primarily inhabit warmer parts of the world, malaria occurs predominantly in the tropics and subtropics. Malaria did not exist in the northern regions of North America prior to settlement by Europeans, but was of great incidence in the South, where appropriate breeding grounds for the mosquito existed. The disease has always been associated with swampy low-lying areas, and the name *malaria* is derived from the Italian words for "bad air."

The life cycle of the malaria parasite is complex and will not be discussed in detail here. One stage of the cycle occurs in humans, with the liberation into the bloodstream of a stage infective for mosquitoes. If a mosquito bites an infected person, the life cycle can be completed in the mosquito, with the formation of a stage infective for humans.

Eradication of malaria   Although there are several drugs (quinine, primaquine), which are effective in humans, because of the obligatory alternation of hosts, control of malaria can be best effected by elimination of the *Anopheles* mosquito. Two approaches to mosquito

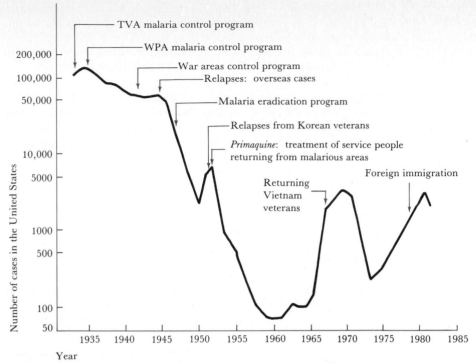

FIGURE 17.18 Dramatic decrease in incidence of malaria in the United States as a result of mosquito eradication programs. Note that the scale is logarithmic, so that the rise during the Vietnam War era is really quite minor. (From Morbidity and Mortality Weekly Report. Annual Summary 1980; issued September, 1981. Centers for Disease Control, Atlanta, Ga.)

control are possible: elimination of the habitat by drainage of swamps and similar areas, and elimination of the mosquito by insecticides. Both approaches have been extensively used. The marked drop in incidence of malaria as a result of mosquito eradication programs in the United States is shown in Figure 17.18. During the 1930s, about 33,000 miles of ditches were dug in 16 southern states, removing 544,000 acres of mosquito breeding area. Millions of gallons of oil were also used to spread on swamps, with the purpose of cutting off the oxygen supply to mosquito larvae. With the discovery of the insecticide dichlorodiphenyltrichloroethane (DDT), chemical control of both larvae and adult mosquitoes was possible. During World War II, the Public Health Service organized an Office of Mosquito Control in War Areas, and because many U.S. military bases were in the southern states, this organization carried out an extensive eradication program in the United States as well as overseas. In 1946, Congress established a 5-year malaria eradication program, which involved treatment of humans and spraying of mosquito areas with DDT to a point where malaria transmission could not occur. As a further indication of the success of this program, in 1935, there were about 4000 deaths from malaria in the United States, and by 1952, the number of deaths had been reduced to 25.

In other parts of the world, eradication has been much more difficult, but the same control measures are used. Despite some of the environmental problems associated with DDT, it is still an effective agent for mosquito control and has been most responsible for the control of malaria in large areas of Africa and South America.

Malaria and biochemical evolution of humans   One of the most interesting discoveries about malaria concerns the mechanism by which human beings in regions of the world where it is endemic acquire resistance to *Plasmodium* infections. Malaria has undoubtedly been endemic in Africa for thousands of years. In West Africans, resistance to malaria caused by *P. falciparum* is associated with the presence in their red cells of hemoglobin S, which differs from normal hemoglobin A only in a single amino acid in each of the two identical halves of the molecule. In hemoglobin S a neutral amino acid, valine, is substituted for an acidic amino acid, glutamic acid. Red cells containing hemoglobin S have reduced affinity for oxygen, and the malaria parasite, having a highly aerobic metabolism, cannot grow as well in these red cells as it can in normal ones. An additional consequence is that individuals with hemoglobin S are less able to survive at high altitudes, where oxygen pressures are lower, but in tropical lowland Africa this disadvantage is not manifested. In West Africans, resistance to another malarial parasite, *P. vivax*, is associated with the presence of another abnormal hemoglobin, hemoglobin E. In certain Mediterranean regions where malaria is endemic, resistance to *P. falciparum* is associated with a deficiency in the red cells of the enzyme glucose-6-phosphate dehydrogenase. The malaria parasite has thus been a factor in the biochemical evolution of human beings. Other microbial parasites have also probably promoted evolutionary changes in their hosts, but in no case do we have such clear evidence as in the case of malaria.

Plague   Pandemic occurrences of **plague** in the past have been responsible for more human deaths than any other infectious disease. Plague is caused by a Gram-negative, facultatively anaerobic rod called *Yersinia pestis* (see Section 19.13). Plague is a natural disease of domestic and wild rodents, but rats appear to be the primary disease reservoir.* Most infected rats die soon after symptoms begin, but a low proportion develop a chronic infection and can serve as a ready source of virulent *Y. pestis*. The majority of cases of human plague in the United States originate in the southwestern states, where the disease is endemic among wild rodents.

Plague is transmitted by the rat flea (*Xenopsylla cheopsis*), which ingests *Y. pestis* cells by sucking blood from an infected animal. Cells multiply in the flea's intestine and can be transmitted to a healthy animal in the next bite. As the disease spreads, rat mortality becomes so great that infected fleas seek new hosts, including humans. Once in humans, cells of *Y. pestis* usually travel to the lymph nodes, where they cause the formation of swollen areas referred to as *buboes*, and for this reason the disease is frequently referred to as **bubonic**

*The disease plague in rats is referred to as *sylvatic* plague.

plague. The buboes become filled with *Y. pestis,* but antiphagocytic surface antigens and a distinct capsule on cells of *Y. pestis* prevent them from being phagocytized. Secondary buboes form in peripheral lymph nodes and cells eventually enter the bloodstream, causing a generalized septicemia. Multiple hemorrhages produce dark splotches on the skin.* If not treated prior to the bacteremic stage, the symptoms of plague, including extreme lymph node pain, prostration, shock, and delirium, usually cause death within 3 to 5 days.

Pneumonic plague occurs when cells of *Y. pestis* are either inhaled directly or reach the lungs during bubonic plague. Symptoms are usually absent until the last day or two of the disease when large amounts of bloody sputum are emitted. Untreated cases rarely survive more than 2 days. Pneumonic plague, as one might expect, is a highly contagious disease, and can spread rapidly via the respiratory route if infected individuals are not immediately quarantined. **Septicemic plague** involves the rapid spread of *Y. pestis* throughout the body without the formation of buboes, and usually causes death before a diagnosis can be made.

Plague can be successfully treated if swiftly diagnosed. Although *Y. pestis* is naturally resistant to penicillin, most strains are sensitive to streptomycin, chloramphenicol, or the tetracyclines. If treatment is begun soon enough, mortality from bubonic plague can be reduced to as few as 1 to 5% of those infected. Pneumonic and septicemic plague can also be treated, but these forms progress so rapidly that antibiotic therapy in the latter stages of the disease is usually too late. Although potentially a devastating disease, an average of fewer than 10 cases of human plague are reported each year in the United States; this is undoubtedly due to improved sanitary practices and the overall control of rat populations.

17.13
Summary

In this chapter we have discussed the general problem of **epidemiology** and **public health,** with emphasis on the way in which an understanding of the spread of disease can lead to control of disease. Although dramatic increases in the health of the population have come about through discovery of cures for infectious disease (personal medicine), it is probably safe to say that for most infectious diseases, the greatest control has come about through public health measures. Although there have been many successes attesting to the value of an epidemiological approach to disease control, the field of epidemiology is much less highly developed than the field of medicine. Probably the main reason for this retarded development is that the application of epidemiological principles to the control of disease involves social problems that are often difficult to solve. There is no better example of this than gonorrhea, a disease for which the means of control have been available since World War II (penicillin), but which is currently of near-epidemic proportions.

Epidemiological control requires to a great extent the infusion of

*These skin splotches probably gave rise to the term "Black Death" used to describe the European pandemic of plague during the mid-1300s.

government regulation, and it is frequently difficult to sort out political from scientific problems. Although we may know how to handle a particular problem, we may find it difficult to obtain the necessary legislation; and even if the legislation is available, its enforcement may present vast areas of difficulty. Frequently, there is an infringement on the rights of the individual when the public welfare is to be protected, and society must often tread a fine line between repression of individual freedom and universal application of a significant health measure. The phenomenon of herd immunity well illustrates this. It is well established that a variety of infectious diseases could be completely controlled in the population if a sufficiently high proportion of the individuals were immunized. It is not essential to immunize everyone, but for some diseases immunity levels of greater than 70 percent are required to prevent the disease from becoming established in the population. Some people object to immunization on personal or religious grounds, yet they are being protected indirectly by the fact that their fellow citizens have been immunized. If, because of sufficiently high immunity in the population, a disease is absent, then motivation to become immunized for this disease falls, and with time a sufficiently large susceptible population builds up so that the disease can become established again. Government regulation could overcome the resistance of the individual to immunization, and everyone could be compelled by law to be immunized, yet in Western society, such government regulation may not find popular acceptance. It is only when government regulation involves the control of nonhuman aspects of disease transmittal, such as water, food, and insect control, that it becomes easily possible to institute appropriate measures.

From one viewpoint, epidemiology is an aspect of the broad field of microbial ecology, as it deals with the environmental relationships of a large group of microorganisms, the pathogens. From an evolutionary viewpoint, it can be understood how microbes have evolved which are able to infect animal hosts and to spread from one host to another. Dispersal problems differ for different microorganisms, and the field of environmental microbiology deals to a great extent with the manner by which pathogenic microorganisms achieve effective transmission from one host to another.

Supplementary Readings

Benenson, A. L., ed. 1981. Control of communicable diseases in man, 13th ed. American Public Health Association, Washington, D.C. A concise and useful reference to the complete spectrum of infectious diseases of humans, with emphasis on diagnostic and public health measures.

Bitton, G. 1980. Introduction to environmental virology. John Wiley & Sons, New York. An excellent source on transmission and survival of pathogenic viruses in the environment.

Centers for Disease Control. Morbidity and Mortality Weekly Report. Atlanta, Ga. Issued weekly. This publication, available in most large library systems, gives a random view of epidemiological problems in the United States. Most, but not all, of the issues relate to infectious disease. **Centers for Disease Control.** Foodborne disease. Annual summary. Atlanta, Ga., and Water-related disease outbreaks. Annual summary. Atlanta, Ga. Issued yearly. Useful, brief discussions of the epidemiology of food- and waterborne diseases, and a detailed presentation of all reported outbreaks for the year. The information given provides insight into principles of disease control for food- and water-borne diseases. **Centers for Disease Control.** Reported morbidity and mortality in the United States. Morbidity and Mortality Weekly Report, annual supplement. Atlanta, Ga. Issued yearly. This publication provides an overview of past and present incidence of infectious disease in the United States. Useful graphs and tables.

Davis, B. D., R. Dulbecco, H. N. Eisen, and **H. S. Ginsberg.** 1980. Microbiology, 3rd ed. Harper & Row, Hagerstown, Md. An in depth discussion of the venereal diseases and respiratory diseases can be found in this standard medical school textbook.

Federal Register, vol. 40, no. 248. 1975. Pages 59566–59588. National interim primary drinking water regulations. U.S. Government Printing Office. Washington, D.C. This material provides some of the rationale and scientific background, as well as the details, of the federal drinking water standards.

Fitzgerald, T. J. 1981. Pathogenesis and immunology of *Treponema pallidum.* Annu. Rev. Microbiol. 35:29–54. A good review of the features of *T. pallidum* thought to be important in the pathogenesis of syphilis.

Fraser, D. W., and **J. E. McDade.** 1979. Legionellosis. Sci. Am. 241:82–99. An exciting description of the successful hunt for the causative agent of Legionnaire's disease.

Freidman, M. J., and **W. Trager.** 1981. The biochemistry of resistance to malaria. Sci. Am. 244:154–164. Biochemistry and genetics of how those with the genetic disorders sickle-cell anemia and thalassemia are protected from malaria.

Hurst, A., and **D. L. Collins-Thompson.** 1979. Food as a bacterial habitat. Pages 79–134 *in* M. Alexander, ed. Advances in microbial ecology, vol. 3. Plenum Press, New York. An excellent review dealing with food as an environment for microbial growth. Emphasis is on growth of microorganisms in food rather than on foodborne diseases.

Kaplan, M. M., and **R. G. Webster.** 1977. The epidemiology of influenza. Sci. Am. 237:88–106. A readily understandable review article which presents a nice integration of the molecular biology, immunology, epidemiology, and medical history of influenza.

Lennette, E. H., A. Balows, W. J. Hausler, Jr., and **J. P. Truant.** 1980. Manual of clinical microbiology, 3rd ed. American Society for Microbiology, Washington, D.C. A compendium of clinically important pathogens.

Mausner, J. S., and **A. K. Bahn.** 1974. Epidemiology, an introductory text. W. B. Saunders, Philadelphia. An excellent, brief text of the principles of epidemiology. Chapter 12 deal specifically with infectious diseases.

Merson, M. H. 1976. Travelers' diarrhea in Mexico. N. Engl. J. Med. 294: 1299–1305. A fascinating study of travelers' diarrhea in a group of physicians attending a medical congress in Mexico City.

Snow on cholera. 1936. A reprint of two papers by John Snow, M.D. The Commonwealth Fund, New York. A reprinting of Snow's 1855 study of cholera, the first epidemiological investigation, and a fascinating detective story. Highly recommended.

18

Bacterial Taxonomy and Identification

In preceding chapters we have discussed the structure and function of procaryotic microorganisms. We learned that a variety of environments exist, and that microorganisms have evolved which are adapted to live in these environments; thus we can anticipate a diversity of organisms to exist. No group shows greater physiological and biochemical diversity than the procaryotes, and the present chapter will deal in an introductory way with this diversity, in anticipation of the discussion of individual groups in the next chapter.

18.1

Bacterial Taxonomy

One of the goals of taxonomy is to discover order in the apparent chaos of biological diversity. Just as the chemist examines the properties of the chemical elements and is able to see relationships that made possible the construction of the periodic table, so the taxonomist attempts to construct a convenient classification scheme for organisms. No implication of evolutionary relationships among the organisms of a single taxonomic group need be raised, although frequently such relationships do appear likely. It is important to distinguish between *taxonomy,* the attempt to find order in diversity, and *identification,* the act of recognizing that an organism under study is similar to previously characterized organisms. Taxonomy is a broad, philosophically oriented activity, whereas identification is a narrow, practically oriented activity. It is also important to distinguish between taxonomy and **nomenclature,** which is the naming of organisms. The name is merely a convenient handle, which expresses in a kind of shorthand the collective properties of the organism. In the first part of this chapter we discuss general aspects of bacterial taxonomy and then discuss the more practical problems of isolation and identification of various groups of bacteria.

Before we begin classifying organisms, we must decide what precisely are the basic units with which we shall be dealing. The taxonomic unit of microbiology is the **clone** or **strain,** which is a population of genetically identical cells derived from a single cell. It is very easy to isolate from nature a large number of clones that reflect the enormous diversity of microbial types. In a taxonomic study one cannot deal with the totality of bacteria, but must select some subgroup for more detailed study. It is most desirable to assemble a collection of strains of the subgroup under study, in order that the variability within the group can be recognized.

Approaches to bacterial taxonomy   There are many ways in which bacterial classification can be approached, depending on the group of organisms to be studied and the facilities available. The first might be called **classical taxonomy** since it is the way bacterial taxonomy has been carried out for over 100 years. A variety of characteristics of different organisms are measured, and these traits are then used in the separation of groups. A group of strains that have most or all characteristics in common will be classified as a single species, and related species are classified in the same genus.

Characteristics of taxonomic value that are widely used can be mentioned here: morphology, nutritional classification (phototroph versus heterotroph versus lithotroph), cell-wall chemistry, cell inclusions and storage products, capsule chemistry, pigments, nutritional requirements, ability to use various carbon, nitrogen, and sulfur sources, fermentation products, gaseous needs, temperature and pH requirements and tolerances, antibiotic sensitivity, pathogenicity, symbiotic relationships, immunological characteristics, and habitat. Some of these characteristics may not be applicable or useful with a particular group, and one must judge how extensive a compilation of data is needed to effect a good classification.

The classical approach is casual and nonsystematic, but it has proved very useful for some bacterial groups, especially those that are morphologically complex—such as the gliding, sheathed, and budding bacteria.

A second approach has been called **numerical taxonomy.** It resembles the classical approach in that a large number of characteristics are measured for the different organisms, but instead of placing more weight on some characteristics than on others and inspecting the data casually, all characteristics are given equal weight and a computer is used to compare the characteristics of each organism with those of others to recognize similarities and differences.

A third and perhaps the most fundamental is **genetic** or **molecular taxonomy,** which aims to ascertain the degree of genetic relatedness of different organisms. The ultimate goal could be to determine the DNA base sequences of the complete genome of the organisms, since the better the sequences of two organisms match, the closer is their genetic relatedness. Although this last goal has not yet been attained, several interesting approaches to it can be carried out, and these will be discussed in more detail below.

Molecular and Genetic Taxonomy

Molecular and genetic taxonomy involves studies designed to show directly or indirectly that the DNA base sequences of two organisms are similar or identical. If two organisms are able to mate and show genetic recombination, they are probably closely related, since recombination requires a close homology between the DNA molecules of the participating genomes. Unfortunately, sexual reproduction occurs infrequently in many procaryotes, so that detailed genetic studies often are not possible. However, several simplifying alternatives are available.

We discussed in Section 9.2 the general structure of DNA, and noted that the DNA base compositions of different procaryotes vary markedly. It is conventional to express the DNA base composition of an organism in terms of the percent of its DNA that contains guanine plus cytosine base pairs (mol % $G + C = \dfrac{G + C}{A + T + G + C} \times 100$).

The DNA base compositions of different organisms range from about 20 to greater than 78 percent guanine plus cytosine $(G + C)$. Despite the degeneracy in the genetic code, it has been calculated that if two organisms differ in $G + C$ content by more than 10 percent, there will be few base sequences in common. Therefore, it would be reasonable to conclude that two organisms having such different $G + C$ contents would not be closely related in the sense that they are direct descendents of a common ancestor. However, this does not mean that these two organisms do not share any common genetic elements, since they could well have plasmids, or even segments of the chromosome, which were identical.

Base compositions of DNA have been determined for a wide variety of bacteria, and several correlations can be observed. (1) Organisms with similar phenotypes often possess similar DNA base ratios (although numerous exceptions have been noted). (2) If two organisms thought to be closely related are found to have widely different base ratios, closer examination usually indicates that these organisms are not so closely related as was supposed. (3) However, two organisms can have identical base ratios and yet be quite unrelated since a variety of base sequences is possible with DNA of the same base composition.

We discussed in Section 9.2 the general approach to nucleic acid homology via hybridization studies. Nucleic acid hybridization provides a much more precise approach than measuring $G + C$ in determining the genetic relatedness among organisms. Both DNA-DNA and DNA-RNA hybridization can be determined, the latter using either messenger RNA made by radioactive labeling, or ribosomal RNA. Studies have shown that organisms that appear to be fairly distantly related by DNA-DNA or DNA-mRNA hybridization, may actually have very similar ribosomal RNAs. This suggests that ribosomal genes have been conserved during evolution, whereas other genes have been more greatly altered (see below). Relatedness of tRNA molecules has also been studied, and the results are similar to those obtained for ribosomal RNA, although the divergences are greater. Data available demonstrate that bacteria as a group are quite diversified, and that many organisms differ from each other by over 95 percent of their genomes. Mammals, on the other hand, are much more closely related as a group since even the most diverse mammals have about 20 percent of their nucleic acid sequences in common.

Nucleic acid sequencing   Procedures are now widely available for determining the nucleotide sequences of relatively small segments of nucleic acids, both RNA and DNA. Ultimately, such an approach may permit detailed characterization of genetic relatedness between organisms even in the absence of genetic recombination. The most thorough attempt to use nucleotide sequence determination in bacterial taxonomy is that involving the sequencing of 16S RNA isolated from the ribosomes of various bacterial species. **Ribosomal 16S RNA** consists of about 1500 nucleotides but can be broken up into fragments of from one to about 15 bases in length using enzymes which attack RNA. By sequencing the short stretches of nucleotides so generated, a "catalog" of sequences can be established for the 16S RNA of any particular species. Using statistical techniques it is possible to compare two sequence "catalogs" and determine the degree of similarity between the two (a perfect match is given a value of 1, while two totally unrelated species will usually show similarity coefficients of less than 0.1).

   Ribosomal RNA sequence studies have proven quite useful as an independent measure of microbial relatedness. By use of this approach, it has been possible not only to classify bacteria, but to give some indication of their **phylogeny** (evolutionary relatedness). Sequence studies have also been attempted using certain proteins as experimental subjects (for example, cytochrome c). Although some useful information has been obtained from protein sequence studies, 16S ribosomal RNA sequencing has the following advantages: (1) all bacteria contain ribosomes and hence, 16S ribosomal RNA, while a specific protein, for example cytochrome c, may or may not be found in any given species; and (2) it appears that ribosomal RNA sequences have been highly conserved and have changed at a very slow rate with time as compared with the rate of mutational change observed in DNA that codes for many proteins. The latter is an important point because it means that although some 16S RNA sequence changes have occurred, extensive changes have not, and thus the genetic record is not scrambled to the point where one 16S ribosomal RNA sequence would bear no resemblance to another. We will see in Section 19.33 how the 16S ribosomal RNA sequencing technique has resulted in a proposal for dividing procaryotes into two major classes, the **eubacteria** and the **archaebacteria,** and how members of the latter group share a number of highly unusual features in common.

Genetic recombination as a taxonomic tool   In organisms that possess mechanisms for genetic recombination, relatedness may be shown through studies on the efficiency of genetic exchange. However, since even closely related organisms may not mate with each other for various reasons, genetic analysis makes possible the recognition of similarities between organisms but not of differences since, if recombination does not occur, it is not usually possible to ascertain whether this is for trivial reasons or because the organisms are unrelated. In eucaryotes the limiting factor for recombination is often the inability of the chromosomes to pair at meiosis. In procaryotes recombination is virtually always a fragmentary process, and hence it is difficult to study the relatedness of whole genomes.

In procaryotes DNA-mediated transformation has proved useful for taxonomic studies. The DNA can be extracted from the strain to be used as a potential donor and can then be used in attempts to transform a competent recipient. This approach has been used extensively in studies on the genus *Bacillus,* using *B. subtilis* as a recipient. The DNAs of *Bacillus* species that have overall base compositions different from that of *B. subtilis* do not transform it, whereas some species that have base compositions like *B. subtilis* transform it at very high frequencies. Others transform it at lower frequencies. DNA from *Bacillus stearothermophilus* does not transform *B. subtilis* at all, even though it has the same base composition. This indicates that a similarity in base composition is necessary but is not sufficient to guarantee interspecies transformation. Presumably, the lack of recombination results from the absence of homologous DNA base sequences.

Transduction has also been used to study the genetic relatedness of the various regions of the genomes of different organisms. Transduction is especially useful in fine-structure genetic analysis, and thus permits determining whether two organisms differ in the structure of particular genes or gene clusters.

18.3

Bacterial Taxonomy and Phylogeny

The primary goal of any classification procedure is to collect related organisms into taxonomic groups called **species.** It is not easy to define a species. In microbiology "species" most commonly refers to a collection of similar strains or clones that differ significantly from other groups of clones. Unfortunately, this definition is so vague that it leaves much room for subjective interpretation. Furthermore, if this definition is followed, species distinctions will continually change as our knowledge of organisms broadens. Ultimately, we may see not several clear-cut species but only a continuum of strains, grading from one extreme to another.

Groups of species are collected in **genera,** and genera are grouped into **families.** Related families are placed in **orders,** orders are placed in **classes,** classes in **phyla,** and phyla in **kingdoms.** The ordering of organisms in this way is usually done with the aim of discerning evolutionary relationships among them. Such ordering is usually referred to as **phylogeny,** and can be done at any level, from genera through kingdoms. However, at our present state of knowledge of microbial evolution no phylogeny can be very informed.

The classification of procaryotes is a difficult matter. It is fairly easy to recognize species and genera, but difficulty arises when an attempt is made to collect related genera into families and orders. A few genera lend themselves well to classification into higher taxa, but most do not. The main problem is the difficulty in deciding whether physiological or morphological characteristics should be paramount in dividing groups.

The most extensive effort to organize the bacteria is that represented by *Bergey's Manual of Determinative Bacteriology,* the first edition of which was published in 1923. This work, whose name has been changed in the latest edition to *Bergey's Manual of Systematic Bacteriology,* is divided into four volumes. Volume I deals with Gram-negative bacteria of medical or industrial significance; Volume II, with Gram-positive bacteria of medical or industrial significance; Volume III, with the remaining Gram-negative bacteria, including the cyanobacteria; and Volume IV covers the actinomycetes. *Bergey's Manual* places emphasis on groups of bacteria at the family level and below, and generally arranges various genera into groups based on either morphological or physiological criteria. An outline of the classification scheme used in *Bergey's Manual* is given in Appendix 4.

In the rest of this chapter we present some of the approaches and techniques used in the isolation and characterization of bacteria with the use of *Bergey's Manual.* This material also can be considered to be an introduction to the more detailed treatment of the bacteria presented in Chapter 19.

18.4

The Classification of Procaryotes

Morphology Although molecular and genetic approaches to taxonomy have cast some doubt on the usefulness of morphological criteria in the classification of procaryotes, morphological properties remain of major importance in the separation of many groups of bacteria. Although not exactly a morphological property, Gram-staining characteristics are generally grouped with morphology, because one can determine Gram stain and some aspects of morphology at the same time. Several of the groups of bacteria in *Bergey's Manual* are separated on the basis of morphology and Gram stain: Gram-positive rods, Gram-positive cocci, Gram-negative rods, Gram-negative cocci, Gram-negative spiral and curved bacteria and so on. The bacterial endospore is a structure of unique characteristics and practical importance, and all endospore-forming bacteria are classified together. A very large group of bacteria possess either a filamentous or coryneform (indeterminate) morphology, and are classified together as the Actinomycetes and related organisms. Several other groups that are separated on the basis of morphology are: the gliding bacteria, rod-shaped or filamentous organisms which move by gliding motility; the sheathed bacteria, generally rod-shaped organisms that grow within long hollow tubes called sheaths; budding or appendaged bacteria, which have stalks, buds, or other protoplasmic extrusions called prosthecae, and which often multiply by budding rather than by simple cell division; the spirochetes, a group of bacteria which are always long spirals or helices, and which move by means of a unique system of fibers, called the axial filament (Figure 18.1). Finally, the mycoplasmas are a group of bacteria that lack a cell wall, and are generally detected by means of the characteristic colonies they make on agar plates, or by their unusual sensitivity to osmotic lysis. Although some mycoplasmas may have been derived from other bac-

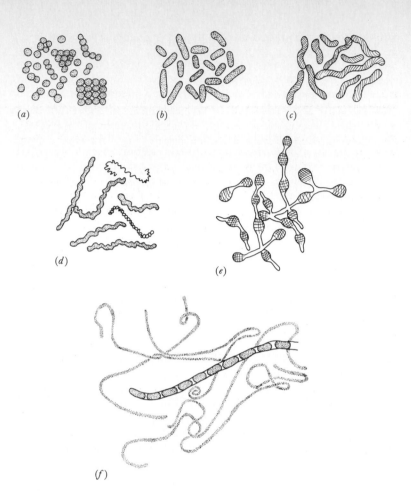

FIGURE 18.1 Representative morphological types of procaryotes: (a) cocci, (b) rods, (c) spirilla, (d) spirochetes, (e) budding bacteria, and (f) filamentous and sheathed-filamentous bacteria.

teria by accidental loss of the ability to form a cell wall, there is good evidence that other mycoplasmas are unique organisms, not directly related to wall-containing bacteria; thus they are grouped together.

One major group of bacteria, the rickettsias, are separated on the basis of their constant and obligate association with animal hosts. The rickettsias are either rod-shaped or coccoid, have typical bacterial cell walls, and are Gram negative. However, the rickettsias appear to be obligate intracellular parasites of warm-blooded animals and insects, and this property seems so fundamental that they are classified as a separate group.

Energy metabolism Several of the major divisions of bacteria are based on energy-generating mechanisms; especially photosynthesis and lithotrophy. The utilization of light energy is considered such a fundamental property that it becomes paramount in the classification scheme. Thus if an organism grows phototrophically and has only photosystem I, it is immediately classified among the purple and green bacteria. If both photosystems I and II are present and the organism is clearly procaryotic, it is placed within the cyanobacteria.

Similarly, if an isolate has the ability to use an inorganic compound as electron donor for energy generation, it is generally classified among the lithotrophs.

Rarely does a natural environment contain only a single type of microorganism. In most cases, a wide variety of organisms are present and it is the task of the bacteriologist to devise methods and procedures which will permit the isolation and culture of organisms of interest. The most common approach to this goal is the **enrichment culture method**. In this method, a medium or set of incubation conditions are used that are *selective* for the desired organism, and that are inhibitory, or counterselective, for undesired organisms. An outline of some successful enrichment culture procedures is given in Table 18.1. This table can be referred to from time to time during the study of Chapter 19.

Successful enrichment requires that an appropriate **inoculum** source be used. The dictum: "Everything is everywhere, the environment selects" has some basis in fact, but it should be understood that a given handful of soil or mud will not automatically contain the organism sought. Thus we begin by going to the appropriate habitat for our inoculum and again, experience provides the clue.

A Bit of History

Students of bacterial diversity owe a great debt to the Delft school of microbiology, which was initiated by Martinus Beijerinck and continued by A. J. Klyuver and C. B. van Niel. Beijerinck, one of the world's greatest microbiologists (among other things, he was the first to characterize viruses), first devised the enrichment culture technique (see below) and used this technique in the isolation and characterization of a wide variety of bacteria. Subsequently, Kluyver and van Niel used the enrichment culture technique to isolate photosynthetic and lithotrophic bacteria, and to show the fascinating physiological diversity among the bacteria. Another important figure from the Dutch school was L. G. M. Baas-Becking, who carried out the first calculations of the energetics of photosynthetic and lithotrophic bacteria, and emphasized the importance of these organisms in geochemical processes. van Niel subsequently came to the United States, and was responsible for the training of a number of general bacteriologists who have carried on this tradition, including R. Y. Stanier, Robert E. Hungate, and M. Doudoroff. In addition, a number of scientists have visited van Niel's laboratory at Pacific Grove, California and learned his methods and approaches. In recent years, the Delft tradition has been carried on in its home country, The Netherlands, by Hans Veldkamp (at Groningen), and in Germany by Norbert Pfennig (at Konstanz), both of whom spent time in van Niel's laboratory.

An excellent example of the application of the enrichment culture technique can be found in the isolation of *Azotobacter,* a nitrogen-fixing bacterium discovered by Beijerinck in 1901 (see Section 19.9). Beijerinck was interested in knowing whether any aerobic bacteria capable of fixing nitrogen existed in the soil. To answer this question he added a small amount of soil to an Erlenmeyer flask containing a thin layer of mineral salts medium devoid of ammonia, nitrate, or any other form of fixed nitrogen, and which contained

TABLE 18.1   Enrichment Culture Methods for Bacteria

Light   Phototrophic bacteria: main C source, CO_2

Aerobic incubation:		Organisms enriched	Inoculum
N_2 as N source		Cyanobacteria	Pond or lake water, sulfide-rich muds, stagnant water, moist decomposing leaf litter, moist soil exposed to light
Anaerobic incubation:			
H_2 or organic acids; N_2 as sole nitrogen source		Rhodospirillaceae ⎫	
H_2S as electron donor		Chromatiaceae ⎬ purple bacteria	
H_2S as electron donor		Chlorobiaceae: green bacteria	

Dark   Lithotrophic bacteria: main C source, CO_2 (medium must lack organic C)

Aerobic incubation:

Electron donor	Electron acceptor	Organisms enriched	Inoculum
NH_4^+	O_2	Nitrosofying bacteria (*Nitrosomonas*)	Soil, mud, sewage effluent
NO_2^-	O_2	Nitrifying bacteria (*Nitrobacter*)	
H_2	O_2	Hydrogen bacteria (various genera)	
H_2S, S^0, $S_2O_3^{2-}$	O_2	*Thiobacillus* spp.	
Fe^{2+}, low pH	O_2	*Thiobacillus ferrooxidans*	
Anaerobic incubation:			
S^0, $S_2O_3^{2-}$	NO_3^-	*Thiobacillus denitrificans*	
H_2	NO_3^- + yeast extract	*Paracoccus denitrificans*	

Dark   Heterotrophic and methanogenic bacteria: main C source, organic compounds

Aerobic incubation:

	Main ingredients	Electron acceptor	Organisms enriched	Inoculum
	Lactate + NH_4^+	O_2	*Pseudomonas fluorescens*	Soil, mud, decaying vegetation, lake sediments; pasteurize inoculum (80°C) for all *Bacillus* enrichments
	Benzoate + NH_4^+	O_2	*Pseudomonas fluorescens*	
	Starch + NH_4^+	O_2	*Bacillus polymyxa*, other *Bacillus* spp.	
Respiration	Ethanol (4%) + 1% yeast extract pH 6.0	O_2	*Acetobacter, Gluconobacter*	
	Urea (5%) + 1% yeast extract	O_2	*Sporosarcina ureae*	
	Hydrocarbons (e.g., mineral oil) + NH_4^+	O_2	*Mycobacterium, Nocardia*	
	Cellulose + NH_4^+	O_2	*Cytophaga, Sporocytophaga*	
	Mannitol or benzoate, N_2 as N source	O_2	*Azotobacter*	

Anaerobic incubation:

	Main ingredients	Electron acceptor	Organisms enriched
	Organic acids	2% KNO_3	*Pseudomonas* ⎫
	Yeast extract	10% KNO_3	*Bacillus* ⎬ Dentrifying species
Anaerobic respiration	Organic acids	Sulfate	*Desulfovibrio, Desulfotomaculum*
	Acetate, propionate, butyrate	Sulfate	Type II sulfate reducers
	H_2	Carbonate	Methanogenic bacteria (lithotrophic species only)
	CH_3OH	Carbonate	*Methanosarcina barkeri*

Anaerobic incubation:

	Main ingredients	Electron acceptor	Organisms enriched	Inoculum
	Glutamate or histidine	None	*Clostridium tetanomorphum*	Mud, lake sediments, rotting plant material, dairy products (lactic and propionic acid bacteria), rumen or intestinal contents
	Starch + NH_4^+	None	*Clostridium* spp.	
	Starch, N_2 as N source	None	*Clostridium pasteurianum*	
	Lactate + yeast extract	None	*Veillonella* spp.	
Fermentation	Glucose + NH_4^+	None	*Enterobacter*, other fermentative organisms	
	Glucose + yeast extract (pH 5)	None	Lactic acid bacteria (*Lactobacillus*)	
	Lactate + yeast extract	None	Propionic acid bacteria	

Translated and adapted from Schlegel, H. G. 1976. Allgemeine Mikrobiologie, Thieme-Verlag, Stuttgart, West Germany.

mannitol as carbon source. Within 3 days a thin film developed on the surface of the liquid and the liquid became quite turbid. Beijerinck observed large, rod-shaped cells which appeared quite distinct from the spore-containing rods of the only other nitrogen-fixing organism known at that time, *Clostridium pasteurianum,* an anaerobe discovered years earlier by Sergei Winogradsky. Beijerinck streaked agar plates containing phosphate and mannitol with the turbid liquid, incubated aerobically and within 48 hours obtained large slimy colonies typical of *Azotobacter.* Pure cultures were obtained by picking and restreaking well-isolated colonies a number of times. Beijerinck assumed his new organism was utilizing N_2 from air as its source of cell nitrogen and later proved this by showing total nitrogen increases in pure cultures of *Azotobacter* grown in the absence of fixed nitrogen.

The advantage of the enrichment culture approach used here should be evident. By omitting from his medium any source of combined nitrogen and incubating aerobically, Beijerinck placed severe constraints on the microbial population. Any organism that developed had to be able to both fix its own nitrogen and tolerate the presence of molecular oxygen. Beijerinck noted that if he placed too much liquid in his flasks, or employed readily fermentable substances such as glucose or sucrose in place of mannitol, his enrichment would frequently turn anaerobic and favor the growth of *Clostridium* rather than *Azotobacter.* The addition of ammonia or nitrate to the original enrichment never resulted in the isolation of *Azotobacter,* but only to a variety of non-nitrogen-fixing bacteria instead. Hence Beijerinck showed that the composition of the growth medium as well as the incubation conditions employed were of paramount importance in the development of an enrichment culture.

18.6

Pure Cultures

Before anything more than a very superficial understanding of specific bacterial characteristics can be obtained, it is essential to undertake **pure culture** studies. With very few exceptions, the morphological properties of procaryotes are so limited that one cannot make an absolute or frequently even an approximate identification of an organism without isolation in pure culture. Also, without pure cultures one cannot determine the characteristics of an organism that are of most general interest, such as nutritional requirements, responses to environment, metabolic products, or pathogenicity (see Section 1.5).

Pure means free of foreign elements or of other living organisms. Hence a pure culture is a culture containing a *single kind* of microorganism. Any culture that contains more than one kind of microorganism is not a pure culture, and is, by definition, a **mixed** (or contaminated) **culture.** In examining a culture microscopically, it is almost impossible to detect a stray contaminant, because the sensitivity of the light microscope is low. With the average oil immersion lens, the field size is such that if the bacterial count is 10^6 cells/ml, there will be on the average only 1 cell per field. This means that if a contaminant numbered 10^4 per milliliter, a not especially small number, one would have to examine 100 fields in order to find *one* organism. Much more sensitive methods involve inoculation of a putative pure culture into a medium which may favor the growth of contaminants; growth of contaminants in such media will quickly indicate that the culture is not pure. On the other hand, absence of growth of contaminants under these conditions does not mean that they are absent, since the culture medium used may not be favorable for their growth. Thus critical determination that a culture is pure is quite difficult, and might even become a small research project in itself.

Pure cultures can be obtained in many ways but the most frequently employed means are the streak plate, the agar shake, or liquid dilution. For organisms that grow well on agar plates the **streak plate** is the method of choice. By repeated picking and restreaking of a well-isolated colony one eventually obtains a pure culture that can then be transferred to a liquid medium. With proper incubation it is possible to purify both aerobes and anaerobes on agar plates via the streak plate method. The **agar shake-tube** method involves the dilution of a mixed culture in tubes of molten agar resulting in colonies embedded *in* the agar rather than on the surface of a plate. The shake tube method has been found useful for purifying particular types of microorganisms (for example, phototrophic sulfur bacteria and sulfate-reducing bacteria; see Section 19.16). For organisms that do not grow well in or on agar, purification can be obtained by successively diluting a cell suspension in tubes of liquid medium until a point is reached beyond which no growth is obtained. By repeating this **liquid dilution** procedure using the highest dilution showing growth as inoculum, it is possible to obtain pure cultures, although the process frequently takes much longer than agar-based purification methods.

18.7

Diagnostic Tests

The detailed characterization of any organism generally depends upon determination of specific physiological and biochemical characteristics. However, it should be emphasized that often what the diagnostic bacteriologist means by "biochemical" is something different from the specific biochemical characteristics that might be deduced from a study of metabolic pathways (such as outlined in Chapter 5 and Appendix 3). In diagnostic work, biochemical tests are generally simple tests that are designed to indicate in a clear-cut fashion the presence or absence of a biochemical characteristic, such as an enzyme. The basis of all of these tests can be ultimately found in the presence or absence in the organism of an enzyme, a group of enzymes, or a whole metabolic pathway. These tests have been developed over the past 75 years, and have undergone numerous modifications, substitutions, and deletions. Many of these tests are now performed in commercially prepared minaturized media kits, which consist of a series of small wells containing various diagnostically useful media or reagents. Most such kits are designed to be inoculated by simply passing a sterile rod containing a portion of a colony through each of the wells. After an appropriate incubation period (usually 30 hours or less) the results of the tests are read by observing a color change or some other easily interpreted reactions. Although diagnostic tests are used most often in the identification of medically important bacteria, they may also be used for similar bacteria of non-medical importance. In Table 18.2, the most important diagnostic tests are given, and the references at the end of this chapter can be used to learn about the much wider array of tests that are also used.

Steps in the Identification of an Unknown Bacterial Culture

The following series of steps will provide a framework within which identification of pure cultures can be accomplished. The information obtained can then be used as an entry into the keys and descriptions in *Bergey's Manual*, with the likelihood that at least a genus name can be put on the isolate, if not a species name.

1. Begin with a pure culture.
2. Determine energy requirements: it can usually be determined by the isolation and culture methods whether the organism is phototrophic, lithotrophic, or heterotrophic.
3. Examine living cells by phase contrast and Gram-stained cells by bright-field microscopy. Determine whether the organism is Gram positive or Gram negative, and describe the morphology: rod, coccus, vibrio, spiral, spirochete, filament, sheath, and so on.
4. Look for spores, holdfasts, stalks, prosthecae, or other identifying characteristics. If necessary, carry out a spore stain. Many endospore-forming bacteria sporulate poorly. Look for spores carefully, especially if the organism is a large Gram-positive rod, and try several different culture media to induce sporulation.
5. Look for motility in wet mounts. Be sure to use freshly grown cultures. From the type of motility, attempt to decide whether the organism is polarly or peritrichously flagellated. If the matter is critical, carry out a flagella stain, or examine negatively-stained preparations under the electron microscope.
6. Examine colonies or mass growth for pigments or other unique characteristics.
7. Test for oxygen requirements. Will the culture grow both aerobically and anaerobically? If the culture has originally been isolated under anaerobic conditions, is it an obligate or facultative anaerobe? Or microaerophilic?
8. If a heterotrophic organism, test for dissimilation of glucose or other simple sugar. Oxidative or fermentative (O/F test).
9. If phototrophic, check to see if it will also grow in the dark aerobically (characteristic of nonsulfur purple bacteria, Rhodospirillaceae).
10. If lithotrophic, usually it has been isolated on an inorganic electron donor (for example, sulfur, ammonia, iron). This provides an immediate entry into the appropriate group in *Bergey's Manual*.
11. After the above, attempt a preliminary identification. From the cluster of genera selected, complete additional tests, if necessary, to narrow the identification.

If, after the steps described above, a clearly positive identification cannot be made, do not despair or assume that you have isolated a new species or genus. It should be remembered that the descriptions

TABLE 18.2 Important Diagnostic Tests for Bacteria

Test	Principle	Procedure	Most Common Use
Carbohydrate fermentation	Acid and/or gas during fermentative growth with sugars or sugar alcohols	Broth medium with carbohydrate and phenol red as pH indicator; inverted tube for gas	Enteric bacteria differentiation (also several other genera or species separations with some individual sugars)
Catalase	Enzyme decomposes hydrogen peroxide, H_2O_2	Add drop of H_2O_2 to dense culture and look for bubbles (O_2) (Figure 8.16)	*Bacillus* ($+$) from *Clostridium* ($-$); *Streptococcus* ($-$) from *Micrococcus/Staphylococcus* ($+$)
Citrate utilization	Utilization of citrate as sole carbon source, results in alkalinization of medium	Citrate medium with bromthymol blue as pH indicator, look for intense blue color (alkaline pH)	*Klebsiella-Enterobacter* ($+$) from *Escherichia* ($-$), *Edwardsiella* ($-$) from *Salmonella* ($+$)
Coagulase	Enzyme causes clotting of blood plasma	Mix dense liquid suspension of bacteria with plasma, incubate, and look for sign of clot in few minutes	*Staphylococcus aureus* ($+$) from *S. epidermidis* ($-$)
Decarboxylases (lysine, ornithine, arginine)	Decarboxylation of amino acid releases CO_2 and amine	Medium enriched with amino acid. Bromcresol purple pH indicator. Alkaline pH if enzyme action, indicator becomes purple	Aid in determining bacterial group among the enteric bacteria
β-Galactosidase (ONPG) test	Orthonitrophenyl-β-galactoside (ONPG) is an artificial substrate for the enzyme. When hydrolyzed, nitrophenol (yellow) is formed.	Incubate heavy suspension of lysed culture with ONPG, look for yellow color	*Citrobacter* and *Arizona* ($+$) from *Salmonella* ($-$). Identifying some *Shigella* and *Pseudomonas* species
Gelatin liquefaction	Many proteases hydrolyze gelatin and destroy the gel	Incubate in broth with 12% gelatin. Cool to check for gel formation. If gelatin hydrolyzed, tube remains liquid upon cooling	To aid in identification of *Serratia*, *Pseudomonas*, *Flavobacterium*, *Clostridium*
Hydrogen sulfide (H_2S) production	H_2S produced by breakdown of sulfur amino acids or reduction of thiosulfate	H_2S detected in iron-rich medium from formation of black ferrous sulfide (many variants: Kliger's iron agar, triple sugar iron agar, also detect carbohydrate fermentation)	In enteric bacteria, to aid in identifying *Salmonella*, *Arizona*, *Edwardsiella*, and *Proteus*
Indole test	Tryptophan from proteins converted to indole	Detect indole in culture medium with dimethylaminobenzaldehyde (red color)	To distinguish *Escherichia* ($+$) from *Klebsiella-Enterobacter* ($-$); *Edwardsiella* ($+$) from *Salmonella* ($-$)
Methyl red test	Mixed-acid fermenters produce sufficient acid to lower pH below 4.3	Glucose-broth medium. Add methyl red indicator to a sample after incubation	To differentiate *Escherichia* ($+$, culture red) from *Enterobacter* and *Klebsiella* (usually $-$, culture yellow)
Nitrate reduction	Nitrate as alternate electron acceptor, reduced to NO_2^- or N_2	Broth with nitrate. After incubation, detect nitrite with α-naphthylamine-sulfanilic acid (red color). If negative, confirm that NO_3^- still present by adding zinc dust to reduce NO_3^- to NO_2^-. If no color after zinc, then $NO_3^- \rightarrow N_2$	To aid in identification of enteric bacteria (usually $+$)

TABLE 18.2 *(continued)*

Test	Principle	Procedure	Most Common Use
Oxidase test	Cytochrome oxidizes artificial electron acceptor: tetramethyl (or dimethyl)-*p*-phenylenediamine	Broth or agar. Oxidase-positive colonies on agar can be detected by flooding plate with reagent and looking for blue or brown colonies	To separate *Neisseria* and *Moraxella* (+) from *Acinetobacter* (−). To separate enteric bacteria (all −) from pseudomonads (+). To aid in identification of *Aeromonas* (+)
Oxidation-fermentation (O/F) test	Some organisms produce acid only when growing aerobically	Acid production in top part of sugar-containing culture tube; soft agar used to restrict mixing during incubation	To differentiate *Micrococcus* (aerobic acid production only) from *Staphylococcus* (acid anaerobically). To characterize *Pseudomonas* (aerobic acid production) from enteric bacteria (acid anaerobically)
Phenylalanine deaminase test	Deamination produces phenylpyruvic acid, which is detected in a colorimetric test	Medium enriched in phenylalanine. After growth, add ferric chloride reagent, look for green color	To characterize the genus *Proteus* and the *Providencia* group
Starch hydrolysis	Iodine-iodide gives blue color with starch	Grow organism on plate containing starch. Flood plate with Gram's iodine, look for clear zones around colonies	To identify typical starch hydrolyzers such as *Bacillus* spp.
Urease test	Urea $(H_2N-\overset{\overset{O}{\|\|}}{C}-NH_2)$ split to $NH_3 + CO_2$	Medium with 2% urea and phenol red indicator. Ammonia release raises pH, intense pink-red color	To distinguish *Klebsiella* (+) from *Escherichia* (−). To distinguish *Proteus* (+) from *Providercia* (−)
Voges-Proskauer test	Acetoin produced from sugar fermentation	Chemical test for acetion using α-napthol	To separate *Klebsiella* and Enterobacter (+) from *Escherichia* (−). To characterize members of genus *Bacillus*.

in *Bergey's Manual* are based on majorities, and the reactions given are those characteristic of the majority of the isolates of that genus or species. If an identification cannot be accomplished, (1) check the culture for its purity; (2) ascertain whether the appropriate tests have been carried out properly; (3) check that the methods have been used properly; and (4) be certain that the appropriate keys and tables have been sought.

It is always a good idea to carry through along with the unknown culture, several known cultures of the same or similar groups. It is especially important to carry out Gram stains on cultures known to be Gram positive and Gram negative, because errors in deciding Gram characteristics are very commonly made, and such errors lead one immediately into the wrong part of the *Manual*. Other common errors occur in deciding morphology (for example, small rods are often hard to tell from cocci; coccobacilli are especially difficult to identify, thus it is best to obtain some authentic cultures, if possible), and in motility, especially in distinguishing polar from peritrichous flagellation.

In most cases, with experience, identification of the correct genus should be easily possible, although identification of species or subspecies may require the assistance of a specialized reference laboratory, or the help of an expert.

We now turn to a consideration of the extensive diversity of procaryotic organisms. As Chapter 19 is covered, it might be profitable, from time to time, to return to some of the concepts outlined in the present chapter.

Supplementary Readings

Buchanan, R. E., and **N. E. Gibbons,** eds. 1974. Bergey's manual of determinative bacteriology, 8th ed. Williams and Wilkins, Baltimore. The standard work on bacterial taxonomy, with extensive descriptions of genera and species; also includes numerous tables, keys, and photographs. See Section 18.3 and Appendix 4 of *Biology of Microorganisms* 4th edition, for details of the new edition of *Bergey's Manual.*

Lennette, E. H., A. Balows, W. J. Hausler, Jr., and **J. P. Truant.** 1980. Manual of clinical microbiology, 3rd ed. American Society for Microbiology, Washington, D.C. A detailed laboratory manual for the identification of pathogenic microorganisms, with precise directions for carrying out various diagnostic tests, and summaries of the diagnostic characteristics shown by various groups of organisms.

Schlegel, H. G. 1965. Anreicherungskultur and Mutantenauslese. Zentralblatt für Bacteriologie, I. Abteilung, Suppl. 1. Gustav Fischer Verlag, Stuttgart, West Germany. Although the title is in German, many of the articles are in English. This is an extremely valuable reference to enrichment culture methods for various groups of microorganisms, primarily bacteria.

Schlegel, H. G. and **H. W. Jannasch.** 1981. Prokaryotes and their habitats. Pages 43–82 *in* M. P. Starr, H. Stolp, H. G. Trüper, A. Balows, and H. G. Schlegel, eds. The prokaryotes, a handbook of habitats, isolation, and identification of bacteria. Springer-Verlag. New York.

Skerman, V. B. D., V. McGowan, and **P. H. A. Sneath,** eds. 1980. Approved lists of bacterial names. Int. J. Sys. Bacteriol. 30:225–420. A source list of all the bacterial names that have been validly published to describe strains of bacteria for which authentic type or neotype cultures exist.

Stolp, H., and **M. P. Starr.** 1981. Principles of isolation, cultivation, and conservation of bacteria. Pages 135–175 *in* M. P. Starr, H. Stolp, H. G. Trüper, A. Balows, and H. G. Schlegel, eds. The prokaryotes, a handbook of habitats, isolation, and identification of bacteria. Springer-Verlag, New York.

Trüper, H. G., and **J. Kramer.** 1981. Principles of characterization and identification of prokaryotes. Pages 176–193 *in* M. P. Starr, H. Stolp, H. G. Trüper, A. Balows, and H. G. Schlegel, eds. The prokaryotes, a handbook of habitats, isolation, and identification of bacteria. Springer-Verlag, New York.

Representative Procaryotic Groups

In the preceding chapter we discussed general approaches to the study of procaryotic diversity. In this chapter we shall describe in some detail the diversity among the procaryotes. Thus, rather than discussing the features procaryotes have in common, we shall consider the ways in which they differ. Our discussion builds on the foundation of information provided by Chapter 2, which dealt with cell structure, and by Chapters 4 through 6, which dealt with cell metabolism. Many of the organisms discussed in the present chapter have already been discussed in an ecological or applied context in Chapters 8 and 13 through 17. Ideally, the present chapter should not be read as a unit, but rather a little at a time and not necessarily in sequence. Perhaps the best approach would be to study selected sections of this chapter at the same time that earlier chapters are covered, in order to illuminate concepts discussed there.

The order in which the groups are presented in this chapter reflects the general sequence of presentation in *Bergey's Manual*. The first of a four-volume new edition (renamed *Bergey's Manual of Systematic Bacteriology*) appeared in 1984 and Appendix 4 lists the genera to be covered in each of the four volumes. There is nothing especially significant about the order of presentation here, and the student is cautioned not to conclude that organisms placed early in this chapter are in any way more primitive than organisms placed later in the chapter. Our knowledge of the taxonomy and evolution of the procaryotes is much too imperfect to construct a convincing scheme for the placement of genera and higher taxa. Molecular approaches to taxonomy may ultimately resolve these phylogenetic questions and, as mentioned in the previous chapter (see Section 18.2), significant strides in this direction have already been taken. The present discussion, however, will emphasize the biochemical, physiological, morphological, and ecological diversity found in the procaryotic world.

Purple and Green (Phototrophic) Bacteria

Bacteria able to use light as an energy source comprise a large and heterogeneous group of organisms, grouped together primarily because they possess one or more pigments called chlorophylls and are able to carry out light-mediated generation of ATP, a process called **photophosphorylation**. Two major groups are recognized, the purple and green bacteria as one group, and the blue-green algae (cyanobacteria) as the other group. The basic distinction between the purple and green bacteria and the cyanobacteria is based on the photopigments and overall photosynthetic process:

Group	Photopigment	Photosynthesis	Photosynthetic O_2 Production
Cyanobacteria	Chlorophyll *a*	Photosystems I and II	Yes
Purple and green bacteria	Bacteriochlorophyll	One photosystem only	No

The biochemistry of photosynthesis has been discussed in some detail in Chapter 6. Green-plant photosynthesis, which is also exhibited by the cyanobacteria, involves two light reactions, and H_2O serves as the electron donor. As a result of the photolysis of water, O_2 is produced. Photosynthesis in the purple and green bacteria, on the other hand, involves only one light reaction, water is not photolysed, and O_2 is not produced. Because the purple and green bacteria are unable to photolyse water, they must obtain their reducing power for CO_2 fixation from a reduced substance in their environment. This can be either an organic compound, a reduced sulfur compound, or H_2. Purple and green bacteria can only grow photosynthetically under anaerobic conditions, probably because only under anaerobic conditions are many of the required reduced electron donors stable. Cyanobacteria, on the other hand, can develop readily under aerobic conditions with no reduced compounds present. Many purple and green bacteria can grow autotrophically with CO_2 as carbon source and a reduced sulfur compound or H_2 as reductant under anaerobic conditions in the light. Under microaerobic conditions, these same reductants serve as electron donors to support lithotrophic growth of many purple bacteria. Phototrophic growth is also possible, however, with light serving as energy source and an organic compound as a carbon source, and under such conditions, CO_2 is only a minor source of cell carbon. Some purple and green bacteria can also grow in the dark using an organic compound as electron donor, but under these conditions a suitable electron acceptor is necessary; O_2 is the most commonly used electron acceptor, but NO_3^-, elemental sulfur, thiosulfate, dimethylsulfoxide, and trimethylamine oxide can be used by certain organisms. A few purple bacteria can also grow in the dark (albeit slowly) by strictly fermentative metabolism. One interesting organism, a *Rhodopseudomonas* species, can even grow anaerobically via the fermentation of carbon monoxide: $CO + H_2O \rightarrow CO_2 + H_2$.

c_s

Pigment	R_1	R_2	R_3	R_4	R_5	R_6**	R_7	Long wave-length absorption maxima (nm) In vivo	Extract (methanol)
Bacterio-chlorophyll a	—C(=O)—CH₃	—CH₃*	—CH₂—CH₃*	—CH₃	—C(=O)—O—CH₃	P/Gg	—H	805 830–890	771
Bacterio-chlorophyll b	—C(=O)—CH₃	—CH₃†	=C(H)—CH₃†	—CH₃	—C(=O)—O—CH₃	P	—H	835–850 1020–1040	794
Bacterio-chlorophyll c	—C(H)(OH)—CH₃	—CH₃	—C₂H₅ / —C₃H₇†† / —C₄H₉	—C₂H₅ / —CH₃(?)	—H	F	—CH₃	745–755	660–669
Bacterio-chlorophyll c_s	—C(H)(OH)—CH₃	—CH₃	—C₂H₅	—CH₃	H	S	—CH₃	740	667
Bacterio-chlorophyll d	—C(H)(OH)—CH₃	—CH₃	—C₂H₅ / —C₃H₇ / —C₄H₉	—C₂H₅ / —CH₃(?)	—H	F	—H	705–740	654
Bacterio-chlorophyll e	—C(H)(OH)—CH₃	—C(H)(=O)	—C₂H₅ / —C₃H₇ / —C₄H₉	—C₂H₅	—H	F	—CH₃	719–726	646

*No double bond between C-3 and C-4; additional H-atoms are in position C-3 and C-4.

†No double bond between C-3 and C-4; an additional H-atom is in position C-3.

**P = phytyl ester (C₂₀H₃₉O—); F = farnesyl ester (C₁₅H₂₅O—); Gg = geranylgeraniol ester (C₁₀H₁₇O—); S = stearyl alcohol (C₁₈H₃₇O—).

††Bacteriochlorophylls c, d, and e consist of isomeric mixtures with the different substituents on R₃ as shown.

From Gloe, A., N. Pfennig, H. Brockmann, and W. Trowitsch. 1975. Archives of Microbiology 102: 103–109, and Gloe, A., and N. Risch. 1978. Archives of Microbiology 118: 153–156.

FIGURE 19.1 Structure of all bacteriochlorophylls. The different substituents present in the positions R_1 to R_7 are given in the accompanying table.

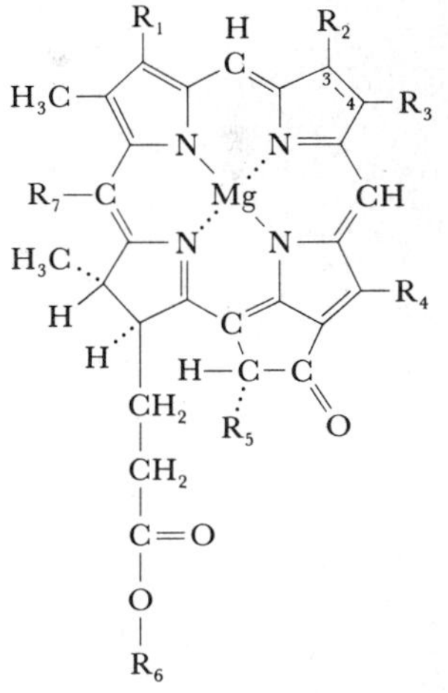

Bacteriochlorophylls We compared a typical **bacteriochlorophyll** with algal chlorophyll in Figure 6.2. A number of bacteriochlorophylls exist, differing in substituents on various parts of the porphyrin ring, and these are outlined in Figure 19.1. The various modifications lead to changes in the characteristic absorption spectra of the bacteriochlorophylls, so that an organism containing a certain bacteriochlorophyll is best able to utilize light of particular wavelengths. The ecological significance of ability to utilize different wavelengths of light was discussed in Section 6.9; it is likely that this selective absorption provides an evolutionary pressure for the development of organisms with various chlorophylls. The long-wavelength absorption maxima of the bacteriochlorophylls are the most characteristic, and these are given in Figure 19.1, as measured in the living cell (in vivo) and in solvent extract. Although the in vivo absorption maximum is of most significance ecologically, from the viewpoint of characterizing the various purple and green bacteria taxonomically, the absorption spectrum in solvent extract is most convenient, because its measurement is easier. Thus to characterize the bacteriochlorophyll of a new isolate of a purple or green bacterium, a simple extract in methanol or ether is made and the absorption spectrum determined. From the long-wavelength maximum listed in Figure 19.1, the likely identification of the bacteriochlorophyll can be made. For confirmation, it is desirable to carry out chromatographic studies on the isolated chlorophyll.

Classification   The purple and green bacteria are a diverse group morphologically, with cocci, rods, vibrios, spirals, budding, and gliding types known. Both polarly and peritrichously flagellated organisms are known in this group. It thus seems likely that the ability to grow phototrophically has developed in a wide variety of bacterial types, and that the only evolutionary similarity among the whole group is the ability to carry out photophosphorylation. Traditionally, these bacteria were classified into three major groups: green sulfur bacteria, purple sulfur bacteria, and purple nonsulfur bacteria. However, research over the past several decades has shown that this classification is oversimplified. Most important, the distinctions between the purple sulfur and the purple nonsulfur bacteria in relation to sulfur metabolism can no longer be maintained. Because of this, only the common names "purple" and "green" are used. Four families among the purple and green bacteria are currently recognized, as follows:

Group	Bacteriochlorophylls	Photosynthetic Membrane Systems (see Figure 19.3)
Purple bacteria: Rhodospirillaceae Chromatiaceae	Bchl *a*, Bchl *b*	Lamellae or tubes, continuous with plasma membrane
Green bacteria: Chlorobiaceae	Bchl *c*, Bchl *d*, Bchl *e*, plus small amounts of Bchl *a*	Vesicles ("chlorosomes"), attached to but not continuous with plasma membrane
Chloroflexaceae	Bchl c_s, plus small amounts of Bchl *a*	

Both purple and green bacteria generally also produce **carotenoid pigments** (Figure 19.2), and the carotenoids of the purple bacteria differ from those of the green bacteria. Carotenoid pigments are responsible for the purple color of the purple bacteria, and mutants lacking carotenoids are blue-green in color. In fact, purple photosynthetic bacteria are frequently not purple, but brown, pink, brown-red, purple-violet, or orange-brown, depending on their carotenoid pigments. Also, many of the "green" bacteria are brown-colored, due to their complement of carotenoids (Figure 19.2). Thus color is not a good criterion for use in recognizing types of purple and green bacteria.

Photosynthetic membrane systems   A major difference between the green and purple bacteria is in the nature of the photosynthetic membrane system. In the purple bacteria, the photosynthetic pigments are part of an elaborate internal membrane system, connected to and

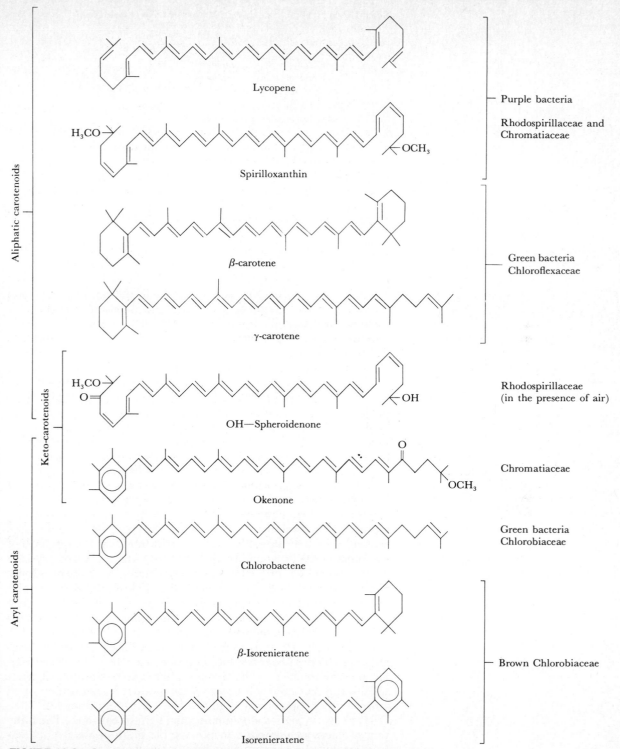

FIGURE 19.2 Carotenoids of the photosynthetic bacteria. This figure gives a few representative structures of carotenoids of the various groups. A number of variants on the above structures also occur and are listed in the accompanying table.

Group	Names	Color of organisms
1	Lycopene, rhodopin, spirilloxanthin	Orange-brown, brownish-red, pink, purple-red
2	Spheroidene, hydroxyspheroidene, spheroidenone, hydroxyspheroi-denone, spirilloxanthin	Red (aerobic) Brownish red to purple (anaerobic)
3	Okenone (Chromatiaceae only), methoxylated keto carotenoids (*Rhodopseudomonas globiformis* only)	Purple-red
4	Lycopenal, lycopenol, rhodopin, rhodopinal, rhodopinol	Purple-violet
5	Chlorobactene, hydroxychloro-bactene, β-isorenieratene, isorenieratene	Green (chlorobactene) Brown (isorenieratene)
6	β-carotene, γ-carotene	Orange-green

Based on Schmidt, K. 1978. Pages 729–750, and S. Liaaen-Jensen, 1978. Pages 233–247 *in* R. K. Clayton and W. R. Sistrom (eds.), *The Photosynthetic Bacteria,* Plenum Press, New York. N.Y.

Figure 19.2 continued

ramifying from the plasma membrane; the membrane often occupies much of the cell interior. In some cases the membrane system is an array of flat sheets called **lamellae** (Figure 19.3*b*), whereas in others it consists of round tubes referred to as **chromatophores** (Figure 19.3*c*). The membrane content of the cell varies with pigment content, which is itself affected by light intensity and presence of O_2. When cells are grown aerobically, synthesis of bacteriochlorophyll is repressed, and the organisms may be virtually devoid of photopigments as well as the internal membrane systems. (This experiment can, of course, only be done in those purple and green bacteria able to grow aerobically, such as members of the Rhodospirillaceae, and members of the genera *Chloroflexus, Thiocapsa, Chromatium,* and *Ectothiorhodospira.*) Consequently, phototrophic growth is only possible under anaerobic conditions, when bacteriochlorophyll synthesis can occur. Superimposed upon this O_2 effect is an effect of light intensity. Even under anaerobic conditions, when synthesis of the photosynthetic apparatus is not repressed, the level of the photopigments and internal membranes is affected by light intensity. At high light intensity, the synthesis of the photosynthetic apparatus is inhibited, whereas when cells are grown at low light intensity the bacteriochlorophyll content is high, and the cells are packed with membranes. (This increase in cell pigment content at low light intensities allows the organism to better utilize the available light.) Usually carotenoid pigment synthesis is coordinately regulated with the bacteriochlorophyll content.

An interesting new group of bacteria which produce bacterio-chlorophyll only under aerobic conditions have been described from marine environments. The aerobic phototrophs resemble Rhodo-spirillaceae, but they will not grow or produce bacteriochlorophyll anaerobically. Significant pigment levels are only observed when cells are grown aerobically in either the light or the dark. The function of bacteriochlorophyll in the aerobic phototrophs is not clear (presumably the pigment functions in photosynthesis but this has not been demonstrated), but it does appear clear that O_2 regulates

Photosynthetic vesicle

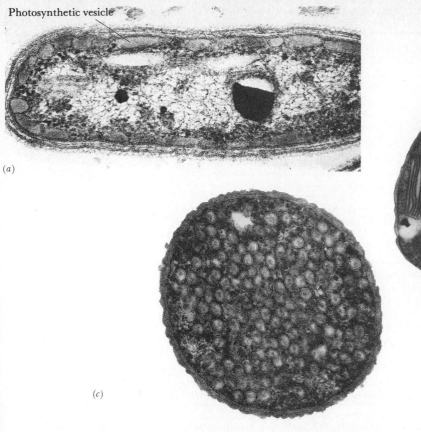

(a)

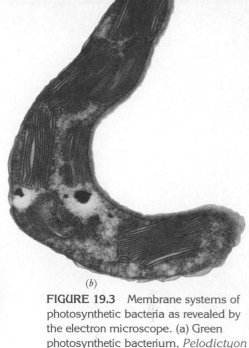

(b)

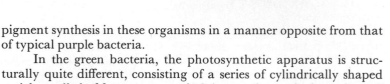

(c)

FIGURE 19.3 Membrane systems of photosynthetic bacteria as revealed by the electron microscope. (a) Green photosynthetic bacterium, *Pelodictyon* sp., showing location of photosynthetic vesicles (chlorosomes). Magnification, 29,200×. (Courtesy of G. Cohen-Bazire.) (b) Purple photosynthetic bacterium, *Ectothiorhodospira mobilis,* showing the photosynthetic lamellae in flat sheets. Magnification, 27,880×. (From Remsen, C. C., S. W. Watson, J. B. Waterbury, and H. G. Trüper, 1968. J. Bacteriol. 95:2374–2392.) (c) *Chromatium* sp., strain D, another purple photosynthetic bacterium, showing the photosynthetic membranes as individual long tubes (chromatophores). Magnification, 76,000×. (Courtesy of Jeffrey C. Burnham and S. C. Conti.)

pigment synthesis in these organisms in a manner opposite from that of typical purple bacteria.

In the green bacteria, the photosynthetic apparatus is structurally quite different, consisting of a series of cylindrically shaped vesicles called **chlorosomes** underlying and attached to the cell membrane (Figure 19.3a). These vesicles are enclosed within a thin membrane that does not have the usual bilayered appearance (it is sometimes called a *nonunit* membrane). Bacteriochlorophylls c, c_s, d, or e (depending on the species) are present inside the chlorosomes, while Bchl a and components of the photosynthetic electron transport chain are located in the cytoplasmic membrane.

Enrichment culture The purple and green bacteria are generally found in anaerobic zones of aquatic habitats, often where H_2S accumulates. These organisms can be enriched by duplicating the habitat in the laboratory (Table 19.1). A basal mineral salts medium is used, to which bicarbonate is added as a source of CO_2. Since many photosynthetic bacteria require vitamin B_{12}, this vitamin is usually added. The optimum pH range is 6 to 8, and the optimum temperature is 20 to 30°C. Incubation is anaerobic, and a small amount of sodium sulfide (0.05 to 0.1 percent $Na_2S \cdot 9H_2O$) is added as photosynthetic electron donor. Selection of appropriate light conditions is important.

Solution 1: 0.83 g $CaCl_2 \cdot 2H_2O$ in 2.5 liters H_2O. For marine organisms, add NaCl, 130 g.

Solution 2: H_2O 67 ml; KH_2PO_4, 1 g; NH_4Cl, 1 g; $MgCl_2 \cdot 2H_2O$, 1 g; KCl, 1 g; 3 ml vitamin B_{12} solution (2 mg/100 ml H_2O); 30 ml trace element solution (1 liter H_2O, ethylene diamine tetraacetic acid, 500 mg; $FeSO_4 \cdot 7H_2O$, 200 mg; $ZnSO_4 \cdot 7H_2O$, 10 mg; $MnCl_2 \cdot 4H_2O$, 3 mg; H_3BO_3, 30 mg; $CoCl_2 \cdot 6H_2O$, 20 mg; $CuCl_2 \cdot 2H_2O$, 1 mg; $NiCl_2 \cdot 6H_2O$, 2 mg; $Na_2MoO_4 \cdot 2H_2O$, 3 mg, pH 3).

Solution 3: Na_2CO_3, 3 g; H_2O, 900 ml. Autoclave in a container suitable for gassing aseptically with CO_2.

Solution 4: $Na_2S \cdot 9H_2O$, 3 g; H_2O, 200 ml. Autoclave in flask containing a Teflon-covered magnetic stirring rod.

Dispense the bulk of solution 1 in 67-ml aliquots in 30 screw-capped bottles of 100 ml capacity, and autoclave with screw caps loosely on. Autoclave the other solutions in bulk. After autoclaving, cool all solutions rapidly by placing the bottles in a cold water bath (to prevent lengthy exposure to air). Solution 3 is then gassed with CO_2 gas until it is saturated (about 30 minutes; pH drops to 6.2). Add this to cooled solution 2 and aseptically place 33 ml of this mixture in each bottle containing solution 1. Solution 4 is partially neutralized by adding dropwise (while stirring on a magnetic stirrer) 1.5 ml of sterile 2 M H_2SO_4. Add 5-ml portions of solution 4 to each bottle, fill the bottles completely with remaining solution 1, and tightly close. The final pH should be between 6.7 and 7.2. Store overnight to consume residual oxygen before using. (For some organisms, the sulfide concentration should be reduced. Use 2.5 ml instead of 5 ml of solution 4 per bottle.) It may be necessary to "feed" the cultures with sulfide (solution 4) from time to time, as the sulfide is used up during growth. To do this, remove 2.5 or 5 ml of liquid from a bottle and refill with an equal amount of sterile neutralized solution 4.

Originally described by Pfennig, N. 1965. Zentralbl. Bakteriol. Parasitenkd. Infektionskr. Hyg. Abt. 1, Supplementh. 1; 179. For an English version, see van Niel, C. B. 1971. Methods Enzymol. 23:3–28.

Light intensities should not be too high, since these bacteria usually live in deep areas of lakes where light is low. Intensities between 10 and 100 footcandles are adequate. The quality of light is also important. Most purple and green bacteria use radiation of the near infrared, 700 to 900 nm, which can be most easily obtained in the laboratory by use of conventional tungsten light bulbs. Fluorescent bulbs are usually deficient in infrared radiation and are not as satisfactory. After inoculation, culture tubes or bottles are incubated for several weeks and examined periodically for signs of visible growth. Enrichment cultures should appear pigmented, and microscopic examination of positive cultures should reveal organisms resembling purple and green bacteria. Pure cultures can be obtained from positive enrichments by conventional plating methods, care being taken, of course, to keep cultures anaerobic throughout. However, the purple and green bacteria are not as sensitive to oxygen as some other anaerobes, so that extreme precautions to maintain anaerobic conditions are usually not necessary. A widely used medium for the culture of purple and green bacteria is given in Table 19.1.

For the Rhodospirillaceae, the sulfide concentration of the medium should be reduced to a much lower level, 0.01 to 0.02 percent $Na_2S \cdot 9H_2O$ (or sulfide can be eliminated altogether), and an organic substance added to provide a carbon source and/or electron donor. Because many of the Rhodospirillaceae have multiple growth-factor requirements, usually one or more B vitamins, addition of 0.01 to 0.02 percent yeast extract as a growth-factor source is recommended. The organic substrate used should be nonfermentable (to avoid enrichment of fermentative organisms such as clostridia); acetate, ethanol, benzoate, isopropanol, butyrate or dicarboxylic acids such as succinate are ideal.

It should be noted that many members of the Chromatiaceae and Chlorobiaceae are incapable of assimilatory sulfate reduction, so that they must be given a source of reduced sulfur. The sulfide added to the medium thus not only uses up the remaining O_2 and serves as an electron donor, but provides this source of reduced sulfur. If an electron donor such as H_2 or an organic compound is to be used, and sulfide must for some reason be avoided, then it is necessary to add another source of reduced sulfur, such as methionine or cysteine (if yeast extract is added, it often serves as a source of reduced sulfur as well as a growth-factor source).

Rhodospirillaceae, the nonsulfur purple bacteria   These bacteria have been called the *nonsulfur purple bacteria* because it had been thought that they were unable to use sulfide as an electron donor for the reduction of CO_2 to cell material, but it is now known that sulfide can be used provided the concentration is maintained at a low level, preferably by use of a chemostat. It appears that levels of sulfide utilized well by the Chlorobiaceae or Chromatiaceae are toxic to the Rhodospirillaceae; hence the previous inability to demonstrate sulfide metabolism. However, the Rhodospirillaceae differ from the other groups of purple and green bacteria in two ways: (1) they are unable to oxidize elemental sulfur to sulfate (although they can oxidize sulfide to sulfate without the intermediate accumulation of elemental sulfur); and (2) they are facultative phototrophs, able to grow aerobically in the dark or anaerobically in the light. (It should be noted that certain species of Chromatiaceae have the capacity to grow aerobically in darkness as well.) Some members of the Rhodospirillaceae can also grow anaerobically in the dark, using fermentative metabolism. Because of the difficulty of providing nontoxic levels of sulfide, these organisms are generally cultured under phototrophic conditions, with an organic compound as major carbon source. Most members of this group also require growth factors, so that yeast extract or some other source of vitamins and other growth factors is usually provided.

The morphological diversity of this group is much less than that of the other purple bacteria and the green bacteria (Table 19.2 and Figure 19.4), although it is clearly a heterogeneous group as it contains both polarly and peritrichously flagellated genera, the latter growing by budding.

Many of the Rhodospirillaceae have the ability to utilize methanol or formate as sole carbon source for phototrophic growth. Two other reduced one-carbon compounds, methylamine and formaldehyde, are not utilized. When growing anaerobically with methanol, some CO_2 fixation is necessary, and the following stoichiometry is found:

TABLE 19.2   Genera and Characteristics of Rhodospirillaceae

Characteristics	Genus	DNA (mol % G + C)
Spirals, polarly flagellated	*Rhodospirillum*	62–66
Rods, ovals, or spheres, polarly flagellated; may bud but no hyphae	*Rhodopseudomonas*	62–72
Ovals, peritrichously flagellated; growth by budding and hypha formation	*Rhodomicrobium*	62–64
Ring-shaped; nonmotile, requires vitamin B_{12}	*Rhodocyclus*	65

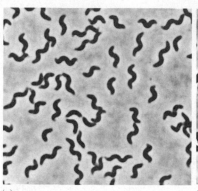

(a)

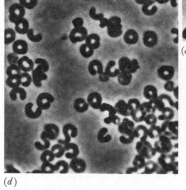

(b)

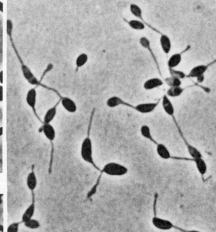

(c)

(d)

FIGURE 19.4   The four morphological types of Rhodospirillaceae (see also Table 19.2.) (a) *Rhodospirillum fulvum.* Magnification, 1600×. (b) *Rhodopseudomonas acidophila.* Magnification, 1620×. (c) *Rhodomicrobium vannielii.* Magnification, 2000×. (Photographs a–c courtesy of Peter Hirsch.) (d) *Rhodocyclus purpureus.* Magnification, 1330×. (Courtesy of Norbert Pfennig.)

$$2CH_3OH + CO_2 \longrightarrow 3(CH_2O) + H_2O$$
Cell material

Apparently, CO_2 is required because methanol is at a more reduced oxidation state than cell material, and the CO_2 serves as an electron sink (electron acceptor).

Enrichments for Rhodospirillaceae can be made highly selective by omitting fixed nitrogen sources such as ammonia or nitrate from the medium, and substituting an ample supply of gaseous nitrogen, N_2. Most Rhodospirillaceae are active N_2 fixers and grow well in a medium in which N_2 is the sole nitrogen source.

Chromatiaceae, the purple sulfur bacteria   Purple bacteria that deposit sulfur and oxidize it to sulfate are morphologically diverse

TABLE 19.3   Genera and Characteristics of Chromatiaceae

Characteristics	Genus	DNA (mol % G + C)
Sulfur deposited externally:		
Spirals, polar flagella	*Ectothiorhodospira*	62–70
Sulfur deposited internally:		
Do not contain gas vesicles		
Ovals or rods, polar flagella	*Chromatium*	48–70
Spheres, diplococci, tetrads, nonmotile	*Thiocapsa*	63–70
Spheres or ovals, polar flagella	*Thiocystis*	62–68
Large spirals, polar flagella	*Thiospirillum*	45
Contain gas vesicles		
Irregular spheres, ovals, nonmotile	*Amoebobacter*	65
Spheres, ovals, polar flagella	*Lamprocystis*	64
Rods; nonmotile; forming irregular network	*Thiodictyon*	65–66
Spheres; nonmotile; forming flat sheets of tetrads; not available in pure culture	*Thiopedia*	

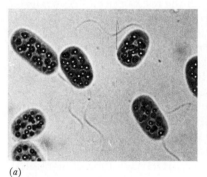

(a)

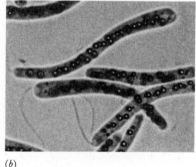

(b)

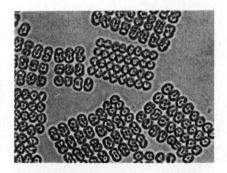

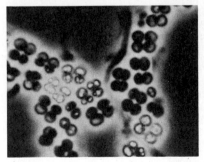

(d)

FIGURE 19.5   Chromatiaceae (purple sulfur bacteria). (a) *Chromatium okenii.* Note the globules of elemental sulfur inside the cells. Magnification, 1450×. (b) *Thiospirillum jenense,* a very large, polarly flagellated spiral. Notice the sulfur globules. Magnification, 1200×. (c) *Thiopedia rosea.* Magnification, 1450×. (d) *Thiocapsa.* Magnification, 1360×. (Photograph a courtesy of Norbert Pfennig; photographs b–d courtesy of Peter Hirsch.)

(Table 19.3). The cell is usually larger than that of the green bacteria, and in sulfide-rich environments may be packed with sulfur granules (Figure 19.5a and b), although in the smaller-celled genera the sulfur granules may not be so obvious (Figure 19.5d). Members of the Chromatiaceae are commonly found in anaerobic zones of lakes as well as in sulfur springs; because of their conspicuous purple color they are often easily seen in large blooms or masses (see Color Plate 6a), and in fact, blooms of purple sulfur bacteria were described in ancient literature. In general, Chromatiaceae grow at somewhat higher pH values than do the green bacteria. The genus *Ectothiorhodospira* is of interest because it deposits sulfur externally but also because it is halophilic, growing at sodium chloride concentrations approaching saturation. It is found in saline lakes, salterns, and other bodies of water high in salt. Members of the Chromatiaceae and Rhodospirillaceae which contain bacteriochlorophyll *b* can be enriched by using infrared radiation instead of visible light, since bacteriochlorophyll *b* has an absorption maximum at 1025 nm, radiation which is invisible to the human eye.

The Chromatiaceae have a limited ability to utilize organic compounds as carbon sources for phototrophic growth. Acetate and pyruvate are utilized by most strains; some strains will use other organic acids, sugars, or ethanol, but methanol is not utilized. A few Chromatiaceae will grow lithotrophically in darkness with thiosulfate as electron donor, and *Thiocapsa* will grow heterotrophically on acetate.

Chlorobiaceae and Chloroflexaceae, the green bacteria At one time considered a relatively small group, it is now known that the green bacteria are quite diverse, including nonmotile rods, spirals, and spheres (Chlorobiaceae), and motile filamentous, gliding forms (Chloroflexaceae, Table 19.4). We discuss the gliding bacteria in some detail in Section 19.3. The gliding green bacteria (Figure 19.6*c* and *d*) could easily be classified with the rest of the gliding bacteria, but it is generally considered that the ability to grow phototrophically is a more fundamental characteristic; thus these gliding organisms are classified with the rest of the phototrophic bacteria. Several other green bacteria have complex appendages called prosthecae, and could be classified with the budding and/or appendaged bacteria (Section 19.5 and Figure 19.23*c*).

The green bacteria that live planktonically in lakes generally possess gas vesicles, whereas the species that live on the bottoms of sulfur and hot springs, or in other benthic habitats, are not gas vesiculate. Members of one genus, *Pelodictyon,* consist of rods that undergo branching, and since the rods remain attached, a three-dimensional network is formed (Figure 19.6*b*).

TABLE 19.4 Genera and Characteristics of Green Photosynthetic Bacteria (Chlorobiaceae and Chloroflexaceae)

Characteristics	Genus	DNA (mol % G + C)
No gas vesicles:		
Straight or curved rods, nonmotile	*Chlorobium* (Figure 19.6*a*)	49–58
Spheres and ovals, nonmotile, forming prosthecae (appendages)	*Prosthecochloris*	50–56
	Ancalochloris (Figure 19.23*c*)	
Filamentous, gliding	*Chloroflexus* (Figure 19.6*c*)	53–55
Filamentous, gliding, large diameter (5 μm)	*Oscillochloris* (Figure 19.6*d* and *e*)	–
Contain gas vesicles:		
Branching nonmotile rods, in loose irregular network	*Pelodictyon* (Figure 19.6*b*)	48–58
Spheres and ovals, nonmotile, forming chains	*Clathrochloris*	–
Filamentous, gliding, large diameter (2–2.5 μm)	*Chloronema*	–

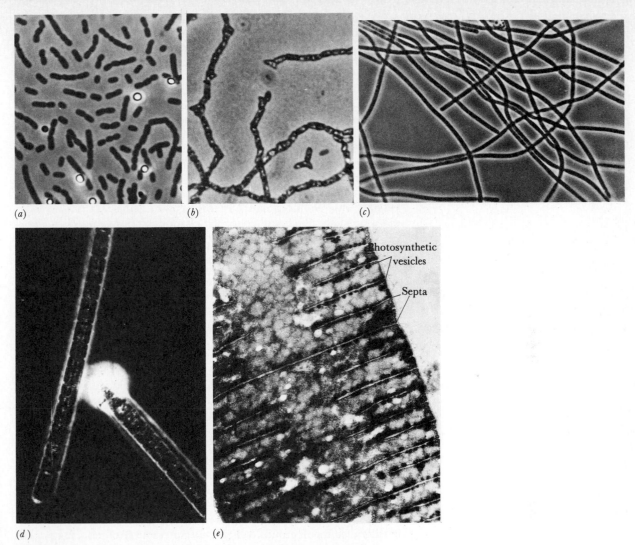

(a)

(b)

(c)

(d)

(e)

FIGURE 19.6 Chlorobiaceae and Chloroflexaceae (green photosynthetic bacteria). (a) *Chlorobium limicola.* Note the sulfur granules deposited extracellularly. Magnification, 1700×. (b) *Pelodictyon clathratiforme,* a bacterium forming a three-dimensional network. Magnification, 1700×. (c) *Chloroflexus aurantiacus,* a filamentous gliding bacterium. Magnification, 1500×. (d) *Oscillochloris,* a large, filamentous, gliding green bacterium. Phase contrast. The brightly contrasting material is the holdfast. Magnification, 788×. (e) Electron micrograph of *Oscillochloris.* The chlorosomes in this preparation are darkly stained. Magnification, 14,800×. (Photographs *a* and *b* courtesy of Norbert Pfennig. Photograph *c* from Madigan, M., and T. D. Brock. 1977. J. Gen. Microbiology. *102:* 279–285. Photographs *d* and *e* courtesy of V. M. Gorlenko, Institute of Microbiology, Moscow, U.S.S.R. From Gorlenko, V. M., and T. A. Pivovarova. 1977. Bulletin of Academy of Sciences of U.S.S.R. Biological Series, No. 3, 396–409.)

All of the Chlorobiaceae are strictly anaerobic and obligately phototrophic, being unable to carry out respiratory metabolism in the dark. Some Chlorobiaceae can assimilate simple organic substances for phototrophic growth, provided that a reduced sulfur compound is present as a sulfur source (since they are incapable of assimilatory sulfate reduction). Organic compounds used by these species include acetate, propionate, pyruvate, and lactate. *Chloroflexus* is much more versatile than most Chlorobiaceae, being able to grow heterotrophically in the dark under aerobic conditions, as well as phototrophically on a wide variety of sugars, amino acids, and organic acids, or photoautotrophically with H_2S or H_2 and CO_2.

The gliding green bacteria are of considerable evolutionary interest because they resemble morphologically the cyanobacteria, yet have only photosystem I. Organisms of the genus *Oscillochloris* look almost exactly like members of the cyanobacterial genus *Oscillatoria* (compare Figures 19.6*d* and *e* and 19.10*d*) suggesting a close evolutionary relationship. Thus the gliding green bacteria may represent a link between those phototrophic bacteria which possess only photosystem I, and those which possess both photosystems I and II and carry out typical oxygenic photosynthesis. However, as discussed below, there is considerable doubt whether the Chlorobiaceae possess the conventional photosynthetic carbon cycle (Calvin cycle) for autotrophic CO_2 fixation (see below). Because of this and because of the presence as part of the photosynthetic apparatus of the unique chlorosomes, another valid viewpoint would be that the Chlorobiaceae are a group quite separate from other phototrophic organisms and are not on an evolutionary line to the oxygenic cyanobacteria.

Physiology of phototrophic growth   The overall picture that emerges from a study of the comparative physiology of the various purple and green bacteria is that light is used exclusively in the generation of ATP, and is not involved in generation of reducing power, as it is in organisms exhibiting oxygenic photosynthesis. There is a good possibility that reducing power is generated by an energy-linked reversal of steps in the electron transport chain (Section 6.4), but this has not been proved for all groups. Whether or not a source of reducing power is needed depends upon the carbon source supplied. If CO_2 is the sole carbon source, then reducing power is needed, and this can come from a reduced sulfur compound or H_2. Reducing power for CO_2 fixation may also come from an organic compound, but if an organic compound is supplied the situation is more complex. This is because the organic compound may serve as a carbon source itself, so that CO_2 need not necessarily be reduced. Whether or not CO_2 is reduced when an organic compound is added will depend at least in part on the oxidation state of the organic compound. Compounds such as acetate, glucose, and pyruvate are at about the oxidation level of cell material and can thus be assimilated directly as carbon sources with no requirement for either oxidation or reduction.* Fatty acids longer than acetate (for example, propionate, butyrate, caprylate) are more re-

*We are speaking here of *net* oxidation or reduction. Some cell constituents are more reduced, but other cell constituents are more oxidized, so that the net oxidation state is zero.

duced than cell material, and some means of disposing of excess electrons is necessary, such as the reduction of CO_2 (as described on page 634 for methanol utilization). Thus the amount of CO_2 fixed by a purple or green bacterium growing with an organic compound will depend upon the oxidation state of the compound, and whether or not inorganic electron donors such as sulfide are also present.

Many purple and green bacteria can grow phototrophically using H_2 as sole electron donor, with CO_2 as carbon source. These organisms have a hydrogenase for activating H_2 and the photosynthetic carbon (Calvin) cycle (ribulose diphosphate carboxylase, see Section 6.7) for fixing CO_2. Since many phototrophic bacteria are incapable of carrying out assimilatory sulfate reduction, it is essential when testing for growth on H_2 to add a small amount of sulfide as a source of reduced sulfur. It may also be necessary to add vitamins for some strains.

In the absence of ammonia as nitrogen source, many purple bacteria produce H_2 in the light. This has been shown to be due to the presence in these organisms of nitrogenase, the enzyme involved in N_2 fixation (see Section 5.13). All nitrogenase enzymes are able to reduce H^+ to H_2, and this reaction runs in competition with N_2 reduction. Since ammonia represses nitrogenase synthesis, production of H_2 by phototrophic bacteria only occurs when ammonia is absent.

Although the presence of the Calvin cycle (see Section 6.7) has been conclusively shown in many phototrophic bacteria, there is considerable doubt about the existence of this cycle in the Chlorobiaceae. Extensive studies have failed to reveal significant amounts of ribulose diphosphate carboxylase, the key enzyme of the cycle, in one strain of *Chlorobium,* and only low levels have been found in one strain of *Chloroflexus.* Also in *Chlorobium,* the pattern of labeling of intermediates when *Chlorobium* is given $^{14}CO_2$ is not consistent with the Calvin cycle. *Chlorobium* contains ferredoxin-linked enzymes, which will cause a reductive fixation of CO_2 into intermediates of the tricarboxylic acid cycle, and it has been proposed that CO_2 is fixed by a reversal of the TCA cycle (which normally functions oxidatively to produce CO_2 from acetate; see Figure 19.7). Since another group of autotrophic bacteria, the methane-producing bacteria (see Section 19.20), apparently do not possess a Calvin cycle for CO_2 fixation, it appears that other mechanisms for autotrophic CO_2 fixation exist in the microbial world. Further elucidation of the pathway of phototrophic CO_2 fixation in the Chlorobiaceae will be of considerable interest.

Sulfur metabolism in the purple and green bacteria Most of the purple and green bacteria are able to oxidize reduced sulfur compounds under anaerobic conditions, with the formation of sulfate. The most common reduced sulfur compounds used are sulfide and thiosulfate. Tetrathionate, which is an oxidation product of thiosulfate, is also used by some organisms, probably after it is reduced to thiosulfate. Elemental sulfur is frequently formed during the oxidation of sulfide or thiosulfate, and is either deposited inside or outside the cells. Elemental sulfur deposited inside the cells is readily available as a further source of reduced sulfur compounds, and externally deposited elemental sulfur may also be utilized, although probably less efficiently

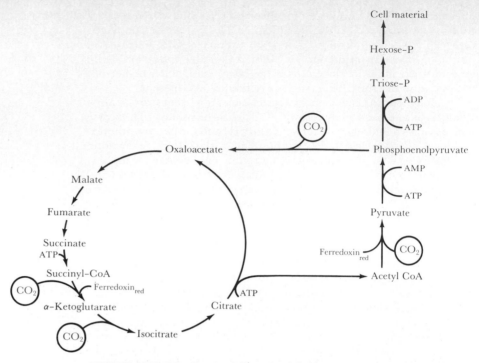

FIGURE 19.7 The Reversed Citric Acid Cycle as a possible mechanism of CO_2 fixation in *Cholorobium*. **Ferredoxin$_{red}$** indicates carboxylation reactions requiring reduced ferredoxin. Starting from acetate, each turn of the cycle results in four molecules of CO_2 being incorporated. Citrate is then split into **acetyl CoA** (C-2) and **oxaloacetate** (C-4).

(because of its high insolubility and relative unavailability when present in the medium instead of inside the cells). The overall pathway of oxidation of reduced sulfur compounds in purple and green bacteria is shown in Figure 19.8. As seen, either sulfide or thiosulfate is oxidized first to sulfite, SO_3^{2-}, and the enzymes adenylsulfate reductase (APS reductase) and ADP-sulfurylase catalyze the oxidation of sulfite to sulfate. Note that a substrate-level phosphorylation occurs at this step, providing for the synthesis of a high-energy phosphate bond (in ADP).

It seems likely that elemental sulfur is not an obligatory intermediate between sulfide and sulfate, but merely a side product. Elemental sulfur is probably a storage product, formed when sulfide concentrations in the environment are high, and under limiting sulfide levels sulfide is oxidized directly to sulfate without formation of elemental sulfur. The formation of elemental sulfur as a storage product has been most clearly shown in *Chromatium,* where it is deposited inside the cells (Figure 19.5*a*). Intracellular elemental sulfur in *Chromatium* can serve as an electron donor for phototrophic growth when sulfide is absent. Additionally, when *Chromatium* is placed in the dark, S^0 can also serve as an electron acceptor, being reduced to sul-

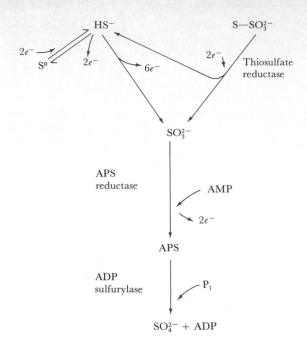

$$\text{Overall: } HS^- + 4H_2O \longrightarrow SO_4^{2-} + H^+ + 8H + 8e^-$$

$$8H + 8e^- + 2CO_2 \longrightarrow 2(CH_2O) + 2H_2O$$

$$\text{Cell material}$$

$$\text{Summation: } HS^- + 2H_2O + 2CO_2 \longrightarrow SO_4^{2-} + 2(CH_2O) + H^+$$

FIGURE 19.8 Pathways of oxidation of reduced sulfur compounds in phototrophic bacteria. Note that a substrate-level phosphorylation occurs via the enzyme ADP sulfurylase.

fide. Under these conditions of dark maintenance metabolism, electrons for reduction of S^0 come from a glycogen storage product; thus *Chromatium* stores both its electron donor (glycogen) and electron acceptor (S^0) as intracellular polymers. If an external electron donor is present, such as acetate, S^0 can also serve as an electron acceptor for dark anaerobic maintenance in the absence of an intracellular storage polysaccharide. Growth of *Chromatium* anaerobically in darkness has not been demonstrated.

Symbiotic associations and mixed culture interactions A two-membered system in which each organism does something for the benefit of the other has been called a **consortium.** An association of two organisms consisting of a large, colorless central bacterium, which is polarly flagellated, surrounded by smaller, ovoid- to rod-shaped green bacteria arranged in four to six rows has been called *Chlorochromatium aggregatum*. The genus and species name are invalid, because they refer not to a single organism, but the association is seen quite commonly in lakes and in muds. The green organism in this association has been cultured and classified as *Chlorobium chlorochromatii*, but the large colorless central organism has not been cultured. A similar association in which the colored organism is a brown-pigmented *Chlorobium* has been called *Pelochromatium roseum*. A clue to the possible role of the colorless central organism comes from the common observation that the sulfide needed by purple and green bacteria can be derived from sulfate- or sulfur-reducing bacteria associated with them. Sulfate- and sulfur-reducing bacteria use organic compounds such as ethanol, lactate, formate, or fatty acids as electron donors and reduce sulfate or sulfur to sulfide. The sulfide produced can then be used by associated phototrophic bacteria, which in the light oxidize the sulfide back to sulfate. If some of the organic matter produced by the phototrophic bacterium is used by the sulfate reducer, a self-feeding system can develop, driven by light energy.

A culture, which formerly was called *Chloropseudomonas ethylica*, can be used as an example of the tightness with which a sulfate reducer and a phototrophic bacterium can be coupled. We now know that this culture was really a mixture of a sulfate reducer and either *Chlorobium limicola* or *Prosthechochloris* spp., but this mixed culture, assumed to be a single organism, was maintained for many years in a number of laboratories, because it was easy to grow. Ethanol was normally added to cultures of *Chloropseudomonas ethylica* (hence the species name), and the ethanol was oxidized and the sulfate reduced by the sulfate-reducing bacterium thus providing sulfide for the growth of the *Chlorobium*. Hence the culture gave the appearance of a green bacterium using ethanol as organic electron donor. Since the discovery of this mixed system, sulfate-reducing bacteria have been isolated from a number of stable enrichment cultures of purple and green bacteria.

The roles of the purple and green bacteria in these stable consortia are probably to remove sulfide that might build up and be toxic to the sulfate reducer, and to function in reformation of sulfate needed as electron acceptor. An additional role might be to synthesize organic substances used by the sulfate reducer as electron donor.

Ecology The greatest number of purple and green bacteria are found in the depths of certain kinds of lakes where stable conditions for growth occur. We discussed the development of thermal stratification in lakes in Chapter 13 (see Figure 13.3); after stratification occurs, stable anaerobic conditions may continue in the deep waters throughout the summer season. If there is a sufficient supply of H_2S and the lake water is sufficiently clear so that light penetrates to the anaerobic zone, a massive layer of purple or green bacteria can develop. This layer, hidden from view of the observer on the surface, can be studied by sampling water at various depths (see Color Plate 6*b*). The bacteria often form a distinct layer just at the depth where H_2S is first present (Figure 19.9). The most favorable lakes for development of these bacteria are those called **meromictic**, which are essentially permanently stratified due to the presence of denser (usually saline) water in the bottom. In such permanently stratified lakes, the bloom of purple or green bacteria may be present throughout the year. Often the photosynthetic activity of these bacteria is sufficiently great so that it is an important source of organic matter to the lake

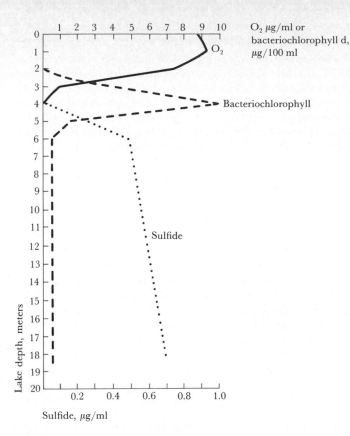

O$_2$ μg/ml or
bacteriochlorophyll d,
μg/100 ml

O$_2$

Bacteriochlorophyll

Sulfide

Lake depth, meters

Sulfide, μg/ml

FIGURE 19.9 Vertical distribution of *Chlorobium* in a stratified lake. The phototrophic bacterium forms a layer just at the top of the anaerobic zone. Bacterial population size is quantified by measuring bacteriochlorophyll d concentration. Figures in the upper abscissa represent both the amount of O$_2$ and H$_2$S. (Unpublished data for Lake Mary, Wisconsin, of T. Parkin and T. D. Brock.)

ecosystem. In some cases more photosynthesis may occur in the bacterial layer than in the surface algal layer. In general, blooms of purple and green bacteria are more common in small lakes and ponds than in large bodies of water, mainly because the stable stratification of large bodies of water is more affected by wind. The ideal lake for observing blooms of purple and green bacteria is a fairly deep small one nestled in a valley protected from strong winds with a fairly large amount of hydrogen sulfide in the bottom waters.

The easiest place to observe purple and green bacteria in nature is in sulfur springs, where massive blooms are often present a few inches below the surface of the water (see Color Plate 6*a*). Along the seacoast, blooms also occur in warm shallow pools of seawater not connected to the open ocean, where the activities of sulfate-reducing bacteria lead to production of large amounts of H$_2$S.

The purple and green bacteria are one of the most diverse groups of bacteria and are of interest for a wide variety of reasons. They have provided extremely useful systems for studying fundamental aspects of the photosynthetic process (see Chapter 6), and are of great evolutionary significance. Their ecological roles may also be important, and their extreme diversity challenges the bacterial taxonomist. Much more work on this interesting group of bacteria would be desirable.

Cyanobacteria (Blue-Green Algae)

The **cyanobacteria** comprise a large and heterogeneous group of phototrophic organisms, studied primarily by phycologists. Classification of cyanobacteria has improved in the last decade, due to the increased availability of pure cultures. Because of the procaryotic nature of these organisms, it has been proposed that they be called the cyanobacteria or blue-green bacteria, instead of blue-green algae, to indicate clearly that they are not eucaryotic algae.

Structure The morphological diversity of the cyanobacteria is considerable (Figure 19.10). Both unicellular and filamentous forms are known, and considerable variation within these morphological types occurs. Recent taxonomic studies divide the cyanobacteria into five major groups: unicellular dividing by binary fission; unicellular dividing by multiple fission (colonial); filamentous containing differentiated cells called heterocysts which function in nitrogen fixation; filamentous nonheterocystous forms; and branching filamentous types (see Figures 19.10 through 19.13). Cyanobacterial cells range in size from those of typical bacterial size (0.5 to 1 μm in diameter) to cells as large as 60 μm in diameter (in the species *Oscillatoria princeps*). The latter has the largest cells known among the procaryotes.

The cyanobacteria differ in fatty acid composition from all other procaryotes. Other bacteria contain almost exclusively saturated and monounsaturated fatty acids (one double bond), but the cyanobacteria frequently contain unsaturated fatty acids with two or more double bonds.

The fine structure of the cell wall of some cyanobacteria is similar to that of Gram-negative bacteria. Many cyanobacteria produce extensive mucilaginous envelopes, or sheaths, that bind groups of cells or filaments together. The photosynthetic lamellar membrane system is often complex and multilayered (see Figure 2.12), although in some of the simpler cyanobacteria the lamellae are regularly arranged in concentric circles around the periphery of the cytoplasm (Figure 19.11). Cyanobacteria have only one form of chlorophyll, chlorophyll a, and all of them also have characteristic biliprotein pigments, **phycobilins** (see Figure 6.10), which function as accessory pigments in photosynthesis. One class of phycobilins, the phycocyanins, are blue, absorbing light maximally at 625 to 630 nm, and together with the green chlorophyll *a*, are responsible for the blue-green color of the bacteria. However, some cyanobacteria produce phycoerythrin, a red phycobilin absorbing light maximally at 570 to 580 nm, and bacteria possessing this pigment are red or brown in color. Even more confusing, the eucaryotic red algae (Rhodophyta) are red because of phycoerythrin, but some species have phycocyanin instead and are blue-green.*

*Several procaryotic phototrophs symbiotic in marine invertebrates (didemnid ascidians) have been shown to lack completely phycobilin pigments, but to contain chlorophyll *b* in addition to chlorophyll *a*. These phototrophs are classified in the genus *Prochloron,* but their precise relationship to cyanobacteria is unclear. (Lewin, R. A. Pages 257–266 *in* M. P. Starr, H. Stolp, H. G. Trüper, A. Balows, and H. G. Schlegel (eds.) The prokaryotes: a handbook on habitats, isolation, and identification of bacteria, Springer-Verlag, New York, 1981.)

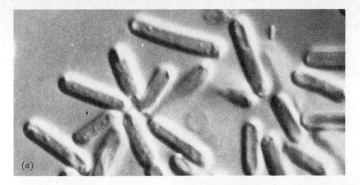

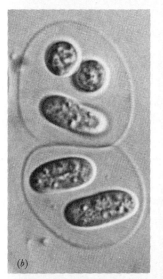

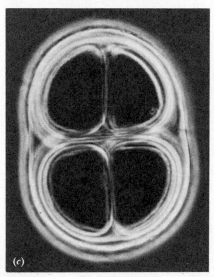

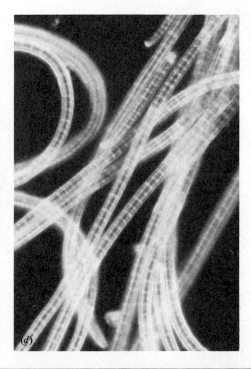

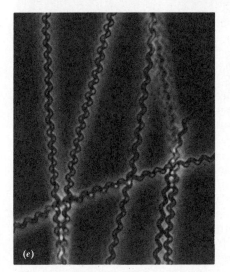

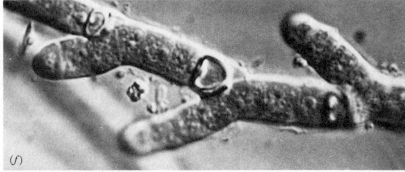

FIGURE 19.10 Cellular diversity among the cyanobacteria. (a) Unicellular. *Synechococcus* sp. Magnification, 2500×. (b) Unicellular colonial, *Gloeocapsa* sp.
Magnification, 1500×. (c) Large unicellular colonial,
Chroococcus sp. Magnification, 500×. (d) Filamentous,
Oscillatoria sp. Magnification, 500×. (e) Spiral filament,
Spirulina sp. Magnification, 500×. (f) Branching filament, *Tolypothrix* sp. Magnification, 1500×. Parts
a, b, and *f* by Nomarski interference contrast, *c* and *e*
by phase contrast; *d* by dark field.

645

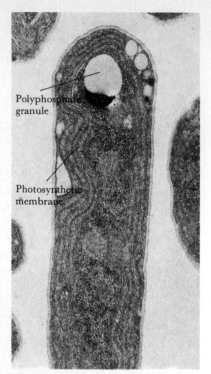

FIGURE 19.11 Electron micrograph of a thin section of the cyanobacterium *Synechococcus lividus*. Magnification, 27,500×. (From Edwards, M. R., D. S. Berns, W. C. Ghiorse, and S. C. Holt. 1968. J. Phycol. 4:283–298.)

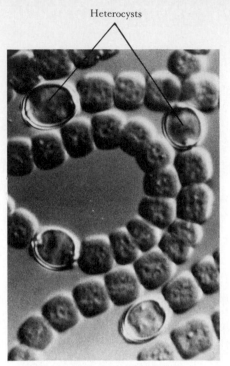

Heterocysts

FIGURE 19.12 Heterocysts of the cyanobacterium *Anabaena* sp. by Nomarski interference contrast. Heterocysts are the main if not sole site of nitrogen fixation in heterocystous cyanobacteria. Magnification, 1500×.

Structural variation   Among the cytoplasmic structures seen in many cyanobacteria are gas vesicles (see Section 2.10), which are especially common in species that live in open waters (planktonic species). Their function is probably to provide the organism with flotation (see Figure 2.36), so that it may remain where there is most light. Some cyanobacteria form heterocysts, which are rounded, seemingly more or less empty cells, usually distributed individually along a filament or at one end of a filament (Figure 19.12). Heterocysts arise from vegetative cells and may be resting cells, although they have rarely been observed to germinate. All heterocystous species of cyanobacteria fix nitrogen, and the heterocysts are the major sites of nitrogen fixation. The heterocysts have intercellular connections with adjacent vegetative cells, and there is mutual exchange of materials between these cells, with products of photosynthesis moving from vegetative cells to heterocysts and products of nitrogen fixation moving from heterocysts to vegetative cells. Heterocysts are low in phycobilin pigments and lack photosystem II, the oxygen-evolving photosystem. Because of the reductive nature of nitrogen fixation and the oxygen lability of the nitrogenase enzyme (see Section 5.13), it seems likely that the

Separation of hormogonium

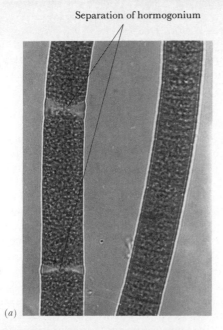

(a)

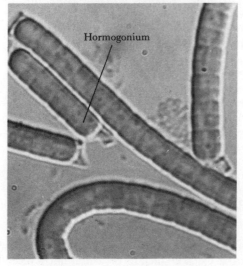

Hormogonium

(b)

FIGURE 19.13 Structural differentiation in cyanobacteria. (a) Initial stage of hormogonium formation in *Oscillatoria*. Notice the empty spaces where the hormogonium is separating from the filament. Magnification, 470×. (b) Hormogonium of a smaller *Oscillatoria* species. Notice that the cells at both ends are rounded. Nomarski interface contrast. Magnification, 1500×. (c) Akinete (resting spore) of *Anabaena* sp. by phase contrast. Magnification, 500×.

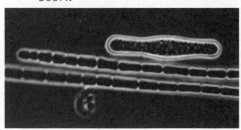

(c)

heterocyst, by maintaining an anaerobic environment, makes possible stabilization of the nitrogen-fixing system in organisms that are not only aerobic but also oxygen-producing. Indeed, some nonheterocystous filamentous cyanobacteria produce nitrogenase and fix nitrogen in normal vegetative cells if they are grown anaerobically. However, a few unicellular cyanobacteria of the sheath-forming *Gloeocapsa* type (Figure 19.10*b*) do not produce heterocysts but nevertheless fix nitrogen aerobically. Marine *Oscillatoria* (*Trichodesmium*) also fix nitrogen without heterocysts, and seem to produce a series of cells in the center of the filament that lack photosystem II activity;

nitrogen fixation apparently occurs in this O_2-nonproducing region.

A structure called the **cyanophycin granule** can be seen in electron micrographs of many cyanobacteria. This structure is a simple polymer of aspartic acid, with each aspartate residue containing an arginine molecule:

$$asp - asp - asp - asp - asp -$$
$$arg \quad arg \quad arg \quad arg \quad arg$$

and can constitute up to 20 percent of the cell mass. It appears that this co-polymer serves as a nitrogen storage product in many cyanobacteria, and when nitrogen in the environment becomes deficient, this polymer is broken down and used. The phycobilin pigments can also constitute a major portion of the cell mass, up to 10 percent, and likewise serve as a nitrogen storage material, being broken down under nitrogen starvation. Because of this, nitrogen-starved cyanobacteria often appear green instead of blue-green in color.

Many, but by no means all, cyanobacteria exhibit gliding motility; flagella have never been found. The rate of gliding varies from so slow that it is not directly observable in the microscope to over 10 μm/s in *Oscillatoria princeps*. Gliding occurs only when the cell or filament is in contact with a solid surface or with another cell or filament. In some cyanobacteria gliding is not a simple translational movement but is accompanied by rotations, reversals, and flexings of filaments. Most gliding forms exhibit directed movement in response to light (phototaxis); it is usually positive although negative movement from bright light may also occur. Chemotaxis (see Section 2.8) may occur as well.

Among the filamentous cyanobacteria, fragmentation of the filaments often occurs by formation of **hormogonia** (Figure 19.13*a* and *b*), which break away from the filaments and glide off. In some species resting spores or **akinetes** (Figure 19.13*c*) are formed, which protect the organism during periods of darkness, drying, or freezing. These are cells with thickened outer walls; they germinate through the breakdown of the outer wall and outgrowth of a new vegetative filament. However, even the vegetative cells of many cyanobacteria are relatively resistant to drying or low temperatures.

Physiology The nutrition of cyanobacteria is simple. Vitamins are not required, and nitrate or ammonia is used as nitrogen source. Nitrogen-fixing species are also common. Most species tested are obligate phototrophs, being unable to grow in the dark on organic compounds. Some cyanobacteria can assimilate simple organic compounds such as glucose and acetate if light is present. Apparently they are unable to make ATP by oxidation of organic compounds, but if ATP is provided by means of photosynthetic phosphorylation, organic compounds can be utilized as carbon sources. However, some species, mainly filamentous forms, can grow in the dark on glucose or other sugars, using the organic material as both carbon and energy source. As we discussed in Section 6.5, a number of cyanobacteria can carry out anoxygenic photosynthesis using only photosystem I, when sulfide is present in the environment.

Several metabolic products of cyanobacteria are of considerable practical importance. Many cyanobacteria produce potent neurotoxins, and during water blooms when massive accumulations of cyanobacteria may develop, animals ingesting such water may succumb rapidly. Fortunately, the massive accumulations needed to cause death do not occur extensively, although subclinical manifestations of cyanobacterial water blooms may be a common, but unobserved, occurrence. Many cyanobacteria are also responsible for the production of earthy odors and flavors in fresh waters, and if such waters are used as drinking water sources, considerable problems may arise. The compound produced is **geosmin** (*trans*-1,10-dimethyl-*trans*-9-decalol). This substance is also produced by many actinomycetes (see the discussion in Section 19.28) and is also responsible for the distinctive "earthy" odor of soil.

Ecology and evolution   Cyanobacteria are widely distributed in nature in terrestrial, freshwater, and marine habitats. In general they are more tolerant to environmental extremes than are eucaryotic algae and are often the dominant or sole photosynthetic organisms in hot springs (Table 8.1), saline lakes, and other extreme environments. Many members are found on the surfaces of rocks or soil. In desert soils subject to intense sunlight, cyanobacteria often form extensive crusts over the surface, remaining dormant during most of the year and growing during the brief winter and spring rains. Other cyanobacteria are common inhabitants of the soils of greenhouses. In shallow marine bays, where relatively warm seawater temperatures exist, cyanobacterial mats of considerable thickness may form. Freshwater lakes, especially those that are fairly rich in nutrients, may develop blooms of cyanobacteria (Figure 19.14). A few of these species are symbionts of liverworts, ferns, and cycads; a number are found as the photosynthetic component of lichens (Section 14.1). In the case of the water fern *Azolla* (see Section 14.10), it has been shown that the cyanobacterial endophyte (a species of *Anabaena*) fixes nitrogen that becomes available to the plant. Some of the photosynthetic symbionts of corals and other invertebrates also are cyanobacteria.

Base compositions of DNA of a variety of cyanobacteria have been determined. Those of the unicellular forms vary from 35 to 71 percent G + C, a range so wide as to suggest that this group contains many members with little relationship to each other. On the other hand, the values for the heterocyst formers vary much less, from 39 to 47 percent G + C. Because of the paucity of physiological characteristics that can be used in classifying cyanobacteria, techniques of molecular taxonomy will probably play a major role in their identification and classification.

The evolutionary significance of cyanobacteria is discussed in Section 19.33, and it is shown that these organisms probably were the first oxygen-evolving photosynthetic organisms and were responsible for the initial conversion of the atmosphere of the earth from anaerobic to aerobic. Fossil evidence of cyanobacteria in the Precambrian is good, and there is evidence that the cyanobacteria occupied vast areas of the earth in those ancient times. Although quantitatively much less significant today, the cyanobacteria still exist in consider-

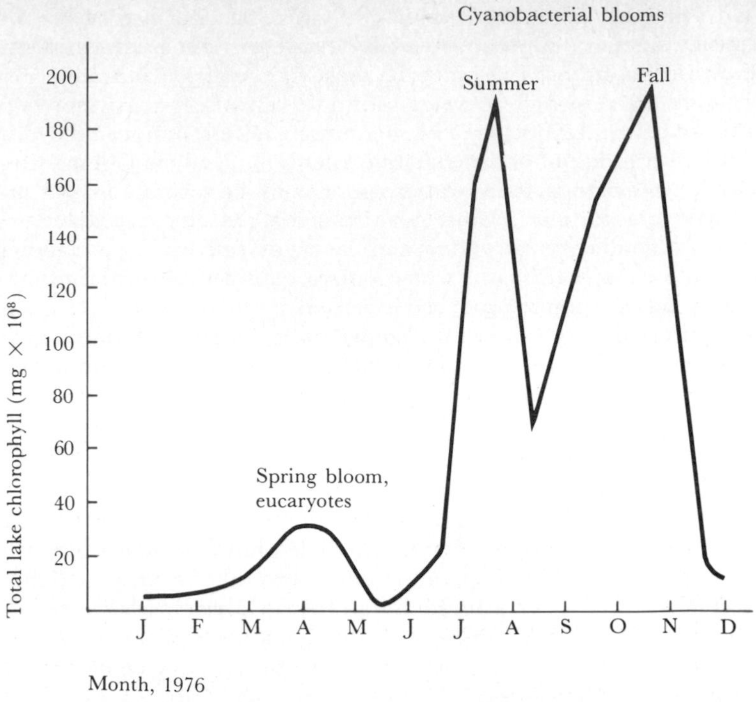

FIGURE 19.14 Development of cyanobacterial blooms in Lake Mendota, Wisconsin, during the 1976 season. Photosynthetic organisms in the lake are quantified by measuring chlorophyll *a*. In this lake, the main reason for the great dominance of cyanobacteria is probably their ability to form gas vesicles and thus remain high in the photic zone (see also Figure 13.4). (Unpublished data of T. D. Brock.)

able numbers and show wide morphological diversity. They are a procaryotic group about which very little solid information is available. Because of their inherent interest, practical implications, and structural beauty and complexity, they are worthy of a major research effort.

19.3

Gliding Bacteria

A variety of bacteria exhibit gliding motility. These organisms have no flagella, but are able in some manner to move when in contact with surfaces. A few of the **gliding bacteria** are morphologically very similar to cyanobacteria and have been considered to be their non-photosynthetic counterparts. Most of the gliding bacteria appear morphologically similar to typical Gram-negative bacteria. One group of gliding bacteria, the fruiting myxobacteria, possesses the interesting property of forming multicellular structures of complex morphology called **fruiting bodies.** Table 19.5 gives a brief ouline of some of the genera of gliding bacteria.

Cytophaga and related genera Organisms of the genus *Cytophaga* are long slender rods, often with pointed ends, which move by gliding. Many digest cellulose, agar, or chitin. They are widespread in the soil and water, often being present in great abundance. The cellulose decomposers can be easily isolated by placing small crumbs of soil on pieces of cellulose filter paper laid on the surface of mineral agar. The bacteria attach to and digest the cellulose fibers, forming transparent spreading colonies that are usually yellow or orange in color. Microscopic examination reveals the bacteria aligned upon the surface of the cellulose fibrils. The cytophagas do not produce soluble, extracellular, cellulose-digesting enzymes (cellulases); the enzymes probably remain attached to the cell envelope, accounting for the fact that the cells must adhere to cellulose fibrils in order to digest them. Organisms of the genus *Sporocytophaga* are similar to *Cytophaga* in morphology and physiology, but form resting spherical structures called *microcysts,* similar to those produced by some fruiting myxobacteria (see below), although they are produced without formation of fruiting bodies. Despite its ability to form microcysts, *Sporocytophaga* is not related to the fruiting myxobacteria; its DNA base composition is similar to that of *Cytophaga,* but far removed from those of the fruiting myxobacteria (Table 19.5). In pure culture, *Cytophaga* can be cultured on agar containing embedded cellulose fibers, the presence of the organism being indicated by the clearing that occurs as the cellulose is digested (Figure 5.6).

Several diseases of freshwater and marine fish are caused by organisms related to *Cytophaga.* Columnaris disease occurs in a variety of kinds of fish, and is usually most prominent when water temperatures rise over 20°C, the disease thus being promoted by thermal pollution of natural waters. The causal agent was formerly classified as *Chondrococcus columnaris,* a fruiting myxobacterium, although since it does not actually produce fruiting bodies, it should probably be placed in the genus *Cytophaga.* It lives saprophytically in natural waters and is able to invade fish only when water temperatures rise. The organism probably enters through small, naturally induced lesions of the skin surface, invading the skin and underlying connective tissue and inducing hyperplasia (abnormal multiplication of tissues). Other

TABLE 19.5 Characteristics of Some Genera of Gliding Bacteria

Characteristics	Genus	DNA (mol % G + C)
Rods and nonseptate filaments:		
Unicellular, rod shaped, heterotrophic; many digest cellulose, chitin, or agar	*Cytophaga* (no microcysts)	28–39
	Sporocytophaga (microcysts formed)	36
Helical or spiral shaped, heterotrophic	*Saprospira*	35–48
Filamentous, phototrophic	*Chloroflexus* (Figure 19.6c)	53–55
Filamentous, heterotrophic	*Microscilla, Flexibacter*	46–48
Septate filaments:		
Filamentous, heterotrophic or lithotrophic, producing S⁰ granules from H₂S	*Beggiatoa*	37–43
Filamentous, heterotrophic; life cycle involving gonidia and rosette formation	*Leucothrix* (Figure 19.16)	46–50
Filamentous, lithotrophic; life cycle like *Leucothrix*	*Thiothrix*	–
Cells in chains or short filaments; heterotrophic; occurs in oral cavity or digestive tract of man and other animals	*Simonsiella, Alysiella*	41–55
Filamentous, heterotrophic	*Vitreoscilla*	44–45
Rods, forming fruiting bodies:		
Unicellular, rod shaped; life cycle involving aggregation, fruiting-body formation, and myxospore formation	Fruiting myxobacteria; *Archangium, Chondromyces, Myxococcus, Polyangium,* etc. (see Table 19.6)	67–71

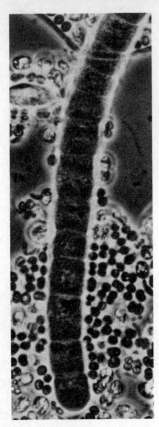

FIGURE 19.15 Phase-contrast photomicrograph of a portion of a filament of a large species of *Beggiatoa* collected from a small pond rich in organic matter. Magnification, 1360×. Notice the resemblance to *Oscillatoria* (Figure 19.12*b*). The small cells around the filament are *Thiocapsa* sp., a photosynthetic bacterium. (Courtesy of Peter Hirsch.)

diseases of fish caused by cytophagalike organisms include cold water disease, peduncle disease, bacterial gill disease, fin rot, and tail rot. Many of these diseases can be serious, especially in hatchery fish, but may be treated with sulfonamides.

Beggiatoa   Organisms of this genus are morphologically very similar to cyanobacteria, resembling especially closely *Oscillatoria*. The filaments of *Beggiatoa* are usually quite long, consisting of many short disklike cells (Figure 19.15). In addition to moving by gliding, they can flex and twist so that many filaments may become intertwined to form a complex tuft. *Beggiatoa* is found in nature primarily in habitats rich in H_2S, such as sulfur springs, decaying seaweed beds, mud layers of lakes, and waters polluted with sewage, and in these habitats the filaments of *Beggiatoa* are usually filled with sulfur granules. It was with *Beggiatoa* that Winogradsky first demonstrated that a living organism could oxidize H_2S to S^0 and then to SO_4^{2-}, leading him to formulate the concept of lithotrophy. However, most pure cultures of *Beggiatoa* so far isolated grow heterotrophically on organic compounds such as acetate, succinate, and glucose, and when H_2S is provided as an electron donor, they still require organic substances for growth. (A few strains of *Beggiatoa* have been reported to be true autotrophs, but this does not seem to be common.) Organisms that can use inorganic compounds or light as energy sources but cannot use CO_2 as sole carbon source have been called **mixotrophs** (see Section 6.11).

An interesting habitat of *Beggiatoa* recently discovered is the rhizosphere of plants (rice, cattails) living in flooded, and hence anaerobic, soils. Such plants pump oxygen down into their roots, so that a sharply defined boundary develops at the root surface between O_2 on the root and H_2S in the soil. *Beggiatoa* (and probably other sulfur bacteria) develops at this boundary, and it has been suggested that *Beggiatoa* plays a beneficial role for the plant by oxidizing and thus detoxifying hydrogen sulfide. The growth of *Beggiatoa* is greatly stimulated by the addition to culture media of the enzyme catalase (which converts hydrogen peroxide into water and oxygen), and since plant roots contain catalase, it has been suggested that the plant promotes the growth of *Beggiatoa* in its rhizosphere via catalase production, thus leading to the development of a loose mutualistic relationship between the plant and the bacterium.

Leucothrix and Thiothrix   These two genera are related in cell structure and life cycle. *Thiothrix,* a lithotroph that oxidizes H_2S, is probably an obligate lithotroph, although this has not been proved with pure cultures. *Leucothrix* is the heterotrophic counterpart of *Thiothrix,* and since its members have been amenable to cultivation, the details of its life cycle and physiology are fairly well established. *Leucothrix* is a filamentous organism that has been found in nature only in marine environments, where it grows most commonly as an epiphyte on marine algae. *Leucothrix* filaments are usually 2 to 5 μm in diameter and may reach lengths of 0.1 to 0.5 cm. The filaments have clearly visible cross walls, and cell division is not restricted to either end but occurs throughout the length of the filament. The free filaments never glide (thus distinguishing them from *Beggiatoa*), although they occasionally wave back and forth in a jerky fashion. Under environmental conditions unfavorable to rapid growth, individual cells of the filaments become round and form ovoid structures called **gonidia,** which are released individually, often from the tips of the filaments (Figure 19.16*a*). The gonidia are able to glide in a jerky manner when they come into contact with a solid surface. They settle down on solid surfaces, synthesize a holdfast, and through growth and successive cell divisions form new filaments. Presumably, in nature the gonidia are elements of dispersal, enabling the organism to spread to other areas. If there are high concentrations of gonidia, individual cells may aggregate, probably because of mutual attraction; they then synthesize a holdfast that causes their ends to adhere in a rosette, and new filaments grow out (Figure 19.16*b*). Rosette formation is found in both *Leucothrix* and *Thiothrix* and is an important means of distinguishing these organisms from many other filamentous bacteria.

Leucothrix is a strict aerobe that grows best under conditions of good aeration. In nature it is most usually found associated with seaweeds, and it probably obtains its organic materials for growth from the alga to which it is attached. *Leucothrix* also attaches to and grows upon the surface of many marine animals. Most strains of *Leucothrix* have no vitamin requirements and can grow on simple sugars or amino acids as sole carbon and energy sources.

Thiothrix is commonly found in flowing sulfide-rich springs, where its ability to attach gives it a selective advantage over *Beggiatoa*. Another common habitat for *Thiothrix,* albeit artificial, is the activated sludge sewage treatment system, especially if the effluent being treated contains hydrogen sulfide. The massive growth of filamentous *Thiothrix* in such activated sludge systems frequently presents practical problems, because the sludge does not settle properly.

Fruiting myxobacteria The fruiting myxobacteria exhibit the most complex behavioral patterns and life cycles of all known procaryotic organisms. The vegetative cells of the fruiting myxobacteria are simple, nonflagellated, Gram-negative rods that glide across surfaces and obtain their nutrients primarily by causing the lysis of other bacteria. Under appropriate conditions a swarm of vegetative cells aggregate and construct "fruiting bodies," within which some of the cells become converted into resting structures called **myxospores.** (A myxospore is defined as a resting cell contained in a fruiting body. A myxospore enclosed in a hard slime capsule is called a **microcyst.**) It is the ability to form complex fruiting bodies that distinguishes the myxobacteria from all other procaryotes. Since the vegetative cells of fruiting myxobacteria look like those of nonfruiting gliding bacteria, it is only through observation of the fruiting bodies that these organisms can be identified (see Table 19.6).

The fruiting bodies of the myxobacteria vary from simple globular masses of myxospores in loose slime to complex forms with a fruiting-body wall and a stalk. The fruiting bodies are often strikingly colored (see Color Plate 8). Occasionally they can be seen with a hand lens or dissecting microscope on pieces of decaying wood or plant material. Fruiting bodies of myxobacteria often develop on dung

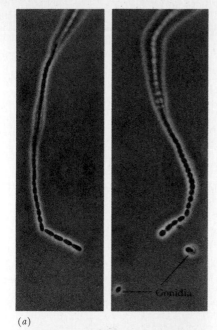

(*a*)

(*b*)

FIGURE 19.16 *Leucothrix mucor.* (a) Filaments showing multicellular nature and release of gonidia. Magnification, 725×. (b) Rosette composed of several multicellular filaments. Nomarski interference contrast. Magnification, 625×.

TABLE 19.6 Classification of the Fruiting Myxobacteria (Order Myxobacterales)

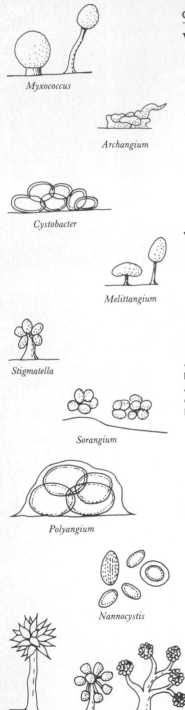

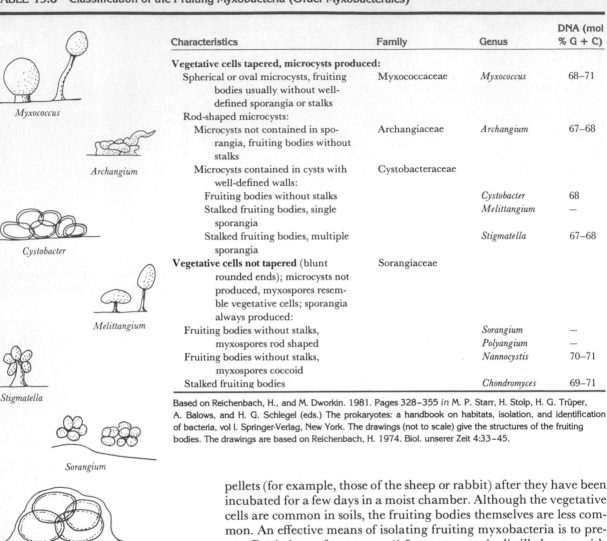

Characteristics	Family	Genus	DNA (mol % G + C)
Vegetative cells tapered, microcysts produced:			
Spherical or oval microcysts, fruiting bodies usually without well-defined sporangia or stalks	Myxococcaceae	*Myxococcus*	68–71
Rod-shaped microcysts:			
Microcysts not contained in sporangia, fruiting bodies without stalks	Archangiaceae	*Archangium*	67–68
Microcysts contained in cysts with well-defined walls:	Cystobacteraceae		
Fruiting bodies without stalks		*Cystobacter*	68
Stalked fruiting bodies, single sporangia		*Melittangium*	—
Stalked fruiting bodies, multiple sporangia		*Stigmatella*	67–68
Vegetative cells not tapered (blunt rounded ends); microcysts not produced, myxospores resemble vegetative cells; sporangia always produced:	Sorangiaceae		
Fruiting bodies without stalks, myxospores rod shaped		*Sorangium*	—
		Polyangium	—
Fruiting bodies without stalks, myxospores coccoid		*Nannocystis*	70–71
Stalked fruiting bodies		*Chondromyces*	69–71

Based on Reichenbach, H., and M. Dworkin. 1981. Pages 328–355 *in* M. P. Starr, H. Stolp, H. G. Trüper, A. Balows, and H. G. Schlegel (eds.) The prokaryotes: a handbook on habitats, isolation, and identification of bacteria, vol I. Springer-Verlag, New York. The drawings (not to scale) give the structures of the fruiting bodies. The drawings are based on Reichenbach, H. 1974. Biol. unserer Zeit 4:33–45.

pellets (for example, those of the sheep or rabbit) after they have been incubated for a few days in a moist chamber. Although the vegetative cells are common in soils, the fruiting bodies themselves are less common. An effective means of isolating fruiting myxobacteria is to prepare Petri plates of water agar (1.5 percent agar in distilled water with no added nutrients) on which is spread a heavy suspension of any of several bacteria that the myxobacteria can lyse and use as a source of nutrients (for example, *Micrococcus luteus* or *E. coli*). In the center of the plate a small amount of soil, decaying bark, or other natural material is placed. Myxobacteria in the inoculum lyse the bacterial cells and use their liberated products as nutrients; as they grow, they swarm out across the plate from the inoculum site. After several days to a week, the plates are examined under a dissecting microscope for myxobacterial swarms or fruiting bodies, and pure cultures are obtained by transfer to organic media of cells from the fruiting bodies or from the edge of the swarm.

The life cycle of a typical fruiting myxobacterium is shown in Figure 19.17. The vegetative cells are typical Gram-negative rods and do not reveal in their fine structure any clue to their gliding motility

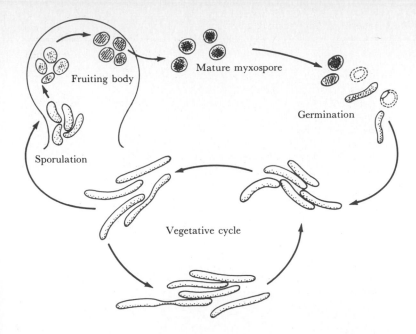

FIGURE 19.17 Life cycle of *Myxo-coccus xanthus.* (Courtesy of Hans Reichenbach and Martin Dworkin.)

Fruiting body

Mature myxospore

Germination

Sporulation

Vegetative cycle

or to their ability to aggregate and form fruiting structures. The vegetative cells of many strains grow poorly or not at all when first dispersed in liquid medium but can often be adapted to growth in liquid by making several passages through shaken liquid growth medium. A vegetative cell usually excretes a slime, and as it moves across a solid surface it leaves a slime trail behind (Figure 19.18a). This trail is preferentially used by other cells in the swarm so that often a characteristic radiating pattern is soon created, with cells migrating along slime trails (Figure 19.18b). The fruiting body ultimately formed (Figure 19.18c) is a complex structure formed by the differentiation of cells in the stalk region and in the myxospore-bearing head. A wide variety of Gram-positive and Gram-negative bacteria, as well as fungi, yeasts, and algae, can be used as food sources. A few fruiting myxobacteria (in the Polyangiaceae) can also use cellulose. Many myxobacteria can be grown in the laboratory on media containing peptone or casein hydrolysate, which provides organic nutrients in the form of amino acids or small peptides; carbohydrates do not ordinarily promote or stimulate growth, and vitamins are usually not required. Inability to utilize carbohydrates reflects the lack of certain enzymes of the Embden–Meyerhof pathway. The organisms are typical aerobes with a well-developed citric acid cycle and cytochrome system.

Fruiting-body formation does not occur so long as adequate nutrients for vegetative growth are present, but upon the exhaustion of amino acids the vegetative swarms begin to fruit. Cells aggregate, possibly through a chemotactic response, with the cells migrating toward each other and forming mounds or heaps (Figure 19.19a). A single fruiting body may have 10^9 or more cells. As the cell mounds become higher, the differentiation of the fruiting body into stalk and head begins (Figure 19.19b and c). Figure 19.19d clearly illustrates

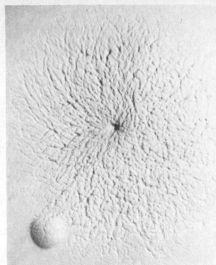

(a)

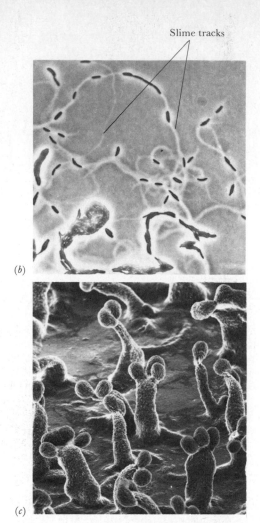

Slime tracks

(b)

(c)

FIGURE 19.18 (a) Photomicrograph of a swarming colony (9-mm diameter) of *Myxococcus xanthus* on agar. (b) Single cells of *M. fulvus* from an actively gliding culture, showing the characteristic slime tracks on the agar. Magnification, 370×. (c) Scanning electron micrograph of a fruiting body of *Stigmatella aurantiaca*. Magnification, 208×. (Photographs *a* and *b* courtesy of Hans Reichenbach. Photograph *c* from Stephens, K., G. D. Hegeman, and D. White. 1982. J. Bacteriol. 149:739–747.)

the differentiation of the fruiting body into stalk and head. The stalk is composed of nonliving slime, within which a few cells may be trapped. The majority of the cells accumulate in the fruiting-body head and undergo differentiation into myxospores. Some genera form encapsulated myxospores sometimes called microcysts (Figure 19.20*b* and Table 19.6). And, in some genera, the myxospores are enclosed in larger walled structures called **cysts.** Compared to the vegetative cell, the microcyst is more resistant to drying, sonic vibration, UV radiation, and heat, but the degree of heat resistance is much less than that of the bacterial endospore. Typically, the microcyst can withstand 58 to 60°C for 10 to 30 minutes, temperatures that would kill the vegetative cell. It seems likely that the main function of the microcyst is to enable the organism to survive desiccation during dispersal or during drying of the habitat. The microcyst germinates by a localized rupture of the capsule, with the growth and emergence of a typical vegetative rod.

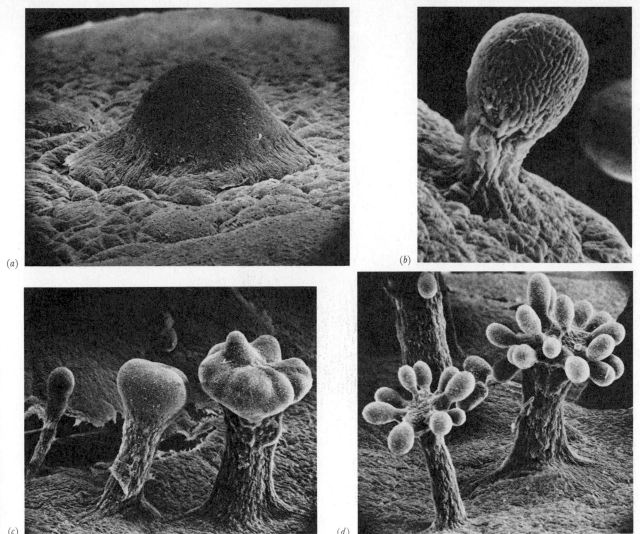

(a)

(b)

(c)

(d)

FIGURE 19.19 Scanning electron micrographs of fruiting body formation in *Chondromyces crocatus.* (a) Early stage, showing aggregation and mound formation. Magnification, 560×. (b) Initial stage of stalk formation. Slime formation in the head has not yet begun so that the cells of which the head is composed are still visible. Magnification, 2590×. (c) Three stages in head formation. Note that the diameter of the stalk also increases. Magnification, 2200×. (d) Mature fruiting bodies. Magnification, 2200×. (From Grillone, P. L., and J. Pangborn. 1975. J. Bacteriol. 124:1558–1565.)

Myxobacteria are usually colored by carotenoid pigments. The main pigments are peculiar carotenoid glycosides, which are esterified in the sugar moiety with fatty acids. Pigment formation is promoted by light, and at least one function of the pigment is photoprotection, as was described for other bacteria in Chapter 8. Since in nature the myxobacteria usually form fruiting bodies in the light, the presence of these photoprotective pigments is understandable. In the genus *Stigmatella,* light greatly stimulates fruiting body formation, and it is thought that light catalyzes production of a pheromone* that initiates the aggregation step. The fruiting myxobacteria are

*A pheromone is a chemical or mixture of chemicals secreted by an organism that acts to elicit a specific behavioral or developmental response from other organisms of the same species. Pheromones are important in the life cycles of various insects, fungi, and ciliates.

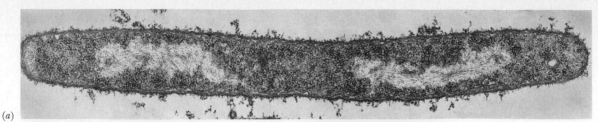

(a)

FIGURE 19.20 (a) Electron micrograph of a thin section of a vegetative cell of *M. xanthus.* Magnification, 33,750×. (b) Myxospore (microcyst) of *M. xanthus,* showing the multilayered outer wall. Magnification, 32,800×. (Courtesy of Herbert Voelz.)

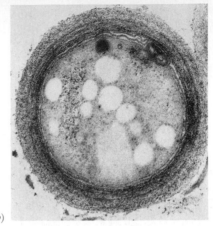

(b)

classified primarily on morphological grounds using characteristics of the vegetative cells, the myxospores, and fruiting body structure (Table 19.6). However, most of the genera outlined in Table 19.6 have not been studied in much detail, and further study may alter ideas on how they should be classified.

The fruiting myxobacteria provide experimental material for the study of a number of interesting problems in developmental microbiology, microbial ecology, and microbial evolution.

19.4

Sheathed Bacteria

Sheathed bacteria are filamentous organisms with a unique life cycle involving formation of flagellated swarmer cells within a long tube or sheath. Under certain (generally unfavorable) conditions, the swarmer cells move out and become dispersed to new environments, leaving behind the empty sheath. Under favorable conditions, vegetative growth occurs within the filament, leading to the formation of long, cell-packed sheaths. The sheathed bacteria are common in freshwater habitats that are rich in organic matter, such as polluted streams, trickling filters, and activated sludge plants, being found primarily in flowing waters. In habitats where reduced iron or manganese compounds are present the sheaths may become coated with a precipitate of ferric hydroxide or manganese oxide. Iron precipitation is strictly nonspecific, but some sheathed bacteria have the specific ability to oxidize manganous ions to manganese oxide. Two genera are currently recognized: *Sphaerotilus,* in which manganese oxidation does not occur, and *Leptothrix,* whose members do oxidize

Mn^{2+}. A single species of *Sphaerotilus* is recognized, *S. natans,* but several species of *Leptothrix* have been discerned, distinguished primarily on size, flagellation of swarmers, and some other morphological characteristics. Most of our discussion will concern *S. natans,* the organism which has been most extensively studied.

The *Sphaerotilus* filament is composed of a chain of rod-shaped cells with rounded ends enclosed in a closely fitting sheath. This thin and transparent sheath is difficult to see when it is filled with cells, but when the filament is partially empty the sheath can easily be seen by phase-contrast microscopy (Figure 19.21a) or by staining. The cells within the sheath divide by binary fission (Figure 19.21b), and the new cells pushed out at the end synthesize new sheath material. Thus the sheath is always formed at the tips of the filaments. The cells are 1 to 2 μm wide by 3 to 8 μm long and stain Gram-negatively. Indi-

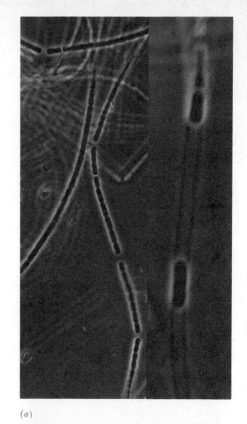

(a)

FIGURE 19.21 *Sphaerotilus natans.* (a) Phase-contrast photomicrographs of material collected from a polluted stream. Active growth stage (left–magnification, 690×) and swarmer cells leaving the sheath (magnification, 1540×). (b) Electron micrograph of a thin section through a filament. Magnification, 25,600×. (c) Electron micrograph of a negatively stained swarmer cell. Notice the polar flagellar tuft. Magnification, 10,290×. (Electron micrographs *b* and *c* from Hoeniger, J. F. M., H. D. Tauschel, and J. L. Stokes. 1973. Canad. J. Microbiol. 19:309–313 and plates I–VII. By permission of The National Research Council of Canada.)

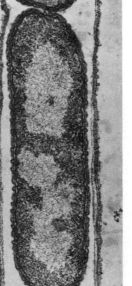

(b)

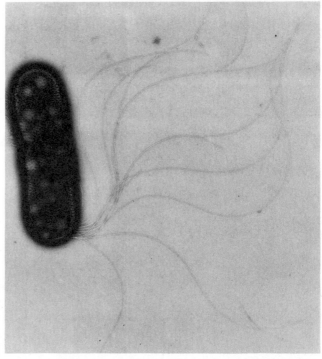

(c)

vidual cells are liberated from the sheaths, probably when the nutrient supply is low. These free cells are actively motile, the flagella being arranged lophotrichously (in a bundle at one pole) (Figure 19.21c). Probably the flagella are synthesized before the cells leave the sheath and, if so, may even aid in their liberation. It is thought that the swarmer cells then migrate, settle down, and begin to grow, each swarmer being the forerunner of a new filament. The sheath, which is devoid of muramic acid or other components of the peptidoglycan cell wall, is a protein-polysaccharide-lipid complex, possibly analogous to the capsules formed by many Gram-negative bacteria but differing in that it forms a linear structure.

Sphaerotilus cultures are nutritionally versatile, able to use a wide variety of simple organic compounds as carbon and energy sources, with inorganic nitrogen sources. Many strains require vitamin B_{12}, a substance frequently needed by aquatic microorganisms. Befitting its habitat in flowing waters, *Sphaerotilus* is an obligate aerobe.

As we noted, *Sphaerotilus* is widespread in nature in aquatic environments receiving rich organic matter. *Sphaerotilus* blooms often occur in the fall of the year in streams and brooks when leaf fall causes a temporary increase in the organic content of the water. Its filaments are the main component of a microbial complex that sanitary engineers call "sewage fungus," which is the fungus-like filamentous slime found on the rocks in streams receiving sewage pollution. In activated sludge plants (see Section 13.15). *Sphaerotilus* growth is often responsible for a detrimental condition called "bulking." The tangled masses of *Sphaerotilus* filaments so increase the bulk of the sludge that it does not settle properly, thus presenting difficulties in sludge clarification.

The oxidation of Fe^{2+} and Mn^{2+} by members of the *Sphaerotilus-Leptothrix* group has been the topic of considerable research and controversy. Because members of this group live at neutral pH in aerobic habitats, where iron and manganese often oxidize spontaneously, it has been difficult to show a specific effect of these organisms on the oxidation process. Although at one time, the *Sphaerotilus-Leptothrix* group was considered to be a specific group of iron bacteria, there is currently no basis for this classification. However, the ability of *Sphaerotilus* and *Leptothrix* to cause precipitation of iron oxides on their sheaths is well established. Such iron-encrusted sheaths are frequently seen in iron-rich waters (Figure 19.22). The process whereby iron deposition occurs is as follows: in iron-rich waters, ferrous iron is often held in solution as a chelate with organic materials such as humic and tannic acids. The sheathed bacteria can take up these soluble chelates, oxidize the organic compound, liberating the ferrous ions, which then oxidize spontaneously and become precipitated, generally in the region of the sheath.

In the case of Mn^{2+}, specific oxidation by *Leptothrix,* but not by *Sphaerotilus,* is known to occur. (Oxidation of Mn^{2+} is not unique to *Leptothrix,* as a number of other bacteria, as well as fungi and yeasts, are able to carry out this process.) At pH values below 8, Mn^{2+} does not oxidize spontaneously, so that a significant biological oxidation is possible. However, even with Mn^{2+}, there is no evidence that *Leptothrix* obtains energy from the process, either lithotrophically or mixo-

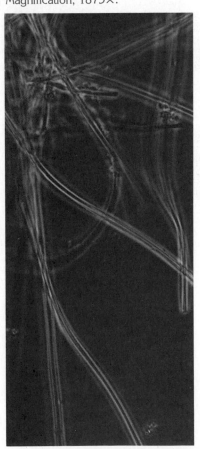

FIGURE 19.22 Phase-contrast photomicrograph of empty iron-encrusted sheaths of *Sphaerotilus* collected from seepage at the edge of a small swamp. Magnification, 1875×.

trophically. *Leptothrix* is unable to grow on completely inorganic media containing Mn^{2+} as sole electron donor for energy generation and when organic compounds are present, careful analysis of growth yields shows no increase in yield due to the added Mn^{2+}. Although *Leptothrix* produces a protein which catalyzes Mn^{2+} oxidation, it is unclear as yet what benefit the organism derives from this oxidation process.

19.5

Budding and/or Appendaged (Prosthecate) Bacteria

This large and rather heterogeneous group contains bacteria which form various kinds of cytoplasmic extrusions: stalks, hyphae, or appendages (Table 19.7). Extrusions of these kinds, which are smaller in diameter than the mature cell and which contain cytoplasm and are bounded by the cell wall, are called **prosthecae** (singular, **prostheca**) (Figure 19.23). However, some bacteria in this group form stalks which are not cytoplasmic extrusions, but are simply excretions of slime or protein outside the cell wall in the form of a definite stalk-like structure. Of considerable interest in this group of bacteria is that

TABLE 19.7 Characteristics of Stalked, Appendaged (Prosthecate), and Budding Bacteria

Characteristics	Genus	DNA (mol % G + C)
Stalked bacteria, stalk an extension of the cytoplasm and involved in cell division	*Caulobacter*	62–67
Stalked fusiform-shaped cells	*Prosthecobacter*	55–60
Stalked, but stalk an excretory product, not containing cytoplasm:		
Stalk depositing iron, cells vibrioid	*Gallionella*	—
Laterally excreted gelatinous stalk, not depositing iron	*Nevskia*	—
Appendaged (prosthecate) bacteria:		
Single or double prosthecae	*Asticcacaulis*	55–61
Multiple prosthecae		
Short prosthecae, multiply by fission	*Prosthecomicrobium*	66–70
Flat, star-shaped cells	*Stella*	60
Long prosthecae, multiply by budding	*Ancalomicrobium*	70–71
Phototrophic	*Prostechochloris*	50–56
With gas vesicles	*Ancalochloris*	—
Budding bacteria:		
Phototrophic	*Rhodomicrobium,* certain	61–64
Heterotrophic, budding without hyphae:	*Rhodopseudomonas* species	
Pear-shaped or globular cells with long, slender true stalks	*Planctomyces*	50
Pear-shaped, stalks lacking	*Pasteuria*	57
Rod-shaped cells	*Blastobacter*	59–64
Heterotrophic, buds on tips of slender hyphae:		
Single hypha from parent cell	*Hyphomicrobium*	59–67
Multiple hyphae from parent cell	*Pedomicrobium*	—

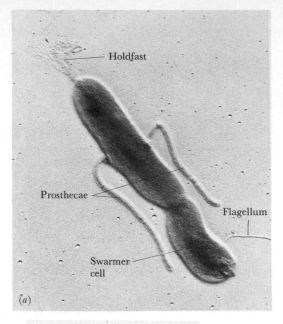

Holdfast

Prosthecae

Flagellum

Swarmer
cell

(a)

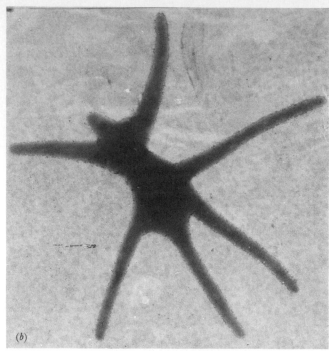

(b)

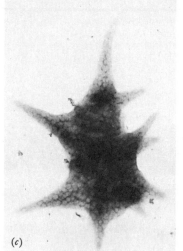

(c)

FIGURE 19.23 (a) Electron micrograph of a shadow-cast preparation of *Asticcacaulis biprosthecum,* illustrating the location and arrangement of the prosthecae. Note also the holdfast material and the swarmer cell in process of differentiation. Magnification, 25,200×. (From Pate, J. L., and E. J. Ordal. 1965. The fine structure of two unusual stalked bacteria. J. Cell Biol. 27:133–150.) (b) Electron micrograph of a negative-stained preparation of a cell of the prosthecate bacterium, *Ancalomicrobium adetum.* The appendages are cellular (i.e., prosthecae) because they are bounded by the cell wall and contain cytoplasm. Magnification, 12,125×. (From Staley, J. T. 1968. *Prosthecomicrobium* and *Ancalomicrobium:* New Prosthecate Freshwater Bacteria. J. Bacteriol. *95:*1921–1942.) Micrograph courtesy of J. T. Staley. (c) Electron micrograph of a whole cell of another prosthecate green bacterium, *Ancalochloris perfilievii.* The structures seen within the cell are gas vesicles. Magnification, 22,500×. (From V. M. Gorlenko, Academy of Sciences, Moscow.)

cell division often occurs as a result of unequal cell growth. In contrast to cell division in the typical bacterium, which occurs by binary fission and results in the formation of two equivalent cells (Figure 19.24), cell division in the stalked and budding bacteria involves the formation of a new daughter cell with the mother cell retaining its identity after the cell division process is completed (Figure 19.24). The genera from Table 19.7, which show this unequal cell-division process, are indicated in Figure 19.24.

The critical difference between these bacteria and conventional types is not the formation of buds or stalks, but the formation of a new cell wall from a single point (polar growth) rather than throughout the whole cell (intercalary growth). Several genera not normally con-

Products of cell division are equal:

Binary fission: conventional bacteria

Products of cell division are unequal:

Simple budding: *Pasteuria, Blastobacter*

Budding from hyphae: *Hyphomicrobium, Rhodomicrobium, Pedomicrobium*

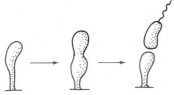

Cell division of stalked organism: *Caulobacter*

Polar growth without differentiation of cell size: *Rhodopseudomonas, Nitrobacter, Methylosinus*

FIGURE 19.24 Contrast between cell division in conventional bacteria and in budding and stalked bacteria.

sidered to be budding bacteria show polar growth without differentiation of cell size (Figure 19.24). An important consequence of **polar cell growth** is that internal structures, such as membrane complexes, do not have to become involved in the cell-division process, thus permitting the formation of more complex structures than with intercalary growth. Several additional consequences of polar growth include: aging of the mother cell occurs; cells are mortal (rather than immortal as in intercalary growth); cell division may be asymmetrical; the daughter cell at division is immature and must form internal or budding structures before it can divide; organisms showing polar growth have a potential for morphogenetic evolution not possible in cells with intercalary growth. Thus some of the most complex morphogenetic processes in the procaryotes are found in the budding and stalked bacteria.

Most of the bacteria in this group are aquatic; in nature many live attached to surfaces, their stalks or appendages serving as attachment sites. Many of the prosthecate forms are free-floating, and it is thought that their appendages serve as absorptive organs, making possible more efficient growth in the nutritionally dilute aquatic environment. Many of these free-floating forms have gas vesicles, presumably an adapation to the planktonic existence. We mentioned the

(a)

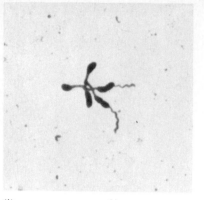

(b)   *(c)*

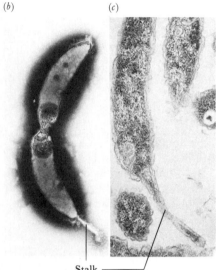

Stalk

FIGURE 19.25 (a) A *Caulobacter* rosette. The five cells are attached by their stalks (prosthecae). Two of the cells have divided and the daughter cells have formed flagella. Magnification, 2000×. (Courtesy of Einar Leifson.) (b, c) Electron micrographs of *Caulobacter* cells. (b) Negatively stained preparation of a cell in division. Magnification, 15,120×. (c) A thin section. Notice that the cytoplasmic constituents are present in the stalk region. Magnification, 31,080×. (From Cohen-Bazire, G., R. Kunisawa, and J. S. Poindexter. 1966. J. Gen. Microbiol. 42:301–308.)

phototrophic forms of this group in Section 19.1; here we discuss the two main heterotrophic groups, the stalked and the budding bacteria.

Stalked bacteria The stalked bacteria comprise a group of Gram-negative, polarly flagellated rods that possess a **stalk**, an organ by which they attach to solid substrates. Most members of this group are classified in the genus *Caulobacter*. Stalked bacteria are frequently seen in aquatic environments attached to particulate matter, plant materials, or other microorganisms; generally they are found attached to microscope slides that have been immersed in lake or pond water for a few days. When many *Caulobacter* cells are present in the suspension, groups of stalked cells are seen attached, exhibiting the formation of rosettes (Figure 19.25a). Electron-microscopic studies reveal that the stalk is not an excretion product but is an outgrowth of the cell, since it contains cytoplasm surrounded by cell wall and plasma membrane (Figure 19.25b and c). The holdfast by which the stalk attaches the cell to a solid substrate is at the tip of the stalk, and, once attached, the cell usually remains permanently fixed. Since the stalk is cytoplasmic, it is also a prostheca. A stalk which is not a prostheca can be seen in an electron micrograph of the budding bacterium *Planctomyces* (Figure 19.26). In this organism the stalk contains no cytoplasm or cell wall and is probably proteinaceous in nature. The stalk functions to anchor the cell to surfaces via a holdfast located at the tip of the stalk.

The *Caulobacter* cell-division cycle (Figure 19.27) is of special interest because it involves a process of unequal binary fission. Cell

FIGURE 19.26 An electron micrograph of a metal-shadowed preparation of *Planctomyces maris*. Note the fibrillar nature of the stalk. Pili are also abundant. Note also the flagella (curly appendages) on each cell and the bud that is developing from the non-stalked pole of one cell. Magnification, 13,950×. (From Bauld, J., and J. T. Staley. 1976. J. Gen. Microbiol. 97: 45–55. Micrograph courtesy of John Bauld and J. T. Staley.)

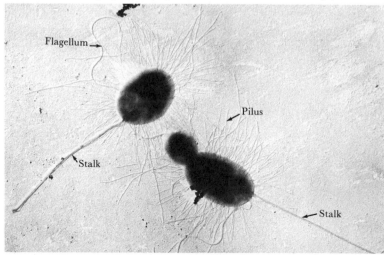

Flagellum

Pilus

Stalk

Stalk

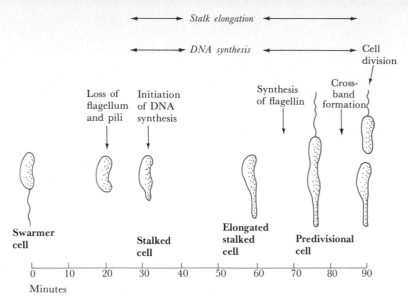

Caulobacter cell cycle beginning with swarmer cell

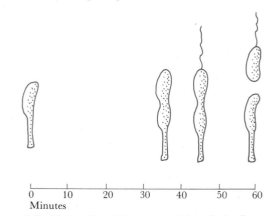

Shortened *Caulobacter* cell cycle beginning with stalked cell

FIGURE 19.27 Stages in the *Caulobacter* cell cycle. (Based on Poindexter, J. S. 1964. Bacteriol. Revs. 28:231–295, and Shapiro, L. 1976. Ann. Rev. Microbiol. 30:377–407.)

division occurs by elongation of the cell followed by fission, a single flagellum forming at the pole opposite the stalk. The flagellated cell so formed, called a "swarmer," separates from the nonflagellated mother cell, swims around, and settles down on a new surface, forming a new stalk at the flagellate pole; the flagellum then disappears (Figure 19.27). Stalk formation is a necessary precursor to cell division and is coordinated with DNA synthesis (Figure 19.27). The time span for the division of the stalked cell is thus shorter than the time span for division of the swarmer cell, owing to the requirement that the swarmer (flagellated) cell must synthesize a stalk before it divides. A crossband is produced in the stalk during or shortly after each cell division (Figure 19.27), so that the age of a stalked cell can be read from the number of stalk cross-bands. The cell-division cycle in *Caulobacter* is thus more complex than simple binary fission since the stalked and swarmer cells have polar differentiation, and the cells themselves are structurally different.

Caulobacters are heterotrophic aerobes; they usually have one or several vitamin requirements, but are able to grow on a variety of organic carbon compounds as sole sources of carbon and energy. Various amino acids serve as nitrogen sources. The enrichment culture for caulobacters makes use of the fact that they occur quite commonly in the organic film that develops at the surface of an undisturbed liquid. If pond, lake, or seawater is mixed with a small amount of organic material such as 0.01 percent peptone, and incubated at 20 to 25°C for 2 to 3 days, a surface film consisting of bacteria, fungi, and protozoa develops, and in this microbial film caulobacters are common. A sample of the surface film is then streaked on an agar medium containing 0.05 percent peptone, and after 3 to 4 days the plates are examined under a dissecting microscope for the presence of microcolonies which are typical of *Caulobacter*. These colonies are then picked and streaked on fresh medium containing a higher concentration of organic matter (for example, 0.5 percent peptone + 0.1 percent yeast extract), and the resulting colonies are examined microscopically for stalked bacteria. Water low in organic matter is a good source of caulobacters; tap or distilled water that has been left undisturbed usually shows good *Caulobacter* development in the surface film. The ability to grow in dilute media is a common property of organisms that attach to solid substrates; in the case of stalked organisms such as *Caulobacter* the stalk itself may also function as an ab-

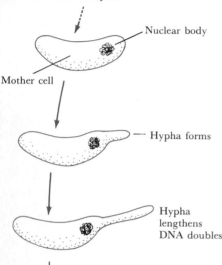

FIGURE 19.28 Stages in the *Hyphomicrobium* cell cycle.

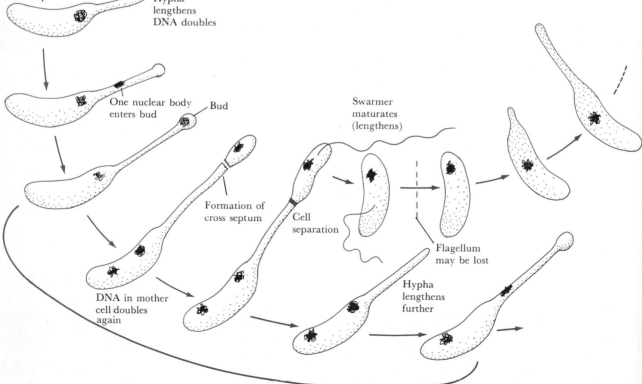

sorptive organ, making it possible for the organism to acquire larger amounts of the restricted supply of organic nutrients than can organisms lacking these appendages.

A stalked organism sometimes classified with the caulobacters is *Gallionella,* which forms a twisted stalk containing ferric hydroxide. However, the stalk of *Gallionella* is not an integral part of the cell but is excreted from the cell surface. It contains an organic matrix on which the ferric hydroxide accumulates. *Gallionella* is frequently found in the waters draining bogs, iron springs, and other habitats where ferrous iron is present, usually in association with sheathed bacteria such as *Sphaerotilus.* In very acidic waters containing iron, *Gallionella* is not present, and acid-tolerant thiobacilli replace it.

Budding bacteria   Certain bacteria are unique in that they multiply by budding rather than by binary fission. The two best studied genera are *Hyphomicrobium,* which is heterotrophic, and *Rhodomicrobium,* which is phototrophic. The process of reproduction in a budding bacterium is illustrated in Figure 19.28. The mother cell, which is often attached by its base to a solid substrate, forms a thin outgrowth that lengthens to become a hyphalike structure, and at the end of the hypha a bud forms. This bud enlarges, forms a flagellum, breaks loose from the mother cell, and swims away. Later, the daughter cell loses its flagellum and after a period of maturation forms a hypha and buds. Further buds can also form at the hyphal tip of the mother cell. Many variations on this cycle are possible. In some cases the daughter cell does not break away from the mother cell but forms a hypha from its other pole. Complex arrays of cells connected by hyphae are frequently seen (Figure 19.29). In some cases a bud begins to form directly from the mother cell without the intervening formation of hypha, whereas in other cases a single cell forms hyphae from each end (Figure 19.29). The hypha is a direct cellular extension of the mother cell (Figure 19.30), containing cell wall, cytoplasmic membrane, ribosomes, and occasionally DNA.

The nuclear events during the budding cycle are of interest (Figure 19.28). The "nuclear body" (region of DNA) located in the mother cell doubles and then divides; once the bud has formed, DNA is moved down the length of the hypha and into the bud. A cross-septum then forms, separating the still developing bud from the hypha and mother cell.

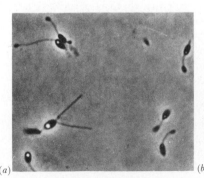
(a)

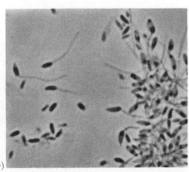
(b)

FIGURE 19.29   Photomicrographs of *Hyphomicrobium.* Magnification, 1360×. (a, b) By phase contrast, showing typical fields. Notice the long hyphae and the occasional budding cell. (Courtesy of Peter Hirsch.)

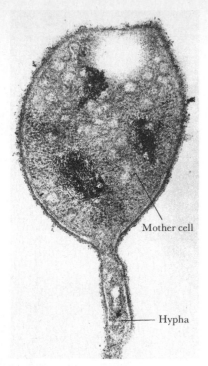

FIGURE 19.30 Electron micrograph of a thin section of a single *Hypho-microbium* cell. Magnification, 51,600×. (From Conti, S. F., and P. Hirsch. 1965. J. Bacteriol. 89:503–512.)

Hyphomicrobium has unique nutritional characteristics. Preferred carbon sources are one-carbon compounds such as methanol, methylamine, formaldehyde, formate, and even cyanide (CN^-). Growth on acetate, ethanol, or higher aliphatic compounds is usually slow, and growth does not occur at all on sugars or most amino acids. Urea, amides, ammonia, nitrite, and nitrate can be utilized as nitrogen sources; no vitamins are required. *Hyphomicrobium* not only is able to grow on very low concentrations of carbon, but will also grow in a liquid medium in the complete absence of any added carbon source, apparently obtaining its energy and carbon from volatile compounds present in the atmosphere. This has frequently led to the erroneous conclusion that *Hyphomicrobium* is an autotroph, but there is no evidence that the enzymes for autotrophic CO_2 fixation are present. *Hyphomicrobium* is widespread in freshwater, marine, and terrestrial habitats. Initial enrichment cultures can be prepared using a mineral-salts medium lacking organic carbon and nitrogen, to which a sample of natural material is added. After several weeks incubation, the surface film that develops is streaked out on agar medium containing methylamine or methanol as a sole carbon source. Colonies are then checked microscopically for the characteristic *Hyphomicrobium* morphology. A fairly specific enrichment procedure for *Hyphomicrobium* uses methanol as electron donor with nitrate as electron acceptor under anaerobic conditions. Virtually the only denitrifying organisms using methanol are *Hyphomicrobium,* so that this procedure selects this organism out of a wide variety of environments.

Hyphomicrobium has been found associated with extensive manganese deposits in water pipelines. Probably, the manganese oxidation occurs spontaneously, and the hyphomicrobia are not involved but are growing on the small amounts of organic matter present in the water. It has been suggested that the function of the budding habit in this case is that it provides a mechanism for the bacteria to escape from the prisonlike manganese deposit, the thin hyphae growing out through the precipitate, so that swarmer cell formation can then take place in the open water.

19.6

Spirochetes

The **spirochetes** are bacteria with a unique morphology and mechanism of motility. They are widespread in aquatic environments and in the bodies of animals. Some of them cause diseases of animals and humans, of which the most important is syphilis, caused by *Treponema pallidum.* The spirochete cell is typically slender, flexuous, helical (coiled) in shape, and often rather long (Figure 19.31). The "protoplasmic cylinder," consisting of the regions enclosed by the cytoplasmic membrane and the cell wall, constitutes the major portion of the spirochetal cell. Fibrils, referred to as **axial fibrils** or **axial filaments,** are attached to the cell poles and wrapped around the coiled protoplasmic cylinder (Figure 19.32). Both the axial fibrils and the protoplasmic cylinder are surrounded by a three-layered membrane called the "outer sheath" or "outer cell envelope" (Figure 19.32). The outer sheath and the axial fibrils are usually not visible by light microscopy,

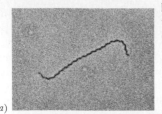

(a)

FIGURE 19.31 Two spirochetes at the same magnification, showing the wide size range in the group.
(a) *Spirochaeta stenostrepta,* by phase-contrast microscopy. Magnification, 1250×. (From Canale-Parola, E., S. C. Holt, and Z. Udris. 1967. Arch. Mikrobiol. 59:41–48.) (b) *Spirochaeta plicatilis.* Magnification, 1220×. (From Blakemore, R. P., and E. Canale-Parola. 1973. Arch. Mikrobiol. 89:273–289.)

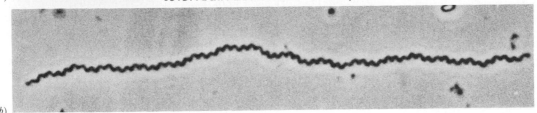

(b)

but are observable in negatively stained preparations or thin sections examined by electron microscopy.

From 2 to more than 100 axial fibrils are present per cell, depending on the type of spirochete. The ultrastructure and the chemical composition of axial fibrils are similar to those of bacterial flagella (see Section 2.7). As in typical bacterial flagella (Figure 2.26), basal hooks and paired disks are present at the insertion end. The shaft of each fibril is composed of a core surrounded by an "axial fibril sheath," so that the spirochete axial fibrils are in a sense analogous to sheathed flagella.

The manner in which the spirochete cell moves depends on the medium through which it is moving. In liquids, locomotion is generally accompanied by rapid rotation around the longitudinal axis, and by lashing, bending, curling, or snakelike contortions of the cell. The coiled configuration is usually retained during locomotion. However, in a viscous medium, such as an agar gel, the helical cell slowly bores itself through the viscous substrate in a screwlike manner. Some spirochetes, such as the large *Spirochaeta plicatilis* (Figure 19.31b), move both by swimming in liquids and by "creeping" on solid surfaces.

The axial fibrils play a significant role in spirochete motility. Each fibril is anchored at one end and extends for approximately two-thirds of the length of the cell. It is thought that the axial fibrils rotate rigidly, as do bacterial flagella (see Section 2.7). Since the protoplasmic cylinder is also rigid, whereas the outer sheath is flexible, if both axial fibrils rotate in the same direction, the protoplasmic cylinder will rotate in the opposite direction, as illustrated in Figure 19.33. If the sheath is not in contact with a surface, it also rotates (see arrow in Figure 19.33). This simple mechanism is all that is needed to generate the wide variety of motions exhibited by spirochetes. If the protoplasmic cylinder is helical (as in most spirochetes), then forward motion will be generated when the sheath is moving through a liquid or semisolid medium by the circumferential slip of the helix through the medium. If the sheath is in contact along its length with a solid surface, the protoplasmic cylinder may not be able to rotate so that the roll of the sheath will cause the cell to slide in a direction nearly parallel to the axis of the helix, generating a "creeping" motility. In

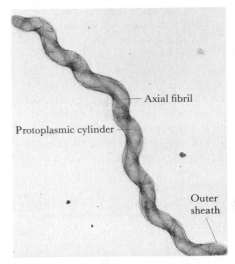

FIGURE 19.32 Electron micrograph of a negatively stained preparation of *Spirochaeta zuelzerae,* showing the position of the axial filament. Magnification, 11,840×. (From Joseph, R., and E. Canale-Parola. 1972. Arch. Mikrobiol. 81:146–168.)

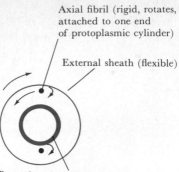

Axial fibril (rigid, rotates, attached to one end of protoplasmic cylinder)

External sheath (flexible)

Protoplasmic cylinder (rigid, generally helical)

FIGURE 19.33 Cross-section of a spirochete cell, showing the arrangement of the protoplasmic cylinder, axial fibrils, and external sheath, and the manner by which the rotation of the rigid axial fibril can generate rotation of the protoplasmic cylinder and (in opposite direction) rotation of the external sheath. If the sheath is free, the cell will rotate about its longitudinal axis and move along it. If the sheath is in contact with a solid surface, the cell will creep forward. See text for details. (Based on Berg, H. C. 1976. How spirochetes may swim. J. Theor. Biol. 56:269–273.)

free liquid, many narrow diameter spirochetes show flexing or lashing motions due to torque exerted at the ends of the protoplasmic cylinder by the twisting axial fibrils.

In this model of motility, the external sheath plays a central role, as do the opposing axial fibrils. Similarly, the rigidity of the protoplasmic cylinder is essential in this mechanism. Interestingly, antibodies directed against the axial fibrils do not cause immobilization of whole cells, whereas antibodies against some component of the external sheath do cause immobilization, perhaps by stiffening the external sheath. It thus appears that despite superficial differences, spirochetes have fundamentally the same motility mechanism as other bacteria, namely the rotation of rigid flagellar fibrils attached in the cell membrane via a basal hook.

Cell division in spirochetes is by transverse fission, the outer sheath of the cell usually being the last structure to separate.

Spirochetes are classified into five genera primarily on the basis of habitat, pathogenicity, and morphological and physiological characteristics. Table 19.8 lists the major genera and their characteristics.

Spirochaeta and Cristispira The genus *Spirochaeta* includes free-living, anaerobic, and facultatively aerobic spirochetes. These organisms are common in aquatic environments, such as the water and mud of rivers, ponds, lakes, and oceans. One species of the genus *Spirochaeta* is *S. plicatilis* (Figure 19.31*b*), a fairly large organism that was the first spriochete to be discovered; it was reported by C. G. Ehrenberg in the 1830s. It is found in freshwater and marine H_2S-containing habitats, and is probably anaerobic. The axial fibrils of *S. plicatilis* are arranged in a bundle that winds around the

TABLE 19.8 Genera of Spirochetes and Their Characteristics

Genus	Dimensions (μm)	General Characteristics	Number of Axial Fibrils	DNA (mol % G + C)	Habitat	Diseases
Cristispira	30–150 × 0.5–3.0	3–10 complete coils; bundle of axial fibrils visible by phase-contrast microscopy	>100	—	Digestive tract of molluscs; has not been cultured	None known
Spirochaeta	5–500 × 0.2–0.75	Anaerobic or facultatively aerobic; tightly or loosely coiled	2–40	50–66	Aquatic, free-living	None known
Treponema	5–15 × 0.1–0.5	Anaerobic, coil amplitude up to 0.5 μm	2–15	38–53	Commensal or parasitic in humans, other animals	Syphilis, yaws
Borrelia	3–15 × 0.2–0.5	Anaerobic; 5–7 coils of approx. 1 μm amplitude	Unknown	46	Humans and other mammals, arthropods	Relapsing fever
Leptospira	6–20 × 0.1	Aerobic; tightly coiled, with bent or hooked ends	2	35–53	Free-living or parasitic of humans, other mammals	Leptospirosis

coiled protoplasmic cylinder. From 18 to 20 axial fibrils are inserted at each pole of this spirochete. Another species, *S. stenostrepta*, has been cultured, and is shown in Figure 19.31*a*. It is an obligate anaerobe commonly found in H_2S-rich, black muds. It ferments sugars via the glycolytic pathway to ethanol, acetate, lactate, CO_2, and H_2. The species *S. aurantia* is an orange-pigmented facultative aerobe, fermenting sugars via the glycolytic pathway under anaerobic conditions, and oxidizing sugars aerobically mainly to CO_2 and acetate.

The genus *Cristispira* (Figure 19.34) contains organisms with a unique distribution, being found in nature primarily in the crystalline style of certain molluscs, such as clams and oysters. The crystalline style is a flexible, semisolid rod seated in a sac and rotated against a hard surface of the digestive tract, thereby mixing with and grinding the small particles of food. Being large spirochetes, the cristispiras can readily be seen microscopically within the style as they rapidly rotate forward and backward in corkscrew fashion. *Cristispira* may occur in both freshwater and marine molluscs, but not all species of molluscs possess them. Unfortunately, *Cristispira* has not been cultured, so that the physiological reason for its restriction to this unique habitat is not known. There is no evidence that *Cristispira* is harmful to its host; in fact, the organism may be more common in healthy than in diseased molluscs.

Treponema Anaerobic, host-associated spirochetes that are commensals or parasites of humans and animals are placed in the genus *Treponema*. Another genus, *Borrelia*, includes those spirochetes that cause relapsing fever, an acute febrile illness of people and animals (see below). *Treponema pallidum*, the causal agent of syphilis, is the best known species of *Treponema*. It differs in morphology from other spriochetes; the cell is not helical, but has a flat wave form. Furthermore, electron microscopy does not show the presence of an outer sheath surrounding both the axial fibrils and the protoplasmic cylinder. Apparently the axial fibrils of *T. pallidum* lie on the outside of the organism. The *T. pallidum* cell is remarkably thin, measuring approximately 0.2 μm in diameter. Living cells are clearly visible in the dark-field microscope or after staining with fluorescent antibody; dark-field microscopy has long been used to examine exudates from suspected syphilitic lesions. In nature *T. pallidum* is restricted to humans, although artificial infections have been established in rabbits and monkeys. It has now been established that virulent *T. pallidum* cells (purified from infected rabbits) contain a cytochrome system, and may in fact be aerobes. This property further separates *T. pallidum* from the other species of *Treponema*, which, as noted, are obligate anaerobes. Meaningful taxonomic studies on this species will have to await its cultivation away from its host.

At present, virulent *T. pallidum* cannot be grown in vitro. Those treponemes that have been cultured are nonpathogenic obligate anaerobes requiring complex media, and all grow slowly, with generation times of 12 to 18 hours. *Treponema pallidum* is quite sensitive to increased temperature, being rapidly killed by exposure to 41.5 to 42.0°C. This was once the basis of fever therapy for

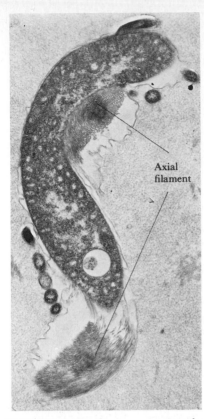

FIGURE 19.34 Electron micrograph of a thin section of *Cristispira*, a very large spirochete. Notice the numerous fibrils in the axial filament. Magnification, 15,500×. (From Ryter, A., and J. Pillot. 1965. Ann. Inst. Pasteur 109:552–562.)

syphilis, that is, increasing the body temperature of the patient in order to kill the spirochete, but the procedure is now supplanted by antibiotic therapy. The heat sensitivity of *T. pallidum* is also reflected in the fact that the organism becomes most easily established in cooler sites of the body, such as the male genital organs, although once established in other areas of the body it will multiply there. Infection of rabbits is also most extensive in cooler sites such as the testicles or skin; indeed, artificial cooling of rabbit skin results in a dramatic increase in the number of lesions. The organism is rapidly killed by drying, and this at least partially explains why transmission between persons is only by direct contact, usually sexual intercourse.

Despite the relative effectiveness of penicillin in curing syphilis, the disease is still common (see Section 17.11), mainly because of the social problems involved in locating and treating sexual contacts of infected individuals.

Other species of the genus *Treponema* are common commensal organisms in the oral cavity of humans and can generally be seen in material scraped from between the teeth and from the narrow space between the gums and the teeth. Various oral spirochetes can be cultivated anaerobically in complex media containing serum. Three species, *T. denticola, T. macrodentium,* and *T. oralis,* have been described, differing in morphology and physiological characteristics. *Treponema denticola* ferments amino acids such as cysteine and serine, forming acetate as the major fermentation acid and CO_2, NH_3, and H_2S. This spirochete can also ferment glucose, but in media containing both glucose and amino acids, the amino acids are used preferentially. The true relationship between *T. pallidum* and the remaining members of the genus *Treponema* may be a distant one, since the G + C base ratio of *T. pallidum* is about 53% while other species of this genus cluster tightly between 38 to 40 percent.

Leptospira The genus *Leptospira* contains strictly aerobic spirochetes that use long-chain fatty acids (for example, oleic acid) as electron donor and carbon sources. With few exceptions these are the only substrates utilized by leptospiras for growth. The leptospira cell is thin, finely coiled, and usually bent at each end into a semicircular hook. At present only one species, *L. interrogans,* is recognized in this genus. Strains of this species are divided into two "complexes": the "parasitic complex" including leptospiras parasitic for humans and animals, and the "biflexa complex" including the free-living aquatic leptospiras. Leptospiras of the parasitic complex are catalase positive, whereas aquatic leptospiras are catalase negative. Different strains are distinguished serologically by agglutination tests, and a large number of serotypes have been recognized. Rodents are the natural hosts of most leptospiras although dogs and pigs are also important carriers of certain strains. In human beings the most common leptospiral syndrome is called "Weil's disease"; in this disorder the organism usually localizes in the kidney. Leptospiras ordinarily enter the body through the mucous membranes or through breaks in the skin. After a transient multiplication in various parts of the body the organism localizes in the kidney and liver, causing nephritis and jaundice. The organism passes out of the body in the urine and infection of another individual is most commonly by contact with infected urine. Therapy with penicillin, streptomycin, or the tetracyclines is possible but may require extended courses to eliminate the organism from the kidney; this is probably because of the slow growth and protected location of the leptospiras. Domestic animals are vaccinated against **leptospirosis** with a killed virulent strain; dogs are usually immunized routinely with a combined distemper-leptospira-hepatitis vaccine. In people, prevention is effected primarily by elimination of the disease from animals. The serotype that infects dogs, *L interrogans* serotype *canicola,* does not ordinarily infect humans, but the strain attacking rodents, *L. interrogans* serotype *icterohaemorrhagiae,* does; hence elimination of rats from human habitation is of considerable aid in preventing the organism from reaching human beings.

Borrelia The majority of species in the genus *Borrelia* are animal or human pathogens. *B. recurrentis* is the causative agent of **relapsing fever** in humans and is transmitted via an insect vector, usually by the human body louse. Relapsing fever is characterized by a high fever and generalized muscular pain which lasts for 3 to 7 days followed by a recovery period of 7 to 9 days. Left untreated, the fever returns in two to three more cycles (hence the name relapsing fever) and causes death in up to 40 percent of those infected. Fortunately, the organism is quite sensitive to tetracycline, and if the disease is correctly diagnosed, treatment is straightforward. Other borrelia are of veterinary importance causing diseases in cattle, sheep, horses, and birds. In most of these diseases the organism is transmitted by ticks.

19.7

Spiral and Curved Bacteria

The division of genera of Gram-negative, motile, rod-shaped, and curved bacteria has presented some difficulties to the organizers of *Bergey's Manual.* A large number of organisms have been studied, and characteristics overlap among groups. Some of the key taxonomic criteria used are cell shape, size, kind of polar flagellation (single or multiple), relation to oxygen (obligately aerobic, microaerophilic, facultative), relationship to plants (as symbionts or plant pathogens) or animals (as pathogens), fermentative ability, and certain other physiological characteristics (nitrogen-fixing ability, ability to utilize methane as electron donor, halophilic nature, thermophilic nature, luminescence). The genera to be covered in the present discussion are given in Table 19.9.

Spirillum, Aquaspirillum, Oceanospirillum and Azospirillum These are helically curved rods, which are motile by means of polar flagella (usually tufts at both poles, see Figure 19.35). The number of turns in the helix may vary from less than one complete turn (in which case the organism looks like a vibrio; see Section 19.14) to many turns. Spirilla with many turns can superficially resemble spirochetes, but differ in that they do not have an outer sheath and axial filaments, but instead contain typical bacterial flagella. Some of the spirilla are

TABLE 19.9   Characteristics of the Genera of Spiral-Shaped Bacteria*

Genus	Characteristics	DNA (mol % G + C)
Spirillum	Cell diameter 1.7 μm; microaerophilic; fresh water	38
Aquaspirillum	Cell diameter 0.2–1.5 μm; aerobic; fresh water	50–65
Oceanospirillum	Cell diameter 0.3–1.2 μm; aerobic; marine (require 3% NaCl)	42–48
Azospirillum	Cell diameter 1 μm; microaerophilic; fixes N_2	69–71
Campylobacter	Cell diameter 0.2–0.8 μm; microaerophilic to anaerobic; pathogenic or commensal in humans and animals; single polar flagellum	29–35
Bdellovibrio	Cell diameter 0.25–0.4 μm; aerobic; predatory on other bacteria; single polar sheathed flagellum	42–50
Microcyclus	Cell diameter 0.5 μm; curved rods forming rings; nonmotile; aerobic; sometimes gas vesiculate	34–68

*All are Gram-negative and respiratory but never fermentative.

very large bacteria, and were seen by early microscopists. It is likely that van Leeuwenhoek first described *Spirillum* in the 1670s, and the genus was first created by the protozoologist Ehrenberg in 1832. The organism seen by these workers is now called *Spirillum volutans,* and is one of the largest bacteria known (Figure 19.35*b*). A phototrophic organism resembling *S. volutans* is *Thiospirillum* (see Figure 19.5*b*). Despite the fact that *S. volutans* has been known for a long time, it resisted cultivation in pure culture until fairly recently. The difficulty turned out to be that the organism is microaerophilic, requiring O_2, but being inhibited by O_2 at normal levels. In mixed culture, contaminants used up much of the oxygen and kept the level low, but once the organism was separated from its contaminants it succumbed to oxygen toxicity (see Section 8.5 for a discussion of O_2 toxicity). The simplest way of achieving microaerophilic growth of *S. volutans* is to use a medium such as nutrient broth in a closed vessel with a large head (gas) space, with an atmosphere of N_2, and to then inject sufficient O_2 from a hypodermic syringe to occupy 1 to 5 percent of the head space volume. Because *S. volutans* is dependent on O_2 for growth, high cell yields cannot be expected in this way, because the O_2 is quickly used up. For cultivation for physiological studies, a continuous stream of 1 percent O_2 in N_2 can be passed through the culture vessel. Another characteristic of *S. volutans* is the formation of prominent granules (volutin granules) consisting of polyphosphate (see Section 2.9).

Azospirillum lipoferum* is a nitrogen-fixing organism, which was originally described and named *Spirillum lipoferum* by Beijerinck in 1922. It has become of considerable interest in recent years because this bacterium has been found to enter into a loose symbiotic relationship with tropical grasses and grain crops (see Section 14.10).

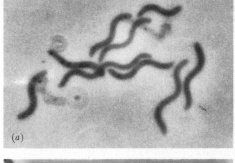

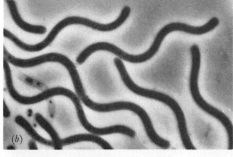

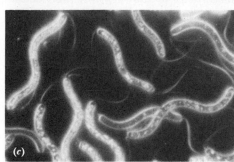

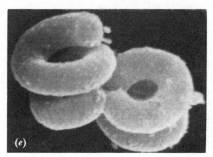

(a)

(b)

(c)

(d)

(e)

FIGURE 19.35 (a) Photomicrograph by phase contrast of *Aquaspirillum peregrinum.* (b) *Spirillum volutans,* the largest spirillum, by phase contrast. (c) *S. volutans,* by dark-field microscopy, showing flagellar bundles and volutin granules. *a–c* magnifications, 1360×. (d) Scanning electron micrograph of an intestinal spirillum. Note the polar flagellar tufts and the spiral structure of the cell surface. Magnification, 26,000×. (e) Scanning electron micrograph of cells of *Microcyclus flavus.* Magnification, 29,000×. (Photographs *a–c* courtesy of Noel Krieg, *d* courtesy of Stanley Erlandsen, *e* courtesy of H. D. Raj.)

Since the publication of the eighth edition of *Bergey's,* the genus *Spirillum* has been redefined so that it includes only a single species, *S. volutans,* characterized by its microaerophilic character, large size, and formation of volutin granules. The small-diameter spirilla (Figure 19.35a), which are not microaerophilic, have been separated into two genera, *Aquaspirillum* and *Oceanospirillum,* the former for freshwater forms and the latter for those living in seawater and requiring NaCl for growth (Table 19.9). At least 16 species of *Aquaspirillum* have been described and 9 species of *Oceanospirillum,* the various species being separated on physiological grounds. The student isolating heterotrophic bacteria from freshwater and marine environments will almost certainly obtain isolates of these genera, as they are among the most common organisms appearing in heterotrophic enrichments, and they are easily purified by streaking on agar. These organisms undoubtedly play an important role in the recycling of organic matter in aquatic environments. Evidence from enrichment culture studies using the chemostat suggests that the aquatic spirilla are adapted to the utilization of organic substrates at very low substrate concentrations, probably because they have very low affinity constants (K_m's) for uptake of organic nutrients.

Bdellovibrio   These small vibrioid organisms have the unique property of preying on other bacteria, using as nutrients the cytoplasmic constituents of their hosts. These bacterial predators are small, highly motile cells, which stick to the surfaces of their prey cells. Because of the latter property, they have been given the name *Bdellovibrio* (*bdello-* is a combining form meaning "leech"). A number of strains of *Bdellovibrio* have been found, and each shows prey specificity, attacking some species of bacteria but not others. In general, Gram-positive bacteria are not attacked. After attachment of a *Bdellovibrio* cell to its prey, the predator penetrates through the prey wall and replicates in the space between the prey wall and membrane (the periplasmic space). The stages of attachment and penetration are shown in Figure 19.36. These processes require the active metabolism of both prey and predator since they will not occur if the prey cells are killed. A schematic representation of the *Bdellovibrio* life cycle is shown in Figure 19.37.

As originally isolated, *Bdellovibrio* cells grow only upon the living prey, but it is possible to isolate mutants that are prey independent

FIGURE 19.36   Stages of attachment and penetration of a prey cell by *Bdellovibrio*. (a) Electron micrograph of a shadowed whole-cell preparation showing *B. bacteriovorus* attacking *Pseudomonas*. Magnification, 10,370×. (From Stolp, H., and H. Petzold. 1962. Phytopath. Z. 45:373.) (b, c) Electron micrographs of thin sections of *Bdellovibrio* attacking *Escherichia coli:* (b) early penetration (magnification, 49,600×); (c) complete penetration (magnification, 64,800×). The *Bdellovibrio* cell is enclosed in a membranous infolding of the prey cell, and replicates in the periplasmic space between wall and membrane. (From Burnham, J. C., T. Hashimoto, and S. F. Conti. 1968. J. Bacteriol. 96:1366–1381.)

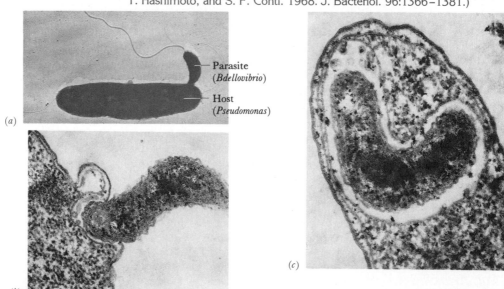

(a)

Parasite
(*Bdellovibrio*)

Host
(*Pseudomonas*)

(b)

(c)

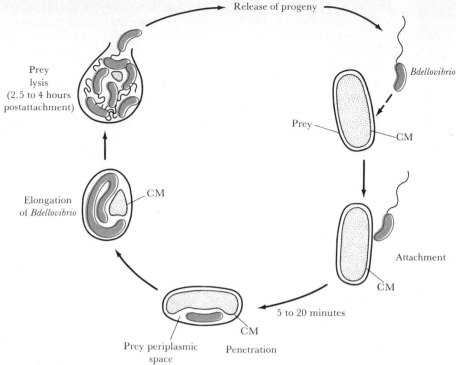

Release of progeny

Prey
lysis
(2.5 to 4 hours
postattachment)

Prey

Bdellovibrio

CM

Elongation
of *Bdellovibrio*

CM

Attachment

CM

5 to 20 minutes

Prey periplasmic
space

Penetration

CM

FIGURE 19.37 Developmental cycle of the bacterial parasite *Bdellovibrio bacteriovorus.* Following primary contact between a highly motile *Bdellovibrio* cell and a Gram negative bacterium, attachment and penetration into the prey periplasmic space occurs. Once inside, *Bdellovibrio* elongates and within 4 hours progeny cells are released. The number of progeny cells released varies with the size of the prey bacterium, for example, 5 to 6 bdellovibrios are released from each infected *E. coli* cell, 20 to 30 for *Spirillum serpens.* CM = prey cytoplasmic membrane.

and are able to grow on complex organic media such as yeast extract-peptone. These strains are unable to utilize sugars as electron donors, but are proteolytic and can oxidize the amino acids liberated by protein digestion. Prey-dependent revertants can be reisolated from the prey-independent mutants by reinfecting a host strain.

    Bdellovibrio is an obligate aerobe, obtaining its energy from the oxidation of amino acids and acetate (via the citric acid cycle). It apparently is unable to utilize sugars as electron donors. Studies have been carried out to determine the growth efficiency of *Bdellovibrio* as indicated by the Y_{ATP} value. As we noted in Section 7.6, most organisms exhibit a Y_{ATP} value of around 10. *Bdellovibrio,* on the other hand, exhibits a Y_{ATP} value of 18 to 26, considerably higher than that of conventional free-living bacteria. This considerably higher efficiency of conversion of substrates into ATP could possibly be due to the more efficient growth process in the periplasmic environment of its host (transport processes are not necessary), and to the fact that *Bdellovibrio* assimilates nucleoside phosphates and fatty acids from its host without first breaking them down. However, these energy-

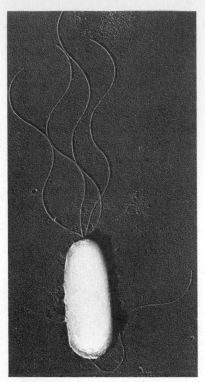

FIGURE 19.38 Typical pseudomonad morphology, *Pseudomonas coronafaciens.* Magnification, 15,860×. (Electron micrograph courtesy of Arthur Kelman.)

sparing effects cannot account for all of the growth efficiency of *Bdellovibrio,* and other reactions must be sought. One idea is that *Bdellovibrio* is much more efficient in coupling energy-generating reactions to biosynthesis, so that energy is not wasted. It is clear that the predatory mode of existence has involved the development in *Bdellovibrio* of interesting and unusual biochemical processes.

Members of the genus *Bdellovibrio* are widespread in soil and water, including the marine environment. Their detection and isolation require methods somewhat reminiscent of those used in the study of bacterial viruses. Prey bacteria are spread on the surface of an agar plate to form a lawn, and the surface is inoculated with a small amount of soil suspension that has been filtered through a membrane filter, which retains most bacteria but allows the small *Bdellovibrio* cells to pass. Upon incubation of the agar plate, plaques analogous to those produced by bacteriophages are formed at locations where *Bdellovibrio* cells are growing. Unlike phage plaques, which continue to enlarge only as long as the bacterial host is growing, *Bdellovibrio* plaques continue to enlarge even after the prey has stopped growing, resulting in large plaques on the agar surface. Pure cultures of *Bdellovibrio* can then be isolated from these plaques. *Bdellovibrio* cultures have been obtained from a wide variety of soils and are thus common members of the soil population. As yet, the ecological role of bdellovibrios is not known, but it seems likely that they play some role in regulating the population densities of their prey.

Microcyclus Members of the genus *Microcyclus* are ring-shaped and nonmotile (Figure 19.35*e*). They resemble very tightly curved vibrios and are widely distributed in aquatic environments. A photosynthetic counterpart to *Microcyclus* has been discovered and placed in the genus *Rhodocyclus* (see Section 19.1).

19.8

The Pseudomonads

All of the genera in this group are straight or slightly curved rods with polar flagella (Figure 19.38). They are heterotrophic aerobes, and never show a fermentative metabolism. (Fermentative organisms with polar flagella are generally classified in the genera *Aeromonas* or *Vibrio,* as indicated in Section 19.14). The most important genus is *Pseudomonas,* discussed in some detail here. Other genera include *Xanthomonas,* primarily a plant pathogen responsible for a number of necrotic plant lesions, and which is characterized by its yellow-colored pigments; *Zoogloea,* characterized by its formation of an extracellular fibrillar polymer, which causes the cells to aggregate into distinctive flocs (this organism is a dominant component of activated sludge, Section 13.15); and *Gluconobacter,* characterized by its incomplete oxidation of sugars or alcohols to acids, such as the oxidation of glucose to gluconic acid or ethanol to acetic acid (this organism is discussed briefly with the other acetic acid bacteria in Section 19.11).

The genus Pseudomonas The distinguishing characteristics of members of the genus *Pseudomonas* are given in Table 19.10. Also given in

TABLE 19.10   The Genus *Pseudomonas*

General characteristics:
Straight or curved rods but not vibrioid; size 0.5–1.0 μm by 1.5–4.0 μm (Figure 19.38); no spores; Gram-negative; polar flagella: single or multiple; no sheaths, appendages or buds; respiratory metabolism, never fermentative, although may produce small amounts of acid from glucose aerobically; use low-molecular-weight organic compounds, not polymers; some are lithotrophic, using H_2 or CO as sole electron donor; some can use nitrate as electron acceptor anaerobically; some can use arginine as energy source anaerobically (arginine dihydrolase, see Figure A3.4)

Minimal characteristics for identification:
Gram-negative, straight or slightly curved; no spores; motile (always); polar flagella (flagellar stain); oxidative-fermentative medium with glucose: open, acid produced; closed, acid not produced; gas not produced from glucose (distinguishes them easily from enteric bacteria and *Aeromonas*); oxidase, almost always positive (enterics are oxidase negative); catalase always positive; photosynthetic pigments absent (distinguishes them from Rhodospirillaceae); indole, negative; methyl red negative; Voges–Proskauer, negative

this table are the minimal characteristics needed to identify an organism as a pseudomonad. Key identifying characteristics are the absence of gas formation from glucose, and the positive oxidase test, both of which help to distinguish pseudomonads from enteric bacteria (Section 19.13). Also, the genus *Rhodopseudomonas* (see Section 19.1) is distinguished from *Pseudomonas* by the presence of photosynthetic pigments when the former is grown anaerobically in the light. (Since *Rhodopseudomonas* will grow aerobically in the dark without producing photosynthetic pigments, it could be mistaken for a pseudomonad.)

The species of the genus *Pseudomonas* are defined on the basis of various physiological characteristics, as outlined in Tables 19.10 and 19.11. Some species (for example, *P. aeruginosa*) are quite homogeneous, so that all isolates fit into a very narrow range of distribution of characteristics, whereas other species are much more heterogeneous. The taxonomy of the genus *Pseudomonas* has been greatly clarified by DNA hybridization studies, and the subgroups given in Table 19.11 are based on DNA homologies.

One of the striking properties of the pseudomonads is the wide variety of organic compounds used as carbon sources and as electron donors for energy generation Some strains utilize over 100 different compounds, and only a few strains utilize fewer than 20. As an example of this versatility, a single strain of *P. aeruginosa* can make use of many different sugars, fatty acids, dicarboxylic acids, tricarboxylic acids, alcohols, polyalcohols, glycols, aromatic compounds, amino acids, and amines, plus miscellaneous organic compounds not fitting into any of the above categories. On the other hand, these organisms are usually unable to break down polymers into their component monomers. The pseudomonads are ecologically important organisms in soil and water and are probably responsible for the degradation of many soluble compounds derived from the breakdown of plant and animal materials.

A few pseudomonads are pathogenic (Table 19.12). Among the fluorescent pseudomonads, the species *P. aeruginosa* is frequently as-

TABLE 19.11 Characteristics of Subgroups and Species
of the Genus *Pseudomonas*

Group	Characteristics	DNA (mol % G + C)
Fluorescent subgroup:	Most produce water-soluble, yellow-green fluorescent pigments; do not form poly-β-hydroxybutyrate; single DNA homology group	
P. aeruginosa	Pyocyanin production, growth at up to 43°C, single polar flagellum, denitrification	67
P. fluorescens	Do not produce pyocyanin or grow at 43°C; tuft of polar flagella	59–61
P. putida	Similar to *P. fluorescens* but do not liquefy gelatin and do grow on benzylamine	62
P. syringae	Lack arginine dihydrolase, oxidase negative, pathogenic to plants	58–60
Acidovorans subgroup:	Nonpigmented, form poly-β-hydroxybutyrate, tuft of polar flagella, do not use carbohydrates; single DNA homology group	
P. acidovorans	Uses muconic acid as sole carbon source and electron donor	67
P. testosteroni	Use testosterone as sole carbon source	62
Pseudomallei-cepacia subgroup:	No fluorescent pigments, tuft of polar flagella, form poly-β-hydroxybutyrate; single DNA homology group	
P. cepacia	Extreme nutritional versatility; some strains pathogenic to plants	67–68
P. pseudomallei	Cause melioidosis in animals; nutritionally versatile	69
P. mallei	Cause glanders in animals; nonmotile; nutritionally restricted	69
Diminuta-vesicularis subgroup:	Single flagellum of very short wavelength, require vitamins (pantothenate, biotin, B_{12})	
P. diminuta	Nonpigmented, do not use sugars	66–67
P. vesicularis	Carotenoid pigment, use sugars	66
Miscellaneous species:		
P. solanacearum	Plant pathogens	66–68
P. saccharophila	Grow lithotrophically with H_2, digest starch	69
P. maltophilia	Require methionine, do not use NO_3^- as N source, oxidase negative	

sociated with infections of the urinary and respiratory tracts in humans. *Pseudomonas aeruginosa* infections are also common in patients receiving treatment for severe burns. *Pseudomonas aeruginosa* is able to grow at temperatures up to 43°C, whereas the nonpathogenic species of the fluorescent group can grow only at lower temperatures. *Pseudomonas aeruginosa* is not an obligate parasite, however, since it can be readily isolated from soil, and as a denitrifier (see Section 13.8) it plays an important role in the nitrogen cycle in nature. As a pathogen

TABLE 19.12   Diseases Caused by *Pseudomonas* Species

Species	Relationship to Disease
P. aeruginosa:	Opportunistic pathogen, especially in hospitals; in patients with metabolic, hematologic, and malignant diseases; hospital-acquired infections from catheterizations, tracheostomies, lumbar punctures, and intravenous infusions; in patients given prolonged treatment with immunosuppressive agents, corticosteroids, antibiotics and radiation; may contaminate surgical wounds, abscesses, burns, ear infections, lungs of patients treated with antibiotics; primarily a soil organism
P. fluorescens:	Rarely pathogenic, as does not grow well at 37°C; may grow in and contaminate blood and blood products under refrigeration
P. maltophilia:	An ubiquitous, free-living organism that is an occasional opportunist in humans
P. cepacia:	Onion bulb-rot; has also been isolated from humans and from environmental sources of medical importance
P. pseudomallei:	Causes melioidosis, a disease endemic in animals and humans in southeast Asia
P. mallei:	Causes glanders, a disease of horses that is occasionally transmitted to humans
P. stutzeri:	Often isolated from humans and environmental sources; may live saprophytically in the body
P. solanacearum:	Plant pathogenic, causing wilts of many cultivated plants (e.g., potato, tomato, tobacco, peanut)

it appears to be primarily an opportunist, initiating infections in individuals whose resistance is low. In addition to urinary infections it can also cause systemic infections, usually in individuals who have had extensive skin damage by burns. The organism is naturally resistant to many of the widely used antibiotics, so that chemotherapy is often difficult. Resistance is frequently due to a resistance transfer factor (R factor; see Section 11.6), which is a plasmid carrying genes coding for detoxification of various antibiotics. *P. aeruginosa* is commonly found in the hospital environment and can easily infect patients receiving treatment for other illnesses. Polymyxin, an antibiotic not ordinarily used in human therapy because of its toxicity, is effective against *P. aeruginosa* and can be used with caution.

Many pseudomonads, as well as a variety of other Gram-negative bacteria, metabolize glucose via the Entner–Doudoroff pathway (Figure A3.5). Two key enzymes of the Entner–Doudoroff pathway are 6-phosphogluconate dehydrase and ketodeoxyglucose-phosphate aldolase. A survey for the presence of these enzymes in a wide variety of bacteria has shown that they are absent from all Gram-positive bacteria (except a few *Nocardia* isolates) and are generally present in bacteria of the genera *Pseudomonas, Rhizobium,* and *Agrobacterium,* as well as in some isolates of several genera of Gram-negative bacteria. It has been suggested that bacteria possessing the Entner–Doudoroff pathway may represent an evolutionary line separate from other bacteria.

Free-Living Aerobic Nitrogen-Fixing Bacteria

A variety of organisms which primarily inhabit the soil are capable of fixing N_2 aerobically (Table 19.13). The genus *Azotobacter* comprises large, Gram-negative, obligately aerobic rods capable of fixing N_2 nonsymbiotically (see Figure 19.39). The first member of this genus was discovered by the Dutch microbiologist M. W. Beijerinck early in the twentieth century, using an enrichment culture technique with a medium devoid of a combined nitrogen source (see "A Bit of History," page 617). Although capable of growth on N_2, *Azotobacter* grows more rapidly on NH_3; indeed, adding NH_3 actually represses nitrogen fixation. Much work has been done in seeking to evaluate the role of *Azotobacter* in nitrogen fixation in nature, especially in comparison to the anaerobic organism *Clostridium pasteurianum* and the symbiotic organisms of the genus *Rhizobium. Azotobacter* is also of interest because it has the highest respiratory rate (measured as the rate of O_2 uptake) of any living organism. In addition to its ecological and physiological importance, *Azotobacter* is of interest because of its ability to form an unusual resting structure called a *cyst* (Figure 19.39*b*).

Azotobacter cells are rather large for bacteria, many isolates being almost the size of yeasts, with diameters of 2 μm or more. Pleomorphism is common, and a variety of cell shapes and sizes have been described. Some strains are motile by peritrichously located flagella. On carbohydrate-containing media, extensive capsules or slime layers are produced. *Azotobacter* is able to grow on a wide variety of carbohydrates, alcohols, and organic acids. The metabolism of carbon compounds is strictly oxidative, and acids or other fermentation products are rarely produced. All members fix nitrogen, but growth also occurs on simple forms of combined nitrogen: ammonia, urea, and nitrate.

Despite the fact that *Azotobacter* is an obligate aerobe, its nitrogenase is as O_2 sensitive as all other nitrogenases (Section 5.13). It is thought that the high respiratory rate of *Azotobacter* (mentioned above) has something to do with protection of nitrogenase from O_2. The intracellular O_2 concentration is kept low enough by metabolism so that inactivation of nitrogenase does not occur.

Like bacterial endospores, *Azotobacter* cysts show negligible endogenous respiration and are resistant to desiccation, mechanical disintegration, and ultraviolet and ionizing radiation. In contrast to endospores, however, they are not especially heat resistant, and they

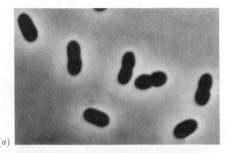

(a)

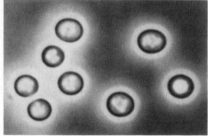

(b)

FIGURE 19.39 *Azotobacter vinelandii:* vegetative cells (a) and cysts (b) by phase-contrast microscopy. Magnification, 2760×. (Courtesy of L. P. Lin and H. L. Sadoff.)

TABLE 19.13   Genera of Free-Living Aerobic Nitrogen-Fixing Bacteria

Genus	Characteristics	DNA (mol % G + C)
Azotobacter	Large rod; produces cysts; primarily found in neutral to alkaline soils	63–66
Azomonas	Large rod; no cysts; primarily aquatic	53–59
Beijerinckia	Pear-shaped rods with large lipid bodies at each end; produces extensive slime; inhabits acidic soils; see Figure 19.40	54–60
Derxia	Rods; form coarse, wrinkled colonies	70

are not completely dormant since they rapidly oxidize exogenous energy sources. Treatment of cysts with metal-binding agents such as citrate results in solubilization of the outer layers, presumably because the structure of the outer layer is stabilized by metals. The central body is then liberated in viable form. The central body does not possess the resistance characteristics of the cyst, thus suggesting that it is the cyst coat that confers resistance. The carbon source of the medium greatly influences the extent of cyst formation, butanol being especially favorable; compounds related to butanol, such as β-hydroxybutyrate, also promote cyst formation.

Considerable research has been done on the role of *Azotobacter* in the nitrogen economy of soils. When large quantities of carbohydrates are incorporated into soil, significant increases in combined nitrogen occur, most likely caused by the action of these carbohydrates in promoting the growth of *Azotobacter*. However, the gains in nitrogen seen are rather small in relation to the amount of carbohydrate added, so that carbohydrate enrichment of soil would not be an economical agricultural practice.

The remaining genera in this group include *Azomonas*, a genus of large rod-shaped bacteria that resemble *Azotobacter* except that they do not produce cysts and are primarily aquatic, and *Beijerinckia* (Figure 19.40) and *Derxia*, two genera that grow well in acidic soils.

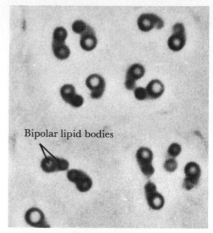

FIGURE 19.40 Phase-contrast photomicrograph of cells of *Beijerinckia indica.* The cells are roughly pear-shaped and contain a large globule of poly-β-hydroxybutyrate at each end. Magnification, 1000×. (Courtesy of Michael K. Ochman.)

19.10

Rhizobium Agrobacterium Group

Agrobacterium and *Rhizobium* are closely related and in some schemes are lumped together in the single genus *Rhizobium*. As currently defined, the genus *Rhizobium* contains those Gram-negative flagellated rods that are capable of entering into symbiosis with legumes by inducing root nodules, and of fixing free nitrogen during this symbiosis. The genus *Agrobacterium* comprises organisms of similar morphology that cause tumorlike growths of plant tissues, resulting in the disease conditions called crown gall and hairy root. In both genera the organisms are small, short rods, motile by means of a few flagella arranged either peritrichously or subpolarly. The DNA base composition varies from 58 to 65 percent G + C. *Rhizobium* and *Agrobacterium* are not only related morphologically and in ability to infect plants, but DNA hybridization studies reveal that homology exists between the genomes of these two genera.

The ability of these bacteria to induce uncontrolled plant growth (tumors and nodules) is of considerable fundamental interest. We discussed in Section 14.10 the process of infection and nodulation of a legume by *Rhizobium*. The crown-gall bacteria induce formation of tumorous growths (Figure 19.41) on a wide variety of plant genera. Thus, in contrast to the root-nodule bacteria, each of which has a fairly narrow host range, the crown-gall bacteria have a very wide host range.

Although plants often form disorganized tissue called *callus* when wounded, the growth induced by *Agrobacterium* is different in that the callus shows uncontrolled growth. It thus resembles tumor growth in animals, and considerable research on **crown gall** has been

FIGURE 19.41 Photograph of tumor on tomato plant caused by crown-gall bacteria of the genus *Agrobacterium*. (Courtesy of H. W. Spurr, A. C. Hildebrandt, and D. J. Riker.)

carried out with the idea that it may provide a model for how cancer occurs in humans. In general, plant growth in tissue culture requires the presence of specific plant hormones, but crown-gall tissue will grow in culture without the addition of such hormones. Interestingly, crown-gall growth in culture can occur even if the bacteria have been removed experimentally from the plant cells. Thus, once *Agrobacterium* has brought about the induction of the tumorous condition, its presence is no longer necessary, and considerable research has been done to find out whether any genetic information from the *Agrobacterium* has been transferred to the plant.

It is now established that a large plasmid called the Ti (for *tumor induction*) **plasmid** must be present in the *Agrobacterium* cells if they are to induce tumor formation, and a restricted portion of this plasmid is transferred via transformation to the plant cells. This plasmid constitutes about 5 percent of the total bacterial DNA and contains not only the trait for tumor formation, but also a number of other traits, of which the most interesting relate to the production and utilization of two unusual amino acids, octopine [N^2-(D-1-carboxyethyl)-L-arginine] and nopaline [N^2-(1,3-dicarboxypropyl)-L-arginine]. Crown-gall tumors contain elevated levels of one or the other of these compounds, and tumor-inducing *Agrobacterium* are able to degrade one or the other of these compounds. Bacterial strains that degrade nopaline induce tumors that synthesize only nopaline, and strains degrading octopine induce tumors synthesizing only octopine. However, despite this correlation, synthesis of one or the other of these amino acids is not essential for tumorigenesis because mutations blocking synthesis of these compounds do not prevent tumorigenesis. After infection, a portion of the plasmid is incorporated into tumor cells, as can be shown by hybridization experiments with tumor mRNA and plasmid DNA. By use of restriction enzymes, it has been shown that only a small portion of the plasmid, equivalent to about 10 to 20 genes, is transcribed by the plant. More research will be necessary to determine the details of the tumorigenic process, but the crown-gall system is clearly an excellent model for understanding at the molecular level the process by which uncontrolled growth is induced. *Agrobacterium* is also now finding considerable use as a vector for introducing foreign DNA into plants by means of recombinant DNA techniques (see Section 12.1).

19.11

Acetic Acid Bacteria

As originally defined, the acetic acid bacteria comprised a group of Gram-negative, aerobic, motile rods that carried out an incomplete oxidation of alcohols, leading to the accumulation of organic acids as end products. With ethanol as a substrate, acetic acid is produced; hence the derivation of the common name for these bacteria. Another property is the relatively high tolerance to acidic conditions, most strains being able to grow at pH values lower than 5. This acid tolerance should of course be essential for an organism producing large amounts of acid. The genus name *Acetobacter* was originally used to encompass the whole group of acetic acid bacteria, but it is now clear that the acetic acid bacteria as so defined are a heterogeneous assemblage, comprising both peritrichously and polarly flagellated organisms. The polarly flagellated organisms are related to the pseudo-monads, differing mainly in their acid tolerance and their inability to carry out a complete oxidation of alcohols. These organisms are now classified in the genus *Gluconobacter* (Table 19.14).

The genus *Acetobacter* now comprises the peritrichously flagellated organisms; these have no definite relationship to other Gram-negative rods. In addition to flagellation, *Acetobacter* differs from *Gluconobacter* in being able to further oxidize the acetic acid it forms to CO_2. This difference in ability to oxidize acetic acid is related to the presence of the citric acid cycle. *Gluconobacter,* which lacks a complete citric acid cycle, is unable to oxidize acetic acid, whereas *Acetobacter,* which has the cycle, can oxidize it. Gluconobacters are sometimes called underoxidizers; the acetobacters, overoxidizers.

The acetic acid bacteria are frequently found in association with alcoholic juices, and probably originally arose in sugar-rich flowers and fruits where a yeast-mediated alcoholic fermentation is common. Acetic acid bacteria can often be isolated from alcoholic fruit juices such as cider or wine. Colonies of acetic acid bacteria can be recognized on $CaCO_3$-agar plates containing ethanol, the acetic acid produced causing a dissolution and clearing of the insoluble $CaCO_3$. Cultures of acetic acid bacteria are used in commercial production of vinegar.

In addition to ethanol, these organisms carry out an incomplete oxidation of such organic compounds as higher alcohols and sugars. For instance, glucose is oxidized only to gluconic acid, galactose to galactonic acid, arabinose to arabonic acid, and so on. This property of underoxidation is exploited in the manufacture of ascorbic acid (vitamin C). Ascorbic acid can be formed from sorbose, but sorbose is difficult to synthesize chemically. It is, however, conveniently obtainable microbiologically from acetic acid bacteria, which oxidize sorbitol (a readily available sugar alcohol) only to sorbose. The use of acetic acid bacteria makes the manufacture of ascorbic acid economically feasible.

Another interesting property of some acetic acid bacteria is their ability to synthesize cellulose. The cellulose formed does not differ significantly from that of plants, but instead of being a part of the cell wall the bacterial cellulose is formed as a matrix outside the wall, and the bacteria become embedded in the tangled mass of cellulose microfibrils. When these species of acetic acid bacteria grow in an unshaken vessel, they form a surface pellicle of cellulose, in which the bacteria develop. Since these bacteria are obligate aerobes, the ability

TABLE 19.14 Differentiation of *Acetobacter, Gluconobacter,* and *Pseudomonas*

Character	*Acetobacter*	*Gluconobacter*	*Pseudomonas*
Flagellation	Peritrichous	Polar	Polar
Growth at pH 4.5	+	+	−
Oxidation of ethanol to acetic acid at pH 4.5	+	+	−
Complete citric acid cycle	Present	Absent	Present
DNA (mol % G + C)	55–64	60–64	58–70

to form such a pellicle may be a means by which the organisms are assured of remaining at the surface of the liquid, where oxygen is readily available.

19.12

MethaFe-Oxidizing Bacteria: Methylotrophs

Methane, CH_4, is found extensively in nature. It is produced in anaerobic environments by methanogenic bacteria (see Section 13.7) and is a major gas of anaerobic muds, marshes, anaerobic zones of lakes, the rumen, and the mammalian intestinal tract. Methane is the major constituent of natural gas and is also present in many coal formations. It is a relatively stable molecule; but a variety of bacteria, the **methane oxidizers,** oxidize it readily, utilizing methane and a few other one-carbon compounds as electron donors for energy generation and as sole sources of carbon. These bacteria are all aerobes, and are widespread in nature in soil and water. They are also of a diversity of morphological types, seemingly related only in their ability to oxidize methane. There has been considerable interest in methane-utilizing bacteria in recent years because of the possibility of using this simple and widely available energy source to produce bacterial protein as a food or feed supplement (single-cell protein).

In addition to methane, a number of other one-carbon compounds are known to be utilized by microorganisms. A list of these compounds is given in Table 19.15. From a biochemical viewpoint, these compounds share a key characteristic: they contain no carbon-carbon bonds, so that all carbon-carbon bonds of the cell must be synthesized de novo. Organisms that can grow using only one-carbon compounds are generally called **methylotrophs.** From the viewpoint of carbon assimilation, the methylotrophs have something in common with autotrophs (Chapter 6), which also use a carbon compound without a carbon-carbon bond, CO_2. The two groups differ, however, in that the methylotrophs require a carbon compound more reduced than CO_2.

TABLE 19.15 Substrates Used by Methylotrophic Bacteria*

Substrates Used for Growth	Substrates Oxidized, but Not Used for Growth (Co-metabolism)
Methane, CH_4	Ammonium, NH_4^+
Methanol, CH_3OH	Ethylene, $H_2C=CH_2$
Methylamine, CH_3NH_2	Chloromethane, CH_3Cl
Dimethylamine, $(CH_3)_2NH$	Bromomethane, CH_3Br
Trimethylamine, $(CH_3)_3N$	Higher hydrocarbons (ethane, propane)
Tetramethylammonium, $(CH_3)_4N^+$	
Trimethylamine N-oxide, $(CH_3)_3NO$	
Trimethylsulfonium, $(CH_3)_3S^+$	
Formate, $HCOO^-$	
Formamide, $HCONH_2$	
Carbon monoxide, CO	
Dimethyl ether, $(CH_3)_2O$	
Dimethyl carbonate, $CH_3OCOOCH_3$	

*A single isolate does not use all of the above, but at least one methylotrophic bacterium has been reported to oxidize each of the listed compounds.

It is important to distinguish between those methylotrophic bacteria that can utilize methane from those that can only use some of the more oxidized one-carbon compounds listed in Table 19.15. A wide variety of bacteria are known which will grow on methanol, methylamine or formate, and these bacteria are members of various genera of heterotrophs: *Hyphomicrobium, Pseudomonas, Bacillus, Vibrio*. In contrast, the methane-oxidizing bacteria are unique in that they can grow not only on some of the more oxidized one-carbon compounds, but also on methane. The methane-oxidizing bacteria possess a specific enzyme system, methane monooxygenase for the introduction of an oxygen atom into the methane molecule, leading to the formation of methanol. It should be noted that although the methane-oxidizing bacteria can also oxidize more oxidized one-carbon compounds such as methanol and formate, initial isolation from nature requires the use of methane as a sole electron donor, since if one of these other one-carbon compounds is used in initial enrichment, a methylotrophic bacterium that does not oxidize methane will almost certainly be isolated. Most methane-oxidizing bacteria appear to be obligate methylotrophs, unable to utilize compounds with carbon-carbon bonds, but bacteria of one genus, *Methylobacterium*, are facultative methylotrophs, also being able to utilize organic acids, ethanol, and sugars.

Methane-oxidizing bacteria are also unique among procaryotes in possessing relatively large amounts of **sterols.** As we noted in Chapter 3, sterols are found in eucaryotes as a functional part of the membrane system, but seem to be absent from most procaryotes. In the methane-oxidizing bacteria, sterols may be an essential part of the complete internal membrane system (see below) that is involved in methane oxidation.

Classification An overview of the classification of methane-oxidizing bacteria is given in Table 19.16. These bacteria were initially distinguished on the basis of morphology and formation of resting stages,

TABLE 19.16   Some Characteristics of Methane-Oxidizing Bacteria

Organism	Morphology	Flagellation	Resting Stage	Internal Membranes*	Carbon Assimilation Pathway[†]	DNA (mol % G + C)
Methylomonas	Rod	Polar	Cystlike body	I	Ribulose monophosphate	50–54
Methylobacter	Rod	Polar	Thick-walled cyst	I	Ribulose monophosphate	50–54
Methylococcus	Coccus	None	Cystlike body	I	Ribulose monophosphate	62
Methylosinus	Rod or vibrioid	Polar tuft	Exospore	II	Serine	62–66
Methylocystis	Vibrioid	None	Poly-β-hydroxy-butyrate-rich cyst	II	Serine	—
Methylobacterium	Rod	Nonmotile	None	II	Serine[‡]	58–66

*Internal membranes: Type I, bundles of disk-shaped vesicles distributed throughout the organism; Type II, paired membranes running along the periphery of the cell. See Figure 19.42.

[†]See Figures 19.43 and 19.44.

[‡]Also grows on glucose (facultative methylotroph).

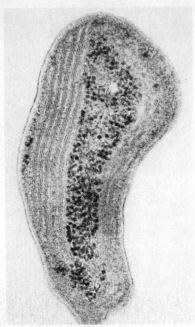

(a)

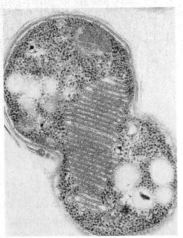

(b)

FIGURE 19.42 Electron micrographs of methane-oxidizing bacteria. (a) *Methanomonas methanooxidans,* illustrating type II membrane system. Magnification, 66,000×. (From Smith, U., and D. W. Ribbons. 1970. Arch. Mikrobiol. 74:116–122.) (b) *Methylococcus capsulatus,* illustrating type I membrane system. Magnification, 35,000×. (From Smith, U., D. W. Ribbons, and D. S. Smith. 1970. Tissue and Cell 2:513–520.)

but it was then found that they could be divided into two major groups depending on their internal cell structure and carbon assimilation pathway. Type I methylotrophs assimilate one-carbon compounds via a unique pathway, the **ribulose monophosphate cycle,** whereas Type II assimilate C-1 intermediates via a **serine pathway.** The requirement for O_2 as a reactant in the initial oxidation of methane explains why all methane oxidizers are obligate aerobes, whereas some organisms using methanol as electron donor can grow anaerobically (with nitrate or sulfate as electron acceptor). The requirement for O_2 also explains why methane gas remains stable indefinitely in natural anaerobic habitats such as in petroleum or coal deposits. Both groups of methane-oxidizers contain extensive internal membrane systems, which appear to be related to their methane-oxidizing ability (the facultative organism *Methylobacterium* loses its internal membranes when grown on glucose or methanol and regains them when transferred back to methane). Type I bacteria are characterized by internal membranes arranged as bundles of disk-shaped vesicles distributed throughout the organism (Figure 19.42b) whereas Type II bacteria possess paired membranes running along the periphery of the cell (Figure 19.42a). Type I methylotrophs are also characterized by a lack of a complete tricarboxylic acid cycle (the enzyme α-ketoglutarate dehydrogenase is absent), whereas Type II organisms possess a complete cycle.

Biochemistry of methane oxidation   Two aspects of the biochemistry of methane oxidizers are of interest, the manner of oxidation of methane to CO_2 (and how this is coupled to ATP synthesis), and the manner by which one-carbon compounds are assimilated into cell material. The overall pathway of methane oxidation involves stepwise, two-electron oxidations

$$CH_4 \rightarrow CH_3OH \rightarrow HCHO \rightarrow HCOOH \rightarrow HCO_3^-$$

	−30 kcal	−46 kcal	−51 kcal	−57 kcal	
Methane	*Methanol*	*Formaldehyde*	*Formate*	*Bicarbonate*

It is now well established that the initial step in the oxidation of methane involves an enzyme called **methane monooxygenase** which is a mixed-function oxygenase. As we discussed in Section 5.5, oxygenase enzymes catalyze the incorporation of oxygen from O_2 into carbon compounds and seem to be widely involved in the metabolism of hydrocarbons. Oxygenases require a source of reducing power, usually NADH, but in *Methylosinus,* electrons come not from NADH but from cytochrome *c.* This cytochrome *c* is involved in recycling of electrons from methanol dehydrogenase (and possibly from the formaldehyde dehydrogenase). In this scheme, ATP synthesis could occur by electron-transport phosphorylation during the reduction of cytochrome *c* by either the methanol or formaldehyde dehydrogenases. This is possible because the reduction potentials of the HCHO/CH_3OH and HCOOH/HCHO couples are −182 and −450 mV, respectively, whereas the reduction potential of the cytochrome *c* is +310 mV. No ATP synthesis occurs during the first step, the oxidation of methane to methanol, and this is consistent with the fact that growth yields of methane-oxidizing bacteria are the same whether methane or methanol is used as substrate. Thus, although considerable energy is potentially available in the oxidation of CH_4 to CH_3OH (about 30 kcal/mole), this energy is not available to organisms, apparently because no biochemical mechanism is available for conserving the energy of methane oxidation.

Biochemistry of one-carbon assimilation   As noted above, all Type I methane oxidizers possess the ribulose monophosphate pathway for carbon assimilation, whereas the Type II methane oxidizers have the serine pathway.

The serine pathway is outlined in Figure 19.43. It is not only present in the Type II methane oxidizers, but is also the pathway in the facultative methylotrophs (*Hyphomicrobium, Pseudomonas,* for example). In this pathway, a two-carbon unit, acetyl-CoA, is synthesized from one molecule of formaldehyde and one molecule of CO_2. The pathway requires the introduction of reducing power and energy in the form of two molecules each of NADH and ATP for each acetyl-CoA sythesized.

The ribulose monophosphate pathway, present in Type I organisms, is outlined in Figure 19.44. It is more efficient than the serine pathway in that all of the carbon atoms for cell material are derived from formaldehyde, and since formaldehyde is at the same oxidation level as cell material no reducing power is needed. The ribulose monophosphate pathway requires the introduction of energy in the form of one ATP for each molecule of 3-phosphoglyceric acid synthesized.

Consistent with the different energy requirements of the two pathways, the cell yield of Type I organisms from a given amount of methane is *higher* than the cell yield of Type II organisms. It seems clear that if a methane-oxidizing bacterium were to be used as a

FIGURE 19.43   The serine pathway for the assimilation of formaldehyde into cell material by Type II methylotrophic bacteria. From acetyl-CoA, three, four, five, and six-carbon compounds are synthesized by reactions of the glyoxylate cycle (see Figure 5.12) and the reversal of glycolysis (see Figure 4.9).

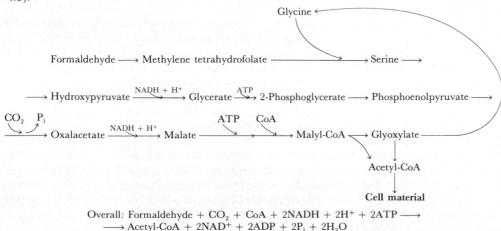

Overall: Formaldehyde + CO_2 + CoA + 2NADH + 2H$^+$ + 2ATP $\longrightarrow$
$\longrightarrow$ Acetyl-CoA + 2NAD$^+$ + 2ADP + 2P$_i$ + 2H$_2$O

(a) 3 Formaldehyde + 3 Ribulose-5-P $\xrightarrow[\text{Synthase}]{}$ 3 Hexulose-6-P $\xrightarrow[\text{Isomerase}]{}$ 3 Fructose-6-P

     (C-1)   (C-5)   (C-6)   (C-6)

(b) 3 Fructose-6-P + ATP $\longrightarrow$ 3 Ribulose-5-P + ADP + Phosphoglyceraldehyde

     (C-6)   (C-5)   (C-3)

                                                           ↓

                                               Cell material

(c) Overall: 3 Formaldehyde + ATP $\longrightarrow$ Phosphogylyceraldehyde + ADP

        (C-1)   (C-3)

FIGURE 19.44 The ribulose monophosphate pathway for assimilation of one-carbon compounds in methylotrophic bacteria. (a) Synthesis of fructose-6-phosphate from formaldehyde and ribulose-5-phosphate. The complete name of this sugar is D-erythro-L-glycero-3-hexulose-6-phosphate. Three formaldehydes are needed to carry the cycle to completion. (b) Regeneration of ribulose-5-phosphate and formation of phosphoglyceraldehyde via the reactions of the pentose phosphate pathway (see Figure 5.4). One ATP is required and the net result is the formation of phosphoglyceraldehyde, which can be incorporated into cell material via the reversal of glycolysis (see Figure 4.9). (c) Overall result of the ribulose monophosphate pathway. The designations C-1, C-5, etc., indicate the number of carbon atoms in each of the compounds.

source of single-cell protein, that an organism with a ribulose monophosphate pathway would be preferable, since cell yields on methane are higher. However, since no energy is gained from the methane to methanol step, it is not certain that there is any advantage of producing single-cell protein from methane as opposed to methanol. Methanol, not being a gas, is much more convenient to use industrially than methane, so that methanol may be the preferred energy source for single-cell protein. Some methane-oxidizing bacteria possess the enzyme ribulose diphosphate carboxylase, the key enzyme in the Calvin cycle for autotrophic CO_2 fixation (see Section 6.7), and probably use an autotrophic pathway for some or all of their carbon assimilation.

Ecology and isolation   Methane-oxidizing bacteria are widespread in aquatic and terrestrial environments, being found wherever stable sources of methane are present. Methane produced in the anaerobic portions of lakes rises through the water column, and methane-oxidizers are often concentrated in a narrow band at the thermocline, where methane from the anaerobic zone meets oxygen from the aerobic zone. There is some evidence that although methane oxidizers are obligate aerobes, they are sensitive to O_2 at normal concentrations, and prefer microaerophilic habitats for development. One reason for this O_2 sensitivity may be that most aquatic methane oxidizers fix N_2, and nitrogenase is O_2 sensitive, so that optimal development occurs where O_2 concentrations are reduced. Methane-oxidizing bacteria play a small but possibly important role in the carbon cycle, converting methane derived from anaerobic decomposition back into cell material (and CO_2).

Most methane oxidizers are moderately sensitive to methanol, and if they overproduce methanol during methane oxidation, their growth may be inhibited. In nature, methane-oxidizing bacteria are often found in close association with methanol oxidizers. It has been suggested that the methanol oxidizer promotes growth of the methane oxidizer by removing the toxic methanol as it is formed.

Although many methane-oxidizing bacteria fix N_2, the nitrogenase of these organisms cannot be assayed in the conventional manner by means of the acetylene reduction technique (see Section 5.13) because acetylene strongly inhibits methane oxidation. In addition, many methane oxidizing bacteria also oxidize ethylene, the product of acetylene reduction by nitrogenase, so that measurement of ethylene production by nitrogenase cannot be accurately done. Nitrous oxide (N_2O) is also a substrate for nitrogenase but does not inhibit methane oxidation, so that N_2O can be used as a convenient substrate for assaying nitrogenase in methane-oxidizing bacteria. Many methane-oxidizing bacteria possess the enzyme hydrogenase, which uses H_2, but are unable to grow on H_2 as sole electron donor for energy generation. It appears that in these organisms hydrogenase functions primarily in N_2 fixation, the electrons from H_2 being used to reduce N_2 to ammonia.

The initial enrichment of methane-oxidizing bacteria is relatively easy, and all that is needed is a mineral salts medium over which an atmosphere of 80 percent methane and 20 percent air is maintained. Once good growth is obtained, purification is carried out by streaking on mineral salts agar plates, which are incubated in a jar with methane-air mixture. Colonies appearing on the plates will be of two types, common heterotrophs growing on traces of organic matter in the medium, which appear in 1 to 2 days, and methane oxidizers, which appear after about a week. The colonies of many methane oxidizers are pink in color. Colonies of methane oxidizers should be picked when small, and purification by continued picking and restreaking is essential. It should be emphasized that cultures of methane oxidizers are often contaminated with common heterotrophs, or with methanol oxidizers, so that careful attention should be paid to the purity of an isolate before any detailed studies are undertaken.

The methane-oxidizing bacteria are able to oxidize ammonia, although they apparently cannot grow lithotrophically using ammonia as sole electron donor. The methane oxygenase also functions as an ammonia oxygenase, and a competitive interaction between the two substrates exists. For this reason, ammonia is generally toxic to methane oxidizers, and the preferred nitrogen source is nitrate. It has been speculated that perhaps the methane-oxidizing bacteria could have arisen from the nitrifying bacteria via a mutation causing the conversion of ammonia oxidase to methane oxidase. The fact that both groups of bacteria have elaborate internal membrane systems (see Section 19.18) encourages such a theory.

Enteric Bacteria

The enteric bacteria, or Enterobacteriaceae, comprise a relatively homogeneous group characterized as follows: Gram-negative, non-sporulating rods, nonmotile or motile by peritrichously occurring flagella, facultative aerobes, oxidase negative with relatively simple nutritional requirements, fermenting sugars to a variety of end products. The characteristics used to separate the enteric bacteria from other bacteria of similar morphology and physiology are given in Table 19.17.

Among the enteric bacteria are many strains pathogenic to humans, animals, or plants as well as other strains of industrial importance. Probably more is known about *Escherichia coli* than about any other bacterial species.

Because of the medical importance of the enteric bacteria, an extremely large number of isolates have been studied and characterized, and a fair number of distinct genera have been defined. Despite the fact that there is marked genetic relatedness between many of the enteric bacteria, as shown by DNA homologies and genetic recombination, separate genera are maintained, largely for practical reasons. Since these organisms are frequently cultured from diseased states, some means of recognizing and designating them is necessary.

One of the key taxonomic characteristics separating the various genera of the enteric bacteria is the range and proportion of end products produced by anaerobic fermentation of glucose. Two broad patterns are recognized, the **mixed-acid fermentation** and the **2,3-butanediol fermentation** (Figure 19.45). In mixed-acid fermentation, three acids are formed in significant amounts—acetic, lactic, and succinic—and ethanol, CO_2, and H_2 are also formed, but not butanediol. In butanediol fermentation, smaller amounts of acids are formed, and butanediol, ethanol, CO_2, and H_2 are the main products. As a result of a mixed-acid fermentation equal amounts of CO_2 and H_2 are produced, whereas with a butanediol fermentation more CO_2 than H_2 is produced. This is because mixed-acid fermenters produce CO_2 only from formic acid by means of the enzyme system formic hydrogenlyase:

$$HCOOH \longrightarrow H_2 + CO_2$$

TABLE 19.17   Defining Characteristics of the Enteric Bacteria

General characteristics: Gram-negative rods; motile by peritrichous flagella, or nonmotile; nonsporulating; facultative aerobes, producing acid from glucose; catalase positive; oxidase negative; usually reduce nitrate to nitrite (not to N_2)

Key tests to distinguish enteric bacteria from other bacteria of similar morphology: oxidase test, enterics always negative—separates enterics from oxidase-positive bacteria of genera *Pseudomonas, Aeromonas, Vibrio, Alcaligenes, Achromobacter, Flavobacterium, Cardiobacterium,* which may have similar morphology; nitrate reduced only to nitrite, (assay for nitrite after growth)—distinguishes enteric bacteria from bacteria that reduce nitrate to N_2 (gas formation detected), such as *Pseudomonas* and many other oxidase-positive bacteria; ability to ferment glucose—distinguishes enterics from obligately aerobic bacteria.

(a) **Mixed acid fermentation,** *e.g., E. coli*

Typical products (molar amounts)

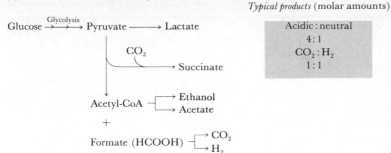

Acidic : neutral
4 : 1
$CO_2 : H_2$
1 : 1

(b) **Butanediol fermentation,** *e.g., Enterobacter*

Typical products (molar amounts)

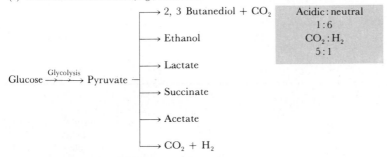

Acidic : neutral
1 : 6
$CO_2 : H_2$
5 : 1

FIGURE 19.45 Distinction between mixed acid and butanediol fermentation in enteric bacteria.

and this reaction results in equal amounts of CO_2 and H_2. The butanediol fermenters also produce CO_2 and H_2 from formic acid, but they produce additional CO_2 during the reactions that lead to butanediol.

A variety of diagnostic tests and differential media are used to separate the various genera in the two broad groups of enteric bacteria, and these are listed in Table 19.18. On the basis of these

TABLE 19.18 Key Differential Media and Tests for Classifying Enteric Bacteria

Triple sugar iron (TSI) agar
   Butt reactions: acid (yellow), gas
   Slant reactions: acid (yellow), alkaline (red)
   H_2S reaction: blackening of butt (in the absence of acid butt reaction)

Urea medium: urea agar with phenol red indicator; look for alkaline reaction due to urease action on urea, liberating ammonia

Citrate medium: growth on citrate as sole energy source

Indole test: peptone broth with high tryptophan content; assay for indole

Voges–Proskauer (VP) test: peptone-containing broth; assay for acetoin

Methyl red (MR) test: buffered glucose-peptone broth; measure pH drop with methyl red indicator

KCN medium: broth with $75\mu g/ml$ potassium cyanide; look for growth

Phenylalanine agar: contains 0.1 percent phenylalanine; after growth, look for phenylpyruvic acid production (indicative of phenylalanine deaminase) by adding ferric chloride solution and looking for green color formation

Utilization of carbon sources: mannitol, tartrate, mucate, acetate, dulcitol, sorbitol, adonitol, inositol

TABLE 19.19 Key Diagnostic Reactions* Used to Separate the Various Genera of Mixed-Acid Fermenters

Organism	H$_2$S(TSI)	Urease	Indole	Motility	Gas from Glucose	β-Galactosidase
Tribe Escherichieae:						
Escherichia	–	–	+	+ or –	+	+
Shigella	–	–	+ or –	–	–	+ or –
Tribe Edwardsielleae:						
Edwardsiella	+	–	+	+	+	–
Tribe Salmonelleae:						
Salmonella	+	–	–	+	+	+ or –
Arizona	+	–	–	+	+	+
Citrobacter	+ or –	–	–	+	+	+
Tribe Proteeae:						
Proteus	+ or –	+	+ or –	+	+ or -	–
Providencia	–	–	+	+	–	–
Tribe Yersinieae:						
Yersinia	–	+	–	†	–	+

Organism	KCN	Citrate	Mucate	Tartrate	Phenylalanine deaminase	DNA (mol % G + C)
Tribe Escherichieae:						
Escherichia	–	–	+	+	–	50–51
Shigella	–	–	–	–	–	50
Tribe Edwardsielleae:						
Edwardsiella	–	–	–	–	–	50
Tribe Salmonelleae:						
Salmonella	–	+ or –	+ or –	+ or –	–	50–53
Arizona	–	+		–	–	50
Citrobacter	+ or –	+	+	+	–	50
Tribe Proteeae:						
Proteus	+	+ or –			+	39–50
Providencia	+	+			+	41
Tribe Yersinieae:						
Yersinia		–			–	45–47

*See Tables 19.17 and 19.18 for the procedures for these diagnostic reactions. Tartrate: utilization as carbon source; mucate: fermentation.

†Motile when grown at room temperature; nonmotile at 37°C.

TABLE 19.20 Key Diagnostic Reactions Used to Separate the Various Genera of 2,3-Butanediol Producers (Tribe Klebsielleae)

Genus	Ornithine Decarboxylase	Gelatin Hydrolysis	Temperature Optimum (°C)	Pigmentation	Motility	Lactose	DNase	Sorbitol	DNA (mol % G + C)
Klebsiella	−	−	37–40	None	−	+	−	+	54–59
Enterobacter	+	slow	37–40	Yellow (or none)	+	+	−	+	52–59
Serratia	+	+	37–40	Red (or none)	+	−	+	−	53–59
Erwinia	−	+ or −	27–30	Yellow (or none)	+	+ or −	−	+	50–58
Hafnia	+	−	35	None	+	−	−	−	52–57

and other tests, the genera can be defined, as outlined in Tables 19.19 and 19.20.

Identification of members of the Enterobacteriaceae presents considerable difficulty because of the large number of strains that have been characterized. Almost any small set of diagnostic characteristics will fail to provide clear-cut distinctions between genera if a large number of strains are tested, because exceptional strains will be isolated. In advanced work in clinical laboratories, identification is now based on computer analysis of a large number of diagnostic tests, carried out using minaturized rapid diagnostic media kits, with consideration being given for variable reactions of exceptional strains. Thus the separations of genera given in Tables 19.19 and 19.20 must be considered to be only approximate, for it will always be possible to isolate a strain that does not possess one or another characteristic normally considered positive for the genus as a whole. With these limitations in mind, an even more simplified separation of the key genera is found in Figure 19.46. This key will permit quick decision on the likely genus in which to place a new isolate.

Escherichia Members of the genus *Escherichia* are almost universal inhabitants of the intestinal tract of humans and warm-blooded animals, although they are by no means the dominant organisms in these habitats. *Escherichia* may play a nutritional role in the intestinal tract by synthesizing vitamins, particularly vitamin K. As a facultative aerobe, this organism probably also helps consume oxygen, thus rendering the large intestine anaerobic. Wild-type *Escherichia* strains rarely show any growth-factor requirements and are able to grow on a wide variety of carbon and energy sources such as sugars, amino acids, organic acids, and so on. Only rarely is *Escherichia* pathogenic, and then usually when host resistance is low. Some strains have been implicated in diarrhea in infants, occasionally occurring in epidemic proportions in children's nurseries or obstetric wards, and *Escherichia* may also cause urinary tract infections in older persons or in those whose resistance has been lowered by surgical treatment or by exposure to ionizing radiation. Enteropathogenic strains of *E. coli* are becoming more frequently implicated in dysenterylike infections and generalized fevers (Section 15.5). As noted, these strains form K antigen, permitting attachment and colonization of the small intestine, and enterotoxin, responsible for the symptoms of diarrhea.

1 MR +; VP − (mixed-acid termenters)   2
   MR −; VP + (butanediol producers)   7
2 Urease +   *Proteus*
   Urease −   3
3 H₂S (TSI) +   4
   H₂S (TSI) −   6
4 KCN +   *Citrobacter*
   KCN −   5
5 Indole +; Citrate −   *Edwardsiella*
   Indole −; Citrate +   *Salmonella*
6 Gas from glucose   *Escherichia*
   No gas from glucose   *Shigella*
7 Nonmotile; Ornithine −   *Klebsiella*
   Motile; Ornithine +   8
8 Gelatin +; DNAse +   *Serratia* (red pigment)
   Gelatin slow; DNAse −   *Enterobacter*

FIGURE 19.46 A very simplified key to the main genera of enteric bacteria. Note that only the most common genera are given. See text for precautions in use of this key. Diagnostic tests for use with this figure are given in Table 19.18. Other characteristics of the genera are given in Tables 19.19 and 19.20.

Shigella The shigellas are very closely related to *Escherichia;* in fact they are so similar that they are able to undergo genetic recombination with each other and are susceptible to some of the same bacteriophages. In contrast to *Escherichia,* however, *Shigella* is commonly pathogenic to humans, causing a rather severe gastroenteritis usually called bacillary dysentery. *S. dysenteriae* is transmitted by food and waterborne routes and is capable of invading intestinal epithelial cells. Once established, it produces both an endotoxin and a neurotoxin that exhibits enterotoxic effects.

Salmonella *Salmonella* and *Escherichia* are quite closely related; the two genera have about 45 to 50 percent of their DNA sequences in common, although the related sequences contain some 13 percent unpaired bases. However, in contrast to *Escherichia,* members of the genus *Salmonella* are usually pathogenic, either to man or to other warm-blooded animals. In humans the most common diseases caused by salmonellas are typhoid fever and gastroenteritis (see Section 17.9 and 17.10). The salmonellas are characterized immunologically on the basis of three cell-surface antigens, the O, or cell-wall (somatic) antigen; the H, or flagellar, antigen; and the Vi (outer polysaccharide layer) antigen, found primarily in strains of *Salmonella* causing typhoid fever. The O antigens are complex lipopolysaccharides that are part of the endotoxin structure of these organisms. We discussed the chemical structure of these lipopolysaccharides in Section 2.5 (see Figure 2.18). The genus *Salmonella* contains over 1000 distinct types having different antigenic specificities in their O antigens. Additional antigenic subdivisions are based on the antigenic specificities of the flagellar H antigens. There is little or no correlation between the antigenic type of a *Salmonella* and the disease symptoms elicited, but antigenic typing permits tracing a single strain involved in a epidemic.

Proteus The genus *Proteus* is characterized by rapid motility and by production of urease. It is a frequent cause of urinary-tract infections in people and rarely may cause enteritis. The species of *Proteus* probably do not form a homogeneous group, as is indicated by the fact that DNA base compositions vary over a fairly wide range (39 to 50 percent $G + C$). Because of the rapid motility of *Proteus* cells, colonies growing on agar plates often exhibit a characteristic **swarming** phenomenon. Cells at the edge of the growing colony are more rapidly motile than are those in the center of the colony; the former move a short distance away from the colony in a mass, then undergo a reduction in motility, settle down, and divide, forming a new crop of motile cells that again swarm. As a result, the mature colony appears as a series of concentric rings, with higher concentrations of cells alternating with lower concentrations.

Although all enteric bacteria can use nitrate as alternate electron acceptor anaerobically, *Proteus* has the additional ability to use several sulfur compounds as electron acceptors for anaerobic growth: thiosulfate, tetrathionate, and dimethylsulfoxide.

Butanediol fermenters The butanediol fermenters are genetically more closely related to each other than to the mixed-acid fermenters, a finding that is in agreement with the observed physiological differences. Their DNA base composition is higher, 50 to 59 percent G + C, and genetic recombination does not occur between the two groups. However, it is possible to transfer plasmids from a mixed-acid fermenter to a butanediol fermenter; the particle replicates in the latter but does not become integrated into the chromosome. The genus designation *Aerobacter,* at one time widely used for one group of the butanediol fermenters, has been abandoned, and now the genus name *Enterobacter* is used. A current classification of this group is outlined in Table 19.20.

One species of *Klebsiella, K. pneumoniae,* occasionally causes pneumonia in humans, but klebsiellas are most commonly found in soil and water. Most *Klebsiella* strains fix N_2 when growing anaerobically, a property not found among other enteric bacteria. The genus *Serratia* is also physiologically related, but differs in certain biochemical properties and in forming a series of red pyrrole-containing pigments called **prodigiosins**. This pigment, a linear tripyrrole, is of interest because it contains the pyrrole ring also found in the pigments involved in energy transfer: porphyrins, chlorophylls, and phycobilins. There is no evidence that prodigiosin plays any role in energy transfer, however, and its exact function is unknown. Species of *Serratia* can be isolated from water and soil as well as from the gut of various insects and vertebrates and occasionally from the intestines of man.

Erwinia is a rather ill-defined genus. One species, *E. amylovora,* causes fire blight of apples and pears, and another, *E. carotovora,* causes soft rot of vegetables by producing a pectinolytic enzyme that causes a breakdown of pectin, the intercellular cement of plant cells. DNA homology studies as well as biochemical properties indicate that the genus *Erwinia* comprises a heterogeneous assemblage of organisms, giving little justification for maintaining a single genus designation.

Yersinia The genus *Yersinia* has been created to accommodate former members of the genus *Pasteurella* that are obviously members of the Enterobacteriaceae. The genus *Yersinia* consists of three species, *Y. pestis,* the causal agent of the ancient and dread disease bubonic plague (see Section 17.12); *Y. pseudotuberculosis,* the causal agent of a tuberculosis-like disease of the lymph node in animals (and rarely in humans); and *Y. enterocolitica,* the causal agent of an intestinal infection (and also occasionally systemic infections) in humans and animals. Although the latter two species are rarely involved in fatal infections, *Yersinia pestis* was responsible for the so-called "Black Death" that ravaged Europe during the fourteenth century killing over one-fourth of the European population. Other pandemics of bubonic plague were recorded in subsequent centuries causing great suffering and countless thousands of deaths. The differentiation of *Yersinia* from other mixed-acid fermenters is given in Table 19.19. In recent years, the prevalence of *Y. enterocolitica* as a waterborne and foodborne pathogen has been realized, and now that methods have been developed for its isolation and recognition, it is being detected with increasing frequency.

19.14

Vibrio and Related Genera

The family Vibrionaceae is placed adjacent to the family Enterobacteriaceae in *Bergey's Manual* because it contains Gram-negative, facultatively aerobic rods, which possess a fermentative metabolism. Most of the members of the family Vibrionaceae are polarly flagellated, although some members are peritrichously flagellated. One key difference between family Vibrionaceae and family Enterobacteriaceae is that members of the former are oxidase positive (see Table 18.2), whereas members of the latter are oxidase negative. Although *Pseudomonas* is also polarly flagellated and oxidase positive, it is not fermentative, and hence can be separated by a simple sugar fermentation test.

Most members of family Vibrionaceae are aquatic, found either in freshwater or seawater habitats, although one important organism, *Vibrio cholerae*, is pathogenic for humans. A separation of some of the genera of the family Vibrionaceae is given in Table 19.21.

Vibrio cholerae is the specific cause of the disease *cholera* in humans; the organism does not normally infect other hosts. Cholera is one of the most common infectious human diseases and one that has had a long history. The organism is transmitted almost exclusively via water, and studies on its distribution in the nineteenth century played a major role in demonstrating the importance of

TABLE 19.21   Distinguishing Characteristics of Genera of the Family Vibrionaceae and Related Genera*

Characteristic	Vibrio	Aeromonas	Photobacterium	Beneckea	Chromobacterium
Morphology	Straight or curved rods, polar flagella	Straight rods, polar flagella	Straight rods, polar flagella	Straight rods, in liquid medium single, polar, sheathed flagellum; on solid medium may have additional unsheathed peritrichous flagella	Straight rods, polar plus lateral flagella
Gas production	−	+	+	−	−
Sensitivity to vibriostat (2,4-diamino-6,7-diisopropyl pteridine)	+	−	+	−	−
Luminescence	−	−	+	+ or −	−
DNA (mol % G + C)	40–50	57–63	39–44	45–54	63–72
Pigment	None	None or brown	None	None	Purple or violet (violacein)

*All are Gram-negative rods, straight or curved, facultative aerobes with a fermentative metabolism (O/F medium), but oxidase positive. They are predominantly aquatic organisms.

water purification in urban areas (see page 591). We discussed the pathogenesis of *V. cholerae* in Section 15.5. *Vibrio cholerae* is capable of good growth at a pH of over 9, and this characteristic is frequently employed in the selective isolation and identification of this organism.

    Vibrio parahemolyticus is a marine organism. It is a major cause of gastroenteritis in Japan (where raw fish is widely consumed) and has also been implicated in outbreaks of gastroenteritis in other parts of the world, including the United States. The organism can be frequently isolated from seawater or from shellfish and crustaceans, and its primary habitat is probably marine animals, with human infection being a secondary development.

Luminescent bacteria   A number of Gram-negative, polarly flagellated rods possess the interesting property of emitting light (luminescence). Most of these bacteria have been classified in the genera *Beneckea* and *Photobacterium* (Table 19.21), but a few *Vibrio* isolates are also luminescent. Most **luminescent bacteria** are marine forms, usually found associated with fish. Some fish possess a special organ in which luminescent bacteria grow (Figure 19.47 *a–d*). Other luminescent marine bacteria live saprophytically on dead fish. A good way of isolating luminescent bacteria is to incubate a dead marine fish for 1 or 2 days at 15 to 20°C; the luminescent bacterial colonies that usually appear on the surface of the fish can be easily seen and isolated (Figure 19.47*e* and *f*). (To see luminescence readily, one should observe the material in a completely dark room after the eyes have become adapted to the dark.)

FIGURE 19.47 Bacterial luminescence. (a) The flashlight fish, *Photoblepharon palpebratus,* photographed at night along a reef in the Gulf of Eilat, Israel, by the light emitted from its own luminescent organ. (b) The same fish photographed with an underwater flash camera, illustrating the location of the luminescent organ under the eye. (c) The same fish with the lid associated with the luminescent organ closed. Luminescence occurs continuously in the organ, and by opening and closing the lid the fish can regulate light emission. (*a–c* from Morin, J. G., A. Harrington, K. Nealson, N. Krieger, T. O. Baldwin, and J. W. Hastings. 1975. Science 190:74–76. Copyright 1975 by the American Association for the Advancement of Science.) (d) Electron micrograph of a thin section through the light-emitting organ of *Photoblepharon palpebratus,* showing the dense array of luminescent bacteria. Magnification, 1820×. (e) Colonies of luminescent bacteria growing on an agar plate, photographed by their own light. (f) The same colonies, photographed by artificial light. (All photographs courtesy of Kenneth Nealson.)

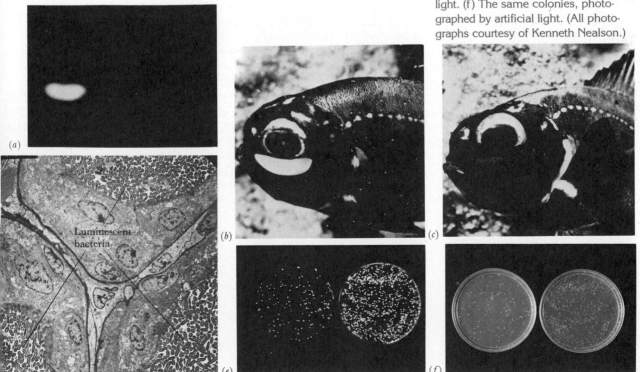

(a)

Luminescent bacteria

(b)

(c)

(e)

(f)

Although *Photobacterium* isolates are facultative aerobes, they are luminescent only when O_2 is present. The amount of O_2 required is quite low, however, and in fact bacterial luminescence can be used as a relatively sensitive way of detecting small amounts of O_2 in a solution. If bacteria are incubated anaerobically for a while and air is then introduced into the culture, a bright flash of light results, following which the light intensity decreases to a low steady-state value. It is thought that under anaerobic conditions a component required for luminescence (probably NADH) accumulates in amounts higher than normal and that its rapid utilization when O_2 is introduced produces the flash.

Two specific components are needed for bacterial luminescence, the enzyme **luciferase,** and a long-chain aliphatic aldehyde (for example, dodecanal); flavin mononucleotide (FMN) and O_2 are also involved. The primary electron donor is NADH, and the electrons pass through FMN to the luciferase. The aldehyde is not required for reduction of the enzyme, but in the absence of the aldehyde the amount of light emitted when the activated enzyme returns to the ground state is low. The light-generating system competes with the normal cytochrome-linked respiratory chain for electrons from NADH, so that if the activity of the cytochrome system is blocked by an inhibitor, the intensity of luminescence is increased.

The enzyme luciferase shows a unique kind of regulatory synthesis called **autoinduction.** The luminous bacteria produce a specific substance, the autoinducer, which accumulates in the culture medium during growth, and when the amount of this substance has reached a critical level, induction of the enzyme occurs. Thus cultures of luminous bacteria at low cell density are not luminous, but become luminous when growth reaches a sufficiently high density so that the autoinducer can accumulate and function. Although autoinduction occurs in all species of luminous bacteria, the autoinducer of one organism is specific for that type, and will not induce enzyme synthesis in another species. Because of the autoinduction phenomenon, it is obvious that a single bacterium existing free in seawater will not be luminous, because the autoinducer could not accumulate, and luminescence only develops when conditions are favorable for the development of high population densities. In symbiotic strains of luminescent bacteria, the rationale for density-dependent luminescence is clear, because luminescence would only develop when high population densities are reached in the light organ of the fish. But it is less clear why luminescence is density-dependent in free-living luminescent bacteria. One idea is that these bacteria are only luminescent when growing saprophytically on decomposing fish or shellfish, where they can grow to high cell density and form luminescent colonies. Such luminescent colonies would serve to attract a higher organism to ingest the material, thus ensuring the transfer of the bacteria (via fecal pellets of the ingesting animal) to another habitat favorable for growth.

The energy involved in the luminescent property is considerable. Obviously, light emission represents wasted energy, as far as the energy metabolism of the organism is concerned. It can be calculated that emission of a single photon of light requires at least

6 ATP, and because of the low efficiency of the luminescent system, up to 60 ATP per photon may be required. Under conditions of maximal luminescence, about 10 to 20 percent of the total cellular O_2 uptake is due to the luciferase system. This shows that a major part of cellular metabolism can be given over to the luminescent system under certain conditions, although obviously at low cell densities, where autoinduction cannot occur, wastage of energy through the luminescent pathway does not occur.

Aeromonas The genus *Aeromonas* is widespread in aquatic environments, and members of this genus have been widely implicated in diseases of aquatic animals such as frogs, turtles, and fish. *Aeromonas* also causes infections of snakes and, rarely, of humans. The organism can be readily isolated from fresh, uncontamined water and is probably an opportunistic pathogen of aquatic animals. *Aeromonas* presents some difficulties in conventional sanitary water analysis by the coliform test, since many *Aeromonas* strains ferment lactose with the production of acid and gas, thus mimicking typical coliforms. However, *Aeromonas* can be easily distinguished from the coliforms by the oxidase test (*Aeromonas* is oxidase positive, the coliforms oxidase negative) and, if motile, by type of flagellation (*Aeromonas* is polarly flagellated, the coliforms peritrichously flagellated).

19.15

The Genus *Zymomonas*

The genus *Zymomonas* consists of large Gram-negative rods (2 to 6 μm long by 1 to 1.4 μm wide), which carry out a vigorous fermentation of sugars to ethanol. Although not all strains are motile, if motility occurs it is by lophotrichous flagella. *Zymomonas* is a common organism involved in alcoholic fermentation of various plant saps, and in many tropical areas of South and Central America, Africa, and Asia, it occupies a position similar to that of *Saccharomyces cerevisiae* (yeast) in North America and Europe. *Zymomonas* has been shown to be involved in the alcoholic fermentation of agave in Mexico, and palm sap in many tropical areas. It also carries out an alcoholic fermentation of sugar cane juice and honey. Although *Zymomonas* is rarely the sole organism involved in these alcoholic fermentations, it is often the dominant organism, and is probably responsible for the production of most of the ethanol (the desired product) in these beverages. *Zymomonas* is also responsible for spoilage of fruit juices such as apple cider and perry. It also may be a constituent of the bacterial flora of spoiled beer, and may be responsible for the production in beer of an unpleasant odor of rotten apples (due to traces of acetaldehyde and H_2S).

Zymomonas is distinguished from *Pseudomonas* by its fermentative metabolism, its microaerophilic to anaerobic nature, oxidase negativity, and its molecular taxonomic characteristics. It resembles most closely the acetic-acid bacteria, specifically *Gluconobacter,* because of its polar flagellation, and it is often found in nature associated with the acetic-acid bacteria. This is of interest because *Zymomonas* ferments glucose to ethanol whereas the acetic acid bacteria ferment ethanol to acetic acid. Thus the acetic-acid bacteria may depend upon the activity of *Zymomonas* for the production of their growth substrate, ethanol. Like the acetic-acid bacteria, *Zymomonas* is quite tolerant of low pH. Unlike yeast, which ferments glucose to ethanol via the Embden–Meyerhof (glycolytic) pathway, *Zymomonas* employs the Entner–Doudoroff pathway. This pathway is active in many pseudomonads as a means of catabolizing glucose. The reactions of the Entner–Doudoroff pathway are outlined in Appendix 3.

19.16

Sulfate-Reducing Bacteria

Sulfate is used as a terminal electron acceptor under anaerobic conditions by a heterogenous assemblage of bacteria which utilize organic acids, fatty acids, and alcohols as electron donors. Many of these organisms possess the enzyme hydrogenase and are thus able to use H_2 as an electron donor as well. Although morphologically diverse, the sulfate-reducing bacteria can be considered a physiologically unified group, in the same manner as the phototrophic or methanogenic bacteria. Eight genera of **dissimilatory sulfate-reducing bacteria** are currently recognized and they are placed in two broad physiological subgroups as outlined in Table 19.22. The genera in group I, *Desulfovibrio* (Figure 19.48a), *Desulfomonas,* and *Desulfotomaculum,* utilize lactate, pyruvate, ethanol, or certain fatty acids as carbon and energy sources, reducing sulfate to hydrogen sulfide. The genera in group II, *Desulfobulbus, Desulfobacter, Desulfococcus, Desulfo-*

TABLE 19.22   Characteristics of Sulfate- and Sulfur-Reducing Bacteria

Genus	Characteristics	DNA (mol % G + C)
Group I sulfate reducers:		
Desulfovibrio	Polarly flagellated, curved rods, no spores; Gram negative; contain desulfoviridin; seven species recognized, one thermophilic	46–61
Desulfomonas	Long, fat rods; nonmotile; no spores Gram negative; contain desulfoviridin; habitat, intestinal tract; one species	66–67
Desulfotomaculum	Straight or curved rods; motile by peritrichous or polar flagellation; Gram negative; desulfoviridin absent; produce endospores; four species, one thermophilic; one species capable of utilizing acetate as energy source	37–46
Group II sulfate reducers:		
Desulfobulbus	Ovoid or lemon-shaped cells; no spores; Gram negative; desulfoviridin absent; if motile, by single polar flagellum; utilizes only propionate as electron donor with acetate + CO_2 as products; one species	59
Desulfobacter	Rods; no spores; Gram negative; desulfoviridin absent; if motile, by single	45

TABLE 19.22 *continued*

Genus	Characteristics	DNA (mol % G + C)
	polar flagellum; utilizes only acetate as electron donor and oxidizes it to CO_2; one species	
Desulfococcus	Spherical cells; nonmotile; Gram negative; desulfoviridin present, no spores; utilizes C_1–C_{14} fatty acids as electron donor with complete oxidation to CO_2; one species	57
Desulfonema	Large filamentous gliding bacteria; Gram positive; no spores; desulfoviridin present or absent; utilizes C_1–C_{12} fatty acids as electron donor with complete oxidation to CO_2; one species	34–41
Desulfosarcina	Cells in packets (sarcina arrangement); Gram negative; no spores; desulfoviridin absent; utilizes C_1–C_{14} fatty acids as electron donor with complete oxidation to CO_2; one species	
Dissimilatory sulfur reducers:		
Desulfuromonas	Straight rods, single lateral flagellum; no spores; Gram negative; does not reduce sulfate; acetate, ethanol or propanol used as electron donor; obligate anaerobe; two species	50–63
Campylobacter	Curved, vibrio-shaped rods; polar flagella; Gram negative; no spores; unable to reduce sulfate but can reduce sulfur, sulfite, thiosulfate, nitrate, or fumarate anaerobically with acetate or a variety of other carbon/electron donor sources; facultative aerobe	40–42
Desulfurococcus	Gram negative cocci; does not reduce sulfate; extreme thermophile, growth optimal at 85°C (pH 6); obligate anaerobe using proteins and small peptides as electron donor with sulfur as electron acceptor; motile; two species	51
Thermoproteus	Gram negative rods to filaments; does not reduce sulfate; extreme thermophile, growth optimal at 90°C (pH 5); obligate anaerobe using glucose and a variety of other organic compounds as electron donor with sulfur as electron acceptor; nonmotile; one species	55
Thermococcus	Cocci, 1 μm in diameter; marine requires 3.5–4.0% NaCl; rest of description as for *Desulfurococcus;* one species	56
Thermofilum	Thin rods to filaments—growth dependent on polar lipid extract of *Thermoproteus;* rest of description as for *Thermoproteus;* one species	57

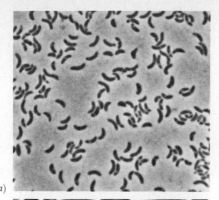

(a)

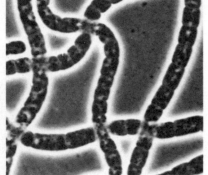

(b)

FIGURE 19.48 Photomicrographs by phase contrast of two representative sulfate-reducing bacteria of vastly different size. (a) *Desulfovibrio desulfuricans,* a Type I sulfate-reducer. Magnification, 1430×. (Courtesy of Norbert Pfennig.) (b) *Desulfonema limicola,* a filamentous, gliding Type II sulfate-reducer. Magnification, 1430×. (From Pfennig, N., F. Widdel, and H. G. Trüper. 1981. The dissimilatory sulfate-reducing bacteria. Pages 926–940 in M. P. Starr, H. Stolp, H. G. Trüper, A. Balows, and H. G. Schlegel, eds. The prokaryotes, a handbook of habitats, isolation, and identification of bacteria. Springer-Verlag, New York, N.Y. Micrographs courtesy of Norbert Pfennig.)

sarcina, and *Desulfonema* (Figure 19.48*b*), specialize in the oxidation of fatty acids, particularly acetate, reducing sulfate to sulfide. The sulfate-reducing bacteria are all obligate anaerobes, and strict anaerobic techniques must be used in their cultivation, although they are not as fastidious with respect to oxygen as the methanogenic bacteria (see Section 19.20).

Sulfate-reducing bacteria are widespread in aquatic and terrestrial environments that become anaerobic due to active microbial decomposition processes. The best known genus is *Desulfovibrio* (Figure 19.48*a*), which is common in aquatic habitats or waterlogged soils containing abundant organic material and sufficient levels of sulfate. *Desulfotomaculum* consists of endospore-forming rods primarily found in soil, and one species is thermophilic. Growth and reduction of sulfate by *Desulfotomaculum* in certain canned foods leads to a type of spoilage called *sulfide stinker*. The remaining genera of sulfate reducers are indigenous to freshwater or marine anaerobic environments; *Desulfomonas* can also be isolated from the intestine.

Although not a sulfate reducer, *Desulfuromonas* is often grouped with the sulfate-reducing bacteria as **dissimilatory sulfur reducers.** Members of the genus *Desulfuromonas* can grow anaerobically by coupling the oxidation of nonfermentable substrates such as acetate to the reduction of elemental sulfur to hydrogen sulfide. However, the ability to reduce elemental sulfur, as well as other sulfur compounds such as thiosulfate, sulfite, or dimethyl sulfoxide (DMSO), is a widespread property of a variety of heterotrophic, generally facultatively aerobic bacteria (for example *Proteus, Campylobacter, Pseudomonas,* and *Salmonella*). *Desulfuromonas* differs from the latter in that it is an obligate anaerobe and utilizes only sulfur as an electron acceptor (see Table 19.22). In addition, certain sulfate-reducing bacteria are capable of substituting sulfur for sulfate as an electron acceptor for anaerobic growth.

Studies of sulfur rich thermal environments have led to the discovery of a host of extremely thermophilic dissimilatory sulfur reducers. Although morphologically distinct and separated into several genera, these organisms are metabolically similar, all carrying out a heterotrophic metabolism with small peptides or glucose as electron donors and elemental sulfur as electron acceptor. Sulfate is not reduced. Since many of these organisms have growth temperature optima near 90°C, little is known of other characteristics of the group. From studies thus far, however, it is clear that these extremely thermophilic anaerobes represent a distinct branch of dissimilatory sulfur reducers, probably sharing little else with organisms like *Desulfuromonas,* except the ability to use sulfur as an electron acceptor.

Physiology The range of electron donors used by sulfate-reducing bacteria is broad. Lactate and pyruvate are almost universally used, and many strains of group I will utilize malate, formate, and certain primary alcohols (for example, methanol, ethanol, propanol, and butanol). Some strains of *Desulfotomaculum* will utilize glucose but this is rather rare among sulfate reducers in general. Group I sulfate reducers oxidize their energy source to the level of acetate and excrete this fatty acid as an end product. Group II organisms differ from those in group I by their ability to oxidize fatty acids, lactate, succinate, and even benzoate in some cases, all the way to CO_2. *Desulfosarcina* and *Desulfonema* (Figure 19.48b) are unique in their ability to grow lithotrophically with H_2 as the electron donor, sulfate as electron acceptor, and CO_2 as sole carbon source.

In addition to growth using sulfate as electron acceptor, many sulfate-reducing bacteria can use certain organic compounds for energy generation by fermentative pathways in the complete absence of sulfate or other terminal electron acceptors. The most common fermentable organic compound is pyruvate, which is converted via a phosphoroclastic reaction to acetate, CO_2, and H_2. With lactate or ethanol, insufficient energy is available from fermentation, so that sulfate is required, although if a H_2-utilizing organism (for instance, a methanogen) is present in a mixed culture, fermentation of these substrates by the sulfate reducer to acetate and H_2 can occur (the H_2 is immediately consumed by the methanogenic component of such a mixed culture to form CH_4). The value of sulfate as an electron acceptor can be readily demonstrated by comparing growth yields on pyruvate with and without sulfate. In the presence of sulfate, the growth yield is considerably higher, because of the much larger amount of energy available when pyruvate utilization is coupled with sulfate reduction.

An interesting energy source utilized by *Desulfotomaculum* is the polymer of phosphoric acid, polyphospate (PP_i). It has been known for many years that the anhydride bond holding together the individual phosphate molecules in polyphosphate contains sufficient free energy to allow the synthesis of ATP, yet no organism has ever been described which was capable of utilizing polyphosphate as an energy source. *Desulfotomaculum,* however, is capable of growing in a medium containing mineral salts, acetate, sulfate, and PP_i, but does not grow in the same medium when PP_i is omitted. It is thought that ATP is produced from polyphosphate by the transfer of a phosphate group to acetate to yield acetyl$\sim$P, followed by the production of ATP via transfer of the phosphate group from acetyl$\sim$P to ADP. It will be interesting to see how widespread the utilization of PP_i is by microorganisms, especially since many bacteria are known to actively store granules of polyphosphate within the cell.

Biochemistry The enzymology of sulfate reduction has been extensively studied. The first step is the formation of **adenosine phosphosulfate (APS)** from ATP and sulfate, as outlined in Section 5.14 and in Figure 19.49. The enzyme APS reductase then catalyzes the reduction of the sulfate moiety to sulfite (see also Figure 19.8). Two mechanisms have been described for the further reduction of sulfite to sulfide, one a direct six-electron reduction via a single enzyme, sulfite reductase, and the second a stepwise reduction of sulfite to sulfide via trithionate and thiosulfate. The second mechanism (Figure 19.49), is more likely to permit stepwise ATP synthesis and is currently favored. (See Section 13.10 for a discussion of the various redox states of sulfur.)

Sulfate-reducing bacteria contain a low reduction potential cytochrome designated cytochrome c_3 (E_0' $-$ 250 to $-$ 300 mV),

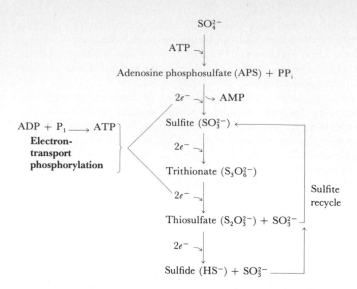

FIGURE 19.49 Scheme for eight-electron reduction of sulfate to sulfide via tetrathionate and thiosulfate. Note that three sulfite ions are needed to form trithionate, but two sulfites are regenerated and recycle. The net result is reduction of one sulfite to sulfide. For a discussion of the oxidation states of sulfur, see Section 13.10.

which functions as an electron carrier, and most sulfate reducers contain cytochromes of the b type as well. *Desulfovibrio* and most other sulfate reducers contain a unique electron carrier, desulfoviridin (absorption peak 630 nm) which can be readily detected by the red fluorescence it exhibits when cells are made alkaline with a few drops of 2.0 N NaOH and then examined under long-wavelength ultraviolet radiation (365 nm). (Some strains, which lack desulfoviridin, contain another pigment, desulforubidin.) Although not all sulfate-reducing bacteria contain desulfoviridin, examination of cultures for its presence (by the fluorescence test) provides a simple technique for detecting typical sulfate-reducing bacteria.

It is now clear that sulfate-reducing bacteria must couple the reduction of sulfate to ATP synthesis via an electron transport chain. The ability of many sulfate reducers to utilize H_2 as an electron donor for energy generation is strong evidence for electron transport involvement, and electron transport in sulfate-reducers probable involves the generation of a membrane potential as in other respiratory schemes (see Section 4.9). The ATP that is consumed in the conversion of sulfate to APS is apparently recovered by electron-transport phosphorylation during the reduction of APS to AMP plus sulfite. A second site of ATP synthesis via electron-transport phosphorylation probably occurs at the step between trithionate and thiosulfate (Figure 19.49), resulting in the net synthesis of one mole of ATP per sulfate reduced.

Isolation and ecology The enrichment of *Desulfovibrio* is relatively easy on a lactate-sulfate medium to which ferrous iron is added. A reducing agent such as thioglycolate or ascorbate is also added. The sulfide formed from sulfate reduction combines with the ferrous iron to form black insoluble ferrous sulfide. This blackening not only serves as an indicator of the presence of sulfate reduction, but the iron ties up and detoxifies the sulfide, making possible growth to higher cell yields. The conventional procedure is to set up liquid enrichments, and after some growth has occurred as evidenced by blackening of the medium, purification is accomplished by streaking onto roll tubes or Petri plates in an anaerobic environment. Although the sulfate reducers are obligate anaerobes, they are not as rapidly inactivated by oxygen as the methane producers, so that streaking on plates can be done in air, provided a reducing agent is present in the medium and plates are incubated in an anaerobic environment. Alternatively, agar "shake tubes" can be used for purification purposes. In the **shake tube method** a small amount of liquid from the original enrichment is added to a tube of molten agar growth medium, mixed thoroughly, and sequentially diluted through a series of molten agar tubes. Upon solidification, individual cells distributed throughout the agar form colonies, which can be removed aseptically and the whole process repeated until pure cultures are obtained. Colonies of sulfate reducing bacteria are recognized by the black deposit of ferrous sulfide, and purified by further streaking.

Sulfate reducers can also be enriched in media containing H_2 and sulfate as electron donor and acceptor, respectively, and acetate as a source of cell carbon. In the absence of H_2, acetate or longer-chain fatty acids can serve as electron donors for enrichment of organisms of group II. Thioglycolate and ascorbate are not added to enrichments for group II organisms, because for unknown reasons these compounds inhibit growth of group II sulfate reducers. For most purposes, it is desirable to add a small amount of an organic supplement to the medium, such as yeast extract, in order to provide required growth factors. Some strains of *Desulfovibrio* have been reported to fix N_2, although this is not yet known to be a general property of the group.

Because of their geochemical importance (Section 13.10), the ecology of the sulfate-reducing bacteria is of considerable importance, but relatively little work has been done in this area. It is known that sulfate-reducing bacteria are common in both freshwater and marine environments, as well as in hypersaline habitats. They are also found in many soils, especially flooded (and hence anaerobic) soils, and their activity in producing sulfide in such habitats may be of importance in affecting plant growth, since most plants are rather sensitive to sulfide. Sulfide buildup in rice paddies can result in restriction of plant growth, although as noted in Section 19.3, oxidation of sulfide around the rice plant roots by *Beggiatoa* may result in detoxification of sulfide. Sulfate-reducing bacteria have also been widely implicated in iron and steel corrosion, through the formation of ferrous sulfide. Pipes buried in sulfate-rich soils or muds (especially in areas where seawater intrusion occurs, since seawater is high in sulfate) are often rapidly corroded as a result of the activities of sulfate-reducing bacteria.

19.17
Gram-Negative Cocci and Coccobacilli

This group comprises a diverse collection of organisms, related by Gram stain, morphology, lack of motility, nonfermentative aerobic metabolism, and similar DNA base composition. The four genera *Neisseria*, *Moraxella*, *Branhamella*, and *Acinetobacter* are distinguished as outlined in Table 19.23. At one time all of these organisms were classi-

Characteristics	Genus	DNA (mol % G + C)
Oxidase-positive; penicillin sensitive:		
Cocci; complex nutrition, utilize carbohydrates	*Neisseria*	46–52
Grow in simple media; do not utilize carbohydrates	*Branhamella*	40–47
Rods, forming cocci in stationary phase; generally no growth factor requirements, generally do not utilize carbohydrates; exhibit "twitching" motility	*Moraxella*	40–47
Oxidize-negative, penicillin resistant; some strains can utilize a restricted range of sugars, and some exhibit "twitching" motility	*Acinetobacter*	40–46

fied in the genus *Neisseria*, but detailed taxonomic work including genetic and DNA homology studies has revealed that several genera are warranted. A major distinction is made on the basis of oxidase reaction. The oxidase-positive organisms are unusually sensitive to penicillin, being inhibited by 1 μg/ml, differing in this way from most other Gram-negative bacteria, including *Acinetobacter*.

In the genus *Neisseria* the cells are always cocci, whereas the cells of the other genera are rod shaped, becoming coccoid in the stationary phase of growth. This has led to designation of these organisms as **coccobacilli**. Organisms of the green *Neisseria* and *Moraxella* are commonly isolated from animals, and some of them are pathogenic, whereas organisms of the genus *Acinetobacter* are common soil and water organisms, although they are occasionally found as parasites of certain animals. Some strains of *Moraxella* and *Acinetobacter* possess the interesting property of **twitching motility,** exhibited as brief translocative movements or "jumps," covering distances of about 1 to 5 μm. We discussed the clinical microbiology of *Neisseria gonorrhoeae* in Section 15.11.

19.18

Lithotrophs: The Nitrifying Bacteria

Bacteria able to grow lithotrophically at the expense of reduced inorganic nitrogen compounds are called **nitrifying bacteria.** Several genera are recognized on the basis of morphology and the particular steps in the oxidation sequences that they carry out (Table 19.24).* No lithotrophic organism is known that will carry out the complete oxidation of ammonia to nitrate; thus **nitrification** of ammonia in nature results from the sequential action of two separate groups of

*Those bacteria oxidizing ammona to nitrite can be called *nitrosofying* bacteria and their generic names all begin with the prefix "nitroso-." Those bacteria oxidizing nitrite to nitrate are the true *nitrifying* bacteria and their generic names all begin with the prefix "nitro-."

TABLE 19.24 Characteristics of the Nitrifying Bacteria

Characteristics	Genus	DNA (mol % G + C)	Habitats
Oxidize ammonia:			
Gram-negative rods, motile (polar flagella) or nonmotile; peripheral membrane systems:			
Short rods, pointed ends	*Nitrosomonas* (group 1)	51	Soil, sewage
Long rods, rounded ends	*Nitrosomonas* (group 2)	47–49	Fresh water
Long rods, rounded ends, chains	*Nitrosomonas* (group 3)	48–49	Marine
Large cocci, motile (peritrichous flagella) or nonmotile; membrane system in center of cell	*Nitrosococcus*	50–51	Soil, marine
Spirals, motile (peritrichous flagella); no obvious membrane system	*Nitrosospira*	54	Soil
Pleomorphic, lobular, compartmented cells; motile (petitrichous flagella)	*Nitrosolobus*	53–56	Soil
Oxidize nitrite:			
Short rods, reproduce by budding, occasionally motile (single subterminal flagellum); membrane system arranged as a polar cap	*Nitrobacter*	60–61	Soil, fresh water, marine
Long, slender rods, nonmotile; no obvious membrane system	*Nitrospina*	57	Marine
Large cocci, motile (one or two subterminal flagella); membrane system randomly arranged in tubes	*Nitrococcus*	61	Marine

organisms, the **ammonia-oxidizing bacteria** and the **nitrite-oxidizing bacteria.** Some heterotrophic bacteria and fungi will oxidize ammonia completely to nitrate, but the rate of the process is much less than that accomplished by the lithotrophic nitrifying bacteria, and may not be significant ecologically. Historically, the nitrifying bacteria were the first organisms to be shown to grow lithotrophically; Winogradsky showed that they were able to produce organic matter and cell mass when provided with CO_2 as sole carbon source.

Many of the nitrifying bacteria have remarkably complex internal membrane systems (Figure 19.50), although not all genera have such membranes.

Biochemistry Ammonia oxidation occurs best at high pH values, and it is inferred that this is because the enzyme involved in the initial step uses the nonionized NH_3 rather than the NH_4^+ ion. Molecular oxygen is required for ammonia oxidation, the initial step involving a mixed-function oxygenase, which uses NADH as electron donor. The first product of ammonia oxidation is hydroxylamine, NH_2OH, and no energy is generated in this step (energy is actually used up via the oxidation of NADH). Hydroxylamine is then oxidized to nitrite and ATP formation occurs at this step via electron-transport phosphorylation through a cytochrome system (see Figure 19.51).

There are some interesting similarities between the ammonia-oxidizing bacteria and the methane-oxidizing bacteria. As noted in Section 19.12, the methane-oxidizing bacteria generally oxidize ammonia to nitrite, and ammonia inhibits methane oxidation. In a similar fashion, certain ammonia oxidizers have been found capable of oxidizing methane, and to incorporate significant amounts of carbon originating from methane into cell material. Methane will not serve as sole carbon and electron donor for growth of ammonia

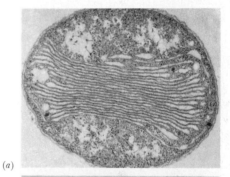

(a)

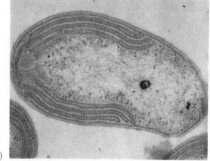

(b)

FIGURE 19.50 Complex internal membranes in the nitrifying bacteria. (a) *Nitrosococcus oceanus.* Magnification, 22,970×. (b) *Nitrobacter winogradskyi.* Magnification, 38,980×. (From Watson, S. W. 1971. Intern. J. System. Bacteriol. 21:254–270.)

FIGURE 19.51 Reactions involved in
the oxidation of inorganic nitrogen
compounds.

Nitrosofying Bacteria

1. $NH_3 + O_2 + NADH + H^+ \longrightarrow NH_2OH + H_2O + NAD^+$
2. $NH_2OH + O_2 \longrightarrow NO_2^- + H_2O + H^+$

Sum: $NH_3 + 2[O_2] + NADH \longrightarrow NO_2^- + 2H_2O + NAD^+$

$\Delta G^{0\prime} = -65$ kcal/mole

Nitrifying Bacteria

$NO_2^- + \frac{1}{2}O_2 \longrightarrow NO_3^- \qquad \Delta G^{0\prime} = -18$ kcal/mole

oxidizers, however. Nitrite oxidizers do not oxidize methane. It appears that the two substrates methane and ammonia have some structural similarities, so that an enzyme which recognizes one can also combine with, and be inhibited by, the other. Another similarity between ammonia oxidizers and methane oxidizers is that both groups generally possess extensive internal membrane systems (compare Figure 19.50 with Figure 19.42).

Only a single step is involved in the oxidation of nitrite to nitrate by the nitrite-oxidizing bacteria (Figure 19.51). This reaction is carried out by a nitrite oxidase system, the electrons being transported to O_2 via cytochromes, with ATP being generated by electron-transport phosphorylation. The energy available from the oxidation of nitrite to nitrate is only 18.1 kcal/mole, which is sufficient for the formation of two ATP. However, careful measurements of molar growth yields suggest that only one ATP is produced per each NO_2^- oxidized to NO_3^-. As we discussed in Section 6.10, the generation of reducing power (NADPH) for the reduction of CO_2 to organic compounds comes from ATP driven reversed electron transport reactions, since the NO_3^-/NO_2^- reduction potential is too high to reduce NADP directly.

Ecology Although rarely present in large numbers, the nitrifying bacteria are widespread in soil and water. They can be expected to be present in highest numbers in habitats where considerable amounts of ammonia are present, such as sites where extensive protein decomposition occurs (ammonification). Nitrifying bacteria develop especially well in lakes and streams, which receive inputs of treated (or even untreated) sewage, because sewage effluents are generally high in ammonia. The classic picture downstream from a sewage outfall, as shown in Figure 13.4, indicates high ammonia concentrations close to the outfall, and falling concentrations downstream as nitrification occurs and nitrate builds up. Because O_2 is required for ammonia oxidation, ammonia tends to accumulate in anaerobic habitats, and in stratified lakes nitrifying bacteria may develop especially well at the thermocline, where both ammonia and O_2 are present. Nitrification results in acidification of the habitat, due to the buildup of nitric acid (the situation is analogous to the buildup of sulfuric acid through the activities of sulfur-oxidizing bacteria). Since nitrous acid can form at acid pH values as a result of ammonia oxidation, the accumulation of this toxic (and mutagenic; see Table 9.6) agent in acidic environments can result in inhibition of further nitrification. In general, nitrification is much more extensive in neutral and alkaline than in acidic habitats, due partly to avoidance of nitrous acid accumulation and partly to the requirement of the non-ionized NH_3 by the ammonia-oxidizing bacteria.

Culture Despite the fact that the nitrifying bacteria have been known since the nineteenth century, work on these organisms has not been extensive, at least in part because of the difficulty of obtaining reproducible pure cultures. Enrichment cultures of nitrifying bacteria are readily obtained by using selective media containing ammonia or nitrite as electron donor, and bicarbonate as sole carbon source. Because of the inefficiency of growth of these organisms, visible turbidity may not develop even after extensive nitrification has occurred, so that the best means of monitoring growth is to assay for production of nitrite (with ammonia as electron donor) or disappearance of nitrite (with nitrite as electron donor). After one or two weeks incubation, chemical assays will reveal whether a successful enrichment has been obtained, and attempts can be made to get pure cultures by streaking on agar plates. The same problem arises when purifying nitrifying bacteria by colony selection on agar that was discussed with the methane-oxidizing bacteria: many common heterotrophs present in the enrichment will grow rapidly on the traces of organic matter present in the medium. Thus purification must be done by repeated picking and streaking, followed by testing to be certain that heterotrophic contaminants have not been taken. In addition, many nitrifying bacteria, especially the ammonia oxidizers, appear to be inhibited by the traces of organic material present in most agar preparations. Culture of these organisms on solid media sometimes requires the use of extensively washed, high purity agar, or the completely inorganic solidifying agent, **silica gel.** The latter compound was first used by Winogradsky to successfully grow the nitrosifyers on solid media. Because most nitrifying bacteria do not grow on organic compounds, the simplest way of checking for contamination is to inoculate a presumed nitrifying isolate into media containing organic matter. If growth occurs, it can be assumed that the isolate was not a nitrifier. In general, better luck is experienced isolating nitrite oxidizers than ammonia oxidizers. With careful attention to detail, it should be possible to isolate a diversity of nitrifying bacteria from most habitats. Most of the nitrifying bacteria appear to be obligate lithotrophs and obligate autotrophs, and have not been successfully grown on organic media. *Nitrobacter* is an exception, however, and is able to grow, although slowly, on acetate as sole carbon and energy source. None of these bacteria require growth factors. Although the group is somewhat heterogeneous morphologically, it seems to be more homogeneous than the sulfur-oxidizing or photosynthetic bacteria, as shown by the fairly narrow range of DNA base composition (Table 19.24) and the similar biochemical properties.

19.19

Lithotrophs: Sulfur- and Iron-Oxidizing Bacteria

The ability to grow lithotrophically on reduced sulfur compounds is a property of a diverse group of microorganisms. Many of these organisms have not, however, been cultured, and precise information on them is minimal. Only four genera, *Thiobacillus, Thiomicrospira, Thermothrix,* and *Sulfolobus* have been consistently cultured so that our discussion here is restricted to these genera. Two broad ecological classes of sulfur-oxidizing bacteria can be discerned, those living at neutral pH and those living at acid pH. Many of the forms living at acid pH also have the ability to grow lithotrophically using ferrous iron as electron donor. We discussed the biogeochemistry of these acidophilic sulfur- and iron-oxidizing bacteria in Sections 13.10 and 13.11.

Thiobacillus The genus *Thiobacillus* contains those Gram-negative, polarly flagellated rods that are able to derive their energy from the oxidation of elemental sulfur, sulfides, and thiosulfate (Table 19.25). Some pseudomonads have been confused with members of the genus *Thiobacillus*. Morphologically, *Thiobacillus* is similar to *Pseudomonas*, but the two differ in that *Thiobacillus* can grow lithotrophically using reduced sulfur compounds. A few thiobacilli can also grow heterotrophically with organic electron donors, and under such conditions resemble pseudomonads. Further, a few pseudomonads (both marine and freshwater) can oxidize thiosulfate to tetrathionate:

$$2S-SO_3^{2-} \longrightarrow {}^{2-}O_3S-S-S-SO_3^{2-} + 2e^-$$

and although they cannot grow solely from the energy obtained in this reaction, they do obtain some slight growth advantage from this process when growing on sugars or other organic compounds. The mechanism by which growth is stimulated during the oxidation of thiosulfate to tetrathionate is not known.

The biochemical steps in the oxidation of various sulfur compounds are summarized in Figure 19.52. Oxidation of sulfide and sulfur involves first the reaction of these substances with sulfhydryl groups of the cell such as glutathione with formation of a sulfide-sulfhydryl complex. The sulfur is then oxidized to sulfite (SO_3^{2-}) by the enzyme sulfide oxidase. There are two ways in which sulfite can be oxidized to produce high-energy phosphate bonds. In one, sulfite is oxidized to sulfate by a cytochrome-linked sulfite oxidase, with

TABLE 19.25 Physiological Characteristics of Sulfur-Oxidizing Bacteria

	Lithotrophic electron donor	Range of pH for growth	DNA (mol % G + C)
Thiobacillus			
species growing poorly in organic media:			
1. *T. thioparus*	H_2S, sulfides, S^0 $S_2O_3^{2-}$	6–8	62–66
2. *T. denitrificans**	H_2S, S^0, $S_2O_3^{2-}$	6–8	63–67
3. *T. neapolitanus*	S^0, $S_2O_3^{2-}$	5–8	55–57
4. *T. thiooxidans*	S^0	2–5	52–57
5. *T. ferrooxidans*	S^0, sulfides, Fe^{2+}	1.5–4	53–60
Thiobacillus			
species growing well in organic media:			
1. *T. novellus*	$S_2O_3^{2-}$	6–8	66–68
2. *T. intermedius*	$S_2O_3^{2-}$	3–7	64
Thiomicrospira†	$S_2O_3^{2-}$, H_2S	6–8	36–44
*Thermothrix**	H_2S, $S_2O_3^{2-}$, SO_3^-	6.5–7.5	—
Sulfolobus	H_2S, S^0	1–5	41

*Facultative aerobes; use NO_3^- as electron acceptor anaerobically.

†One of its species is capable of using NO_3^- anaerobically.

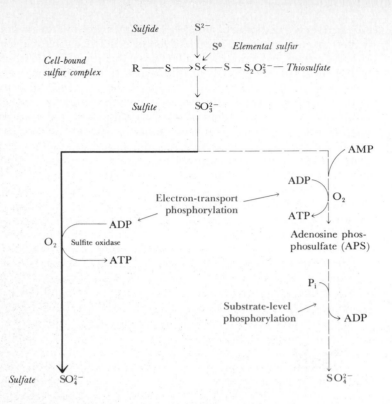

FIGURE 19.52 Steps in the oxidation of different compounds by thiobacilli. The sulfite oxidase pathway is thought to account for the majority of sulfite oxidized.

the formation of ATP via electron-transport phosphorylation. This pathway is universally present in thiobacilli. In the second, sulfite reacts with AMP; two electrons are removed, and adenosine phosphosulfate (APS) is formed. The electrons removed are transferred to O_2 via the cytochrome system, leading to the formation of high-energy phosphate bonds through electron-transport phosphorylation. In addition, a substrate-level phosphorylation occurs, APS reacting with P_i and being converted to ADP and sulfate. With the enzyme adenylate kinase, two ADP can be converted to one ATP and one AMP. Thus oxidation of two sulfite ions via this system produces three ATP, two via electron-transport phosphorylation and one via substrate-level phosphorylation. The significance of the APS pathway in thiobacilli in general is unclear, however, since it has been found only in a few *Thiobacillus* species.

Thiosulfate ($S_2O_3^{2-}$), which can be viewed as a sulfide of sulfite (SSO_3^{2-}), is split into sulfite and sulfur. The sulfite is oxidized to sulfate with production of ATP, and the other sulfur atom is converted into insoluble elemental sulfur. Thus, when they oxidize thiosulfate, the thiobacilli produce elemental sulfur but when they oxidize sulfides they do not. The elemental sulfur produced can itself be oxidized later when the thiosulfate supply is exhausted. If thiosulfate is low, elemental sulfur does not accumulate, probably being oxidized as soon as it is formed.

Enrichment cultures of thiobacilli are quite easy to prepare. Sulfur or thiosulfate is added to a basal salts medium with NH_4^+ as a nitrogen source and bicarbonate as a carbon source, and the medium is then inoculated with a sample of soil or mud. After aerobic incubation at room temperature for a few days, the liquid should appear turbid owing to the growth of thiobacilli. If thiosulfate has been used, droplets of amorphous sulfur will also be present. If elemental sulfur is used, many of the bacteria may be attached to the insoluble sulfur crystals, and the edges of such crystals should be examined for presence of bacteria. The bacteria also attach to the crystals of metal sulfides such as PbS, HgS, or CuS. From thiosulfate enrichment cultures, pure cultures might be obtained by streaking onto a solidified medium of the same composition. However, pure cultures are fairly difficult to obtain, since growth is usually slow and heterotrophic contaminants grow on small amounts of organic matter released by the thiobacilli. Another problem in obtaining pure cultures of thiobacilli that grow best at neutral pH, such as *T. thioparus,* is that sulfur oxidation leads to sulfuric acid production and a drop in pH, resulting in the death of the culture. Thus highly buffered media (2 to 10 g/liter of a mixture of K_2HPO_4 and KH_2PO_4) and frequent transfers of the culture are necessary. On the other hand, isolation of the acidophilic *T. thiooxidans* is relatively easy, since this organism is resistant to the acid it produces and most heterotrophic contaminants cannot grow at the low pH values (2 to 5) where *T. thiooxidans* thrives.

Many isolates of acidophilic thiobacilli can also oxidize ferrous iron. The geochemical aspects of iron oxidation were discussed in Section 13.11. At acid pH, ferrous iron is not readily oxidized spontaneously, and the acid-tolerant thiobacilli are the main agents in nature for the oxidation of ferrous iron in acidic environments. Those isolates which oxidize both iron and sulfur compounds are currently classified as the species *T. ferrooxidans,* the species *T. thiooxidans* being restricted to those acid-tolerant thiobacilli which cannot oxidize iron.

Sulfolobus Another genus of sulfur-oxidizing organisms is *Sulfolobus.* Members of this genus are not only acidophilic but also thermophilic, growing over the pH range from 1 to 5 (optimum 2 to 3) and temperature range of 60 to 85°C (optimum 70 to 80°C). *Sulfolobus* lives primarily in sulfur-rich geothermal habitats, such as hot springs, fumaroles, and hot, acid soils. Hydrogen sulfide emitted from the source is oxidized by *Sulfolobus* to elemental sulfur and then further oxidized to sulfuric acid, and therefore this organism is responsible for the acidity of sulfur-rich geothermal habitats.

Sulfolobus has a generally spherical shape and forms distinct lobes (Figure 19.53). The cells adhere tightly to sulfur crystals, where they can be visualized microscopically by use of fluorescent dyes (Figure 6.17*b*). *Sulfolobus* will grow either lithotrophically on elemental sulfur as the sole electron donor, or heterotrophically on certain amino acids or sugars. Sulfur oxidation can also be coupled to the reduction of ferric iron, but growth anaerobically under these conditions does not occur since *Sulfolobus* is an obligate aerobe. *Sulfolobus* will also grow lithotrophically on sulfide minerals, and it has been shown that it can be used for leaching of low-grade ores under

FIGURE 19.53 *Sulfolobus acidocaldarius.* Electron micrograph of a thin section. Magnification, 48,970×. (From Brock, T. D., K. M. Brock, R. T. Belly, and R. L. Weiss. 1972. Arch. Mikrobiol. 84:54–68.)

thermophilic conditions. The cell-wall structure of *Sulfolobus* is unusual (Figure 19.53), and the cell membrane is also unusual, being constructed chemically so that it can withstand hot, acidic conditions. In contrast to the membrane of most other organisms, the *Sulfolobus* membrane is not a phospholipid bilayer, but a monolayer consisting of a long-chain hydrocarbon (actually an isoprenoid) connected at both ends by ether linkages (instead of the usual ester linkages) to glycerol residues. This structure is unusually acid and heat stable, but has the conventional hydrophilic outside and inside surfaces (the glycerol and phosphate residues) separated by a hydrophobic interior (the hydrocarbon chain) typical of membranes in general. A similar type of cell membrane has also been found in another acidophilic thermophile, *Thermoplasma*, as well as the halophilic bacteria of the genus *Halobacterium* (see Section 19.33 for additional discussion of *Thermoplasma*, *Halobacterium* and *Sulfolobus*).

***Thiomicrospira* and *Thermothrix*.**   Members of the genera *Thiomicrospira* and *Thermothrix* are thiosulfate-oxidizing bacteria which grow at neutral pH. *Thiomicrospira*, as its name implies, is a tiny, spiral-shaped bacterium. In fact, this organism is so small that it can be selectively enriched by inoculating a thiosulfate- or hydrogen sulfide-containing medium with the filtrate remaining from filtering a mud slurry through a membrane filter with a pore size of just 0.22 μm. Such a filter would retain organisms the size of *Thiobacillus*. The genus *Thiomicrospira* consists of two species; both are obligate lithotrophs and one species can grow anaerobically with nitrate as electron acceptor. *Thermothrix* is a filamentous bacterium which inhabits hot sulfur springs having a neutral or slightly acidic pH. Cultures of *Thermothrix* grow aerobically or anaerobically (with nitrate as electron acceptor) at temperatures between 55 and 85°C (optimum at 70°C). Like *Sulfolobus, Thermothrix* is capable of heterotrophic growth, and utilizes a variety of organic compounds, including glucose, acetate, and amino acids as electron donors.

19.20

Methane-Producing Bacteria: Methanogens

We have described the overall process of methanogenesis in Section 13.7 and noted that this process was carried out by a unique and specialized group of bacteria, the **methanogenic bacteria**. Methane formation occurs to a significant extent only in strictly anaerobic environments. Eight substrates have been shown to be converted to methane by one or another methanogenic bacterium: acetate (CH_3-COO^-), formate ($HCOO^-$), methanol (CH_3OH), methylamine (CH_3-NH_2), dimethylamine [$(CH_3)_2NH$], trimethylamine [$(CH_3)_3N$], carbon monoxide (CO), and CO_2. In the case of acetate, methane formation involves a cleavage of the molecule with the formation of CH_4 from the CH_3-group and CO_2 from the carboxyl group of acetate. The details of this process are not understood, and very few of the typical methanogenic bacteria are able to carry out this process. In the case of CO_2, methane formation occurs by a reductive process, with the electrons usually coming from H_2. Methane formation from H_2 and CO_2 can be viewed as a type of anaerobic respiration in which CO_2 serves as the electron acceptor, but biochemical studies indicate that the mechanism of electron transport to CO_2 is quite different from that of other anaerobic respiratory processes, such as nitrate and sulfate reduction, in that a conventional electron-transport system involving cytochromes, flavins, and quinones is not present in methanogens. Despite the absence of these electron carriers, evidence is good that ATP synthesis occurs via some sort of electron-transport phosphorylation process. One electron-transport component that has been identified in the methanogenic bacteria is **factor 420 (F_{420})**, molecular weight 630, a blue-green fluorescing compound, which exhibits a strong absorption at 420 nm when in the oxidized form.* Other coenzymes found in methanogens, include F_{430}, which is a nickel-containing tetrapyrrole, and F_{342}, which is a pterin called **methanopterin**. F_{430} serves as the prosthetic group of methylcoenzyme M reductase in the terminal step of methanogenesis (see following sections on coenzyme M and biochemistry of methanogenesis), and F_{342} is thought to be the carrier of C-1 units during the reduction of CO_2 to the hydroxymethyl (-CH_2OH) level. In addition, a component referred to as the "carbon dioxide reduction factor" is required for the initial steps of CO_2 plus H_2-mediated methanogenesis. In the case of methanol and formate, electrons for their reduction to methane can also come from H_2, but in the absence of H_2, methane formation can still occur from both of these substrates. Under such conditions, some molecules of the substrate serve as electron donor and are oxidized to CO_2, whereas other molecules are reduced and serve as electron acceptor, a fermentation process. In the case of methanol, the overall reaction to methane has the following stoichiometry:

$$4CH_3OH \longrightarrow 3CH_4 + CO_2 + 2H_2O$$

Diversity A variety of morphological types of methanogenic bacteria have been isolated in recent years, and an in-depth study of their physiology and ribosomal RNA sequences has resulted in the classification of this group into four families containing a total of twelve genera (Table 19.26 and Figures 19.54 and 19.55). It is of interest that despite this marked morphological diversity, and the widely divergent DNA base compositions found (30 to 61 percent G + C), all methanogenic bacteria seem remarkably similar in their physiology and biochemistry. Analyses of cell-wall chemistry of a number of the methanogenic bacteria has failed to find the presence of peptidoglycan, and it seems likely that the cell wall of methanogens is constructed on a different principle than that of most other bacteria. The lack of peptidoglycan is especially interesting when it is noted (Table 19.26) that many of the methanogenic bacteria

*This characteristic fluorescence can be used in the identification of methanogenic bacteria. Upon introduction of air, colonies of methanogenic bacteria show a blue-green fluorescence when examined with blue light. With a fluorescence microscope, even the fluorescence of individual cells can be seen. See Color Plate 7c and d.

TABLE 19.26   Characteristics of Methane-Producing Bacteria

Genus and Species	Morphology	Gram Stain	Substrates for Methanogenesis	Special Characteristics	DNA (mol % G + C)
Methanobacterium:					
formicicum	Long rod/filament	+ or −	$H_2 + CO_2$, formate	Nonmotile	40–42
bryantii	Long rod	+ or −	$H_2 + CO_2$	Requires B vitamins; nonmotile	32–38
thermoautotrophicum	Long rod/filament	+	$H_2 + CO_2$	Thermophile; nonmotile	49–52
Methanobrevibacter:					
ruminantium	Short rods, often in chains	+	$H_2 + CO_2$, formate	Requires acetate, several amino acids, 2-methylbutyrate, and coenzyme M	30
smithii	Short rods, often in chains	+	$H_2 + CO_2$, formate	Nonmotile; requires acetate	31–32
arboriphilus	Short rods	+	$H_2 + CO_2$	Found in wet wood of living trees; nonmotile	27–31
Methanomicrobium:					
mobile	Short rods with polar flagella	−	$H_2 + CO_2$, formate	Highly motile; requires rumen fluid	48
paynteri	Rods	−	$H_2 + CO_2$	Marine, requires NaCl	45
Methanogenium:					
cariaci	Small irregular cocci	−	$H_2 + CO_2$, formate	Marine, requires NaCl, acetate, and yeast extract	51
marisnigri	Small irregular cocci	−	$H_2 + CO_2$, formate	Marine, requires NaCl and trypticase	61
olentangyi	Small irregular cocci	−	$H_2 + CO_2$	Requires acetate	54
thermophilicum	Small irregular cocci	−	$H_2 + CO_2$, formate	Marine, requires NaCl, trypticase, and vitamins; thermophile, grows optimally at 55°C	59
Methanospirillum:					
hungatei	Short to long wavy filament	−	$H_2 + CO_2$, formate	Motile	45–46
Methanococcus:					
vannielii	Irregular, lobed cocci	−	$H_2 + CO_2$, formate	Motile, requires selenium or tungsten	31
deltae	Irregular cocci	−	$H_2 + CO_2$, formate	Nonmotile, requires 3% NaCl	40
voltae	Irregular, lobed cocci	−	$H_2 + CO_2$, formate	Motile	30
mazei	Cocci, in clusters	+ or −	Methanol, acetate, methylamines, $H_2 + CO_2$, (weak)	Undergoes life cycle; requires NaCl; not yet in pure culture	—
thermolithotrophicus	Cocci	−	$H_2 + CO_2$, formate	Thermophile; requires NaCl; found in volcanic areas on sea floor	31
Methanosarcina:					
barkeri	Large cocci in packets	+	$H_2 + CO_2$, formate, methanol, methylamines, acetate	Most metabolically versatile of all methanogens; one strain thermophilic; one strain contains gas vesicles	38–51
Methanoplanus:					
limicola	"Plate-shaped" with angular corners	−	$H_2 + CO_2$, formate	Acetate required for growth but not for methanogenesis	47

TABLE 19.26 (*continued*)

Genus and Species	Morphology	Gram Stain	Substrates for Methanogenesis	Special Characteristics	DNA (mol % G + C)
Methanothrix:					
soehngenii	Filamentous, which break easily into small rods	−	Acetate only	Unable to use H_2 + CO_2 for methanogenesis	52
Methanothermus:					
fervidus	Rods	+	H_2 + CO_2	Grows optimally at 83°C, temperature range, 60–97°C; requires yeast extract	33
Methanolobus:					
tindarius	Lobed-shaped cocci	−	Methanol, methylamines	Marine, unable to use H_2 + CO_2, acetate, or formate for methanogenesis; motile; yeast extract does not stimulate growth	46
Methanococcoides					
methylutens	Irregular cocci	−	Methanol and methylamines only	Marine, unable to use H_2 + CO_2, acetate, or formate for methanogenesis; nonmotile; colonies yellow pigmented; yeast extract stimulates growth	42

(a)

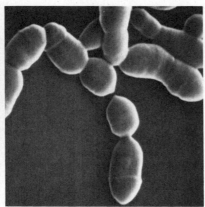

(b)

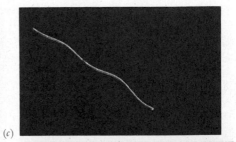

(c)

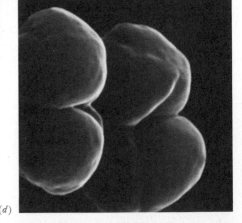

(d)

FIGURE 19.54 Scanning electron micrographs of whole cells of methanogenic bacteria, showing the considerable morphological diversity.
(a) *Methanobrevibacter ruminantium.*
(b) *Methanobacterium* strain AZ.
(c) *Methanospirillum hungatii.*
(d) *Methanosarcina barkeri.* All magnifications, 11,220×. (Courtesy of Alexander Zehnder.)

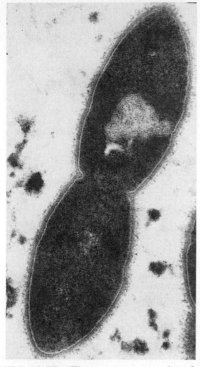

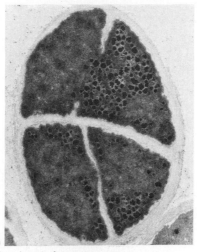

(a)   (b)

FIGURE 19.55 Electron micrographs of methanogenic bacteria.
(a) *Methanobacterium ruminantium.* Magnification, 42,100×. (b) *Methano-sarcina barkeri,* showing the thick cell wall and the manner of cell segmentation and cross-wall formation. Magnification, 21,000×. (From Zeikus, J. G., and V. G. Bowen. 1975. Comparative ultrastructure of methanogenic bacteria. Canad. Jour. Microbiol. 21:121–129. Reproduced by permission of the National Research Council of Canada.)

stain Gram-positive. Another feature indicating the molecular similarity of methanogenic bacteria, which distinguishes them from other bacteria, is shown by studies on the structure of their 16S ribosomal RNAs. Detailed studies of the 16S ribosomal RNAs of a wide variety of organisms have shown that sequence similarities and differences can be used to construct an overall classification of procaryotic and eucaryotic organisms (see Section 19.33). The methanogenic bacteria are found to be in a class quite distinct from most other procaryotes, based on this feature. Interestingly, several other bacteria that also lack a typical peptidoglycan cell wall, *Halobacterium, Sulfolobus,* and *Thermoplasma,* are also found to be related to the methanogenic bacteria, and distinct from other procaryotes, based on studies of 16S ribosomal RNA (see Section 19.33).

Coenzyme F_{420} As noted, coenzyme F_{420} is a unique coenzyme found in methanogenic bacteria and functions as a low potential electron carrier. The structure of F_{420} is shown below.

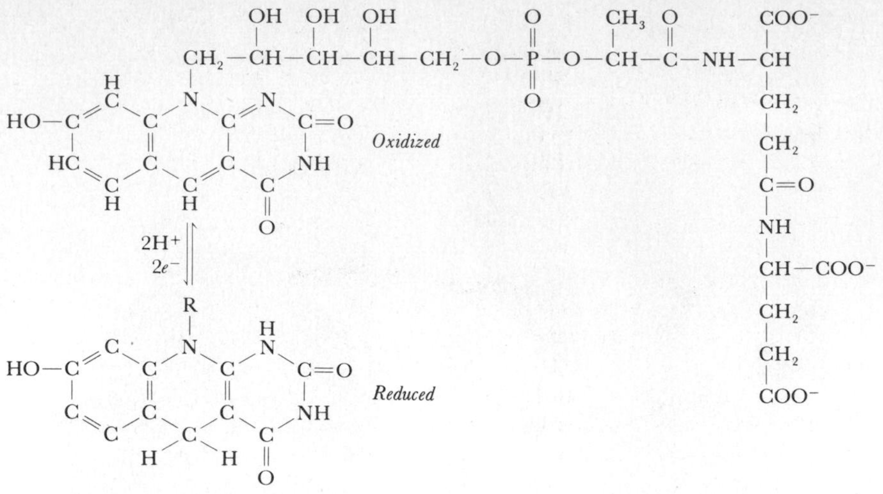

OH OH OH   O   CH₃ O   COO⁻

CH_2—CH—CH—CH—CH_2—O—P—O—CH—C—NH—CH

Oxidized

2H⁺
2e⁻

R

Reduced

The structure of F_{420} has some similarity to the structure of a flavin (Figure 4.13), but lacks a nitrogen in the middle ring, as well as the methyl groups at positions 7 and 8. In contrast to conventional flavins, coenzyme F_{420} is confined to donating two electrons at a time rather than one, since it cannot form a semiquinone. The reduction potential of F_{420} (oxidized/reduced) is about -380 millivolts, making it considerably lower than the reduction potential of the flavins (Table A1.2).

Coenzyme M All of the methanogenic bacteria which have been examined possess a unique coenzyme, coenzyme M, which is known to be involved in at least the final step in methane formation. Coenzyme M has the chemical name 2-mercaptoethanesulfonic acid and has the following simple structure: HS—CH_2—CH_2—SO_3H. This coenzyme is the carrier of the methyl group, which is reduced to methane by the methyl-CoM reductase–F_{430} complex in the final stage in methanogenesis:

$$CH_3-S-CoM + 2H \xrightarrow{\text{ATP}} HS-CoM + CH_4$$

Despite the simplicity of coenzyme M, it is highly specific in the methylreductase reaction. A number of closely related analogs of coenzyme M have been found to be inactive (even the propane analog of coenzyme M, which differs by addition of only a single CH_2 group, is inactive). The only compound which shows any activity with the methylreductase enzyme is ethyl-CoM, which is converted to ethane. However, since methanogenic bacteria do not form ethane, this reaction must be considered only a laboratory artifact.

Methanobrevibacter ruminantium has the interesting property of requiring coenzyme M as a growth factor. Thus coenzyme M can also be considered to be a vitamin. Some methanogenic bacteria excrete coenzyme M, and this is apparently the source of coenzyme M in the rumen to *M. ruminantium*. Coenzyme M is so active as a vitamin for *M. ruminantium,* that the organism shows a growth response at concentrations as low as 5 nanomolar (5×10^{-9} M).

A potent analog of coenzyme M is bromoethanesulfonic acid, $Br-CH_2-CH_2-SO_3H$. This compound causes 50 percent inhibition of methylreductase activity at a concentration of 10^{-6} M, and it also inhibits growth of methanogenic bacteria. Because coenzyme M is restricted to methanogenic bacteria, the bromo analog can be used to specifically inhibit methanogenesis in natural environments, permitting the dissection of various stages in the anaerobic breakdown of organic matter.

In addition to the bromo analog of coenzyme M, a number of other compounds are known to partially or completely inhibit methanogenesis. These include acetylene ($HC \equiv CH$), ethylene ($H_2C = CH_2$), chloroform ($CHCl_3$), carbon tetrachloride (CCl_4), and the insecticide, DDT [1,1,1-trichloro-2,2-bis(*p*-chlorophenyl)ethane]. These compounds probably all function as analogs of the methyl group (the central trichloro group of DDT resembles chloroform).

Biochemistry of methane bacteria Methane bacteria carry out two distinct reductive processes with CO_2, one leading to methane formation, the other involving the synthesis of cell carbon. Growing cells convert about 90 to 95 percent of the CO_2 provided into methane, and the rest to cell carbon. Despite the fact that many methanogenic bacteria grow completely autotrophically on H_2 plus CO_2, there is no evidence of a conventional Calvin cycle involving ribulose diphosphate carboxylase. It seems likely that methanogenic bacteria have a unique mechanism for autotrophic CO_2 fixation. Recent evidence suggests that the pathway of CO_2 fixation in methanogens involves the reduction of two molecules of CO_2 to form acetyl $\sim$ CoA (by an unknown enzymatic pathway), with a further carboxylation yielding pyruvate. Pyruvate can be converted to phosphoenolpyruvate and hence to hexoses, or further carboxylated to give the four- and five-carbon intermediates of the citric acid cycle.

The reduction of CO_2 to methane is usually H_2-dependent and all methanogens capable of growing on $H_2 + CO_2$ possess a nickel-containing hydrogenase which couples the uptake of H_2 to the reduction of $NADP^+$ to NADPH via the intermediate electron carrier, F_{420}. The reduction of CO_2 to CH_4 is thought to occur via several intermediates, most likely with oxidation states at the levels of formate, formaldehyde, and methanol, but it is likely that at all of these stages, the carbon atom is combined with a carrier. The current picture of the steps involved in methanogenesis beginning with $H_2 + CO_2$ is as follows:

1. H_2 is activated by hydrogenase and used to reduce F_{420}.
2. Reduced F_{420} reduces $NADP^+$ to NADPH.
3. CO_2 is bound by F_{342} (methanopterin, MP) to give $MP-C{\overset{\displaystyle O}{\underset{\displaystyle OH}{\big\langle}}}$.
4. $MP-C{\overset{\displaystyle O}{\underset{\displaystyle OH}{\big\langle}}}$ is reduced by an unknown electron donor to

   $MP-C{\overset{\displaystyle O}{\underset{\displaystyle H}{\big\langle}}}$ in the presence of the "carbon dioxide reduction factor."

5. The $-C{\overset{\displaystyle O}{\underset{\displaystyle H}{\Big\langle}}}$ group, bound either to MP or to an unidentified carrier, is reduced to the level of $-CH_2OH$. The source of electrons is probably reduced F_{420} or NADPH.

6. The $-CH_2OH$ group is transferred to $SH-CoM$ to give $HOCH_2-S-CoM$.

7. $HOCH_2-S-CoM$ is reduced to $H_3C-S-CoM$, probably by reduced F_{420}.

8. $H_3C-S-CoM$ is reduced to $HS-CoM$ and CH_4 by the methylreductase–F_{430} complex. The source of electrons is probably F_{420}.

These steps are summarized in Figure 19-56.

When *Methanosarcina barkeri* is grown on methyl-containing compounds (for example, methanol) in place of $H_2 + CO_2$, a vitamin B_{12}-containing protein is synthesized and may be involved in methanogenesis at or near the terminal step of the process. In *Methanosarcina barkeri* grown on methanol, CH_3-B_{12} (the origin of which is unclear) donates its methyl group to CoM-SH to produce CoM-CH_3. Methane is made from the latter via the methyleductase–F_{430} system described for cells grown on $H_2 + CO_2$. Interestingly, however, $(H_2 + CO_2)$-grown cells lack the B_{12}-containing enzyme system; hence vitamin B_{12} probably plays a role only when *Methanosarcina* grows on methylated substrates.

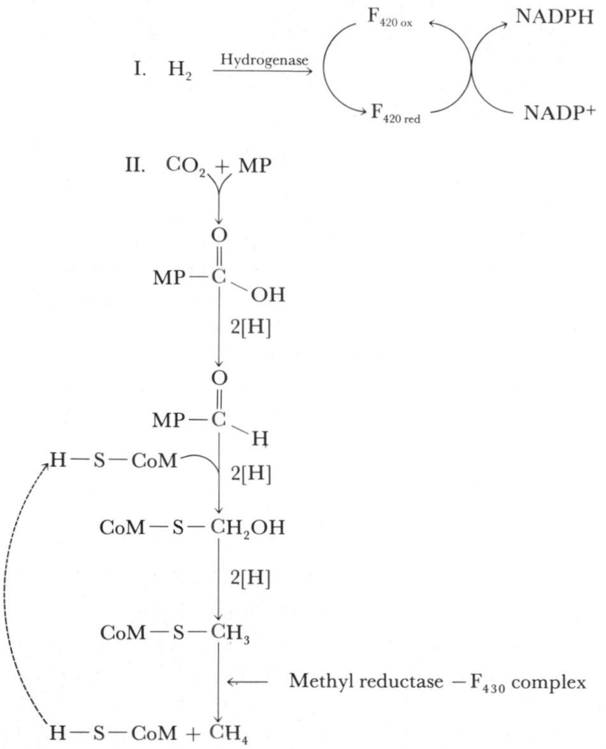

FIGURE 19.56 Biogenesis of methane from $CO_2 + H_2$. F_{420} = factor 420; F_{430} = factor 430; MP = methanopterin (F_{342}); HS-CoM = reduced coenzyme M. The source of [H] may be NADPH, $F_{420reduced}$, or some other hydrogen donor.

On theoretical grounds, it appears that ATP synthesis in the methanogenic bacteria occurs via electron-transport phosphorylation, since there is no possibility of substrate-level phosphorylation during growth on acetate, methanol, or CO_2 and H_2. Under standard conditions, the free energy change of the reduction of CO_2 to CH_4 with H_2 is -31 kcal/mole. However, concentrations of H_2 in methanogenic habitats are usually quite low, no higher than 1 micromolar, and because of the influence of concentration of reactants on free energy change (see Appendix 1), the free energy of formation of CH_4 by methanogenic bacteria in their natural habitat is much lower, about -15 kcal/mole. Thus no more than one ATP will be formed during CO_2 reduction to CH_4, and this agrees with the molar growth yields which have been obtained for methanogenic bacteria growing on H_2 plus CO_2. Energy generation during the conversion of acetate to CH_4 and CO_2 is a real mystery at this point, and much remains to be done to elucidate the energetic mechanisms involved in methane formation from methylated compounds.

Methane-producing bacteria may have been among the first autotrophic organisms to evolve on this planet, since their electron donor, H_2, and their electron acceptor, CO_2, would both have been present on the primitive earth, and since these bacteria grow under the anaerobic conditions that would have prevailed at those times. The evolutionary and biochemical relationships of this interesting group of bacteria should provide a fertile field for study (see Section 19.33).

Acetogenic Bacteria   A second group of organisms, distinct from the methanogens, are capable of growing lithotrophically on H_2 + CO_2, but the product of CO_2 reduction in this case is acetate instead of methane. Organisms such as *Acetobacterium woodii* or *Clostridium aceticum* (an endospore-former), carry out a "homoacetic acid" metabolism through the conversion of H_2 + CO_2 to acetate. The reaction involved is as follows:

$$2CO_2 + 4H_2 \longrightarrow CH_3COOH + 2H_2O$$

Unlike the methanogens, however, **acetogenic bacteria** are also able to grow heterotrophically on sugars. Fructose is fermented by these organisms with acetic acid being the sole fermentation product. Biochemical studies have shown that the acetogens lack the unique series of coenzymes found in the methanogens, and it appears that they resemble typical fermentative anaerobes much more than they do the methanogens. They are included in the discussion of the methanogens because they use H_2 as an electron donor (anaerobically) with CO_2 as electron acceptor.

19.21

Gram-Positive Cocci

The Gram-positive cocci are not a natural taxonomic grouping, since they include bacteria with widely differing physiological characteristics. The two most commonly seen genera are *Staphylococcus* and *Micrococcus,* and our discussion here is restricted to these genera. (The genus *Streptococcus* is discussed in Section 19.23 with the other lactic acid bacteria.) *Staphylococcus* and *Micrococcus* are both aerobic organisms with a typical respiratory metabolism. They are catalase positive (see Table 18.2), and this test permits their distinction from *Streptococcus* and some other genera of Gram-positive cocci.

As we discussed in Section 8.2, the Gram-positive cocci are relatively resistant to reduced water potential, and tolerate drying and high salt fairly well. Their ability to grow in media with high salt provides a simple means for isolation. If an inoculum is spread on an agar plate containing a fairly rich medium containing about 6 to 7.5 percent NaCl, and the plate incubated aerobically, Gram-positive cocci will often form the predominant colonies. Often, these organisms are pigmented, and this provides an additional aid in selecting Gram-positive cocci.

The two genera *Micrococcus* and *Staphylococcus* can easily be separated based on the oxidation/fermentation test. *Micrococcus* is an obligate aerobe, and produces acid from glucose only aerobically, whereas *Staphylococcus* is a facultative aerobe, and produces acid from glucose both aerobically and anaerobically. Their DNA base compositions are also widely different: *Micrococcus,* 66 to 73 percent G + C; *Staphylococcus,* 30 to 38 percent G + C.

Staphylococci are common parasites of humans and animals, and occasionally cause serious infections. In humans, two major forms are recognized, *S. epidermidis,* a nonpigmented, nonpathogenic form that is usually found on the skin or mucous membranes, and *S. aureus,* a yellow pigmented form that is most commonly associated with pathological conditions, including boils, pimples, pneumonia, osteomyelitis, meningitis, and arthritis. We listed the exotoxins of *S. aureus* in Table 15.1. One of the significant exotoxins is **coagulase,** an enzymelike factor that causes fibrin to coagulate and form a clot. Strains of *S. aureus* are generally coagulase positive, whereas *S. epidermidis* is coagulase negative. We discussed the possible role of the yellow carotenoid pigment of *S. aureus* in resistance to phagocytosis in Section 15.7 and staphylococcal food poisoning was discussed in Section 17.10.

Certain strains of *S. aureus* have been implicated as the agents responsible for the so-called **toxic shock syndrome** (TSS), a severe infection characterized by high fever, vomiting, and diarrhea. Toxic shock occurs most frequently in a small percentage of menstruating women who use tampons. Blood and mucus in the vagina become colonized by hemolytic *S. aureus* from the skin, and the presence of a tampon may concentrate this material, creating ideal microbial growth conditions. The symptoms of TSS are thought to be due to an exotoxin released by *S. aureus,* which causes a considerable blood pressure drop followed shortly by the aforementioned symptoms. A small number of cases of TSS have been fatal. Women can markedly reduce their risk of TSS by either not using tampons or by alternating tampons with other products such as sanitary napkins or shields.

The micrococci are not associated with pathological conditions, and because their DNA base compositions are so different from those of the staphylococci, it is unlikely that the two groups are very closely related. Micrococci are widespread in soil and become dispersed through the air, often being isolated as contaminants on agar plates.

They are usually pigmented, either yellow or pink, and such pigment may help these organisms to survive destruction by sunlight during airborne dispersal (Section 8.6).

The structure, mode of formation, and heat resistance of the bacterial endospore have been discussed in Sections 2.11 and 8.1. Several genera of endospore-forming bacteria have been recognized, distinguished on the basis of morphology, relationship to O_2, and energy metabolism (Table 19.27). The two genera most frequently studied are *Bacillus,* the species of which are aerobic or facultatively aerobic, and *Clostridium,* which contains the strictly anaerobic species. (Heat-resistant spores are also produced by certain thermophilic actinomycetes, but we shall not discuss these forms here.) The genera *Bacillus* and *Clostridium* consist of Gram-positive or Gram-variable rods, which are usually motile, possessing peritrichous flagella. Members of the genus *Bacillus* produce the enzymes catalase and superoxide dismutase. Clostridia do not produce catalase and produce only low levels of superoxide dismutase, and it is thought that one reason they are obligately anaerobic is that they have no way of getting rid of the toxic H_2O_2 and O_2^- produced from molecular oxygen (see Section 8.5). Molecular taxonomic studies of the genus *Bacillus* have shown that it is a heterogeneous group and can hardly be considered an assemblage of closely related organisms. The DNA base composition of sundry *Bacillus* species vary from about 30 to 50 percent G + C, and studies on nucleic acid homologies by hybridization and genetic transformation also suggest considerable genetic heterogeneity. Although less molecular work has been done with members of the genus *Clostridium,* data from DNA base compositions suggest less genetic heterogeneity in this group; values from 23 to 54 percent G + C have been reported, but most species cluster between 25 to 30 percent G + C. Because of the complex series of enzymatic steps involved in sporulation, it seems reasonable to hypothesize that the ability to form endospores arose only once during evolution and that a primitive spore former was, by evolutionary divergence, the forerunner of the variety of spore-forming bacteria known today.

Even though they are not closely related genetically, all the spore-forming bacteria are ecologically related since they are found

TABLE 19.27 The Genera of Endospore-Forming Bacteria

Characteristics	Genus	DNA (mol % G + C)
Rods:		
Aerobic or facultative, catalase produced	*Bacillus*	32–68
Microaerophilic, no catalase	*Sporolactobacillus*	39
Anaerobic:		
Sulfate-reducing	*Desulfotomaculum*	42–46
Do not reduce sulfate, fermentative	*Clostridium*	23–54
Cocci (usually arranged in packets)	*Sporosarcina*	40–43

in nature primarily in the soil. Even those species that are pathogenic to humans or animals are primarily saprophytic soil organisms, and infect hosts only incidentally. Spore formation should be advantageous for a soil microorganism because the soil is a highly variable environment. Although at some times nutrient supply is in excess, at other times it is deficient. Soil temperatures can be quite high in summer, especially at the surface. Thus a heat-resistant dormant structure should offer considerable survival value in nature. On the other hand, the ability to germinate and grow quickly when nutrients become available is also of value as it enables the organism to capitalize on a transitory food supply. The longevity of bacterial endospores is noteworthy. Records of spores surviving in the dormant state for over 50 years are well established.

Bacillus Members of the genus *Bacillus* are easy to isolate from soil or air and are among the most common organisms to appear when soil samples are streaked on agar plates containing various nutrient media. Spore formers can be selectively isolated from soil, food, or other material by exposing the sample to 80°C for 10 to 30 minutes, a treatment that effectively destroys vegetative cells while many of the spores present remain viable. When such pasteurized samples are streaked on plates and incubated aerobically, the colonies that develop are almost exclusively of the genus *Bacillus*. Bacilli usually grow well on synthetic media containing sugars, organic acids, alcohols, and so on, as sole carbon sources and ammonium as the sole nitrogen source; some isolates have vitamin requirements. Many bacilli produce extracellular hydrolytic enzymes that break down polysaccharides, nucleic acids, and lipids, permitting the organisms to use these products as carbon sources and electron donors. Many bacilli produce antibiotics, of which bacitracin, polymyxin, tyrocidin, gramicidin, and circulin are examples. In most cases antibiotic production seems to be related to the sporulation process, the antibiotic being released when the culture enters the stationary phase of growth and after it is committed to sporulation. An outline of the subdivision of the genus *Bacillus* is given in Table 19.28.

A feature of the genus *Bacillus* is the great diversity of organisms which it encompasses. When we discussed the enteric bacteria (Section 19.13), we mentioned that genera in that group have been created based on rather minor physiological differences. Similar physiological differences exist in the genus *Bacillus,* yet separate genera have not been pulled out. This is not a scientific matter, but a matter of scientific psychology: the spore-forming characteristic is so distinctive, and so overriding, that there has been a reluctance on the part of taxonomists to create a number of separate genera. A number of types of energy generation have been recognized in the genus *Bacillus:*

1. *B. coagulans* carries out a typical homolactic fermentation, as described for the lactic acid bacteria (Figure 19.60).
2. *B. subtilis* and *B. cereus* carry out a 2,3-butanediol fermentation (see Figure 19.45 in the enteric bacteria section); these organisms also produce glycerol as a fermentation product.

TABLE 19.28 Subdivision of the Genus *Bacillus* and Representative Species

Characteristics	Species
Spores oval or cylindrical; fermentative;	
casein and starch usually hydrolyzed:	
Sporangia not swollen; spore wall thin;	
Gram positive:	
Thermophiles and acidophiles	*B. coagulans, B. acidocaldarius*
Mesophiles	*B. licheniformis, B. cereus* (and *B. anthracis*), *B. megaterium, B. subtilis*
Insect pathogen	*B. thuringiensis*
Sporangia distinctly swollen; spores oval;	
spore wall thick; Gram variable:	
Thermophile, grows at 65°C	*B. stearothermophilus*
Mesophiles	*B. polymyxa, B. macerans, B. circulans*
Insect pathogens	*B. larvae, B. popilliae*
Spores spherical; nonfermentative; casein	
and starch not hydrolyzed:	
Sporangia swollen; nutritional require-	
ments complex; Gram variable	*B. sphaericus, B. pasteurii*

3. *B. polymyxa* produces 2,3-butanediol, and also produces ethanol and H_2; it is also a N_2 fixer.

4. *B. macerans* ferments sugars to ethanol, acetone, acetic, and formic acids; it also fixes N_2.

5. *B. schlegelii* is a lithotroph, able to use H_2 as sole electron donor.

In addition to the taxonomic criteria listed in Table 19.28, additional tests that permit distinction of species of the genus *Bacillus* include: acid and/or gas production from glucose; Voges–Proskauer test (for butanediol producers); nitrate reduction; ability to grow anaerobically; and motility.

The *B. cereus* group consists of the species *B. cereus, B. anthracis,* and *B. thuringiensis.* The first is a common soil organism; the latter two are similar to *B. cereus* morphologically and genetically but are distinguished because they are pathogenic, *B. anthracis* causing anthrax in animals and humans and *B. thuringiensis* causing a variety of diseases in insects (see below). DNA homology studies have shown that all three of these species are related, with *B. anthracis* and *B. cereus* being the more similar. In fact, *B. anthracis* cannot be distinguished from *B. cereus* by any criterion except pathogenicity.

A number of bacilli, most notably *B. larvae, B. popilliae* and *B. thuringiensis* are insect pathogens, and in recent years there has been considerable interest in these organisms because of their potential use in the biological control of insect infestations of plants. These insect pathogens form a crystalline protein during sporulation, called the **parasporal body,** which is deposited within the sporangium but outside the spore (Figure 19.57). These crystal-forming bacilli cause fatal diseases of moth larvae such as the silkworm, the cabbage

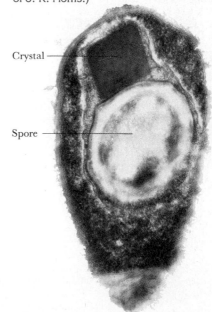

FIGURE 19.57 Formation of the toxic parasporal crystal in the insect pathogen *Bacillus thuringiensis.* Electron micrograph of a thin section. Magnification, 44,000×. (Courtesy of J. R. Norris.)

Crystal

Spore

worm, the tent caterpillar, and the gypsy moth, due to the action of this toxic substance.

Clostridium   The clostridia lack a cytochrome system and a mechanism for electron-transport phosphorylation, and hence they obtain ATP only by substrate-level phosphorylation. A wide variety of anaerobic energy-yielding mechanisms are known in the clostridia; indeed, the separation of the genus into subgroups is based primarily on these properties and on the nature of the electron donors used (Table 19.29).

A number of clostridia ferment sugars, producing as a major end product butyric acid. Some of these also produce acetone and butanol, and at one time the acetone-butanol fermentation by clostridia was of great industrial importance as it was the main commercial source of these products. Today, however, the chemical synthesis of acetone and butanol from petroleum products has mostly replaced the microbiological process. Some clostridia of the acetone-butanol

TABLE 19.29   Characteristics of Some Groups of the Genus *Clostridium*

Key Characteristics	Other Characteristics	Species	DNA (mol % G + C)
Ferment carbohydrates:			
Ferment cellulose	Fermentation products: acetic acid, lactic acid, succinic acid, ethanol, CO_2, H_2	*C. cellobioparum* *C. thermocellum*	25–28 38–39
Ferment sugars, starch, and pectin	Fermentation products: acetone, butanol, ethanol, isopropanol, butyric acid, acetic aid, propionic acid, lactic acid, CO_2, H_2; some fix N_2	*C. butyricum* *C. acetobutylicum* *C. pasteurianum* *C. perfringens*	27–28 28–29 26–28 24–27
Ferment sugars primarily to acetic acid	Total synthesis of acetate from CO_2; cytochromes present in some species	*C. aceticum* *C. thermoaceticum* *C. formicoaceticum*	33 54 34
Ferment proteins or amino acids	Fermentation products: acetic aid, fatty acids, NH_3, CO_2, sometimes H_2; some also ferment sugars to butyric and acetic acids; may produce exotoxins	*C. sporogenes* *C. tetani* *C. botulinum* *C. histolyticum* *C. tetanomorphum*	26 25–26 21–28 — —
Purine fermenters	Ferment uric acid and other purines, forming acetic acid, CO_2, NH_3	*C. acidiurici*	27
Ethanol fermentation to fatty acids	Produces butyric acid, caproic acid, and H_2; does not attack sugars, amino acids, or purines	*C. kluyveri*	35

type fix N_2; the most vigorous N_2 fixer is *C. pasteurianum,* which probably is responsible for most anaerobic nitrogen fixation in the soil. One group of clostridia ferments cellulose with the formation of acids and alcohols, and these are the main organisms decomposing cellulose anaerobically in soil. There is considerable industrial interest in the production of ethanol (an automotive fuel) by the clostridial fermentation of cellulose, and studies are underway to increase the yield of ethanol and reduce the formation of acidic fermentation products, the goal being to use waste cellulose as a motor fuel.

The biochemical steps in the formation of butyric acid and butanol from sugars are well understood (Figure 19.58). Glucose is converted to pyruvate via the Embden–Meyerhof pathway, and pyruvate is split to acetyl-CoA, CO_2, and reducing equivalents (reduced ferredoxin) by a phosphoroclastic reaction (see Section 4.7). Acetyl-CoA is then reduced to fermentation products using the NADH derived from glycolytic reactions. The proportions of the various products are influenced by the duration and the conditions of the fermentation. During the early stages, butyric and acetic acids are the predominant products, but as the pH of the medium drops, synthesis of the acids ceases and acetone and butanol begin to ac-

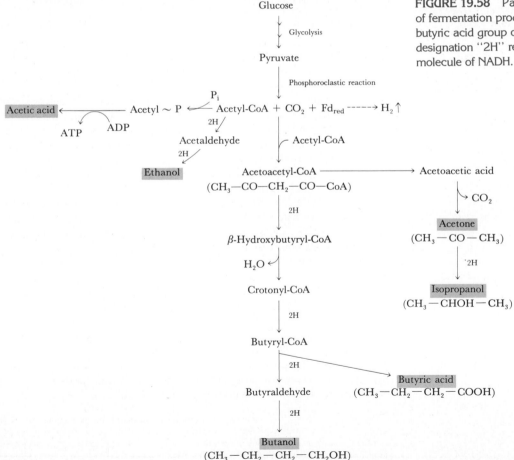

FIGURE 19.58 Pathway of formation of fermentation products from the butyric acid group of clostridia. The designation "2H" represents one molecule of NADH.

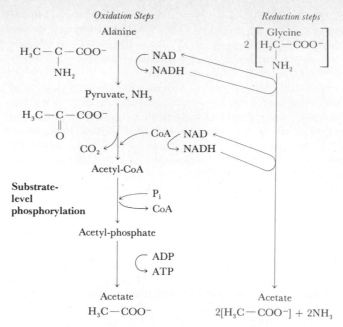

FIGURE 19.59 Coupled oxidation-reduction reaction (Stickland reaction) in *Clostridium sporogenes* between alanine and glycine. In addition to the site of substrate-level phosphorylation shown, there is also a possible site of electron-transport phosphorylation to the glycine/acetate reaction.

Overall: Alanine + 2 Glycine + ADP + $P_i \longrightarrow$ 3 Acetate + CO_2 + $3NH_3$ + ATP

TABLE 19.30 Amino Acids Participating in Coupled Fermentations (Stickland Reaction)

Amino acids oxidized:
Alanine
Leucine
Isoleucine
Valine
Histidine

Amino acids reduced:
Glycine
Proline
Hydroxyproline
Ornithine
Tryptophan
Arginine

cumulate. If the medium is kept alkaline with $CaCO_3$, very little of the neutral products are formed and the fermentation products consist of about 3 parts butyric and 1 part acetic acid.

Another group of clostridia obtain their energy by fermenting amino acids. Some strains do not ferment single amino acids, but when two separate amino acids are present in the medium, one functions as the electron donor and is oxidized, while the other acts as the electron acceptor and is reduced. This type of coupled decomposition is known as the **Stickland reaction.** For instance, *C. sporogenes* will attack a mixture of glycine and alanine, as outlined in Figure 19.59.*

Various amino acids that can function as either oxidants or reductants are listed in Table 19.30. The products of the oxidation are always NH_3, CO_2, and a carboxylic acid with one less carbon atom than the amino acid which is oxidized.

Some amino acids can be fermented singly, rather than in a Stickland-type reaction. These are alanine, cysteine, glutamate, glycine, histidine, serine, and threonine. It is usually found that each group of clostridia is specific in the kinds of substances it can ferment; usually either sugars or amino acids are utilized, although there are strains that can ferment both. Many of the products of amino acid fermentation by clostridia are foul-smelling substances, and the odor that results from putrefaction is a result mainly of clostridial action. In addition to butyric acid, other odoriferous compounds produced are isobutyric acid, isovaleric acid, caproic acid, hydrogen sulfide,

*The glycine reductase is a membrane-bound selenium-containing enzyme, so that growth on this mixture only occurs if selenium is present in the medium. Tap water media generally contain sufficient selenium for growth, but distilled water media probably will not. Selenium can be added in the form of sodium selenite.

methylmercaptan (from sulfur amino acids), cadaverine (from lysine), putrescine (from ornithine), and ammonia.

The main habitat of clostridia is the soil, where they live primarily in anaerobic "pockets," made anaerobic primarily by facultative organisms acting upon various organic compounds present. A number of clostridia have adapted to the anaerobic environment of the mammalian intestinal tract. Also, as was discussed in Section 15.3, several clostridia that live primarily in soil are capable of causing disease in humans under specialized conditions. Botulism is caused by *C. botulinum,* tetanus by *C. tetani,* and gas gangrene by *C. perfringens* and a number of other clostridia, both sugar and amino acid fermenters. These pathogenic clostridia seem in no way unusual metabolically, but are distinct in that they produce specific toxins or, in those causing gas gangrene, a group of toxins (Table 15.1). An unsolved ecological problem is what role these toxins play in the natural habitat of the organism. Many gas-gangrene clostridia also cause diseases in domestic animals, and botulism occurs in sheep and ducks, and a variety of other animals.

Endospore formation   The differences between the endospore and the vegetative cell are profound (Table 19.31), and sporulation involves a very complex series of events. Bacterial sporulation does not occur

TABLE 19.31   Differences Between Bacterial Endospores and Vegetative Cells

Characteristic	Vegetative Cells	Spores
Structure	Typical Gram-positive cell	Thick spore cortex Spore coat Exosporium (some species)
Microscopic appearance	Nonrefractile	Refractile
Chemical composition:		
Calcium	Low	High
Dipicolinic acid	Absent	Present
PHB	Present	Absent
Polysaccharide	High	Low
Protein	Lower	Higher
Parasporal crystalline protein (some species)	Absent	Present
Sulfur amino acids	Low	High
Water content	High	Low
Enzymatic activity	High	Low
Metabolism (O_2 uptake)	High	Low or absent
Macromolecular synthesis	Present	Absent
mRNA	Present	Low or absent
Heat resistance	Low	High
Radiation resistance	Low	High
Resistance to chemicals and acids	Low	High
Stainability by dyes	Stainable	Stainable only with special methods
Action of lysozyme	Sensitive	Resistant

when cells are dividing exponentially, but only when growth ceases owing to the exhaustion of an essential nutrient. For instance, if a culture growing on glucose as an electron donor exhausts the glucose in the medium, vegetative growth ceases and several hours later spores begin to appear. If more glucose is added to the culture just at the end of the growth period, sporulation is inhibited. Glucose probably prevents sporulation through catabolite repression, inhibiting the synthesis of the specific enzymes involved in forming the spore structures. Glucose is not the only substance repressing spore formation; many other electron donors also can do this. We thus see that growth and sporulation are opposing processes. It seems reasonable to assume that the elaborate control mechanisms in endospore-forming bacteria ensure that in nature sporulation will occur only when conditions are no longer favorable for growth.

19.23

Lactic Acid Bacteria

The lactic acid bacteria are characterized as Gram-positive, usually nonmotile, nonsporulating bacteria that produce lactic acid as a major or sole product of fermentative metabolism. Members of this group lack porphyrins and cytochromes, do not carry out electron-transport phosphorylation, and hence obtain energy only by substrate-level phosphorylation.* All of the lactic acid bacteria grow anaerobically. Unlike many anaerobes, however, most of these are not sensitive to O_2 and can grow in its presence as well as in its absence; thus they are **aerotolerant anaerobes.** Some strains are able to take up O_2 through the mediation of flavoprotein oxidase systems, producing H_2O_2, although most strains lack catalase and most dispose of H_2O_2 via alternative enzymes referred to as peroxidases. No ATP is formed in the flavoprotein oxidase reaction, but the oxidase system can be used for reoxidation of NADH. Most lactic acid bacteria can obtain energy only from the metabolism of sugars and related compounds, and hence are usually restricted to habitats in which sugars are present. They usually have only limited biosynthetic ability, and their complex nutritional requirements include needs for amino acids, vitamins, purines, and pyrimidines.

Traditionally, the lactic acid bacteria were considered a single group containing both cocci and rods. Some workers prefer to separate the group; however, we continue to keep both rods and cocci together in the present text, recognizing the obvious physiological similarities of the group as a whole.

Homo- and heterofermentation One important difference between subgroups of the lactic acid bacteria lies in the nature of the products formed during the fermentation of sugars. One group, called **homofermentative,** produces virtually a single fermentation product, lactic acid, whereas the other group, called **heterofermentative,** produces

*If heme (the parent ring structure of cytochromes) is added to the growth medium, certain lactic acid bacteria will make cytochromes of the *a* and *b* types. Cells containing these cytochromes are apparently capable of synthesizing ATP aerobically via electron transport phosphorylation.

other products (especially ethanol and CO_2) as well. Abbreviated pathways for the fermentation of glucose by a homo- and a hetero-fermentative organism are shown in Figure 19.60. The differences observed in the fermentation products are determined by the presence or absence of the enzyme **aldolase,** one of the key enzymes in glycolysis (see Figure 4.6). The heterofermenters, lacking aldolase, cannot break down hexose diphosphate to triose phosphate. Instead, they oxidize glucose-6-phosphate to 6-phosphogluconate and then

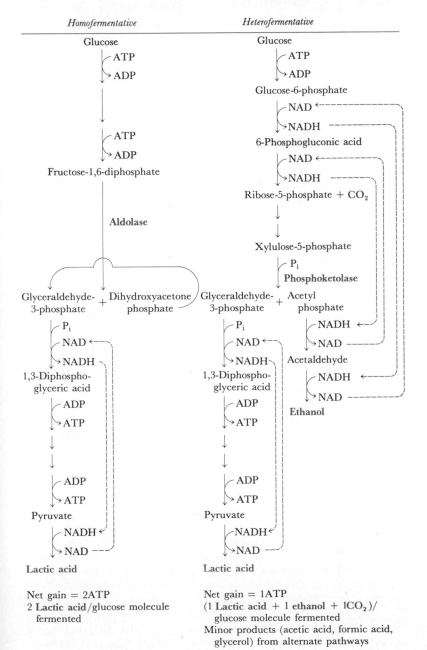

Homofermentative

Glucose

ATP
ADP

ATP
ADP

Fructose-1,6-diphosphate

Aldolase

Glyceraldehyde-3-phosphate + Dihydroxyacetone phosphate

P_i
NAD
NADH

1,3-Diphospho-glyceric acid

ADP
ATP

ADP
ATP

Pyruvate

NADH
NAD

Lactic acid

Net gain = 2ATP
2 **Lactic acid**/glucose molecule fermented

Heterofermentative

Glucose

ATP
ADP

Glucose-6-phosphate

NAD
NADH

6-Phosphogluconic acid

NAD
NADH

Ribose-5-phosphate + CO_2

Xylulose-5-phosphate

P_i
Phosphoketolase

Glyceraldehyde-3-phosphate + Acetyl phosphate

P_i NADH
NAD   NAD

1,3-Diphospho-   Acetaldehyde
glyceric acid   NADH
ADP   NAD
ATP

Ethanol

ADP
ATP

Pyruvate

NADH
NAD

Lactic acid

Net gain = 1ATP
(1 **Lactic acid** + 1 ethanol + 1CO_2)/ glucose molecule fermented
Minor products (acetic acid, formic acid, glycerol) from alternate pathways

FIGURE 19.60 The fermentation of glucose in homofermentative and heterofermentative lactic acid bacteria.

Genus	Cell Form and Arrangement	Fermentation	DNA (mol % G + C)
Streptococcus	Cocci in chains	Homofermentative	33–40
Leuconostoc	Cocci in chains	Heterofermentative	38–41
Pediococcus	Cocci in tetrads	Homofermentative	38
Lactobacillus	(1) Rods, usually in chains	Homofermentative	34–50
	(2) Rods, usually in chains	Heterofermentative	36–53

decarboxylate this to pentose phosphate, which is broken down to triose phosphate and acetylphosphate by means of the enzyme **phosphoketolase.** Triose phosphate is converted ultimately to lactic acid with the production of 1 mole of ATP, while the acetylphosphate accepts electrons from the NADH generated during the production of pentose phosphate and is thereby converted to ethanol without yielding ATP. Because of this, heterofermenters produce only 1 mole of ATP from glucose instead of 2 moles as are produced by homofermenters. This difference in ATP yield from glucose is reflected in the fact that homofermenters produce twice as much cell mass as heterofermenters from the same amount of glucose. Because the heterofermenters decarboxylate 6-phosphogluconate, they produce CO_2 as a fermentation product, whereas the homofermenters produce little or no CO_2; therefore one simple way of detecting a heterofermenter is to observe production of CO_2 in laboratory cultures. At the enzyme level, heterofermenters are characterized by the lack of aldolase and the presence of phosphoketolase. Many strains of heterofermenters can use O_2 as an electron acceptor with a flavoprotein serving as electron donor. In this reaction half of the NADH generated from the oxidation of glucose to ribose is transferred to a flavin and on to O_2. Acetylphosphate can then be converted to acetate instead of being reduced to ethanol, and an additional ATP is synthesized.

The various genera of the lactic acid bacteria have been defined on the basis of cell morphology and type of fermentative metabolism, as is shown in Table 19.32. Members of the genera *Streptococcus, Leuconostoc,* and *Pediococcus* have fairly similar DNA base ratio compositions; in addition, there is very little variation from strain to strain. The genus *Lactobacillus,* on the other hand, has members with widely diverse DNA compositions and hence does not constitute a homogeneous group.

Streptococcus and other cocci   The genus *Streptococcus* (Figure 19.61) contains a wide variety of members with quite distinct habitats, whose activities are of considerable practical importance to humans. Some members are pathogenic to people and animals. As producers of lactic acid, certain streptococci play important roles in the production of buttermilk, silage, and other fermented products.

The genus *Streptococcus* is subdivided into a number of groups of related species on the basis of a series of characteristics that are enumerated in Table 19.33. Hemolysis on blood agar is of consider-

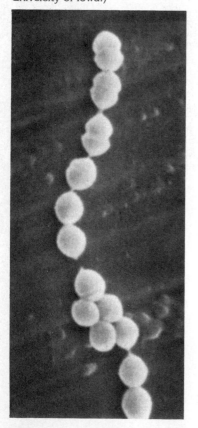

FIGURE 19.61 Scanning electron micrograph showing typical streptococcal morphology. Magnification, 9680×. (Courtesy of Bryan Larsen, University of Iowa.)

TABLE 19.33 Differential Characteristics of Streptococci

Group	Antigenic (Lancefield) Groups	Representative Species	Type of Hemolysis on Blood Agar	Good Growth at 10°C	Good Growth at 45°C
Pyogenes	A, B, C, F, G	*S. pyogenes*	Lysis (β)	−	−
Viridans	Not grouped	*S. mutans*	Greening (α)	−	+
Fecal (enterococci)	D	*S. faecalis*	Lysis (β), greening (α), or none	+	+
Lactic	N	*S. cremoris*	None	+	−

Group	Optimum Temp. (°C)	Survive 60°C for 30 Min	Growth in 6.5% NaCl Broth	Growth in Broth at pH 9.6	Growth in Milk with 0.1% Methylene Blue	Growth in Broth with 40% Bile	Habitat
Pyogenes	37	−	−	−	−	−	Respiratory tract, systemic
Viridans	37	−	−	−	−	−	Mouth, intestine
Fecal (enterococci)	35–37	+	+	+	+	+	Intestine, vagina, plants
Lactic	25	+	−	−	+	+	Plants, dairy products

able importance in the subdivision of the genus. Colonies of those strains producing streptolysin O or S are surrounded by a large zone of complete red blood cell hemolysis, a condition called β **hemolysis.** On the other hand, many streptococci that do not produce hemolysins cause the formation of a greenish or brownish zone around their colonies, which is due not to true hemolysis but to discoloration and loss of potassium from the red cells. This type of reaction has classically been referred to as α hemolysis. The streptococci are also divided into immunological groups based on the presence of specific carbohydrate antigens. These antigenic groups (or **Lancefield groups** as they are commonly known, named for Rebecca Lancefield, a pioneer in *Streptococcus* taxonomy), are designated by letters; A through O are currently recognized. Those β-hemolytic streptococci found in human beings usually contain the group A antigen, which is a cell-wall polymer containing N-acetylglucosamine and rhamnose. The fecal streptococci contain the group D antigen, a glycerol teichoic acid containing glucose side chains. Group B streptococci are usually found in association with animals and are a cause of mastitis in cows. Streptococci that are found in milk, the so-called lactic streptococci, are of antigen group N.

Placed in the genus *Leuconostoc* are cocci that are morphologically similar to streptococci but are heterofermentative. Strains of *Leuconostoc* also produce the flavoring ingredients diacetyl and acetoin by breakdown of citrate and have been used as starter cultures in dairy fermentations, but their place has now been taken by *S. diacetilactis*. Some strains of *Leuconostoc* produce large amounts of dextran polysaccharides (α-1,6-glucan) when cultured on sucrose (see Figure 5.11). Dextrans produced by *Leuconostoc* have found some medical use as plasma extenders in blood transfusions. Other strains of *Leuconostoc* produce fructose polymers called "levans."

Lactobacillus   In contrast to the streptococci, the genus *Lactobacillus* contains a more heterogeneous assemblage of organisms. Most species are homofermentative, but some are heterofermentative. The genus has been divided into three major subgroups (Table 19.34).

Lactobacilli are often found in dairy products, and some strains are used in the preparation of fermented products. For instance, *L. bulgaricus* is used in the preparation of yogurt, *L. acidophilus* in the production of acidophilus milk, and other species are involved in the production of sauerkraut, silage, and pickles. The lactobacilli are usually more resistant to acidic conditions than are the other lactic acid bacteria, being able to grow well at pH values around 5. Because of this, they can be selectively isolated from natural materials by use of carbohydrate-containing media of acid pH, such as tomato juice-peptone agar. The acid resistance of the lactobacilli enables them to continue growing during natural lactic fermentation when the pH value has dropped too low for the other lactic acid bacteria to grow, and the lactobacilli are therefore responsible for the final stages of lactic acid fermentations. The lactobacilli are rarely or never pathogenic.

TABLE 19.34 Characteristics of Subgroups in the Genus *Lactobacillus*

Characteristics	Species	DNA (mol % G + C)
Homofermentative:		
Lactic acid the major product (>85%) from glucose		
No gas from glucose; aldolase present		
(1) Grow at 45°C but not at 15°C; long rods; glycerol teichoic acid	*L. delbrueckii*	50
	L. leichmanii	50
	L. lactis, L. bulgaricus	50
	L. acidophilus	36
(2) Grow at 15°C, variable growth at 45°C; short rods and coryneforms; ribitol and glycerol teichoic acids	*L. casei, L. plantarum*	45–46
	L. curvatus	43
Heterofermentative:		
Produce about 50% lactic acid from glucose; produce CO_2 and ethanol; aldolase absent; phosphoketolase present; long and short rods; glycerol teichoic acid	*L. fermentum*	53
	L. cellobiosus	53
	L. brevis, L. buchneri	42–46

19.24

The Actinomycetes and Related Genera

An extremely large variety of bacteria falls under this heading, as evidenced by the entire volume of the new edition of *Bergey's Manual,* which is devoted to this group. There are considerable difficulties in drawing clear-cut distinctions between various genera. All of these organisms show a few common features: They are Gram-positive, rod-shaped to filamentous, and generally nonmotile in the vegetative phase (although motile stages are known). A continuum exists from simple rod-shaped organisms, to rod-shaped organisms which occasionally grow in a filamentous manner, to strictly filamentous forms. DNA analyses (mol % G + C) reflect a considerable diversity with values as low as 36 and as high as 78 reported. An attempt has been made in Table 19.35 to provide a complete overview of this group. In the following sections, we discuss some of the more interesting and important genera.

19.25

Coryneform Bacteria

The coryneform bacteria are Gram-positive, aerobic, nonmotile, rod-shaped organisms that have the characteristic of forming irregular-shaped, club-shaped, or V-shaped cell arrangements during normal growth. V-shaped cell groups arise as a result of a snapping movement that occurs just after cell division (called post-fission snapping movement or snapping division) (Figure 19.62). **Snapping division** has been shown to occur in one species because the cell wall consists of two layers; only the inner layer participates in cross-wall formation, so that after the cross-wall is formed, the two daughter cells remain attached by the outer layer of the cell wall. Localized rupture

TABLE 19.35 Actinomycetes and Related Genera (All Gram-positive)

Major Groups	DNA (mol % G + C)
Coryneform group of bacteria: Rods, often club-shaped, morphologically variable; not acid-fast or filamentous; snapping cell division	
Genus I. *Corynebacterium:* irregularly staining segments, sometimes granules; club-shaped swellings frequent	51–72
Genus II. *Arthrobacter:* coccus-rod morphogenesis	60–76
Genus III. *Cellulomonas:* coryneform morphology; cellulose digested; facultative aerobe	71–73
Genus IV. *Kurthia:* rods with rounded ends occurring in chains; coccoid later	36–38
Corynebacterium: animal and plant pathogens; also soil saprophytes	
Arthrobacter: soil organisms	
Propionic acid bacteria: anaerobic to aerotolerant; rods or filaments, branching	
Genus I. *Propionibacterium:* nonmotile; anaerobic to aerotolerant; produce propionic acid and acetic acid	65–67
Genus II. *Eubacterium:* obligate anaerobes; produce mixture of organic acids, including butyric, acetic, formic, and lactic	—
Propionibacterium: dairy products (Swiss cheese); skin, may be pathogenic	
Eubacterium: intestine, infections of soft tissue, soil; may be pathogenic; probably the predominant member of the intestinal flora	
Actinomycetes: filamentous, often branching; highly diverse; facultative aerobes	
Family I. Actinomycetaceae: not acid-alcohol-fast; facultatively aerobic; mycelium not formed; branching filaments may be produced; rod, coccoid, or coryneform cells	
Genus *Actinomyces:* anaerobic to facultatively aerobic; filamentous microcolony, but filaments transitory and fragment into coryneform cells; may be pathogenic for humans or animals; teeth	59–63
Genus *Bifidobacterium:* smooth microcolony, no filaments; coryneform and bifid cells common; found in intestinal tract of breast-fed infants.	59–66
Other genera: *Arachnia, Bacterionema, Rothia, Agromyces*	
Family II. Mycobacteriaceae: acid-alcohol fast, filaments transitory	
Genus *Mycobacterium* (only genus in family): pathogens, saprophytes; obligate aerobes; lipid content of cells and cell walls high; waxes, mycolic acids; simple nutrition; growth slow; tuberculosis, leprosy, granulomas; avian tuberculosis; also soil organisms; hydrocarbon oxidizers	63–70
Family III. Frankiaceae: nitrogen-fixing symbionts of plants; true mycelium produced	
Genus *Frankia:* forms nodules of two types on various plant roots; probably microaerophilic; grows slowly; fixes N_2	—
Family IV. Actinoplanaceae: true mycelium produced; spores formed, borne inside sporangia	
Genera *Actinoplanes, Streptosporangium*	69–71
Family V. Dermatophilaceae: mycelial filaments divide transversely, and in at least two longitudinal planes, to form masses of motile, coccoid elements; aerial mycelium absent; occasionally responsible for epidermal infections	
Genera *Dermatophilus, Geodermatophilus*	—
Family VI. Nocardiaceae: mycelial filaments commonly fragment to form coccoid or elongate elements; aerial spores occasionally produced; sometimes acid-alcohol-fast	
Genus *Nocardia:* common soil organisms; obligate aerobes; many hydrocarbon utilizers	61–72
Genus *Rhodococcus:* soil saprophytes, also common in gut of various insects	59–69
Family VII. Streptomycetaceae: mycelium remains intact, abundant aerial mycelium and long spore chains	
Genus *Streptomyces:* Nearly 500 recognized species, many produce antibiotics	69–75
Other genera (differentiated morphologically): *Streptoverticillium, Sporichthya, Microellobosporia, Kitasatoa, Chainia*	67–73
Family VIII. Micromonosporaceae: mycelium remains intact; spores formed singly, in pairs, or short chains; several thermophilic; saprophytes found in soil, rotting plant debris; one species produces endospores	
Genera: *Micromonospora, Thermoactinomyces, Thermomonospora*	54–79

of this outer layer on one side results in a bending of the two cells away from the ruptured side (Figure 19.63), and thus development of V-shaped forms.

The main genera of coryneform bacteria are *Corynebacterium* and *Arthrobacter*. The genus *Corynebacterium* consists of an extremely diverse group of bacteria, including animal and plant pathogens as well as saprophytes. The genus *Arthrobacter*, consisting primarily of soil organisms, is distinguished from *Corynebacterium* on the basis of a cycle of development in *Arthrobacter* involving conversion from rod to sphere and back again (Figure 19.64). However, some corynebacteria are pleomorphic and form coccoid elements during growth, so that the distinction between the two genera on the basis of life cycle is more quantitative than qualitative. The *Corynebacterium* cell frequently has a swollen end, so that it has a club-shaped appearance (hence the name of the genus: *koryne* is the Greek word for "club"), whereas *Arthrobacter* is less commonly club shaped.

Organisms of the genus *Arthrobacter* are among the most common of all soil bacteria. They are remarkably resistant to desiccation and starvation, despite the fact that they do not form spores or other resting cells. (The coccoid and rod-shaped forms seem equally resistant to desiccation and starvation.) Arthrobacters are a heterogeneous group that have considerable nutritional versatility, and strains have been isolated that decompose herbicides, caffeine, nicotine, and other unusual compounds.

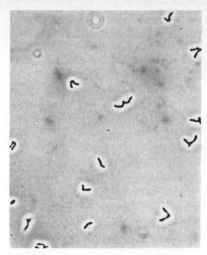

FIGURE 19.62 Photomicrograph of characteristic V-shaped cell groups in *Arthrobacter crystallopoietes,* resulting from snapping division. Magnification, 500×. (From Krulwich, T. A., and J. L. Pate. 1971. J. Bacteriol. 105:408–412.)

FIGURE 19.63 Electron micrograph of cell division in *Arthrobacter crystallopoietes,* illustrating how snapping division and V-shaped cell groups arise. Magnification, 36,720×. (a) Before rupture of the outer cell-wall layer. (b) After rupture of the outer layer on one side. (From Krulwich, T. A., and J. L. Pate. 1971. J. Bacteriol. 105:408–412.)

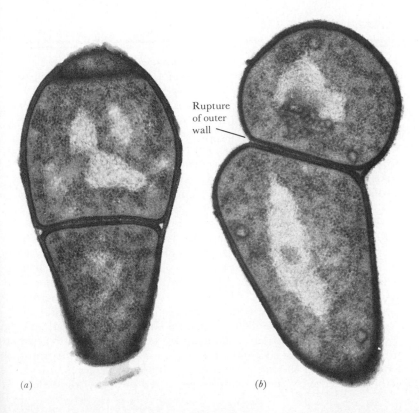

Rupture of outer wall

(a)   (b)

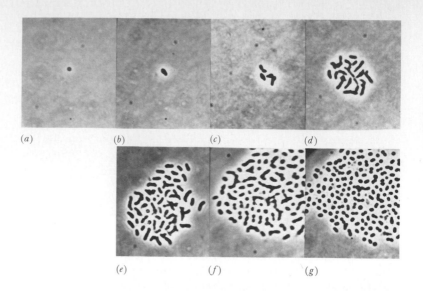

FIGURE 19.64 Stages in the life cycle of *Arthrobacter globiformis* as observed in slide culture; (a) single coccoid element; (b–e) conversion to rod and growth of microcolony consisting predominantly of rods; (f–g) conversion of rods to coccoid forms. Magnification, 1180×. (From Veldkamp, H., G. van den Berg, and L. P. T. M. Zevenhuizen. 1963. Antonie van Leeuwenhoek 29:35–51.)

(a) (b) (c) (d)

(e) (f) (g)

One way in which *Corynebacterium* and *Arthrobacter* differ markedly is in cell-wall structure. *Corynebacterium* has a cell wall generally containing arabinose-galactose polysaccharide complexes linked to the peptidoglycan, and in this way is similar to *Mycobacterium* (see Section 19.27) whereas *Arthrobacter* does not have these sugars in its cell wall.

19.26

Propionic Acid Bacteria

The propionic acid bacteria (genus *Propionibacterium*) were first discovered as inhabitants of Swiss (Emmentaler) cheese, where their fermentative production of CO_2 produces the characteristic holes; the presence of propionic acid is at least partly responsible for the unique flavor of the cheese. Although this acid is produced by some other bacteria, its production by the propionic acid bacteria is a distinguishing characteristic of the genus. The bacteria in this group are Gram-positive, pleomorphic, nonsporulating rods, nonmotile and anaerobic. They ferment lactic acid, carbohydrates, and polyhydroxy alcohols, producing propionic acid, succinic acid, acetic acid, and CO_2. Their nutritional requirements are complex, and they usually grow rather slowly. In some taxonomic schemes, the facultatively aerobic coryneforms are also classified as propionic acid bacteria.

The enzymatic reactions leading from glucose to propionic acid are of interest (Figure 19.65). The initial catabolism of glucose to pyruvate follows the Embden–Meyerhof pathway as in the lactic acid bacteria, but the NADH formed is reoxidized as one part of a cycle in which propionic acid is formed. Pyruvate accepts a carboxyl group from methylmalonyl-CoA by a transcarboxylase reaction, leading to the formation of oxalacetate and propionyl-CoA. The latter substance reacts with succinate in a step catalyzed by a CoA transferase, producing succinyl-CoA and propionate. The succinyl-CoA is then isomerized to methylmalonyl-CoA, and the cycle is complete.

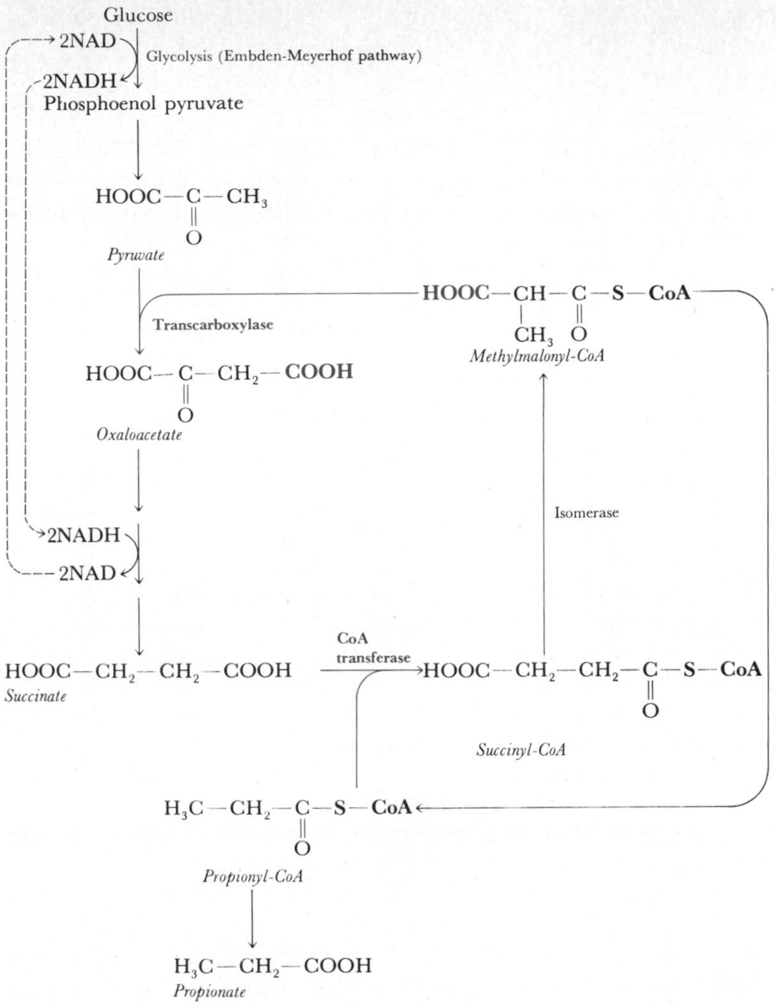

Glucose

2NAD

Glycolysis (Embden-Meyerhof pathway)

2NADH

Phosphoenol pyruvate

HOOC—C—CH$_3$
||
O

Pyruvate

Transcarboxylase

HOOC—CH—C—S—CoA
| ||
CH$_3$ O

Methylmalonyl-CoA

HOOC—C—CH$_2$—COOH
||
O

Oxaloacetate

Isomerase

2NADH

2NAD

CoA transferase

HOOC—CH$_2$—CH$_2$—COOH → HOOC—CH$_2$—CH$_2$—C—S—CoA
                                                    ||
                                                    O

Succinate

Succinyl-CoA

H$_3$C—CH$_2$—C—S—CoA ←
              ||
              O

Propionyl-CoA

H$_3$C—CH$_2$—COOH

Propionate

FIGURE 19.65 The formation of propionic acid by *Propionibacterium*.

Reoxidation of NADH occurs in the steps between oxalacetate and succinate, and the oxidation-reduction balance is restored.

Most propionic acid bacteria also ferment lactate with the production of propionate, acetate, and CO_2. The anaerobic fermentation of lactic acid to propionate is of interest because lactic acid itself is an end product of fermentation for many bacteria. The propionic acid bacteria are thus able to obtain energy anaerobically from a substance that other bacteria are producing.

It is the fermentation of lactate to propionate that is important in Swiss cheese manufacture. The starter culture consists of a mixture of homofermentative streptococci and lactobacilli, plus propionic acid bacteria. The initial fermentation of lactose to lactic acid during formation of the curd is carried out by the homofermentative organisms. After the curd (protein and fat) has been drained, the propionic acid bacteria develop rapidly and usually reach numbers of 10^8 per gram by the time the cheese is 2 months old. Swiss cheese "eyes" are formed by the accumulation of CO_2, the gas diffusing through the curd and gathering at weak points.

19.27

Mycobacterium

The genus *Mycobacterium* consists of rod-shaped organisms, which at some stage of their growth cycle possess the distinctive staining property called **acid-alcohol fastness.** This property is due to the presence on the surface of the mycobacterial cell of unique lipid components called **mycolic acids** and is found only in the genus *Mycobacterium*. First discovered by Robert Koch during his pioneering investigations on tuberculosis, this unique staining property permitted the identification in tuberculous lesions; it has subsequently proved to be of great taxonomic use in defining the genus *Mycobacterium*.

Acid-alcohol fastness   In the staining procedure (Ziehl-Neelsen stain) a mixture of the dye basic fuchsin and phenol is used in the primary staining procedure, the stain being driven into the cells by slow heating of the microscope slide to the steaming point for 2 to 3 minutes. The role of the phenol is to enhance penetration of the fuchsin into the lipids. After washing in distilled water, the preparation is decolorized with acid alcohol (3 percent HCl in 95 percent ethanol); the fuchsin dye is removed from other organisms, but is retained by the mycobacteria. After another wash in water, a final counterstain of methylene blue is used. Acid-alcohol-fast organisms on the final preparation appear red whereas the background and nonacid-alcohol-fast organisms appear blue.*

As noted, the key component necessary for acid-alcohol fastness is a unique lipid fraction of mycobacterial cells called mycolic acid. Mycolic acid is actually a group of complex branched-chain hydroxy lipids with the overall structure shown in Figure 19.66a. The carboxylic acid group of the mycolic acid must be free (unesterified), and it reacts on a one-to-one basis with the fuchsin dye (Figure 19.66b). The mycolic acid is complexed to the peptidoglycan of the mycobacterial wall, and this complex somehow prevents approach of the acid-alcohol solvent during the decolorization step. It was also first demonstrated by Koch that disruption of cellular integrity destroys the acid-alcohol-fast property; thus cellular integrity is a necessary prerequisite of this property.

The mycobacteria are not readily stained by the Gram method, because of the high surface lipid content, but if the lipoidal portion of the cell is removed with alkaline ethanol (1 percent KOH in absolute ethanol), the intact cell remaining is non-acid-alcohol fast but Gram-positive. Thus *Mycobacterium* can be considered to be Gram-positive. If the lipoidal portion is not removed, then the cells are resistant to decolorization by the Gram procedure even when stained

FIGURE 19.66   Structure of (a) mycolic acid and (b) basic fuchsin, the dye used in the acid-alcohol-fast stain. The fuchsin dye probably combines with the mycolic acid via ionic bonds between COO^- and NH_2^+.

(a)   Mycolic acid; R_1 and R_2 are long-chain aliphatic hydrocarbons

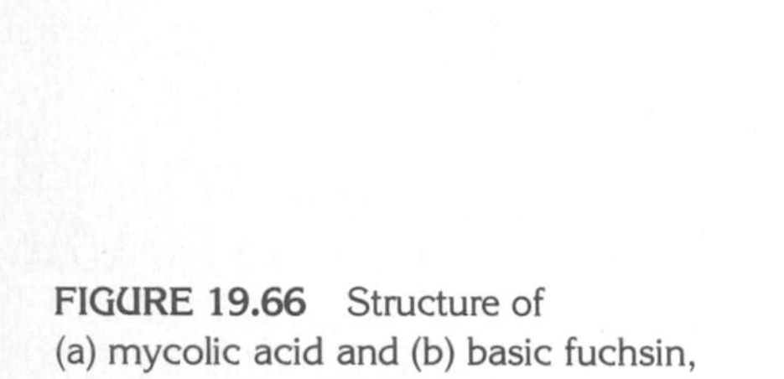

(b)   Basic fuchsin

*In some older acid-fast staining procedures, the acid was dissolved in water instead of ethanol. A number of corynebacteria and nocardias grown on media containing glycerol are resistant to decolorization with aqueous acid, but are decolorized by acid-alcohol. Thus, although the property is sometimes called acid fastness, acid-alcohol fastness is preferred, since it is more specific for mycobacteria.

with crystal violet alone, in the absence of iodine, whereas iodine is essential for the conventional Gram-staining procedure (Figure 2.2).

Stage of growth influences the acid-alcohol fastness of mycobacteria. In an actively growing population, many cells may appear nonacid-alcohol fast, whereas a higher proportion of stationary phase cells will exhibit this staining characteristic. Thus it is important to carry out the acid-alcohol procedure on fully grown cultures.

Characteristics of mycobacteria Mycobacteria are generally rather pleomorphic, and may undergo branching or filamentous growth. However, in contrast to the actinomycetes, filaments of the mycobacteria become fragmented into rods or coccoid elements upon slight disturbance; a true mycelium is not formed. In general, growth is slow or very slow, and easily visible colonies are produced from dilute inoculum only after days to weeks of incubation. (The reason Koch was successful in first isolating *M. tuberculosis* was that he waited long enough after inoculating media.) When growing on solid media, mycobacteria generally form tight, compact, often wrinkled colonies, the organisms piling up in a mass rather than spreading out over the surface of the agar (Figure 19.67a). This formation is probably due to the high lipid content and hydrophobic nature of the cell surface. The characteristic slow growth of most mycobacteria is probably also due, at least in part, to the hydrophobic character of the cell surface, which renders the cells strongly impermeable to nutrients; species having less lipid grow considerably more rapidly.

For the most part, mycobacteria have relatively simple nutritional requirements. Growth often occurs in simple mineral salts medium with ammonium as nitrogen source and glycerol or acetate as sole carbon source and electron donor. Growth of *M. tuberculosis* is stimulated by lipids and fatty acids, and egg yolk (a good source of lipids) is often added to culture media to achieve more luxuriant growth. A glycerol-whole egg medium (Lowenstein-Jensen medium) is often used in primary isolation of *M. tuberculosis* from pathological materials. Perhaps because of the high lipid content of its cell walls, *M. tuberculosis* is able to resist such chemical agents as alkali or phenol for considerable periods of time, and this property is used in the selective isolation of the organism from sputum and other materials that are grossly contaminated. The sputum is first treated with 1 N NaOH for 30 minutes, then neutralized and streaked onto isolation medium.

A characteristic of many mycobacteria is their ability to form yellow carotenoid pigments. Based on pigmentation, the mycobacteria can be divided into three groups: nonpigmented (including *M. tuberculosis, M. bovis*); forming pigment only when cultured in the light, a property called **photochromogenesis** (including *M. kansasii, M. marinum*); and forming pigment even when cultured in the dark, a property called **scotochromogenesis** (including *M. gordonae, M. paraffinicum*). The property of photochromogenesis is of some interest and has been extensively studied. This property is not unique to mycobacteria, as it also occurs in a number of fungi. Photoinduction of carotenoid formation involves short-wavelength (blue) light, and only occurs in the presence of O_2. The evidence indicates that the

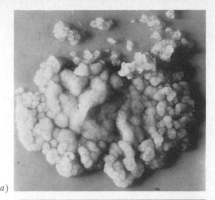

(a)

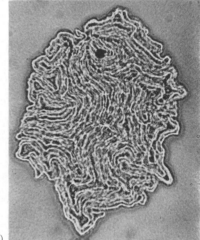

(b)

FIGURE 19.67 (a) Characteristic colony appearance of a mycobacterium culture, *M. tuberculosis,* showing the compact, wrinkled appearance of the colony. The colony is about 7 mm in diameter. (Courtesy of N. Rist, Pasteur Institute, Paris.) (b) A colony of virulent *M. tuberculosis* at an early stage, showing the characteristic cordlike growth. (From Lorian, V. 1968. Am. Rev. Resp. Dis. 97:1133–1135.)

critical event in photoinduction is a photooxidation event (see Section 8.6 for a discussion of photooxidation), and it appears that one of the early enzymes in carotenoid biosynthesis is photoinduced. As with other carotenoid-containing bacteria (see Section 8.6), it has been suggested that carotenoids protect mycobacteria against photooxidative damage involving singlet oxygen.

The cell walls of mycobacteria contain a peptidoglycan that is covalently bound to an arabinose-galactose-mycolic acid polymer, and it is this lipid-polysaccharide-peptidogylcan complex which confers the hydrophobic character to the mycobacterial cell surface. In addition to this lipid component, mycobacteria form a wide variety of other lipids, providing the chemist studying lipids with a fascinating amount of material.

Pathogenesis Virulence of mycobacteria is related in some way to the surface properties. A characteristic of *M. tuberculosis* infection, known since the time of Koch, is the induction of a delayed-type (cell-mediated) hypersensitivity (see Section 16.6). Injection of mycobacterial cells, or cell fractions from mycobacteria, together with an antigen of another source, strongly stimulates antibody formation against the antigen. Nonspecific materials that stimulate antibody formation against antigens are called **adjuvants,** and one of the best adjuvants (sometimes called *Freund's adjuvant*) is a mixture of dead *M. tuberculosis* cells in a water-in-oil emulsion. Because of the great clinical importance of adjuvants, considerable research has been done on the mechanism of action of mycobacterial adjuvants. There is now some evidence that the active fraction is the mycolic acid-arabinogalactan-peptidoglycan structure of the mycobacterial cell wall (which must be introduced in a water-in-oil emulsion to be effective). The mechanism of mycobacterial adjuvant action is probably related in some way to the ability of mycobacterial components to stimulate cell division of lymphocytes or other cells involved in the development of the immune response.

Despite the marked stimulation of the immune response by mycobacteria and their products, organisms such as *M. tuberculosis* can enter into long-term and potentially serious relationships with the host. Capable of surviving inside macrophages, *M. tuberculosis* thus sets up an intracellular infection (see Section 15.8) of long duration. Due to the slow growth of virulent *M. tuberculosis,* the infection may develop only over many years, and be very difficult to control. Although the prime site of infection of *M. tuberculosis* is the lungs, the organism can also cause infection of the bones, spleen, meninges, and skin. Initial infection is almost always via the respiratory route into the lungs; depending on virulence of the parasite and resistance of the host, the organism may cause a minor or extensive infection of lungs. Spread of the organism from lungs to other parts of the body is rare, and is greatly influenced by nonspecific host resistance and previous exposure to *M. tuberculosis.*

Virulence of *M. tuberculosis* cultures has been correlated with the formation of long cordlike structures (Figure 19.67) on agar or in liquid medium, due to side-to-side aggregation and intertwining

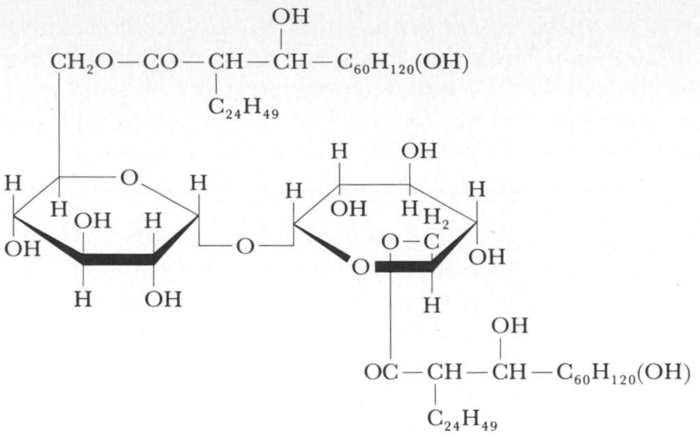

FIGURE 19.68 Structure of "cord factor," a mycobacterial glycolipid: 6,6'-dimycolyltrehalose.

of long chains of bacteria. Growth in cords reflects the presence on the cell surface of a characteristic lipid, the **cord factor,** which is a glycolipid (Figure 19.68).

Chemotherapy of tuberculosis has been a major factor in control of the disease and has resulted in the virtual elimination of the tuberculosis sanatoriums, which once were commonly found in the countryside. The initial success in chemotherapy occurred with the introduction of streptomycin, but the real revolution in tuberculosis treatment came with the discovery of **isonicotinic acid hydrazide** (isoniazid, INH), an agent virtually specific for mycobateria. This agent is not only effective and free from toxicity, but inexpensive and readily absorbed when given orally. Although the mode of action of INH is not completely understood it is known that it affects in some way the synthesis of mycolic acid. Treatment of organisms with very small amounts of INH (as little as 5 picomoles per 10^9 cells) results in complete inhibition of mycolic acid synthesis, and continued incubation results in a complete loss of outer membrane areas of the cell, a loss of cellular integrity, and death. Following treatment with INH, mycobacteria lose their acid-alcohol fastness, in keeping with the role of mycolic acid in this staining property (see above).

Other species of the genus *Mycobacterium* also are pathogenic for man and other animals. *M. bovis* enters humans via the intestinal tract typically from the ingestion of raw milk. A localized intestinal infection eventually spreads to the respiratory tract and is followed shortly by the classic symptoms of tuberculosis. *M. leprae* is the causative agent of the ancient disease leprosy. Unfortunately, *M. leprae* has never been grown on artificial media, and mice (where the typical human symptoms of leprosy are not observed) or armadillos (where symptoms are apparent) must be used as sources of the organism for further study.

Many mycobacteria are saprophytic and can be readily isolated from soil or water. Many of the saprophytic strains will grow using aliphatic hydrocarbons as sole carbon source, and enrichments using mineral oil, ethane, decane, or other hydrocarbons will often result in the isolation of some of these mycobacteria. The close relationship between the various species of mycobacteria is revealed by their DNA base compositions, most of which fall in a narrow range around 65 to 67 percent G + C. Clearly, this is a group of considerable evolutionary and ecological interest, and deserves much further work.

19.28

Actinomycetes

The actinomycetes are a large group of filamentous bacteria, usually Gram-positive, which form branching filaments. As a result of successful growth and branching, a ramifying network of filaments is formed, called a **mycelium** (Figure 19.69). Although it is of bacterial dimensions, the mycelium is in some ways analogous to the mycelium formed by the filamentous fungi. Most actinomycetes form spores; the manner of spore formation varies and is used in separating subgroups, as outlined in Table 19.34. The genus *Mycobacterium*, members of which often show a tendency to form branches, is also placed in the actinomycetes by some taxonomists. The DNA base compositions of all members of the actinomycetes fall within a relatively narrow range of 63 to 78 percent G + C. Organisms at the upper end of this range have the highest G + C percentage of any bacteria. In the present discussion we concentrate on the genus *Streptomyces*.

Streptomyces   *Streptomyces* is a genus represented by a large number of species and varieties. *Streptomyces* filaments are usually 0.5 to 1.0 μm in diameter and of indefinite length, and often lack cross walls in the vegetative phase. Growth occurs at the tips of the filaments and is often accompanied by branching so that the vegetative phase consists of a complex, tightly woven matrix, resulting in a compact convoluted colony. As the colony ages, characteristic aerial filaments (sporophores) are formed, which project above the surface of the colony and give rise to spores (Figure 19.70). *Streptomyces* spores, usually called **conidia,** are not related in any way to the endospores of *Bacillus* and *Clostridium* since the streptomycete spores are produced simply by the formation of cross walls in the multinucleate aerial filaments followed by separation of the individual cells directly into spores (Figure 19.71). The surface of the conidial wall often has convoluted projections, the nature of which is characteristic of each species. Differences in shape and arrangement of aerial filaments and spore-bearing structures of various species are among the fundamental features used in separating the *Streptomyces* groups (Figure 19.72). The conidia and aerial filaments are often pigmented and contribute a characteristic color to the mature colony; in addition, pigments sometimes are produced by the substrate mycelium and contribute to the final color of the colony. The dusty appearance of the mature colony, its compact nature, and its color make detection of *Streptomyces* colonies on agar plates relatively easy.

Ecology and isolation of Streptomyces   Although some streptomycetes can be found in fresh waters and a few inhabit the ocean, they are primarily soil organisms. In fact, the characteristic earthy odor of soil is caused by the production of a series of streptomycete metabolites called **geosmins.** These substances are sesquiterpenoid compounds, unsaturated ring compounds of carbon, oxygen, and hydro-

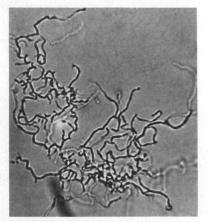

FIGURE 19.69 A young colony of an actinomycete, *Nocardia corallina,* showing typical filamentous cellular structure (mycelium). Magnification, 700×. (Courtesy of Hubert and Mary P. Lechevalier.)

(a)

Chain of conidia

Sporophores

(b)

FIGURE 19.70 Photomicrographs of several spore-bearing structures of actinomycetes. (a) *Streptomyces,* a monoverticillate type. Magnification, 270×. (Courtesy of Peter Hirsch.) (b) *Streptomyces,* a spiral type. Magnification, 680×. (Courtesy of Hubert and Mary P. Lechevalier.)

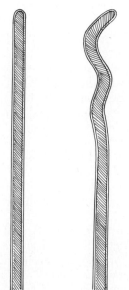
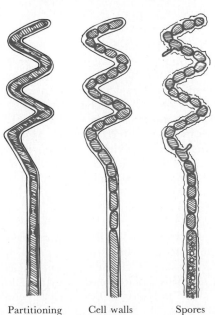

Growth phase   Tip curls   Partitioning of tip   Cell walls thicken and constrict   Spores mature

FIGURE 19.71 Diagram of stages in the conversion of a streptomycetes aerial hypha into spores (conidia). (From Wildermuth, H., and D. A. Hopwood. 1970. J. Gen. Microbiol. 60:51–58. Used by permission of Cambridge University Press.)

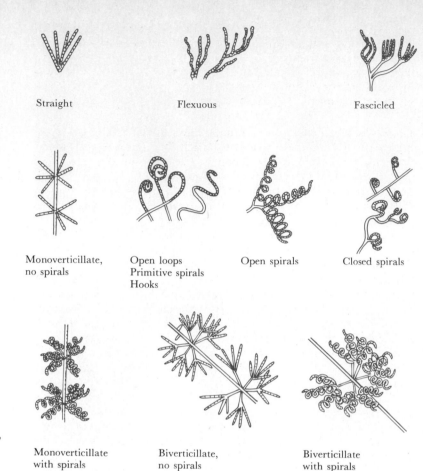

Straight   Flexuous   Fascicled

Monoverticillate,   Open loops   Open spirals   Closed spirals
no spirals   Primitive spirals
    Hooks

Monoverticillate   Biverticillate,   Biverticillate
with spirals   no spirals   with spirals

FIGURE 19.72 Various types of spore-bearing structures in the streptomycetes. (From Pridham, T. G., C. W. Hesseltine, and R. G. Benedict. 1958. Appl. Microbiol. 6:52–79.)

gen. The geosmin first discovered has the chemical name *trans*-1,10-dimethyl-*trans*-9-decalol (see also Section 19.2).

Alkaline and neutral soils are more favorable for the development of *Streptomyces* than are acid soils. Isolation of large numbers of *Streptomyces* from soil is relatively easy: a suspension of soil in sterile water is diluted and spread on selective agar medium, and the plates are incubated at 28 to 30°C. Media often selective for *Streptomyces* contain the usual inorganic salts to which starch, asparagine, or calcium malate is added as a carbon source and undigested casein as a nitrogen source. After incubation for 2 to 7 days the plates are examined for the presence of the characteristic *Streptomyces* colonies and spores of interesting colonies can be streaked and pure cultures isolated.

Nutritionally, the streptomycetes are quite versatile. Growth-factor requirements are rare, and a wide variety of carbon sources, such as sugars, alcohols, organic acids, and amino acids, can be utilized. Most isolates produce extracellular enzymes that permit utilization of polysaccharides (starch, cellulose, hemicellulose), proteins, and fats, and some strains can use hydrocarbons, lignin, tannin, or rubber. A single isolate may be able to break down over 50 distinct carbon sources. Streptomycetes are strict aerobes, whose growth in liquid culture is usually markedly stimulated by forced aeration. Sporulation usually takes place not in liquid culture but only when the organism is growing on the surface of agar or another solid substrate; it can occur, however, when organisms form a pellicle on the surface of an unshaken liquid culture.

Antibiotics of Streptomyces   Perhaps the most striking property of the streptomycetes is the extent to which they produce **antibiotics.** Evidence for antibiotic production is often seen on the agar plates used in the initial isolation of *Streptomyces:* adjacent colonies of other bacteria show zones of inhibition (Figure 19.73). In some studies close to 50 percent of all *Streptomyces* isolated have proved to be antibiotic producers. Because of the great economic and medical importance of many streptomycete antibiotics, an enormous amount of work has been done on these producers. Over 500 distinct antibiotic substances have been shown to be produced by streptomycetes, and a large number of these have been studied chemically. Some organisms produce more than one antibiotic, and often the several kinds produced by one organism are not even chemically related. The same antibiotic may be formed by different species found in widely scattered parts of the world. A change in nutrition of the organism may result in a change in the nature of the antibiotic produced. The organisms are usually resistant to their own antibiotics, but they may be sensitive to antibiotics produced by other streptomycetes.

More than 50 streptomycete antibiotics have found practical application in human and veterinary medicine, agriculture, and industry. Some of the more common antibiotics of *Streptomyces* origin are listed in Table 19.36. They are grouped into classes based on the chemical structure of the parent molecule. The search for new strep-

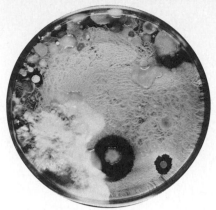

FIGURE 19.73 Antibiotic action of soil microorganisms on a crowded plate. The smaller colonies surrounded by inhibition zones are streptomycetes; the larger, spreading colonies are *Bacillus* sp. (Courtesy of Eli Lilly & Co.)

TABLE 19.36   Some Common Antibiotics Synthesized by Species of *Streptomyces*

Chemical Class	Common Name	Produced by	Active Against*
Aminoglycosides	Streptomycin	*S. griseus*	Most Gram negatives
	Spectinomycin	*Streptomyces spp.*	*M. tuberculosis,* penicillinase-producing *N. gonorrhoeae*
	Neomycin	*S. fradiae*	Broad spectrum, usually used in topical applications due to toxicity
Tetracyclines	Tetracycline	*S. aureofaciens*	Broad spectrum, Gram positives, Gram negatives, rickettsias and chlamydias, *Mycoplasma*
	Chlortetracycline	*S. aureofaciens*	As for tetracycline
Macrolides	Erythromycin	*S. erythreus*	Most Gram positives, frequently used in place of penicillin, *Legionella*
	Clindamycin	*S. lincolnensis*	Effective against obligate anaerobes, especially *Bacteroides fragilis*
Polyenes	Nystatin	*S. noursei*	Fungi, especially *Candida* infections
	Amphocetin B	*S. nodosus*	Fungi
None	Chloramphenicol	*S. venezuelae*	Broad spectrum; drug of choice for typhoid fever

*Most antibiotics are effective against several different bacteria. The entries in this column refer to the most frequent clinical application of a given antibiotic.

tomycete antibiotics continues, since many infectious diseases are still not adequately controlled by existing antibiotics. Also, the development of antibiotic-resistant strains requires the continual discovery of new agents. Despite the extensive work on antibiotic-producing streptomycetes, their ecological relationships are poorly understood.

19.29

Rickettsias

The rickettsias are small bacteria that have a strictly intracellular existence in vertebrates, usually in mammals, and are also associated at some point in their natural cycle with blood-sucking arthropods such as fleas, lice, or ticks. Rickettsias cause a variety of diseases in humans and animals, of which the most important are typhus fever, Rocky Mountain spotted fever, scrub typhus (tsutsu-gamushi disease), and Q fever. Rickettsias take their name from Howard Ricketts, a scientist of the University of Chicago, who first provided evidence for their existence and who died from infection with the rickettsia that causes typhus fever, *Rickettsia prowazekii*. Most rickettsias have not been unequivocally cultured in nonliving media and hence must be considered **obligate intracellular parasites,** although there is nothing so unusual about their physiology as to suggest that they cannot eventually be grown in vitro. (The causal agent of trench fever, *R. quintana,* has been cultured in cell-free medium.) Rickettsias have been cultivated in laboratory animals, lice, mammalian tissue cultures, and the yolk sac of chick embryos. In animals, growth takes place primarily in phagocytic cells.

The rickettsias are Gram-negative, coccoid or rod-shaped cells in the size range of 0.3 to 0.7 μm wide by 1 to 2 μm long. Electron micrographs of thin sections show profiles with a normal bacterial morphology (Figure 19.74); both cell wall and cell membrane are visible. The cell wall contains muramic acid and diaminopimelic acid. Both RNA and DNA are present, and the DNA is in the normal double-stranded form, with a G + C content varying from 30 to 33 percent in various species of the genus *Rickettsia* and 43 percent for *Coxiella burnetii* (the causal agent of Q fever). The rickettsias divide by normal binary fission, with doubling times of about 8 hours. The penetration of a host cell by a rickettsial cell is an active process, requiring both host and parasite to be alive and metabolically active. Once inside the phagocytic cell, the bacteria multiply primarily in the cytoplasm and continue replicating until the host cell is loaded with parasites (Figure 15.12), at which time the host cell bursts and liberates the bacteria into the surrounding fluid.

Much attention has been directed to the metabolic activities and biochemical pathways of rickettsias, in an attempt to explain why they are obligate intracellular parasites. Since biochemical studies must be done with large populations of cells, and since these populations can be obtained only by growing the parasites in animal cells, much effort has been expended on devising methods for purifying rickettsias and for separating them from any contaminating host tissues that might confuse metabolic or biochemical studies. Many rickettsias possess a highly distinctive energy metabolism, being able to oxidize only one amino acid, glutamate, and being unable to

FIGURE 19.74 Electron micrograph of cells of *Rickettsiella melolonthae* within a blood cell of its host, the beetle *Melolontha melolontha*. Notice that the bacteria are growing within a vacuole within the host cell. Magnification, 25,200×. (From Devauchelle, G., G. Meynadier, and C. Vago. 1972. J. Ultrastructure Res. 38:134–148.)

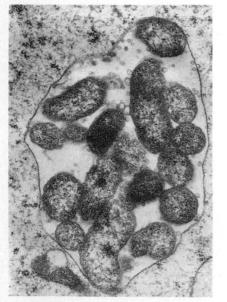

oxidize glucose, glucose-6-phosphate, or organic acids. However, *Coxiella burnetii* is able to utilize both glucose and pyruvate as electron donors. Rickettsias possess a complete cytochrome system and are able to carry out electron-transport phosphorylation, using as the electron donor NADH. They are also able to synthesize at least some of the small molecules needed for macromolecular synthesis and growth, while they obtain the rest of their nutrients from the host cell. There is some suggestion that the host also provides some key coenzymes, such as NAD and CoA. Such large coenzymes do not usually penetrate readily into bacteria that live independently, and there is evidence that the rickettsial membrane is looser and more "leaky" than those of other bacteria. For a fastidious organism capable of infecting other cells rich in nutrients and cofactors, such a permeable membrane would be advantageous. A summary of the biochemical properties of rickettsias is given in Table 19.37.

If the membranes of the rickettsias are indeed unusually leaky, the organisms may die quickly when out of their hosts, and this may explain why they must be transmitted from animal to animal by arthropod vectors. When the arthropod obtains a blood meal from an infected vertebrate, rickettsias present in the blood are inoculated directly into the arthropod, where they penetrate to the epithelial cells of the gastrointestinal tract, multiply, and appear later in the feces. When the arthropod feeds upon an uninfected individual, it then transmits the rickettsias either directly with its mouthparts or by contaminating the bite with its feces. However, the causal agent of Q fever, *C. burnetii,* can also be transmitted to the respiratory system by aerosols. *Coxiella burnetii* is the most resistant of the rickettsias to physical damage, which probably explains its ability to survive in air.

The relationship of the rickettsias to other bacteria is unknown,

TABLE 19.37 Comparison of Biochemical Properties of Rickettsias, Chlamydias, and Viruses

Property	Rickettsias	Chlamydias	Viruses
Structural:			
Nucleic acid	RNA and DNA	RNA and DNA	Either RNA or DNA, never both
Ribosomes	Present	Present	Absent
Cell wall	Muramic acid, DAP	Muramic acid, D-alanine	No wall
Structural integrity during multiplication	Maintained	Maintained	Lost
Biosynthetic:			
Macromolecular synthesis	Carried out	Carried out	Only with use of host machinery
ATP-generating system	Present	Absent	Absent
Sensitivity to antibacterial antibiotics	Sensitive	Sensitive	Resistant

although it is frequently postulated that rickettsias evolved from other bacteria by progressive loss of function as a result of mutation, a process called "degenerate evolution." Although this would explain the obligate intracellular habitat, it would not indicate how rickettsias are able to survive and grow within phagocytic cells, since this environment is hostile to most other bacteria. Another point is that the rickettsias we currently know are those which cause pathological changes, but many arthropods possess other intracellular rickettsia-like organisms that do not seem to be related to any disease either in the arthropod or in an alternate host and could be intracellular symbionts. Such intracellular symbionts may be, like rickettsias, representatives of a much larger group of intracellular bacteria that arose from free-living forms by degenerate evolution. It does not necessarily follow, that all rickettsias arose from the same free-living bacterium. The evolutionary steps may have occurred independently in many quite unrelated free-living bacteria.

19.30

The Psittacosis Group: Chlamydias

The psittacosis group, which comprises the genus *Chlamydia*, probably represents a further stage in degenerate evolution from that discussed above for the rickettsias, since the chlamydias are obligate parasites in which there has been an even greater loss of metabolic function. In fact, for many years the chlamydias were considered to be large viruses rather than bacteria, and only since the nature of virus replication has been well understood has the bacterial nature of the chlamydias been firmly established. Many chlamydias are smaller than some of the true viruses, such as the smallpox virus, but the chlamydias divide by binary fission, and do not replicate in the manner of viruses. Of the diseases caused by chlamydias, **psittacosis,** an epidemic disease of birds that is transmitted occasionally to humans, is one of the most important. In addition, **trachoma,** a debilitating disease of the eye characterized by vascularization and scarring of the cornea is caused by *C. trachomitis*. Other strains of *C. trachomitis* infect the genitourinary tract, and chlamydial infections are one of the leading sexually transmitted diseases in society today (see Section 17.11). Trachoma is the leading cause of blindness in humans.

It is not as disease entities that the chlamydias are of most interest, however; they are intriguing because of the biological and evolutionary problems they pose. The bacterial nature of the chlamydias was first suspected when it was discovered that, unlike viruses, they were susceptible to penicillin and other antibiotics whose action is restricted to the bacteria. When the specificity of penicillin for the bacterial cell wall was understood, the apparent viral nature of the chlamydias was further refuted. Biochemical studies showed that the chlamydias have typical bacterial cell walls, and they have both DNA and RNA. Electron microscopy of thin sections of infected cells shows forms that clearly are undergoing binary fission. The biosynthetic capacities of the chlamydias are very restricted, however, even more so than the rickettsias. This raised the interesting question of the limits to which evolutionary loss of function can be pushed while independence of macromolecular function is still retained.

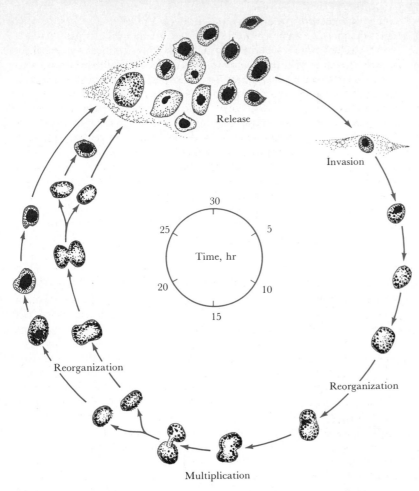

FIGURE 19.75 Cell-division cycle of a member of the psittacosis group. (From Moulder, J. W. 1964. The psittacosis group as bacteria. John Wiley & Sons, Inc., New York.)

Release

Invasion

30

25   5

Time, hr

20   10

15

Reorganization

Reorganization

Multiplication

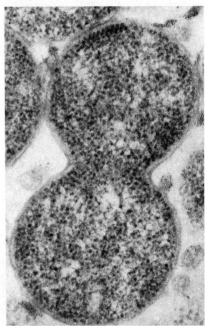

FIGURE 19.76 Electron micrograph of a thin section of a dividing cell of *Chlamydia psittaci,* a member of the psittacosis group, within a cell of a mouse tissue culture. Magnification, 128,800×. (Courtesy of Robert R. Friis.)

Although no convincing evidence exists that all the chlamydias are closely related, they are currently classified in a single genus; but the name *chlamydia* should be viewed more as a convenience than as a taxonomic entity.

The life cycle of a typical member of the genus *Chlamydia* is shown in Figure 19.75. Two cellular types are seen in a typical life cycle: a small, dense cell, which is relatively resistant to drying and is the means of dispersal of the agent, and a larger, less dense cell, which divides by binary fission and is the vegetative form. Unlike the rickettsias just discussed above, the chlamydias are not transmitted by arthropods but are primarily airborne invaders of the respiratory system—hence the significance of resistance to drying of the small cells. When a virus infects a cell, it loses its structural integrity and liberates nucleic acid. When a chlamydia enters a cell, however, although it changes form, it remains a structural unit and enlarges and begins to undergo binary fission. A dividing form of a chlamydia is seen in Figure 19.76. After a number of divisions, the vegetative cells are converted into small, dense cells that are released when the host cell disintegrates and can then infect other cells. Generation times of 2 to 3 hours have been reported, which are considerably faster than those found for the rickettsias.

As we noted, chlamydia cells have a chemical composition similar to that of other bacteria. Both RNA and DNA are present. The DNA content of a chlamydia corresponds to a molecular weight of approximately 4×10^8, about twice that of vaccinia virus and one-tenth that of *Escherichia coli*. The base composition is about 29 percent G + C. At least some of the RNA is in the form of ribosomes, and, like those of other procaryotes, the ribosomes are 70S particles composed of one 50S and one 30S unit.

The small, dense chlamydial form has a typical rigid bacterial cell wall, and muramic acid has been shown to be present. When the small form reorganizes to the large form it loses its cell-wall rigidity, and there is some evidence that the cross-linking of the peptidoglycan is reduced in extent.

The metabolic properties of chlamydias purified from infected cells have been studied by methods similar to those used with the rickettsias. The biosynthetic capacities of the chlamydias are much more limited than are those of the rickettsias (see Table 19.37). Although macromolecular syntheses occur in the chlamydias, no energy-generating system is present and the cells obtain their ATP from the host. They are able to oxidize glucose to pentose by means of enzymes of the pentose phosphate pathway, but they do not obtain energy from this oxidation. Chlamydias synthesize their own folic acid, D-alanine, lysine, and probably a number of other small molecules. One interesting feature of chlamydias is that their proteins are deficient in the amino acids arginine and histidine, which are found in most other organisms as well as in viruses. In keeping with this lack, neither amino acid is required for multiplication. From the limited biosynthetic capacities of chlamydias it is easy to see why they are obligate parasites. Their inability to manufacture ATP means that at the least they are energy parasites of their hosts; probably they also obtain from these hosts a variety of other co-enzymes, as well as low-molecular-weight building blocks. The chlamydias have the simplest biochemical abilities known among cellular organisms.

19.31

Mycoplasmas

The mycoplasmas are organisms without cell walls that do not revert to walled organisms. They are probably the smallest organisms capable of autonomous growth and are of special evolutionary interest because of their extremely simple cell structure.

The lack of cell walls in the mycoplasmas has been proved by electron microscopy and by chemical analysis, the latter showing that the key wall components, muramic acid and diaminopimelic acid, are missing. In Chapter 2 we discussed protoplasts and showed how these structures can be formed when cell-wall-digesting enzymes act on cells that are in an osmotically protected medium, and that when the osmotic stabilizer is removed, protoplasts take up water, swell, and burst. The mycoplasmas resemble protoplasts in their lack of a cell wall, but they are more resistant to osmotic lysis and are able to

TABLE 19.38 Cell-wall-less Bacteria

Genus	Properties	DNA (mol % G + C)
Mycoplasma	Many pathogenic; require sterols; facultative aerobes	24–40
Acholeplasma	Do not require sterols; facultative aerobes	29–35
Anaeroplasma	May or may not require sterols; obligate anaerobes; degrade starch, producing acetic, lactic, and formic acids plus ethanol and CO_2; inhibited by thallium acetate	29 (sterol-requiring strains) 40 (non-sterol-requiring strains)
Spiroplasma	Spiral to corkscrew-shaped cells; associated with various phyto-pathogenic (plant disease) conditions	25–31
Ureaplasma	Coccoid cell; occasional clusters and short chains; growth optimal at pH 6; strong urease reaction; associated with certain urinary tract infections in humans; inhibited by thallium acetate	26–29
Thermoplasma	Thermophilic, acidophilic; found in heated coal refuse sites; no sterol requirement; DNA contains histones	46
Methanoplasma	Methanogen	—

survive conditions under which protoplasts lyse. This ability to resist osmotic lysis is at least partially determined by the nature of the mycoplasma cell membrane, which is more stable than that of other procaryotes. In one group of mycoplasmas, the membrane contains **sterols** that seem to be responsible for stability, whereas in other mycoplasmas, carotenoids or other compounds may be involved instead. Those mycoplasmas possessing sterols in their membrane do not synthesize them but require them preformed in the culture medium. This sterol requirement is a current basis for separating some of the mycoplasmas into two genera, *Mycoplasma,* which requires sterols, and *Acholeplasma,* which does not. Members of a third genus, *Thermoplasma,* also do not require sterols but are distinguished from acholeplasmas by the fact that thermoplasmas are thermophilic. The only species described, *T. acidophilum,* has a temperature optimum around 55°C and a pH optimum around 2. The current taxonomic status of the cell-wall-less bacteria is shown in Table 19.38.

Mycoplasma cells are usually fairly small, and they are highly pleomorphic, a consequence of their lack of rigidity. A single culture may exhibit small coccoid elements, larger, swollen forms, and fila-mentous forms of variable lengths, often highly branched (Figure 19.77). It is from the production of filamentous, funguslike forms that the name *Mycoplasma* (*myco* means "fungus") derives. A common growth form is seen in cultures that divide by budding; division occurs with the cells remaining either directly attached or connected by thin hyphae (Figure 19.78).

FIGURE 19.77 Electron micrograph of a metal-shadowed preparation of *Mycoplasma mycoides.* Note the coccoid and hyphalike elements. Magnification, 8600×. (Courtesy of Alan Rodwell.)

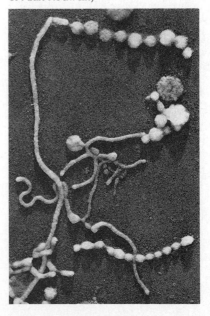

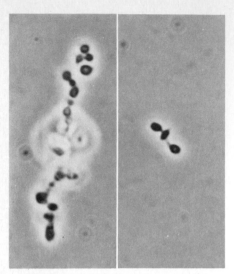

FIGURE 19.78 Photomicrograph by phase-contrast microscopy of a mycoplasma culture, showing typical cell arrangement. Magnification, 1200×.

FIGURE 19.79 Typical "fried-egg" appearance of mycoplasma colonies on agar. The colonies are around 1 mm in diameter.

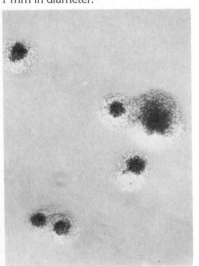

The small coccoid elements (0.2 to 0.3 μm in size) are the smallest mycoplasma units capable of independent growth. Because of flexibility due to lack of a cell wall, mycoplasma cells pass through filters with pore sizes smaller than the true diameter of the cells, and this has led to erroneous estimates of the minimum cell size capable of growth. Cellular elements of diameters close to 0.1 μm exist in mycoplasma cultures, but these are not capable of growth. Even so, the minimum reproductive unit of 0.2 to 0.3 μm probably represents the smallest free-living cell. Additionally, the genome size of mycoplasmas is also smaller than that of most procaryotes, between 4×10^8 and 1×10^9 in molecular weight, which is comparable to that of the obligately parasitic chlamydia and rickettsia, and about one-fifth to one-half that of *Escherichia coli*.

The mode of growth differs in liquid and agar cultures. On agar there is a tendency for the organisms to grow so that they become embedded in the medium, and the fibrous nature of the agar gel seems to affect the division process, perhaps by promoting separation of units from the growing mass. Colonies of mycoplasmas on agar exhibit a characteristic "fried-egg" appearance because of the formation of a dense central core, which penetrates downward into the agar, surrounded by a circular spreading area that is lighter in color (Figure 19.79). Growth of mycoplasmas is not inhibited by penicillin, cycloserine, or other antibiotics that inhibit cell wall synthesis, but the organisms are as sensitive as other bacteria are to antibiotics that act on targets other than the cell wall. Use is made of the natural penicillin resistance of mycoplasmas in preparing selective media for their isolation from natural materials. The culture media used for the growth of most mycoplasmas have usually been quite complex. Growth is poor or absent even in complex yeast extract-peptone-beef heart infusion media unless fresh serum or ascitic fluid is added. The main constituents provided by these two adjuncts are unsaturated fatty acids and sterols. Some mycoplasmas can be cultivated on relatively simple media, however, and synthetic media have been developed for some strains. Most mycoplasmas use carbohydrates as energy sources, and they require a range of vitamins, amino acids, purines, and pyrimidines. The energy metabolism of mycoplasmas is not unique. Some species are oxidative, possessing the cytochrome system and making ATP by electron-transport phosphorylation. Other species resemble the lactic acid bacteria in being strictly fermentative, producing energy by substrate-level phosphorylation and yielding lactic acid as the final product of sugar fermentation. Members of the genus *Anaeroplasma* are obligate anaerobes which ferment glucose or starch to a variety of acidic products. *Methanoplasma* is a methanogenic bacterium which lacks a cell wall.

Once appropriate culture media and isolation techniques were devised, it was possible to show that mycoplasmas are widespread in nature. Some strains can be routinely isolated from normal and pathological material of warmblooded animals, while other strains can be isolated from plants, insects, sewage, soil, compost, and other natural materials. One of the most effective means of isolating mycoplasmas is to pass the material through a filter small enough to hold back most other types of bacteria, and then to streak out the filtrate on a complex agar medium containing serum and penicillin. Plates are incubated for 1 to 2 weeks and examined with a dissecting microscope for colonies with the typical fried-egg appearance. For some purposes, preliminary filtration may not be desirable, and the material can be streaked directly onto plates of the selective agar medium. Another inhibitor frequently added to isolation media is thallium acetate (usually added at 250 μg/ml), a substance to which mycoplasmas are more resistant than are other organisms.

Many isolates are saprophytes or harmless commensals, but quite a few are pathogenic. Bovine contagious pleuropneumonia, caused by *M. mycoides,* has been a serious problem to the cattle industry throughout the world (the mycoplasma group used to be referred to as the PPLO group, for *p*leuro*p*neumonia-*l*ike *o*rganisms). Another important disease is agalactia of sheep and goats, a disease that primarily affects the mammary glands. In humans, mycoplasmas have been implicated in pneumonia, urethritis, kidney infections, and infections of the oral cavity.

The genus *Spiroplasma* consists of pleomorphic cells, spherical or slightly ovoid, which can become helical or branched nonhelical filaments. Although they lack a cell wall and flagella, they are motile by means of a rotary (screw) motion or a slow undulation. Despite the absence of a cell wall, *Spiroplasma* is reported to be Gram-positive. The organism has been isolated from the leaves of citrus plants, where it causes a disease called "citrus stubborn disease" and from corn plants suffering from "corn stunt disease." A number of other mycoplasma-like bodies have been detected in diseased plants by electron microscopy, which indicates that there may be a large group of plant-associated mycoplasmas. In addition, spiroplasmas have been found in the gut of ticks and honeybees, and are associated (apparently without causing disease) with the surfaces of various flowers.

19.32

New Frontiers

The student experiencing a first course in microbiology is often overwhelmed by what at first seems to be an endless list of different bacteria. If organisms are thought of not as individuals, but as members of various groups unified by a peculiar physiological or morphological property, it is much easier to keep the scope of microbial diversity within bounds. By no means, however, should the material in this chapter be interpreted to mean that all procaryotes have been discovered and characterized. On the contrary, most microbiologists would agree that only a small percentage of the organisms of the procaryotic world have ever been grown in the laboratory.

That this is the case can be illustrated by a perusal of the recent microbiological literature, where every month several new bacteria are formally described and characterized as new species or even new genera. As examples, Figures 19.80*a* and *b* and 19.81 show photo- or electron micrographs of three rather unusual bacteria isolated within the last few years. The square bacterium (Figure

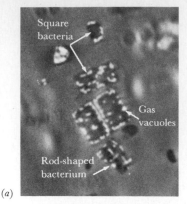

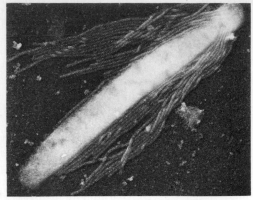

(a)

(b)

FIGURE 19.80 Photo and electron micrographs of recently discovered unusual procaryotes. (a) A square bacterium found in hypersaline pools in the Sinai region of the Middle East. The cells measure anywhere from 1.5 to 11 μm on a side and are just 0.2–0.5 μm thick. The light areas in the cell are gas vesicles. Magnification, 2000×. (b) A rod-shaped bacterium containing a tuft of very thick flagella isolated from an African snail. Magnification, 18,700×. (Photograph a courtesy of A. E. Walsby. Photograph b from Cole, R. M., C. S. Richards, and T. J. Pophin. 1977. J. Bacteriol. 132:950–966.)

19.80a), which also contains gas vesicles, was found growing in hypersaline pools in the Sinai region of the Middle East (see Figure 2.5 for an electron micrograph of the square bacterium); the long rod containing a tuft of sheathed, polar flagella (Figure 19.80b) was found in the tissues of the African snail; and magnetic bacteria, which contain particles of magnetite (Fe_3O_4) that act as internal compasses to orient the cells in a particular direction (Figure 19.81) have been found in both freshwater and marine sediments. Hence many procaryotes are out there just waiting to be discovered—the procaryotic world is vast indeed!

19.33

Summary: Evolution of Microbial Diversity

We have learned in this chapter, and throughout this book, of the enormous diversity of microorganisms. How did such diversity arise? It should be clear by now that the broad morphological, physiological, and ecological characteristics of the various groups of microorganisms are ultimately controlled by the genetic constitution of the organisms, interacting with the environments of which they are a part. Thus microbial diversity arises as a result of mutation and genetic recombination processes, working in an environment which constantly selects. Microbial diversity reflects the diversity of habitats on earth suitable for life.

The age of the earth is about $4\frac{1}{2}$ billion years, and there is good fossil evidence that microbial life has existed for at least the past 3 billion years. In rocks younger than 2 billion years, the variety and morphological diversity of fossil microorganisms is considerable. Figure 19.82 shows some photomicrographs of extremely thin sections

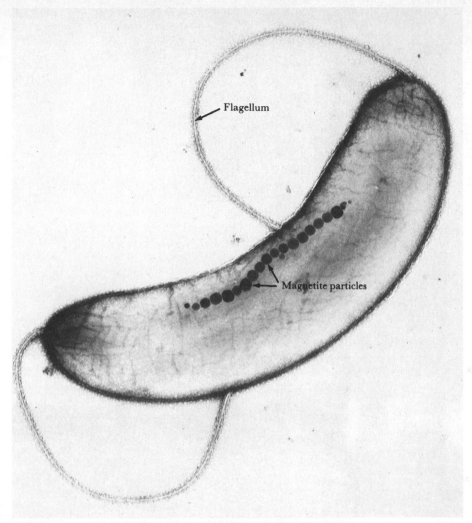

Flagellum

Magnetite particles

FIGURE 19.81 Negatively stained electron micrograph of a magnetotactic bacterium. This bacterium contains particles of Fe_3O_4 (magnetite) arranged in a chain; the particles serve to align the cell along geomagnetic lines. The organism was isolated from a water treatment plant in Durham, New Hampshire. Magnification, 41,600×. (Micrograph courtesy of R. Blakemore and N. Blakemore. From Blakemore, R. P., and R. B. Frankel. 1981. Scientific American. 245:58–65.)

of rocks containing structures remarkably similar to certain modern filamentous bacteria. Structures such as those in Figure 19.82a are surprisingly similar to certain of the filamentous green photosynthetic bacteria (Figure 19.6) or the cyanobacterium *Oscillatoria* (Figure 19.10d), and provide strong evidence that the procaryotes as a group had evolved an impressive cellular diversity longer than one billion years ago. More recent evidence pushes this back even further, since organic microstructures thought to be cellular organisms have been

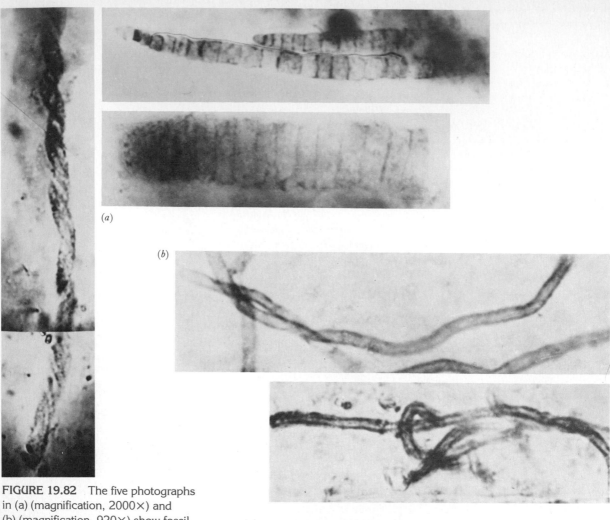

(a)

(b)

FIGURE 19.82 The five photographs in (a) (magnification, 2000×) and (b) (magnification, 920×) show fossil procaryotic microorganisms found in the Bitter Springs formation, a rock formation in central Australia about a billion years old. These forms bear a striking resemblance to modern cyanobacteria or their colorless counterparts. The two photographs in (c) (magnification, 2000×) show fossils possibly of a eucaryotic alga. The cellular structure is remarkably similar to that of certain modern chlorophyta, such as *Chlorella* sp. These are from the same rock formation as are the procaryotic organisms. (From Schopf, J. W. 1968. J. Paleontol. Pt. I 42:651–688.)

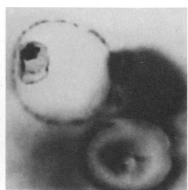

(c)

discovered in rocks from Australia known to be more than $3\frac{1}{2}$ billion years old!

Primitive environments   The original environment of the earth was devoid of significant amounts of O_2 and hence was a reducing environment. A variety of gases and other key biological elements existed at that time, including CH_4, CO_2, CO, N_2, NH_3, H_2O, and H_2S. It is now well established that the synthesis of biologically important molecules can occur if reducing atmospheres containing such gases are subjected to intense energy sources. Of the energy sources available on the primitive earth, the most important was ultraviolet (UV) radiation from the sun, but lightning discharges, radioactivity, and thermal energy from volcanic activity were also available. If gaseous mixtures resembling those thought to be present on the primitive earth are irradiated with UV or subjected to electric discharge, a wide variety of biochemically important molecules can be made, such as sugars, amino acids, purines, pyrimidines, nucleotides, and fatty acids. It has also been shown that under prebiological conditions some of these biochemical building blocks could have polymerized, leading to the formation of polypeptides, polynucleotides, and the like. We can therefore imagine that on the primitive earth a rich mixture of organic compounds accumulated, but in the absence of living organisms these compounds would have been stable and should have persisted for countless years. Thus, with time, there should have been an extensive accumulation of organic materials, and the stage was set for biological evolution.

Primitive organisms   A possible course of biological evolution is illustrated in Figure 19.83. From an organic soup of small molecules and macromolecules to a primitive living organism is a giant step. There are two basic features that primitive organisms must have had: (1) **metabolism,** that is, the ability to accumulate, convert, and transform nutrients and energy; and (2) a **hereditary mechanism,** that is, the ability to replicate and produce offspring. Both of these features require the development of a cellular structure. Such structures probably arose through the spontaneous coming together of lipid and protein molecules to form membranous structures, within which were trapped polynucleotides, polypeptides, and other substances. This step may have occurred countless times to no effect, but just once the proper set of constituents could have become associated, and a primitive organism arose. This primordial organism was probably structurally simple (that is, procaryotic) and would have found itself surrounded by a rich supply of organic materials usable as nutrients for energy metabolism and growth.

The first primitive organism was probably anaerobic and heterotrophic. Its closest living relative is probably a nonsporulating, anaerobic organism without a cell wall, most likely resembling a mycoplasma. Such an organism was probably simple, in the sense of possessing very few enzymes. No cytochrome or photosynthetic system would have been present, no cell wall, no flagella, no ability to sporulate, no citric acid cycle. It would have had a variety of growth-factor requirements. As time went on, however, growth and

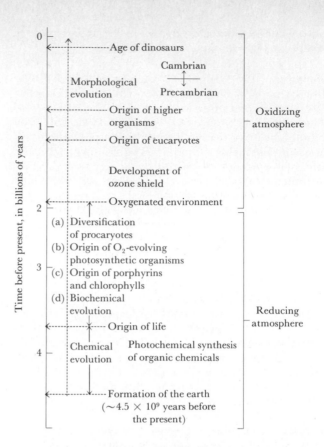

FIGURE 19.83 Possible course of biological evolution. The positions of the various stages are only approximate.

reproduction of this organism would have depleted the organic nutrients in its environment. Mutation and selection could then have resulted in the appearance of new organisms with greater biosynthetic capacities, which hence were better adapted to the changing environment. To make survival in osmotically varying environments possible, a cell wall probably appeared next. A key step would have been the development of the first cytochrome-like porphyrin since this could have led to the construction of an electron-transport particle able to carry on electron-transport phosphorylation, therefore opening up the possibility of utilizing a variety of energy sources not available from substrate-level phosphorylation. In an environment in which preformed organic compounds were being depleted, this development no doubt provided a distinct selective advantage. Since conditions were still anaerobic, an electron acceptor other than oxygen was needed.

One possible electron acceptor is CO_2, present in primordial gases, which would have been reduced to CH_4, and could have permitted evolution of methanogenic bacteria, a group still widespread today in anaerobic environments. Another possible anaerobic electron acceptor is sulfate, a substance synthesized chemically by reaction of H_2S and ozone, which could have permitted the evolution of organisms such as the sulfate-reducing bacteria. Although the methane bacteria lack cytochromes and use other unique electron carriers, the sulfate-reducing bacteria do possess a primitive type of cytochrome called cytochrome c_3.

Once iron-containing porphyrins had evolved, the next step might conceivably have been substitution of magnesium for iron, producing chlorophyll. With photosynthesis then possible, a great explosion of life could have occurred because of the availability of the enormous amount of energy from the visible light of the sun. The first photosynthetic organism was no doubt anaerobic, probably using light only for ATP synthesis and using reduced compounds from its environment—such as H_2S—as sources of reducing power. Such an organism may have resembled *Chlorobium, Chromatium,* or one of the other photosynthetic bacteria.

The next step probably involved development of the second light reaction of green-plant photosynthesis, making it possible to use the plentiful supply of H_2O as an electron donor. Since both ATP and reduced pyridine nucleotides (NADH, NADPH) could now be made photosynthetically, light energy could be used more efficiently. Clearly such an organism would have considerable competitive advantage over other photosynthetic organisms. The first of this type was probably similar to one of the present-day cyanobacteria, some of which can grow completely anaerobically, although they produce O_2 photosynthetically. The evolution of an oxygen-producing photosynthetic apparatus had enormous consequences for the environment of the earth since, as O_2 accumulated, the atmosphere changed from a reducing to an oxidizing type. With O_2 available as an electron acceptor, aerobic organisms evolved; these were able to obtain much more energy from the oxidation of organic compounds than could anaerobes (see Chapter 4). More energy was made available, and higher population densities could develop, increasing the chances for the appearance of new types of organisms. There is good evidence from the fossil record that, at about the time that the earth's atmosphere became oxidizing, there was an enormous burst in the rate of evolution, leading to the appearance of eucaryotic microorganisms and from them to higher animals and plants.

Another major consequence of the appearance of O_2 was that it is the major source of **ozone** (O_3), a substance that provides a barrier preventing the intense ultraviolet radiation of the sun from reaching the earth. When O_2 is subject to short-wavelength ultraviolet radiation, it is converted to O_3, which strongly absorbs the wavelengths up to 300 nm. Until this ozone shield developed, evolution could only have continued in environments protected from direct radiation from the sun, such as under rocks or in the deeper parts of the oceans. After the photosynthetic production of O_2 and development of an ozone shield, organisms could have ranged generally over the surface of the earth, permitting evolution of a greater diversity of living organisms.

Origin of the eucaryotes As we have emphasized frequently in this book, there are vast differences between procaryotic and eucaryotic organisms, which override their fundamental similarities at the biochemical level. Because of the increased structural complexity of eucaryotes, coupled with the evidence from the fossil record, it seems reasonable to propose that eucaryotes arose later than, and presumably from, procaryotic ancestors. It seems likely that the eucaryotic nucleus and mitotic apparatus arose as a necessity for ensuring the replication and orderly partitioning of DNA once the genome size had increased to the point where replication as one molecule was no longer feasible. The widespread occurrence of plasmids in bacteria suggests that even in procaryotes there are evolutionary advantages for segregation of genetic information into more than one DNA molecule. It is possible to imagine how separate chromosomes might have arisen in a procaryote from plasmids and become segregated within the cell into a membrane-enclosed nucleus. Probably spindle fibers and the mitotic apparatus would also have had to evolve at the same time. There is no obvious reason why this primitive eucaryote would have needed other typical eucaryotic organelles, and these could have arisen later.

Many students of microbial evolution believe that the modern eucaryotic cell evolved in steps through the incorporation into a primitive eucaryote of heterotrophic and phototrophic symbionts. This theory, referred to as the **endosymbiotic theory** of eucaryotic evolution, has, through the years, gathered increasing experimental support (see also Section 3.3 and supplementary readings for Chapter 3). The theory postulates that an aerobic bacterium established residency within the cytoplasm of a primitive eucaryote and supplied the eucaryotic partner with energy in exchange for a stable, protected environment, and a ready supply of nutrients. This aerobic bacterium would represent the forerunner of the present eucaryotic mitochondrion. In similar fashion, the endosymbiotic uptake of a cyanobacterium would have made the primitive eucaryote photosynthetic and no longer dependent on organic compounds for energy production. The phototrophic endosymbiont would then be considered the forerunner of the chloroplast. Studies of the nucleic acids and ribosomes of eucaryotic cells and their organelles have built an impressive case for the theory of endosymbiosis. Small amounts of DNA exist within mitochondria and chloroplasts and this DNA is arranged in a closed circular fashion typical of procaryotes. The ribosomes of mitochondria and chloroplasts are of bacterial size (70S) instead of the 80S ribosomes observed in the cytoplasm of eucaryotic cells. Antibiotics that specifically inhibit protein synthesis by 70S ribosomes inhibit the activity of mitochondrial and chloroplast ribosomes as well, but do not affect the larger ribosomes of the eucaryotic cytoplasm. Finally, molecular sequencing data have shown that the 16S ribosomal RNA of mitochondrial and chloroplast ribosomes contains nucleotide sequences typical of certain bacterial and cyanobacterial 16S RNA (see Section 18.2). These sequences are not related to the corresponding nucleotide sequences in the equivalent RNA species (18S) from eucaryotic cytoplasmic ribosomes. Whether the endosymbiotic theory is correct or not, it is clear that by some process or another, a division of labor was developed within the primitive eucaryotic cell, which was beneficial and made further evolution possible.

However, in order to evolve to the eucaryotic stage, a cell necessarily had to sacrifice certain procaryotic features such as genetic plasticity, structural simplicity, and the ability to adapt rapidly to new environments. Also, the greater complexity of the eucaryotic cell meant that it would face difficulty in adapting to life in extreme environments, such as thermal areas, where the procaryote is pre-eminent. For these reasons the evolution of the eucaryote did not toll the death bell for the procaryote. Both types of cells continued to evolve, serving as the forerunners of the various species we know today.

A point that deserves emphasis here is that none of the organisms living today are primitive. They are all modern organisms, well adapted to, and successful in, their ecological niches. Certain of these organisms may indeed be very similar to primitive organisms and may represent stems of the evolutionary tree that have not changed for millions of years; in this respect they are related to primitive organisms, but they are not themselves primitive. Further-more, we must distinguish between primitive organisms and simple organisms, that is, those whose cell structures and biochemical potentialities are uncomplicated. The latter may represent organisms that evolved late and became simple through adaptive processes which selected for the reduction and loss of properties possessed by their ancestors. The best example of this is probably the psittacosis group, which consists of obligate intracellular parasites with very limited biochemical potentialities (see Section 19.30). These organisms grow only within the cells of vertebrates, and hence could not have evolved until the appropriate host existed. Once they were established within the vertebrate cell, certain functions would no longer have been necessary and conceivably could have been lost through mutation, without affecting the success of the organisms.

Archaebacteria   Biologists for many years have accepted the assertion that life has evolved around two different cellular plans, eucaryotic and procaryotic. The picture that is emerging from molecular sequencing studies using ribosomal RNAs shows that three groups of cellular organisms may have evolved from the ancestral cell, two of these groups being procaryotic, and the remaining one eucaryotic. The two procaryotic groups have been designated the **eubacteria,** consisting of the majority of Gram-positive and Gram-negative bacteria plus the cyanobacteria, and the **archaebacteria,** a group of unusual procaryotes including the methanogenic bacteria, the extremely halophilic bacteria, and certain thermoacidophilic bacteria.

Archaebacteria as a group share little in terms of their basic physiology; methanogens are strict anaerobes, halophiles such as *Halobacterium* are strict aerobes and require high concentrations of NaCl for growth, and thermoacidophiles such as *Sulfolobus* inhabit highly acidic thermal habitats. Considering these facts, how can it be that these organisms are related? Based on the 16S ribosomal RNA sequencing studies mentioned earlier (see Section 18.2), these organisms form a group which share related 16S ribosomal RNA sequences—sequences that differ drastically from those of other pro-caryotes. It should be noted, however, that the archaebacteria them-selves are likely to represent a rather broad group. This is shown by the fact that 16S ribosomal RNA sequence relationships between certain archaebacteria are no stronger than between archaebacteria and eubacteria (for example, *Sulfolobus* and *Thermoplasma,* both thermoacidophilic archaebacteria, are no more closely related via 16S ribosomal RNA sequence criteria than *Methanobrevibacter,* an archaebacterium, is to *Bacillus,* a eubacterium). Nevertheless, many microbiologists favor the idea of three cellular kingdoms, and *Bergey's Manual* now deals with archaebacteria as a unit despite this diversified physiology.

Archaebacteria are undeniably procaryotic cells since they lack a nucleus and the cellular organelles of eucaryotes and are of the same microscopic dimensions as the eubacteria. In addition, archaebacterial ribosomes are the same size as eubacterial ribosomes (70S; but with vastly different ribosomal RNA sequences). Archaebacteria differ from eubacteria in the chemistry of their membrane lipids [archaebacterial lipids are ether (C—O—C)-linked while

$$\overset{\text{O}}{\overset{\|}{}}$$

eubacterial and eucaryotic lipids are ester (C—O—)-linked], in the fact that they do not contain peptidoglycan in their cell walls (archaebacterial cell walls consist of protein, glycoprotein, or polysaccharide, while all eubacteria that possess cell walls contain peptidoglycan), in their spectrum of antibiotic sensitivity (most eubacteria are sensitive to chloramphenicol or kanamycin while archaebacteria are not), and in certain details of the translational (protein synthesizing) machinery. In the latter connection it has been found that the amino acid carried by the initiator tRNA in archaebacteria is methionine instead of formylmethionine. Curiously, methionine also serves an initiator function in the eucaryotic translational process. Study of the secondary structure of the 5S ribosomal RNA from archaebacteria and the 5S counterpart in eucaryotes also shows strong structural similarities.

Thus it appears that archaebacteria are a group distinct from the eubacteria and eucaryotes, but which have certain features in common with both groups. Although the name *archaebacteria* connotes an ancient origin, it has been by no means proven that these organisms originated before the eubacteria. Answers to such phylogenetic questions hopefully will come from further molecular sequencing studies and a better understanding of bacterial physiology, biochemistry, and genetics. It is hoped that future studies of archaebacteria and eubacteria will be able to fill the gaps in the geological time scale shown in Figure 19.83, and ultimately help us understand how the procaryotic world diversified through the course of microbial evolution.

Supplementary Readings
General References

Buchanan, R. E., and N. E. Gibbons, eds. 1974. Bergey's manual of determinative bacteriology, 8th ed. Williams & Wilkins, Baltimore. Complete coverage of all recognized genera of bacteria, exclusive of cyanobacteria. A good place to begin a literature survey on a specific group. The 1st edition of *Bergey's Manual of Systematic Bacteriology* will be published in four volumes, the first of which was published in 1984. The remaining volumes will appear at approximately two-year intervals. The cyanobacteria will be covered in the new *Bergey's Manual* (Volume 3).

Carlile, M. J., J. F. Collins, and B. E. B. Moseley (eds.). 1981. Molecular and cellular aspects of microbial evolution. Soc. Gen. Microbiol. Symp. Volume 32. Cambridge University Press, New York. A series of articles dealing with various aspects of microbial evolution.

Fox, G. E., E. Stackebrandt, R. B. Hespell, J. Gibson, J. Maniloff, T. A. Dyer, R. S. Wolfe, W. E. Balch, R. S. Tanner, L. H. Magium, L. B. Zablen, R. Blakemore, R. Gupta, L. Bonen, B. J. Lewis., D. A. Stahl, K. R. Luehrsen, K. N. Chen, and C. R. Woese. 1980. The phylogeny of prokaryotes. Science 209:457–463. A proposal for a bacterial phylogeny based on 16S rRNA sequence analysis.

Starr, M. P., H. Stolp, H. G. Trüper, A. Balows, and H. G. Schlegel. 1981. The prokaryotes—a handbook on habitats, isolation, and identification of bacteria. Published in two volumes, 2284 pages, 169 chapters, Springer-Verlag, New York. The most complete reference on the characteristics of bacteria. Also includes formulation of various media and isolation procedures for virtually every procaryotic group known. Extremely well illustrated.

Woese, C. R. 1981. Archaebacteria. Sci. Am. 244:98–122. An overview of ribosomal RNA sequencing studies and how they have changed our view of cellular evolution. Highly recommended.

Phototrophic Bacteria

Pfennig, N. 1978. General physiology and ecology of photosynthetic bacteria. Pages 3–18 in Clayton, R. K., and W. R. Sistrom, eds. The photosynthetic bacteria. Plenum Press, New York. A general overview of the physiology and ecology of photosynthetic bacteria.

Stanier, R. Y., N. Pfennig, and H. G. Trüper. 1981. Introduction to the phototrophic prokaryotes. Pages 197–211 in The prokaryotes (see Starr et al., General References section).

Trüper, H. G., and N. Pfennig. 1981. Characterization and identification of the anoxygenic phototropic bacteria. Pages 299–312 in The prokaryotes (see Starr et al., General References section).

Van Niel, C. B. 1971. Techniques for the enrichment, isolation, and maintenance of the photosynthetic bacteria. Methods Enzymol. 23:3–28. An excellent review of methods for culturing these organisms, plus a broad overview of the ecology of the group. Highly recommended.

Cyanobacteria

Rippka, R., J. D. Waterbury, and R. Y. Stanier. 1981. Isolation and purification of cyanobacteria: some principles. Pages 212–220 in The Prokaryotes (see Starr et al., General References section).

Rippka, R., J. D. Waterbury, and R. Y. Stanier. 1981. Provisional generic assignments for cyanobacteria in pure cultures. Pages 247–258 in The prokaryotes (see Starr et al., General References section). A simplified taxonomy of the cyanobacteria.

Stanier, R. Y., and G. Cohen-Bazire. 1977. Phototrophic prokaryotes: the cyanobacteria. Annu. Rev. Microbiol. 31:225–274. An extensive review of the properties of cyanobacteria, with emphasis on evolutionary relationships and biochemical diversity.

Walsby, A. E. 1981. Cyanobacteria: Planktonic gas-vacuolate forms. Pages 224–235 in The prokaryotes (see Starr et al., General References section). Descriptions of the common gas vacuolate cyanobacteria and how to culture them.

Gliding and Sheathed Bacteria

Burchard, R. P. 1981. Gliding motility of prokaryotes: ultrastructure, physiology, and genetics. Annu. Rev. Microbiol. 35:497–529. A detailed review of the characteristics of gliding motility and gliding bacteria.

Mulder, E. G., and **M. H. Deinema.** 1981. The sheathed bacteria. Pages 425–440 *in* The prokaryotes (see Starr et al., General References section). Covers the general properties of this group.

Reichenbach, H. 1981. Taxonomy of the gliding bacteria. Annu. Rev. Microbiol. 35:339–364. A short review giving an overview of the gliding bacteria.

Reichenbach, H., and **M. Dworkin.** 1981. Introduction to the gliding bacteria. Pages 315–327 *in* The prokaryotes (see Starr et al., General References section). A well-illustrated introduction to the biology of the gliding bacteria.

Stalked, Budding, and Appendaged (Prosthecate) Bacteria

Moore, R. L. 1981. The biology of *Hyphomicrobium* and other prosthecate, budding bacteria. Annu. Rev. Microbiol. 35:567–594. A short review concentrating on morphological and biochemical events occurring during budding and growth of *Hyphomicrobium*.

Poindexter, J. S. 1964. Biological properties and classification of the *Caulobacter* group. Bacteriol. Rev. 28:231–295. The major work on the caulobacters and the starting point for all subsequent work.

Staley, J. T., **P. Hirsch,** and **J. M. Schmidt.** 1981. Introduction to the budding and/or appendaged bacteria. Pages 451–455 *in* The prokaryotes (see Starr et al., General References section). A concise overview of the budding/prosthecate group.

Spirochetes, Spiral, and Curved Bacteria

Canale-Parola, E. 1977. Physiology and evolution of spirochetes. Bacteriol. Rev. 41:181–204. Excellent general review of the biology of spirochetes, with emphasis on the nonpathogenic forms.

Canale-Parola, E. 1978. Motility and chemotaxis of spirochetes. Annu. Rev. Microbiol. 32:69–99. A detailed review which considers the unique motility mechanisms of the spirochetes.

Holt, S. C. 1978. Anatomy and chemistry of spirochetes. Microbiol. Rev. 42:114–160. A useful review of the structure of the spirochete wall and axial fibrils, with many electron micrographs.

Krieg, N. R. 1981. The genera *Spirillum, Aquaspirillum* and *Oceanospirillum*. Pages 595–608 *in* The prokaryotes (see Starr et al., General References section). A short review of the nonpathogenic spirilla.

Lennette, E. H., **A. Balows, W. J. Hausler, Jr.,** and **J. P. Truant.** 1980. Manual of clinical microbiology, 3rd ed. American Society for Microbiology, Washington, D.C. Chapters 32 through 34 of this text deal with laboratory study of the pathogenic spirochetes.

Free-Living, Nitrogen-Fixing Bacteria

Becking, J. H. 1981. The family Azotobacteraceae. Pages 795–817 *in* The prokaryotes (see Starr et al., General References section). A review of the taxonomy of the heterotrophic, aerobic nitrogen fixers, with many excellent photomicrographs.

Gordon, J. K. 1981. Introduction to the nitrogen-fixing prokaryotes. Pages 782–794 *in* The prokaryotes (see Starr et al., General References section). A basic introduction to the nitrogen-fixing bacteria.

Rhizobium-Agrobacterium Group

Lippincott, J. A., **B. B. Lippincott,** and **M. P. Starr.** 1981. The genus *Agrobacterium*. Pages 842–855 *in* The prokaryotes (see Starr et al., General References section). An overview of the biology of this interesting group of plant pathogens.

Thomashow, M. F., C. G. Panagopoulou, M. P. Gordon, and **E. W. Nester.** 1980. Host range of *Agrobacterium tumefaciens* is determined by the Ti plasmid. Nature (Lond.) 283:794–796. Research paper showing that the tumor induction (Ti) plasmid dictates which host plants are susceptible to tumorigenesis.

Haber, C. L., L. N. Allen, S. Zhao, and **R. S. Hanson,** 1983. Methylotrophic bacteria: biochemical diversity and genetics. Science. 221: 1147–1153. An overview of the genetics of methylotrophs with emphasis on recombinant DNA techniques that can be used for constructing industrially and environmental useful methylotrophs.

Hanson, R. S. 1980. Ecology and diversity of methylotrophic organisms. Adv. Appl. Microbiol. 26:3–39. An overview of methylotrophs and of methods for measuring their activities in nature.

Higgins, I. J., D. J. Best, R. C. Hammond, and **D. Scott.** 1981. Methane-oxidizing microorganisms. Microbiol. Rev. 45:556–590. A detailed consideration of the physiology and biochemistry of methylotrophs.

Baumann, P., and **L. Baumann.** 1981. The marine Gram-negative eubacteria: genera *Photobacterium, Beneckea, Alteromonas,* and *Alcaligenes.* Pages 1302–1351 *in* The prokaryotes (see Starr et al., General References section). A thorough treatment of the taxonomy of marine Gram-negative rods and vibrios.

Hastings, J. W., and **K. H. Nealson.** 1981. The symbiotic, luminous bacteria. Pages 1332–1345 *in* The prokaryotes (see Starr et al., General References section). A nice summary of the habitats of luminescent bacteria and techniques for isolating these interesting prokaryotes.

Laanbroek, H. J., and **N. Pfennig.** 1981. Oxidation of short chain fatty acids by sulfate-reducing bacteria in freshwater and in marine sediments. Arch. Microbiol. 128:330–335. A research paper dealing with the physiology and ecology of sulfate reducers.

Pfennig, N., F. Widdel, and **H. G. Trüper.** 1981. The dissimilatory sulfate-reducing bacteria. Pages 926–947 *in* The prokaryotes (see Starr et al., General References section). A good summary of the taxonomy of sulfate-reducing bacteria.

Thauer, R. K., K. Jungermann, and **K. Decker.** 1977. Energy conservation in chemotrophic anaerobic bacteria. Bacteriol. Rev. 41:100–180. Contains a good section on biochemistry of sulfate-reducing bacteria.

Bøvre, K., and **N. Hagen.** 1981. The Family Neisseriaceae: rod-shaped species of the genera *Moraxella, Acinetobacter, Kingella,* and *Neisseria,* and the *Branhamella* group of cocci. Pages 1506–1529 *in* The prokaryotes (see Starr et al., General References section).

Kelley, D. P. 1981. Introduction to the chemolithotrophic bacteria. Pages 997–1004 *in* The prokaryotes (see Starr et al., General References section). A nice overview of lithotrophic metabolism.

Smith, A. J., and **D. S. Hoare.** 1977. Specialist phototrophs, lithotrophs, and methylotrophs: a unity amongst a diversity of procaryotes? Bacteriol. Rev. 41:419–448. This review compares and contrasts lithotrophs with photosynthetic and methylotrophic bacteria.

Methane-Oxidizing Bacteria

Vibrio and Related Bacteria

Sulfate-Reducing Bacteria

Gram-Negative Cocci

Lithotrophs

Methane-producing Bacteria

Balch, W. E., G. E. Fox, L. J. Magrum, C. R. Woese, and **R. S. Wolfe.** 1979. Methanogens: reevaluation of a unique biological group. Microbiol. Rev. 43:260–296. The new taxonomy of the methanogenic bacteria.

Smith, M. R., S. H. Zinder, and **R. A. Mah.** 1980. Microbiol methanogenesis from acetate. Process. Biochem. 15:34–39. A short review emphasizing the biochemistry and energetics of acetate conversion to methane.

Wolfe, R. S., and **I. L. J. Higgins.** 1979. Microbial biochemistry of methane: a study in contrasts. Int. Rev. Biochem. 21:267–353. An excellent review of the physiology and biochemistry of methanogens.

Zehnder, A. J. B. 1978. Ecology of methane formation. *In* **R. Mitchell,** ed. Water pollution microbiology, vol. II. John Wiley, New York. Emphasizes energetics and nutritional relationships of ecological significance.

Zeikus, J. G. 1980. Chemical and fuel production by anaerobic bacteria. Annu. Rev. Microbiol. 34:423–464. A detailed review article covering the production of methane, liquid fuels, and biomass by microorganisms.

Zeikus, J. G. 1983. Metabolism of one-carbon compounds by chemotrophic anaerobes. Adv. Microbial Physiol. 24:215–289. An overview of the physiology and biochemistry of methanogenesis and of organisms that utilize one-carbon compounds. Emphasizes the methanogens.

Lactic Acid Bacteria

London, J. 1976. The ecological and taxonomic status of the lactobacilli. Annu. Rev. Microbiol. 30:279–301. Good review of the ecology, physiology, and genetics of the lactobacilli.

Actinomycetes

Lechevalier, H. A., and **M. P. Lechevalier.** 1981. Introduction to the order Actinomycetales. Pages 1915–1922 *in* The prokaryotes (see Starr et al., General References section). A summary of the genera included in the Actinomycetes group.

Mycobacterium

Barksdale, L., and **K. S. Kim.** 1977. *Mycobacterium.* Bacteriol. Rev. 41:217–372. An extensive and extremely valuable review on the basic biology, as well as the pathogenesis, of the mycobacteria. Extensively illustrated. The bibliography contains over 1,300 citations.

Chlamydia and Rickettsia

Baca, O. G., and **D. Paretsky.** 1983. Q-fever and *Coxiella burnettii:* a model for host-parasite interactions. Microbiol. Rev. 47:127–149. An excellent review of a typical rickettsial organism. Considers the biological and biochemical aspects of the organism as well as the pathology and immunology of the disease Q-fever.

Page, L. A. 1981. Obligately intracellular bacteria: the genus *Chlamydia.* Pages 2210–2222 *in* The prokaryotes (see Starr et al., General References section).

Shachter, J., and **H. D. Caldwell.** 1980. Chlamydiae. Annu. Rev. Microbiol. 34:285–309. A consideration of the major chlamydial species.

Mycoplasma

Masover, G., and **L. Hayflick.** 1981. The genera *Mycoplasma, Ureaplasma,* and *Acholeplasma,* and associated organisms (Thermoplasmas and Anaeroplasmas). Pages 2245–2270 *in* The prokaryotes (see Starr et al., General References section). A detailed consideration of the major groups of cell wall-less bacteria.

Whitcomb, R. F. 1981. The biology of Spiroplasmas. Annu. Rev. Entomol. 26:397–425. Review focusing on the diseases caused by spiroplasmas.

Blakemore, R. P. 1982. Magnetotactic bacteria. Annu. Rev. Microbiol. 36:217–238. A short review describing some aspects of the ecology, physiology, and fascinating discovery of these unusual procaryotes.

Kessel, M., and **Y. Cohen.** 1982. Ultrastructure of square bacteria from a brine pool in southern Sinai. J. Bacteriol. 150:851–860. Electron microscopy of the square bacteria.

Walsby, A. E. 1980. A square-bacterium. Nature (Lond.) 283:69–71. First description of extremely thin, square-shaped procaryotes.

Yayanos, A. A., A. S. Dietz, and **R. VanBoxtel.** 1981. Obligately barophilic bacterium from the Mariana Trench. Proc. Natl. Acad. Sci. USA 78:5212–5215. Describes the basic properties of procaryotes that grow only at elevated pressures.

Energy Calculations

Appendix

1

Definitions

1. ΔG^0 = standard free energy change of the reaction, at 1 atm pressure and 1 M concentrations; ΔG = free energy change under the conditions used; $\Delta G^{0'}$ = free energy change under standard conditions at pH 7.
2. Calculation of ΔG^0 for a chemical reaction from the free energy of formation, G_f^0, of products and reactants:

    $$\Delta G^0 = \Sigma\, \Delta G_f^0 \text{(products)} - \Sigma\, \Delta G_f^0 \text{(reactants)}$$

    That is, sum the ΔG_f^0 of products, sum the ΔG_f^0 of reactants, and subtract the latter from the former.
3. For energy-yielding reactions involving H^+, converting from standard conditions (pH 0) to biochemical conditions (pH 7):

    $$\Delta G^{0'} = \Delta G^0 + m\, \Delta G_f'\,(H^+)$$

    where m is the net number of protons in the reaction (m is negative when more protons are consumed than formed), and $\Delta G_f'\,(H^+)$ is the free energy of formation of a proton at pH 7 = -9.55 kcal at 25°C.
4. Effect of concentrations on ΔG: with soluble substrates, the concentration ratios of products formed to exogenous substrates used are generally equal to or greater than 10^{-2} at the beginning of growth, and equal to or less than 10^{-2} at the end of growth. From the relation between ΔG and equilibrium constant (see item 8 below), it can be calculated that ΔG for the free energy yield in practical situations differs from free energy yield under standard conditions by at most 2.8 kcal, a rather small amount, so that for a first approximation, standard free energy yields can be used in most situations. However, with H_2 as a product, H_2-consuming bacteria present may keep the concentration of H_2 so low that the free energy yield is significantly affected. Thus in the fermentation of ethanol to acetate and H_2, the $\Delta G^{0'}$ at 1 atm H_2 is $+2.3$ kcal,

773

but at 10^{-4} atm H_2 it is -8.6 kcal. With H_2-consuming bacteria present, therefore, the ethanol fermentation becomes useful. (See also item 9 below.)

5. Reduction potentials: by convention, electrode equations are written in the direction, oxidant $+ ne^- \rightarrow$ reductant (that is, as reductions), where n is the number of electrons transferred. The standard potential (E_0) of the hydrogen electrode, $2H^+ + 2e^- \rightarrow H_2$ is set at 0.0 mV at 1.0 atm of H_2 gas and 1.0 M (molar) H^+, at 25°C. E_0' is the standard reduction potential at pH 7. See also Table A1.2.

6. Relation of free energy to reduction potential:
$$\Delta G^{0'} = -nF\,\Delta E_0'$$
where n is the number of electrons transferred, F is the faraday (23,062 cal), and $\Delta E_0'$ is the E_0' of the electron-accepting couple minus the E_0' of the electron-donating couple.

7. Equilibrium constant, K. For the generalized reaction
$$a\text{A} + b\text{B} \rightleftharpoons c\text{C} + d\text{D}$$
$$K = \frac{[\text{C}]^c[\text{D}]^d}{[\text{A}]^a[\text{B}]^b}$$
where A, B, C, and D represent reactants and products; $a, b, c,$ and d represent number of molecules of each; and brackets indicate concentrations.

8. Relation of equilibrium constant, K, to free energy change. At constant temperature and pressure,
$$\Delta G = \Delta G^0 + RT \ln K$$
where R is a constant (1.98 cal $\cdot$ mol$^{-1} \cdot$ °K^{-1}) and T is the absolute temperature (°K).

9. Effect of concentration on reduction potential is given in Figure A1.1. The curves present the observed potentials for three redox pairs; the standard potentials, E_0', are given by the midpoint values, which represent the potentials when equal amounts of oxidized and reduced forms are present. The graph shows that two substances can be coupled in a redox reaction even if the standard potentials are unfavorable, provided that the concentrations are appropriate. Thus, normally the reduced form of A would donate electrons to the oxidized form of B. However, if the concentration of the reduced form of A were low and the concentration of the reduced form of B were high, it would be possible for the reduced form of B to donate electrons to the oxidized form of A. Thus the redox couple would function in the reverse direction from that predicted from standard potentials. A practical example of this is the utilization of H^+ as an electron acceptor to produce H_2. Normally, H_2 production in fermentative bacteria is not extensive because H^+ is a poor electron acceptor; the midpoint potential of the $2H^+/H_2$ pair is -0.42 V. However, if the concentrations of H_2 is kept low by continually removing it (a process done by methane bacteria, which use $H_2 + CO_2$ to produce methane, CH_4), the potential will be more positive and then H^+ will serve as a suitable electron acceptor.

FIGURE A1.1 Effect of concentration on reduction potential.

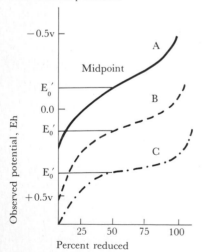

1. The oxidation state of an element in an elementary substance (i.e., H_2, O_2) is zero.
2. The oxidation state of the ion of an element is equal to its charge (i.e, $Na^+ = +1$, $Fe^{3+} = +3$, $O^{2-} = -2$).
3. The sum of oxidation numbers of all atoms in a neutral molecule is zero. Thus H_2O is neutral because it has two H at $+1$ each and one O at -2.
4. In an ion, the sum of oxidation numbers of all atoms is equal to the charge on that ion. Thus, in the OH^- ion, $O(-2) + H(+) = -1$.
5. In compounds, the oxidation state of O is virtually always -2, and that of H is $+1$.
6. In simple carbon compounds, the oxidation state of C can be calculated by adding up the H and O atoms present and using the oxidation states of these elements as given in item 5, since in a neutral compound the sum of all oxidation numbers must be zero. Thus the oxidation state of carbon in methane, CH_4, is -4 (4H at $+1$ each $= -4$); in carbon dioxide, CO_2, the oxidation state of carbon is $+4$ (2O at -2 each $= -4$).
7. In organic compounds with more than one C atom, it may not be possible to assign a specific oxidation number to each C atom, but it is still useful to calculate the oxidation state of the compound as a whole. The same conventions are used. Thus the oxidation state of carbon in glucose, $C_6H_{12}O_6$, is zero (12H at $+1 = 12$; 6O at $-2 = -12$) and the oxidation state of carbon in ethanol, C_2H_6O, is -4 (6H at $+1. = +6$; one O at -2).
8. In all oxidation-reduction reactions there is a balance between the oxidized and reduced products. To calculate an oxidation-reduction balance, the number of molecules of each product is multiplied by its oxidation state. For instance, in calculating the oxidation-reduction balance for the alcoholic fermentation, there are two molecules of ethanol at $-4 = -8$ and two molecules of CO_2 at $+4 = +8$, so that the net balance is zero. When constructing model reactions, it is useful to calculate redox balances to be certain that the reaction is possible.

Calculating Free Energy Yields for Hypothetical Reactions

Energy yields can be calculated either from differences in reduction potentials of electron-donating and electron-accepting partial reactions or from free energy of formations of the reactants and products (Table A1.1). The procedure using free energy values will be described here and the procedure using reduction potentials is given in Table A1.2.

1. *Balancing reactions* In all cases, it is essential to ascertain that the coupled oxidation-reaction is balanced. Balancing involves three things: (a) the total number of each kind of atom must be identical on both sides of the equation; (b) there must be an ionic balance, so that when positive and negative ions are added up on the right

TABLE A1.1 Free Energies of Formation (kcal/mole)

Carbon Compound	Metal	Nonmetal	Nitrogen Compound
CO, -32.78	Cu^+, $+12.0$	H_2, O	N_2, O
CO_2, -94.25	Cu^{2+}, $+15.5$	H^+, 0 at pH 0	NO, $+20.7$
CH_4, -12.13	CuS, -11.7	-1.36 per pH	NO_2, $+12.4$
H_2CO_3, -148.94	Fe^{2+}, -20.3	O_2, O	NO_2^-, -8.9
HCO_3^-, -140.26	Fe^{3+}, -2.5	OH^-, -37.6 at pH 14	NO_3^-, -26.6
CO_3^{2-}, -126.17	FeS_2, -36.0	-47.2 at pH 7	NH_4^+, -19.0
Formate, -83.9	$FeSO_4$, -198	-56.7 at pH 0	N_2O, $+24.9$
Methanol, -41.9	PbS, -22.1	H_2O, -56.7	
Acetate, -88.29	Mn^{2+}, -54.4	H_2O_2, -32.05	
Alanine, -88.8	Mn^{3+}, -19.6	PO_4^{3-}, -245	
Aspartate, -167	MnO_4^{2-}, -120.9	Se^0, O	
Butyrate, -84.3	MnO_2, -109	H_2Se, -18.4	
Citrate, -279	$MnSO_4$, -228	SeO_4^{2-}, -105	
Ethanol, -43.4	HgS, -11.7	S^0, O	
Formaldehyde, -31.2	MoS_2, -53.8	SO_3^{2-}, -116	
Glucose, -219	ZnS, -47.4	SO_4^{2-}, -178	
Glutamate, -167		$S_2O_3^{2-}$, -122	
Glycerol, -117		H_2S, -6.5	
Lactate, -124		HS^-, $+2.88$	
Malate, -202		S^{2-}, $+20.5$	
Pyruvate, -113			
Succinate, -165			
Urea, -48.7			

Values for free energy of formation of various compounds can be found in Dean, J. A. 1973. Lange's handbook of chemistry, 11th ed, McGraw-Hill, New York; Garrels, R. M., and C. L. Christ. 1965. Solutions, minerals, and equilibria. Harper & Row, New York; Burton, K. 1957. Appendix *in* Krebs, H. A., and H. L. Kornberg, Energy transformations in living matter, Ergebnisse der Physiologie. Springer-Verlag, Berlin; and Thauer, R. K., K. Jungermann, and K. Decker. 1977. Bacteriol. Rev. 41:100–180.

side of the equation, the total ionic charge (whether positive, negative, or neutral) exactly balances the ionic charge on the left side of the equation; and (*c*) there must be an oxidation-reduction balance, so that all of the electrons removed from one substance must be transferred to another substance. In general, when constructing balanced reactions, one proceeds in the reverse of the three steps listed above. Usually, if steps (*c*) and (*b*) have been properly handled, step (*a*) becomes correct automatically.

2. *Examples* (*a*) What is the balanced reaction for the oxidation of H_2S to SO_4^{2-} with O_2? First, decide how many electrons are involved in the oxidation of H_2S to SO_4^{2-}. This can be most easily calculated from the oxidation states of the compounds, using the rules given in item 5 of the list on page 775. Since H has an oxidation state of $+1$, the oxidation state of S in H_2S is -2. Since O has an oxidation state of -2, the oxidation state of S in SO_4^{2-} is $+6$ (since it is an ion, using the rules given in items 4 and 5 on page 775). Thus the oxidation of H_2S to SO_4^{2-} involves an eight-electron transfer (from -2 to $+6$). Since each O atom can accept two electrons (the oxidation state of O in O_2 is zero, but in H_2O is -2), this means that two molecules of molecular oxygen O_2,

TABLE A1.2 Reduction Potentials of Some Redox Pairs of Biochemical and Microbiological Importance*

Redox Pair	$E_0'(V)$	Redox Pair	$E_0'(V)$
SO_4^{2-}/HSO_3^-	-0.52	HSO_3^-/HS^-	-0.11
CO_2/formate	-0.43	Menaquinone ox/red	-0.075
H^+/H_2	-0.41	APS + AM/HSO_3^-	-0.060
Flavodoxin ox/red[†]	-0.37	Rubredoxin ox/red	-0.057
Ferredoxin ox/red	-0.38	Acrylyl-CoA/propionyl-CoA	-0.015
$S_2O_3^{2-}/HS^- + HSO_3^-$	-0.36	Glycine/acetate$^-$ + NH_4^+	-0.010
NAD/NADH	-0.32	$S_4O_6^{2-}/S_2O_3^{2-}$	$+0.025$
Cytochrome c_3 ox/red	-0.29	Fumarate/succinate	$+0.030$
CO_2/acetate$^-$	-0.29	Cytochrome b ox/red	$+0.030$
S^0/HS^-	-0.27	Ubiquinone ox/red	$+0.11$
SO_4^{2-}/HS^-	-0.25	Cytochrome c_1 ox/red	$+0.23$
CO_2/CH_4	-0.24	$S_2O_6^{2-}/S_2O_3^{2-} + HSO_3^-$	$+0.23$
FAD/FADH	-0.22	NO_2^-/NO	$+0.36$
$HSO_3^-/S_3O_6^{2-}$	-0.20	Cytochrome a_3 ox/red	$+0.385$
Acetaldehyde/ethanol	-0.20	NO_3^-/NO_2^-	$+0.43$
Pyruvate$^-$/lactate$^-$	-0.19	Fe^{3+}/Fe^{2+}	$+0.77$
FMN/FMNH	-0.19	O_2/H_2O	$+0.82$
Dihydroxyacetone phosphate/		NO/N_2O	$+1.18$
glycerolphosphate	-0.19	N_2O/N_2	$+1.36$
Flavodoxin ox/red[†]	-0.12		

*Calculating energy yields in oxidation-reduction reactions: the amount of energy that can be released when two half-reactions are coupled can be calculated from the differences in reduction potentials of the two reactions and from the number of electrons transferred. The further apart the two half-reactions are, and the greater the number of electrons, the more energy released. The conversion of potential difference to free energy is gven by the formula $\Delta G^0 = nF \Delta E_0'$, where n is the number of electrons, F is the faraday constant (23 kcal) and $\Delta E_0'$ is the difference in potentials. Thus the $2H^+/H_2$ redox pair has a potential of -0.41 and the $\frac{1}{2}O_2H_2O$ pair has a potential of $+0.82$, so that the potential difference is 1.23, which (since two electrons are involved) is equivalent to a free energy yield (ΔG^0) of -57 kcal/mole. On the other hand, the potential difference between the $2H^+/H_2$ and the NO_3^-/NO_2^- reactions is less, 0.84 V, which is equivalent to a free energy yield of -39 kcal/mole. Because many biochemical reactions are two-electron transfers, it is often useful to give energy yields for two-electron reactions, even if more electrons are involved. Thus the SO_4^{2-}/H_2 redox pair involves eight electrons, and complete reduction of SO_4^{2-} with H_2 would require $4H_2$ (equivalent to eight electrons). From the reduction potential difference between $2H^+/H_2$ and SO_4^{2-}/H_2S (0.16 V), a free-energy yield of -29 kcal/mole is calculated, or -7 kcal/mole per two electrons. By convention, reduction potentials are given for conditions in which equal concentrations of oxidized and reduced forms are present. In actual practice, the concentrations of these two forms may be quite different. As discussed in this appendix (item 9, page 774), it is possible to couple half-reactions even if the potential difference is unfavorable, providing the concentrations of the reacting species are appropriate.

[†]Separate potentials are given for each electron transfer in this potentially two-electron transfer.

Data from Thauer, R. K., K. Jungermann, and K. Decker, 1977. Bacteriol. Rev. 41:100–180.

will be required to provide sufficient electron-accepting capacity. Thus, at this point, we know that the reaction requires one H_2S and $2O_2$ on the left side of the equation, and one SO_4^{2-} on the right side. To achieve an ionic balance, we must have two positive charges on the right side of the equation to balance the two negative charges of SO_4^{2-}. Thus two H^+ must be added to the right side of the equation, making the overall reaction

$$H_2S + 2O_2 \longrightarrow SO_4^{2-} + 2H^+$$

By inspection, it can be seen that this equation is also balanced in terms of the total number of atoms of each kind on each side of the equation.

(b) What is the balanced reaction for the oxidation of H_2S to SO_4^{2-} with Fe^{3+} as electron acceptor? We have just ascertained that the oxidation of H_2S to SO_4^{2-} is an eight-electron transfer. Since the reduction of Fe^{3+} to Fe^{2+} is only a one-electron transfer, $8Fe^{3+}$ will be required. At this point, the reaction looks like this:

$$H_2S + 8Fe^{3+} \longrightarrow 8Fe^{2+} + SO_4^{2+} \quad \text{(not balanced)}$$

We note that the ionic balance is incorrect. We have 24 positive charges on the left and 14 positive charges on the right (16 + from Fe, 2 − from sulfate). To equalize the charges, we add $10H^+$ on the right. Now our equation looks like this:

$$H_2S + 8Fe^{3+} \longrightarrow 8Fe^{2+} + 10H^+ + SO_4^{2+} \quad \text{(not balanced)}$$

To provide the necessary hydrogen for the H^+ and oxygen for the sulfate, we add $4H_2O$ to the left and find that the equation is now balanced:

$$H_2S + 4H_2O + 8Fe^{3+} \longrightarrow 8Fe^{2+} + 10H^+ + SO_4^{2-} \quad \text{(balanced)}$$

In general, in microbiological reactions, ionic balance can be achieved by adding H^+ or OH^- to the left or right side of the equation, and since all reactions take place in an aqueous medium, H_2O molecules can be added where needed. Whether H^+ or OH^- is added will depend upon whether the reaction is taking place in acid or alkaline conditions.

3. *Calculation of energy yield for balanced equations from free energies of formation* Once an equation has been balanced, the free energy yield can be calculated by inserting the values for the free energy of formation of each reactant and product from Table A1.1, and using the formula in item 2 on page 773.

For instance, for the equation

$$H_2S + 2O_2 \longrightarrow SO_4^{2-} + 2H^+$$

ΔG values $\longrightarrow \quad -6.5 \quad 0 \qquad\qquad -177 \quad 2 \times -9.1$

(assuming pH 7)

$\Delta G^{0\prime} = -188.7$ kcal/mole

The values on the right are summed and subtracted from the values on the left, taking care to ensure that the signs are correct. From the data in Table A1.1, a wide variety of free energy yields for reactions of microbiological interest can be calculated.

Energetics of Light-Driven Reactions

The energy of a quantum of light can be calculated from the equation

$$E \text{ (energy of quantum)} = \frac{h \text{ (Planck's constant)} \times c \text{ (velocity of light)}}{\lambda \text{ (wavelength)}}$$

$$h = 6.62 \times 10^{-27} \text{ erg sec}$$

$$c = 2.99 \times 10^{10} \text{ cm/sec}$$

$$\lambda \text{ (for red light of 680)} = 6.8 \times 10^{-5} \text{ cm}$$

Thus for light of 680 nm, $E = 2.9 \times 10^{-12}$ erg per quantum. A mole of quanta is equivalent to an Avogadro's number of quanta. Avoga-

TABLE A1.3 Energy Expenditure for Polymer Synthesis

Substance	Approximate Dry Weight (percent)	Monomer Units in Polymer		ATP Required	
		Average Molecular Weight	Micromoles/ 100 mg Cells	Per Monomer*	Micromoles ATP/ 100 mg Cells
Protein	60	110	545	5	2725
Nucleic acid	20	300	67	5	335
Lipid	5	262	19	1	19
Polysaccharide	5	166	30	2	60
Peptidoglycan	10	1000	10	10	100
Total					or 31 g of cell material per mole ATP

*Calculation of ATP required per monomer polymerized:

Protein: amino acid activation: 2ATP; ribosome function: 2GTP = 2ATP; mRNA turnover: 1ATP.

Nucleic acid: nucleotide formation from free base: 3ATP; polymerization: 2ATP.

Lipid: formation of glycerol ester from fatty acyl-CoA and glycerophosphate: 1ATP.

Polysaccharide: formation of uridine diphospho sugar: 2ATP.

Peptidoglycan: activation of five amino acids: 5ATP; UDP-muramic acid 2ATP; UDP-N-acetylglucosamine: 2ATP; lipid carrier: 1ATP.

dro's number is 6×10^{23}. Thus a mole of quanta of light of 680 nm is $2.9 \times 10^{-12} \times 6 \times 10^{-23} = 17.4 \times 10^{11}$ ergs. Since there are 4.185×10^7 ergs/cal, we divide 17.4×10^{11} by 4.185×10^7 and obtain 4.1×10^4 cal. Thus 1 mole of quanta of red light of 680 nm is equivalent to 4.1×10^4 cal, or 41 kcal.

Appendix 2

The Mathematics of Growth and Chemostat Operation

Exponential growth Although many analyses of microbial growth can be done graphically, as discussed in Chapter 7, for some purposes it is convenient to use a differential equation that expresses the quantitative relationships of growth:

$$\frac{dX}{dt} = \mu X \tag{1}$$

where X may be cell number or some specific cellular component such as protein* and μ is the *instantaneous growth-rate constant*. If this equation is integrated, we obtain a form which reflects the activities of a typical batch culture population:

$$\ln X = \ln X_0 + \mu(t) \tag{2}$$

where ln refers to the natural logarithm (logarithm base e), X_0 is cell number at time 0, X is cell number at time t, and t is elapsed time during which growth is measured. This equation fits experimental data from the exponential phase of bacterial cultures very well, such as those in Table 7.1. Taking the antilogarithm of each side gives

$$X = X_0 e^{\mu t} \tag{3}$$

This equation is useful because it allows prediction of population density at a future time by knowing the present value and μ, the growth-rate constant. As discussed in Section 7.1, an important constant parameter for an exponentially growing population is the doubling (or generation) time. Doubling of the population has occurred when $X/X_0 = 2$. Rearranging and substituting this value into equation 3 gives

$$2 = e^{\mu(t_{\text{gen}})} \tag{4}$$

---

*In specific experiments it is sometimes desirable to measure cell mass, cell protein, cell DNA content, or some other growth parameter. All may be accounted for by Equation 1. For the sake of simplicity, X will be referred to as cell number in the rest of this discussion, although the alternatives must be kept in mind.

Taking the natural logarithm of each side and rearranging gives

$$\mu = \frac{\ln 2}{t_{gen}} = \frac{0.693}{t_{gen}} \tag{5}$$

The generation time t_{gen} may be used to define another growth parameter, k, as follows:

$$k = \frac{1}{t_{gen}} \tag{6}$$

where k is the growth-rate constant for a batch culture. Combining Equations 5 and 6 shows that the two growth-rate constants, μ and k, are related:

$$\mu = 0.693k$$

It is important to understand that μ and k are both reflections of the same growth process of an exponentially increasing population. The difference between them may be seen in their derivation above: μ is the *instantaneous* rate constant and k is an *average* value for the population over a finite period of time. (See Table A2.1 for calculation of k values from experimental results.) This distinction is more than a mathematical point. As was emphasized in Chapter 7, microbial growth studies must deal with population phenomena, not the activities of individual cells. The constant k reflects this averaging assumption. However, the constant μ, being instantaneous, is a closer approximation of the rate at which individual activities are occurring. Further, the instantaneous constant μ allows us to consider bacterial growth dynamics in a theoretical framework separate from the traditional batch culture.

Mathematical relationships of chemostats An especially important application of the instantaneous growth-rate constant μ is, of course, the chemostat, a culture device (Section 7.5) in which population size and growth rate may be maintained at constant values of the experimenter's choosing, over a wide range of values. As we saw in Section 7.4, the rate of bacterial growth is a function of nutrient concentration (Figure 7.6b). This figure represents a saturation process which may be described by the equation

$$\mu = \mu_{max}\frac{S}{K_s + S} \tag{8}$$

where μ_{max} is the growth rate at nutrient saturation, S is the concentration of nutrient,* and K_s is a saturation constant which is numerically equal to the nutrient concentration at which $\mu = \frac{1}{2}\mu_{max}$. Equation 8 is formally equivalent to the Michaelis–Menten equation used in enzyme kinetic analyses. Estimates of the unknown parameters μ_{max} and K_s can be made by plotting $1/\mu$ on the y axis versus $1/s$ on the x axis. A straight line will be obtained which intercepts the y axis, and the value at the intercept is $1/\mu_{max}$. The slope of the line is K_s/μ_{max}, and

---

TABLE A2.1   Calculation of k Values

Bacterial population densities are expressed in scientific notation with powers of 10, so Equation 3 may be converted to terms of logarithm base 10 and k substituted for the instantaneous constant μ:

$$k = \frac{\log_{10}X_t - \log_{10}X_0}{0.301t}$$

Example 1:
$X_0 = 1000\ (= 10^3)$ $\log_{10}$ of 1000 = 3
$X_t = 100{,}000\ (= 10^5)$ $\log_{10}$ of 100,000 = 5
$t = 4$ hours

$$k = \frac{5 - 3}{(0.301)4} = \frac{2}{1.204}$$

$k = 1.66$ doublings per hour
$t_{gen} = 0.60$ hour (36 minutes) for population to double

Example 2:
$X_0 = 1000\ (= 10^3)$ $\log_{10}$ of 1000 = 3
$X_t = 100{,}000{,}000\ (= 10^8)$ $\log_{10}$ of 10^8 = 8
$t = 120$ hours

$$k = \frac{8 - 3}{(0.301)120} = \frac{5}{36.12}$$

$k = 0.138$ doubling per hour
$t_{gen} = 7.2$ hours (430 minutes) for population to double

---

*It is usual to speak of a *limiting nutrient,* that is, some component of the growth medium (often the electron donor) which, if increased in concentration, causes an increase in growth rate. One must be careful with this terminology, however, for, as Figure 7.6b shows, there is usually observed a saturation level for each nutrient. At saturation values the particular nutrient under consideration no longer limits growth and some other component or factor becomes limiting.

since μ_{max} is known, K_s can then be calculated. Values for K_s as a rule are very low. A few examples are: 2.1×10^{-5} M for glucose (*Escherichia coli*), 1.34×10^{-6} M for oxygen (*Candida utilis*), 3.5×10^{-8} M for phosphate (*Spirillum* spp.), and 4.7×10^{-13} M for thiamin (*Cryptococcus albidus*).

The key to the operation of the chemostat is that, in a steady-state population, the nutrient concentration is very low, usually very near 0. Therefore, each drop of fresh medium entering the chemostat supplies nutrient which is consumed almost instantaneously by the population and the experimenter has direct control over the growth rate: if nutrient is added more rapidly, μ increases; if it is added more slowly, μ decreases. In a chemostat the rate at which medium enters is equal to the rate at which spent medium and cells exit through the overflow. This rate is called the dilution rate, D, and is defined:

$$D = \frac{F}{V} \tag{9}$$

where F is the flow rate (in units of volume per time, usually ml/hour) and V is chemostat volume (in ml). Therefore, D has units of time^{-1} (usually hour^{-1}) as does μ. In fact, at steady state, $\mu = D$. That μ must equal D may be seen by consideration of Equations 8 and 9 and the discussion above. When S is low, μ is a direct function of S. But since nutrient is instantly consumed upon addition, the value of S at any moment is controlled by the rate of medium addition, that is, the dilution rate D. A further crucial consequence of the relation $\mu = D$ is that a chemostat is a *self-regulating* system within broad limits of D. Consider a steady-state chemostat functioning with $\mu = D$. If μ were to increase momentarily due to some variation in the culture, S would necessarily decline, as the increased growth caused increased nutrient consumption. The lower value of S would in turn cause μ to decrease back to its stable level. Conversely, if μ were to decrease momentarily, then S would increase as "extra" nutrient went unconsumed. This rise in S would lead to higher values of μ until steady state is again reached. This biological feedback system functions well in practice, as it does in theory. Chemostat populations may be maintained at constant growth rates (μ) for long periods of time: months or even years.

There is another important aspect to chemostat studies and that is population density. The relation between cell growth and nutrient consumption may be defined:

$$Y = \frac{dX}{dS} \tag{10}$$

where Y is the yield constant for the particular organism on the particular medium and dX/dS is the amount of cell increase per unit of substrate consumed. The yield constant is in units of

$$Y = \frac{\text{weight of bacteria formed}}{\text{weight of substrate consumed}}$$

and is a measure of the efficiency with which cells convert nutrient to more cell material. The composition of the medium in the reservoir is central to the successful steady-state operation of the chemostat. All components in this medium are present in excess, except one, the growth-limiting substrate. The symbol for the concentration of that substrate in the reservoir is S_R, whereas that for the same substrate in the culture vessel is S. Taking into account Equations 1, 8, and 10, the following considerations can be made.

When a chemostat is inoculated with a small number of bacteria, and the dilution rate does not exceed a certain critical value (D_c), the number of organisms will start to increase. The increase is given by

$$\frac{dX}{dt} = \text{growth} - \text{output} \qquad \text{or} \qquad \frac{dX}{dt} = \mu X - DX \qquad (11)$$

since initially $\mu > D$, dX/dt is positive. However, as the population density increases, the concentration of the growth-limiting substrate decreases, causing a decrease of μ (Equation 8). As discussed above, if D is kept constant, a steady state will be reached in which $\mu = D$ and $dX/dt = 0$.

Steady states are possible only when the dilution rate does not exceed a critical value D_c. This value depends on the concentration of the growth-limiting substrate in the reservoir (S_R):

$$D_c = \mu_{\max}\left(\frac{S_R}{K_s + S_R}\right) \qquad (12)$$

Thus, when $S_R \gg K_s$ (which is usually the case), steady states can be obtained at growth rates close to $\mu_{\max}$. The change in S is given by

$$\frac{dS}{dt} = \text{input} - \text{output} - \text{consumption}$$

Since consumption is growth divided by yield constant (see Equation 10), we have

$$\frac{dS}{dt} = S_R D - sD - \frac{\mu X}{Y} \qquad (13)$$

When a steady state is reached, $dS/dt = 0$.

From Equations 11 and 13, the steady-state values of organism concentration $(\overline{X})$ and growth-limiting substrate $(\overline{S})$ can be calculated:

$$\overline{X} = Y(S_R - S) \qquad (14)$$

$$\overline{S} = K_s\left(\frac{D}{\mu_{\max} - D}\right) \qquad (15)$$

These equations have the constants S_R, Y, K_s, and $\mu_{\max}$. Once their values are known, the steady-state values of $\overline{X}$ and $\overline{S}$ can be predicted for any dilution rate. As can be seen from Equation 15, the steady-state substrate concentration $(\overline{S})$ depends on the dilution rate (D) applied, and is independent of the substrate concentration in the reservoir (S_R). Thus D determines $\overline{S}$, and $\overline{S}$ determines $\overline{X}$. A specific example of the use of the equations to predict chemostat behavior can be obtained by study of Figure 7.9. In that figure, the constants are $\mu_{\max} = 1.0$ hour^{-1}; $Y = 0.5$; $K_s = 0.2$ g/liter; $S_R = 10$ g/liter. With these constants, the curves in Figure 7.9 can be generated.

Experimental uses of the chemostat One of the major theoretical advantages of a chemostat is that this device allows the experimenter to control growth rate (μ) and population density $(\overline{X})$ independently of each other. Over rather wide ranges (see above and Figure 7.9) any desired value of μ can be obtained by alteration of D. In practice this means changing the flow rate, since volume is usually fixed (Equation 9). Similarly, the population density $\overline{X}$ may be determined by varying nutrient concentration in the reservoir (Equation 14). This independent control of these two crucial growth parameters is not possible with conventional batch cultures, as discussed in connection with Figure 7.6*a*.

A practical advantage to the chemostat is that a population may be maintained in a desired growth condition (μ and $\overline{X}$) for long periods of time. Therefore, experiments can be planned in detail and performed whenever most convenient. Comparable studies with batch cultures are essentially impossible, since specific growth conditions are constantly changing.

Mathematics of the stationary phase The differential equation (Equation 1) expressing the exponential growth phase can be modified to include the trend toward the stationary phase, which occurs at high population density. A second term is added to the equation that expresses the maximum population attainable for that organism under the environmental conditions specified, here called X_m. The equation then assumes a form often called the *logistics equation:*

$$\frac{dX}{dt} = \mu X - \frac{\mu}{X_m} X^2 \tag{16}$$

The second term in this equation essentially expresses self-crowding effects, such as nutrient depletion or inhibitor buildup. In integrated form, this equation will graph as a typical microbial growth curve, showing exponential and stationary phases. The length of the exponential phase will be determined by the size of X_m. When X is small, the second term will have little effect, whereas as X approaches X_m, the growth rate (dX/dt) will approach zero. The logistics equation is widely used by ecologists studying higher organisms to express the growth rate and carrying capacity of the habitat for animals and plants. The equation is equally applicable to microbial situations.

Mathematics of disinfection The equation expressing the death rate of unicellular organisms upon treatment with a disinfectant or other lethal agent is sometimes called *Chick's law,* named after H. Chick, who first analyzed death rates in 1908. If X_d represents the number of dead cells at time t, then $X_0 - X_d$ will represent the survivors. The death rate is proportional to the survivors:

$$\frac{dX}{dt} = k(X_0 - X_d) \tag{17}$$

Upon integration, this equation gives a typical first-order relationship:

$$kt = \ln\frac{X_0}{X_0 - X_d} \tag{18}$$

and when the logarithm of $X_0 - X_d$ (that is, the number of survivors) is plotted against time, a straight line will be obtained. This equation is useful when determining the length of time necessary to allow a disinfectant to completely sterilize something (for example, the contact time for chlorine gas in a water purification plant).

TABLE A2.2 Five-Tube Most-Probable-Number (MPN) Table

See Section 7.2 for a discussion of MPN counts.

Five culture tubes were inoculated with 1-ml aliquots from each of three successsive tenfold dilutions. After the cultures were incubated, the number of tubes showing growth at each dilution was recorded. The table gives the MPN per milliliter of the inoculum taken from the first dilution. Two examples of the use of this table appear at right.

10^0	10^{-1}	10^{-2}	MPN	10^0	10^{-1}	10^{-2}	MPN
0	1	0	0.18	5	0	0	2.3
1	0	0	0.20	5	0	1	3.1
1	1	0	0.40	5	1	0	3.3
2	0	0	0.45	5	1	1	4.6
2	0	1	0.68	5	2	0	4.9
2	1	0	0.68	5	2	1	7.0
2	2	0	0.93	5	2	2	9.5
3	0	0	0.78	5	3	0	7.9
3	0	1	1.1	5	3	1	11.0
3	1	0	1.1	5	3	2	14.0
3	2	0	1.4	5	4	0	13.0
4	0	0	1.3	5	4	1	17.0
4	0	1	1.7	5	4	2	22.0
4	1	0	1.7	5	4	3	28.0
4	1	1	2.1	5	5	0	24.0
4	2	0	2.2	5	5	1	35.0
4	2	1	2.6	5	5	2	54.0
4	3	0	2.7	5	5	3	92.0
				5	5	4	160.0

*For a brief discussion of how MPN tables are calculated, and a more detailed table, see de Man, J. C. 1975. The probability of most probable numbers. Eur. J. Appl. Microbiol. 1:67–78. Another source for MPN tables is: Standard methods for the examination of water and wastewater, 15th ed. 1980. American Public Health Association, Washington, D.C.

Example 1:

Five 1-ml aliquots were inoculated from 10^0, 10^{-1}, and 10^{-2} dilutions of a pasteurized milk sample. After incubation, five tubes showed growth at 10^0, three tubes at 10^{-1}, and no tubes at 10^{-2}. The MPN from Table A2.3 is 7.9 viable cells per milliliter of milk.

Example 2:

Five 1-ml aliquots were inoculated from 10^{-3}, 10^{-4}, and 10^{-5} dilutions of a raw milk sample. After incubation, four tubes showed growth at 10^{-3}, two tubes at 10^{-4}, and oen tube at 10^{-5}. The MPN from Table A2.3 for the 10^{-3} dilution is 2.6 cells/ml and the viable count per milliliter of milk is therefore 2.6×10^3.

Chick, H. 1930. The theory of disinfection. Pages 179–207 *in* A system of bacteriology, vol. 1. Medical Research Council, London. A still useful discussion of quantitative aspects of the killing process.

Kubitschek, H. E. 1970. Introduction to research with continuous cultures. Prentice-Hall, Englewood Cliffs, N.J. A very lucid account of the mathematics of continuous cultures.

Rahn, O. 1948. Mathematics in bacteriology. Burgess, Minneapolis, Minn. Although quite old-fashioned, this book provides a useful summary of mathematical treatments of the bacterial growth curve.

Supplementary Readings

3

Biochemical Pathways

In this appendix, further details are given of biochemical reactions, which were discussed in outline form in Chapters 4 through 6. Some additional reactions of unusual microbiological interest are also included here.

Notes on the enzyme reaction for glycolysis, as illustrated in Figure A3.1:

1. *Hexokinase* The reaction catalyzed is essentially irreversible. The enzyme will also react with sugars other than glucose: fructose, mannose, glucosamine.
2. *Glucose-6-phosphate isomerase* The reaction catalyzed is freely reversible.
3. *Phosphofructokinase* The reaction catalyzed is irreversible. The activity of the enzyme is inhibited by ATP but stimulated by ADP or AMP. Therefore, when ATP levels are high, glycolysis is inhibited, but when ATP levels fall, glycolysis is stimulated. In this way, the level of ATP in the cell is regulated. Phosphofructokinase is the most important control point in the glycolytic sequence.
4. *Aldolase* The reaction catalyzed is reversible, although in the direction written it has a positive $\Delta G^{0\prime}$. However, because the products formed are quickly utilized, the reaction is pulled in the forward direction. Aldolase is a key enzyme in glycolysis; bacteria (for example, heterolactics) which lack this enzyme, ferment hexose sugar by a different pathway, with reduced energy yield.
5. *Triose phosphate isomerase* As a result of the aldolase reaction, two products are formed, dihydroxyacetone phosphate and glyceraldehyde phosphate. Only the latter can be degraded further. Triose phosphate isomerase catalyzes the interconversion of

TABLE A3.1 Energy-Rich Compounds Involved in Substrate-Level Phosphorylation

Name	Structure	Free Energy of Hydrolysis,* $-\Delta G^{0\prime}$ (kcal/mole)
Acetyl-CoA	$CH_3-\overset{\overset{\displaystyle O}{\|\|}}{C}-O-CoA$	8.5
Propionyl-CoA	$CH_3-CH_2-\overset{\overset{\displaystyle O}{\|\|}}{C}-O-CoA$	8.5
Butyryl-CoA	$CH_3-CH_2-CH_2-\overset{\overset{\displaystyle O}{\|\|}}{C}-O-CoA$	8.5
Succinyl-CoA	$-OOC-CH_2-CH_2-\overset{\overset{\displaystyle O}{\|\|}}{C}-O-CoA$	8.4
Acetylphosphate	$CH_3-\overset{\overset{\displaystyle O}{\|\|}}{C}-O-PO_3H_2$	10.7
1,3-Diphosphoglycerate	$H_2PO_3OC-\overset{\overset{\displaystyle H}{\|}}{C}-\overset{\overset{\displaystyle H}{\|}}{\underset{\underset{\displaystyle OH}{\|}}{C}}-OPO_3H_2$	12.4
Carbamyl phosphate	$H_2N-\overset{\overset{\displaystyle O}{\|\|}}{C}-OPO_3H_2$	9.6
Phosphoenolpyruvate	$H_2C=\underset{\underset{\displaystyle COOH}{\|}}{C}-OPO_3H_2$	12.3
Adenosine-phosphosulfate	$Adenosine-O-PO_3-O-SO_3$	21
N^{10}-formyltetrahydrofolate	$Tetrahydrofolate-O-\overset{\overset{\displaystyle O}{\|\|}}{C}$	5.6

*From Table 4, Thauer, R. K., K. Jungermann, and K. Decker. 1977. Bacteriol. Rev. 41:100–180.

these two substances. The equilibrium lies in the direction of dihydroxyacetone phosphate, but as glyceraldehyde phosphate is consumed, the reaction is pulled in the direction of the latter.

6. *Glyceraldehyde phosphate dehydrogenase* This is one of the most important reactions in glycolysis, as it leads to *substrate-level phosphorylation*. Inorganic phosphate is converted into a high-energy phosphate bond, and NAD is reduced to NADH. The high-energy phosphate bond is shown on the figure.

7. *Phosphoglycerate kinase* This is the first reaction in which ATP is synthesized. Although reversible, the reaction proceeds most favorably towards the right, and hence serves to pull the preceding reaction.

8. *Phosphoglyceromutase* This enzyme catalyzes the migration of the remaining phosphate ester from the 3 to the 2 position. This reaction is freely reversible.

9. *Enolase* In this important reaction, the low-energy phosphate ester of 2-phosphoglycerate is converted into a high-energy phosphate bond in phosphoenolpyruvate. Although, as written, the reaction appears to involve simply the removal of water, it can also be considered an intramolecular oxidation-reduction in

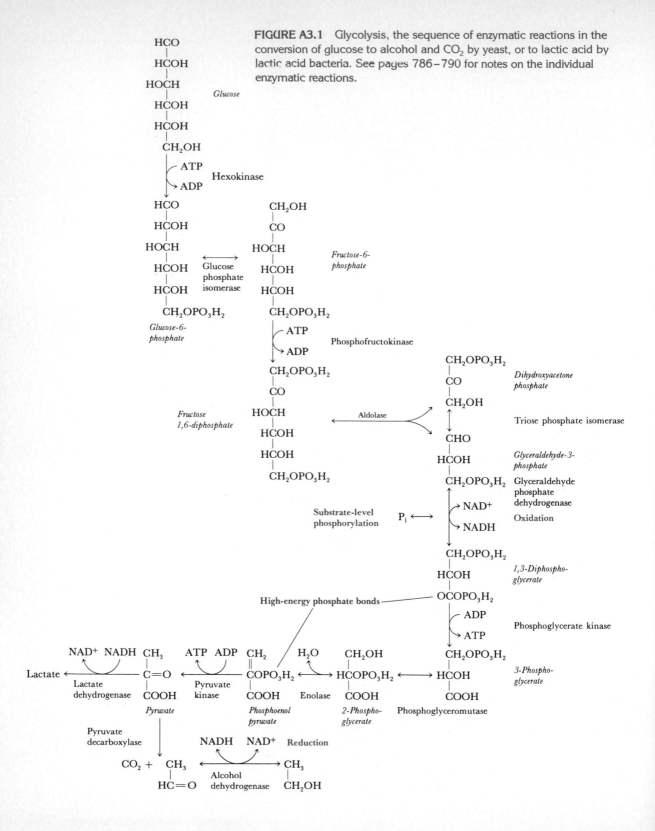

FIGURE A3.1 Glycolysis, the sequence of enzymatic reactions in the conversion of glucose to alcohol and CO_2 by yeast, or to lactic acid by lactic acid bacteria. See pages 786–790 for notes on the individual enzymatic reactions.

which carbon atom 2 becomes more oxidized and carbon atom 3 becomes more reduced. The free energy of hydrolysis of the high-energy phosphate bond in phosphoenolpyruvate is one of the highest, -14.8 kcal/mole, and this compound is an important intermediate in a variety of biosynthetic reactions, as well as in the succeeding steps of glycolysis.

10. **Pyruvate kinase** The synthesis of the second ATP occurs at this step. The reaction proceeds with a large negative $\Delta G^{0\prime}$ and is essentially irreversible.

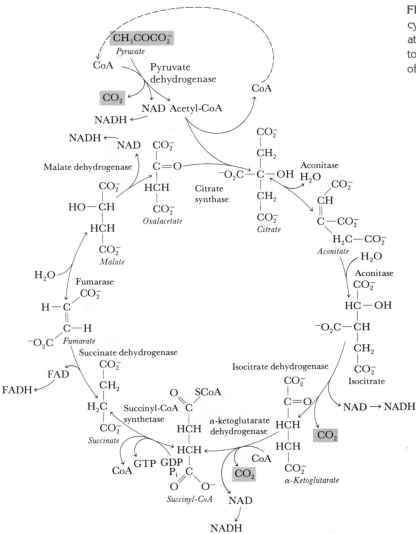

FIGURE A3.2 The tricarboxylic acid cycle. The overall reaction is shown at the bottom. Pyruvate is oxidized to 3 molecules of CO_2 with production of 15 molecules of ATP.

Overall reaction: Pyruvate + 4NAD + FAD $\longrightarrow$ 3CO_2 + 4NADH + FADH

GDP + phosphate $\longrightarrow$ GTP
GTP + ADP $\longrightarrow$ GDP + ATP

Electron-transport phosphorylation: 4NADH $\equiv$ 12ATP ⎤ 15ATP
FADH $\equiv$ 2ATP ⎦

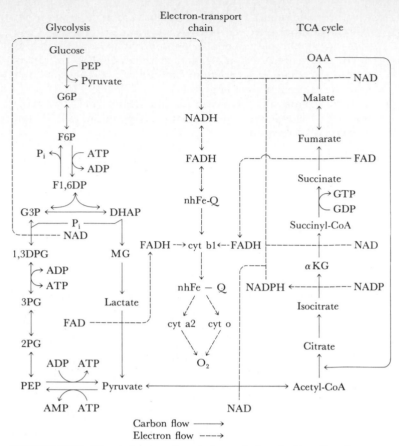

FIGURE A3.3 The pathways of central metabolism in *E. coli* and *S. typhimurium*. The following abbreviations are used: glucose-6-P (G6P), fructose-6-P (F6P), fructose-1,6-diP (F1,6DP), glyceraldehyde-3-P (G3P), 1,3-diP-glycerate (1,3DPG), 3-P-glycerate (3PG), 2-P-glycerate (2PG), P-enolpyruvate (PEP), dihydroxyacetone-P (DHAP), methyl glyoxal (MG), non-heme iron-coenzyme Q complex (nh Fe-Q), cytochrome (cyt), oxaloacetate (OAA), and α-ketoglutarate (αKG). (From Bochner and Savageau. 1977. Applied and environmental microbiology. 33:442.)

11. ***Pyruvate decarboxylase***   This enzyme reaction involves a tightly bound coenzyme, thiamine pyrophosphate, a derivative of vitamin B_1. The reaction is essentially irreversible.

12. ***Alcohol dehydrogenase***   It is at this step that the NADH formed earlier is reoxidized, maintaining the oxidation-reduction balance. Although called a dehydrogenase, in the present context the enzyme is acting in the reverse direction, and adding hydrogen to acetaldehyde.

13. ***Lactic dehydrogenase***   In the alternate glycolytic sequence which occurs in muscle, lactic acid bacteria, and many other microorganisms, pyruvate is converted to lactate by this enzyme, using NADH as electron donor. The redox balance is thus maintained.

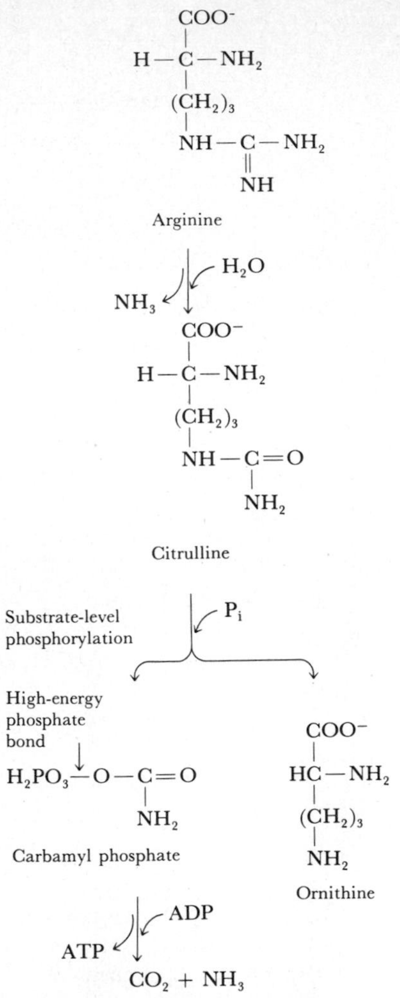

FIGURE A3.4 The arginine dihydrolase system. Generation of
ATP by substrate-level phosphorylation
and carbamyl phosphate formation.
This system is used for anaerobic
energy generation by normally
nonfermentative pseudomonads,
as well as by some lactic acid bacteria,
some mycoplasmas, some clostridia,
and many bacilli. Note that energy
generation by this system does not
involve an oxidation-reduction, but a
hydrolysis.

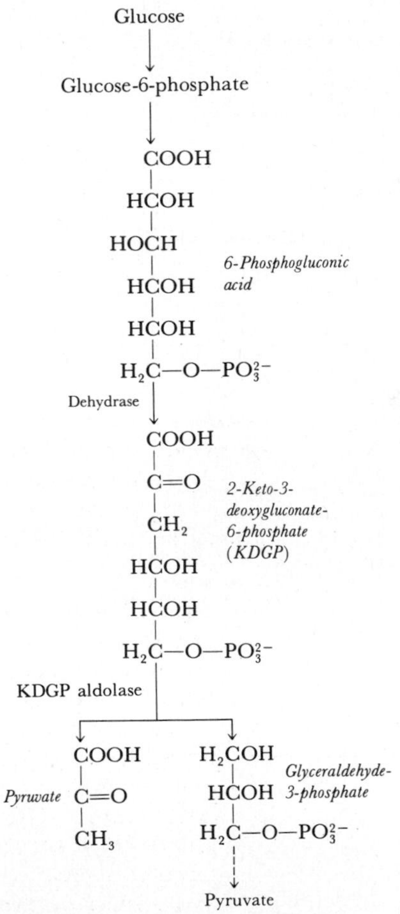

FIGURE A3.5 Entner-Doudoroff
pathway.

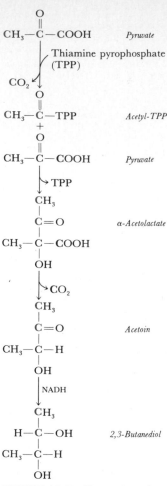

FIGURE A3.7 Formation of 2,3-butanediol from two molecules of pyruvate. In the first two steps, the coenzyme thiamin pyrophosphate (TPP) is involved and two molecules of pyruvate condense to form α-acetolactate, with the loss of one CO_2.

FIGURE A3.6 Pentose-phosphate pathway (hexose-monophosphate) with relations to other cellular processes. Compounds and reactions in color are not part of the pathway but rather represent connections to other metabolic activities.

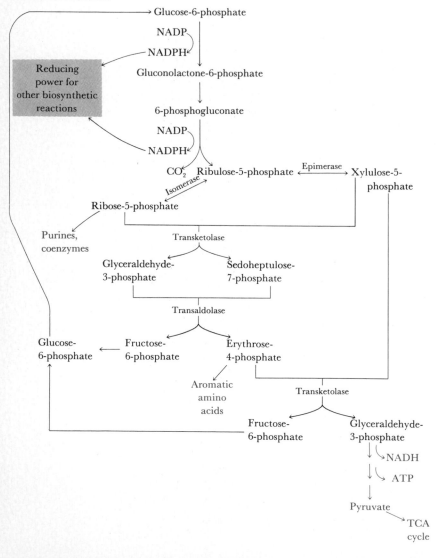

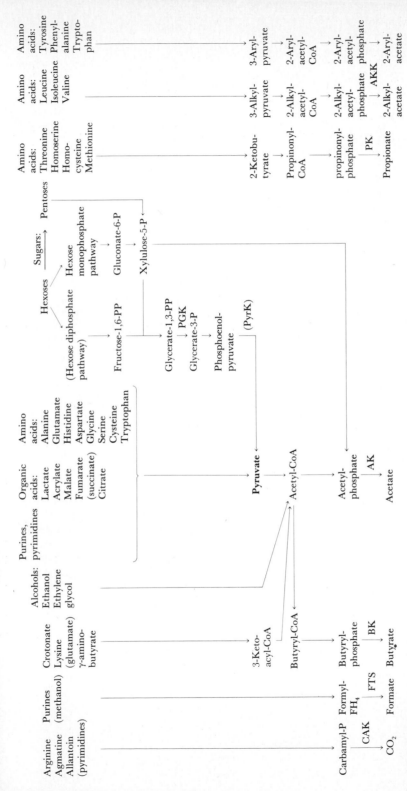

FIGURE A3.8 Pathways for the anaerobic breakdown of various fermentable substances. The sites of substrate-level phosphorylation are shown by the abbreviations: CAK, carbamyl phosphate kinase; FTS, formyltetrahydrofolate synthetase; AK, acetate kinase; PK, propionate kinase; BK, butyrate kinase; AKK, alkyl (aryl) acetate kinase; PGK, phosphoglycerate kinase; PyrK, pyruvate kinase. (From Decker, K., K. Jungermann, and R. K. Thauer. 1970. Energy production in anaerobic organisms. Angew. Chem. Int. Ed. Engl. 9:138–158.)

Basic amino acid structure:

$$R-\underset{\underset{NH_2}{|}}{\overset{\overset{H}{|}}{C}}-COOH$$

R groups

Glycine   H—

Alanine CH_3—

Valine $\underset{CH_3}{\overset{CH_3}{}}CH-$

Leucine $\underset{CH_3}{\overset{CH_3}{}}CH-CH_2-$

Isoleucine $CH_3-CH_2-\underset{CH_3}{\overset{|}{CH}}-$

Serine $OH-CH_2-$

Threonine $CH_3-\underset{\overset{|}{H}}{\overset{\overset{OH}{|}}{C}}-$

Cysteine $HS-CH_2-$

Methionine $CH_3-S-CH_2-CH_2-$

Phenylalanine   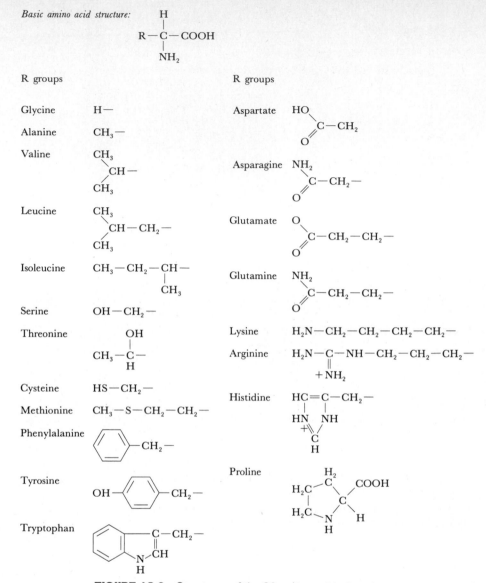$-CH_2-$

Tyrosine $OH-$⬡$-CH_2-$

Tryptophan

R groups

Aspartate $\underset{O}{\overset{HO}{}}C-CH_2$

Asparagine $\underset{O}{\overset{NH_2}{}}C-CH_2-$

Glutamate $\underset{O}{\overset{O}{}}C-CH_2-CH_2-$

Glutamine $\underset{O}{\overset{NH_2}{}}C-CH_2-CH_2-$

Lysine $H_2N-CH_2-CH_2-CH_2-CH_2-$

Arginine $H_2N-\underset{+NH_2}{\overset{||}{C}}-NH-CH_2-CH_2-CH_2-$

Histidine $HC=C-CH_2-$
            $\quad | \quad |$
            $HN \quad NH$
            $\quad +\backslash//$
            $\quad\quad C$
            $\quad\quad H$

Proline

FIGURE A3.9 Structures of the 20 amino acids found in proteins.

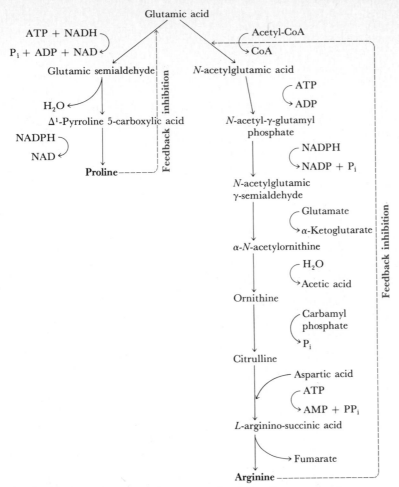

FIGURE A3.10 Synthesis of amino acids of the glutamate family, proline, and arginine. Note that each product is a feedback inhibitor of the first enzyme in the pathway for its own synthesis.

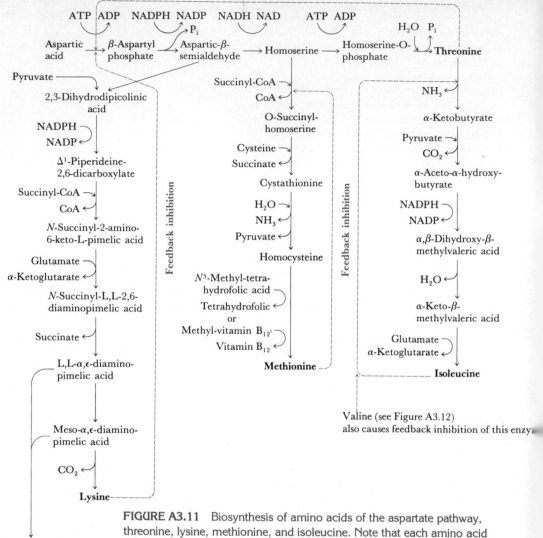

FIGURE A3.11 Biosynthesis of amino acids of the aspartate pathway, threonine, lysine, methionine, and isoleucine. Note that each amino acid is a feedback inhibitor of the first enzyme in its own pathway. In some fungi, a completely different pathway for lysine biosynthesis is present, called the aminoadipic acid pathway, but the one given here is that found in procaryotes, higher plants, and animals.

FIGURE A3.12 Biosynthesis of two branched-chain amino acids, valine and leucine. The enzymes of the valine pathway also catalyze the analogous steps in the isoleucine pathway (Figure A3.11, α-Ketobutyrate to isoleucine). Because valine is a feedback inhibitor of the enzyme catalyzing both α-Ketobutyrate to α-Aceto-α-hydroxybutyrate (in the isoleucine pathway) and pyruvate to α-Acetolactate (in its own pathway), valine inhibits the synthesis of isoleucine. However, this inhibition is reversed by isoleucine, thus providing a finely tuned regulatory system for the biosynthesis of these amino acids. Leucine is a feedback inhibitor of the first enzyme specific to its own pathway, and thus does not affect valine or isoleucine biosynthesis. In addition to the feedback inhibitions, these amino acids also cause repression of the enzymes involved in their biosynthesis (as do many of the other amino acids whose pathways are presented in this appendix).

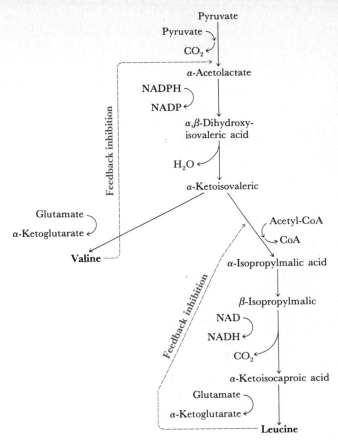

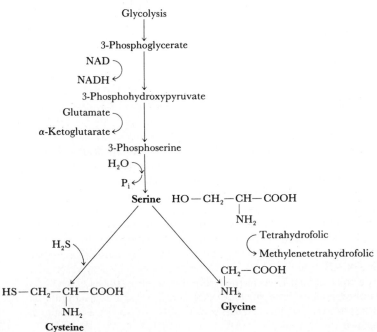

APPENDIX A3.13 Biosynthesis of the amino acids of the serine pathway, serine, glycine, and cysteine. In addition to the direct formation of cysteine from serine + H_2S shown here, an additional mechanism exists via O-acetylserine, which is found in some organisms.

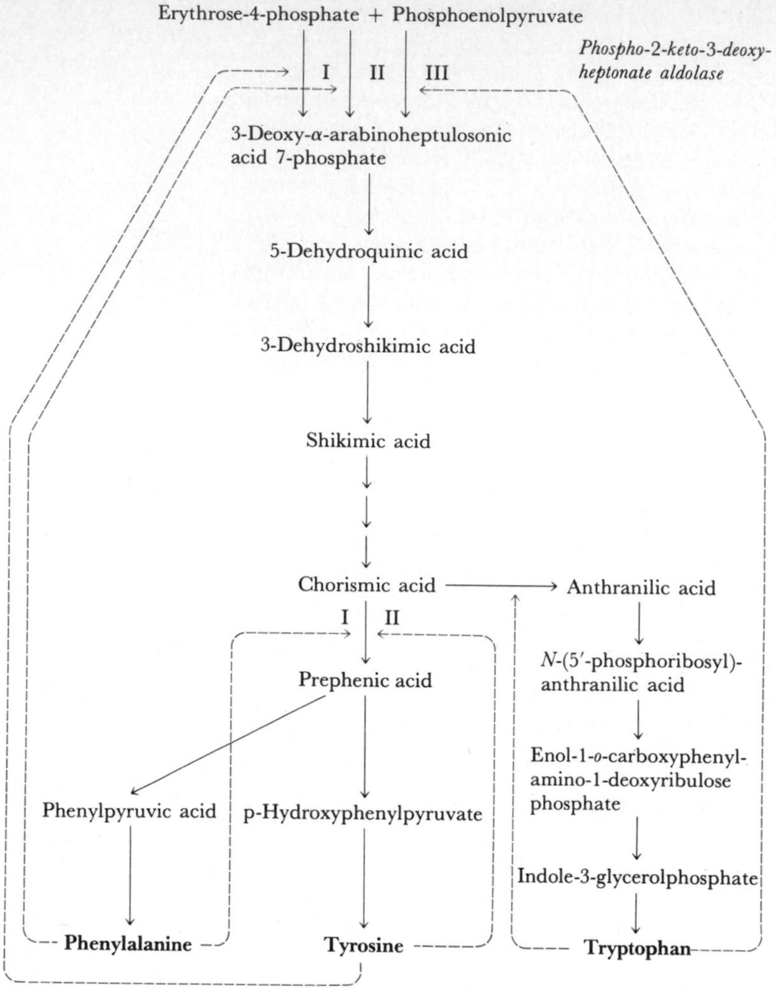

Erythrose-4-phosphate + Phosphoenolpyruvate

I   II   III

Phospho-2-keto-3-deoxy-heptonate aldolase

3-Deoxy-α-arabinoheptulosonic acid 7-phosphate

5-Dehydroquinic acid

3-Dehydroshikimic acid

Shikimic acid

Chorismic acid ⟶ Anthranilic acid

I   II

Prephenic acid

N-(5′-phosphoribosyl)-anthranilic acid

Enol-1-*o*-carboxyphenyl-amino-1-deoxyribulose phosphate

Phenylpyruvic acid   p-Hydroxyphenylpyruvate

Indole-3-glycerolphosphate

Phenylalanine   **Tyrosine**   **Tryptophan**

FIGURE A3.14 Aromatic amino acids, biosynthesis and regulation. Not all of the intermediate steps are shown. Each of the three aromatic amino acids is a feedback inhibitor (lines in color) of the first step in the pathway, catalyzed by the enzyme phospho-2-keto-3-deoxy-heptonate aldolase. Three distinct isozymes of this enzyme exist (I, II, III), each inhibited by a different aromatic amino acid. In addition, two isozymes (I, II) exist for the conversion of chorismic acid to prephenic acid, each of which is inhibited by a separate amino acid, phenylalanine or tyrosine. Tryptophan is also a feedback inhibitor of the enzyme catalyzing the first step specific for its synthesis, chorismic acid to anthranilic acid.

FIGURE A3.15 (opposite) Biosynthesis of the amino acid histidine. The five-membered imidazole ring of histidine is derived from three components (see lower left hand corner of diagram): two carbons from the ribose of phosphoribosyl-pyrophosphate, one nitrogen from the amide group of glutamine, and a nitrogen and carbon from the purine ring of ATP. During the dismemberment of the purine ring of ATP to obtain the nitrogen and carbon, another imidazole derivative is formed, AICAR (see lower left), which participates in the resynthesis of the purine ring of ATP (see Figure A3.17, purine ring synthesis). The amino acid side chain of histidine is formed from three carbon atoms of the ribose of phosphoribosylpyrophosphate (see right hand side of diagram). The final product, histidine, is a feedback inhibitor of the first enzyme in the pathway.

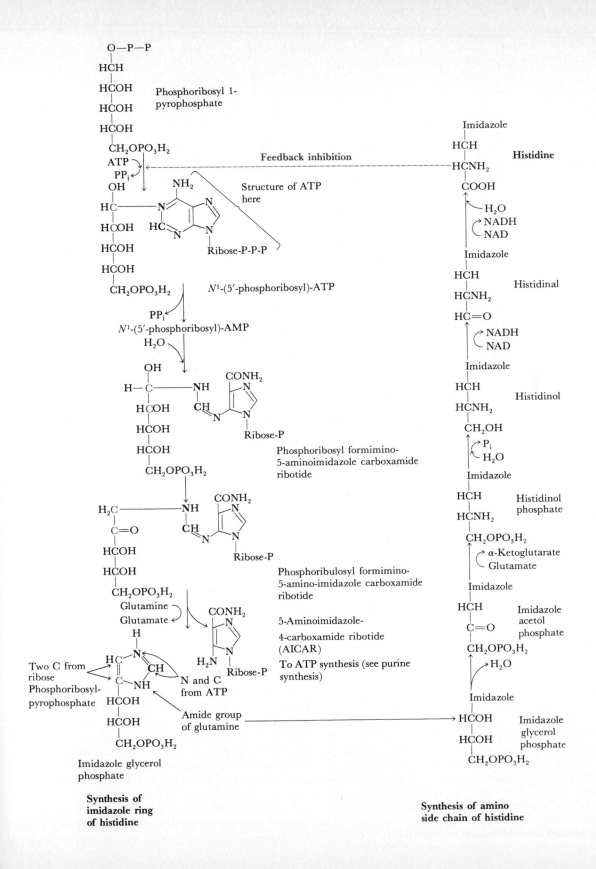

Phosphoribosyl 1-pyrophosphate

Feedback inhibition

Structure of ATP here

Ribose-P-P-P

N^1-(5'-phosphoribosyl)-ATP

N^1-(5'-phosphoribosyl)-AMP

Ribose-P

Phosphoribosyl formimino-5-aminoimidazole carboxamide ribotide

Ribose-P

Phosphoribulosyl formimino-5-amino-imidazole carboxamide ribotide

5-Aminoimidazole-4-carboxamide ribotide (AICAR)

To ATP synthesis (see purine synthesis)

N and C from ATP

Two C from ribose Phosphoribosyl-pyrophosphate

Amide group of glutamine

Imidazole glycerol phosphate

Synthesis of imidazole ring of histidine

Histidine

Histidinal

Histidinol

Histidinol phosphate

Imidazole acetol phosphate

Imidazole glycerol phosphate

Synthesis of amino side chain of histidine

799

FIGURE A3.16 Structure of the purine and pyrimidine nucleotides.

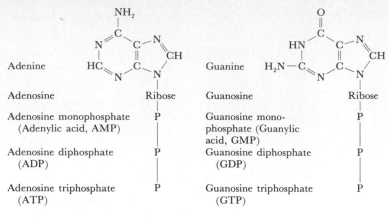

Adenine		Guanine	
Adenosine	Ribose	Guanosine	Ribose
Adenosine monophosphate (Adenylic acid, AMP)	P	Guanosine monophosphate (Guanylic acid, GMP)	P
Adenosine diphosphate (ADP)	P	Guanosine diphosphate (GDP)	P
Adenosine triphosphate (ATP)	P	Guanosine triphosphate (GTP)	P

Structure of pyrimidine nucleotides

Uracil	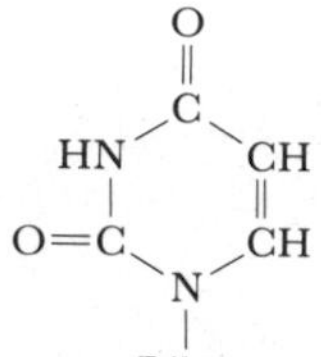	Cytosine	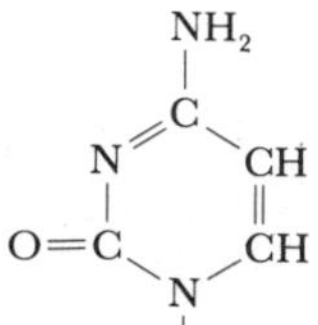
Uridine	Ribose	Cytidine	Ribose
Uridine monophosphate (Uridylic acid, UMP)	P	Cytidine monophosphate (Cytidylic acid, CMP)	P
Uridine diphosphate (UDP)	P	Cytidine diphosphate (CDP)	P
Uridine triphosphate (UTP)	P	Cytidine triphosphate (CTP)	P

The deoxynucleotides are the same as the ribonucleotides except deoxyribose is substituted for ribose. One pyrimidine, thymine (methyl-uracil), is found only as a deoxynucleotide.

Thymine

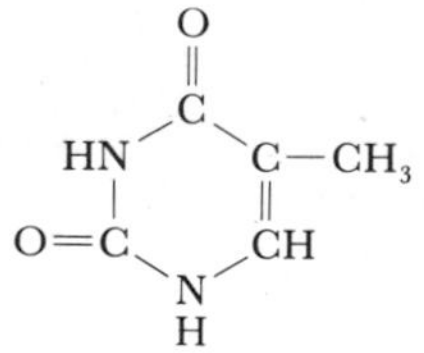

FIGURE A3.17 (opposite) Biosynthesis of the purine ring. The first purine synthesized is inosinic acid, from which the purines adenylic and guanylic acids are derived (see Figure A3.18). There are ten steps in the synthesis of the purine ring. The ring is built up on a ribose-phosphate carrier. (1) Attachment of an amino group to the ribose-phosphate carrier. This amino group is derived from the amide nitrogen of glutamine and will become the nitrogen in the 9 position of the purine ring. (2) The amino acid glycine is attached intact to the amino group. (3) A formyl group is attached, using tetrahydrofolic acid as carrier. This formyl group is derived either from formate or from the 3 carbon of the amino acid serine. (4) Addition of an amino group, derived from the amide of glutamine. (5) Closing of the imidazole ring. (6) Addition of a carboxyl group, derived from CO_2. (7) Formation of a peptide bond between this carboxyl group and the amino group of aspartate. (8) Retention of the amino group from aspartate, with the rest of this moiety being split off as fumarate. This leads to the formation of AICAR, a compound also formed as a byproduct of histidine biosynthesis (see Figure A3.15). (9) Formylation of AICAR by a formyl group carried by tetrahydrofolic acid coenzyme. (10) Closure of the pyrimidine portion of the purine ring, forming the first purine in the pathway, inosinic acid. The derivation of each atom of the purine ring is shown at the top of the diagram.

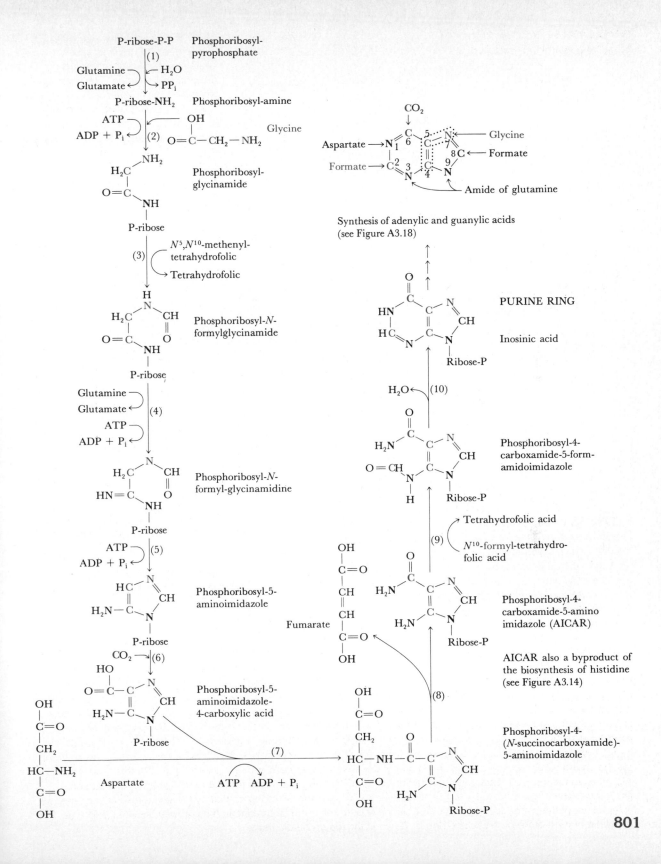

P-ribose-P-P   Phosphoribosyl-pyrophosphate

(1)
Glutamine → ← H₂O
Glutamate ← → PPᵢ

P-ribose-NH₂   Phosphoribosyl-amine

ATP →
ADP + Pᵢ ←   (2)   OH   Glycine
O=C—CH₂—NH₂

NH₂
H₂C   Phosphoribosyl-glycinamide
O=C
NH
P-ribose

(3)   N⁵,N¹⁰-methenyl-tetrahydrofolic
→ Tetrahydrofolic

H
N
H₂C   CH   Phosphoribosyl-N-formylglycinamide
O=C   O
NH
P-ribose

Glutamine →
Glutamate ←   (4)
ATP →
ADP + Pᵢ ←

N
H₂C   CH   Phosphoribosyl-N-formyl-glycinamidine
HN=C   O
NH
P-ribose

ATP →
ADP + Pᵢ ←   (5)

N
HC   CH   Phosphoribosyl-5-aminoimidazole
H₂N—C   N
P-ribose

CO₂ →   (6)
HO
O=C—C   N   Phosphoribosyl-5-aminoimidazole-4-carboxylic acid
H₂N—C   CH
N
P-ribose

OH
C=O
CH₂
HC—NH₂
C=O
OH
Aspartate   (7)   ATP   ADP + Pᵢ

CO₂
C   5   N   Glycine
N 1   6   C   7   Formate
Aspartate →   8 C ←
Formate →   C 2   3   C   9
N   4   N
Amide of glutamine

Synthesis of adenylic and guanylic acids (see Figure A3.18)

O
C   N   PURINE RING
HN   C   CH
HC   C   N   Inosinic acid
N   N
Ribose-P

H₂O ←   (10)

O
C   N   Phosphoribosyl-4-carboxamide-5-form-amidoimidazole
H₂N   C   CH
O=CH   N   C   N
H   Ribose-P

(9)   Tetrahydrofolic acid
N¹⁰-formyl-tetrahydro-folic acid

OH
C=O   O
CH   C   N   Phosphoribosyl-4-carboxamide-5-amino imidazole (AICAR)
CH   H₂N   C   CH
Fumarate   CH   H₂N   C   N
C=O   Ribose-P
OH

AICAR also a byproduct of the biosynthesis of histidine (see Figure A3.14)

OH
C=O   O
CH₂   C   N   Phosphoribosyl-4-(N-succinocarboxyamide)-5-aminoimidazole
(8)   HC—NH—C—C   CH
C=O   C   N
OH   H₂N   N
Ribose-P

APPENDIX A3.18 Biosynthesis of adenylic and guanylic acids, containing the key purine rings, adenine and guanine. The starting material is inosinic acid; the biosynthesis of inosinic acid was outlined in Figure A3.17. For the regulation of purine biosynthesis, see Figure A3.19.

GTP

GDP + P$_i$

Inosinic acid

Aspartate

COOH
|
HC—NH$_2$
|
CH$_2$
|
COOH

HOOC—CH$_2$—CH—COOH
|
NH

NAD

NADH

Xanthylic acid

Ribose-P

Adenylosuccinic acid

Ribose-P

COOH
|
CH
‖
Fumarate CH
|
COOH

Glutamine

Glutamate

ATP

AMP + PP$_i$

Ribose-P

Adenylic acid (AMP)

Ribose-P

Guanylic acid (GMP)

Phosphoribosyl-pyrophosphate

FIGURE A3.19 Regulation of purine biosynthesis. Feedback inhibition occurs at the first step in the pathway leading to the purine ring, by both adenine and guanine nucleotides. In addition, AMP and GMP inhibit the first steps specific to their own syntheses. A further control exists because ATP is required for one of the stages of GMP synthesis and GTP is required for one of the stages of AMP synthesis (see Figure 3.18), so that build-up of one of the two purine nucleotides would be expected to increase the rate of synthesis of the other. In this way, a balance would be maintained between the levels of the two nucleotides, both of which are required in about equal amounts for nucleic acid synthesis.

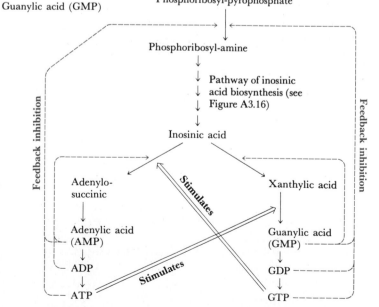

Feedback inhibition

Phosphoribosyl-amine

Pathway of inosinic acid biosynthesis (see Figure A3.16)

Inosinic acid

Adenylo-succinic

Stimulates

Xanthylic acid

Adenylic acid (AMP)

Guanylic acid (GMP)

ADP

GDP

Stimulates

ATP

GTP

Feedback inhibition

Bergey's Classification of Bacteria

KINGDOM: PROCARYOTAE
Division I: The Cyanobacteria
Division II: The Bacteria

Part 1: Phototrophic bacteria
Order 1: Rhodospirillales
Family I: Rhodospirillaceae
  Genus I: *Rhodospirillum*
  Genus II: *Rhodopseudomonas*
  Genus III: *Rhodomicrobium*
Family II: Chromatiaceae
  Genus I: *Chromatium*
  Genus II: *Thiocystis*
  Genus III: *Thiosarcina*
  Genus IV: *Thiospirillum*
  Genus V: *Thiocapsa*
  Genus VI: *Lamprocystis*
  Genus VII: *Thiodictyon*
  Genus VIII: *Thiopedia*
  Genus IX: *Amoebobacter*
  Genus X: *Ectothiorhodospira*
Family III: Chlorobiaceae
  Genus I: *Chlorobium*
  Genus II: *Prosthecochloris*
  Genus III: *Chloropseudomonas*
  Genus IV: *Pelodictyon*
  Genus V: *Clathrochloris*

Part 2: Gliding bacteria
Order I: Myxobacterales
Family I: Myxococcaceae
  Genus I: *Myxococcus*
Family II: Archangiaceae
  Genus I: *Archangium*
Family III: Cystobacteraceae
  Genus I: *Cystobacter*
  Genus II: *Melittangium*
  Genus III: *Stigmatella*
Family IV: Polyangiaceae
  Genus I: *Polyangium*
  Genus II: *Nannocystis*
  Genus III: *Chondromyces*

Order II: Cytophagales
Family I: Cytophagaceae
  Genus I: *Cytophaga*
  Genus II: *Flexibacter*
  Genus III: *Herpetosiphon*
  Genus IV: *Flexithrix*
  Genus V: *Saprospira*
  Genus VI: *Sporocytophaga*
Family II: Beggiatoaceae
  Genus I: *Beggiatoa*
  Genus II: *Vitreoscilla*
  Genus III: *Thioploca*

Family III: Simonsiellaceae
  Genus I: *Simonsiella*
  Genus II: *Alysiella*
Family IV: Leucothrichaceae
  Genus I: *Leucothrix*
  Genus II: *Thiothrix*
Families and genera of uncertain affiliation
  Genus: *Toxothrix*
Family: Achromatiaceae
  Genus: *Achromatium*
Family: Pelonemataceae
  Genus: *Pelonema*
  Genus: *Achroonema*
  Genus: *Peloploca*
  Genus: *Desmanthos*

Part 3: Sheathed bacteria
  Genus: *Sphaerotilus*
  Genus: *Leptothrix*
  Genus: *Streptothrix*
  Genus: *Lieskeella*
  Genus: *Phragmidiothrix*
  Genus: *Crenothrix*
  Genus: *Clonothrix*

Genus III: *Ochrobium*
Genus IV: *Siderococcus*

Part 13: Methane-producing bacteria
Family I: Methanobacteriaceae
Genus I: *Methanobacterium*
Genus II: *Methanosarcina*
Genus III: *Methanococcus*

Part 14: Gram-positive cocci
a. Aerobic and/or facultatively anaerobic
Family I: Micrococcaceae
Genus I: *Micrococcus*
Genus II: *Staphylococcus*
Genus III: *Planococcus*
Family II: Streptococcaceae
Genus I: *Streptococcus*
Genus II: *Leuconostoc*
Genus III: *Pediococcus*
Genus IV: *Aerococcus*
Genus V: *Gemelia*
b. Anaerobic
Family III: Peptococcaceae
Genus I: *Peptococcus*
Genus II: *Peptostreptococcus*
Genus III: *Ruminococcus*
Genus IV: *Sarcina*

Part 15: Endospore-forming rods and cocci
Family I: Bacillaceae
Genus I: *Bacillus*
Genus II: *Sporolactobacillus*
Genus III: *Clostridium*
Genus IV: *Desulfotomaculum*
Genus V: *Sporosarcina*
Genus of uncertain affiliation
Genus: *Oscillospira*

Part 16: Gram-positive, nonsporing, rod-shaped bacteria
Family I: Lactobacillaceae
Genus I: *Lactobacillus*
Genera of uncertain affiliation
Genus: *Listeria*
Genus: *Erysipelothrix*
Genus: *Caryophanon*

Part 17: Actinomyces and related organisms
Coryneform group of bacteria
Genus I: *Corynebacterium*
a. Human and animal pathogens
b. Plant-pathogenic corynebacteria
c. Nonpathogenic corynebacteria
Genus II: *Arthrobacter*
Genera incertae sedis
(*Brevibacterium*)
(*Microbacterium*)
Genus III: *Cellulomonas*
Genus IV: *Kurthia*
Family I: Propionibacteriaceae
Genus I: *Propionibacterium*
Genus II: *Eubacterium*

Order I: Actinomycetales
Family I: Actinomycetaceae
Genus I: *Actinomyces*
Genus II: *Arachnia*
Genus III: *Bifidobacterium*
Genus IV: *Bacterionema*
Genus V: *Rothia*
Family II: Mycobacteriaceae
Genus I: *Mycobacterium*
Family III: Frankiaceae
Genus I: *Frankia*
Family IV: Actinoplanaceae
Genus I: *Actinoplanes*
Genus II: *Spirillospora*
Genus III: *Streptosporangium*
Genus IV: *Amphosporangium*
Genus V: *Ampullariella*
Genus VI: *Pilimelia*
Genus VII: *Planomonospora*
Genus VIII: *Planobispora*
Genus IX: *Dactylosporangium*
Genus X: *Kitasatoa*
Family V: Dermatophilaceae
Genus I: *Dermatophilus*
Genus II: *Geodermatophilus*
Family VI: Nocardiaceae
Genus I: *Nocardia*
Genus II: *Pseudonocardia*
Family VII: Streptomycetaceae
Genus I: *Streptomyces*
Genus II: *Streptoverticillium*
Genus III: *Sporichthya*
Genus IV: *Microellobospora*

Family VIII: Micromonosporaceae
Genus I: *Micromonospora*
Genus II: *Thermoactinomyces*
Genus III: *Actinobifida*
Genus IV: *Thermomonospora*
Genus V: *Microbiospora*
Genus VI: *Micropolyspora*

Part 18: The Rickettsias
Order I: Rickettsiales
Family I: Rickettsiaceae
Tribe I: Rickettsieae
Genus I: *Rickettsia*
Genus II: *Rochalimea*
Genus III: *Coxiella*
Tribe II: Ehrlichieae
Genus IV: *Ehrlichia*
Genus V: *Cowdria*
Genus VI: *Neorickettsia*
Tribe III: Wolbachieae
Genus VII: *Wolbachia*
Genus VIII: *Symbiotes*
Genus IX: *Blattabacterium*
Genus X: *Rickettsiella*
Family II: Bartonellaceae
Genus I: *Bartonella*
Genus II: *Grahamelia*
Family III: Anaplasmataceae
Genus I: *Anaplasma*
Genus II: *Paranaplasma*
Genus III: *Aegyptionella*
Genus IV: *Haemobartonella*
Genus V: *Eperythrozoon*
Order II: Chalamydiales
Family I: Chlamydiaceae
Genus I: *Chlamydia*

Part 19: The Mycoplasmas
CLASS I: MOLLICUTES
Order I: Mycoplasmatales
Family I: Mycoplasmataceae
Genus I: *Mycoplasma*
Family II: Acholeplasmataceae
Genus I: *Acholeplasma*
Genus of uncertain affiliation
Genus: *Thermoplasma*
Mycoplasmalike bodies in plants

Bergey's Manual of Systematic Bacteriology

Bergey's Manual has recently been revised, and the material will be published in four volumes. Volume 1 was published in 1984—subsequent volumes will be published in approximately two year intervals. The name of the work has been changed from *Bergey's Manual of Determinative Bacteriology* to *Bergey's Manual of Systematic Bacteriology,* and the new *Manual* will reflect the enormous body of new taxonomic information published since the early 1970s. The contents of each volume is broken into several sections, with each section containing a number of genera. The bacteria have been partitioned among the four volumes as follows:

Volume I. **Gram-negative bacteria of medical and commercial importance:** spirochetes, spiral and curved bacteria, Gram-negative aerobic and facultatively aerobic rods, Gram-negative obligate anaerobes, Gram-negative aerobic and anaerobic cocci, rickettsias, mycoplasmas.

Volume II. **Gram-positive bacteria of medical and commercial importance:** Gram-positive cocci, Gram-positive endospore-forming and nonsporing rods, and the non-filamentous actinomycetes.

Volume III. **Remaining Gram-negative bacteria:** phototrophic, gliding, sheathed, and appendaged bacteria, cyanobacteria, lithotrophic bacteria, and the archaebacteria (methanogens, extreme halophiles, and the thermoacidophiles).

Volume IV. **Filamentous actinomycetes and related bacteria.**

Microscopy

A compound microscope (Figure A5.1) has two lenses, the *objective lens,* placed close to the object to be viewed, and the *ocular lens,* placed close to the eye. The primary enlargement of the object is produced by the objective lens, and the image so produced is transmitted to the ocular, where the final enlargement occurs.

The total magnification of a compound microscope is the product of the magnifications of its objective and ocular lenses. With an objective magnifying 40X and an ocular magnifying 10X, the total magnification is 400X. The compound microscope is therefore able to achieve considerably greater powers of magnification than a microscope constructed of only a single lens. The latter, called a *simple microscope,* is used mainly in hand lenses and magnifying glasses.

An important property of a microscope is its *resolving power*—its ability to show as distinct and separate two points that are close together. The greater the resolving power, the greater the definition of an object; hence microscopes with high resolving power are especially good for viewing small structures. Resolving powers in a compound microscope depends on the wavelength of light used and an optical property of the objective lens known as its *numerical aperture.* The formula for calculating resolving power is: diameter of smallest resolvable object (resolving power) = 0.5(wavelength of light/numerical aperture of objective). Since the wavelength of light is usually fixed, the resolution of an object is in practice a function of the numerical aperture: the larger the numerical aperture, the smaller the object resolved (Table A5.1). There is a rough correspondence between the magnification of an objective lens and its numerical aperture: lenses with higher magnification usually have higher numerical apertures. (The value of the numerical aperture is printed on the side of the lens.) But the medium through which the light passes also affects numerical aperture. As long as the objective is separated from the object by air, its numerical aperture can never be greater than 1.0;

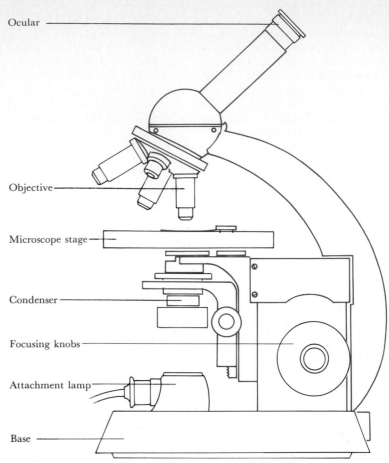

Ocular

Objective

Microscope stage

Condenser

Focusing knobs

Attachment lamp

Base

FIGURE A5.1 Diagram of a modern light microscope. The light source can be either built-in (as in the diagram) or external.

TABLE A5.1 Optical Data for Commonly Used Objective Lenses*

	Objective Magnification		
	10×	40×	90×[†]
Magnification with 10× eyepiece	100×	400×	900×
Numerical aperture	0.25	0.76	1.30
Resolving power, μm	2.0	0.45	0.27
Approximate depth of field, μm	7.0	1.3	0.5
Approximate area of field with 10× eyepiece, mm	1.5	0.35	0.17

*The values given are for average lenses; different manufacturers produce lenses of various specifications.

[†]Oil immersion.

to achieve values greater than this the objective must be immersed in a medium of higher refractive index than that of air. Oils of various kinds are used, and lenses designed for use with oil are called *oil-immersion lenses.* The numerical aperture of a high-quality oil-immersion lens is between 1.2 and 1.4. Although oil-immersion lenses are usually of higher magnification than are lenses designed for dry use, they need not be so. An oil-immersion lens used in air presents a very unsatisfactory image, however; the lens should never be employed without the use of oil.

According to optical theory, the highest resolution possible in a compound light microscope will permit the visualization of an object whose diameter is about 0.2μm (micrometer), or 0.0002 mm (1 μm is 0.001 mm). If an object 0.2μm in diameter is magnified 1,000 times, it will appear to the eye as an object 0.2mm in diameter, which can easily be seen. With the compound microscope, magnifications greater than 1,000 are attainable by choice of appropriate oculars, but increased magnification does not lead to increased resolution, although it may make an object easier to see. High magnification that increases the size of an object but does not increase its detail is called *empty magnification.*

Most microscopes used in microbiology have oculars that magnify about $10\times$ and objectives of $10\times$, $40\times$, and $90\times$ or $100\times$ (oil immersion) (Table A5.1). The low-power lens is used for scanning the specimen to locate objects of interest; the $40\times$ lens permits the detailed visualization of large microorganisms such as algae, protozoa, and fungi: and the $90\times$ or $100\times$ lens is used for viewing bacteria and small eucaryotic microorganisms.

The illumination system of a microscope is also of considerable importance, especially when high magnifications are used. The light entering the system must be focused on the specimen, and a *condenser lens system* is used for this purpose. By raising or lowering the condenser the plane of focus of the light can be altered and a position can be chosen that leads to precise focus. The condenser lens system also has an *iris diaphragm,* which controls the diameter of the circle of light as it leaves the condenser system. The purpose of the iris diaphragm is not to control the intensity of light falling on the object but to ensure that the light leaving the condenser system just fills the objective lens. If the iris diaphragm is too large, some of the light will pass not into the objective but around it and will cause glare, thus reducing the clarity of the image. If the light is too bright, it should be reduced not by altering the position of the condenser or the iris diaphragm but by using neutral density filters or by decreasing the voltage to the lamp itself. Specimens vary in degree of contrast, and the light must be adjusted carefully for each object being examined. It cannot be too strongly emphasized that the proper adjustment of the light is crucial to good microscopy, especially at higher magnifications.

Two other factors related to the lens system used are depth of field and area of field (Table A5.1). *Depth of field* means the thickness of the specimen in focus at any one time, and is greater with low power than with high power. Under oil immersion the depth of field is very shallow, usually less than 1 μm. The *area of field* is represented by the diameter of the specimen that is within view. Area of field is greater at low power than at high power, and it is for this reason that low-power lenses are useful for scanning a slide.

Phase-contrast microscopy   The phase-contrast microscope makes it possible to see small cells easily even without staining. Cells differ in refractive index from their surrounding medium, and this difference can be used to create an image with a much higher degree of contrast than can be obtained with the normal light microscope. In Figure A5.2 the light path of the phase-contrast microscope is compared with that of the conventional bright-field microscope. In the phase-contrast microscope the source of illumination is a hollow cone of light coming through an annular (ring) stop in the condenser lens. In the objective lens a phase ring shifts the phase of the light going through it by a quarter of a wavelength. Light passing unretarded through the slide passes through the phase ring and is seen as normal white light by the eye. Light passing through a specimen of refractive index different from the medium is retarded and has a longer light path, thus arriving at the ocular out of phase. Interference between

FIGURE A5.2   Light paths of (a) bright-field, (b) phase-contrast, and (c) dark-field microscopes. Images seen by these three optical systems are shown in Figure 2.1.

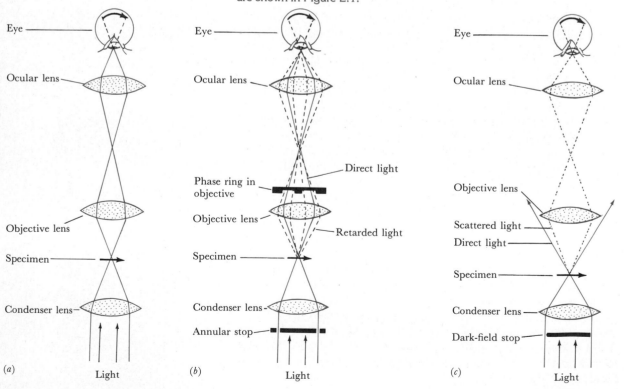

retarded and unretarded light produces an image of the specimen. In most phase-contrast microscopes the image appears dark on a light background, although with a different phase ring in the objective lens the image can appear bright on a dark background. Since the image depends on differences in refractive index between the specimen and the surrounding medium, the image disappears if the specimen is immersed in a liquid of identical refractive index. To achieve best results with the phase-contrast microscope, it is essential that the optical system be accurately aligned, and a special device, the centering telescope, is used to adjust the annular stop in relation to the phase ring. One point of caution: because of the phase plate, the phase-contrast objective is less suitable than a normal objective for observing colored material, so that bright-field objectives are still necessary and should always be used for observing stained specimens. Because a lot of the available light is eliminated by the annular stop, a fairly intense light source is necessary for effective phase microscopy. This is one of the main reasons phase microscopes are not more commonly used in teaching situations.

Dark-field microscopy The dark-field microscope relies on the scattering of light by the specimen to render it visible. The dark-field condenser has an opaque circle in its center so that none of the direct light reaches the specimen (Figure A5.2). The light which reaches the specimen comes from the outside of this opaque circle and passes at an angle through the specimen, so that it does not reach the objective at all. When looking through a dark-field microscope in the absence of a specimen, nothing is seen, only blackness. Specimens scatter some of the light which strikes them, and this scattered light does reach the objective. The specimen thus appears as a bright object surrounded by a black background (Figure 2.1). Contrast in dark-field microscopy is extremely good, but only the outlines of the specimen can be clearly seen, interior structures generally being invisible. Considerable care must be taken to ensure that the condenser is adjusted properly, otherwise the specimen may not be seen at all. Resolution by dark-field microscopy is quite high, and objects can be seen that are even smaller than the theoretical resolution of the light microscope. This is because the scattering of light leads to an enlargement of the object size. Even large virus particles can be seen by dark-field microscopy, although they are normally invisible in the light microscope. Because mineral particles and other forms of tiny dirt will also scatter light, it is important that the material being examined be suspended in clean, preferably filtered, water. Because most of the light does not reach the objective, a very intense light source is necessary for dark-field microscopy, especially at high magnifications.

Electron microscopy The principles of electron microscopy were discussed in Section 2.1. The layout and function of the conventional electron microscope (sometimes called the transmission electron microscope) and the scanning electron microscope are given in Figures A5.3 through A5.5.

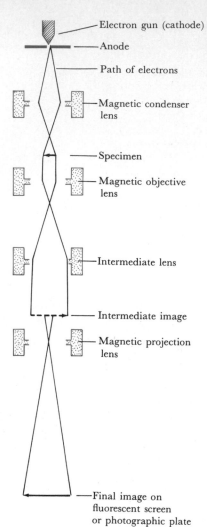

Electron gun (cathode)

Anode

Path of electrons

Magnetic condenser lens

Specimen

Magnetic objective lens

Intermediate lens

Intermediate image

Magnetic projection lens

Final image on fluorescent screen or photographic plate

FIGURE A5.3 Path of electrons in an electron microscope. The whole system is operated in a high vacuum. Compare with the path of light through an optical microscope in Figure A5.2.

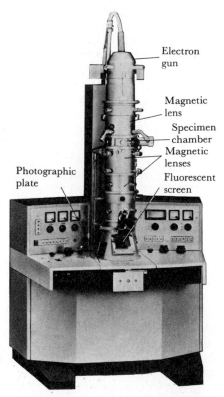

Electron gun

Magnetic lens

Specimen chamber

Magnetic lenses

Fluorescent screen

Photographic plate

FIGURE A5.4 A modern electron microscope. (Courtesy of Siemens Corporation.)

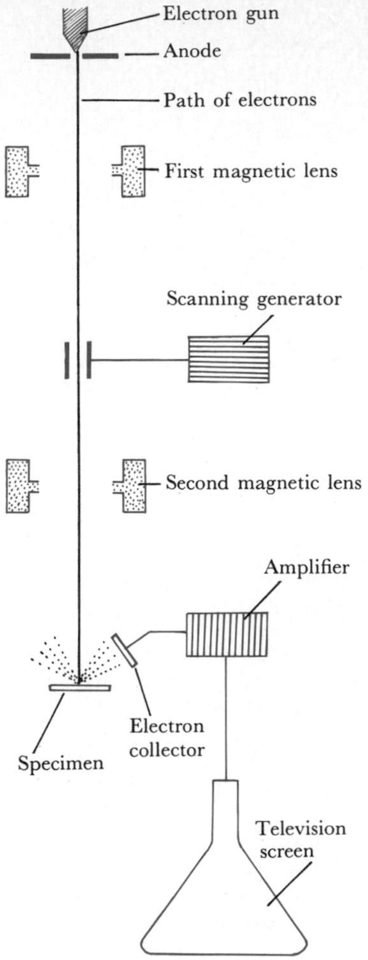

Electron gun
Anode
Path of electrons
First magnetic lens

Scanning generator

Second magnetic lens

Amplifier

Specimen
Electron collector

Television screen

FIGURE A5.5 Path of electrons in a scanning electron microscope.

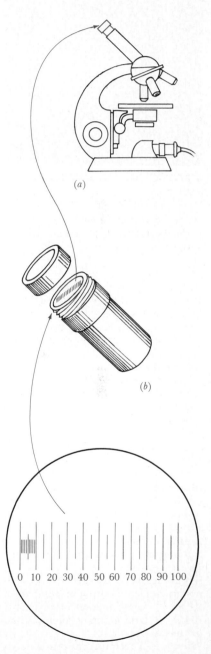

(a)

(b)

FIGURE A5.6 Ocular micrometer disk (c) used for measuring cell size. The disk fits into the ocular (b), which in turn is positioned in the light microscope (a). The cell is placed under the microscope so that the dimension to be measured extends along the length of the micrometer rule. The ocular micrometer must be calibrated for each microscope and for each objective of the microscope; once calibrated, the lines on the ocular micrometer will be equivalent to micrometers or parts of micrometers. Size is now measured by moving the cell so that it is optically adjacent to the rule of the ocular, and its size read.

0 10 20 30 40 50 60 70 80 90 100

Glossary

Numbers within parentheses indicate the pages on which defined words are first used in the text. If a term is not here, consult the index.

Accessory pigments Colored compounds other than chlorophyll which absorb and transfer light energy to chlorophyll (193).

Acetogenic bacteria Organisms capable of reducing CO_2 to acetic acid or converting sugars quantitatively into acetate (723).

Acid-alcohol fastness A staining property of *Mycobacterium* species where cells stained with hot carbolfuschin will not decolorize with acid-alcohol (742).

Activation energy Energy needed to make substrate molecules more reactive; enzymes function by lowering activation energy (98).

Active immunity An immune state achieved by self-production of antibodies. Compare with passive immunity (554).

Active site The portion of an enzyme which specifically combines with the substrate (99).

Aerobe An organism that grows in the presence of O_2; may be facultative or obligate (260).

Aerosol Suspension of particles in airborne water droplets.

Aerotolerant Of an anaerobe, not being inhibited by O_2 (261).

Agglutination Reaction between antibody and cell-bound antigen resulting in clumping of the cells (530).

Akinete A resting spore of a cyanobacterium (648).

Algae A large group of chlorophyllous eucaryotic microorganisms.

Allergy A harmful antigen-antibody reaction, usually caused by a foreign antigen in food, pollen, or chemicals; immediate-type hypersensitivity (550).

Allosteric Characteristic of some proteins, especially enzymes, in which a compound combines with a site on the protein other than the active site. The result is a change in conformation at the active site; useful in regulating activity (159).

Amphibolic pathway A biochemical pathway that serves the dual functions of catabolism and anabolism (131).

Anabolism The biochemical processes involved in the synthesis of cell constituents from simpler molecules, usually requiring energy (130).

Anaerobe An organism that grows in the absence of oxygen (261).

Anaerobic respiration Use of an electron acceptor other than O_2 in an electron-transport oxidation. Most common anaerobic electron acceptors are nitrate, sulfate, and carbonate (123).

Anaphylaxis (anaphylactic shock) A violent allergic reaction caused by an antigen-antibody reaction (550).

Anaplerotic reaction A reaction that replenishes intermediates that are removed from a biochemical cycle for biosynthesis (131).

Anoxygenic photosynthesis Use of light energy to synthesize ATP by cyclic photophosphorylation in green and purple bacteria (184).

Antibiotic A chemical agent produced by one organism that is harmful to other organisms (226).

Antibody A protein present in serum or other body fluid that combines specifically with antigen (522).

Anticodon A sequence of three purine and pyrimidine bases in transfer RNA that is complementary to the codon in messenger RNA (288).

Antigen A substance, usually macromolecular, that induces specific antibody formation (521).

Antimicrobial Harmful to microorganisms by either killing or inhibiting growth (226).

Antiseptic An agent that kills or inhibits growth, but is not harmful to human tissue (226).

Antiserum A serum containing antibodies (556).

Antitoxin An antibody active against a toxin (528).

Archaebacteria A group of unusual procaryotes, including the methanogenic, extremely halophilic, and thermoacidophilic bacteria (765).

Ascospore The sexual spore of an Ascomycete fungus (82).

Ascus A saclike structure in Ascomycete fungi within which ascospores are borne (82).

Aseptic technique Laboratory procedures that prevent contamination (6).

Asexual Reproduction, either vegetative or involving spore formation, which occurs without sexual processes.

Attenuation Selection from a pathogen of nonvirulent strains still capable of immunizing (498). Also: a process that plays a role in the regulation of enzymes involved in amino acid biosynthesis (299).

Autoclave A device for sterilizing by heat under steam pressure (248).

Autoimmunity Immune reactions of a host against its own self constituents (551).

Autolysis Spontaneous lysis (31).

Autotroph Organism able to utilize CO_2 as sole source of carbon (181).

Auxospore A spore in diatom algae that leads to reformation of an enlarged vegetative cell (77).

Auxotroph A mutant that has a growth factor requirement. Contrast with a prototroph (304).

Axenic Pure, uncontaminated; an axenic culture is a pure culture.

Axial fibrils Flagella-like structures involved in motility in spirochetes. Each axial fibril is inserted near the end of the protoplasmic cylinder and is intertwined with the cylinder, beneath the outer sheath.

Axial filament See *axial fibrils*.

Bacteremia The presence of bacteria in the blood (487).

Bacteria The whole group of procaryotic microorganisms.

Bactericidal Capable of killing bacteria (226).

Bacteriophage A virus that infects bacteria (316).

Bacteriostatic Capable of inhibiting bacterial growth without killing (226).

Bacteroid A swollen, deformed *Rhizobium* cell, found in the root nodule (476).

Barophile An organism able to live optimally at high hydrostatic pressure (257).

Barotolerant An organism able to tolerate high hydrostatic pressure, although growing better at normal pressures (257).

Basidiospore A sexual spore of a Basidiomycete (84).

Basidium A structure of a Basidiomycete fungus that bears spores called basidiospores (84).

Biliprotein Accessory pigment for photosynthesis; phycobilin (193).

Biodegradable Capable of being broken down by living organisms; usually used in reference to artificial organic compounds such as pesticides (450).

Biogeochemical cycle The conversion of an element from oxidized to reduced form (or vice versa) by microorganisms (404).

Bioluminescence Living organisms' light production (699).

Biotechnology The discipline dealing with all phases of the production of useful products by means of microorganisms (381).

Budding A process of cell division in which the mother cell retains its identity, the daughter cell forming by growth of a new cell upon one part of the mother (661).

Calorie A unit of heat or energy; that amount of heat required to raise the temperature of 1 gram of water by $1°C$ (96).

Calvin cycle A series of reactions used by autotrophs to fix carbon dioxide into organic compounds (196).

Capsid The protein coat of a virus (317).

Capsomere An individual protein subunit of the virus capsid (317).

Capsule A compact layer of polysaccharide exterior to the cell wall in some bacteria (47). See also *Glycocalyx* and *Slime layer*.

Carcinogen A substance that causes the initiation of tumor formation. Frequently, a mutagen (311).

Carrier An individual that continually releases infective organisms but does not show symptoms of disease (567).

Catabolism The biochemical processes involved in the breakdown of organic compounds, usually leading to the production of energy (95).

Catalyst A substance that promotes a chemical reaction without itself being changed in the end (98).

Cell The fundamental unit of living matter, an entity isolated from other cells by a membrane (1).

Cell-mediated immunity An immune response generated by the activities of non-antibody-producing cells. Compare with *Humoral immunity* (520).

Chelator A compound that combines with a metal and keeps it in solution (168).

Chemiosmosis The use of ion gradients across membranes, especially proton gradients, to generate ATP See *Proton-motive force*.

Chemoautotroph An autotrophic organism obtaining energy from the oxidation of inorganic compounds.

Chemolithotroph An organism that uses an inorganic compound as energy source.

Chemoorganotroph See *Heterotroph*.

Chemoprophylaxis The use of chemicals or antibodies to prevent future infection or disease (517).

Chemostat A continuous culture device controlled by the concentration of limiting nutrient (222).

Chemotaxis Movement toward or away from a chemical (42).

Chemotherapy Treatment of infectious disease with chemicals or antibiotics (512).

Chlorination The practice of adding small amounts of chlorine to drinking water to ensure microbiological safety (593).

Chloroplast A green, chlorophyll-containing organelle in eucaryotes; site of photosynthesis (61).

Chlorosomes The photosynthetic apparatus of the green bacteria. Contain bacteriochlorophylls c, c_s, d, or e (631).

Chromatophores Intracellular membranes of purple photosynthetic bacteria. Contain bacteriochlorophylls and carotenoids (630).

Chromogenic Producing color; a chromogenic colony is a pigmented colony.

Chromosome The structure that contains the DNA in eucaryotes, usually complexed with histones (68).

Cilium Short, filamentous structure that beats with many others to make a cell move (63).

Clone A population of cells all descended from a single cell (304); a number of copies of a DNA fragment obtained by allowing the inserted DNA fragment to be replicated by a phage or plasmid (381).

Cloning vector A DNA molecule that is able to bring about the replication of foreign DNA fragments (381).

Coccoid Sphere-shaped.

Coccus A spherical bacterium (21).

Codon A sequence of three purine and pyrimidine bases on DNA or messenger RNA that codes for a specific amino acid (273, 302).

Coenobium A group of cells that have remained together after division, either in an irregular or regular array (74).

Coenocytic A growth form in which several nuclei and protoplasmic components mix freely; common to fungi (82).

Coenzyme A low-molecular-weight chemical which participates in an enzymatic reaction by accepting and donating electrons or functional groups. Examples: NAD, FAD. (100).

Coliforms Gram-negative, nonsporing, facultative rods that ferment lactose with gas formation within 48 hours at 35°C (588).

Colony A population of cells growing on solid medium, arising from a single cell (9).

Cometabolism The degradation of a compound only in the presence of other organic material which serves as the primary energy source (449).

Commensalism A relationship between two organisms in which one of the organisms is benefited and the other is unaffected (452).

Competence Ability to take up DNA and become genetically transformed (352).

Complement A complex of proteins in the blood serum that acts in concert with specific antibody in certain kinds of antigen-antibody reactions (534).

Complementary Nonidentical but related genetic structures that show precise pairing (275).

Concatamer A DNA molecule consisting of two or more separate molecules linked end-to-end to form a long linear structure (334).

Conidiophore In a fungus, a hyphal structure bearing conidia (82).

Conidium An asexual spore formed by Zygomycete fungi or actinomycete bacteria. Synonym: conidiospore (82).

Conjugation In eucaryotes, the process by which haploid gametes fuse to form a diploid zygote (69); in procaryotes, transfer of genetic information from one cell to another by cell-to-cell contact (360).

Consortium A two-membered bacterial culture (or natural assemblage) in which each organism benefits from the other (641).

Constitutive enzyme One synthesized continuously during growth, not subject to induction or repression (301).

Contagious Of a disease, transmissible (7).

Cortex The region inside the spore coat of an endospore, around the core (51).

Crista Inner membrane in a mitochondrion, site of respiration (61).

Cryophile Psychrophile (241).

Culture A particular strain or kind of organism growing in a laboratory medium (9).

Cutaneous Relating to the skin.

Cyanobacteria Blue-green bacteria (formerly blue-green algae); photosynthetic procaryotes which perform oxygenic photosynthesis (2).

Cyclic photophosphorylation Light-driven ATP synthesis in which no external electron donor or acceptor is involved (185).

Cyst A resting stage formed by some bacteria and protozoa in which the whole cell is surrounded by protective layer; not the same as spore.

Cytochrome Iron-containing porphyrin rings complexed with proteins, which act as electron carriers in the electron-transport system (115).

Cytokinesis Separation of cytoplasm into daughter cells following nuclear division (70).

Cytoplasm Cellular contents inside the plasma membrane, excluding the nucleus.

Decimal reduction time Time at a particular temperature to reduce the viable population by 90 percent (248).

Decomposers Organisms that break down dead organic material (403).

Degenerate code The genetic code is said to be degenerate because many amino acids are coded for by more than one codon (307).

Deletion A removal of a portion of a gene (307).

Denitrification Conversion of nitrate into nitrogen gases under anaerobic conditions, resulting in loss of nitrogen from ecosystems (123, 422).

Deoxyribonucleic acid (DNA) A polymer of nucleotides connected via a phosphatedeoxyribose sugar backbone; the genetic material of the cell (275).

Desiccation Drying.

Diauxic growth   Growth occurring in two separate phases between which a temporary lag occurs (298).

Differential stain   A staining procedure in which different organisms or parts of organisms are stained in different ways (16).

Diploid   In eucaryotes, an organism or cell with two chromosome complements, one derived from each haploid gamete (69).

Direct count   Measurement of cell concentration by use of a microscope and a calibrated counting chamber (216).

Disinfectant   An agent that kills microorganisms, but may be harmful to human tissue (226).

Dispersal   Processes involved in spread of organisms through the environment.

Doubling time   The time needed for a population to double (214). See also *Generation time.*

Ecosystem   The total community of living organisms, together with their physical and chemical environment (403).

Electron acceptor   A substance that accepts electrons during an oxidation-reduction reaction. An electron acceptor is an oxidant (101).

Electron donor   A compound that donates electrons in an oxidation-reduction reaction. An electron donor is a reductant (101). See also *Primary electron donor.*

Electron-transport phosphorylation   Synthesis of ATP involving a membrane-associated electron transport chain and the creation of a proton-motive force. Previously termed oxidative phosphorylation (108). See also *Chemiosmosis.*

Endemic   A disease that is constantly present in low numbers in a population (570). Compare with *Epidemic.*

Endergonic   A chemical reaction requiring input of energy to proceed (97).

Endocytosis   A process in which a particle such as a virus is taken intact into an animal cell. Phagocytosis and pinocytosis are two kinds of endocytosis (324).

Endoplasmic reticulum   An extensive array of internal membranes in eucaryotes (60).

Endospore   A bacterial spore formed within the cell and extremely resistant to heat as well as to other harmful agents (51).

Endosymbiosis   The hypothesis that mitochondria and chloroplasts are the descendants of ancient procaryotic organisms (62).

Endotoxin   A toxin not released from the cell; bound to the cell surface or intracellular (493). Compare with *Exotoxin.*

Enrichment culture method   A technique for isolating organisms from nature using media and growth conditions highly selective for a particular type (617).

Enteric   Intestinal.

Enterotoxin   A toxin affecting the intestine (494).

Entner-Doudoroff pathway   A reaction sequence used by some bacteria to catabolize glucose (791).

Enzyme   A protein functioning as the catalyst of living organisms, which promotes specific reactions or groups of reactions (98).

Epidemic   A disease occurring in an unusual number of individuals in a community at the same time (570). Compare with *Endemic.*

Epidemiology   The study of the incidence and prevalence of disease in populations (564).

Epilimnion   The upper region of a lake, subject to wind mixing. Above the thermocline (409).

Epiphyte   An organism living on the surface of a plant.

Epitheca   The outer half wall of a diatom cell (78).

Eucaryote   A cell or organism having a true nucleus (2).

Eurythermal   Growing over a wide temperature range (240).

Eutrophication   Nutrient enrichment of natural waters, usually from artificial sources, which frequently leads to excessive algal growth (425).

Exergonic reaction   A chemical reaction that proceeds with the liberation of heat (97).

Exon   The coding sequences in a split gene. Contrasted with *Introns,* the intervening noncoding regions (274).

Exotoxin   A toxin released extracellularly (487). Compare with *Endotoxin.*

Exponential phase   A period during the growth cycle of a population in which growth increases at an exponential rate (214).

Expression   The ability of a gene to function within a cell in such a way that the gene product is formed (386).

Facultative   A qualifying adjective indicating that an organism is able to grow either in the presence or absence of an environmental factor (e.g., "facultative aerobe," "facultative psychrophile").

Feedback inhibition   Inhibition by an end product of the biosynthetic pathway involved in its synthesis (159).

Fermentation   Catabolic reactions producing ATP in which organic compounds serve as both primary electron donor and ultimate electron acceptor (108).

Fermentation (industrial)   A large-scale microbial process.

Ferredoxin   An electron carrier of low reduction potential; small protein containing iron (113).

Filamentous   In the form of very long rods, many times longer than wide (21).

Fimbria (plural fimbriae) Short filamentous structure on a bacterial cell; although flagella-like in structure, generally present in many copies and not involved in motility. Plays role in adherence to surfaces and in the formation of pellicles (46). See also *Pilus.*

Flagellum   An organ of motility (37, 62).

Flavoprotein   A protein containing a derivative of riboflavin, which acts as electron carrier in the electron-transport system (114).

Fluorescent   Having the ability to emit light of a certain wavelength when activated by light of another wavelength.

Fluorescent antibody   Immunoglobulin molecule which has been coupled with a fluorescent dye so that it exhibits the property of fluorescence (530).

Formal reaction An equation describing an electrochemical reaction in which electrons are included as reactants (101). See also *Half-reaction*.

Fractional sterilization A process in which some material is made sterile by several brief applications of heat over a period of time (7). Synonym; *Tyndallization*.

Frame shift Since the genetic code is read three bases at a time, if reading begins at either the second or third base of a codon, a faulty product usually results. This is called a frame shift (the reading frame refers to the pattern of reading) (308).

Free energy Energy available to do useful work (97).

Fruiting body A macroscopic reproductive structure produced by some fungi (e.g., mushrooms) and some bacteria (e.g., Myxobacteria). Fruiting bodies are distinct in size, shape, and coloration for each species (87, 650).

Fungi A large group of colorless eucaryotic microorganisms.

Gametangia Hyphal portions of Zygomycetes which fuse to form zygospores in sexual recombination (82).

Gametes In eucaryotes, haploid cells analogous to sperm and egg, which conjugate and form a diploid (69).

Gas vesicle A gas-filled structure in certain procaryotes that confers ability to float (49). Sometimes called *gas vacuole*.

Gene A unit of heredity; a segment of DNA specifying a particular protein or polypeptide chain.

Generation time Time needed for a population to double (214). See also *Doubling time*.

Genetic engineering Procedures involved in the isolation, manipulation, and expression of genetic material, either for basic research or in the development of industrial processes (379).

Genetic marker Any mutant gene useful in genetic analysis.

Genome The complete set of genes present in an organism.

Genotype The genetic complement of organisms. Compare with *Phenotype*.

Genus A group of related species (2, 614).

Geosmin A compound produced by many cyanobacteria and streptomycetes which causes the "earthy" odor of soil (649).

Germ free Devoid of microorganisms.

Germ-free animals Animals that have been raised since birth in a sterile environment and which contain no microbial flora (466).

Germicide A substance that inhibits or kills microorganisms (226).

Gliding motility A property found in many bacteria of motility without the agency of flagella or other exterior organs of motility (650).

Gluconeogenesis Synthesis of glucose from smaller molecules (139).

Glycocalyx General term for polysaccharide components outside the bacterial cell wall (47). See also *Capsule* and *Slime layer*.

Glycolysis Reactions of the Embden-Meyerhof pathway in which glucose is oxidized to pyruvate (109).

Glyoxylate cycle A pathway which is a branch from the tricarboxylic acid cycle; useful in replenishing oxalacetate for further TCA cycle activity (142).

Gnotobiotic A system, usually animal, in which the composition of the microbial flora is known (467). Compare to *Germ free*.

Growth factor An organic compound required in very small amounts as a nutrient (171).

Growth rate The rate at which growth occurs, usually expressed as the generation time (213).

Half-reaction A reduction or oxidation written alone. Two half-reactions may be combined to form a complete electrochemical reaction (101). See also *Formal reaction*.

Halophile Organism requiring salt (usually NaCl) for growth (253).

Haploid In eucaryotes, an organism or cell containing one chromosome complement and the same number of chromosomes as the gametes (69).

Hapten A substance not inducing antibody formation but able to combine with a specific antibody (522).

Haustorium A special absorptive hypha of a fungus (453).

Hemagglutination Agglutination of red blood cells (530).

Hemolysis Lysis of red blood cells (146, 488).

Herbicide A chemical used to control growth of undesirable plants such as weeds (448).

Heterocyst Specialized cell in filamentous cyanobacteria frequently associated with nitrogen fixation (646).

Heteroduplex A double-stranded DNA in which one strand is from one source and the other strand from another, usually related, source (280).

Heterofermentation Fermentation of glucose or other sugar to a mixture of products (732).

Heterologous antigen An antigen reacting with an antibody other than the one it induced (522).

Heterothallic Sexual recombination requiring two different mating types (82).

Heterotroph Organism obtaining carbon from organic compounds (181).

Hexose monophosphate pathway (HMP) Reaction sequence used primarily to synthesize pentoses from hexoses (136, 792).

Holdfast An adhesive material at a localized position on a cell, enabling the cell to attach to a surface (664).

Homofermentation Fermentation of glucose or other sugar leading to virtually a single product, lactic acid (732).

Homologous antigen An antigen reacting with the antibody it had induced (522).

Homothallic Sexual recombination requiring only one mating type (82).

Hormogonium A motile segment of a filamentous cyanobacterium, usually involved in dispersal (648).

Host An organism capable of supporting the growth of a virus or parasite.

Humoral immunity An immune response involving the activities of antibodies (520).

Hybridoma The fusion of a malignant cell with a single B-lymphocyte to produce a malignant lymphocyte producing monoclonal antibody (546).

Hydrolysis Breakdown of a polymer into smaller units, usually monomers, by addition of water; digestion (137).

Hypersensitivity An immune reaction, usually harmful to the animal, caused either by antigen-antibody reactions (see *Allergy*) or cellular-immune processes.

Hypha A single filament of a fungus (78).

Hypolimnion The lower region of a stratified lake, not subject to wind mixing during at least part of the year. Below the thermocline (409).

Hypotheca Inner half wall of a diatom cell (78).

Icosahedron A geometrical shape occurring in many virus particles, with 20 triangular faces and 12 corners (318).

Immune Able to resist infectious disease (521).

Immunization Induction of specific immunity by injecting antigen or antibodies (553).

Immunoglobulin Antibody (522).

Immunological Refers to processes of antibody formation and antigen-antibody reactions, whether or not immunity to infectious disease results (521).

Induced enzyme An enzyme subject to induction (295).

Induction The process by which an enzyme is synthesized in response to the presence of an external substance, the inducer (295).

Infection Growth of an organism within the body (483).

Infection thread Primary colonization and inward movement of *Rhizobium* cells within the root hair of a leguminous plant (476).

Inflammation Characteristic reaction to foreign particles and stimuli, resulting in redness, swelling, heat, and pain (504).

Informational macromolecule A macromolecule that plays a role in the transfer or expression of genetic information (271).

Inhibition Prevention of growth or function.

Inoculum Material used to initiate a microbial culture.

Insertion A genetic phenomenon in which a piece of DNA is inserted into the middle of a gene (308).

Insertional inactivation Loss of function as a result of integration of a DNA vector into a site conferring a property such as antibiotic resistance (382).

Insertion element Specific nucleotide sequences that are involved in the transfer and integration of pieces of DNA (308, 365).

In situ In place; generally used in microbiology to refer to the study of microorganisms in their natural environments, such as soil, water, the animal body.

Integration The process by which a DNA molecule becomes incorporated into the genome (340).

Interferon A protein produced as a result of virus infection which interferes with virus replication (347).

Intron The intervening noncoding regions in a split gene. Contrasted with *Exons,* the coding sequences (274).

Invasiveness Degree to which an organism is able to spread through the body from a focus of infection (497).

In vitro In glass, in culture.

In vivo In the body, in a living organism.

Isozymes Enzymes that have identical catalytic function but different structures, differing usually in regulatory control (159).

Lag phase The period after inoculation of a population before growth begins (218).

Lamella A flat layer (usually a membranous layer; e.g., "photosynthetic lamella") (61).

Latent virus A virus present in a cell, yet not causing any detectable effect.

Leaching The use of bacteria to extract commercially valuable metals from ores (439).

Leader A specific amino acid sequence involved in the attenuation process (299).

Leghemoglobin A protein containing heme which functions to bind oxygen in leguminous root nodules (473).

Leukocidin A substance able to destroy phagocytes (501).

Leukocyte A white blood cell, usually a phagocyte (499).

L form A wall-less procaryote derived from an organism with a wall.

Lichen A regular association of an alga or cyanobacterium with a fungus, usually leading to the formation of a plantlike structure (452).

Lipid Water-insoluble molecules important in structure of cell membrane and (in some organisms) cell wall (25). See also *Phospholipid.*

Lipopolysaccharide (LPS) Complex lipid structure containing unusual sugars and fatty acids found in many Gram-negative bacteria, and constituting the chemical structure of the outer layer (32).

Lithotroph An organism that can obtain its energy from oxidation of inorganic compounds (204). Equivalent to *Chemolithotroph.*

Lophotrichous Having a tuft of polar flagella (37).

Luciferase The enzyme that catalyzes light formation in luminescent bacteria (700).

Luminescence Production of light (699).

Lymphocyte A white blood cell involved in antibody formation or cellular immune response (538).

Lymphokines Substances secreted from T-lymphocytes which stimulate the activity of macrophages (547).

Lysin An antibody that induces lysis.

Lysis Rupture of a cell, resulting in loss of cell contents (31).

Lysogeny The hereditary ability to produce virus (337). See also *Temperate virus.*

Lysosome A cell organelle containing digestive enzymes (60).

Macrocyst A thick-walled structure of multicellular origin in the cellular slime molds, which may be involved in sexual reproduction (89).

Macromolecule A large molecule formed from the connection of a number of small molecules. A polymer. Informational macromolecules play a role in transfer of genetic information (271).

Meiosis In eucaryotes, reduction division, the process by which the change from diploid to haploid occurs (70).

Membrane Any thin sheet or layer. See especially *Plasma membrane* (24).

Meromictic lake A lake in which complete mixing to the bottom does not occur. Generally has a permanently anaerobic bottom layer (642).

Mesophile Organism living in the temperature range around that of warm-blooded animals (240).

Messenger RNA (mRNA) An RNA molecule containing a base sequence complementary to DNA; directs the synthesis of protein (284).

Metabolism All biochemical reactions in a cell, both anabolic and catabolic (95).

Metachromatic granule Reserve of inorganic phosphate stored within the cell and stainable by basic dyes; also called *volutin* (49).

Methanogen A methane-producing bacterium (715).

Methylotroph An organism that can grow on organic compounds containing no carbon-carbon bonds (686).

Microaerophilic Requiring O_2 but at a level lower than atmospheric (260).

Microbiological assay A procedure to measure small concentrations of vitamins or other nutrients by measuring bacterial growth (221).

Microcyst A myxospore of a myxobacterium enclosed in a hard slime capsule (653).

Micrometer One-millionth of a meter, or 10^{-6} m (abbreviated μm), the unit used for measuring microbes (17). Formerly called *micron*.

Micron See *Micrometer*.

Microtubules Tubes that are the structural entity for eucaryotic flagella, have a role in maintaining cell shape, and function as mitotic spindle fibers (62).

Mitochondrion Eucaryotic organelle responsible for processes of respiration and electron-transport phosphorylation (59).

Mitosis A highly ordered process by which the nucleus divides in eucaryotes (69).

Mixotroph An organism able to assimilate organic compounds as carbon sources while using inorganic compounds as electron donors (209).

Monera Taxonomic classification of bacteria as a separate kingdom (3).

Monoclonal antibody A monospecific antibody (546).

Monotrichous The state of having a single polar flagellum (37).

Morbidity Incidence of disease in a population, including both fatal and nonfatal cases (566).

Mortality Incidence of death in a population (566).

Most probable number (MPN) A statistical expression providing a measure of cell number in a population (217).

Motility The property of movement of a cell under its own power.

Murein See *Peptidoglycan*.

Mutagen An agent that induces mutation, such as radiation or chemicals (306).

Mutant A strain differing from its parent because of mutation (304).

Mutation A sudden inheritable change in the phenotype of an organism (304).

Mycelium A mass of hyphae in a fungus (78).

Mycolic acid Long-chain, branched fatty acids characteristic of members of the genus *Mycobacterium* (742).

Mycology The study of fungi.

Mycorrhiza Symbiotic association between plant roots and fungi (479).

Myxospore The resting cell of a myxobacterium contained within a fruiting body (653).

Negative stain A procedure in which the background is stained whereas the specimen is not (16).

Nitrification The conversion of ammonia to nitrate (424).

Nitrogen fixation Reduction of nitrogen gas to ammonia (420).

Nodule A tumorlike structure produced by the roots of symbiotic nitrogen-fixing plants. Contains the nitrogen-fixing microbial component of the symbiosis. (473).

Noncyclic photophosphorylation Light-driven ATP synthesis in which water is the electron donor and NADP the electron acceptor (189). See also *Oxygenic photosynthesis*.

Nonsense mutation A mutation that changes a normal codon into one which does not code for an amino acid (303).

Nonseptate The absence of apparent cross-walls in filamentous organisms at the light-microscope level of resolution (84).

Nucleic acid A polymer of nucleotides. See *Deoxyribonucleic acid* and *Ribonucleic acid*.

Nucleic acid hybridization Joining together of a strand of nucleic acid (DNA or RNA) from one organism with a strand from another by base sequence homology (280).

Nucleoid The region in a procaryotic cell where the DNA is located. Although frequently diffuse, sometimes the DNA is sufficiently contracted so that a defined region can be detected (37).

Nucleolus A structure seen within the nondividing nucleus, having a high RNA content; the site of ribosomal RNA synthesis (68).

Nucleosome A compact, highly folded unit of a eucaryotic chromosome which contains about 200 base pairs (283).

Nucleus A membrane-enclosed structure containing the genetic material (DNA) organized in chromosomes (67).

Nutrient A substance taken by a cell from its environment and used in catabolic or anabolic reactions (132).

Obligate A qualifying adjective referring to an environmental factor always required for growth (e.g., "obligate anaerobe").

Oligotrophic Describing a body of water in which nutrients are in low supply (425).

Operator A specific region of the DNA at the initial end of the gene, where the repressor protein attaches and blocks mRNA synthesis (296).

Operon A cluster of genes whose expression is controlled by a single operator (296).

Opsonization Promotion of phagocytosis by a specific antibody in combination with complement (528).

Organelle A membrane-enclosed body specialized for carrying out certain functions (57).

Osmophilic Requiring an environment with increased solute concentration (253).

Osmosis Diffusion of water through a membrane from a region of low solute concentration to one of higher concentration (34).

Outer layer A thin membrane lying outside the peptidoglycan layer in Gram-negative bacteria; consists of lipopolysaccharide attached to the peptidoglycan layer (32).

Oxidation A process by which a compound gives up electrons, acting as an electron donor, and becomes oxidized (101).

Oxidation-reduction (redox) reaction A coupled pair of reactions, in which one compound becomes oxidized, while another becomes reduced and takes up the electrons released in the oxidation reaction (101).

Oxidative phosphorylation See *Electron-transport phosphorylation.*

Oxygenic photosynthesis Use of light energy to synthesize ATP and NADPH by noncyclic photophosphorylation with the production of O_2 from water (187).

Palindrome A nucleotide sequence on a DNA molecule in which the same sequence is found on each strand, but in the opposite direction, leading to the formation of a repetitious inversion (278).

Pandemic A worldwide epidemic (570).

Parasite An organism able to live on and cause damage to another organism (483).

Passive immunity Immunity resulting from transfer of antibodies from an immune to a nonimmune individual (556).

Pasteurization A process using mild heat to reduce the microbial level in heat-sensitive materials (250).

Pathogen An organism able to inflict damage on a host it infects (483).

Pentose-phosphate pathway See *Hexose monophosphate pathway.*

Peptidoglycan The rigid layer of bacterial walls, a thin sheet composed of *N*-acetylglucosamine, *N*-acetylmuramic acid, and a few amino acids (29).

Periplasmic space The area between the plasma membrane and the cell wall, containing certain enzymes involved in nutrition (33).

Peritrichous flagellation Having flagella attached to many places on the cell surface (37).

Permuted sequence The termini of DNA sequences on different viral particles of the same virus which are nonidentical but overlapping (334).

Peroxisome Microbody within which photorespiration occurs, glycolate being oxidized (60).

Pesticide A chemical that kills undesired insects or animals (448).

pH An expression indicating the hydrogen-ion concentration of a solution: the negative logarithm of the hydrogen-ion concentration (257).

Phage See *Bacteriophage.*

Phage conversion A process in which a normal temperate phage modifies the phenotype of a host cell in a way unrelated to the phage immunity system (357).

Phagocyte A body cell able to ingest and digest foreign particles (498).

Phagocytosis Ingestion of particulate material such as bacteria by protozoa and phagocytic cells of higher organisms (91, 500).

Phenotype The characteristics of an organism observable by experimental means. Compare with *Genotype.*

Phospholipid Lipids containing a substituted phosphate group and two fatty acid chains on a glycerol backbone (24).

Phosphoroclastic reaction Breakdown of pyruvate in reactions involving inorganic phosphate and coenzyme A and resulting in the formation of acetylphosphate (112).

Phosphorylation Addition of phosphate to a molecule, often to activate it prior to oxidation.

Photoautotroph An organism able to use light as its sole source of energy and CO_2 as sole carbon source.

Photochromogen An organism (e.g., a *Mycobacterium*) forming pigment only when exposed to light (743). Compare with *Scotochromogen.*

Photoheterotroph An organism using light as a source of energy and organic materials as carbon source and/or electron donor (208).

Photophosphorylation Synthesis of high-energy phosphate bonds as ATP, using light energy.

Photosynthesis Light-driven ATP synthesis (182). See also *Anoxygenic photosynthesis* and *Oxygenic photosynthesis.*

Phototaxis Movement toward light (202).

Phototroph An organism that obtains energy from light.

Phycology The study of algae.

Phylogeny The ordering of species into higher taxa based on evolutionary relationships (614).

Phylum A taxonomic entity embodying a group of other taxonomic entities of lesser rank (e.g., the bacteria comprise one phylum) (614).

Phytoplankton Algae that exist floating or suspended freely in a body of water (407).

Pilus A fimbria-like structure that is present on fertile cells, both Hfr and F+, and is involved in DNA transfer during conjugation. Sometimes called *Sex pilus* (47). See also *Fimbria*.

Pinocytosis In protozoa, the uptake of macromolecules into a cell by a drinking type of action (91).

Plankton Organisms existing floating or suspended freely in a body of water (407).

Plaque A localized area of virus lysis on a lawn of cells (318).

Plasma The noncellular portion of blood (499).

Plasma membrane The thin structure enclosing the cytoplasm, composed of phospholipid and protein, in a bimolecular leaflet structure (24).

Plasmid An extrachromosomal genetic element not essential for growth (357).

Plasmodium A multinucleate mass of protoplasm formed by the acellular slime molds (88).

Plasmolysis Collapse of the protoplast as a result of dehydration, resulting, for example, from the cell's being in a medium high in solute (34).

Polar flagellation Condition of having flagella attached at one end or both ends of the cell (37).

Poly-β-hydroxybutyric acid (PHB) A polymer of β-hydroxybutyric acid; an energy-storage compound in procaryotes (49).

Positive stain Procedure in which specimen is stained while background is not (16).

Post-translational modification Modification of a protein after it has been synthesized by enzymatic action (294).

Precipitation A reaction between antibody and soluble antigen resulting in a visible mass of antibody-antigen complexes (528).

Primary electron donor The first molecule to donate electrons in an electrochemical sequence (104).

Procaryote A cell or organism lacking a true nucleus, usually having its DNA in a single molecule (2).

Promoter The binding site for RNA polymerase, near the operator (286).

Prophage The state of a temperate virus when it is integrated into the host genome (337).

Prophylactic Treatment, usually immunologic, designed to protect an individual from a future attack by a pathogen (517).

Prostheca A cytoplasmic extrusion from a cell such as a bud, hypha, or stalk (661).

Prosthetic group The tightly bound, nonprotein portion of an enzyme; not the same as *Coenzyme* (100).

Protista Taxonomic classification including all microorganisms (3).

Proton-motive force An energized state of a membrane created by expulsion of protons through action of an electron transport chain (118). See also *Chemiosmosis*.

Protoplasm The complete cellular contents, plasma membrane, cytoplasm, and nucleus; usually considered to be the living portion of the cell, thus excluding those layers peripheral to the plasma membrane.

Protoplasmic cylinder The central structure of a spiro-chete, consisting of the cytoplasm and regular wall units. Outside the protoplasmic cylinder are the axial fibrils and the outer envelope (668).

Protoplast A cell from which the wall has been removed (34).

Prototroph The wild-type parent from which an auxotrophic mutant has been derived. Contrast with *Auxotroph* (304).

Protozoa Eucaryotic microorganisms with animal affinities (90).

Psychrophile An organism able to grow at low temperatures (241).

Psychrotroph A facultative psychrophile; an organism able to grow at 0°C but also able to grow at temperatures of 25 to 30°C (242).

Pure culture An organism growing in the absence of all other organisms (9).

Pyogenic Pus-forming; causing abscesses (501).

Pyrogenic Fever-inducing.

Quarantine The limitation on the freedom of movement of an individual, to prevent spread of a disease to other members of a population (579).

Radioimmunoassay An immunological assay employing radioactive antibody for the detection of certain antigens in serum (532).

Reading-frame shift See *Frame shift*.

Recombination Process by which genetic elements in two separate genomes are brought together in one unit.

Redox See *Oxidation-reduction reaction*.

Reduction A process by which a compound accepts electrons to become reduced (101).

Reduction potential The electrode potential for a formal reaction written as a reduction (101).

Regulation Processes that control the rates of synthesis of proteins. Induction and repression are examples of regulation (274).

Replication Conversion of one double-stranded DNA molecule into two identical double-stranded DNA molecules (271).

Repression The process by which the synthesis of an enzyme is inhibited by the presence of an external substance, the repressor (295).

Respiration Catabolic reactions producing ATP in which either organic or inorganic compounds are primary electron donors and inorganic compounds are ultimate electron acceptors (113).

Reticuloendothelial system The total system of phagocytic cells able to remove particles or microorganisms from the bloodstream (499).

Rhizosphere A region around the plant root immediately adjacent to the root surface where microbial activity is usually high (472).

Ribonucleic acid (RNA) A polymer of nucleotides connected via a phosphate-ribose backbone, involved in protein synthesis (284).

Ribosome A cytoplasmic particle composed of RNA and

protein, which is part of the protein-synthesizing machinery of the cell (36, 290).

Rumen A special organ found in herbivorous mammals that subsist solely on plant material. Contains an extensive microbial flora, many of which are cellulolytic (468).

Satellite DNA DNA in eucaryotes that is present in large numbers of repetitive copies (283).

Scotochromogen An organism (e.g., *Mycobacterium*) forming pigment both in the presence and absence of light (743). Compare with *Photochromogen*.

Self-assembly The process by which a number of protein molecules associate and form a specific structure; for example, virus protein coats and bacterial flagella (330).

Septate The state of a filament in which cross-walls (septa) are present.

Septicemia Invasion of the bloodstream by microorganisms; bacteremia (507).

Septum A cross-wall of a fungus hypha (82).

Serology The study of antigen-antibody reactions in vitro (528).

Serum Fluid portion of blood remaining after the blood cells and materials responsible for clotting are removed (499).

Sex pilus A specific pilus that is present in fertile cells, both Hfr and F$^+$, and is involved in DNA transfer during conjugation (360).

Sheath A filamentous structure external to the wall (658).

Signal sequence A specific sequence of 15 to 20 amino acids that is involved in the movement of a secretor protein through the plasma membrane (293).

Siphonaceous Of certain filamentous algae, a state in which the filaments are nonseptate (74).

Slime layer A diffuse layer of polysaccharide exterior to the cell wall in some bacteria (47). See also *Capsule* and *Glycocalyx*.

Species A collection of closely related strains (2, 614).

Spheroplast A spherical, osmotically sensitive cell derived from a bacterium by loss of some but not all of the rigid wall layer. If all the rigid wall layer has been completely lost, the structure is called a *Protoplast* (34).

Sporangiospores Asexual reproductive spores of Zygomycetes (82).

Sporangium An envelope in which spores are formed: applies to asexual spores of Zygomycetes and endospores of bacteria (82).

Spore A general term for resistant resting structures formed by many bacteria and fungi (7, 51, and 80).

Stalk An elongate structure, either cellular or excreted, which anchors a cell to a surface (661).

Stationary phase The period during the growth cycle of a population in which growth ceases (220).

Stenothermal Growing over a narrow temperature range (240).

Sterile Free of living organisms (235).

Sterilization Treatment resulting in death of all living organisms and viruses in a material (6, 235, and 250).

Strain A population of cells all descended from a single cell; a clone (304).

Substrate The compound undergoing reaction with an enzyme (98).

Substrate-level phosphorylation Synthesis of high-energy phosphate bonds through reaction of inorganic phosphate with an activated (usually) organic substrate (108).

Suppressor A mutation that restores wild-type phenotype without affecting the mutant gene, usually arising by mutation in another gene (308).

Swarming Spread of a motile organism over the surface of a semisolid medium due to movement of individual cells in groups (45).

Symbiosis A relationship between two organisms (452).

Systemic Not localized in the body; an infection disseminated widely through the body is said to be systemic.

Taxis Movement toward or away from a stimulus.

Taxon Any taxonomic entity, such as a species, genus, or phylum.

Taxonomy The science of classification (610).

Teichoic acid Acidic polysaccharide containing either glycerol or ribitol, connected by phosphate diester bonds. Found in the walls of Gram-positive bacteria (33).

Temperate virus A virus which upon infection of a host does not necessarily cause lysis but may become integrated into the host genetic material (337). See *Lysogeny*.

Tetrasporal Describing an arrangement of cells in certain algae in which the individual cells are embedded in a mucilage (74).

Thallus The plantlike structure of a fungus or alga, usually macroscopic in size (78).

Thermal death time Time at a particular temperature required to kill all cells in a population (249).

Thermocline The region in the depth of a lake where the temperature undergoes a sharp drop (409).

Thermophile An organism living at high temperature (240).

Thylakoid Photosynthetic lamella; the internal membrane of chloroplasts, with which chlorophyll is associated (61, 184).

Toxigenicity The degree to which an organism is able to elicit toxic symptoms (497).

Toxin A microbial substance able to induce host damage (488).

Toxoid A toxin modified so that it is no longer toxic but is still able to induce antibody formation (554).

Transcription Synthesis of a messenger RNA molecule complementary to one of the two double-stranded DNA molecules (273).

Transduction Transfer of genetic information via a virus particle (354).

Transfer RNA (tRNA) A type of RNA involved in the

translation process; each amino acid is combined with one or more specific transfer RNA molecules (287).

Transformation Transfer of genetic information via free DNA (351).

Translation The process during protein synthesis in which the genetic code in messenger RNA is translated into the polypeptide sequence in protein (274).

Transposon A genetic element that can move from place to place; contains an insertion element at each end (366).

Transposon mutagenesis Insertion of a transposable element within a gene generally leads to inactivation of that gene, resulting in the appearance of a mutant phenotype (366).

Tricarboxylic acid cycle (citric acid cycle, Krebs cycle) A series of steps by which pyruvate is oxidized completely to CO_2, also forming NADH, which allows ATP production (115).

Trichome In bacteriology, generally equivalent to a *filament.*

Tyndallization See *Fractional sterilization.*

Vaccine Material used to induce antibody formation resulting in immunity (554).

Vacuole A small space in a cell containing fluid and surrounded by a membrane. In contrast to a vesicle, a vacuole is not rigid.

Vector An agent, usually an insect or other animal, able to carry pathogens from one host to another. Also: A genetic element able to incorporate DNA and cause it to be replicated in another cell (381).

Venereal disease Any of a number of different sexually transmitted diseases (599).

Vesicle A body constructed as a bladder; a small, rigid, thin-walled structure. See *Gas vesicle.*

Viable Alive; able to reproduce.

Viable count Measurement of the concentration of live cells in a microbial population (216).

Virion A virus particle; the virus nucleic acid surrounded by protein coat (316).

Viroid A small RNA molecule that has viruslike properties (327).

Virulence Degree of pathogenicity of a parasite (483).

Virus A genetic element containing either DNA or RNA that is able to alternate between intracellular and extracellular states, the latter being the infectious state (316).

Volutin See *Metachromatic granule.*

Zoogloea A growth form (especially *Zoogloea ramigera*) in which individual cells are enclosed in an extensive slime (446).

Zoonoses Diseases primarily of animals which are occasionally transmitted to humans (567).

Zoospore A motile spore of a fungus (77).

Zygospore A sexual spore formed by Zygomycetes (82).

Zygote In eucaryotes, the single diploid cell resulting from the conjugation of two haploid gametes (69).

Index